Exploring Agriculture, Food, and Natural Resources

D. Barry Croom

Kevin Jump

Melissa Riley

Ashley Yopp

Publisher
The Goodheart-Willcox Company, Inc.
Tinley Park, IL
www.g-w.com

Preface

Middle school is the perfect time to introduce students to the many leadership development and career opportunities available in agriculture. Our authors, all long-time educators, have had their lives enriched by participating in agriculture activities, teaching agriculture to students, and training agriculture teachers. They believe in the power of agriculture to create and give direction to bright futures.

Exploring Agriculture, Food, and Natural Resources captures this excitement and sense of possibility for middle school students. In building a product from the ground up with middle schoolers in mind, the authors set several goals for themselves. First, the text's coverage needed to be directed specifically to middle school students. The content needed to be presented clearly, directly, at the right level, and in the right amount of detail. Second, coverage needed to span the breadth of agriculture. If the goal of a middle school agriculture education course is to promote interest in agriculture and CTE, it needed to go well beyond farming to cover all the career pathways and interests that agriculturalists represent. Third, it had to provide flexibility and ease of use for agriculture teachers, a dedicated group of professionals long on passion and short on time. Developing a text in terms of small lessons allows for easy customization to different schedules, interests, and grade levels. The presence of activities and higher-level thinking questions in all lessons adds further resources for teachers. Lastly, the authors wanted to make this text fun, because agriculture is fun. Agriculture is a place for curious minds, hands-on learning, and active participation. Students of all backgrounds and experiences are welcomed into the wonderful world of agriculture and encouraged to stay.

In creating the text, the author team drew on their experience as educators and subject matter experts, providing pedagogical soundness and accuracy. They were guided by a diverse panel of teacher-reviewers, who critiqued each lesson and made helpful suggestions for improvement.

Special care has been taken to design a product that is visually stimulating and pleasing to our readers. We are excited about the way this has come together, and we hope that you will be, too!

About the Authors

D. Barry Croom has more than thirty-five years of experience in agricultural and extension education. His career interest developed as a high school junior, when he coached his first dairy judging team for the local junior high school FFA program. Dr. Croom began his professional career as a high school agricultural education teacher. While teaching full-time, he earned a Master of Education in Agricultural Education. During this time, he was also selected by the National FFA Organization to develop and present in-service workshops to teachers across the United States. His students earned state and national recognition for academic achievement. He was recognized as the Sampson County Schools Teacher of the Year and has earned both the American FFA Degree and the Honorary American FFA Degree from the National FFA Organization.

In 1996, Dr. Croom became the state FFA coordinator responsible for the statewide FFA Program in North Carolina. After completing his doctoral degree, he transitioned to teacher education, rising through the ranks to become an Alumni Distinguished Undergraduate Professor at North Carolina State University. He has served as an agriculture teacher, teacher educator, extension associate, state education staff member, professor, and department head in agricultural education and agricultural sciences.

As a professor in Agricultural Leadership, Education and Communication at the University of Georgia, his research and instructional focus is on effective teaching, career and technical education policy, and diversity in agricultural and extension education.

Kevin Jump taught high school agriculture for 16 years and has served for 12 years as a state staff member for Georgia Agricultural Education. As a high school teacher, Mr. Jump taught Basic Agricultural Science, Agricultural Mechanics, Forestry, Natural Resources, Horticulture, Animal Science, and Crop Production. His current position as an Area Teacher in Georgia's Central Region focuses on providing resources for agriculture teachers in Agricultural Mechanics. In addition, he coordinates with other area teachers, state staff members, and industry partners to produce Career Development Events (CDEs), middle school curriculum, high school curriculum, and adult education curriculum for Georgia Agricultural Education. Mr. Jump is also the co-author on G-W's *Agricultural Mechanics and Technology Systems* Lab Workbook.

Melissa Riley works with the Georgia Agricultural Education State Staff as an Area Horticulture Teacher. She has served in this role since 2010. With a BS and MS in agriculture education, Mrs. Riley creates and teaches horticulture and agriscience curriculum for agriculture educators across the state. She also helps teachers with their school greenhouses and other land lab areas. Mrs. Riley coordinates and facilitates several different CDEs for Georgia FFA and serves as the FFA Floriculture CDE Superintendent for the National FFA Organization. In 2022, the Georgia Vocational Agriculture Teachers Association (GVATA) named her Outstanding State Staff Member of the Year.

Before joining state staff, Mrs. Riley spent eight years as a middle school agriculture education teacher. While in the classroom, Mrs. Riley instructed the general agriculture curriculum, and utilized a school greenhouse, livestock barn, and outside garden area for classroom instruction. She presented educational workshops for the community and held fall and spring plant sales each year. Mrs. Riley's efforts earned her the GVATA Young Professional of the Year award in 2006 and Outstanding Agriculture Program in 2009. She was also named the school and system-wide Teacher of the Year in 2009.

Ashley Yopp is the Bureau Chief for Adult and Career Education at the Florida Department of Education, where she leads the administration and coordination of educational programs designed to help adult learners gain the skills needed to enter the workforce and obtain access to postsecondary education. She is a native of North Carolina and worked in secondary and post-secondary education as an agriculture teacher, teacher educator, and research scientist. Dr. Yopp received a BA in Political Science from East Carolina University, a MS in Agricultural Education from North Carolina A&T State University, and a PhD in Agricultural Leadership, Education, and Communication from Texas A&M University.

As a teacher educator and scientist at the University of Georgia, Dr. Yopp's work in the STEM Integration and Research Development Laboratory focused on understanding workforce needs to build interdisciplinary capacities for the future of work in rural communities. Additionally, her experiences include designing and delivering teacher education programs in agricultural education and in-service professional development as well as facilitating instruction in organizational ethics, leadership, and change, strategic communications, and research methodology. Dr. Yopp credits her experience as a student and FFA Member in Agricultural Education at Triton High School as the catalyst for her life and career as an advocate for Career and Technical Education.

Reviewers

The authors and publisher wish to thank the following industry and teaching professionals for their valuable input into the development of *Exploring Agriculture, Food, and Natural Resources*:

Hannah P. Adams
Midway ISD
Waco, TX

Skyler Alexander
Monroe County Middle School
Forsyth, GA

Adeline Amador
Delhi Middle School
Delhi, CA

Morgan Anderson
Cook Middle School
Sparks, GA

Haley Andrews
Dr. Kirk Lewis Career and
Technical High School
Pasadena, TX

Crystal Aukema
Marathon Central School
Marathon, NY

Staci Bartos
Manor Middle School
Manor, TX

Carmen Bennett
Jenkins County Middle
High School
Millen, GA

John Bergin
Mission Valley High School
Eskridge, KS

Paryce Black
J. Frank Hillyard Middle School
Broadway, VA

Johanna Bossard
Hamilton Central School
Hamilton, NY

Gary Bruns
Wilton Community School
District
Wilton, IA

Debora Cahan
Granville Junior/Senior
High School
Granville, NY

Jacqueline M. Cervantes
East Jackson Middle School
Commerce, GA

Lisa Clement
Willowcreek Middle School
Lehi, UT

Brenda Collins
Archdale-Trinity Middle School
Trinity, NC

Shelby C. Conley
Echols County Schools
Echols, GA

Katie Cooper
Salmon Junior/Senior High School
Salmon, ID

April Davis
Commerce Middle School
Commerce, GA

Audrey Davis
Thomas County Middle School
Thomasville, GA

Ron Davis
Foothill Middle School
Walnut Creek, CA

Jessica Duncan
Moyock Middle School
Moyock, NC

Megan Fern
Burke County Middle School
Burke County, GA

Deeanna Fischer
Lady Bird Johnson Middle
School
Irving, TX

Brittany Gibbs
Willie J. Williams Middle School
Moultrie, GA

Kelly Goff
Tomlin Middle School
Plant City, FL

Susan S. Harrison
Robert E. Aylor Middle School
Stephens City, VA

Dani Hodges
Appomattox Middle School
Appomattox, VA

Emma Huber
Tomah Middle School
Tomah, WI

Ryan Inman
EW Grove School
Paris, TN

Teresa Lindberg
Edward W. Wyatt Middle School
Emporia, VA

Kimberly Martin
Miller Grove High School
Cumby, TX

Jenna McCarty
Muenster ISD
Muenster, TX

Jenna McIntire
Big Spring Middle School
Newville, PA

Robin C. McLean
Northern Burlington County
    Regional Middle School
Columbus, NJ

Tara Meade
Lakeside Middle School
Lakeside, CA

Tina Miner-James
Delaware Academy High School
Delhi, NY

Kelsey A. Myers
Aspermont ISD
Aspermont, TX

Kasey Naylor
Sam Rayburn High School
Ivanhoe, TX

Amber Nead
North East Carolina Preparatory
    School
Tarboro, NC

Linda Nichols
Bryant Middle School
Dos Palos, CA

Carrie Owen
DeLand Middle School
DeLand, FL

Melvin Phelps
Lowville Academy and Central
    School
Lowville, NY

Abby Readshaw
Patrick Henry Local Schools
Hamler, OH

Renee Reed
Fall River Junior Senior High
    School
McArthur, CA

Scott Robison
Rolesville Middle School
Rolesville, NC

Eve Rogers
Eighth Street Middle School
Tifton, GA

Brooke Scott
Sixth Ward Middle School
Thibodaux, LA

Cindy Fannon Vance
Stuarts Draft Middle School
Stuarts Draft, VA

Samantha Wagner
Parlier Junior High School
Parlier, CA

Debra Wallace
Mount Olive Middle School
Mount Olive, NC

Tiffany Walters
Rossville Middle School
Rossville, GA

Dr. Susan Weiss
Stewart Magnet Middle School
Tampa, FL

Claudia Zimmer
Dr. Edward Roberson Middle
    School
Houston, TX

Beth Zuilhof
Midway ISD
Waco, TX

Acknowledgments

The authors and publisher would like to thank the following companies, organizations, and individuals for their contribution of resource material, images, or other support in the development of *Exploring Agriculture, Food, and Natural Resources.*

Lucas Brock
Rebecca Carter, National FFA
    Organization
Olivia Cook
Audrey Davis

Tyler Ertzberger
Jessica Fife
Sara Beth Fulton
Cheralyn Keily
Stormy Knight

Abigail Norton
Andy Paul
Christine White, National FFA
    Organization

Getting Started

The instructional design of this text includes student-focused learning tools to help students succeed. This visual guide highlights the features designed for the textbook.

Characters are featured throughout the text to guide students on their journey.

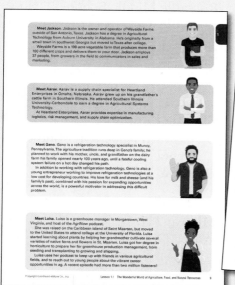

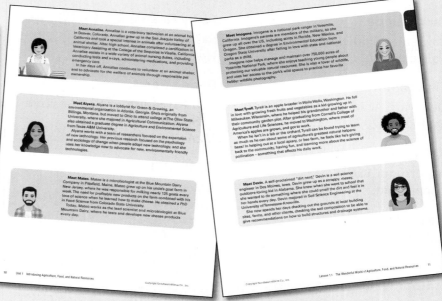

SAE for ALL Profiles open every Unit and provide relevant examples and insight into opportunities made possible with successful student SAE projects.

Words to Know lists the key terms to be learned in the chapter.

Learning Outcomes clearly identify the knowledge and skills to be obtained when the chapter is completed.

Essential Questions provoke deep thought, lively discussion, and new understanding.

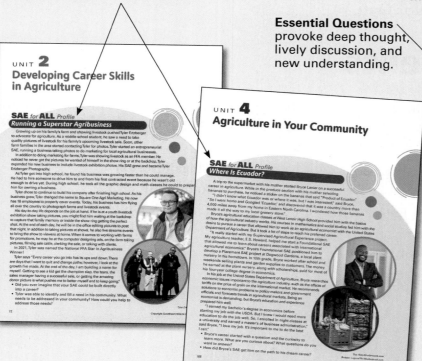

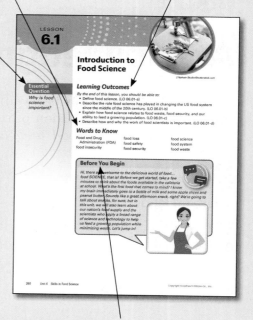

Before You Begin activities at the beginning of each lesson encourage student engagement and reflection on the subjects covered in the lesson.

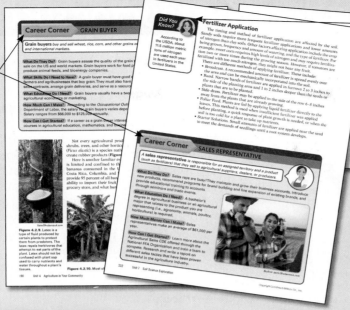

Career Corner features make real-life connections to a wide variety of potential career options.

Did You Know? features highlight fun and memorable facts to help students grasp key lesson takeaways.

Hands-On Activity features are placed in each lesson to provide memorable and fun activities for students to learn key lesson concepts.

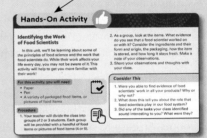

Making Connections features placed throughout the text connect agriculture to other academic subject areas, encouraging students to make connections to their other classes and interests.

Let's Apply What We Know questions and activities allow students to demonstrate concept comprehension in longer forms.

What Did We Learn? summary feature provides additional review and reinforces key learning objectives.

Let's Check and See What We Know questions allow students to demonstrate knowledge, identification, and comprehension of chapter material.

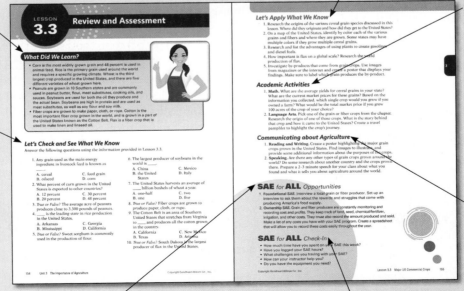

Academic Activities are provided in areas including science, technology, engineering, math, social science, and language arts to extend learning and aid the application of knowledge.

Communicating about Agriculture questions and activities help integrate reading, writing, listening, and speaking skills.

SAE for ALL Opportunities help students make real-life connections to a variety of new and interesting SAE opportunities.

SAE for ALL Check-In reminds students to review and record progress on their SAEs.

TOOLS FOR STUDENT AND INSTRUCTOR SUCCESS

Student Tools

Student Text

Exploring Agriculture, Food, and Natural Resources is a comprehensive text that provides an exciting, energetic, full-color learning resource. The text is available in a print or online version.

Lab Workbook

The Lab Workbook that accompanies *Exploring Agriculture, Food, and Natural Resources* includes instructor-created activities to help students recall, review, and apply concepts introduced in the textbook.

G-W Companion Website

- For digital users, e-flash cards and vocabulary exercises allow interaction with content to create opportunities to increase achievement.

Online Learning Suite

The Online Learning Suite provides the foundation of instruction and learning for digital and blended classrooms. An easy-to-manage shared classroom subscription makes it a hassle-free solution for both students and instructors. An online student text and a lab workbook, along with rich supplemental content, bring digital learning to the classroom. All instructional materials are found on a convenient online bookshelf and are accessible at home, at school, or on the go.

- Videos enrich learning: New videos introduce students to agriculture in action! Meet real middle school students, learn about SAE projects, and explore career paths in videos that supplement and support the concepts introduced in the text.
- Virtual Toolbox: helps students practice identifying tools and includes assessment that instructors can use to test mastery.

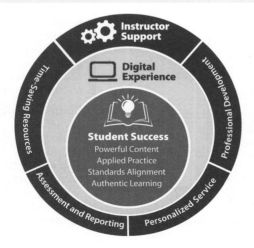

Instructor Tools

LMS Integration

Integrate Goodheart-Willcox content within your Learning Management System for a seamless user experience for both you and your students. LMS-ready content in Common Cartridge® format facilitates single sign-on integration and gives you control of student enrollment and data. With a Common Cartridge integration, you can access the LMS features and tools you are accustomed to using and G-W course resources in one convenient location—your LMS.

G-W Common Cartridge provides a complete learning package for you and your students. The included digital resources help your students remain engaged and learn effectively:

- **Digital Textbook**
- Online **Lab Workbook** content
- **Videos**
- **Drill and Practice** vocabulary activities
- **Virtual Toolbox**

When you incorporate G-W content into your courses via Common Cartridge, you have the flexibility to customize and structure the content to meet the educational needs of your students. You may also choose to add your own content to the course.

For instructors, the Common Cartridge includes the Online Instructor Resources. QTI® question banks are available within the Online Instructor Resources for import into your LMS. These prebuilt assessments help you measure student knowledge and track results in your LMS gradebook. Questions and tests can be customized to meet your assessment needs.

Online Instructor Resources (OIR)

- The **Instructor Resources** provide instructors with time-saving preparation tools such as answer keys, editable lesson plans, and other teaching aids.
- **Instructor's Presentations for PowerPoint®** are fully customizable, richly illustrated slides that help you teach and visually reinforce the key concepts from each lesson.
- Administer and manage assessments to meet your classroom needs using **Assessment Software with Question Banks**, which include hundreds of matching, completion, multiple choice, true/false, and short answer questions to assess student knowledge of the content in each chapter.

See https://www.g-w.com/exploring-agriculture-food-natural-resources-2024 for a list of all available resources.

Professional Development

- Expert content specialists
- Research-based pedagogy and instructional practices
- Options for virtual and in-person Professional Development

See www.g-w.com/pd for more information.

Brief Contents

Contents

Edmund Lowe Photography/Shutterstock.com

FFA.org

Sasina/Shutterstock.com

FenliaQ/Shutterstock.com

kung_tom/Shutterstock.com

Uryupina Nadezhda/Shutterstock.com

UNIT **6**
Skills in Food Science 258

Volodymyr_Shtun/Shutterstock.com

UNIT **7**
Soil Science Exploration . . .296

Maria Sbytova/Shutterstock.com

UNIT **8**
Plant Science Exploration . . **334**

Baronb/Shutterstock.com

UNIT **9**
Animal Science Exploration . **416**

NOOR RADYA BINTI MD RADZI/Shutterstock.com

UNIT 10
Wildlife and Natural Resources Management . . . 536

DedMityay/Shutterstock.com

UNIT **11**
Agricultural Engineering . . .668

UNIT 12
Connecting Producers and Consumers through Ag Marketing 764

eddie-hernandez.com/Shutterstock.com

Career Corner

Hands-On Activity

Making Connections

SAE for ALL Profile

Introducing Agriculture, Food, and Natural Resources

SAE for **ALL** Profile
Learning about Agriculture through an SAE

Meet Brylee Ferre from Spanish Fork, Utah. Brylee did not grow up knowing much about agriculture. One of her first experiences was watching a pig race with a youth group at the county fair. The activity sparked her interest in working with animals, and she enrolled in agriculture classes at her school. The following year, an animal welfare organization protested against the pig races. The county decided not to proceed with pig races in the future. In response, Brylee created a Research SAE on pig races and other types of agricultural entertainment and the community's view on these activities.

Brylee became more involved in FFA and decided to expand her animal knowledge by raising a pig in the school animal lab. She enjoyed this work and created another Research SAE, this one looking at the impact that animal labs have on students and local livestock shows. Brylee was able to use her Research SAEs as projects for the National FFA Agriscience Fair.

Brylee's investigations allowed her to examine other people's perspectives about agriculture while growing her agricultural knowledge. Through her Research SAEs, Brylee feels like she learned a lot about the many aspects of agriculture. She also learned more about communication skills and problem-solving skills.

Brylee said, "I have always had a fear of presenting things in front of people. I needed a script to be able to say what I thought needed to be said. [These projects] really taught me how to better communicate in presentations, interviews, and just life in general." Brylee uses these skills in her current job working as an executive assistant in the insurance industry. Brylee attributes many of her workplace skills to what she learned while in FFA and during these projects.

Completing these projects taught Brylee how to find solutions to problems. The two projects she did were based on her questions and thoughts about animal welfare. She learned how to create surveys and interpret the information compiled from the surveys.

- What questions do you have about how agriculture connects to your life?
- Does an SAE project sound like something that might be fun or interesting?
- How can you create a research project that answers your questions about agriculture, food, or natural resources?

Top: Brylee Ferre
Bottom: jakelv7500/Shutterstock.com

Edmund Lowe Photography/Shutterstock.com

Fotokostic/Shutterstock.com

The Wonderful World of Agriculture, Food, and Natural Resources

Essential Question

What is AFNR, and why is it important to understand it?

Learning Outcomes

By the end of this lesson, you should be able to:
- Identify the foundational components of agriculture, food, and natural resources (AFNR). (LO 01.01-a)
- Describe the value of hands-on experiential learning in career development. (LO 01.01-b)
- Discuss agricultural education as a part of career and technical education (CTE). (LO 01.01-c)

Words to Know

agricultural education
agriculture
agronomist

career and technical education (CTE)
domesticate

experiential learning
livestock
natural resources

Before You Begin

Hello, and welcome to AFNR! I'm Ms. Lopez, an agriculture teacher and your tour guide into the wonderful world of agriculture, food, and natural resources. Together, we'll explore the art and science of producing our food and fiber while thinking critically about our roles as stewards of the environment and natural resources. We'll also meet some interesting people along the way as we imagine the future and think about exciting careers. Did I mention I'm glad you're here? We're going to have a blast!

*Now, before we jump right in, I bet you already know a lot about agriculture, food, and natural resources. Personally, food is my favorite topic! Find a sheet of paper and take two minutes to write down every word that comes to mind when you hear the word **agriculture**. What does it mean to you? What about food? Natural resources? I bet you have quite a few things listed already. At the end of our lesson, circle back to look at your list. What can you add? What did you miss? Agriculture is a pretty big word, and it covers a lot. Let's get started!*

Figure 1.1.1. Agriculture, food, and natural resources create endless opportunities for the American workforce in more than 300 diverse career areas. *Where might you fit in?*

AFNR in Our World

Did you know that by the year 2050, the world will be home to nearly nine billion people and the task of feeding a growing population rests in the hands of people working in agriculture? Agriculture in the United States alone accounts for more than 23 million jobs. This means that about 17 percent of the total workforce in this country is involved in some aspect of agriculture, from the farm to your plate (**Figure 1.1.1**). Nearly endless career opportunities exist within agriculture, food, and natural resources, but each requires a critical foundation in problem solving, ethics, critical thinking, integrity, teamwork, technology, leadership, creativity, communications, and innovation.

Agriculture

What is agriculture? You may be familiar with the word *farming*, which is often used instead. **Agriculture** includes the growing and harvesting of crops and raising domesticated animals, or **livestock**, for human use. For the past 11,000 years, agricultural practices have provided food and materials humans need to survive.

Unlike earlier hunting and gathering, agriculture allowed humans to **domesticate**, or tame, plants and animals with the use of primitive tools. It wasn't until later that humans began to raise livestock animals such as cattle, pigs, goats, and poultry, with the first cow setting foot on American soil in the early 1600s. In the coming lessons, you will learn more about American agriculture's development since then, including some of the most revolutionary machines in modern history (**Figure 1.1.2**).

Did You Know?

Did you know fire was one of the earliest tools used in farming? Fire helps clear large areas of land for planting while making the soil more fertile for crops.

Figure 1.1.2: Agriculture has changed drastically over time, with innovation driving the work of farmers and ranchers around the globe. The work once completed by a steel plow and mule can now be automated. *How do you think this has changed what is possible?*

In addition to food for humans and animals, agriculture produces a wide variety of other useful goods, including lumber, fertilizers, leather, sugar, and chemicals such as ethanol, starch, alcohols, and plastics. If you're wearing blue jeans today, a farmer grew at least two cotton plants to make the raw material for each pair! Farmers are responsible for producing fibers for clothing and other materials, such as cotton, wool, hemp, and flax. Farmers grow many other plants for their beauty, such as flowers, shrubs, or nursery plants.

You will learn about many elements of agriculture in this book, including the various methods farmers use when producing crops like cotton, corn, soybeans, wheat, potatoes, peanuts, and rice. These plants are referred to as *row crops*, because each is typically grown in widely spaced rows to provide room for watering and space for machines, like tractors and sprayers, to drive without crushing them (**Figure 1.1.3**). This practice, like with many others, was developed by **agronomists**, or scientists focused on producing efficient and high-quality food and fiber crops while also managing and maintaining the soil.

Fotokostic/Shutterstock.com

Figure 1.1.3. Large machinery like tractors and sprayers have revolutionized the agriculture industry, helping producers manage thousands of acres of food and fiber crops each year. *Can you imagine what it would have been like to do this work without a machine?*

margouillat photo/Shutterstock.com

Figure 1.1.4. Foods such as cheese, ice cream, yogurt, and butter are value-added products that employ farmers, food scientists in laboratories across the country, and plenty of others in between. *Can you think of what some of these other jobs and roles might be?*

Food

There is a distinction in agriculture between plants and animals grown for food, and those produced for clothing and shelter. When thinking about food production, there are opportunities to learn about every aspect, from the development of seeds, plants, and animal breeds for cost-effective production of tasty, high-quality food, to the packaging, processing, sales, and marketing required to move our food from the farm to our plate. This also includes the research and safety measures used to protect and maintain a healthy and nutritious food supply.

Animals raised for food, such as cattle, hogs, goats, sheep, and poultry, provide most of the protein needed in our diets. In fact, globally, these animals supply nearly 350 million pounds of meat for consumption each year. In addition, animals are raised to produce dairy products such as milk, yogurt, butter, ice cream, and cheese (**Figure 1.1.4**). In the weeks ahead, you will learn more about how dairy animals are raised, as well as how cheese and ice cream are crafted and perfected by artisans and scientists around the world.

 **Did You Know?** Did you know that an average American eats about 31 pounds of cheese every year?

Plants cultivated and grown for food include fruits and vegetables, such as tomatoes, cucumbers, apples, watermelons, peanuts, corn, cabbage, and mushrooms. The production of these plants takes place in orchards, on family farms, in greenhouses, and sometimes in your own backyard. In the pages ahead, you will get a peek into laboratories where the use of cutting-edge science and innovative technology continues to improve the quality and quantity of our food supply.

Do you like bugs? One topic concerning our food involves a few special guests of the wiggly variety. Worms are vital to helping improve the health and composition of fertile soil. Pollinators, such as butterflies, bees, and birds, are responsible for the reproduction of more than 80 percent of the world's flowering plants, so creating spaces to protect them is crucial to the future of agriculture (**Figure 1.1.5**).

Natural Resources

Last for us to consider, but certainly not least, are natural resources. In AFNR, we are concerned with the production of natural resources such as timber, paper, and biofuels, and the protection of clean air, healthy water, and healthy soils needed for thriving ecosystems. The field of natural resources is responsible for more than a million American jobs in areas such as environmental conservation, timber management, wildlife biology and protection, and ecological restoration.

Natural resources are items originally provided by the natural world and come from both *renewable* and *nonrenewable* sources. Different regions of the country provide access to different natural resources. For example, fertile soils found in the central and Great Plains area of the United States provide the perfect environment to produce an abundant food supply, while canals in Arizona support the convenient transportation of goods throughout the Southwest.

One of the United States' most important and increasingly scarce resources is water. Three-fourths of our planet is covered in water, but most of that is saltwater found in Earth's oceans, which isn't usable for drinking. Only three percent of available water is fresh and drinkable, but two-thirds of that is frozen in ice caps and glaciers. That means that only *one* percent of available water is drinkable and accessible on Earth's surface in rivers and lakes, or underground between soil and rocks. The conservation of this valuable resource is an essential link to the future production of food and fiber around the world.

Sunlight is an increasingly important natural resource used in the production of solar energy. Additionally, wind generates power through offshore and land-based turbines. The constant expansion, management, and protection of these natural resources create career opportunities for anyone interested in innovation and conservation in the agriculture industry.

Figure 1.1.5. Monarch butterflies and other pollinators are a vital link in the production of food and fiber! Each fall, millions of monarchs migrate across the United States to spend the winter in warmer climates. *How do you think you can you help them on their journey?*

How Will We Explore AFNR?

There is a lot for us to learn! Don't worry, though; we'll be together every step of the way. I'm excited to introduce you to some of my friends who work in agriculture, food, and natural resources. My hope is that they will help you get a better picture of job opportunities and of the knowledge, skills, and experiences you may need to build a future career in AFNR. There is no better way to learn than from experience, so let's look at our roadmap for the journey ahead.

Experiential Learning

Hands-on learning, or *experiential learning*, is at the center of everything you will learn in AFNR. In fact, you'll soon be "learning by doing" in this class, gaining new skills through active exploration of the world around you. One day you might be outside working in a garden or tending to livestock on your school's campus, and another day you may be extracting DNA from strawberries in a science lab. When you learn this way, not only do you have fun, you also commit new information to memory using all of your senses. This is so much more effective than just memorizing information for a test! Turning information into useful knowledge is easier and more exciting when you learn in a hands-on way.

Agricultural Education

Agricultural education combines classroom lessons about agriculture, food, and natural resources with hands-on experiential learning with the help of agriculture teachers. As you explore the many topics in AFNR, agriculture teachers help you develop a wide variety of skills, including science, math, communications, leadership, management, and technology. Agricultural education has a comprehensive vision for learning that includes three interconnected types of experiences (**Figure 1.1.6**). Along with nearly 800,000 other students and 12,000 teachers from all 50 states, Puerto Rico, Guam, and the Virgin Islands, you will spend time learning and growing in the classroom and educational laboratory; with supervised experiences outside the classroom; and through participation in student organizations like the National FFA Organization. Each area is vital to the total student experience in agricultural education and will prepare you for leadership, personal growth, and career success.

Classroom/laboratory instruction

TOMORROW'S LEADERS

Leadership (FFA)

Experiential learning (SAE)

NAAE

Figure 1.1.6. Agricultural education connects classroom learning to valuable experiences outside the classroom and leadership development through the National FFA Organization. *How is this model different from how you learn in other classes?*

Career and Technical Education

Agricultural education is also unique in that it has a strong focus on career development. As one of eight pathways in *career and technical education (CTE)*, or the practice of teaching specific career skills to middle school and high school students, agriculture, food, and natural resources is preparing you for high-wage, high-skill, and high-demand jobs in the future.

Let's take a quick look into a few future career opportunities by meeting some of my friends. These folks built careers in agriculture, food, and natural resources and are putting years of knowledge and experiences to work every day. You'll see them throughout this book, as they help us to explore the career pathways available to students who are interested in AFNR. Let's meet them!

Meet Jackson. Jackson is the owner and operator of Wayside Farms outside of San Antonio, Texas. Jackson has a degree in Agricultural Technology from Auburn University in Alabama. He's originally from a small town in southwest Georgia but moved to Texas after college.

Wayside Farms is a 198-acre vegetable farm that produces more than 100 different crops and delivers them to your door. Jackson employs 37 people, from growers in the field to communicators in sales and marketing.

Meet Aarav. Aarav is a supply chain specialist for Heartland Enterprises in Omaha, Nebraska. Aarav grew up on his grandfather's cattle farm in Southern Illinois. He attended Southern Illinois University-Carbondale to earn a degree in Agricultural Systems Technology.

At Heartland Enterprises, Aarav provides expertise in manufacturing logistics, risk management, and supply chain optimization.

Meet Geno. Geno is a refrigeration technology specialist in Muncy, Pennsylvania. The agriculture tradition runs deep in Geno's family; he planned to work with his mother, uncle, and grandfather on the dairy farm his family opened nearly 100 years ago, until a fateful cooling system failure on a hot day changed his path.

In addition to working with refrigeration technology, Geno is also a young entrepreneur working to improve refrigeration technologies at a low cost for developing countries. His love for milk and cheese (and his family's past), combined with his passion for expanding opportunities across the world, is a powerful motivator in addressing this difficult problem.

Meet Luisa. Luisa is a greenhouse manager in Morgantown, West Virginia, and host of the AgriNow podcast.

She was raised on the Caribbean island of Saint Maarten, but moved to the United States to attend college at the University of Florida. Luisa started learning about plants by helping her grandmother cultivate several varieties of native ferns and flowers in St. Maarten. Luisa got her degree in horticulture to prepare her for greenhouse production management, from seeding and transplanting to growing and shipping.

Luisa uses her podcast to keep up with friends in various agricultural fields, and to reach out to young people about the vibrant career opportunities in ag. A recent episode had more than two million listeners!

Meet Annalise. Annalise is a veterinary technician at an animal hospital in Denver, Colorado. Annalise grew up in the San Joaquin Valley of California and took a special interest in animals after volunteering at an animal shelter. After high school, Annalise completed a certification in Veterinary Assisting at the College of the Sequoias in Visalia, California. Annalise assists in a wide variety of animal nursing duties, including conducting tests and x-rays, administering medications, and providing emergency care.

In her days off, Annalise continues to volunteer at an animal shelter, and to advocate for the welfare of animals through responsible pet ownership.

Meet Aiyana. Aiyana is a lobbyist for Green & Growing, an environmental organization in Atlanta, Georgia. She's originally from Billings, Montana, but moved to Ohio to attend college at The Ohio State University, where she majored in Agricultural Communication. Aiyana also obtained a graduate degree in Agriculture and Environmental Science from Texas A&M University.

Aiyana works with a team of researchers focused on the expansion of new technology. Her previous research focused on the psychology and sociology of change when people adopt new technology, and she uses her knowledge now to advocate for new, environmentally friendly technologies.

Meet Mateo. Mateo is a microbiologist at the Blue Mountain Dairy Company in Pittsfield, Maine. Mateo grew up on his uncle's goat farm in New Jersey, where he was responsible for milking nearly 125 goats every week. The need for profitable new products on the farm combined with his love of science when he learned how to make cheese. He obtained a PhD in Food Science from Colorado State University.

Today, Mateo works as the lead scientist and microbiologist at Blue Mountain Dairy, where he tests and develops new cheese products every day.

Meet Imogene. Imogene is a national park ranger in Yosemite, California. Imogene's parents are members of the military, so she grew up all over the US, including stints in Florida, New Mexico, and Oregon. She obtained a degree in Environmental Education from Oregon State University after falling in love with state and national parks as a child.

Imogene now helps manage and maintain over 750,000 acres of Yosemite National Park, where she enjoys teaching young people about protecting our valuable natural resources. She is also a lover of wildlife, and uses her access to the park's wild spaces to practice her favorite hobby: wildlife photography.

Meet Tyrell. Tyrell is an apple breeder in Walla Walla, Washington. He fell in love with growing fresh fruits and vegetables as a kid growing up in Milwaukee, Wisconsin, where he helped his grandmother and father with their community garden plot. After graduating from Cornell's College of Agriculture and Life Sciences, he moved to Washington, where most of America's apples are grown, and got to work.

When he isn't in a lab or the orchard, Tyrell can be found trying to learn as much as he can about some of agriculture's greatest natural helpers: bees! In helping out at a local apiary, or bee farm, he feels like he's giving back to the community, having fun, and learning more about the science of pollination – something that affects his daily work.

Meet Devin. A self-proclaimed "dirt nerd," Devin is a soil science engineer in Des Moines, Iowa. Devin grew up as a scrappy, messy, outdoors-loving kid in Alabama. She knew when she went to school that she wanted to do something where she could smell the dirt and feel it in her hands every day. Devin majored in Soil Science Engineering at the University of Tennessee-Knoxville.

She now spends her days checking out the grounds at local building sites, farms, and other clients, checking the soil composition to be able to give recommendations on how to build structures and drainage systems.

What Did We Learn?

- Agriculture is an industry that includes the growing and harvesting of crops and raising animals for the purpose of food, fiber, or shelter.
- Food production covers all points and steps between creating, modifying, or planting seeds or breeding animals, and getting food onto your dinner table.
- The natural resources segment includes the production of timber, paper, and biofuels (among other products), and the protection of clean air, healthy water, and healthy soils needed for thriving ecosystems.
- Experiential learning is a method used to provide hands-on, active exploration for students learning in agricultural education and CTE. Experiential learning helps commit information to memory by engaging the five senses.
- Agricultural education is comprised of three parts: teaching agriculture, food, and natural resources concepts through traditional classroom and laboratory instruction; participation in Supervised Agricultural Experiences; and participation in the National FFA Organization.
- Agricultural education is one pathway (of eight) in career and technical education. CTE is focused on teaching career skills to middle and high school students as a way of preparing students for future careers.

Let's Check and See What We Know

Answer the following questions using the information provided in Lesson 1.1.

1. *True or False?* Career and Technical Education (CTE) is only for students who plan to attend college.
2. The natural resources segment includes the production and/or protection of all of the following, *except*:
   A. biofuels
   B. air
   C. livestock
   D. timber
3. Experiential learning helps students by engaging in active exploration through:
   A. sight, sound, taste, touch, and smell
   B. light, flavor, music, art, and technology
   C. outdoor activities
   D. books and musicals

4. Butterflies, bees, and birds are helpful in food production because of their roles as:
   A. livestock
   B. row crops
   C. bugs
   D. pollinators
5. What type of scientist is focused on producing efficient and high-quality food and fiber crops while also managing and maintaining the soil?
   A. agronomist
   B. genetic engineer
   C. chemist
   D. microbiologist

Let's Apply What We Know

1. What is experiential learning, and why is it used in agricultural education?
2. List possible uses of livestock animals, other than meat.
3. When were the earliest agricultural practices used?
4. What is career and technical education? How does it differ from, say, the instruction in your history class?

Academic Activities

1. **Math.** We learned in this lesson that there are 23 million people working in agriculture in the US. That is 17 percent of the workforce! If this is 17 percent, how big is the US workforce?

Communicating about Agriculture

1. **Oral History.** The average American teenager is roughly three generations removed from living on a farm. That means that if any of your ancestors lived on a farm, they were likely your great-great-grandparents. How might that affect the way you think about producing food? Do you think older family members feel the same way? Create a list of questions to ask your family at dinner tonight about where their food comes from. Consider asking other adults like your teachers, youth leaders, coaches, and mentors to start having conversations about food.

SAE for ALL Opportunities

1. **Foundational SAE.** Think of your favorite teacher. What makes them your favorite? Do they make information come to life? Do they motivate you to try new things? Great teachers enjoy helping others learn and succeed. If you enjoy helping others, maybe you should consider a future career in agriculture education. Make a list of questions that you can ask your teacher about their training, education, and experiences to help you get more information.

Exploring Opportunities in Agriculture, Food, and Natural Resources

only_kim/Shutterstock.com

Learning Outcomes

By the end of this lesson, you should be able to:
- Identify career opportunities in agriculture, food, and natural resources. (LO 01.02-a)
- Describe the eight career pathways in agricultural education. (LO 01.02-b)
- Summarize what knowledge and skills are needed to attain a career in one of the eight career pathways. (LO 01.02-c)

Words to Know

agribusiness

agricultural engineering

animal science

biotechnology

career pathway

environmental science

food science

plant science

Before You Begin

Think back to when you were a bit younger. What did you want to be when you grew up? I wanted to be an astronaut! And then a fire fighter, marine biologist, and even a florist. The funny thing is, though, that I rarely thought about what I would need to do to prepare for those careers. No one can magically become an astronaut overnight, right? It takes years of training and preparation. You can be anything you want to be, but making your dreams real takes planning. So let's start now! What's your dream job? **(Figure 1.2.1)** *What skills will you need? Education? Experiences? Draw a picture of yourself in that future career and list everything you think you will need to accomplish that goal. At the end of this lesson, circle back to your exercise and see if you have anything to add or change.*

Africa Studio/Shutterstock.com

Planning for a future career might seem like a daunting task, but exploration is a great first step. You can find opportunities for your future in agriculture, food, and natural resources by taking a closer look at diverse *career pathways*, or groups of careers that involve similar knowledge and skills.

In **Lesson 1.1**, you learned about agricultural education as a part of career and technical education. In this lesson, you will take your understanding one step further by exploring the eight distinct CTE career pathways in agricultural education and identify careers within each one. You will also consider your personal interests, talents, and characteristics and how to align those with potential careers in those pathways: Agribusiness Systems, Animal Systems, Biotechnology Systems, Environmental Service Systems, Food Products and Processing Systems, Natural Resources Systems, Plant Systems, and Power, Structural, and Technical Systems.

Agricultural Education Career Pathways

Agriculture, as discussed previously, makes up a huge part of the American and global economies. As populations grow and resource levels change, producing and providing for the world will require innovation, creativity, collaboration, and hard work. Nearly endless career opportunities exist within each of the eight pathways moving forward, but each requires different preparation, skills, and interests. Let's take a closer look at each pathway.

Consider each pathway and think about what areas might align with your current interests while also considering areas that, although new, might engage your talents in the future.

Agribusiness Systems

The Agribusiness Systems career pathway prepares students to work in teams, using technology to understand the best ways to efficiently profit from agricultural products. People who work in *agribusiness* may manage their own business providing agricultural goods or services. They may also work in law, accounting, education, or sales. Do any of these areas sound interesting to you? Maybe you're good at math, enjoy graphic design, or find yourself thinking about social media content. If so, consider courses on the Agribusiness Systems pathway.

Earlier, you met Aarav, who works in logistics as a supply chain specialist for Heartland Enterprises. Aarav has always loved data, but more than that, he figured out at an early age that he had a talent for figuring out the quickest, most efficient way to complete almost any task. Now, his work at Heartland requires he develop the fastest, safest way to get crops from the field to the market for sale. He also works to reduce the risk associated with moving large quantities of food and resources across the country.

You also met Luisa. As you might recall, she enjoys talking to and understanding people who work in agriculture. Her podcast, AgriNow, digs into the everyday lives of people working in agriculture, food, and natural resources while uncovering the newest issues facing the industry. After only two years, her podcast has grown to over 400,000 regular listeners in 87 countries around the world! Might you enjoy streaming your own podcast? What might you talk about?

Other careers in agribusiness might include positions in marketing, finance, or rural development (**Figure 1.2.2**). Depending on the job and educational requirements, these professionals could take home a salary between $32,000 and $154,000 a year.

Animal Systems

The Animal Systems career pathway prepares students to work for the care, health, and safety of animals of all sizes, and with the products they provide. This area of study is also commonly referred to as *animal science*. Careers in this pathway include the management of swine, poultry, and dairy cows, the production and maintenance of fish hatcheries, the medical care of small and large animals, and the protection of biological ecosystems (**Figure 1.2.3**). Do you have an animal at home you enjoy caring for? If so, maybe you should learn more about Animal Systems! Many students enjoy working with nonprofit groups with different specializations. For example, there are groups that help people with special needs through therapeutic horseback riding. Might a similar opportunity sound interesting to you?

StockMediaSeller/Shutterstock.com

Figure 1.2.2. The Agribusiness Systems career pathway employs millions of people to ensure the most up-to-date practices and products keep agriculture efficient and profitable. *Does this differ from how you have thought about agriculture in the past?*

santypan/Shutterstock.com

Figure 1.2.3. People working in the Animal Systems career pathway ensure the health and wellness of animals such as livestock, but also help us manage the care of small animals like dogs and cats too. *What might make someone a good fit for this pathway?*

In high school, Annalise volunteered at an animal clinic where she was responsible for the feeding, walking, and general well-being of small animals. Now, after completing a certification in Veterinary Assisting, Annalise conducts tests and x-rays, attends to emergency care, and administers medications and vaccinations. It may be amazing to think about how her career started as a volunteer job on the weekends, but stories like hers are not uncommon!

Other careers in Animal Systems, such as swine managers, poultry scientists, and horse trainers, require various educational experiences including on-the-job trainings and apprenticeships, two- or four-year degrees, and/or professional degrees and medical training. Salaries in this field range from $22,000 to $175,000 per year.

Biotechnology Systems

The Biotechnology Systems career pathway prepares students for careers working to enhance the growth and value of plants, animals, and microorganisms through genetic modification. This area of science is called **biotechnology**. Careers in this area include experimentation conducted in laboratory settings (**Figure 1.2.4**), development of biofuels, ecological monitoring, and genetically engineering plants and animals. Do you enjoy working in the science lab? If so, you should learn more about the Biotechnology Systems career pathway!

Earlier, you met Mateo, a microbiologist at the Blue Mountain Dairy Company. Mateo's love for cheese started on his uncle's goat farm, where he was responsible for milking a herd of 125 goats each week. His interest in science led him to combine his experience on the farm with

Budimir Jevtic/Shutterstock.com

Figure 1.2.4. *Do you enjoy chemistry or working in the science lab?* Careers in Biotechnology Systems help develop the future of our food supply in both quality and quantity. Maybe you will develop the next variety of potato, tomato, or even peanut! *What is a question you might be interested in answering?*

growing microorganisms used to make some of his favorite cheeses. After studying food science in college and getting a PhD in Microbiology, Mateo now works as lead scientist and microbiologist where he tests and modifies new cheese products daily. Can you think of a better job than tasting and creating cheese? I know Mateo can't!

Other careers in Biotechnology Systems may include the genetic engineering of crops such as peanuts for drought resistance, the management of pests through the development of herbicides, and the modification of animal genetics to produce more heat-tolerant livestock. Think here about Tyrell, who we also met earlier—he spends his days trying to breed new strains of apples to be receptive to different growth conditions, or resistant to different pests, or to have a particular taste or texture. If that sounds interesting to you, Biotechnology Systems may be a fit for you.

Salaries in Biotechnology Systems vary based on educational background and experience and range from $28,000 to $122,000 per year.

Environmental Service Systems

The Environmental Service Systems career pathway prepares students for the management of human activities and their related impacts on our environment. This study is often referred to as ***environmental science***. Careers in this area may include the study of soils and water; meteorology and the impacts of weather on agriculture, food, and natural resources; and the biology of aquatic environments.

Aiyana works as a lobbyist for an environmental organization in Atlanta. As a trained social scientist, her work combines a deep understanding of the environment with the psychology and sociology of people and their behaviors. More specifically, Aiyana is interested in helping the people of Georgia adopt new technologies, such as solar panels, to help provide energy for their homes and businesses. Solar panels can be costly additions to homes, and although having these panels will save money in the long term, homeowners must consider the burden this might place on their families. Aiyana wants to help increase the use of sustainable energy sources without hurting families and businesses in the process. She also loves talking to and helping people, which helps her better connect and interpret human behavior. Do you also enjoy talking to and meeting new people? While social skills are not required to work in all environmental science areas, working with people is a cornerstone to this pathway.

Other careers in Environmental Service Systems include water treatment specialists, environmental engineers, and sustainability regulators (**Figure 1.2.5**). Salaries in Environmental Service Systems range from $42,000 to $107,000 per year.

lovelyday12/Shutterstock.com

Figure 1.2.5. *Have you seen solar panels or wind turbines around your community?* The use of wind and solar power create alternative sources for energy to power our homes, businesses, and schools. *Do you think of technology when you think of agriculture?*

Food Products and Processing Systems

The Food Products and Processing Systems career pathway prepares students to enhance the flavor and nutritional value of foods while also ensuring safe and efficient practices. This career path is also referred to as *food science*. Careers in this area may include the development of new ice cream flavors, the regulation of food imports and exports for transportation and safety (**Figure 1.2.6**), and the development of more cost-effective practices in areas where technology is difficult to come by.

Geno is a refrigeration specialist who works in Pennsylvania by day and an entrepreneur by night. His love for milk and cheese obviously started on the farm, but a trip to Ethiopia sparked his passion to help improve the refrigeration of dairy and other products in places electricity is either not available or the cost is too high for many families. Geno is in his final phase of developing a small, portable cooler that mimics the heating and cooling of the human body without the use of any electricity. Each of these small coolers also costs less than $3 to build and can be constructed from simple materials that are easily accessible all over the world. Something as simple as a refrigerator is not so simple in many places! Inventions and new practices developed by people like Geno help extend the shelf life of many foods and provide a safe food supply to people across the globe.

Additional career opportunities in Food Products and Processing Systems include culinary arts, food service management, communications and marketing, finance, and public safety. Salaries range from $24,750 to $123,000 per year, depending on the job, educational experience, and training.

Dusan Petkovic/Shutterstock.com

Figure 1.2.6. Food scientists and quality assurance specialists take precautions to make sure our food supply is safe and secure. *Why is this a crucially important role?*

Natural Resources Systems

The Natural Resources Systems career pathway prepares students for the protection and management of the natural environment in both rural and urban settings. Careers may include law enforcement and protection of fish and wildlife; maintaining local, state, and national parks; timber and forest management; and the surveying of land. Have you ever visited a state or national park?

Ranger Imogene enjoys teaching others about the beauty, history, and preservation of Yosemite National Park in California. As a national park ranger, she is partially responsible for the management of nearly 750,000 acres of land, hundreds of species of wildlife, and the safety of over four million visitors each year. Additionally, Imogene monitors the constantly changing topography of the park, including the shifts and changes occurring to Half Dome, a natural rock formation and popular site for hiking and rock climbing.

Devin, a soil science engineer, also works in this pathway. Her work, making sure that the soil of a client's site is suitable for a proposed purpose (building, planting) helps to protect both her clients' investments in new buildings or crops and the topography and water of the local area. Does that sound like a fun challenge to you?

Kowit Lanchu/Shutterstock.com

Figure 1.2.7. Butterflies and bees are vital partners in pollinating our food supply. Without their help, many of our fruits and vegetables would never flower or produce food for us to eat. *How do you think people can work to help them?* People are working on this in the Natural Resources Systems pathway right now!

Careers in the Natural Resources Systems pathway also include conservation scientists, timber cruisers, foresters, civil engineers, and social scientists (**Figure 1.2.7**). Salaries range from $25,750 to $109,000 per year.

Plant Systems

The Plant Systems career pathway prepares students for the study of plants and how they impact our world (also referred to as ***plant science***). More specifically, this pathway helps feed our growing population in a safe, nutritious, and economical way (**Figure 1.2.8**). Careers in Plant Systems may include the development of insect-resistant crops, monitoring of food safety and quality, conservation of natural resources, and constant improvement of production practices.

Luisa, a greenhouse manager in Morgantown, West Virginia, is responsible for every phase of seeding, transplanting, growing, and shipping large shrubs and trees to homes and businesses around the country. Luisa's love for working in the greenhouse started at home on the Dutch Caribbean island of St. Maarten, where she helped her grandmother cultivate several varieties of native ferns and flowers for sale at the local market. A college degree in horticulture helped her develop the additional knowledge and skills necessary for a successful career growing plants. Is there a greenhouse located at your school? If not, consider walking around one at a home improvement store or plant nursery. It's a great place to learn about new and interesting plant varieties!

Additional careers in the Plant Systems pathway include landscape design, greenhouse management, agronomy, turf and weed science, commodity marketing, and grounds management. Salaries range from $41,000 to $94,000 per year based on training and experience.

Figure 1.2.8. Many greenhouses have developed systems to allow for the growth and harvest of fruits and vegetables all year. Even in the coldest places, people in horticulture and plant science are developing opportunities to increase our food supply, regardless of weather conditions. *Would you be interested in working in this environment?*

Azaddin Evliyaogullari/Shutterstock.com

Power, Structural, and Technical Systems

The Power, Structural, and Technical Systems career pathway prepares students to innovate, design, and build solutions for projects in electronics, power, hydraulics, and engineering. This area of focus is also referred to as *agricultural engineering*. Careers in this pathway may also include robotics, precision agriculture, global positioning systems (GPS), geographic information systems (GIS), drones, welding and fabrication, and heavy equipment mechanics. Do you enjoy large machinery, designing construction projects, or tinkering with robots? Might you enjoy thinking about all the components of a project, drawing them out to scale, assembling materials, and constructing the design—even if they do not always come out the way you envisioned? Sometimes the process of understanding what *doesn't* work is the most important part!

Jackson knows all about the design process. As the owner and operator of Wayside Farms, he develops the master plans for all projects on the farm. Recently, he implemented a large-scale project using drones to monitor crop growth using sensors and digital imaging. The work that once took him three weeks to complete on his 198-acre farm can now be accomplished in three days with the use of technology. It's a fascinating process that makes his vegetable production more efficient and cost-effective.

In addition to the use of drone technology at Wayside Farms, Jackson trains two dozen of his employees on the safe and proper use of heavy machinery and equipment, including tractors, cultivators, and sprayers. He also works with three of his employees in the routine maintenance and repair of this equipment.

Salaries for jobs in Power, Structural, and Technical Systems vary between $34,000 to $125,000 per year, depending on your educational background and area of interest. For example, on Jackson's farm, welders make $53,500 per year, while wind turbine technicians start at a salary of $61,000 (**Figure 1.2.9**).

Everyonephoto Studio/Shutterstock.com

Figure 1.2.9. *Do you enjoy building new things with your hands?* Career opportunities in Power, Structural, and Technical Systems allow you to work with technology in designing and building new structures.

What Did We Learn?

- Agricultural education is comprised of eight career pathways: Agribusiness Systems, Animal Systems, Biotechnology Systems, Environmental Service Systems, Food Products and Processing Systems, Natural Resources Systems, Plant Systems, and Power, Structural, and Technical Systems.
- The Agribusiness Systems pathway prepares students to work in teams, using the newest technology to understand the best ways to be efficient while making a profit from agricultural products.
- The Animal Systems pathway prepares students to work for the care, health, and safety of animals of all sizes, and with the products they provide.
- The Biotechnology Systems pathway prepares students to enhance the growth and value of plants, animals, and microorganisms through genetic modification.
- The Environmental Service Systems pathway prepares students for the management of human activities and their related impacts on the environment.
- The Food Products and Processing Systems pathway prepares students to enhance the flavor and nutritional value of foods while also ensuring safe and efficient practices.
- The Natural Resources Systems pathway prepares students to protect and manage the natural environment in both rural and urban settings.
- The Plant Systems pathway prepares students for the study of plants and how they impact our world.
- The Power, Structural, and Technical Systems pathway prepares students to innovate, design, and build solutions for projects in electronics, power, hydraulics, and engineering.

Let's Check and See What We Know

Answer the following questions using the information provided in Lesson 1.2.

1. *True or False?* A water treatment specialist would prepare for their career by taking courses in the Food Products and Processing pathway.

2. Which career pathway includes careers in turf grass management, landscape design, and agronomy?
   A. Environmental Service Systems
   B. Food Products and Processing
   C. Animal Systems
   D. Plant Systems

3. Which career pathway's main focus is the health and welfare of small and large animals?
   A. Plant Systems
   B. Power, Structural, and Technical Systems
   C. Animal Systems
   D. Biotechnology Systems

4. *True or False?* People working in the field of agricultural engineering design and build projects utilizing a wide array of technology such as controls, hydraulics, and electronics.

5. *True or False?* Careers in Natural Resources Systems are solely focused on the management of urban environments.

6. Careers in biotechnology include all of the following, *except:*
   A. ecologist
   B. park ranger
   C. microbiologist
   D. genetic engineer

7. Which of the following career pathways may include the use of robotics, heavy machinery, or GPS technology?
   A. Environmental Service Systems
   B. Power, Structural, and Technical Systems
   C. Natural Resources Systems
   D. Animal Systems

8. A timber cruiser surveys and estimates the quantity of lumber in a given area. This career is most likely within which career pathway?
   A. Agribusiness Systems
   B. Food Products and Processing
   C. Power, Structural, and Technical Systems
   D. Natural Resources Systems

9. *True or False?* Plant Systems careers focus on the impacts of weather on agriculture, food, and natural resources and the biology of aquatic environments.

10. *True or False?* A veterinarian is one possible career in the Animal Systems pathway.

11. Which career pathway focuses on the biology of aquatic environments?
    A. Environmental Service Systems
    B. Power, Structural, and Technical Systems
    C. Animal Systems
    D. Biotechnology Systems

12. Which of the following career pathways would focus on the flavor and nutritional value of food?
    A. Environmental Service Systems
    B. Food Products and Processing
    C. Animal Systems
    D. Biotechnology Systems

13. Careers in the Natural Resources Systems career pathway include all of the following *except*:
    A. forester
    B. swine manager
    C. conservation scientist
    D. civil engineer

14. Which career pathway would employ food service technicians, refrigeration specialists, and food scientists?
    A. Plant Systems
    B. Power, Structural, and Technical Systems
    C. Animal Systems
    D. Food Products and Processing Systems

15. *True or False?* A meteorologist for the National Weather Service would be considered a possible career in the Power, Structural, and Technical Systems pathway.

Let's Apply What We Know

1. What types of careers are available in Plant Systems? Provide three examples.
2. List the possible areas of concentration for someone interested in Environmental Service Systems.
3. What duties might a veterinary assistant complete at a veterinary office?
4. Find the range of salaries for four career pathways that interest you.
5. Which career pathway includes genetic modification to enhance the growth and value of plants and animals? Why does it fit in this pathway?

Academic Activities

1. **Social Science.** With more than four million visitors each year, the continuous management and maintenance of Yosemite National Park is critical to preserving one of our nation's most precious tracts of land. Research the human impact that visitors have on our country's 419 national parks each year and describe three ways young people can help preserve these areas for future generations to enjoy.

Communicating about Agriculture

1. **Speech.** Research the most interesting career you learned about in this lesson and develop a two-minute video describing that career with at least four interesting facts, educational requirements, and a possible salary range.

SAE for ALL Opportunities

1. **Entrepreneurship SAE.** Many times, the best solutions are created out of great need. For example, earlier we learned of an invention Geno created after a visit to Ethiopia inspired him to help provide low-cost refrigeration for those in need. What inspires you to create new things? Might some of your ideas be solutions for everyday problems? Brainstorm ways you can take ideas and turn them into a small business opportunity.

Understanding Agricultural Education

Juice Verve/Shutterstock.com

1

Essential Question

How can you use your time in agriculture class to accomplish personal goals?

Learning Outcomes

By the end of this lesson, you should be able to:
- Explain the relevance of agricultural education. (LO 01.03-a)
- Describe the three-component model of agricultural education. (LO 01.03-b)
- Discuss the importance of classroom instruction, FFA, and SAEs. (LO 01.03-c)

Words to Know

classroom and laboratory instruction

hands-on learning

National FFA Organization

Smith-Hughes Act

Supervised Agricultural Experience (SAE)

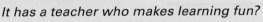

Before You Begin

Let's take a moment to think about your favorite class. What makes that class your favorite? Is the class your favorite because:

It has a teacher who makes learning fun?

It involves experiments or activities?

It involves plenty of field trips?

As you think about why you like your favorite class, I have another question: Would you enjoy a class that combines interactive learning and a virtual or class field trip? How about a class that allows you to apply your knowledge to real-world situations and gives you opportunities to get involved outside of the classroom?

If you said yes, well, guess what? You are in luck! The type of class that was just described does exist, and you are already enrolled. All you must do is take advantage of the opportunities your agricultural education teacher gives you. It's that simple! Are you ready to learn more about this unique and fun class?

Figure 1.3.1. Agricultural education can take place outside of the normal classroom setting. Agriculture teachers often use school gardens, greenhouses, livestock barns, and other locations to demonstrate the concepts students are learning in class. *Where would you like to take a class?*

BearFotos/Shutterstock.com

Did You Know? The Smith-Hughes Act expanded vocational education throughout the United States. A short time later (15–20 years), further efforts to provide career training intensified, as national interest in economic development grew during the Great Depression and World War II.

What Is Agricultural Education?

Agricultural education, as you learned back in **Lesson 1.1**, is a form of teaching that uses hands-on learning to help students learn about the global agriculture industry, which includes food, fiber, and natural resources systems. Agricultural education can also teach many other skills. Agricultural education will help you develop skills in science, math, communications, leadership, and technology, because each of these areas are part of agriculture. Students enrolled in agricultural education classes are provided opportunities for leadership development, personal growth, and career success. Agricultural education programs are typically offered in middle and high schools. Each program is unique based on its location and the students it serves (**Figure 1.3.1**).

The Beginnings of Agricultural Education

Formalized agricultural education began in 1917 when Congress passed the Smith-Hughes Act. The ***Smith-Hughes Act*** provides federal aid to states to fund high school courses in agriculture, industry, and consumer and family sciences. US Senator Hoke Smith and US Representative Dudley M. Hughes (**Figure 1.3.2**), both of Georgia, drafted and promoted the bill known as the National Vocational Education Act, which was later renamed the Smith-Hughes Act.

Figure 1.3.2. Hoke Smith (left) served as Governor of Georgia from 1907–1911, before serving in the US Senate. Dudley M. Hughes (right), before running for political office, was instrumental in bringing railroads to rural Georgia to help farmers sell and trade their crops. *How does their work on the Smith-Hughes Act affect your life today?*

Library of Congress

Library of Congress

The Smith-Hughes Act successfully expanded career education to high school students across the country. Today, more than 800,000 students participate in agricultural education classes throughout the United States and its territories.

Career Corner — Agricultural Education Teacher

If this is interesting to you, you might want to explore being an **agricultural education teacher**, *like me! Let's meet a real ag teacher and learn a bit more about what it takes to succeed.*

Cheralyn Keily is a middle school agricultural education teacher in Bonaire, Georgia. She has been teaching middle school for the last 15 years. She grew up in Wisconsin near her family's dairy farm. Mrs. Keily came to Georgia in college, loved the area, and decided to make it her new home. Mrs. Keily has enjoyed sharing the wide world of agriculture with a diverse student population that is spread across urban and suburban areas.

Her students share what they learn about agriculture, in and out of the classroom, with their local community through outreach and community service projects. Mrs. Keily and her students were recognized by the National FFA Organization for their advocacy efforts and were named the 2021 Outstanding Middle School FFA Chapter.

When asked what keeps her motivated to stay in this career field, Mrs. Keily says, "I enjoy getting to share something I am passionate about with students who share that passion. Not every student enrolled in my class will be passionate about agriculture, but every one of them will need agriculture. It is important for them to understand its importance and how it impacts their life. There are times that are challenging and sometimes even discouraging, as there will be in every job, but when I see a student find joy in what they are learning, and success in their endeavors, I am motivated to keep providing them with those opportunities through agricultural education and FFA."

Cheralyn Keily

If you might be interested in being an agricultural education teacher, here's a bit more information for you (and some questions to consider):

What Do I Need to Be Good at This? The best teachers have a wide variety of skills, which may (or may not) include organization, communication, curiosity, and energy.

What Education Do I Need? You need to have a bachelor's degree in agriculture or agricultural science, along with teacher certification, to become an agriculture teacher. Some teachers have additional credentials. Mrs. Keily's advice is to be sure to work hard in all your classes and seek help in subjects that are difficult for you. Teaching agricultural education will require a comprehensive knowledge of science, math, language arts, and social studies. Secondly, take any opportunity to gain hands-on experience in all different areas of agriculture.

The National Cooperative Extension Service

On May 8, 1914, Hoke Smith and Asbury Lever proposed the Smith-Lever Act, which founded the National Cooperative Extension Service. Asbury Lever was a US Representative from South Carolina; Hoke Smith, as you remember from earlier, was a Senator from Georgia. The two legislators developed the Smith-Lever Act to create a partnership between the United States Department of Agriculture (USDA) and state land-grant universities.

The Cooperative Extension Service provides a nationwide system of community-based training, educational events, and outreach to help farmers, homeowners, and youth use the latest university research to improve their lives. State land-grant universities employ several people to serve as cooperative extension agents. The local extension agents offer classes and training to help educate Americans about topics that are relevant to their local area. Often, these extension agents live and work within the town and community that they serve. The Cooperative Extension Service has been instrumental in helping increase American agricultural productivity.

In its early years, the Extension Service played a critical role helping farmers grow enough wheat and other crops to meet the needs of the nation during World War I. Today, the Extension Service continues to adapt as rural and urban communities and their needs change.

Cooperative extension agents offer classes to keep the local community informed of emerging trends in agriculture. Agents also help homeowners troubleshoot problems they are having with their lawns, gardens, landscaping, pests, and an array of other topics.

Consider This

1. Is there a land-grant university near you? Research land-grant universities and summarize what they are and how they were established.
2. With your teacher's permission, contact the extension agent who serves your community and invite them to speak to your class. Brainstorm with your classmates and write questions for the agent before they present to your class.

Alf Ribeiro/Shutterstock.com

The Three-Component Model

Agricultural education is shared with students in three interconnecting components (**Figure 1.3.3**):

- Classroom and laboratory instruction
- National FFA Organization
- Supervised Agricultural Experiences (SAEs)

Classroom and Laboratory Instruction

Classroom and laboratory instruction refers to learning that takes place in the classroom, but can also extend to instruction in a laboratory, greenhouse, or outdoor setting. Classroom and laboratory instruction for agricultural education is based on both natural and social sciences, including environmental science, natural resources, plant and animal science, food science, aquaculture, and agribusiness.

One great, unique thing about agriculture classrooms is that students have lots of opportunities to take in information and then apply it in hands-on, real-world experiences. ***Hands-on learning*** is instruction with activities that help you apply knowledge, such as interacting with materials, objects, equipment, machinery, or animals. For example, when learning how seeds develop, agriculture students plant seeds and watch them grow, or when learning how a chicken develops agriculture students candle eggs in a laboratory.

National FFA Organization

The ***National FFA Organization*** is a nationwide American student organization that is an integral part of agricultural education. FFA membership is open to middle and high school students in grades seven through twelve. FFA promotes the importance of agricultural education through realistic, hands-on learning opportunities, and competitions outside of the classroom.

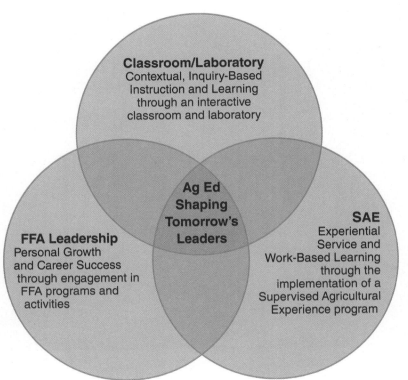

Figure 1.3.3. The three-component model shows that the best agricultural instruction takes place where the three different components intersect. *How do these components make for effective learning?*

Goodheart-Willcox Publisher

Figure 1.3.4. Many students participate in FFA Career Development Events (CDEs) and Leadership Development Events (LDEs). Students participating in these events demonstrate the knowledge they are learning in their agriculture class in a competitive environment. Students compete against other FFA chapters. The student pictured is competing in an Agricultural Technology and Mechanical Systems CDE. There are CDEs that are specific to high school and middle school students. If you have an interest in a particular area of agriculture, talk with your advisor about possible CDEs.

If this is your first time exploring agriculture, you might assume that FFA members are not very diverse, and you would be wrong! FFA members come from rural areas, small towns, and large cities across the United States. FFA helps students prepare for career success, no matter their background or career goals (see **Figure 1.3.4**).

Supervised Agricultural Experience (SAE)

Supervised Agricultural Experiences, or (SAEs), are specialized, personalized projects for agricultural education students, organized according to a given format. These are another way that agricultural education students learn by doing. SAE programs involve practical, hands-on agricultural activities that are completed by students outside of their regular class time (**Figure 1.3.5**).

All students enrolled in an agriculture course are expected to participate in an SAE. SAEs are designed to help students explore careers and develop organizational, financial, and employability skills.

The successful combination of these three components—classroom instruction, FFA, and SAEs—creates a strong agricultural education program. These programs produce well-rounded students who are prepared to be future leaders in the agriculture industry.

Figure 1.3.5. Student SAE projects can vary widely, depending on the interests of the student. Projects range from working with companion animals to creating computer software to help farmers to running a business. The possibilities are endless. *What might you enjoy doing?*

Hands-On Activity

Developing a Personal Three-Component Model

Create a personal mural explaining how the three-component model relates to your own life.

You will need the following materials:

- Large sheets of paper
- Markers

Here are your directions:

1. Draw three circles that intersect like the three-component model (see Figure 1.3.3).
2. Label the circles Classroom Instruction, FFA, and SAE.

3. List things in each ring that you are currently doing that relate to that specific area. If you aren't yet involved in FFA or an SAE, make note of two or three things you might be interested in exploring in each of those areas.
4. List two goals in each circle that you would like to achieve to help yourself grow.

Consider This

1. Share your personal model with two or three classmates. How are your goals different? How are they alike?
2. All students need support and encouragement to help them reach their goals. How can you support your classmates to help them reach their goals?

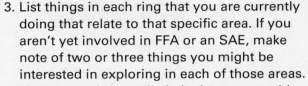

LESSON 1.3 Review and Assessment

What Did We Learn?

- Agricultural education teaches students about the global agriculture industry, including food, fiber, and natural resources systems.
- Agricultural education is shared with students in three interconnecting components: classroom and laboratory instruction, FFA participation, and Supervised Agricultural Experiences, or SAEs.
- Classroom and laboratory instruction consists of learning that takes place in a classroom or specialized setting, such as a laboratory, greenhouse, or outdoors. It consists of teacher-led curriculum related to many different natural and social sciences, as they pertain to agriculture.
- The National FFA Organization, or FFA, is a nationwide student organization that is an integral part of any agricultural education program. FFA emphasizes the importance of classroom lessons through realistic, hands-on opportunities and competitions outside of the classroom.
- Supervised agricultural experience, or SAE, programs involve practical, hands-on, self-directed agricultural activities that are completed by students outside of their regular class time.

Let's Check and See What We Know

Answer the following questions using the information provided in Lesson 1.3.

1. Agricultural education teaches you about global agriculture, as well as _____ system(s).
   - A. food
   - B. natural resources
   - C. fiber
   - D. All of these.

2. Students taking an agriculture class learn additional skills, such as _____.
   - A. communication
   - B. science and math
   - C. leadership
   - D. All of these.

3. The Smith-Hughes Act created _____ in the United States.
   - A. land-grant universities
   - B. agricultural education
   - C. FFA
   - D. extension offices

4. *True or False?* Agricultural education classes are offered outside of the 50 states.

5. *True or False?* There are fewer than 50,000 students involved in agriculture classes nationwide.

6. The three components of the agricultural education model are _____.
   - A. FFA, SAE, and classroom and laboratory instruction
   - B. leadership, personal growth, and career success
   - C. plants, animals, and food
   - D. shop, classroom, and greenhouse

7. FFA membership is open to students in grades _____.
   - A. 1–5
   - B. 6–7
   - C. 7–10
   - D. 7–12

8. *True or False?* Living on a farm is a requirement to be involved in FFA.

9. *True or False?* SAEs take place outside of classroom instruction time.

10. *True or False?* All students enrolled in an agriculture class are expected to have an SAE.

Let's Apply What We Know

Review the pictures below and answer the following questions.

kyslynskahal/Shutterstock.com

goodluz/Shutterstock.com

FFA Image - 128FAL14SUB039

1. How do each of the pictures relate to the three-component model of agricultural education?
2. Why is it important to have all three components together?
3. What do each of the pictures have in common?

Academic Activities

1. **Science.** When following the scientific method, there are three variables in an experiment: controlled, dependent, and independent. Controlled variables are items that remain constant. Which part of the agricultural education model would you relate to a controlled variable?

2. **Science.** Independent variables are items that are changed by the scientist to run an experiment. In which part of the agricultural education model do you think students have the most control over how much they grow and learn?

3. **Science.** There is a direct link between independent and dependent variables. The dependent variable is the effect that occurs from changing the independent variable. Which part of the agricultural education model do you think is most affected by the other two components?

Communicating about Agriculture

1. **Reading and Speaking.** First, choose an area of agriculture that interests you, such as horticulture, greenhouse technology, or farm machinery. After you decide on a subject, select a time period that you would like to research. Using at least three resources, research the history of your chosen subject during that era and write a short summary. Include major breakthroughs and the effects the breakthroughs had on the industry. Present your report to the class using visuals.

2. **Listening and Speaking.** In small groups, review the Words to Know listed at the beginning of this lesson. For each term, discuss the meaning of term and describe the term in simple, everyday language. Record your group's initial description, and then make suggestions to improve your description. Compare your descriptions with those of the other groups in a classroom discussion.

SAE for ALL Opportunities

1. **Foundational SAE.** Begin brainstorming ideas for your own SAE project. Think about things you are currently doing that really interest you. For example, if you raise small animals, such as rabbits, you could turn this hobby into an SAE. You could sell the rabbits and vermicompost made with the rabbit droppings. What topics in agriculture interest you? If you are stumped, browse the Table of Contents to come up with some topics you want to learn more about. Create a list of possible projects based on the topics on your list. Think about relatively simple projects that have the potential to grow over time.

The History, Mission, and Membership Benefits of the FFA

FFA

Learning Outcomes

By the end of this lesson, you should be able to:
- Summarize the history of the FFA. (LO 01.04-a)
- Describe the purpose and mission of the FFA. (LO 01.04-b)
- List the benefits of being an active FFA member. (LO 01.04-c)
- Identify opportunities available to middle school FFA members. (LO 01.04-d)

Words to Know

Career Development Event (CDE)
career success
creed

Discovery Degree
emblem
Leadership Development Event (LDE)

mission statement
motto
personal growth
premier leadership

Before You Begin

Middle school is a great time to get involved and learn more about yourself. Have you gotten involved in any organizations or clubs here at school already? Let's make a list of the organizations and clubs at this school as a class. Write what each of these school organizations do or represent. Have your classmates help you with clubs you are not familiar with, if necessary.

Now that we know what is out there, do any of these sound interesting? Place a star or mark next to the clubs or organizations that you are already a part of or want to learn more about.

The National FFA Organization, or FFA, is a leadership organization for middle and high school students who are taking agricultural education courses. FFA helps students develop leadership skills, provides recognition for their accomplishments, gives back to local communities, and is full of fun opportunities to boot! The opportunities FFA provides are almost endless. Your level of commitment, dedication, and willingness to step out of your comfort zone and try new things will determine how far you go with FFA.

In this lesson, we will explore the FFA's history and the benefits of becoming an FFA member.

The History of the FFA

FFA began in 1928 in Kansas City, Missouri, as the Future Farmers of America—an organization for young white high school men who were studying to become farmers. Leslie Applegate, a young man from New Jersey, was elected the first President of the Future Farmers of America. Over the years, many things have changed about the FFA, including the name and the types of students who were welcomed into its membership.

Around this same time, the New Farmers of America was organized in Tuskegee, Alabama. Founded in 1935 as an organization for young African American men, the NFA served black students interested in studying agriculture and farming in Southern states where schools were segregated. This organization served its members for 30 years, providing experiences in both agricultural education and leadership. It merged with the Future Farmers of America in 1965, and black men were allowed to join moving forward (**Figure 1.4.1**). At the time of the merger, the NFA had just over 1,000 chapters in 12 states and consisted of more than 50,000 members. In 1969, FFA membership opened to female agriculture students.

FFA

Figure 1.4.1. Fred McClure, from Texas, was the first black man to serve as a National FFA Officer. McClure, seen here with President Gerald Ford, was elected as National FFA Secretary in 1973. McClure also served as a special advisor to President George H.W. Bush.

In 1988, the Future Farmers of America changed its name to the National FFA Organization to represent the growing diversity of the agricultural industry. Also in 1988, seventh and eighth grade agriculture students were allowed to become FFA members.

When FFA was founded, annual membership dues were $0.10. In 1965, delegates voted to increase the membership dues to $0.50. As of 2022, National FFA dues are $7.00 per year, per student. FFA members may also have to pay additional membership fees for local and state dues. Membership dues are used for many different things, such as funding FFA events (contests and conferences) on the local, state, and national levels.

FFA Colors

The FFA colors were adopted in 1929, the year after FFA was founded. The National FFA colors are national blue and corn gold.

Did You Know? National blue and corn gold were selected to represent important values to the FFA. The blue was taken from the blue field of the American flag, as FFA is a national organization. The gold represents golden fields of ripened corn, which is one of the few crops grown in all 50 states. This represents the unity of the organization.

THE FFA CREED

I BELIEVE in the future of agriculture, with a faith born not of words but of deeds – achievements won by the present and past generations of agriculturalists; in the promise of better days through better ways, even as the better things we now enjoy have come to us from the struggles of former years.

I BELIEVE that to live and work on a good farm, or to be engaged in other agricultural pursuits, is pleasant as well as challenging; for I know the joys and discomforts of agricultural life and hold an inborn fondness for those associations which, even in hours of discouragement, I cannot deny.

I BELIEVE in leadership from ourselves and respect from others. I believe in my own ability to work efficiently and think clearly, with such knowledge and skill as I can secure, and in the ability of progressive agriculturalists to serve our own and the public interest in producing and marketing the product of our toil.

I BELIEVE in less dependence on begging and more power in bargaining; in the life abundant and enough honest wealth to help make it so – for others as well as myself; in less need for charity and more of it when needed; in being happy myself and playing square with those whose happiness depends on me.

I BELIEVE that American agriculture can and will hold true to the best traditions of our national life and that I can exert an influence in my home and community which will stand solid for my part in that inspiring task.

The creed was written by E.M. Tiffany, and adopted at the 3rd National Convention of the FFA. It was revised at the 38th Convention and the 63rd Convention.

FFA

Figure 1.4.2. The FFA Creed is a strong statement of the beliefs and values of the FFA organization. *What do you think of these values? What values would be in your personal creed?*

FFA Creed

The FFA Creed was written in 1930 by Erwin Milton (E.M.) Tiffany and adopted at the third National FFA Convention that same year. A *creed* is a statement of beliefs that guide the actions of the organization's members. The FFA Creed consists of five paragraphs, each beginning with the words "I Believe" (**Figure 1.4.2**). The FFA Creed notes the importance of agricultural life, hard work, leadership, determination, and citizenship.

FFA Emblem

An *emblem* is an image or series of images that serve as a symbol representing a nation, organization, or family. The FFA emblem is made up of six items that each represent a part of the FFA. The first FFA emblem dates to 1926. Henry Groseclose, who is often referred to as the Father of FFA, designed the emblem. The emblem has not changed much over the years, as you can see in **Figure 1.4.3**. The last update to the emblem occurred in 2015. The items in the emblem are:

- A cross section of an ear of corn, which serves as the foundation of the emblem and as a symbol of unity, as corn is grown in all 50 states
- The rising sun, which is a symbol of progress and a promise that tomorrow will bring a new day
- The plow, which is a symbol of labor and tillage of the soil
- The eagle, which is an American symbol that serves as a reminder of our freedom and heritage
- The owl, which is a symbol of wisdom and the knowledge required to be successful in the agriculture industry
- The words *Agricultural Education* and *FFA*, which indicate the combination of learning and leadership necessary for developing agriculture

Figure 1.4.3. The FFA emblem inspires pride, reflection, and motivation inside millions of past and present FFA members. The FFA emblem has stood the test of time, a symbol that goes back to the roots of the organization and still embodies the FFA's mission today.

FFA

FFA Official Dress

The first official dress, or uniform, was adopted in 1930. It consisted of a dark blue shirt, blue or white pants, a blue cap, and a yellow tie. FFA members are required to wear official dress at official FFA functions.

In 1933, the blue corduroy jacket was adopted as official dress. This is the same blue corduroy jacket worn today. The FFA emblem is displayed on both the front and back of the jacket. You can learn a lot about a member from their FFA jacket. The member's name, home state, and local FFA chapter are printed on the member's jacket. Members are also encouraged to wear up to three FFA pins on the front left chest of their jackets. Members receive pins for offices held, degrees earned, and awards received.

In 1973, the standards for FFA official dress were adopted. The standards stated that the FFA jacket should only be worn by FFA members. It also provided guidelines for proper use of the jacket. The standards were updated in 2002 and again in 2017, allowing girls to wear black dress pants instead of a skirt as official dress. Today, as you can see in **Figure 1.4.4**, official dress consists of:

- An official FFA jacket zipped to the top
- Black slacks and black socks/nylons or black skirt and black nylons
- A white, collared blouse or shirt
- An official FFA tie or scarf
- Black dress shoes with closed heel and toe

FFA Motto

A *motto* is a short sentence or phrase that captures the beliefs or ideals of an organization. The FFA motto describes the ways learning occurs through agricultural education and the value of serving those around you. The FFA motto serves as a guide for members to follow and strive for every day. The four lines of the motto are:

> *Learning to Do,*
> *Doing to Learn,*
> *Earning to Live,*
> *Living to Serve*

Did You Know?

Some FFA events require students to wear attire other than official dress. For example, the Meat Evaluation CDE requires students to wear lab coats and hard hats. The Veterinary Science CDE requires students to wear scrubs that would typically be worn in a veterinary office.

FFA

Figure 1.4.4. Students are required to wear official dress to FFA functions. *Is it ever acceptable for students to wear other clothing to an FFA function? If so, which function?*

FFA Opening and Closing Ceremonies

Official FFA ceremonies are a source of pride, identity, and tradition among FFA members. Opening and closing ceremonies emphasize the purpose of the meetings, the duties of the officers, and the significance of recognition given to individuals and guests. All rules and etiquette for official FFA ceremonies and meetings are to be memorized, rehearsed, and conducted with pride by the FFA members presiding over the meeting. Opening ceremonies should be conducted at the beginning of any official FFA meeting or convention. Meetings and conventions should conclude with closing ceremonies before everyone is dismissed.

FFA Pledge to the Flag

The Pledge of Allegiance to the flag of the United States is the official salute of the National FFA Organization. The Pledge of Allegiance is part of the closing ceremony at all FFA meetings and conventions (**Figure 1.4.5**).

The Mission of the National FFA

FFA helps students develop communication, social, and time management skills. It also helps students build character and promotes being an active member of your community through volunteering. FFA gives students opportunities to travel and to learn by doing, all while preparing students for future careers.

FFA Mission Statement

A *mission statement* is a summary of the aims and values of a group. The FFA mission statement was created by some of the organization's founding members. It explains what FFA strives to provide for each of its members. The FFA mission statement reads:

> *FFA makes a positive difference in the lives of students by developing their potential for premier leadership, personal growth, and career success through agricultural education.*

Figure 1.4.5. The 2020-21 National FFA Officer team salutes the flag during closing ceremonies at the 94th National FFA Convention.

FFA

Premier Leadership

Who comes to mind when you think of a great leader? A president, an athlete, a family member, your school principal, or maybe even your agricultural education teacher? What does it mean to be a premier leader? *Premier leadership* is the quality of inspiring, empowering, guiding, and teaching others using skills that are developed and improved over the course of your life. These skills promote self-awareness, focus, and accountability in all areas of life.

FFA is a great place to begin developing leadership skills. In FFA, students are exposed to organizational skills, teamwork, and time management, as well as self-discipline. FFA can provide opportunities for public speaking, communicating as a team, conducting formal business, using correct parliamentary procedure, and building a healthy competitive spirit.

Hands-On Activity

Developing a Personal Leadership Plan

To be successful, you must set goals for yourself. It is helpful to develop a personal leadership plan and reevaluate your goals every few years to make sure you are staying on track with your plans.

For this activity, you will need the following:

- Notebook or loose-leaf paper
- Pen or pencil
- Computer and printer (optional)

Use the following steps to write your personal leadership plan:

1. Make a list of three goals you hope to accomplish by the time you graduate from high school. These goals can range from small, such as earning a specific grade in a specific class, to larger goals, such as being accepted into a particular university, or serving as the next student body president.
2. Beneath each of your goals, write leadership traits that you have that will help you reach those goals. Being a good listener, hard worker, or dedicated student are examples of leadership traits.
3. Next, list additional skills that you will need to gain or improve over the next few years to help you reach your goals.

4. Review your list of current skills and things you need in the future to help you reach your goals. Make a list of potential roadblocks or things that might make achieving your goals difficult.
5. Next to each of your roadblocks, write something you can do to help yourself overcome it or avoid it all together.
6. Keep this paper and refer to it often, maybe at the beginning of every semester. You may want to post it where you see it every day. This will help you stay focused and dedicated your personal leadership plan.

Consider This

1. Having support is important when trying to achieve a goal. Make a list of people who could support and help you reach your three goals. People who can give you support include family members, teachers, friends, and coaches.
2. Talk to one or two of these individuals about the goals you have set for yourself. Set up a time to check in with them regarding your progress toward your goals. This could be monthly, quarterly, or annually.

National FFA Agriscience Fair

The scientific method can be applied to any science, including agricultural science. The National FFA Agriscience Fair recognizes students who gain hands-on skills in agricultural projects and is open to middle school students just like you. Participation begins at the local level and progresses to state and national levels. The participants must conduct a scientific research project pertaining to a part of the agricultural industry and present their findings to a panel of judges with a display and a report.

Agriscience Categories

- Animal systems
- Environmental services/Natural resources systems
- Food products and processing systems
- Plant systems
- Power, structural, and technical systems
- Social systems

Steps to think about before starting your own agriscience project:

1. Brainstorm an important agricultural issue, question, or principle you want to investigate.
2. Outline the specific scientific details you will be researching.
3. List the steps needed to complete the project and items you will need.
4. Research how you will use the scientific process to collect and analyze data.
5. Decide your level of commitment or the amount of time you can dedicate.
6. Talk to your agriculture teacher about your ideas and seek feedback to help you develop the best project you can.

FFA

Leadership Development Events

Being involved in your local FFA chapter can help you learn about yourself and increase your self-confidence. Members can participate in Leadership Development Events, or LDEs. *Leadership Development Events (LDEs)* are FFA events in which students use critical thinking skills, effective decision-making skills, and communication skills while celebrating fair competition and individual achievement (**Figure 1.4.6**). LDEs develop FFA members' self-confidence and contribute to their advancement in the FFA degree program, which is addressed later in this lesson. LDEs also equip FFA members with strong communication skills and give them an opportunity to advocate for the agriculture industry. National FFA LDE events that are offered each year to FFA members in middle school include Conduct of Chapter Meetings and the FFA Creed Speaking Event.

FFA

Figure 1.4.6. Students participating in FFA competitions, like reciting the FFA Creed from memory during the Creed Speaking LDE, gain confidence, learn from others, and build their résumés.

Career Corner REBECCA CARTER, National FFA Organization

Associate Director of Awards, Recognition, Grants, and Scholarship

*If you're interested in **FFA**, maybe you'd like to work with them! Let's talk to someone who has made it her mission to bring the benefits of FFA to students like you.*

Rebecca Carter believes in the FFA Motto: Learning to Do, Doing to Learn, Earning to Live, and Living to Serve. Mrs. Carter started working for the National FFA organization because she wanted to help students be the best versions of themselves through agricultural education. After teaching agriculture for 13 years, she now manages two teams that facilitate all of the national awards, grants, scholarships, and competitive events for FFA.

Mrs. Carter's main responsibilities include:

- Working closely with other managers as a part of the Programs and Events division;
- Leading strategic initiatives for the organization;
- Developing and managing sponsor/donor relationships; and
- Managing the revision process that ensures all FFA programs' alignment to industry standards.

Rebecca Carter

Mrs. Carter's advice to middle schoolers starting to explore career options is to "be a continuous learner and take advantage of experiences, as you never know where they may lead. This means: serve on committees, conduct a great SAE, participate in as many FFA activities and school activities as you can. Be curious about things and ask lots of questions to learn more."

If coordinating events like this sounds interesting to you, here are some things to consider:

What Do I Need to Be Good at This? An Associate Director would need to enjoy fast-paced work with new challenges each day. Strong organizational and communication skills will help prepare you for success!

What Education Do I Need? In order to work in this or a similar career, you will need to have a bachelor's degree in agriculture or agricultural science, along with background or experience organizing events.

Photographee.eu/Shutterstock.com

Figure 1.4.7. Everyone has the potential to achieve career success. With hard work, dedication, and motivation to achieve your goals, anything is possible. *What are some of your career goals? How do you think FFA can help you reach your career goals?*

Career Success

Talking about careers in middle school may seem odd. However, the skills you learn now will help you succeed in college and in your dream job one day! ***Career success*** is continuously demonstrating the qualities, attributes, and skills necessary to succeed in or prepare for a chosen profession (**Figure 1.4.7**).

Career Development Events

Being involved in FFA will introduce you to many career fields. The agricultural industry is an advanced technological industry, encompassing computers, plants, livestock, companion animals, forestry, woodworking, and many other areas. FFA has many Career Development Events (CDEs) that introduce you to these career fields. ***Career Development Events (CDEs)*** are events in which students demonstrate career-related knowledge and skills they have learned in the agriculture classroom. CDEs are competitive events that encourage teamwork and individual achievement. You can compete in CDEs on many topics, including agricultural sales, agronomy (the science of crop production), floriculture (flower cultivation), food science and technology, livestock evaluation, and veterinary science.

Personal Growth

FFA encourages students to embrace personal growth. ***Personal growth*** is changing and learning from continual effort to improve yourself intellectually, morally, and physically. FFA provides opportunities to strengthen your personal relationships and social skills. These opportunities include serving as a committee member, serving as a committee chair, or maybe even a chapter or state officer.

FFA also helps you build your character through volunteerism and community involvement. Members who serve on CDE and livestock show teams also learn the importance of teamwork and cooperation.

FFA Opportunities

Becoming active middle school FFA members gives students a sense of belonging, fun, and of being part of something bigger than themselves. It also gives students the chance to discover and explore current issues and career opportunities related to agriculture. The following programs are open to middle school FFA chapters and their members:

- Agriscience Fair
- Career and Leadership Development Events, such as Creed Speaking and Conduct of Chapter Meetings
- Discovery FFA Degree
- National Chapter Award Program
- Middle School Models of Excellence
- Partners in Active Learning Support (PALS) (**Figure 1.4.8**)
- SAE Grants through the National FFA Organization

MISTER DIN/Shutterstock.com

Figure 1.4.8. The FFA PALS program is a mentoring program designed to help FFA members build their interpersonal, human relations, and leadership skills while working with elementary students.

Discovery Degree

The FFA *Discovery Degree* recognizes students enrolled in seventh and eighth grade agricultural science classes who are making steps toward productive engagement in the FFA. To receive a Discovery Degree, members must meet the following requirements:

- Enroll in an agriculture class for at least a portion of a school year while in grades seven or eight
- Become a dues-paying FFA member at the chapter, state, and national levels
- Participate in at least one FFA chapter activity outside of scheduled class time
- Demonstrate an understanding of agriculture-related careers and entrepreneurship opportunities
- Become familiar with the local FFA chapter's Program of Activities (POA)
- Submit a written application for the degree

Once an FFA member reaches high school, they can begin to work toward their Greenhouse, Chapter, and American Degrees.

Did You Know? The American FFA Degree is one of the most prestigious awards bestowed by the National FFA Organization. It recognizes the effort FFA members apply toward their Supervised Agricultural Experience and the outstanding leadership abilities and community involvement they exhibit throughout their FFA career. Students typically earn their American FFA Degree within three years of graduating high school.

LESSON 1.4
Review and Assessment

What Did We Learn?

- FFA is a national leadership organization for middle and high school students who are taking agricultural education courses.
- The FFA was founded in 1928 as the Future Farmers of America, an organization for white men who were interested in studying agriculture. In 1935, the New Farmers of America was founded for African American men. The two organizations merged in 1965, and was retitled as the National FFA Organization in 1988. Women were able to join in 1969.
- The mission of the FFA is to make a positive difference in the lives of students by developing their potential for premier leadership, personal growth, and career success.
- Middle school students in FFA are able to participate in chapter meetings, the Agriscience Fair, and some CDEs and LDEs, like Creed Speaking.
- The Discovery Degree recognizes seventh and eighth grade members that are engaged in their local FFA chapter.

Let's Check and See What We Know

Answer the following questions using the information provided in Lesson 1.4.

1. The Future Farmers of America was founded in ____.
   - A. 1925
   - B. 1928
   - C. 1930
   - D. 1965

2. In 1969, ____ were allowed to join the FFA.
   - A. middle school students
   - B. African American students
   - C. female students
   - D. college students

3. When FFA began, membership dues were ____.
   - A. $0.10
   - B. $0.50
   - C. $5.00
   - D. $7.00

4. *True or False?* The official FFA colors are navy blue and corn gold.

5. The writer of the FFA Creed was ____.
   - A. Henry Groseclose
   - B. George Washington
   - C. E.M. Tiffany
   - D. Leslie Applegate

6. The FFA Creed is made up of ____ paragraphs, each beginning with "I Believe."
   - A. 2
   - B. 3
   - C. 4
   - D. 5

7. Henry Groseclose helped to create the FFA ____.
   - A. Creed
   - B. emblem
   - C. colors
   - D. salute

8. The FFA jacket was adopted as official dress in ____.
   - A. 1928
   - B. 1933
   - C. 1965
   - D. 1969

9. How many pins is an FFA member allowed to wear on the front of their jacket?
   - A. 1
   - B. 2
   - C. 3
   - D. 4

10. *True or False?* The FFA Discovery Degree can only be received by middle school FFA members.

Let's Apply What We Know

1. Memorize the FFA Creed and present it to your class.
2. Make a list of personal goals you would like to achieve as an FFA member.
3. Prepare a one- to two-minute speech about the meaning of premier leadership. Record yourself giving the speech or recite the speech to your class.
4. Create an FFA time line highlighting major FFA events that took place in each decade since its founding. Supplement what you learned in this chapter with your own research. Hang your time line in the classroom, hallway, or on a bulletin board to share with your classmates.
5. Create your own personal motto using only 12 words, like the FFA motto.
6. Talk with your school's FFA Advisor and make a list of all the CDEs and LDEs that your local middle and high school FFA chapter compete in every year, or research more about CDEs and LDEs on the National FFA website (www.ffa.org). Challenge yourself to compete in at least one FFA event this year!

Academic Activities

1. **Language Arts.** Create an informational pamphlet about the FFA. Research FFA history and opportunities for premier leadership, personal growth, and career success. Make sure to include images in your pamphlet. Present your finished design to the class.
2. **Communications.** Find videos of people giving speeches on YouTube. Make notes about the speaker's voice, volume, tone, gestures, and expressions. Identify effective and distracting gestures, patterns of speech, and expressions. Keep these in mind as you write a speech that follows the prepared public speaking guidelines for the National FFA Organization. Choose a topic that interests you and make sure your speech has a well-defined outline. Using a computer, cell phone, or other recording device, record yourself giving your speech. Review your recordings and note places where you need improvement.

Communicating about Agriculture

1. **Writing and Speaking.** Interview a former FFA member or high school member. Write a short essay on the impact FFA had or continues to have on their life. Make sure to include their favorite FFA activities and any leadership roles they enjoyed. If time permits and the interviewee agrees, record and edit a video of the interview and present it at an FFA chapter meeting.
2. **Listening and Writing.** In a small group, develop personal mission statements. Each person should share two to four goals. Write down these goals and help each other develop mission statements using them as a starting point. Print your mission statement and hang it where you will see it each day.

SAE for ALL Opportunities

1. **Foundational SAE.** Explore careers related to an agricultural field of your choice. Develop a career plan of experiences and education to prepare someone for a successful career in services related to your chosen subject. Include areas such as education requirements, job outlook, general responsibilities, traveling requirements, and salary range.
2. **Foundational SAE.** Select two career areas of interest and interview people working in those careers. Prepare a list of questions before you hold the interviews. If it is not possible to perform the interview in person, set up a video interview that you can record and show to your classmates.

The Supervised Agricultural Experience Program

Belish/Shutterstock.com

Learning Outcomes

By the end of this lesson, you should be able to:
- Explain the purpose of a Supervised Agricultural Experience (SAE) program. (LO 01.05-a)
- List the benefits of having an SAE program. (LO 01.05-b)
- Describe the types of SAE programs. (LO 01.05-c)
- Explain how a student would go about designing an SAE program. (LO 01.05-d)
- Describe why it is important to maintain current, thorough SAE records. (LO 01.05-e)

Words to Know

Foundational SAE

Ownership SAE
  (also called an
  Entrepreneurship SAE)

Placement SAE
  (also called an
  Internship SAE)

recordkeeping

Research SAE

School-Based Enterprise
  SAE

Service Learning SAE

Before You Begin

Do you have any hobbies? Write down three things you enjoy doing in your spare time. I'll give you 60 seconds. Ready, set, go!

Now, who has their list and wants to share with the class? I enjoy traveling, gardening, and, okay, this one is kind of funny (and a little embarrassing as a teacher), but playing apps and games on my phone, especially those farming games. They are the best!

Have you ever thought about how you could take your hobbies and kick them up a notch? Not sure what I mean? Well, let's dive into this lesson and learn how you can take your favorite hobby and grow it into something that can build your leadership and soft skills (and maybe even earn some money). Who doesn't like that?

Supervised agricultural experiences (SAEs) take lessons that you have learned in your agriculture classes and expand them into hands-on lessons that take place outside of the classroom (**Figure 1.5.1**). The great thing about SAE programs is they can be tailored to fit the individual student. Each SAE program is unique. An SAE may be growing plants in the school greenhouse, raising livestock on a family farm, working at a local feed store, doing observation hours in a stockyard, or conducting research on reindeer in Alaska. No matter where you live or what you enjoy, there is an SAE for you!

What Is a Supervised Agricultural Experience Program, and Why Do You Need One?

Supervised Agricultural Experiences (SAEs) are student-led, instructor-supervised, work-based learning experiences that result in measurable outcomes. Agricultural education teachers provide supervision and guidance for the students' programs (**Figure 1.5.2**). Parents and employers are often also included. While it is not necessary that an SAE take place on a farm, ranch, or other agricultural setting, the experience should connect the agricultural experience with classroom instruction and a student's interests. What is important is that SAE programs take place in a real-world environment or simulated workplace environment outside of the classroom, such as a virtual program, school lab, or greenhouse. SAEs may even occur on the school campus, during the school day, outside of class time.

SAEs are designed to grow and change with you as you develop your skills and identify new interests. The end goal of an SAE is to prepare you for a productive career, no matter what path you take to get there. Why not learn the skills you want while also having meaningful experiences (and possibly earning money) along the way?

Everett Collection/Shutterstock.com

Figure 1.5.1. Successful SAE projects involve activities students are already doing at home. What are some activities that you already do that might be relevant?

Jacob Lund/Shutterstock.com

Figure 1.5.2. One of the best things about an SAE is the *supervised* part. Your agricultural education teacher is here to help. Ask questions and seek their advice as you make decisions about your SAE program.

Benefits of SAE Participation

You might be wondering how an SAE program is going to benefit you, outside of getting a good grade in your agriculture course. The most important benefits are the things you learn and experiences you have. You can take the things you learn in your agriculture class and apply them in real-world situations.

Some other benefits to having a successful SAE program, though, are:
- Exploring different agriculture careers
- Learning about things that interest you
- Gaining work experience
- Developing employability and soft skills
- Honing decision-making skills
- Raising your self-confidence and pride
- Practicing responsibility and recordkeeping
- Learning time management skills
- Developing independence

Types of Supervised Agricultural Experience Programs

Foundational SAE

There are many types of SAE programs. When you are a new agricultural education student, National FFA recommends that you begin with a Foundational SAE. *Foundational SAEs* are short and introduce students to the skills needed to expand or start a more in-depth SAE that they can carry through their middle and high school careers.

In Foundational SAEs, students research and explore career opportunities in the agriculture, food, and natural resources industries. They learn skills needed to succeed in college and in their careers, including responsibility, communication, innovation, critical thinking, and collaboration.

Students also learn how personal financial practices like budgeting, saving, and appropriate use of credit lead to financial independence. The goal of a Foundational SAE is to give students an understanding of the breadth of the agricultural industry and to consider the role that agriculture plays in society, the environment, and the economy.

Once you have learned the fundamentals of an SAE program, you can expand your SAE into one on the five types of Immersion SAE programs:
- Placement (or *Internship*) SAE
- Ownership (or *Entrepreneurship*) SAE
- Research SAE
- School-Based Enterprise SAE
- Service Learning SAE

ALPA PROD/Shutterstock.com

Figure 1.5.3. Placement SAE programs are one of the most common types of SAE because students can work or volunteer in any agriculture-related business. The student here is working at a local flower shop. *Can you think of any agricultural businesses in your community that would make a good SAE program?*

Placement SAE

In a *Placement or Internship SAE*, students can gain knowledge and understanding in a chosen field as paid employees or unpaid volunteers (**Figure 1.5.3**). Students complete the duties assigned by the employer. Students are evaluated by the employer under the guidance of the agriculture instructor.

There is usually no investment of money with these types of SAEs, only the time spent working. Students keep records of the hours they work, skills they learn, and any money they earn.

Did You Know? Internships can be paid or unpaid opportunities for students. If you are uncertain about your interest in a certain area, you can volunteer in that career field until you gain a better understanding. Internships can be a great résumé builder while you are in school. You can work for a temporary period (such as summer, when school is out) to give you some work experience before graduation.

Ownership SAE

If you want to be your own boss, an Ownership SAE might be a good fit. In an ***Ownership (or Entrepreneurship) SAE***, students create, own, and operate a business that provides goods and/or services to the marketplace (**Figure 1.5.4**). There is typically a financial investment, in addition to time spent, with this type of SAE. Students keep records of the time spent on the project, the skills learned, the money spent, and the money earned.

RaksyBH/Shutterstock.com

Figure 1.5.4. Exhibiting livestock is a great Ownership SAE program. Students own the animals and are responsible for the costs that go along with raising an animal, such as feed, treatments, and show fees. *Do you or any of your friends exhibit livestock through your local FFA or 4-H program?*

Making Connections #STEM

Calculating Salaries

Research information on the average hourly pay in your area for three agriculture-related occupations. Assume that each occupation provides work for 40 hours a week and gives their employees one week of paid vacation each year.

1. Calculate what a person in each of these occupations would earn (before taxes) in:
   A. one week.
   B. one month.
   C. one year.
2. If every employee earned a three percent pay raise after every year with the company, how much money would they earn in _____?
   A. two years
   B. three years
   C. five years
3. Create a table or graph to show your results.

Consider This

1. Looking over the rate of pay for each of the careers you selected, can you draw any conclusions from the results?
2. Do the results change your views on your ideal career?

Dean Drobot/Shutterstock.com

Research SAE

In a **Research SAE**, students choose a research question and work through the scientific method to learn new information or find information that supports existing research (**Figure 1.5.5**). Agriculture teachers can help students choose a reasonable topic for this type of SAE. Research SAEs usually require very little financial investment, but do require an investment of time. Students should keep records of the time spent on the research, as well as the procedures, data collected, and conclusions from completing the program.

KOBE611/Shutterstock.com

Figure 1.5.5. Students choosing a Research SAE conduct experiments on agricultural products using the scientific method. A Research SAE can continue through high school by broadening your research and modifying your hypothesis. *What questions might you be interested in answering?*

Did You Know? You can take the research you are doing with an SAE and turn it into an FFA Agriscience Fair project. Middle and high school students can compete at the state and national levels in a wide variety of topic areas related to agriculture. Ask your teacher, or search "Agriscience Fair" on the National FFA website (www.ffa.org).

School-Based Enterprise SAE

A **School-Based Enterprise SAE** might be the best fit for students who enjoy working in partnership with other students, and who would benefit from the school's or FFA chapter's resources. In a School-Based Enterprise SAE, students lead business enterprises that provide goods or services. These types of SAE programs use school facilities, equipment, and other resources provided by the agricultural education department (**Figure 1.5.6**). These SAEs can include things such as managing and selling plants from the school greenhouse, or creating and selling metal art projects from the school welding shop. School-Based Enterprise SAEs do not require the student to invest money. However, any profit made from selling the products goes back to the school or FFA chapter. Students keep records of the hours they work, skills they learn, and items that they created and sold.

Figure 1.5.6. School-Based Enterprises are a great way to give back to your school and FFA chapter. Students spend hours outside of normal class time working in areas such as the agriculture lab or school greenhouse. *How can you give back and help your agriculture program?*

Gorodenkoff/Shutterstock.com

Service Learning SAE

Students interested in volunteering in their community might enjoy a Service Learning SAE. In a *Service Learning SAE*, students (individually or as a small team) plan, conduct, and evaluate a project that is designed to provide a service to the community (**Figure 1.5.7**). The project must benefit an organization, a group, or individuals other than the FFA chapter. These types of SAEs may vary in terms of financial requirements, depending on the project. Sometimes students can get materials donated, and other times there might some expenses. Students keep records of hours worked and the skills they learn, as money is not earned with service learning projects.

Dmytro Zinkevych/Shutterstock.com

Figure 1.5.7. Service Learning SAE programs are a true passion for many students. Do you enjoy volunteering and helping others? Service Learning programs can be big or small, ranging from running a school canned food drive to starting a coat drive, or even picking up trash. *What needs do you see around your community?*

Developing an SAE Program

Now that you understand the types of SAE programs that are available, you can start thinking about the best SAE for you. One of the greatest benefits of an SAE is that it can be tailored to fit your needs. The best type of SAE is one that matches your interests, time lines, and resources, so that you can successfully complete it.

SAE Interests

One of the best things about an SAE is that you have complete control over what you want to learn and how you conduct your SAE program. Think about your personal interests and see if a program comes to mind from one of the following agriculture categories (**Figure 1.5.8**):

- Animal Systems: Includes programs working with all types of animals, from livestock to exotic to companion animals
- Agribusiness: Includes programs that work with the financial side of agriculture, ranging from lending to accounting to advertising
- Leadership, Education, and Communication: Includes programs that work with producers and/or consumers to help develop an understanding about agriculture
- Natural Resources Systems: Includes programs that work to preserve or reclaim our natural resources, including minerals, water, soil, air, and wildlife
- Food Products and Processing: Includes programs that work with the processing or regulating of our food supply
- Power, Structural, and Technical Systems: Includes programs that deal with the design, fabrication, and maintenance of agriculture equipment or structures
- Plant Systems: Includes programs that include growing plants for food or ornamental use
- Biotechnology: Includes programs that enhance the natural world through technological advancements or engineering

SAE Ideas by Category

Sector	Placement/ Internship	Ownership/ Entrepreneurship	Research	School-Based Enterprise	Service Learning
Agribusiness	Intern with a commodity broker	Own a business managing records for local agricultural companies	Analyze consumer trends in spending at the local farmer's market	Sponsor a farmer's market at your school	Raise funds to help purchase crop insurance for producers in need
Animal	Work at a local veterinary office	Raise and sell puppies	Conduct a feed trial to determine palatability	Have a school dog grooming service	Organize a supply drive for a local pet shelter
Biotechnology	Work for a biotechnology laboratory	Produce animals using embryo transfer	Examine yields for GMO vs. non-GMO crops	Grow GMO crop trials on school land	Present information about the use of biotechnology to local elementary students
Education and Communication	Work for a local agricultural TV or radio station	Start a livestock photography company	Experiment with different instructional methods to teach agricultural concepts	Develop flyers and marketing materials for the school greenhouse	Host a local agricultural experience night for at-risk youth
Environmental Service	Obtain an internship with a local water testing company	Own a composting business	Test different methods for preventing erosion	Install and run a biodigester at your school	Coordinate a local highway cleanup project
Food Products and Processing	Find a position with a local meat processor	Start a company that makes and sells jams and jellies	Analyze different methods for preserving food products	Create and sell a product from produce grown in a school garden	Organize a food drive and supply food to the local food pantry
Natural Resources	Get a summer internship with the US Forest Service	Cut and sell firewood or wooden fence posts	Conduct an analysis of local natural resources use	Raise fish at your school with the state fish and wildlife service to stock local lakes and streams	Coordinate a waterway cleanup day
Plants	Find a position at a local nursery/ landscape company	Raise and sell produce at the local farmer's market	Determine the most appropriate amounts of fertilizer for a specific crop	Work in the school greenhouse or garden area	Help plant flowers in your local downtown area
Power, Structural, and Technical	Intern in the repair shop for an agricultural machinery company	Build and sell custom welding projects	Study the effects of different lubrication fluids	Help build projects in the local ag shop	Use your knowledge of structural systems to build and install handicapped accessible rails on a local building

Goodheart-Willcox Publisher

Figure 1.5.8. Use this chart to spark some ideas for your own SAE program. *Do you see an SAE example in the table that you find interesting?*

Selecting Your SAE

Before you select your SAE program, consider the following questions:
- What agricultural area(s) interest you?
- How much time do you have to invest in this program?
- Will you need money or a sponsor to complete your SAE program?
- What kinds of equipment and facilities will you need?

Once you determine your area of interest and the resources that are available, it is time to decide on your SAE program. Use the decision tree in **Figure 1.5.9** to help narrow your options. Select an SAE program that interests you and that fits into your schedule and financial situation. Next, you should finalize the specific details of your project. From there, all you need is a personal commitment, willingness to learn new skills, and to be open to new experiences.

SAE Recordkeeping

An essential part of having an SAE is recordkeeping. **_Recordkeeping_** is creating the written history of your activities by collecting and entering data. You must keep accurate records of your project in a timely manner, so you do not forget things that happened with your SAE program. Learning recordkeeping skills at an early age helps you in many areas of your life and career. Can you imagine how this might be a useful skill?

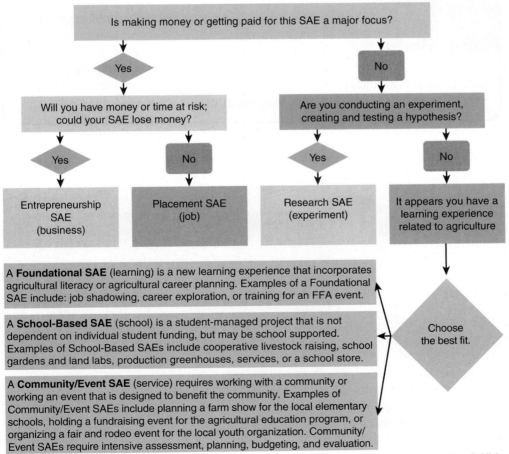

Figure 1.5.9. This decision tree is designed to help narrow your SAE options. To use the decision tree, begin by answering the question at the top and then follow the arrows to questions based on your answers. *What were your results?*

Is making money or getting paid for this SAE a major focus?

Yes → Will you have money or time at risk; could your SAE lose money?
- Yes → Entrepreneurship SAE (business)
- No → Placement SAE (job)

No → Are you conducting an experiment, creating and testing a hypothesis?
- Yes → Research SAE (experiment)
- No → It appears you have a learning experience related to agriculture → Choose the best fit.

A **Foundational SAE** (learning) is a new learning experience that incorporates agricultural literacy or agricultural career planning. Examples of a Foundational SAE include: job shadowing, career exploration, or training for an FFA event.

A **School-Based SAE** (school) is a student-managed project that is not dependent on individual student funding, but may be school supported. Examples of School-Based SAEs include cooperative livestock raising, school gardens and land labs, production greenhouses, services, or a school store.

A **Community/Event SAE** (service) requires working with a community or working an event that is designed to benefit the community. Examples of Community/Event SAEs include planning a farm show for the local elementary schools, holding a fundraising event for the agricultural education program, or organizing a fair and rodeo event for the local youth organization. Community/Event SAEs require intensive assessment, planning, budgeting, and evaluation.

Goodheart-Willcox Publisher

Figure 1.5.10. Recordkeeping is one of the most important parts of your SAE program. Keeping accurate records helps you complete FFA award applications when you continue your SAE program in high school. There are various methods for recordkeeping, including digital records. Talk with your agriculture teacher about which method they prefer that you use.

Keeping accurate records helps you measure the success of your SAE. Many schools or states have specific requirements or record books that agriculture students use to keep records on their SAE programs. Your instructor can help you determine the best method of keeping records for your program (**Figure 1.5.10**).

You need to keep track of the time spent and experiences you have with your SAE program. Keep track of the skills that you learn, such as operating a new power tool, using a new computer software program, learning to ear notch pigs, or propagating a new plant. Records of the money spent or earned with your SAE program also need to be created. Your records provide the whole picture of your SAE program. If someone picks up your record book, they should be able to tell exactly what you did, how you did it, and what you learned. Keeping accurate records also helps with FFA awards when you are in high school.

Did You Know? Recordkeeping may be one of the hardest parts of an SAE program, but it is also one of the most important. Many students use apps on their phone or tablet, such as AET, to keep online records that can be updated quickly. This makes recordkeeping convenient and allows you to input hours and expenses on the spot.

Hands-On Activity

Creating an SAE Time Line

Developing a successful SAE can be very challenging for a beginning agriculture student. The program can seem overwhelming, and you might not be sure where to start. Have you ever heard the question, "How do you eat an elephant?" One bite at a time! You probably get the idea. Breaking your SAE into weekly assignments will help you stay on task and not get behind.

Things you will need:
- SAE program due date, given to you by your agriculture teacher
- Monthly calendar or planner
- Pencil or pen
- SAE notebook or record book

Continued

Creating an SAE Time Line *Continued*

Let's break your SAE program down into manageable pieces.

- **Phase 1: These items should be done in the first few weeks of starting your SAE.**
- Select a date to have your project topic. This should be within the next two weeks.
- Make a list of aspects of the agricultural world that interest you or that you want to learn more about. Small animal care, horseback riding, and woodworking are examples. Make a list of objects or places you have access to or own, such as a lawn mower, family store, or vegetable garden. Read through the SAE for All Profiles located in this book.
- Create SAE program goals. Goals are personal; there is no right or wrong answer. What are three things you hope to gain from completing this program? Maybe you want to earn a certain amount of money or learn how to complete a specific task. Your goals are specific to your SAE.
- Complete an inventory list of what you need to get started or what items you will use with your SAE program. Make note of who owns the items you will be using. Do you own them, or are they owned by family members or friends? You might also want to note the value of each inventory item.
- Create a plan of action. How will you get started? Who do you need to contact to start? Whose help will you need, and when will you notify them about the SAE program you are starting? For example: If I want to volunteer at the local vet office in my town, I will need to talk to my family members about my plans and gain permission from the office to help. Or if I want to start cutting lawns around my neighborhood, and I do not own my own mower, I will need to find a lawn mower I can borrow, determine how I will get it from yard to yard, and create a plan to gain customers.
- **Phase 2: You will complete these items throughout your SAE.**
- Logging SAE Activity. Select a day of the week to log your SAE activities and collect pictures of your SAE program. As a busy student, sometimes it is hard to log your activities each day, but waiting until the SAE is due to log your experiences will lead to incomplete records of hours, expenses, and accomplishments. At least one day each week, you will need to log the days you worked, what you accomplished, and how many hours you spent. You will also need to collect photos of you working on your SAE. Is there anything else you think might be important to log?
- Create a detailed income and expenses log. Keep a record of any money you spend or earn while working on your SAE. Make sure to hold onto receipts and other documents that will help you when it is time to turn in your SAE.
- Take photographs. Pictures are a great way to document the work you do for your SAE. Action shots that show you completing the task are best. Make sure you are wearing proper safety equipment if it is needed.
- **Phase 3: You're almost done! Complete your record book or class project.**
- Skills Learned. Reflect back on your SAE. What are the three most important skills you learned over the course of your project? Give a brief description of each skill and how you acquired it.
- Overall Summary. Write up an overall SAE summary, describing what you learned, how your SAE experience will impact your future, and how you would improve your SAE if you were to start it over. Include any other details you think are important about your SAE program.
- Log/Record book. Look over your class SAE record book if you have one. Update any information that is missing and record any hours that have not been logged.
- Include Pictures. Make sure to include pictures and add captions to your photos that explain the importance of what you are doing in the photo.

Consider This

1. Research different proficiency awards that are available through the National FFA Organization (https://www.ffa.org/participate/awards/proficiencies/). Read the description for each of the category areas.
2. Did any of these categories spark an interest or idea about your SAE program? List the proficiency area that you think best fits your personal SAE program idea.

What Did We Learn?

- Supervised Agricultural Experiences (SAEs) are student-led, instructor-supervised, work-based learning experiences that result in measurable outcomes.
- The purpose of a Supervised Agricultural Experience program is to create career-ready students, regardless of their intended career path.
- All students in an agriculture education course must have an SAE.
- There are many benefits to having an SAE program, including exploring different agricultural careers and gaining work experience.
- Foundational SAEs allow beginning agriculture students a chance to familiarize themselves with agriculture. These are typically short, introductory projects.
- There are five types of Immersion SAE programs: Placement, Ownership, Research, School-Based Enterprise, and Service Learning.
- Students should select their SAE program based on their own personal interest and their available resources.
- Recordkeeping is a vital part of an SAE program. Students should record things, such as time invested, money earned or spent, and skills learned, in a timely manner.

Let's Check and See What We Know

Answer the following questions using the information provided in Lesson 1.5.

1. *True or False?* Your SAE project idea is assigned to you by your agriculture teacher; you have no input in the type of project you complete.

2. _____ SAEs are short and introduce students to the skills needed to expand or start a more in-depth SAE.
   A. Placement
   B. Ownership
   C. Foundational
   D. School-Based

3. *True or False?* All SAE programs must be related to the agriculture industry in some way.

4. Students who own their programs and are responsible for all management and financial decisions are taking part in a(n) _____ SAE program.
   A. Placement
   B. Ownership
   C. Service Learning
   D. School-Based

5. A student who uses school facilities and/or resources to complete an SAE is participating in a(n) _____ SAE program.
   A. Placement
   B. Ownership
   C. Service Learning
   D. School-Based

6. A student using their SAE to give back to their community through volunteering or providing a service is participating in a(n) _____ SAE program.
   A. Placement
   B. Ownership
   C. Service Learning
   D. School-Based

7. *True or False?* Considering your own personal interests and hobbies is *not* typically a part of SAE programs.

8. *True or False?* Before selecting your SAE program, you should consider the amount of time and money you have to devote to this program.

9. The written history of one's activities through collecting and entering data is known as _____.
   A. gaining work experiences
   B. recordkeeping
   C. balancing
   D. volunteering

10. Which of the following needs to be recorded in your SAE records?
   A. Skills learned
   B. Money earned or spent
   C. Time invested
   D. All of these.

Let's Apply What You Know

1. Think about your dream career. Explain how the skills you learn from your SAE program can help you with your future career plans.
2. Why do you think it is important to keep accurate records? List five things you think should be tracked with your SAE program.
3. How do Foundational SAEs differ from the other five categories of SAE programs? Identify the type(s) you would most likely use for your SAE program, and explain why you would or would not use each type.

Academic Activities

1. **Language Arts.** Create an informational flyer about SAE programs. Use the flyer to encourage other agriculture students to have outstanding SAE programs. Highlight different types of SAE programs and opportunities available to agriculture students. Make sure you include images in your flyer. Present your finished design to the class.

Communicating about Agriculture

1. **Speaking and Writing.** Interview an older student who has completed at least one year of an SAE program. Ask the student what makes their SAE successful, the types of challenges they have faced, and what they would do differently if they did it over. If time permits, discuss your SAE and ask for advice on how you can ensure it is successful. Write a short summary on what you learned.
2. **Listening and Speaking.** Using the SAE video library on the National FFA website, watch five student SAE stories. Working in a small group, use the FFA videos as inspiration to create videos of your own SAE projects. If you do not yet have an SAE, think about what you would like to do and use your inspiration for your video. Prepare your videos for presentation to your class. Think about using this as an ongoing activity in which you continue to record your activities throughout the course of your SAE program(s).

SAE for ALL Opportunities

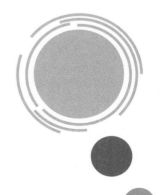

1. **Foundational SAE.** Identify an agriculture career that interests you. Spend the day with a career mentor to learn more about what their job responsibilities are each day. Write up a summary of your experience, making sure to include the pros and cons of having an SAE project in this career field.
2. **Recordkeeping.** Getting in the habit of recording data can be difficult, but recordkeeping is important to any SAE project. Start developing your skills by recording what you ate and drank for a day. Record time, amount, and any other details you think are important. Share your food log with a friend. Find out what questions they have about the data you recorded to help you fill in the gaps for future recordkeeping.

Proper Procedure for Conducting Meetings

Billion Photos/Shutterstock.com

Essential Question

How can using parliamentary procedure keep formal meetings on task and running smoothly?

Learning Outcomes

By the end of this lesson, you should be able to:
- Explain the purpose of running meetings using correct parliamentary procedure. (LO 01.06-a)
- Describe the tasks associated with running a meeting properly. (LO 01.06-b)
- Identify the types of motions that may be used in a meeting, and when it is appropriate to use them. (LO 01.06-c)
- Demonstrate the ability to make motions, discuss, and vote on topics in a formal meeting setting. (LO 01.06-d)

Words to Know

agenda	main motion	privileged motion
amendment	minutes	quorum
chair	motion	second
gavel	order of business	subsidiary motion
incidental motion	parliamentary procedure	

Before You Begin

Have you ever been in a group and had a difficult time making a decision? Maybe it was deciding where the group should go to eat or whose house the group should visit. Maybe it was deciding who should be a team captain, or what movie to see. Maybe there was a time you wanted to do something different from everyone else in the group, and didn't feel like your voice was heard. Making decisions in a group is difficult!

How did you come to a decision that was fair for everyone? Did you play Rock, Paper, Scissors? Did you arm wrestle? Did you break out into a dance-off? I doubt you did any of those things (mostly because they're a bit silly). Did you do something that worked better? Share with a partner some of your best ideas for coming to a fair decision in a group.

1

A ny time you are part of a large group, there will be different opinions and ideas about what is best for the group. This is why many formal meetings are conducted with a set of rules and guidelines, called *Robert's Rules of Order*. *Robert's Rules of Order* is a handbook for running meetings successfully and professionally. The principles included in the handbook are appropriate for any decision-making organization (**Figure 1.6.1**). The set of guidelines outlined in *Robert's Rules of Order* is known as ***parliamentary procedure***. Using parliamentary procedure is an important part of participation in FFA. This lesson will focus on why parliamentary procedure is important, the correct way to use it, and what it means to make a motion.

Purpose of Parliamentary Procedure

The main purpose of parliamentary procedure is to conduct meetings effectively. There are four main rules that ensure meetings run smoothly. They are:

- Handle one item at a time.
- Extend courtesy to everyone.
- Ensure that the rights of the smaller population are protected.
- Ensure that decisions are made with the benefit of a large group in mind.

mark reinstein/Shutterstock.com

Figure 1.6.1. The US Congress, state, and local governments use the rules of parliamentary procedure to conduct business and vote on important issues. *Why might it be important to understand how this works?*

Parliamentary procedure is used in more than 200 countries to run meetings. These include the British Parliament, the Union of South Africa, and the US Senate. It is also used by local and state governments, at civic meetings, and in school organizations, including FFA.

What does this have to do with you? Taking the time to learn parliamentary procedure will allow you to participate in and lead formal meetings, in any organization, in an effective way. This is an opportunity to improve your personal leadership skills and make you an asset to any professional organization.

Conducting a Meeting

Before a formal meeting can begin, it must be determined whether quorum is met. **Quorum** is the number of members that must be present in a meeting for formal business to occur. Quorum is usually half the membership plus one person. For example, if an organization has 30 members, at least 16 people would need to be present at the meeting for any new decisions to be made. The main reason a quorum must be present is to prevent a small number of people in a group from making decisions that might not represent the ideas or best interests of the whole group. Quorum protects the entire membership (**Figure 1.6.2**).

Every meeting must also have a *chair*. A **chair** is the person responsible for conducting meetings. This might be the president, the CEO, an executive on the committee, or an appointed member. The chair must have good knowledge of parliamentary procedures and be able to follow the order of business. The **order of business** is the order in which items should be presented in a meeting. This is also known as the meeting **agenda**.

The chair uses the gavel to begin and end the meeting and to call the group to order. A **gavel** is a hammer-like tool that the chair uses to direct the membership (**Figure 1.6.3**). Different gavel taps mean different things:

- One tap of the gavel signifies that the members need to be seated or that the meeting is over.
- Two taps of the gavel signify the beginning of a meeting.
- Three taps of the gavel signify that members should stand.
- A rapid series of taps signify that members should come to order or stop talking.

Iakov Filimonov/Shutterstock.com

Figure 1.6.2. Taking accurate attendance at meetings ensures that a quorum is met. Establishing quorum guarantees that business cannot take place or votes cannot be taken unless most the membership is present. *How are votes counted with large groups, such as the US Congress?*

everything possible/Shutterstock.com

Figure 1.6.3. *Have you ever seen a gavel being used on TV?* What was the setting of the show? A gavel in parliamentary procedure is used by the chair to keep order with the members and run a meeting effectively.

Every meeting must also have a secretary or recorder to keep the minutes of the meeting (**Figure 1.6.4**). The *minutes* are a complete, written record of the events of the meeting. These minutes are usually presented to the members or the board for review and, once approved, are available to members for years to come. Things that may be included in the minutes are the date and time the meeting began, a record of all motions made and voting that took place, and the time the meeting was adjourned.

Types of Motions

Have you ever heard someone say, "I move to…," or "I make a motion to…"? What they are doing is introducing their idea to the group or trying to get their idea accepted by the group through a motion. A *motion* is a formal proposal by a member, which, if approved by the membership, will result in a certain action. By making a motion, you are bringing something before the group that you want to be decided. A new motion cannot be presented to the members until the preceding motion is finalized. There are five types of motions a person can make:

- Privileged motions
- Incidental motions
- Subsidiary motions
- Main motions
- Motions that bring a question before the assembly

The table in **Figure 1.6.5** identifies the precedence (order) of motions.

Privileged Motions

A *privileged motion* pertains to the rights and needs of the organization. These motions include concerns with the beginning, stopping, pausing, or refocusing the purpose of the meeting. Privileged motions concern urgent matters that have priority over regular business. For example, let us say that a news channel wants to take a video of a meeting. A member may raise a question of privilege and get permission from the members before allowing the recording to occur.

Incidental Motions

Incidental motions are related to parliamentary procedure rules and procedures and not with the business at hand. They include motions that ask questions, correct mistakes, ensure an accurate vote, or change/modify the rules. For example, if someone is not following correct parliamentary procedure and there is an error in the procedure, a member may call for a point of order. This requires the chair to make a ruling on the error.

Figure 1.6.4. Minutes are an important record of what takes place during a meeting. The secretary or recorder usually takes the minutes of the meeting. The minutes are kept on file for use by the members. Board members may also refer to the minutes if they have a question about a past motion or discussion.

Motions in Order of Precedence

Classification	Motion
Privileged	Adjourn Recess Question of Privilege Orders of the Day
Incidental	Appeal Point of Order Parliamentary Inquiry Suspend the Rules Withdraw Divide the Question Division of Assembly
Subsidiary	Lay on Table Previous Question Limit or Extend Debate Postpone Definitely Refer to a Committee Amend Postpone Indefinitely
Main Motion	Main Motion
Motions that bring a question again before the assembly	Take from the Table Reconsider Rescind

Goodheart-Willcox Publisher

Figure 1.6.5. Use this chart as a guide to understand which motions go with each classification. The motions are also listed in order of precedence, which means the ones at the top of the chart come before the motions listed farther down.

Subsidiary Motions

A *subsidiary motion* is a motion that deals with managing other motions. These motions set aside a current motion, modify the amount of time a motion can be debated, or change the motion in some way. For example, if debate is taking a long time and the members have differing opinions about the motion, a subsidiary motion would allow them to refer the matter to a committee and the vote would be postponed.

Main Motion

A *main motion* is a motion that brings a new topic or idea before the membership. For example, if your FFA chapter wanted to have an ice cream social at the next chapter meeting, this new business could be brought before the members for a vote. Because it introduces a new topic, it cannot be made when there is already an item of business being discussed. It can be introduced after the current business is finished.

Motions That Bring an Old Question before the Assembly

This last type of motion brings a question or idea before the members for a second time. These motions allow the members to reconsider a vote, change a previous vote, or discuss a motion on which they had already voted. For example, if the FFA members voted to have the ice cream social and it was later revealed there was not enough money to cover the cost, the motion could be brought before the members to reconsider the previous vote.

Basics of Handling Motions

The steps that you must follow if you want to bring new business before the membership are as follows:

1. Bring the motion before the membership.
2. Debate or discuss the motion as a group.
3. Amend, or make changes, to the motion if needed.
4. Vote on the motion.

Monkey Business Images/Shutterstock.com

Figure 1.6.6. Before someone can make a motion, they must be recognized by the chair of the meeting. After another member seconds the motion, the individual will have a chance to explain their motion to the membership, and the reasoning for making that motion.

Bringing Motions before the Membership

If you have an idea or item you want to discuss with your organization, you must present it as a motion. To bring a motion before the membership, you need to be recognized by the chair (**Figure 1.6.6**). To get the chair's attention, you should stand and say "Mr. or Ms. Chair" and wait for them to call on you. You may then state your motion. All motions begin with "I move...." For example, if you attend an FFA meeting and want to share your idea of participating in the canned food drive at your school, you would need to get the attention of the chair of the meeting. After you are recognized, you would say, "I move that our FFA chapter participate in the school canned food drive next week."

Second

After you present your idea, you must have a second to get your motion to the membership to vote. A **second** is when another member states that they would also like to discuss a motion with the membership. The person who seconds the motion does not have to be in favor of the motion. They can share an opposing point of view with the membership. What a second means is that someone else sees importance in discussing the issue.

Once the motion has been made and seconded, the chair will restate the motion to ensure that all the members heard and understand the motion. The motion is now up for discussion, amendment, or voting.

Making Connections # Communication

Argumentative Writing

Using correct parliamentary procedure is a great way to discuss a controversial topic and ensure that multiple sides of an argument are heard. For this activity, write an argumentative paper discussing your opinion about a current agricultural issue.

1. Select one of the agriculture topics below for your argumentative paper:
   - Rural broadband issues
   - GMO crops
   - Immigration and agriculture labor
   - Antibiotics in meat and dairy products
   - Food safety
   - Wild horse and burro management
   - Climate change
   - Choose an agriculture issue of your own
2. Research your chosen topic and decide how you feel about the issue.
3. Write an argumentative paper that addresses why you feel the way you do about the issue. Make sure your paper has:
   - An introductory paragraph with a thesis statement (one sentence summary of your main point and your claim)
   - At least three body paragraphs that explain the reasons why you support your thesis. Include facts and figures from your research to help support your opinion.
   - A conclusion that restates your thesis and summarizes all the arguments you made supporting your position

4. Talk with your classmates and find other students that also wrote about your topic. Have a class discussion using correct parliamentary procedures about the agricultural topic.

MBI/Shutterstock.com

Consider This

1. There are often many opinions when it comes to controversial topics. Did this activity help you get a better understanding of others' opinions, especially those that are different from your own?
2. Have you ever encountered a situation where you felt like your voice or opinion was not heard? How did you handle the situation?

Debating and Amending Motions

All main motions and amendments to main motions are debatable. This means that the motion can be discussed before it is voted on. When debating a motion, it is common for members to share their thoughts about the motion, both for it and against it. The following parliamentary rules must be followed when debating main motions:

- The person making the motion has the right to debate the motion first.
- Each member has the right to speak twice per motion.
- Debates are typically limited to 10 minutes per motion.

A member may also change certain aspects of the motion. This is known as making an *amendment*. As an example, let us say that a motion is made that FFA members will read agriculture books to second graders at a local elementary school during Agriculture Awareness Week. The motion can be amended to say that FFA members will read to students in kindergarten through third grade instead. Motions cannot be amended more than twice.

The three ways that you may amend a main motion are inserting, striking out, and striking out and inserting.

- **Inserting** allows a word or information to be inserted into the motion. The only exception is that you cannot insert the word *no* or *not* into the motion and change the goal of the motion. For example, if the motion is for "all FFA officers to attend the state FFA convention," it cannot be changed to, "all FFA officers will *not* attend the state FFA convention."
- **Striking out** allows sections of the motion to be deleted.
- **Striking out and inserting** allows a portion of the motion to be deleted, and new information to be added to the motion. See **Figure 1.6.7**.

Both debating and amending motions are great ways to practice public speaking and enhance your personal leadership skills.

Parliamentary Procedure Motions Guide

Privileged Motions				
Motion	**Debatable?**	**Amendable?**	**Vote**	**What the Motion Does**
Adjourn	N	N	Majority	Ends the meeting
Recess	N	Y	Majority	Allows for a break in the meeting
Question of Privilege	N	N	Chair Decision	Allows a member to ask privilege to perform a personal need
Call for the Orders of the Day	N	N	None*	Guides the members to the items that the meeting was intended to discuss

*There are additional rules that apply to these statements; refer to the most recent parliamentary resource for additional information.

Goodheart-Willcox Publisher

Figure 1.6.7. Use this chart as a guide to the possible motions that can take place during a meeting. Practice with your friends or classmates until you feel comfortable using the motions correctly. (*Continued*)

Incidental Motions

Motion	Debatable?	Amendable?	Vote	What the Motion Does
Appeal	N	Sometimes	Sometimes	Allows the floor to vote to overturn the chair's decision
Point of Order	N	N	None	Corrects a parliamentary mistake
Parliamentary Inquiry	N	N	None	Allows members to ask a question about parliamentary law (answered by the chair)
Suspend the Rules	N	N	2/3*	Allows organizational rules to be changed
Withdraw a Motion	N	N	Majority*	–
Object to the Consideration	N	N	2/3*	Allows a member to stop a motion from being considered
Division of the Assembly	N	N	None	Gives a counted vote after the results of a voice vote are unclear

*There are additional rules that apply to these statements; refer to the most recent parliamentary resource for additional information.

Subsidiary Motions

Motion	Debatable?	Amendable?	Vote	What the Motion Does
Lay on the Table	N	N	Majority	Sets aside a motion (does not give a specific time)
Previous Question	N	N	2/3	Takes a vote to close debate and proceed immediately to voting
Limit or Extend Debate	N	Y	2/3	Changes the number or length of allowed debates
Postpone Definitely	Y	N	Majority	Postpones a motion to a specific time
Refer to Committee	Y	Y	Majority	Sends a motion to a committee to gather more information
Amend	Y	Y	Majority	Modifies a motion
Postpone Indefinitely	Y	N	Majority	Sends the motion away; kills the motion

Main Motion

Motion	Debatable?	Amendable?	Vote	What the Motion Does
Main Motion	Y	Y	Majority	Introduces a new item of business to the group

Motions That Bring a Question Again before the Assembly

Motion	Debatable?	Amendable?	Vote	What the Motion Does
Reconsider	Y*	N	Majority	Allows the group to bring up a motion that was previously voted on to discuss and vote on again
Rescind	Y	Y	Majority or 2/3*	Allows the group to completely remove record of a motion that was made
Take from the Table	N	N	Majority	Brings back up a motion that was postponed with lay on the table

*There are additional rules that apply to these statements; refer to the most recent parliamentary resource for additional information.

Goodheart-Willcox Publisher

Figure 1.6.7. *Continued.*

Voting on a Motion

After the members finish debating and amending the motion (or 10 minutes have passed and debate is ended), it is time to vote on the motion. There are several ways to vote:

- A **voice vote** requires the chair to ask everyone in favor of the motion to respond with an "aye" or "yes." The chair would then ask those voting against the motion to respond with "nay" or "no." Voice votes are the most common vote and are used to vote on motions that have not previously sparked much discussion or argument, such as a vote to adjourn.
- A **rising vote** requires the chair to ask all those in favor of the motion to stand or raise their hand. Once votes are counted, the chair will do the same for those against the motion. Usually, in a rising vote, the chair will announce the number of votes for and against and present the decision. A rising vote can take place if the votes are close and an exact count is needed, but not necessarily a list of which member votes for or against. Often, a chair will use a rising vote on a motion that has had a lot of discussion or debate from the members and the membership has strong opinions on the topic.
- A **roll call vote** requires the chair to call each member by name and ask for his or her vote, one at a time. A roll call vote is often used when a recording needs to be made on how each member voted. Roll call votes are what is used when both state and federal House and Senate members are voting on new laws and bills.
- When voting with **ballots**, anonymous votes are written, collected, tallied, and recorded. Ballots are often used when voting takes place over many days, such as during government elections. They can also be used when there are a lot of voting members and an exact count is needed. Ballots allow business to move forward in the meeting while the ballots are being counted.

Motions usually require either a majority vote or a two-thirds vote to pass. The exact number of votes a motion needs to pass depends on the type of motion. Motions that require a two-thirds vote means that two-thirds of the members must vote yes for the motion to pass.

Examples of motions that require a two-thirds vote include:

- To close, limit, or extend debate
- To suspend the rules
- To amend the constitution and bylaws
- To close nominations
- To remove an officer or expel a member
- To object to the consideration of a motion

A majority vote requires that one more than half of the membership vote in favor of the motion. Examples of motions that require a majority vote include:

- To adjourn a meeting
- To recess a meeting
- To suspend further consideration of a motion
- To postpone consideration of a motion
- To amend a motion
- To vote on a ruling by the chair

Hands-On Activity

Practicing Parliamentary Procedure

Learning the basics of parliamentary procedure is important, but knowing how to apply them correctly is more so. Practice makes perfect, so put your parliamentary procedure knowledge to the test. Hold a mock classroom meeting using correct parliamentary procedures to decide on an agriculture class president.

For this activity, you will need the following:

- Paper
- Pens
- Candidate backgrounds

Instructions for using parliamentary procedure:

1. Decide on a chair for the agriculture class meeting.
2. Decide who will serve as the secretary and keep meeting minutes.
3. Decide who will make the motion for a new agriculture class president and who will second that motion.
   A. Make a list of people interested in running for agriculture class president.
   B. Write a few talking points about each of the people interested in running so you are ready to discuss the motion.
4. Have the chair begin the meeting and follow through with correct parliamentary procedure until a new agriculture class president has been elected.
5. Take turns serving as the chair, secretary, person making the motion, and the person seconding the motion.

Consider This

1. Was the class able to elect a new agriculture class president without any parliamentary procedure mistakes?
2. What type of voting was used to elect the new agriculture class president?
3. How can you improve your parliamentary procedure skills?

Additional Practice

1. Practice writing and amending motions. Write five motions on a piece of paper. Write each of your motions exactly how you would state them to the chair. Trade papers with a partner and amend your partner's motions.
2. As a class or in a small group, practice the different methods of taking and counting votes.

Iconic Bestiary/Shutterstock.com

Parliamentary Leadership Opportunity

If you enjoy speaking, debating, or conducting meetings, you may enjoy the FFA Conduct of Chapter Meetings Leadership Development Event. The Conduct of Chapter Meetings LDE allows FFA members to use their parliamentary procedure skills as they demonstrate how to conduct efficient meetings.

Designed for FFA members in seventh, eighth, and ninth grades, this LDE challenges a team of seven students from one chapter to assume officer positions (president, vice president, secretary, treasurer, reporter, sentinel, and advisor) and demonstrate correct use of FFA opening and closing ceremonies.

Career Corner ▶ PARLIAMENTARIAN

If this sounds interesting to you, perhaps a job as a **parliamentarian** *is for you! Let's learn more.*

Parliamentarians act as facilitators for bylaws, consultants to manage meetings, and mentors for members on parliamentary procedure.

What does this job entail? Parliamentarians assist the chair to manage meetings and advise on parliamentary procedure, review and revise unit bylaws as necessary, and arrange information on nominations and oversee the election process.

What do I need to be good at this? Know and understand parliamentary procedure, have good communication skills, and ensure justice and courtesy for all and favoritism for none.

What education do I need? To become a certified parliamentarian, you must pass a certification exam.

How much money can I make? It varies widely, and depends almost entirely what organization you are working with.

How can I get started? Serve in leadership positions within your school, serve on a chapter officer team, volunteer at a government office, or attend a community meeting where formal business is taking place.

Intellistudies/Shutterstock.com

What Did We Learn?

- Parliamentary procedure is a set of rules and regulations for conducting an effective meeting.
- You must meet quorum before any official business may take place. Minutes are kept on official business that takes place at the meeting. The minutes must be approved by the members.
- The five types of motions used in parliamentary procedure are privileged, incidental, subsidiary, main, and motions that bring an old question before the assembly.
- All motions require a second before they can be voted on or discussed.
- There are three ways to amend a motion: inserting a word or phrase, striking out a word or phrase, or striking out and inserting a word or phrase.
- There are four types of voting methods: voice vote, rising vote, roll call vote, and ballots.

Let's Check and See What We Know

Answer the following questions using the information provided in Lesson 1.6.

1. *True or False?* One tap of the gavel signifies that members need to be seated.
2. *True or False?* Two taps of the gavel indicate that the meeting is ending.
3. All motions begin with the phrase, "____."
   - A. I will
   - B. I move
   - C. I demand
   - D. I agree
4. Quorum is usually set at ____ of the organization's members plus one.
   - A. half
   - B. one-third
   - C. two-thirds
   - D. one-fourth
5. The person responsible for conducting the meeting is the ____.
   - A. chair
   - B. secretary
   - C. president
   - D. teacher
6. A motion that brings up a new idea for the membership is a(n) ____ motion.
   - A. privileged
   - B. incidental
   - C. subsidiary
   - D. main

7. *True or False?* The person who made the motion has the right to be the first person to debate the motion before the membership.
8. Votes that are counted by members standing when directed by the chair are known as ____.
   - A. voice votes
   - B. rising votes
   - C. roll call votes
   - D. ballots
9. A written record of the events of the meeting are the meeting ____.
   - A. minutes
   - B. motions
   - C. orders
   - D. business
10. Votes that are counted by calling each member by name are known as ___.
   - A. voice votes
   - B. rising votes
   - C. roll call votes
   - D. ballots

Let's Apply What You Know

1. For each of the following meeting scenarios, name the type of motion you could use to manage the situation and achieve your goal.
   A. The meeting has been going on for a while and you believe the members need a short break.
   B. You notice the chair has skipped over an item on the agenda by accident.
   C. The chair voted a motion down, but there were many people for and against the motion when the voice vote was taken. You would like the vote to be clarified.
   D. You do not feel as though you have enough information to vote on a motion. However, you think if some research were done by a few members, you could vote with confidence in your decision.
   E. You noticed that the chair is moving forward with discussion on a motion before a second is made.

Academic Activities

1. **Math.** Calculating the number of members required for a vote is an important part of parliamentary procedure. For each of the following groups, calculate the number required for a majority vote and a two-thirds vote.

   A. 25 members   D. 215 members
   B. 40 members   E. 410 members
   C. 150 members

2. **Speaking.** In groups of five or six, ask your agriculture instructor to give you practice problems from the local area or state Conduct of Chapter CDEs. Work as a team to select a chair and practice your skills in parliamentary procedure. Make sure that you can make motions, debate, and come to a resolution of the problem. Have a class contest to determine which team is the best.

Communicating about Agriculture

1. **Reading and Writing.** Research how Congress uses parliamentary procedure to conduct its meetings. Determine if they have additional rules. If so, what are they and why are they used? Look for additional interesting information, such as how they count the votes, how much time a person has to speak, and how members extend the meeting time when it is to their advantage. Determine if the President must follow the rules when he is in Congress or if he has special privileges. Present your findings in a short memo to the class.

2. **Writing and Speaking.** Create a set of cards for a Jeopardy or quiz bowl game. First, make a list of parliamentary procedure questions and answers. This can be done as a class or group activity. Write one question and its answer on the same side of each card. Students can play the game as teams or individuals. You will need someone to keep score, set the timer, and to determine which team raised a hand first. Use a timer to limit response times. If you are playing with teams, allow the team members a set amount of time to confer and present their answer. If you are playing Jeopardy, answers should be given in the form of a question (e.g. "What is …(answer)?").

SAE for ALL Opportunities

1. **Foundational SAE.** Things run more smoothly when there is an order of business or plan of action. Create an order of business for your SAE program. Lay out the items you need to complete from start to finish. Start by numbering the main details or tasks. Then add sub details under your main headings. Indent the sub details and label them consecutively with capital letters.

2. **Foundational SAE.** Agricultural law deals with the laws pertaining to agricultural infrastructure, including those that deal with production/use, marketing, seed distribution, water, fertilizer, pesticide, land use, and environmental issues. People working in careers relating to agricultural law must be familiar with parliamentary procedure. Agricultural lawyers provide services to a variety of clients in the agriculture industry, including being an advocate in both state and national capitols for the rights of the agricultural workforce. Research a piece of current agriculture legislation in your state that could affect the agriculture industry. How do you think this legislation would affect your state?

SAE for ALL Check-In

- How much time have you spent on your SAE this week?
- Have you logged your SAE hours?
- What challenges are you having with your SAE?
- How can your instructor help you?
- Do you have the equipment you need?

UNIT 2
Developing Career Skills in Agriculture

SAE for **ALL** Profile
Running a Superstar Agribusiness

Growing up on his family's farm and showing livestock pushed Tyler Ertzberger to advocate for agriculture. As a middle school student, he saw a need to take quality pictures of livestock for his family's upcoming livestock sale. Soon, other farm families in the area started contacting Tyler for photos. Tyler started an entrepreneurial SAE, running a business taking photos to do marketing for local agricultural businesses.

In addition to doing marketing for farms, Tyler was showing livestock as an FFA member. He noticed he never got the pictures he wanted of himself in the show ring or at the backdrop. Tyler expanded his new business to include livestock exhibition photos. His SAE grew and became Tyler Ertzberger Photography.

As Tyler got into high school, he found his business was growing faster than he could manage. He had to hire someone to drive him to and from his first contracted event because he wasn't old enough to drive yet. During high school, he took all the graphic design and math classes he could to prepare him for owning a business.

Tyler chose to continue to build his company after finishing high school. As his business grew, Tyler changed the name to Square One Agri Marketing. He now has 18 employees to properly cover events. Today, his business has him flying all over the country to photograph farms and livestock events.

His day-to-day life depends on the job at hand. If he is at a youth livestock exhibition show taking pictures, you might find him waiting at the backdrop to capture that family memory or inside the show ring getting the perfect shot. At the end of each day, he will be in the office editing pictures to post that night. In addition to taking pictures at shows, he also live streams events to bring the show to viewers at home. When it comes to working with farms for promotions, he may be at the computer designing ads, on the farm taking pictures, filming sale cattle, clerking the sale, or talking with clients.

In 2021, Tyler was named the National FFA Star in Agribusiness Winner!

Tyler says "Every career you go into has its ups and down. There are days that I want to quit and change paths; however, I look at the growth I've made. At the end of the day, I am building a name for myself. Getting to see a kid get the champion slap, the tears, the sales manager having a successful sale, or getting the amazing donor picture is what pushes me to better myself and to keep going."

- Did you ever imagine that your SAE could be built directly into a career?
- Tyler was able to identify and fill a need in his community. What needs to be addressed in your community? How could you help to address those needs?

Tyler Ertzberger

Practicing Positive Character

PRASANNAPIX/Shutterstock.com

Learning Outcomes

By the end of this lesson, you should be able to:
- Explain why strong character is important for living a happy and successful life. (LO 02.01-a)
- Identify character traits needed for success at home, work, and school. (LO 02.01-b)
- Describe behaviors commonly associated with building positive character. (LO 02.01-c)
- Understand the importance of good character in building relationships. (LO 02.01-d)
- Evaluate personal areas for improvement. (LO 02.01-e)

Words to Know

character	gratitude	optimism
creed	grit	respect
empathy	honesty	role model
fairness	kindness	

Before You Begin

Looking at the title of this lesson, you might be wondering what this lesson is about, and why it matters. Your future career, happiness, and well-being are tied to your development as a person of strong character. Take the next three minutes to think back to a time you treated someone unfairly or behaved in a way that you regret. Consider the possible impacts that your actions had on those around you, and reflect on a few core questions: How do you treat other people? What is fair? Would others consider you trustworthy? Who do you look up to as a positive example? Do you have values or beliefs that help guide your thoughts and behaviors? Carry these thoughts with you as we head into this lesson.

Whether you are aware of it or not, everyone—including you!—has values and beliefs that shape their thoughts and actions. These can be shaped by an understanding of your personal experiences and the events going on around you. Have you ever thought about what you value or believe? Developing a personal *creed*, or set of values and beliefs, can help as a guide to your development as an individual of good character. In Unit 1, you learned about the values that guide the National FFA Association, including the FFA Creed. Each of the five paragraphs of the FFA Creed written by E.M. Tiffany in 1928 begin with "I believe" and outline core values of the organization. Today, nearly a century later, FFA members continue to look to the FFA Creed for guidance, including "I believe in leadership from ourselves and respect from others," and "[I believe] in being happy myself and playing square with those whose happiness depends on me." With this as background, let's ask again: what guides you? What would your personal creed include?

Cultivating a Life of Good Character

Once you have your own personal creed, it's time to think about putting it into action. *Character*, or the attributes and traits that make up and distinguish your individual nature, is developed over time, with your personal creed serving as a reference to guide your behavior. The work of developing positive traits and attributes is never done, with your character being constantly tested in different environments and situations. Your ability to consistently act in ways that are kind, trustworthy, honest, and fair, to name a few possible traits, defines how others will think of you (**Figure 2.1.1**). Obviously, good character is important!

Hands-On Activity

My Personal Creed

Having a personal creed can help you figure out what you value and believe.

For this activity, you will need the following:

- Paper
- Pen or pencil

Here are your directions:

1. Take 15 minutes to think and write about the people, experiences, situations, and other influences in your life that have made you who you are.
2. Then, think and write about what you value, or what is important to you.

Consider This

1. Reflection on these two prompts are just the beginning to the development of your own personal creed. What else should it include?

dumayne/Shutterstock.com

Figure 2.1.1. Like constructing the outer structure of a house, building good character can be hard work. Be thoughtful about developing good character by practicing behaviors that align with your personal values and beliefs.

Hands-On Activity

Character Trait Inventory: Part One

Cultivating better character starts with knowing where we need to grow! Let's take some time to start figuring this out. Before diving deeper into this topic, let's pause to take an inventory of your character traits.

For this activity, you will need the following:

- Paper
- Pen or pencil

Here are your directions:

1. On a sheet of paper, create two columns. Label the first column "I got this" and the second column "This could be better."

2. Next, take a look at the words below and place each in the appropriate column, based on how your feel about your ability in that area.

Honest	Forgiving
Empathetic	Grateful
Trustworthy	Fair
Optimistic	Persistent
Kind	Respectful

Consider This

1. What does this tell you about where you need to start?
2. How can you start improving in these areas? Make a quick note or two next to each area you want to improve.

Kamira/Shutterstock.com

Figure 2.1.2. Even with a firm foundation, the job of building good character is never done. Maintaining good character requires consistent attention and improvement. *When might you run into situations that require extra work in developing positive character traits?*

Character education is probably not a new concept for you. As a society, we begin learning terms like honesty, respect, integrity, and fairness at a young age. You may remember seeing each of those words blown up into huge letters and placed along the walls of your elementary school. You may have similar posters and words around your school today. Character becomes more important as you get older and begin to interact more directly with the world. As you make more decisions about your life, independently from your parents or adult family members, this foundation of words and definitions helps you connect the traits of good character with daily behavior. Character education helps define the ways you navigate friendships, professional relationships, and the decisions you make (**Figure 2.1.2**).

Cornerstones of Strong Character

What is respect and why is it important? ***Respect*** describes the positive way you treat and give attention to others, especially when considering their feelings, rights, wishes, and traditions.

Treating other people with respect makes the world a better place to live in, and it's relatively easy to do. Here are a few ways to practice showing respect for others:

- Listen attentively when other people speak.
- Value the opinions of others, even if you do not agree.
- Be considerate of others.
- Don't insult, mock, or tease people.
- Don't engage in gossip or talk about people behind their back.
- Be sensitive to the feelings of others.

Can you think back to the last time you told a fib, maybe about something small? What was it about, and why was it easier than telling the truth? Sometimes **honesty**, or the quality of speaking and acting openly and truthfully, is more difficult than telling a seemingly harmless lie, but dishonesty breeds more dishonesty. Practicing honesty may involve removing yourself from situations that you should not be in or distancing yourself from groups of people who encourage you to lie (or engage in behavior you do not want to be honest about). It is important to practice honesty every day and create situations where it is easy to do the right thing. Honesty is a key to building trust with your friends, parents, teachers, and future employers. Being honest makes life easier, and every day creates a new opportunity to start fresh. What steps can you take each day to choose the honest path forward?

Another cornerstone of strong character is **fairness**, or the ability to recognize your judgements and biases while making decisions that treat others equally. Keeping your personal feelings or biases from getting in the way of your decision-making is a difficult process that requires logic and consideration of multiple perspectives. Can you see things from someone else's perspective?

Do you think it is cool to be kind? Why or why not? If you turn on the TV or hop on social media, it may seem like **kindness**, or the sincere act of being gentle, caring, and helpful toward others, has taken a back seat in our society. Being kind and helpful to others can be as simple as giving your time and attention to a friend who is going through a difficult time or mowing the lawn of a neighbor who has recently been sick (**Figure 2.1.3**). Sometimes you can simply sit with a friend in silence and provide all the kindness they need in that moment. Really being there for someone in need is the best way to show kindness or compassion. How can you cultivate more kindness in your school? Jot down a few ideas of how you can show kindness this week. Being kind benefits you too, and you may be surprised at how great you feel!

Figure 2.1.3. One rewarding way to show kindness is by helping others. *What can you do today to show kindness in your home, school, and community?*

Melinda Nagy/Shutterstock.com

Opportunities for New Growth

Now that we have covered the basics of building strong and unshakable character, consider a few new spaces to practice, cultivate, and grow. **Gratitude**, or the ability to feel and express appreciation, involves voicing a genuine sense of thankfulness to others. Being thoughtful about showing gratitude can have lasting effects on your emotional health. So, when should you show gratitude? The most obvious time to show gratitude is when someone makes a thoughtful gesture or gives you a gift. You can also practice gratitude by thanking the people who bring joy or care to your life, like teachers, pastors, friends, and family.

Feeling grateful doesn't require speaking or showing appreciation to others in the form of a gift or card. Instead, it may be how you feel on a sunny day or when you get spend time with your family. Close your eyes and think of all the people, places, animals, experiences, or things that you feel lucky to have in your life. Do you feel a sense of peace, fulfillment, and well-being? That's gratitude!

Think about the last time someone hurt your feelings. How did you feel? What did you want the other person to know about how you felt? Why did you want them to understand your feelings? In this situation and others, it's important to show **empathy**, or the ability to understand and share the feelings or emotions of another person. Showing empathy is important in almost all areas of life, as it helps you understand the thoughts, feelings, and perspectives of people around you. Empathy also helps you connect with others, develop friendships, and work well as a member of a team. People who lack empathy might act in some of the following ways:

- Make hurtful comments about someone's appearance or clothing
- Fail to recognize when someone else is hurting, especially when that person loses a pet or loved one
- Refuse to hear someone else's point of view
- Have a hard time making friends

Given that these actions show a lack of empathy, how can you practice showing empathy for others? How might you act differently in these situations?

You have probably heard this old saying somewhere: "When life give you lemons, make lemonade." Why do you think people say that? What does it mean? Popular wisdom like this reminds you to make the best of what may feel like a bad situation. It's a statement that reminds you to be hopeful or optimistic. **Optimism** is the act of being purposefully hopeful and confident about the future. Sometimes it's hard to make the best of difficult situations; how can you see the bright side when everything seems dark? Sometimes it is hard to reframe a situation in a positive way and stay hopeful that things will work out for the best. Optimism and positive thinking do not mean you ignore the bad things that come your way, but instead that you approach hardships or obstacles in a more productive way. Practice shifting your focus to a more positive place when times get tough. Take a deep breath and think about what you are grateful for, or maybe a good joke you heard at lunch. Then come back to the situation and consider new ways of thinking. Instead of stopping with all the things you can't do, consider all the things you can. Optimism is not always easy to show, but with practice, it can help you remain grounded, strong, and capable in any situation.

Last (but not least) is grit. **Grit** describes persistence, willingness to fail, and the strength and courage to move forward and try again. Although at times it can feel like climbing the ladder to meet your goals is impossible, continuing to climb that ladder in the face of setbacks is the important part. There are no shortcuts in life, and talent, experience, and luck are not always parceled out equally. Your ability to keep moving toward your goals defines your character (**Figure 2.1.4**).

eufrith/Shutterstock.com

Figure 2.1.4. We often think the most successful people must have the most talent in a particular job or skill, but that's not always the case. In fact, the biggest indicator of one's success is not talent at all, but how hard you're willing to work to meet your goal. *Does this surprise you? Why or why not?*

To better understand grit, ask yourself these questions:
- How do I react to setbacks?
- Do I give up easily? What keeps me on track?
- Do I usually finish the things I start? Why or why not?
- Am I a hard worker? Would other people agree with my answer?

Models of Strong Character

Practicing good character should be a part of your daily routine. Having people in our lives with strong character and surrounding ourselves with people who inspire us to be better and to make good choices can be an important part of this process. Who do you look up to? A friend, coach, or maybe a member of your family? Why? Is it because they are exceptionally skilled in some way, or because of the way they treat others? Finding a *role model*, or a person to look up to or imitate because of their good character, is one way to start building character today.

Building strong character requires you to be intentional with your actions. Creating a plan, even if it's just for tomorrow, is important when practicing a new skill. It is also important that you continue to nurture the positive character traits you already show. Be sure to ask yourself about those, too. You never stop learning and growing, and good character is a critical part of living a happy and successful life (**Figure 2.1.5**).

MPH Photos/Shutterstock.com

Figure 2.1.5. Take a good look in the mirror. *How can you take small steps today to improve your character?* Let's make a plan!

Hands-On Activity

Revisiting Your Character Inventory

Think back to the beginning of this lesson, when you created your initial list of character traits.
1. After reading this lesson, how do you feel about your abilities in these areas? Take a couple of minutes to make any updates.

2. Next, think about how you start building stronger character. For each of the words you placed in the column labeled, "This could be better," ask yourself the following questions:
   A. How have I shown _____ today?
   B. How have I not shown _____ today?
   C. How can I show _____ tomorrow?

What Did We Learn?

- Behaving according to our values and beliefs helps us to build strong relationships with others, contribute to our communities, and do good work.
- Developing a personal creed, or set of values and beliefs, can help guide our progress in building good character.
- Fairness, respect, honesty, and kindness are cornerstones of good character, and foundational to leading a life of happiness and success.
- Exercising traits such as empathy, optimism, grit, and gratitude provide new opportunities to learn and grow at home, school, and work.
- Role models serve as important examples in building and improving our character.
- Building strong character is a process, and it requires consistently asking how we can show positive traits and evaluating our progress honestly.

Let's Check and See What We Know

Answer the following questions using the information provided in Lesson 2.1.

1. *True or False?* Character is a fixed skill that does *not* need to be practiced daily.
2. What trait describes the act of being hopeful for the future?
   A. Respect
   B. Kindness
   C. Optimism
   D. Negativity
3. What can we create to help guide our development as individuals of good character?
   A. A game
   B. A creed
   C. A list of career choices
   D. A list of jokes
4. *True or False?* Keeping personal feelings or biases from affecting the decisions we make is an element of fairness.
5. *True or False?* Positive role models can set examples of good character for us to follow.

6. Showing respect to others may include all of the following behaviors, *except*:
   A. being considerate when others sit at your lunch table.
   B. paying attention when another person is speaking.
   C. helping someone by opening the door for them.
   D. picking on the person who sits behind you in class.
7. What word describes our ability to share and understand the feelings or emotions of others?
   A. Empathy
   B. Integrity
   C. Optimism
   D. Trustworthiness
8. *True or False?* Grit is defined as the act of being purposefully hopeful and confident about the future.

9. All of the following are examples of showing empathy, *except*:
   A. asking a new student if they would like to join you for lunch.
   B. ignoring a classmate's point of view.
   C. recognizing that a friend is sad when they lose a loved one.
   D. caring for your sibling when they are feeling sick.

10. Which of the following terms describes the quality of speaking and acting openly and truthfully?
    A. Optimism
    B. Fairness
    C. Grit
    D. Honesty

Let's Apply What We Know

1. Why is it important to practice positive character traits?
2. List three different ways you can show respect to others in your daily life.
3. What is a personal creed? How can it help you develop good character?
4. Why is empathy an important component of character?

Academic Activities

1. **Social Studies.** Our nation's history is full of examples of positive character. Describe a historical figure and/or event where character was central to success. What role did positive character play in the outcome of the event and/or life of the historical figure?

Communicating about Agriculture

1. **Communication Styles.** Role models are important to our development as leaders in agriculture. When seeking out role models, pay specific attention to how people communicate with the words they use online, in print, and in person. Are they demonstrating positive elements of character? Are they honest? Trustworthy? Do they use empathy toward others? Positive communication is a strong indicator of good character and a trait worth seeking out in a role model. Can you describe the traits you think are most important in showing character through communication, and how you can see your role models demonstrating those traits?

SAE for ALL Opportunities

1. **Placement SAE.** Sometimes the best way to help others learn is through creative expression. Think of three character traits you feel you could communicate well to others through writing, art, video, or another creative platform. If you enjoy sharing information on different topics in new ways, consider talking to your agriculture teacher about SAE opportunities in agricultural education.

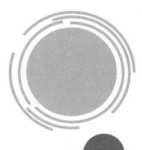

SAE for ALL Check-In

- How much time have you spent on your SAE this week?
- Have you logged your SAE hours?
- What challenges are you having with your SAE?
- How can your instructor help you?
- Do you have the equipment you need?

Building Leadership Skills through Agriculture

bodnar.photo/Shutterstock.com

Learning Outcomes

By the end of this lesson, you should be able to:
- Explain the importance of developing personal leadership skills. (LO 02.02-a)
- Describe the characteristics of a good leader. (LO 02.02-b)
- Identify opportunities for personal leadership. (LO 02-02-c)

Words to Know

accountability	long-term goals	responsibility
followership	nonverbal communication	short-term goal
leader		team
leadership	personal leadership plan	verbal communication

Before You Begin

If I were to ask each member of your class to define "leadership," do you think everyone in the class would give the same definition? One funny thing about leadership is that our personal experiences play into what actions we consider acts of leadership. We often appreciate others who make us feel valued, and who take time to hear our opinions. These people often treat everyone fairly and with respect. We can all also probably think about a time when we felt hurt, let down, left out, mad, or confused by someone else's actions. In those moments, did you see others exhibiting good leadership?

Take out a piece of paper and draw a line down the middle. Label one side of the paper "Good Leadership" and the other "Poor Leadership." I am going to give you a few moments to come up with five examples or characteristics for each, and then we will figure out how your examples fit into this lesson.

Have you ever heard someone say that another person has good leadership skills? What makes someone a good leader? Do you think a person is born a good leader, or do you think they learn what it takes to be a good leader? What are leadership skills and how do you "get" them?

What Is Leadership?

Leadership is the personal quality of being able to guide or direct others. A *leader* is someone who guides or directs a group or organization. There are many opinions about what makes someone a good leader.

Personal Leadership

How many times have you taken a leadership role? Being the captain of a sports team, an officer with a school organization, or mentor at your church can all be examples of personal leadership. Personal leadership is the ability of a person to represent the traits of a good leader and work towards becoming a better leader. Learning more about yourself and understanding the personal strengths that you possess helps you become a better leader. You must set goals for yourself and develop a plan that helps you reach those goals (**Figure 2.2.1**). It is also important to avoid people and places that would prevent you from reaching your goals.

Personal Leadership Plan

A *personal leadership plan* is a strategy for how you will accomplish your goals and grow as a leader. When laying out your first plan, set one or two goals for yourself. Think about challenges or positives that will help you reach these goals; then, set out a plan for reaching the goals. Having a plan to follow will improve your ability to reach your goals.

airdone/Shutterstock.com

Figure 2.2.1. A personal leadership plan is a strategy for how you will accomplish your goals. Take your strengths and weaknesses into account, as well as how you can improve yourself, and reach your goals.

Goals

Having clear goals is an important part of building your personal leadership plan or growing your personal leadership skills. Setting goals helps you to stay focused on what you want to accomplish. Both long- and short-term goals are important for people of all ages.

Short-term goals are tasks you want to accomplish in the near future. The near future can mean today, this week, this month, or even this year. These can be simple goals, such as passing an upcoming test or making a school sports team.

A **long-term goal** is a task you want to accomplish further in the future. Long-term goals require more time and planning. They are not something you can complete in a week or month. Long-term goals usually take 12 months or more to achieve. Examples of long-term goals may include graduating from high school or buying your first car with money you save.

Did You Know? Specific, realistic goals work best when it comes to making a change or taking action. For example, having a goal of studying at least 30 minutes every day is a more measurable and effective goal than saying you are going to study to bring up your grade. When? How can you tell you achieved that goal? Specific, realistic goals are easier to track, understand, and stick with over time.

digitalskillet/Shutterstock.com

Figure 2.2.2. Think of someone you have encountered in a leadership role, such as a teacher, a coach, a pastor, or an older sibling. *Were your experiences with this person positive or negative? What made them a good or a bad leader?*

Characteristics of a Good Leader

How do good leaders handle difficult times in their lives? Do you think they learn how to remain positive through difficult times? What personal traits allow them to remain positive when things go wrong? Leadership takes discipline and a combination of skills or traits that allow you to handle unique situations or challenges (**Figure 2.2.2**). Some of the characteristics leaders possess may include:

- Character: personal traits that show good moral and ethical values. We learned about character, and how to develop it, in **Lesson 2.1**.
- Passion: a person's desire to find meaningful motivation
- Resourcefulness: the ability to obtain physical resources and to place people (human resources) in positions that suit them well
- Accountability: a person's willingness to answer for their actions or decisions
- Focus: a person's ability to remain or stay in contact with a person's or group's goals and desired results

Personal Traits

Think about a fellow classmate who stands out as a leader. Why did you pick this person? What was the first thing about them that came to mind? Can you identify the person's personal traits that contribute to their leadership skills? How would you describe their character? There are many things that you can do to help you stand out as a leader or become a better leader. Think about starting with the following list and expand it by adding other actions. How do you think doing these things might make you a better leader?

- Telling the truth
- Serving others and being helpful
- Encouraging those around you
- Working hard
- Arriving to class/appointments on time or early
- Showing up to class or work prepared
- Learning from your mistakes
- Seeking knowledge

Pursuit of Passion

Almost every success story begins with someone finding something that excites them or awakens great passion. It is difficult to become motivated to do something you do not enjoy. Think about professional athletes. Do you think a professional athlete would be successful if they did not enjoy playing their sport? What about a rancher? Would they be able to run a successful ranch or herd if they did not care about their employees and cattle? What about farmers? They are passionate about providing safe food for all of us.

To be a good leader, you must find a passion, and to find your passion, you must be open to new opportunities. The fellow classmate you thought about earlier, are they involved in activities, such as sports, school clubs, or school government? Sometimes, to find your passion, you must step out of your comfort zone and take new chances. Challenge yourself to try something new at your school or in your community this year.

Listen and Learn

How often has a teacher said, "You need to listen," or "Make sure you are listening"? Those teachers are giving you good advice that can help you become a better leader (**Figure 2.2.3**). Being a good listener helps you build trusting relationships. As a leader, it is difficult to know what people are thinking, what is troubling them, or how to help them—unless you take the time to listen to them. Listening goes well beyond being quiet and giving someone your full attention. It requires you to be aware of body language, facial expressions, and moods.

To help you work on your listening skills, give the person talking your full attention and make eye contact while they are talking. Let the speaker know you are listening by showing interest and reacting to what they are saying. If your mind starts to wander, ask questions to

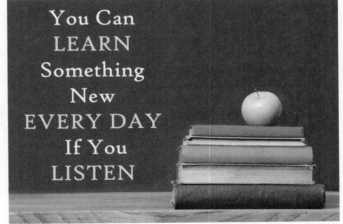

northallertonman/Shutterstock.com

Figure 2.2.3. You can learn a lot when you are a good listener. A wise person knows there is something to learn from everyone, if you just listen. *Can you think of something you learned today by listening?*

help you stay focused during the conversation or repeat things to ensure you understand. Listening provides you with new ideas and possible solutions you could not have thought of on your own. (**Lesson 2.4** talks more about listening, if you want to brush up on this more!)

Effective Communicators

To be a good leader, you must know how and when to be an effective communicator. Communication takes place in two ways: verbal and nonverbal. *Verbal communication* is the use of words to exchange information. Good leaders can communicate on many levels, from speaking to someone one on one or speaking to a whole group of people. However, good leaders also know how to communicate properly through phone calls, emails, text messages, and social media. In order to communicate your ideas and thoughts effectively, you must express yourself clearly. If people do not understand what you are trying to communicate, the message may be misunderstood or not received at all.

Hands-On Activity

Practicing Active Listening and Communication Skills

Listening is an important leadership skill, but is often overlooked. Listening helps build healthy relationships, builds trust and respect, and gives you better understanding of any situation.

Working in small groups, collect the following items:

- Blindfold
- Book
- Chair
- Pencil or pen
- Piece of paper
- Timer (optional)

Here are your directions:

1. Set these items up in an obstacle course. For example, each participant could step over the pencil, spin around on the piece of paper, jump over the book, and sit in the chair.
2. Each person in the group will take turns going through the obstacle course. Each participant will be blindfolded while going through the course.

BearFotos/Shutterstock.com

3. Other students must guide the blindfolded student through the course using voice commands, such as "take three steps forward," "take one step backward," or "take one jump forward." Set a time limit for finishing the course.
4. Take turns being both the communicator and the listener.

Consider This

1. How many listeners finished the course within the time limit?
2. Did the communicators get their message or directions across clearly? What happens if the message is unclear?
3. How can you improve your listening skills?

Nonverbal communication is the exchange of information without words. For example, a speaker can use hand gestures to help get a point across. How we sit, stand, and hold our hands are all part of nonverbal communication. Even our facial expressions can communicate our thoughts to others (**Figure 2.2.4**). Think about a time you heard some exciting news. Do you think your facial expressions told the other person that you were excited?

Accountability and Responsibility

You may be familiar with the words accountability and responsibility, and have some idea what they mean. Do you know the difference between them? The main difference between the two is that responsibility can be shared with other people, while accountability cannot. *Accountability* is answering for one's actions or decisions. *Responsibility* is being reliable or

Koldunov/Shutterstock.com

Figure 2.2.4. *What is this person's expression telling you?* Nonverbal communication is often people's first impression of us. How you stand or the attitude you send to others can determine how approachable you are when someone wants to talk or share something with you. *What are some actions you could take to seem welcoming?*

Making Connections # Leadership

Identifying Communication Styles

Think about the people in your closest friend group. Do all of your friends communicate the same way? Are some loud? Are some quiet? Do some always share ideas or things to do?

These are all different ways to communicate, and each of us communicates using different styles. Use the following chart, based on the Platinum Rule by Dr. Tony Alessandra, to determine which communication style you relate to the most.

Rawpixel.com/Shutterstock.com

Syle	Description
The Socializer	Relationship-oriented; appear to need the approval of others; move, act, and speak quickly; avoid details when possible; risk-taker; enjoy the spotlight; persuasive
The Director	Task-oriented; move, act, and speak quickly; want to be in charge; seek results through others
The Thinker	Task-oriented; move, act, and speak slowly; enjoy solitary, intellectual work; greatly concerned with accuracy; cautious decision-makers; demonstrate good problem-solving skills
The Relator	Relationship-oriented; move, act, and speak slowly; avoid risk; seek tranquility and peace; enjoy teamwork; show good counseling skills

Consider This

1. What communication style fits you best? What about your closest friends?
2. Do you think there would be problems if everyone communicated the exact same way? Why?
3. How do you make sure your opinion is heard when communicating with people who have a different communication style than you?

Diane Garcia/Shutterstock.com

Figure 2.2.5. Practice is a big part of keeping your focus. It is easy to get frustrated and lose focus when things are not going as you planned. Practice helps you feel prepared and confident when it comes time to perform.

dependable for something in one's control or management. If you are working on a group assignment with several classmates, each of you is responsible for completing your assigned parts. You are accountable if you do not complete your part on time, not your classmates. Good leaders know the importance of being responsible *and* accountable for their actions. Leaders who practice responsibility and accountability do not make excuses when they make a mistake; they lead by example.

Focus

Leaders are often responsible for developing a mission or goals for the group they are leading. It can be easy to fall short of these goals and ideals if you stop paying attention to them. It is easy to get distracted! Staying focused on your goals is important if you are going to reach them. How can a leader help their team to focus? You can make sure that the goals you're working toward are compelling, and that communication about them is clear. You can help yourself (and your group) stay focused by reducing distractions, discouraging multitasking, taking short breaks, and continuing to practice (**Figure 2.2.5**).

Team Leadership

A **team** is defined as a group of people who come together to achieve a common goal. An example of a common goal could be winning a competition, having a successful fundraiser, or completing a community service project. Almost everyone has been on a team at some point. Families can even be considered a type of team. Research shows that an effective team can accomplish more working together than by working individually and combining their individual efforts (**Figure 2.2.6**). To ensure that a team is displaying good team leadership, they need the following:

- A shared mission or vision
- The ability to work with the strengths of everyone on the team
- Ground rules and mutual respect for all teammates
- The ability to resolve conflicts when they arise

Figure 2.2.6. A baseball team is a good example of team leadership. Each person on the team needs to play their role for the whole team to be successful. *What do you think would happen if team members only looked out for themselves and did not work to help the team? How many games do you think they would win?*

sirtravelalot/Shutterstock.com

Conflict Resolution

It is natural for team members to have disagreements. When people get together with different personalities, thoughts, and opinions, disagreements are bound to happen. It is how you respond to those disagreements that makes you a good leader. A few tips to help you and your teammates rise above a disagreement are:

- Take responsibility for your own actions and role in the conflict. Apologize if needed.
- Wait until everyone involved in the conflict has his or her emotions under control before trying to talk through the problem.
- Be open and flexible to other people's feelings and ideas. Try to work toward a solution that benefits most of the group.
- Keep the final goal in mind. Seek a solution that allows the group to be successful.

Followership

Some people naturally find themselves attracted to a vocal, up-front leadership role in any group they are a part of, while others would not like to take on such a role. It is normal and okay to not want to take on vocal leadership roles, whether you have more reserved personality or you want to learn and experience more before stepping into a leadership role. As long as you are still participating in the group, you have a role to play. The reality is that not everyone can be an up-front, vocal leader at the same time. At some point, everyone must be a follower (**Figure 2.2.7**).

Tom Franks/Shutterstock.com

Figure 2.2.7. Do you know why Canada geese fly in a V formation? By having a clear leader, they can fly for longer periods without stopping. The leader can change in the formation when it gets tired and needs to rest. The remaining birds follow just slightly above the leader, which reduces the wind resistance and makes the lead bird's job easier. Being a good follower can be just as important as being a good leader. *Do you see yourself as a good follower? Do your actions help or hurt the leader in charge?*

Followership is the ability to follow a leader effectively. You might be confused. What does that even mean? Well, it varies. Not everyone approaches followership in the same way. To be a good follower, try practicing the following things:

- Communicate with the leader. Ask how you can be helpful to the group. Do not expect them to assign you a task. Take initiative!
- Share your ideas. One of the great things about working in a group is the different ideas each member can contribute.
- Be committed to the shared vision or goals of the group.
- Use your skills and talents in a way that contributes to the group's success. For example, if you are knowledgeable about graphic designing on computers, volunteer to make flyers, or if you have many followers on social media, offer to spread awareness of the group's mission in a positive way through social media.
- Be willing to hold true to your values when they disagree with the direction of the leader. Do not be afraid to stand up for yourself if you feel like the group values and morals are not the same as your own.

Opportunities to Lead

School is a great place to find opportunities to lead. By helping to lead a student organization, you can have a big impact on not only your school but also your community and other students that follow you as a leader. Even if you already consider yourself a leader in your school, be open to new opportunities that help you grow your leadership skills.

Opportunities to serve in a leadership role include the following:

- Serving as an officer in a school club or organization such as FFA, FBLA, FCCLA, or DECA
- Competing or participating in public speaking events (**Figure 2.2.8**)
- Serving on school, club, or organization committees
- Serving as a member of student government
- Leading team sports
- Participating in community organizations, such as 4-H

Figure 2.2.8. Public speaking is a great way to practice your leadership skills. Try starting small by speaking in front of small groups or a single class. When you have built up your confidence, try competing in a public speaking event or contest.

Overdose Studio/Shutterstock.com

Career Corner
ANDY PAUL, Leadership Program Specialist, Georgia FFA Association

As a former National FFA President, Andy Paul has a deep passion for the National FFA Organization and its members. He got involved in the FFA his eighth-grade year and quickly fell in love with the blue and gold. Attending his first state convention, he was floored at seeing all the lights, production, and how effective the state FFA officer team was at speaking. He had always enjoyed public speaking and knew that he wanted to be on that stage as a state officer eventually. In 2013, he was able to experience firsthand all the opportunities that Georgia FFA had to offer as the state FFA President. Mr. Paul went on to serve as the National FFA President from 2014–2015. Since serving as an officer, he knew that he wanted to coach others on how to become the best presenters and communicators they could be, and this position as Leadership Program Specialist is the prime opportunity to do just that.

What would it take to be successful in this career?

What does this job entail? Leadership program specialists develop, implement, and support leadership development, coaching, and mentoring programs. They also facilitate in-person and virtual leadership programs and sustainment events. They manage all design and operational aspects of leadership programs, including budget forecasting and tracking for multiple curriculums.

What do I need to be good at this? You need to be good at building relationships, and enjoy encouraging and motivating others. You need to be adaptable, innovative, and creative.

What education do I need? To be a leadership program specialist, you need a bachelor's degree in human resources, agricultural education, or agricultural business.

How much money can I make? Positions like this pay an average salary of $60,000.

How can I get started? Serve in leadership positions within your school, serve on a chapter officer team, volunteer in your community, or job shadow decision makers in your local community.

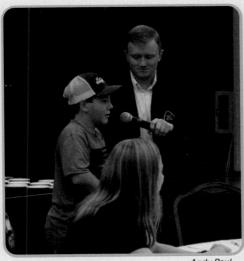
Andy Paul

Andy Paul

What Did We Learn?

- A leader is someone who guides or directs a group or organization. Leadership is the personal quality of being able to guide or direct others.
- Many personal traits display leadership ability. These traits include a pursuit of passion, being a good listener, an interest in learning, effective communication, accountability and responsibility, focus, listening, and teamwork.
- Working to become more aware of your own leadership traits will help you build your personal leadership plan.
- When making a personal leadership plan, you need both long- and short-term goals.
- Your school is a great place to look for opportunities to improve your leadership skills.

Let's Check and See What We Know

Answer the following questions using the information provided in Lesson 2.2.

1. Leadership is the personal qualities one needs to help them ____ others.
   A. boss
   B. befriend
   C. guide
   D. anger

2. Which of the following is a characteristic of a good leader?
   A. Passion
   B. Accountability
   C. Focus
   D. All of these.

3. Using words in a written form, such as an email, is an example of ____ communication.
   A. verbal
   B. nonverbal
   C. classical
   D. social

4. Facial expressions and hand gestures are examples of ways people share their thoughts through ____ communication.
   A. verbal
   B. nonverbal
   C. classical
   D. social

5. Not making excuses when you make a mistake is an example of being ____ for your actions.
   A. verbal
   B. forgiving
   C. tolerant
   D. accountable

6. A(n) ____ is a group of people working together toward a common goal.
   A. community
   B. class
   C. team
   D. audience

7. *True or False?* Followership is the ability to actively follow and work with the leader of a group.

8. An individual who is working toward being a better leader is working on ____ leadership.
   A. followership
   B. ownership
   C. verbal
   D. personal

9. Making an A on an upcoming test is an example of a ____ goal.
   A. team
   B. long-term
   C. short-term
   D. school-based

10. A ____ allows you to plan how you will achieve the goals you have set for yourself.
    A. personal leadership plan
    B. long-term goal
    C. short-term goal
    D. team

Let's Apply What We Know

1. Do you think leaders are born or made? Explain your answer.
2. What are five leadership characteristics you feel you already have?
3. Create a personal leadership plan using both long- and short-term goals.
4. Think of a person you admire as a great leader. Explain what makes them a great leader.
5. Make a list of five leadership traits; then, write how you think they correspond with behaviors one might display on the job. For example: Responsibility on the job means showing up to work on time.

Academic Activities

1. **Language Arts.** Serve as a reading mentor for students in earlier grades. Research some appropriate agriculture-related books you would like to read with the students. Check them out from the library or borrow them. Arrange to read one on one with a student or do a read aloud for the entire class and practice your public speaking skills.
2. **Language Arts.** Trace the outline of a classmate on a large piece of paper, such as bulletin board paper. Dress your person in clothes you think a leader would wear. Write words around your person that you think describe leadership traits your person should have.
3. **Social Science.** Family crests are not as common as they were in medieval times, but they still serve as a visual representation of a clan's strengths. It is something that represents the family. Research family crests to learn the meaning of the symbols used to represent the clan's background. Apply that same idea to leadership and create a crest that represents the values, beliefs, and ideas of a great leader. You may also create a crest that represents you and your strengths.
4. **Social Science.** Find a popular personality assessment (some examples include the Myers-Briggs Type Indicator, Strengthfinders, Enneagram Personality Test, or the Kolb Learning Styles Inventory). Have the members of your class take the assessment to see what it says about their leadership styles. Create a graph or chart about your findings and make predictions about how these findings will influence the way your class handles class discussion and decisions.

Communicating about Agriculture

1. **Reading and Speaking.** Select a motivational leadership quote. Find a partner and share your quote. Explain why you chose it. Listen actively as they share their quote. Switch partners and share your quote and the quote from your previous partner. Discuss the meaning of each quote.
2. **Speaking.** Working in small groups, practice your presentation skills with five-minute speeches. Record each other and review your presentations. Offer each other constructive criticism and critically review your own speech. Be kind and note each other's good points, as well as areas that need improvement. Practice your speech for a few minutes. Repeat the exercise, but apply what you learned after reviewing the first presentation.

SAE *for* ALL *Opportunities*

1. **Foundational SAE.** Make a list of ways your SAE program can help you develop new leadership skills. Are you learning skills that will help you in the future? Learning how to communicate better? Think about what leadership skills you'd like to learn, and then adjust your SAE plan to let you improve your leadership skills.
2. **Foundational SAE.** Interview someone who is a leader in your SAE program area. For example, interview a veterinarian or poultry farmer or owner of a carpentry shop. Find out what steps they took to develop their leadership skills. Write up a summary of what you learned from your interview.

SAE *for* ALL *Check-In*

- How much time have you spent on your SAE this week?
- Have you logged your SAE hours?
- What challenges are you having with your SAE?
- How can your instructor help you?
- Do you have the equipment you need?

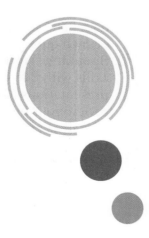

Cultivating Attitudes and Habits for Career Success

Bannafarsai_Stock/Shutterstock.com

Essential Question

How do we develop healthy long-term habits?

Learning Outcomes

By the end of this lesson, you should be able to:
- Describe the value of making a good first impression. (LO 02.03-a)
- Identify the characteristics of a strong work ethic. (LO 02.03-b)
- Demonstrate time management techniques. (LO 02.03-c)
- Understand the rewards for completing quality work. (LO 02.03-d)

Words to Know

extrinsic reward	motivation	self-discipline
first impression	professionalism	time management
intrinsic reward	punctuality	work ethic

Before You Begin

*Did you know that you are already preparing for your future career every day? The choices we make when completing tasks at school or taking on additional chores at home help us develop our attitude about, and ability to, work. Take the next few minutes to jot down all the projects, assignments, and responsibilities you would like to complete this week. Then, separate those into two categories: Home and School. What do your lists look like? Sometimes lists help us see what needs to be done. By writing to-do lists and organizing them into categories like Home and School, we can begin to tackle tasks in a way that feels manageable (***Figure 2.3.1***). Do you think this would help you to get more done? Getting in the habit of completing tasks in a timely and organized fashion will help you build skills for the future.*

Respiro/Shutterstock.com

Figure 2.3.1. Sometimes, responsibilities can seem overwhelming! Take a couple minutes each day to think through and write down the tasks you need to complete. It is much easier to determine our priorities and fulfill our responsibilities when we can see everything out in front of us. *What are other tools you might use to stay organized?*

Are you dependable? Can people count on you to get a job done? Would your friends consider you to be a hard worker? These questions are important to consider when developing positive habits. The workplace has changed a great deal in the past twenty years and is almost certain to change again before you take your first job. What is unlikely to change is your **work ethic**, or the values, attitudes, and behaviors you bring to every job or task you complete. People with a strong work ethic are driven to be dependable and responsible as they tackle tasks and assignments. How can you practice a good work ethic today?

In this lesson, you will learn the value of making a good first impression and identify the characteristics of people who exhibit a strong work ethic. You will also demonstrate good time management techniques and begin to understand the types of rewards received for quality work. It's time to flex your muscles and begin cultivating the attitudes and habits needed for success!

Making a Good First Impression

The impression you make on others can have an impact on the friends you have, the type of job you get, and the success you achieve. The image, or impression, you project to the world may be seen as a true reflection of your work habits and behavior. How can you provide others with a positive image from the start?

Have you ever heard the phrase, "You never get a second chance to make a first impression"? What do you think it means? *First impressions* are the thoughts and opinions people have about you after your first meeting (**Figure 2.3.2**).

Figure 2.3.2. It is important to consider the elements of a positive first impression in starting on the right track at work or school. *What will you be sure to remember next time you meet someone new?*

George Rudy/Shutterstock.com

These impressions provide others with a mental picture of you that may or may not truly reflect who you are as a person. It only takes thirty seconds to make a good first impression, but almost *twenty minutes* to undo a bad impression. At what times in your life might it be most important to ensure you make a good first impression?

Creating and sharing a positive image is critical when first meeting other people, including potential employers who make hiring decisions. How can you make sure you put your best foot forward? Managing your personal appearance is an obvious place to begin thinking about first impressions and includes factors such as wearing appropriate clothing and grooming properly. How you look does not define who you are, but it is important to understand that other people use your appearance to gather information and make assumptions about you. It makes sense to do what you can to try and make that impression positive! Considering these factors may seem silly, but not putting in any effort may prevent others from getting to know more about you.

Here are a few other ways to help you make a good first impression:

1. Smile when meeting others and greet them with a firm handshake.
2. Make eye contact and frequently use their name with appropriate titles (Miss, Mr., Dr., etc.).
3. Listen carefully and show interest in conversation.
4. Ask questions and respond thoughtfully.
5. Share your experiences and skills, when appropriate, without bragging.

Hands-On Activity

Shake It Out!

Find a partner to practice your best professional handshake, and a few silly ones too!

1. Begin by introducing yourself and making up a silly handshake or greeting. Here are a few examples of handshakes to get you started:
   - Mork and Mindy: Stick all of your fingers together and then make a V between your middle and ring fingers. "Shake" by angling your V to stick into your partner's V.
   - Rock and Roll: Place your pinky and pointer finger straight up while curling the others under. Connect hands at the knuckles.
   - Yo Bro: Grip hands around the thumbs, release, and then bump fists.
   - 'Bow to 'Bow: Bend arms at the elbow and touch your partner's alternate elbow twice.
2. Once you've got the hang of some silly ways to greet others, practice a professional handshake by firmly grasping your partner's hand, making eye contact, and introducing yourself. Listen to your partner and ask questions, too!

Dan Kosmayer/Shutterstock.com

3. Switch partners to practice until you get the hang of the perfectly professional handshake.
4. Continue to practice this new skill by introducing yourself to people outside of class. Just be sure to wash your hands often!

Consider This

1. Why do we shake hands when first meeting others?
2. How can a poor handshake make a bad first impression?
3. Did you feel more comfortable introducing yourself at the end compared to the beginning of this activity? Why or why not?

WAYHOME studio/Shutterstock.com

Figure 2.3.3. Have you ever waited, and waited, and waited for a pizza delivery? Sometimes being punctual can mean the difference between a hot and cheesy dinner and cold reheated pie. Timeliness is important in so many areas of life! *When else might being timely make a difference?*

Characteristics of a Good Work Ethic

Employers consider work ethic to be one of the most important qualities of a good employee. What are some characteristics or behaviors exhibited by people with a strong work ethic? Consider the following as habits and characteristics you can address daily to develop a strong work ethic for the future.

Practice Punctuality

Punctuality is the characteristic of completing a task or meeting an obligation before a previously designated time (**Figure 2.3.3**). Punctuality displays respect for other people and their time. You might not think much of running late or missing a deadline, but being late usually gives the impression that you do not value others' time and responsibilities, and can disrupt others' work. Practice the habit of being punctual by getting to class ahead of schedule. This creates an opportunity to talk with your teacher or get mentally prepared for learning. If you're taking classes online, stick with your study schedule, hitting the books at the time you planned. Pay attention to the clock. Set alarms if you need to. Practicing this habit will put you ahead of the game when you get to the workplace.

Develop Professional Habits

Professionalism, or how you conduct yourself in a professional or work setting, goes beyond what you wear to include behavior and attitudes. Practice being positive by leaving your bad attitude at the door. Don't participate in gossip or talk poorly about your classmates. Develop a reputation of integrity by being honest and consistent in your words and actions.

Cultivate Self-Discipline

Anything worth achieving takes ***self-discipline***, or the ability to stay focused on a task or goal without being easily sidetracked. Train yourself to push through on projects and expect excellence from yourself and others. Creating an organized plan to manage your tasks, activities, and other responsibilities can help you keep track of everything and make progress.

Use Your Time Wisely

Have you ever heard an adult say, "Don't put off until tomorrow what you can do today"? If you can do something now, you should! Build good habits by developing good ***time management***, or a process for organizing and planning your day by dividing your time between activities. Good time management helps you to work smarter, not harder. You can often get more done, in less time, when you stay on task and make effective use of your time.

Skills That Pay the Bills

Do you ever wish you had more hours in the day? Time management is a difficult skill to master at any age, especially when you have so many different activities and responsibilities on your plate. You might like to think about your time each week in the same way you would consider a budget. Thinking about your priorities will help you consider the larger (or more expensive) items before allotting time to smaller activities. Let's practice our time management skills by considering your "big ticket" time budget items.

1. First, think about your weekly time budget by calculating the number of minutes in a week.
2. Now, subtract approximately eight hours (in minutes), per day, from the total number of minutes in a week to determine how many minutes you are awake. This will be your weekly time budget.
3. On a separate sheet of paper, create two columns, one titled "How I Spend My Time" and the other "Time Budget."
4. In the "How I Spend My Time" column, jot down as many of your normal weekly activities and responsibilities as you can in three minutes.
5. What are your "big ticket" items or urgent priorities each week? How much time does each usually require? Add time to each in the "Time Budget" column and subtract the total from your weekly minute budget.

Aysezgicmeli/Shutterstock.com

6. Now, consider the littler things. Add time to each in the "Time Budget" column and continue to subtract from your weekly budget.
7. How much time, if any, did you have left over? What would you want to do with this time? Discuss your weekly budget with your neighbor and discuss the value of your time.

Consider This

1. How much time was left over each week? What do you want to do with it? If you didn't have any time left over, what does that tell you about how you spend your time?
2. Were you surprised by how quickly the little things began to add up?
3. What type of items can you reduce each week to make time for the big or important things that need your attention?

Find Balance

Having a good work ethic doesn't mean keeping your eyes glued to a book or computer screen. It requires a healthy balance between work, school, and pleasurable activities (**Figure 2.3.4**). While it is important to stay engaged and complete tasks on time, it is also important to participate in activities away from school and work. A healthy balance between work and play ensures you have time to disconnect, recharge, relax, and enjoy life. Finding that balance can be difficult but is an essential component to maintaining a strong work ethic over the long term.

Ground Picture/Shutterstock.com

Figure 2.3.4. Do you participate in hobbies or activities outside of school and work that help you recharge and relax? *How can these activities keep you balanced at work and school?*

Understanding Motivation

Sometimes work ethic is defined by your ability to stay motivated toward a goal. **Motivation** is the influence that makes you want to do things and finish them, even if those things are difficult, and it is key to staying focused. What motivates you? How do you motivate yourself?

Types of Motivation

Can you think of a time you worked really hard to accomplish something for yourself? Maybe practicing for a softball team tryout or a band audition? What was motivating you? Did you want to try something new, or please the parents or others who were encouraging you? There are many different reasons or motivators that keep you working toward your goals, but they primarily fall into two groups: *intrinsic rewards* and *extrinsic rewards*.

Intrinsic rewards speak to who you are as an individual and what you personally need, want, and value. Intrinsic rewards also address meeting your basic needs, such as food, water, clothing, and shelter—things you would be highly motivated to have in your life, regardless of your interests. Intrinsic rewards can motivate you to improve your self-image, work toward a specific achievement, or simply fulfill your curiosity. In addition to meeting needs, these rewards can give you pride in your work, satisfaction in your abilities, or the pleasure of a sense of responsibility.

Extrinsic rewards are different in that they motivate by providing a physical reward or validation from another person. These rewards can come in the form of good grades, money, recognition, and praise. Extrinsic motivators may come in the form of avoiding punishment or penalties. In the workplace, extrinsic rewards could also mean a new promotion or a higher salary.

Consider this: If motivation comes in these two different forms, could you be motivated by both? For example, you may be working hard to learn your lines to get a part in the school play because you enjoy acting (intrinsic) and want to improve your skills (intrinsic). You may also be motivated because your parents will be proud to watch you on stage (extrinsic). Another example might be what motivates you to help landscape outside this weekend. Will you be receiving $30 to mulch, weed, and help your grandmother plant new flowers in her garden? That might be a great extrinsic reward, but maybe you also really love the intrinsic reward of spending time with your grandmother, being outside, and helping her pick out flowers at the local garden center.

Most people are motivated by some mix of intrinsic and extrinsic awards. It is absolutely normal to use extrinsic rewards to keep you motivated, but sometimes you have to push forward with little in the way of praise, fanfare, or fancy new gadgets. Building and focusing on your intrinsic motivations will help you to keep focused and on task even when no one is cheering you on (**Figure 2.3.5**).

matimix/Shutterstock.com

Figure 2.3.5. *What keeps your eyes on the prize?* Consider the things that keep you motivated toward a goal and build them into your routine. Understanding your own motivation can keep you on target.

What Did We Learn?

- Creating a positive image is important when first meeting other people, including potential employers. Smiling, making eye contact, listening carefully, asking questions, and sharing your experiences can help ensure you make a good first impression.
- Having a strong work ethic means bringing your best values, attitudes, and behaviors to every job or task you complete. It is vital to success at home, work, and school.
- Organizing and planning for your day by creating a list of your responsibilities and activities can help you manage time wisely.
- Understanding the key components of motivation, including intrinsic and extrinsic rewards, can help us keep focused on completing tasks and reaching goals.

Let's Check and See What We Know

Answer the following questions using the information provided in Lesson 2.3.

1. *True or False?* Although clothing does not make us who we are, personal appearance and proper grooming are important because they affect the way we are perceived by others.

2. What term describes the ability to stay focused on a task or goal without being easily sidetracked?
   A. Motivation
   B. Eye contact
   C. Reward
   D. Self-discipline

3. Which of the following are *not* elements of a strong work ethic?
   A. Arriving to a meeting 10 minutes late
   B. Completing an assigned task on time
   C. Pushing through to finish a project
   D. Volunteering to help clean the greenhouse after school

4. *True or False?* First impressions provide others with a mental image of us that may or may not be true about who we are as people.

5. *True or False?* When meeting someone for the first time, you should avoid eye contact.

6. Extrinsic rewards include all of the following *except*:
   A. a high test grade.
   B. a pizza party.
   C. pride in your work.
   D. $20.

7. Which of the following would *not* be considered a professional habit?
   A. Leaving your bad attitude at the door
   B. Gossiping in the hallway about a team member
   C. Being honest and consistent in your words and actions
   D. Working collaboratively to meet goals

8. Which of the following is *not* an intrinsic reward?
   A. Food
   B. Salary or wages for completing a job
   C. Personal feelings of accomplishment
   D. Shelter

9. How long can it take to reverse or undo a bad first impression?
   A. Twenty minutes
   B. One week
   C. Three seconds
   D. Eight years

10. Which of the following describes an essential component to developing a good work ethic?
    A. Avoiding responsibility
    B. Turning in projects past due
    C. Not properly grooming before an interview
    D. Developing balance

Let's Apply What We Know

1. What is a strong work ethic? Provide examples.
2. How can punctuality display respect for others in the workplace?
3. What is motivation and why is it important to achieving our goals?
4. List four ways to make a great first impression.
5. Explain the difference between intrinsic and extrinsic rewards. Give examples of each.
6. How do we cultivate a positive, accurate image of ourselves for other people?

Academic Activities

1. **Social Science.** Interview working adults in your community to gain their perspectives on work ethic and professionalism. Ask questions to better understand who they are, what motivates them, and tips on cultivating career-ready habits for the future.

Communicating about Agriculture

1. **Social Media.** With the permission of your ag teacher, share your weekly newsletter (from the SAE for ALL Opportunities section below) via your school's or chapter's social media accounts.
2. **Social Media.** Brainstorm a list of accounts or hashtags on social media that might highlight professionalism and healthy habits. Share two with your class, and explain why they might be a quality resource.

SAE for ALL Opportunities

1. **Placement SAE.** Start a weekly newsletter highlighting tips and tricks for developing career skills such as professionalism, punctuality, and time management. Ask other FFA chapter members or classmates to contribute and include upcoming events, funny pictures, and stories of interest.

SAE for ALL Check-In

- How much time have you spent on your SAE this week?
- Have you logged your SAE hours?
- What challenges are you having with your SAE?
- How can your instructor help you?
- Do you have the equipment you need?

Budimir Jevtic/Shutterstock.com

Engaging in Effective Communication

Learning Outcomes

By the end of this lesson, you should be able to:
- Describe elements of effective verbal, nonverbal, visual, and written communication. (LO 02.04-a)
- Identify effective listening strategies. (LO 02.04-b)
- Discuss the importance of giving and receiving constructive feedback. (LO 02.04-c)

Words to Know

active listening
communication
constructive

effective communication
feedback
visual communication

written communication

Before You Begin

How have you communicated with other people today? Since you got out of bed this morning, did you post on any social media accounts? Have you sent a text message? Did you talk to your family about an upcoming field trip or an assignment? On a sheet of paper, brainstorm the ways you have communicated so far today. Identify the forms of communication you used as you read the lesson.

Figure 2.4.1. It only takes a few seconds to send an email to a friend overseas, but did you know it takes approximately 16 hours to fly from Atlanta, Georgia to New Delhi, India? That's 7,976 miles at about 500 mph. Technology has rapidly increased our ability to communicate with people around the world!

ioat/Shutterstock.com

Have you ever been amazed by the speed at which you can send a message to someone halfway around the world? Have you thought about how, in a matter of seconds, you can share a photo with friends and family thousands of miles away? Maybe you have not taken the time to appreciate this before, but if you ask your teacher, you may learn something about how this has not always been possible. How long do you think it would take to get the same message or photo to these people if you did not have the internet and cell service? See **Figure 2.4.1**.

The process of exchanging thoughts, ideas, messages, or information between two or more people is called *communication*. On a daily basis, you use several methods of communication, including:

- Verbal
- Nonverbal
- Written
- Visual

In this lesson, you will identify the elements of effective communication and consider the best methods to use at home, school, or in the workplace. You will also discuss the best listening strategies and begin to understand appropriate workplace interactions.

What Is Effective Communication?

Communication serves as the foundation of every interaction at school and in the workplace, but the ways in which we communicate are changing rapidly. While society once used face-to-face communication for most communication, we now use written words in the form of texts, emails, and posts online. With so much of our communication lacking the understanding gained when we can see each other in person, it's important to understand how to communicate effectively. **Effective communication** ensures clear, concise, and consistent information is shared, received, and understood by all involved. When we take time to ensure our messages are shared with others effectively, we can avoid unnecessary issues and misunderstandings with teachers and classmates, at home, and in the workplace. How often do you believe misunderstandings occur when information is lost or taken the wrong way? Can you imagine ways poor communication could go wrong or prove costly to a group or organization (**Figure 2.4.2**)?

Consider several strategies when engaging in effective communication, including making eye contact with the person or group sending or receiving verbal communication, actively listening in meetings, and responding to or providing feedback. Something as simple as looking another person in the eye can convey the complexity of a message and build trust between people. Similarly, good listening skills can improve how we perceive information and understand all components of a project or issue. Considering who you are speaking to is very important when conveying messages, as well as the tone you take when speaking. Appropriate tone can amplify your message, help provide clarity to the topic, and prevent any misunderstandings. Failing to consider these strategies may cause

Olena Yakobchuk/Shutterstock.com

Figure 2.4.2. Have you ever tried to talk to someone on a walkie talkie? While it's fun, it can also be easy to miss words or hard to understand a message. *Would you use a walkie talkie in a professional or educational setting? Why or why not?*

confusion, cause conflict, and hurt others' feelings. Can you think of a time when you did not consider effective communication strategies and unintentionally caused issues with a friend or family member? What happened and how might you have avoided the issue with effective communication?

Types of Communication

The methods of communication you use at home, work, and school are verbal, nonverbal, written, and visual. Our primary default method of communication, and what you probably think of when you think of communication, is verbal. *Verbal communication* is communication using spoken words and sounds. As noted in **Lesson 2.2**, written communication is also a form of verbal communication, as it uses words to convey meaning. *Nonverbal communication* uses body language, facial expressions, and posture to convey information without words. Visual communication is a form of nonverbal communication. Much of your daily interaction, maybe more than you realize, involves nonverbal communication. Can you judge a friend's mood from across the lunchroom without even talking, or maybe know what your parent is about to say just by their posture or facial expression? Many times, thoughts and feelings are conveyed effectively without conscious verbal or written cues. Instead, your brain gathers information received through eye contact, gestures, or tone of voice to communicate.

Can you think of any ways that visual communication is used in your daily life? Here's a hint: You may have used it when you were on your phone this morning or between classes. Today, adding emojis to text messages is a popular way to share information about our feelings and emotional states (**Figure 2.4.3**).

Complex messages are often conveyed with simple symbols, such as a smiley face or a thumbs-up. Have you ever sent a string of emojis that were not taken as you intended? To many people, emojis are not a professional means of communication and special attention should be paid to the use of text messaging and emojis in the workplace or a professional situation. Sometimes, visual communications used in text messages or emails can oversimplify complex thoughts and be misunderstood by the recipient.

Did You Know?

Visual communication is nonverbal communication that uses signs, symbols, or pictures. The ancient Egyptians used *hieroglyphics*, or picture words, on their monuments. The Mayans and Inca also used symbols. Cave paintings in Europe and Asia show that the inclination to use symbols and pictures to communicate was widespread, if not universal, in ancient civilizations.

Abrilla/Shutterstock.com

popicon/Shutterstock.com

Figure 2.4.3. Thousands of years ago, the ancient Egyptians combined symbols representing alphabet-like sounds together with images to represent ideas. Today, we use emojis to express emotion and communication in a similar way. *What do you think your great-great-great-grandchildren may be using to communicate in 150 years?*

Communication in professional settings relies on your ability to express yourself or transmit information with **written communication**, or messages exchanged via composed, written words, such as email, assignments, social media, and blogs (**Figure 2.4.4**). It is important that your language is professional and just as clear and concise in written form as it would be verbally.

Effective Listening Strategies

Have you ever talked to a friend while they scrolled through messages on their phone? Did you feel as though your friend were giving you their full attention? Did you like that feeling? Many times, simple eye contact can be the first step to **active listening**, or the act of concentrating on the person speaking, understanding their message, and responding appropriately. In addition to helping you focus on the topic, eye contact conveys interest or concern to the person speaking. Eye contact also helps draw people together. Focus on your behavior and attempt to keep eye contact, even if the other person is not returning the same listening cues (**Figure 2.4.5**). Being a good listener and modeling good active listening behavior will help the people around you to do better, too!

Keeping an open mind to the message being relayed can also help you improve your listening skills. Listen without jumping to conclusions. Remember: the speaker is attempting to express their thoughts or feelings, from their point of view. Listening to the words as they are being said and picturing them as you hear them may help you become a more effective listener.

With the rapid pace of life, you may often feel the need to convey messages as urgently as possible. However, there are risks to this. Sometimes, when you or someone you are communicating with is in a hurry, you can miss key facts or pieces of information that are important to the conversation. To prevent missing something important or misunderstanding, wait for the speaker to pause before adding new information or changing the topic. Ask questions if you do not understand the message, or if you are not certain. There is no harm in asking for clarification. It may be helpful to keep notes in some circumstances and summarize what is said to ensure you recognize relevant information.

panitanphoto/Shutterstock.com

Figure 2.4.4. Written communication can be done with a pencil or pen, but encompasses much more than that. *How many forms of written communication can you name?*

logoboom/Shutterstock.com

Figure 2.4.5. Eye contact is such a crucial part of communication that it's one of the first things that babies learn to do to gather information. *How do you think it makes a difference when you make eye contact?*

The Importance of Constructive Feedback

Practice makes perfect. When you are playing basketball or the guitar, practice helps you to focus your behaviors and improve your skills. Communication is no different! Good communication can help you gather input from others and build on your abilities at home, school, and work. Effective communication can make tasks easier, improve relationships, and save you time and effort. How do you think that practicing good communication can make your life easier and better?

One of the more difficult tasks in communication is learning how to give and receive *feedback*, or any information from others about your performance. Have you ever had a coach or teacher pull you to the side and give you advice on how to do better? Sometimes, depending on the circumstances, information, and emotions involved, it can be hard to listen to information or feedback that feels negative. It is important to see the feedback as **constructive**, or helpful to your ability to get better. Similarly, have you ever tried to talk to a teammate, classmate, or sibling about something they could improve upon? It can be just as hard to give constructive feedback to others, who may or may not want to hear it, as it is to hear it yourself. Learning how to give and receive constructive feedback will help improve your communication and leadership skills and increase healthy communication with those around you (**Figure 2.4.6**).

Rawpixel.com/Shutterstock.com

Figure 2.4.6. Having the right attitude about getting feedback—being open, thoughtful, and eager to learn—can make a huge difference in your ability to improve on the field, in the classroom, or at a job. *Have you ever gotten feedback that was hard to hear? What did you do with it?*

Hands-On Activity

Say It, Draw It

Stretch your active listening skills with a friend. You will need:

- Paper
- Pen or pencil

Ask your friend to draw an object by providing verbal step-by-step instructions. For example, instructions might include:

1. Draw a small 2-inch square in the bottom-left corner of your paper.
2. Draw a triangle inside the square.
3. Draw a line that divides the triangle into two parts.

Continue to make the activity more difficult by adding instructions for your partner. One simple misinterpretation could change the outcome, so it is important to listen carefully and gain understanding. Switch partners and compare shapes or objects.

Consider This

1. Did different partners' interpretations of the directions differ? Why?
2. How important was effective listening to this activity?
3. Why is it important to ask clarifying questions when communicating with others?

Receiving constructive feedback can be difficult, but considering the positive intent behind it can help reframe your thinking. Consider these strategies next time you are getting advice or feedback from a teacher, coach, parent, or friend:

1. Acknowledge what was said. Sometimes it is helpful to repeat back what you heard from the person providing feedback to ensure you understood the message correctly.
2. Accept the information. This can be difficult, especially if you do not agree, but it is important to accept that someone else's perception of the situation is valid.
3. Ask questions. Gain a clearer picture of what you're doing and how you can get better by asking questions like, "Do you have any suggestions for me?" or "What else should I do?" This will ensure you have advice, information, and concrete actions to take to improve, in addition to the information.
4. Create a plan to address the behavior described in the feedback. How can you help improve the situation by changing your behavior? By developing a plan, you are making a commitment to make things right.
5. Take action. Make strides to improve. Changing behavior doesn't always happen quickly, but taking steps to follow your plan is important.
6. Stay positive. Keep a positive attitude and continuously practice taking constructive feedback.

Career Corner   POULTRY SCIENTIST

Hey, there! If your communication skills are top notch, then maybe being a **Poultry Scientist** *is a career for you to explore!*

What Do They Do? Poultry scientists work with other researchers to help improve the genetic profiles and reproduction practices of poultry at production facilities around the world. They also work with individual production facilities to implement new, research-based techniques and technologies, and to help reduce the spread of infectious diseases.

What Do I Need to Be Good at This? Poultry scientists love science, animals, and solving problems. They are also really good listeners! Clear communication between scientists and farmers could be the difference between a healthy flock ready for market, and widespread disease that affects thousands of animals. This important work relies on clear communication and healthy relationships.

What Education Do I Need? You can become a poultry scientist with a bachelor's degree in agricultural science or animal science; some advanced positions may require you to have a graduate degree.

How Much Money Can I Make? Poultry scientists in the US earn between $29,500 and $69,000 annually, with an average salary around $58,000.

How Can I Get Started? You can build an SAE around animal science. Can you find anyone in your area raising chickens? You could interview them to learn how they got started and what challenges they face. Can you research how much it would cost to start raising your own chickens, and what supplies you might need, or work with your school to see if you are able to incubate some chicks? Now is the time to get started!

What Did We Learn?

- Effective communication ensures clear, concise, and consistent information is shared, received, and understood by everyone involved.
- Various methods of communication, both verbal and nonverbal, help us convey information with the use of language, symbols, and gestures.
- Making eye contact, drawing mental pictures, and asking follow-up questions can improve verbal communication skills and help prevent misunderstanding.
- Learning to take and give constructive feedback can help us seek advice from others and improve our abilities, while improving communication with those around us.

Let's Check and See What We Know

Answer the following questions using the information provided in Lesson 2.4.

1. The first step to effective listening is to ____.
   A. seek feedback
   B. send an email
   C. make eye contact
   D. ask questions

2. *True or False?* Using emojis to convey information in an email is appropriate in the workplace.

3. ____ includes the use of signs, symbols, or pictures.
   A. Constructive feedback
   B. Visual communication
   C. Active listening
   D. Verbal communication

4. Which of the following is a method of exchanging thoughts, ideas, messages and information between two or more people?
   A. Email
   B. Text message
   C. Conversation
   D. All of these.

5. *True or False?* Giving a friend a thumbs-up from across the room is one method of nonverbal communication.

Let's Apply What We Know

1. Consider the following message: "Your room is a mess! Clean it immediately!"
   A. In what ways might this message be conveyed?
   B. How would you clearly and concisely convey the same message without the use of verbal communication?
   C. Take a few moments to draw pictures, symbols, or emojis that convey the same message without words. Now, think about the difficulty one might have decoding your string of symbols, pictures, or emojis without reading the original message. How could you make this message clearer? Is this the best way to get this message across?
   D. Using this exercise as an example, why is it important to use appropriate forms of communication?
2. Why is it important to provide constructive feedback?
3. Explain the difference between verbal and nonverbal communication. Give examples of each.
4. Why is it important to practice active listening?
5. How do we use symbols to communicate? Give examples.
6. What strategies can we use to communicate clearly and effectively?

Academic Activities

1. **Social Studies.** What historical figure do you most admire and why? Research their background and major accomplishments, with specific attention given to communication. Did this person deliver a speech or share any of their written work? What communication skills did they model that might be helpful to you as an effective communicator?

Communicating about Agriculture

1. **Writing and Social Media.** Create a blog or vlog to help educate others on agricultural topics of interest to your school and community. As you create it, think about how you can use good communication skills to share this information with others.

SAE for ALL Opportunities

1. **Foundational SAE.** What makes someone a good communicator in agriculture? Talk to your ag teacher about what skills they use to communicate information effectively to students. What challenges do they face? What other examples of communicators can you find in agriculture? Write up a report covering effective and creative communication strategies in getting relevant, quality information to middle school students.

SAE for ALL Check-In

- How much time have you spent on your SAE this week?
- Have you logged your SAE hours?
- What challenges are you having with your SAE?
- How can your instructor help you?
- Do you have the equipment you need?

Growing through Curiosity and Creativity

pixelrain/Shutterstock.com

2

Essential Question

How do curiosity and creativity help prepare us for successful careers?

Learning Outcomes

By the end of this lesson, you should be able to:
- Describe how curiosity functions as a part of the learning process. (LO 02.05-a)
- Explain the problem-solving process and its value in exploring career opportunities. (LO 02.05-b)
- Describe how creativity is related to problem-solving. (LO 02.05-c)
- Apply creative thinking to solve scientific problems. (LO 02.05-d)
- Understand the importance of critical thinking to solving problems. (LO 02.05-e)

Words to Know

creativity	data	problem-solving
critical thinking	hypothesis	process
curiosity	observation	scientific method

Before You Begin

Before starting this next lesson, let's think back over what we've learned in this unit. Take a few minutes to skim through the earlier lessons. What were the major takeaways? Why is it important to nurture good character? How can you build leadership abilities, and why are they important? What attitudes and habits will help build success in your future career? How can you maintain effective communication at home, work, and school? Today's lesson will build on what we've learned so far by considering how additional characteristics, such as curiosity and creativity, help us to grow as problem-solvers.

Tinkering with Curiosity

What gets you excited about learning something new? Not just in terms of learning at school, but at home, church, with a friend, or on vacation? What is it that leaves you asking questions, searching for more information, and really trying to figure something out? This feeling is called *curiosity*, or the strong desire to know or learn something new. For many people, curiosity is the force behind their desire or motivation to explore new topics or seek solutions to problems (**Figure 2.5.1**). You can think of it like the spark that ignites the fuel in a gasoline engine. One tiny spark gets the whole engine running! Of course, you won't be curious about every idea or new piece of information that comes your way, and that's okay. When something does spark, though, watch out: Your curiosity supercharges your interest and enthusiasm and gives new meaning to learning. It makes learning more fun!

A little bit of curiosity goes a long way when it comes to your motivation to work hard. In fact, scientists believe curiosity and hard work are as important to your success as intelligence. They may even be more valuable to your future career! Understanding curiosity requires you to consider the gap between what you know and what you don't know. You might be wondering how you're supposed to figure out what you don't know when you don't know it; even that question sounds pretty difficult. One of the many amazing things about the human brain is its ability to naturally notice a gap in your understanding when you are presented with new information. It's your curiosity about the gap that sends you into problem-solving mode! *Problem-solving* is the process of searching for a solution to a question or problem. The key word here is *process*, or a series of actions or steps followed on the way to achieving a goal.

Think about your other subjects for a minute. Do you remember the word *process* cropping up there? What other processes do you remember? For instance, do you remember what the letters **PEMDAS** mean? Here we go: parentheses, exponents, multiplication, division, addition, and subtraction. Those letters are a way of remembering the order of operations (a process) to solve math problems. Similarly, when you are presented with a new problem, topic, or question, curiosity alerts the portion of your brain that wants to fill the gap and connect with new information. Harnessing your curiosity through problem-solving can help you build a future filled with the excitement of learning.

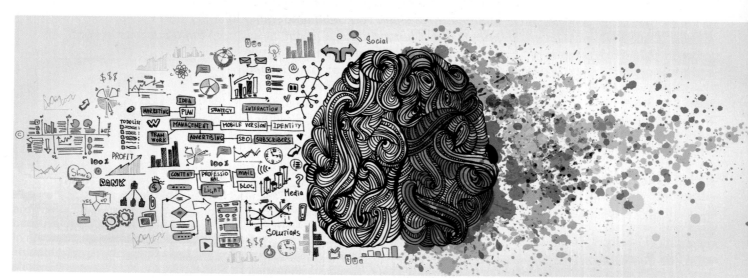

Figure 2.5.1. Curiosity is a powerful tool when learning new things! *What topics get your brain moving?*

Thinking Differently

You may remember reading (or listening to) Dr. Seuss books as a kid. In *Oh, The Thinks You Can Think,* he advises students to think left and right, and low and high. What does he mean? This line celebrates the endless possibilities your imagination can offer. More importantly, it helps you consider how you put that imagination to work with creativity. *Creativity* is the capacity to use your imagination to generate or recognize new ideas and possibilities or make something original that is of value (**Figure 2.5.2**). The item or idea does not have to be tangible, but might be a piece of art, a new invention, or a poem. In what ways are you creative? Take a few minutes to jot down all the different ways you stretch your imagination daily. What's on your list? Share your notes on creativity with a neighbor.

Focusing Curiosity and Creativity with Scientific Inquiry

Sometimes it can be difficult to connect processes or subjects that seem different. In what ways are subjects like art and biology alike? At first glance they may seem different, but both artists and scientists search for answers and approach problems with open-mindedness, curiosity, and creativity. While their approaches may look completely different, the methods and procedures may be structured in the same way. For example, to researchers and scientists, *data*, or information collected through scientific observation, serves as a product gathered during investigation into understanding a problem. It's the same type of scientific *observation*, or process used to gain information through detailed examination and experimentation, used in the development of a new work of art. Both art and science are attempts to understand and describe the world around us, and both welcome failure as part of the creative process.

Figure 2.5.2. What do you notice about this photo of Earth? Sometimes, seeing the world differently is what we need when seeking answers to problems. *How can we change our perspectives to solve old problems?*

canbedone/Shutterstock.com

Creativity Isn't Just Good for Problem-Solving

Studies show that creativity is great for your brain's health. Participating in creative activities eases symptoms of stress, anxiety, and depression. Over the course of the coronavirus pandemic, many people have had extra time at home or alone. Creativity is one way to deal with the stresses this time has caused for many people. Creativity connects you to your emotions and thoughts, provides an outlet to process the things we feel, and can allow us to connect with others on a deep level.

Consider This

1. Does it surprise you that creativity and mental health are related?
2. How do you process your emotions?
3. Think about what sort of creative outlets you participate in. Do you have hobbies or passions that fall under the creative umbrella? If not, what might you be interested in trying?

Do you consider yourself more of a scientist or an artist? Maybe both? Consider problem-solving through the lens of science using the *scientific method*, or the specific step-by-step process that scientists use to think about and make sense of complex questions. These steps may help to simplify your thinking, while giving you some practice using concrete methods of investigation.

Step 1: Ask Questions and Seek Information

What do you wonder about? What would you like to know? Once you have a question in your mind, seek out information to understand the question better. You might start by looking at reliable sources online or at the library. Understanding what is already known about this question before diving in is necessary when thinking of solutions.

Step 2: Make an Informed Guess

A *hypothesis*, or informed guess, is a possible explanation or answer to your question, based on what you already know about the topic. You don't have to be right! A hypothesis is a starting point to work from as you continue to investigate. You can begin to test your hypothesis with further study.

Step 3: Test Your Ideas

Design a way to better understand your hypothesis. Can it be tested? If so, how? Create a step-by-step method of investigation. Make observations and track your results. To be certain of your findings, repeat your investigation several times and record your results each time.

Step 4: Analyze Information and Build Conclusions

Now that you've recorded your observations, what do they mean? Take time to consider and analyze this information. Was your hypothesis correct? Why or why not? How do you know? Did you follow your methods appropriately? Explain your experiment to someone else and review any lingering questions you still have. Many times, this process should be evaluated, adjusted, and/or repeated based on the results. Keep building on your findings and let failure inform your next experiment!

In what ways can you use scientific thinking to help structure your creative thoughts and lingering curiosities? Why is this structure so important (**Figure 2.5.3**)?

Thinking about Thinking

Components of the scientific method can be used far beyond a science lab or project. In fact, *critical thinking*, or the ability to think deeply in a clear and informed way, uses the same components and steps of gathering information, forming hypotheses, testing ideas, analyzing new information, and developing informed conclusions. The key to critical thinking is constant training. By consistently practicing your problem-solving abilities, you can build your ability to think critically. The neat thing

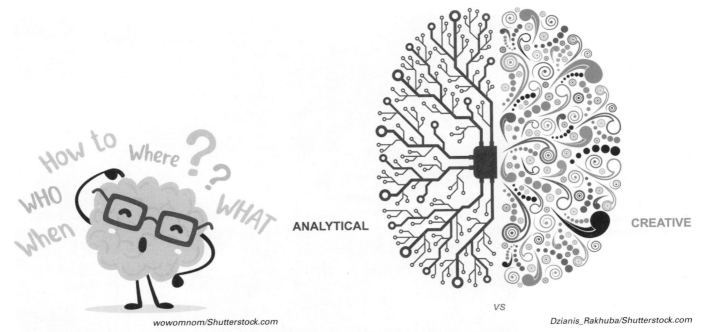

bangoland/Shutterstock.com

Figure 2.5.3. You don't have to be in a lab to use the scientific method. *What ways can you use the scientific method to help think about and solve everyday questions and problems?*

about critical thinking (well, one of them) is the way that your brain quickly filters through information on its own, the more practice it has. Eventually, your brain quickly recognizes the difference between useful and useless information as it solves problems. Think of it as a tiny doorway between the two sides of your brain. The more you think critically and intensely about a problem and the many ways to solve it, the bigger the doorway gets, and the faster messages or possible solutions can run through it (**Figure 2.5.4**).

ANALYTICAL   CREATIVE

VS

wowomnom/Shutterstock.com   *Dzianis_Rakhuba/Shutterstock.com*

Figure 2.5.4. The brain is a powerful organ that works like a central computer to control your body's functions. It's difficult to imagine, but millions of messages are sent and received through the nervous system every day, even when you're not aware it's happening. *In what ways can you focus some of those messages on learning through critical thinking?*

The brain is a complex organ, so it's not exactly that simple, but it's a good way to consider the active process of thinking about a problem and then tinkering with possible solutions.

Critical thinking and problem-solving are valued by employers and important in every career, at every level. When employers discuss these skills, they are often referring to the ability to handle difficult or unexpected situations in the workplace. By placing yourself in new situations, you will encounter a variety of new and different problems to solve. Taking advantage of opportunities to explore new careers and life situations will help you stretch, expand, and improve your abilities to learn, think critically, and solve problems. The more practice you have, the better prepared you will be for your future career!

Employers may also use the term *critical thinking* to refer to the need for technical skills that are industry or job specific. For example, a veterinary assistant will need to use effective communication skills, such as active listening, when understanding a pet owner and the medical needs of their pet. They will also need specific technical knowledge related to animal illnesses and medications to better understand and help solve the problem. Participating in SAEs and career development activities related to your areas of interest will help prepare you with the skills you need to impress employers.

Sometimes, the process of problem-solving seems to slow down your creativity. It can be frustrating to search for an answer to your most curious questions and hit roadblocks along the way, but it is important to stay optimistic and persevere toward a goal. Some of the world's most brilliant scientists experienced years of failed experiments before getting results. You must stick with it and grow from failure (**Figure 2.5.5**). Trusting a new process may be difficult, but it gets easier over time. Test it out yourself!

Figure 2.5.5. Walt Disney, the creator of Mickey Mouse and the Walt Disney Company, once said, "We keep moving forward, opening new doors, and doing new things, because we're curious and curiosity keeps leading us down new paths." His commitment to following his imagination and never giving up is now responsible for decades of entertainment and innovative technology known around the world. *Make a list of other leaders, creators, or famous folks who show creativity in their daily lives. What differences have their contributions made in our world?*

Hands-On Activity

Creativity and Solving Scientific Problems

It might sound strange to you to use creativity to solve scientific problems, but it's not a difficult thing to do, once you get the hang of it! Let's give it a try together and have a little fun.

For this activity, you will need:

- Paper
- Pen or pencil

Your agriculture teacher will divide your class into groups of 3 or 4 students. This will be your problem-solving team.

Each group will come up with an answer to this question: How can our greenhouse grow bigger, brighter, and more beautiful flowers? The goal for each group is to come up with the most fun, creative solution to this problem.

Each group will work together to work through the steps of the scientific method.

What information do we need to answer this question? Make a list of any relevant information you can find in five minutes, or any previous knowledge.

What informed guesses can you make about this question? Come up with as many guesses as you can as a group. Choose the most creative, interesting, out-of-left-field guess that your team comes up with.

How can you test this idea? You won't be doing this experiment, but how could you figure out if this wild guess was true? Let your minds go wild and have some fun in designing an experiment. As long as it would get you the information you need, your ideas are fair game! Make a list of the items you would need, and the steps you would follow to test your idea.

Share your guess and experiment design with the class.

Consider This

1. Were you surprised by any of the ideas that other teams shared? Did any of them make you laugh, or give you any new ideas?
2. If you were going to adapt your experiment to be something you did in class, how would you do it? What changes would you make, if any? Why?
3. These ideas were supposed to be creative and interesting. Did any of them cross the line from creative to ridiculous? Where is that line?

What Did We Learn?

- Curiosity describes our desire to learn something new. It leads us to understand the gaps between what we know and what we do not know.
- Problem-solving is a process that leads you through a series of steps or actions when searching for a solution to a problem or question.
- Focusing our creativity and curiosity with problem-solving methods, such as the scientific method, is helpful when considering complex questions or ideas.
- The scientific method is the step-by-step process that scientists use to test new ideas, and it can help in solving other problems, too. To use the scientific method, you ask a question and seek information, make an informed guess, test your ideas, and analyze the results to draw conclusions.
- Critical thinking is the active process of thinking about a problem and possible solutions and testing your ideas and assumptions to see if they make sense in light of the available information.

Let's Check and See What We Know

Answer the following questions using the information provided in Lesson 2.5.

1. Which of the following items are products of creativity?
   A. A painting of three dogs
   B. An original short story on the adventures of two pigs and a goat
   C. Song lyrics about the weather
   D. All of these.

2. Which of the following terms describes information collected through scientific observation?
   A. Opinion
   B. Data
   C. Communication
   D. Hypothesis

3. *True or False?* Curiosity is the desire to know or learn something new.

4. *True or False?* The key to critical thinking is constant training and exercise of the brain.

5. Which of the following terms describes an informed guess?
   A. Conclusion
   B. Hypothesis
   C. Research
   D. Creativity

Let's Apply What We Know

1. List and describe the steps of the scientific method. Think of a question that you'd like to answer (it can be a fun one) and use that question to help fill in examples in your description.

2. In what ways are art and science similar? How about English and math?
3. How is critical thinking valuable when solving problems?
4. How can following a process help improve the output of your creative thought?
5. Explain how exercising creativity helps you be a better problem-solver.

Academic Activities

1. **Social Studies.** It can be difficult to separate fact from opinion, especially when watching television or reading information on the internet or social media. How do you know what is true? How can you filter out misinformation, or determine what is false? Take the following statements: 1) Oranges contain both calcium and vitamin C; and 2) Manatees are the ugliest of all creatures in the sea. Which of these statements is a fact? Which is an opinion? Ask a friend to tell you something (it can be true or false, opinion or fact). Consider the following questions as you think about the statement, and explain how you come to your conclusions.
   - Can this statement be proven beyond a doubt? How?
   - Does it have a bias?
   - Is there anything misleading about this information?
   - Does this statement rely on assumptions? How can I tell?
   - Just because I agree with this, does that mean it is a fact? If I disagree with it, does that mean it is untrue?
   - Does this information appeal to my emotions?
   - Where else can I go to verify this statement?

Communicating about Agriculture

1. **Understanding Communication.** Curiosity and creative problem-solving lead to significant breakthroughs in agriculture, but it is just as important that agriculturalists inform the public about the issues, problems, and discoveries they make. Progress can be slower when we fail to help others understand our work in relatable ways. In what ways might you communicate the topics learned in this lesson to others? How can you leverage creativity in being an effective communicator?

SAE for ALL Opportunities

1. **Service Learning SAE.** We don't have to search very far to find problems to solve, and they don't have to be big or difficult for us to lend a hand. Take a few hours this week being curious about the problems you can help solve at home or in your community. Then, take action! There's no limit to the problems we can solve and people we can help when we put our mind to it!

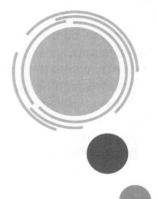

SAE for ALL Check-In

- How much time have you spent on your SAE this week?
- Have you logged your SAE hours?
- What challenges are you having with your SAE?
- How can your instructor help you?
- Do you have the equipment you need?

UNIT 3
The Importance of Agriculture

SAE for ALL Profile
Taking Over the Family Farm

Scotty Huddleston always knew that one day he would take over his family farm. To him, there is no better smell than that of a freshly plowed field or cut corn silage. As a small child he loved nothing more than climbing up on the big farm equipment or riding around the field with his father. Scotty said, "One of my favorite things to do as a high school FFA member was to drive my tractor to school in the FFA tractor parade."

Scotty had a Placement SAE, working with his dad and uncle in forage production. A forage production SAE produces and markets crops for forage such as alfalfa, clover, brome grass, orchard grass, grain forages, corn and grass silages, and sorghum (not used for grain). The forage produced through Scotty's SAE went into cattle feed. Scotty and his family own and lease almost 3,000 acres of silage production.

Scotty says the most rewarding part of his SAE was seeing something through from start to finish. He was the one who plowed the field, planted the seed, fertilized, managed the growth, and harvested the corn. He found this same satisfaction in his high school horticulture classes when working in the school greenhouse, getting ready for the annual spring plant sale. Being able to help local community members pick just the right plants for their own yard and being able to explain how to best take care of the plants was extremely satisfying to him.

Although a high school degree is all Scotty would have needed to continue working on his family farm, he attended a local community college where he earned a two-year agribusiness degree. He plans to use this degree to help him run and manage the farm one day. He is also working with his dad to expand the farm to offer more agritourism opportunities for the local community members, such as pick-your-own pumpkin patches and farm hayrides in the fall.

- Scotty wants to stick with a career in agriculture for many reasons, and he finds the work fulfilling. Can you imagine feeling fulfilled in an agricultural career?
- Scotty has a business degree, even though it isn't a requirement to run a farm. How will this help him to run the farm? What benefits do you see to his decision?

Top photo: HQuality/Shutterstock.com
Bottom photo: Smereka/Shutterstock.com

Sasina/Shutterstock.com

The Effects of Agriculture on Daily Life

Aleksandr Rybalko/Shutterstock.com

Learning Outcomes

By the end of this lesson, you should be able to:
- Describe why agriculture is important. (LO 03.01-a)
- List items provided by agriculture that you use each day. (LO 03.01-b)
- Explain the relationship between individual people and the agriculture industry. (LO 03.01-c)

Words to Know

consumer
ethanol
fiber
producer

United States Department of Agriculture (USDA)
wool

Before You Begin

Who likes to play games? Are you competitive like me? Let's play a game right now. Make a list of all the things that you think you would still be able to do or that you would still have or use if agriculture didn't exist. Take your best shot: What are all of the things you have in your life that are not related to agriculture in any way? The person with the most accurate list at the end of two minutes wins! Ready? Go!

Was it harder than you thought? It was for me. Goodness, this is hard! Agriculture provides food, so I can't write that. It provides the materials to build my shelter (house) too, so that is out. I couldn't walk my dog (no domesticated animals or animal care), there wouldn't be any grass (growing plants in nurseries) or concrete (using natural inputs to create new by-products). No clothing (Yikes! That could be embarrassing). No playing football with my friend—we wouldn't have a football without the leather that comes from raising cattle. I couldn't drive anywhere, as there wouldn't be any gas. I wouldn't even have the paper or pencil I'm using to make this list! There must be something agriculture doesn't provide.

Did you come up with anything?

Whether you are pouring milk onto your morning cereal or taking a bite out of your peanut butter and jelly sandwich, most people know that the food we eat is provided to us by agriculture. This food comes from hardworking farmers and ranchers across our country. You might not realize that many other, less obvious things that you use every day are also made from products provided by agriculture.

Agriculture's Effects on Daily Life

Did you know that you start your day with agriculture? Toothpaste contains corn sugars known as dextrose. Dextrose is a water-soluble bulking agent that makes the toothpaste thick, but also dissolvable when mixed with water. Corn is also used in items as diverse as makeup, shampoo, and conditioner.

Agriculture touches all kinds of aspects of your daily life that you may not be aware of! One aspect of agriculture that most people do not think about is fuel. **Ethanol** is a corn-based fuel that is often added to gasoline. Cornstarch is used in the production of tires by keeping the rubber in the tires from sticking to the molds. Agriculture is also in the things we use for entertainment. The paper in your favorite book or comic is made from trees raised for that purpose, and the ink is likely made from soybeans (**Figure 3.1.1**). Stringed instruments, such as violins and guitars, are made from wood, and the strings may be made from animal intestines. Ew! Who knew?

SherSor/Shutterstock.com

Figure 3.1.1. Soybeans are one of the most versatile crops produced. Nearly all soybeans are processed for their oil. The oil may be refined for cooking and other edible uses such as salad dressings or margarine. Soybean oil can also be used in industrial manufacturing to make products like biodiesel, carpet cleaners, and candles.

Did You Know? Cattle, corn, and soybeans are the top three US farm products. Unsurprisingly, there are many products created from cattle, corn, and soybeans (and their by-products), but you might be surprised by what some of these products are!

Do you think sports depend on agriculture? In basketball, the hardwoods that make up the floors come from trees. The football that is thrown around the field each fall is made from cowhide. Soccer and golf would not be the same without the turf grass and greens. Baseball is included too, from the leather used for the ball to the trees used to make the bat. How are tennis balls tied to agriculture (**Figure 3.1.2**)?

School would not be as much fun without agriculture. Did you know that more than 170,000 pencils can be made from one tree? Crayons are made from soybeans, and an acre of soybeans can make over 82,000 crayons. Glue is made from animal by-products. There would also be no paper (from trees), textbooks (again, paper from trees), or classroom desks without agriculture.

Ellie.tuang/Shutterstock.com

Figure 3.1.2. Tennis balls are also made from agricultural products. The bounce in a tennis ball comes from the natural rubber center, which is harvested from rubber trees found mainly in Thailand. The outside of the ball is covered in woolen fibers. *Where do those come from?*

Hands-On Activity

Create a Daily Agricultural Products Collage

Have you ever thought about how many products you use in a single day? From your soap and toothpaste in the morning to the car or bus to get you to school to your pillow and sheets at night, you touch a tremendous number of different items each day. How are the products you use on a regular basis related to agriculture?

For this activity, you will need the following:

- Colored construction paper
- Markers
- Old magazines or internet images
- Scissors
- Glue or tape
- Device for internet research

Here are your directions:

1. Find images of items you use on a regular basis, such as clothes or a cell phone. You can cut these out of an old magazine or print them off a computer.
2. Research the agricultural products used to create those items, such as trees, plants, or animals.
3. Attach your pictures to your construction paper with glue or tape.
4. Beside each item write in marker the agricultural product it is made from, such as *animal hair*, beside a picture of a hairbrush.

Consider This

1. What everyday product had the most surprising origin to you? Why?
2. Do you think most people realize the impact agriculture has on them every day? How could you spread awareness to people in your friend group, family, or school?

Did You Know?

Farm and ranch families make up less than two percent of the US population.

Agriculture's Impact on the Food Supply

Most people are aware that farmers and ranchers grow or raise almost all the food we eat. Food is any material or substance that is used for human consumption. Farmers work long hours to ensure that Americans have food available at grocery stores across the country every day. Food production goes beyond the farmers growing and harvesting the food or raising the livestock we eat, though. It also includes the processing, distribution, marketing, and sale of all products used for human consumption. Food production is the broadest area of agriculture (**Figure 3.1.3**).

Figure 3.1.3. There are five US states, California, Iowa, Nebraska, Texas, and Minnesota, that produce over one-third of the food Americans eat every day.

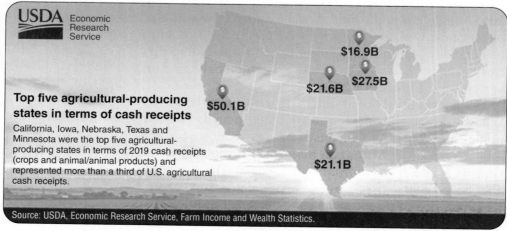

USDA Economic Research Service

Top five agricultural-producing states in terms of cash receipts

California, Iowa, Nebraska, Texas and Minnesota were the top five agricultural-producing states in terms of 2019 cash receipts (crops and animal/animal products) and represented more than a third of U.S. agricultural cash receipts.

$16.9B
$27.5B
$21.6B
$50.1B
$21.1B

Source: USDA, Economic Research Service, Farm Income and Wealth Statistics.

USDA

The United States government ensures that the food we eat is safe through the **United States Department of Agriculture (USDA)**. The USDA is a department of the United States government that manages various programs related to food, agriculture, natural resources, rural development, and nutrition. The USDA writes and enforces regulations that farmers and producers must follow to sell their products to the public. The USDA also keeps information on the

Table 13-3.—Consumption: Per capita consumption of major food commodities, United States, 2012–2016 [1]

Commodity	2012	2013	2014	2015	2016
	Pounds	*Pounds*	*Pounds*	*Pounds*	*Pounds*
Red meats [2]	98.0	98.0	95.0	98.6	100.5
Beef	54.5	53.6	51.5	51.4	52.9
Veal	0.3	0.3	0.2	0.2	0.2
Lamb and mutton	0.6	0.7	0.7	0.7	0.8
Pork	42.6	43.5	42.6	46.3	46.6
Fish	14.3	14.5	15.5	14.9	16.1
Canned	3.7	3.3	3.7	3.3	3.7
Fresh and frozen	10.4	10.9	11.5	11.3	12.0
Cured	0.3	0.3	0.3	0.3	0.3
Poultry	69.3	70.3	71.3	75.2	76.3
Chicken	56.7	57.7	58.8	62.6	63.2
Turkey	12.6	12.6	12.5	12.6	13.1
Eggs	32.7	33.2	34.2	33.0	35.0
Dairy products [3]					
Total dairy products	615.1	607.3	616.0	630.1	645.7
Fluid milk and cream	188.6	184.6	178.6	174.7	172.0
Plain and flavored whole milk	46.5	46.2	45.9	47.6	49.8
Plain reduced fat and light milk (2%, 1%, and 0.5%)	84.7	82.6	79.8	76.8	74.2
Plain fat free milk (skim)	24.8	22.6	20.0	17.7	15.6
Flavored lower fat free milk	12.5	12.3	12.0	12.1	12.3
Buttermilk	1.5	1.6	1.5	1.6	1.6
Eggnog	0.4	0.4	0.4	0.4	0.4
Yogurt (excl. frozen)	14.0	15.0	14.9	14.4	13.8
Sour cream and dip	4.1	4.1	4.1	4.1	4.3
Cheese (excluding cottage) [4]	33.3	33.4	34.2	35.1	36.4
American	13.3	13.4	13.7	14.0	14.4
Cheddar	9.6	9.6	9.8	10.2	10.4
Italian	13.8	13.8	14.2	14.5	15.2
Mozzarella	10.7	10.7	11.2	11.3	11.7
Cottage cheese	2.3	2.1	2.1	2.1	2.1
Condensed and evaporated milk	7.3	7.2	6.8	7.7	7.4
Ice cream	13.2	13.0	12.5	12.9	12.9
Butter	5.5	5.5	5.5	5.6	5.7
Fruits and vegetables [5]	636.0	636.6	636.6	631.9	651.7
Fruits	244.6	254.6	250.1	252.7	256.4
Fresh	131.8	136.3	136.3	136.1	142.3
Citrus	23.5	23.9	23.3	22.7	24.0
Noncitrus	108.3	112.4	113.1	113.4	118.3
Processing	112.8	118.2	113.8	116.6	114.1
Citrus	52.6	55.1	52.6	52.7	51.9
Noncitrus	60.2	63.1	61.2	63.8	62.2
Vegetables	391.4	382.1	386.5	379.2	395.3
Fresh	188.7	184.6	186.4	186.4	199.5
Processing	202.7	197.5	200.1	192.9	195.8
Flour and cereal products	174.1	174.7	174.4	172.9	171.9
Wheat flour [6]	134.3	135.0	134.7	133.0	131.7
Corn products	33.9	33.9	34.0	34.0	34.2
Oat products	4.7	4.5	4.5	4.6	4.6
Barley and rye products	1.1	1.2	1.2	1.3	1.4
Caloric sweeteners (dry weight basis)	129.3	128.2	129.1	129.1	128.1
Sugar (refined)	66.6	68.0	68.4	69.1	69.7
Corn sweeteners [7]	61.0	58.4	58.5	57.8	56.5
Honey and edible syrups	1.8	1.9	2.1	2.1	1.9
Others					
Coffee (green bean equivalent)	9.7	9.9	10.0	10.2	NA
Cocoa (chocolate liquor equivalent) [8]	4.2	4.2	4.1	4.3	NA
Tea (dry leaf equivalent)	0.9	0.9	0.9	0.9	NA
Peanuts (shelled)	6.7	6.9	7.0	7.4	7.2
Tree nuts (shelled)	4.2	4.0	4.1	4.1	4.9

[1] Quantity in pounds, retail weight unless otherwise shown. Totals may not add due to rounding. [2] Boneless, trimmed weight equivalent. [3] Total dairy products reported on a milk-equivalent, milkfat basis. All other dairy categories reported on a product weight basis. [4] Natural equivalent of cheese and cheese products. [5] Farm weight. [6] White, whole wheat, semolina, and durum flour. [7] High fructose, glucose, and dextrose. [8] Chocolate liquor is what remains after cocoa beans have been roasted and hulled; it is sometimes called ground or bitter chocolate.
ERS, Food Economics Division, (202) 694-5400. Historical consumption and supply-disappearance data for food may be found at https://www.ers.usda.gov/data-products/food-availability-per-capita-data-system.

USDA

Figure 3.1.4. Think about some of your favorite foods and see where they fall on the list. What about foods like pizza that aren't listed? Many foods, like pizza, cover many different categories from wheat for the crust, vegetables for the sauce, dairy for the cheese, and pork for the pepperoni.

Tada Images/Shutterstock.com

Figure 3.1.5. When visiting the grocery store, you might notice USDA labels on many of the food products. These labels mean that the farmer followed the guidelines given by the USDA in the production of that item. This label tells the buyer that the apples are organic, which means the farmer followed all the organic rules and regulations given by the USDA when growing and harvesting.

Did You Know?

Every baseball requires 150 yards of wool.

types of products Americans consume each year and how much Americans spend on their food supply. According to the USDA, in 2018 more than 325 million people in the United States depended on American agriculture for food and fiber. Use **Figure 3.1.4** to determine how many pounds of food the average American consumes in a single year.

Did You Know?

On any given day, one in eight Americans will eat pizza. Yum!

Lesson 3.3 takes a closer look at the crops that are used to make our food supply.

Agriculture's Impact on Clothing

In addition to producing our food, farmers in the United States also provide the fiber necessary to produce our clothes. Have you ever looked at the labels of your clothes to see what materials were used to make them? Those tags tell you which fiber was used to produce the clothing. *Fiber* is a thread or filament used to create woven or composite material. The two main types of fibers used to produce clothing or other fabric products are *natural fibers* and *synthetic fibers*. Natural fibers are found in plants or animals. Synthetic fibers, such as polyester or nylon, are made from petroleum products and other natural resources.

Making Connections  **#STEM**

Agriculture Youth Programs

Does learning more about the world of agriculture interest you? There are lots of youth programs offered around the nation that encourage students to learn and explore a wide variety of agricultural topics. The United States Department of Agriculture (USDA) offers a free two-week program in the summer specifically for middle and high school students called AgDiscovery. AgDiscovery is a unique opportunity for students to explore

agricultural sciences and gain knowledge about careers in animal and plant sciences, wildlife management, veterinary medicine, biotechnology, entomology, food safety, food production, agribusiness, forestry, and many more. AgDiscovery is hosted nationwide at many different universities.

Research more about possible youth programs on the internet. Talk to your teacher or guardian(s) about getting involved if it excites you!

Commonly produced natural fibers include cotton, linen, and wool. The largest fiber crop produced in the United States is cotton (**Figure 3.1.6**). One bale of cotton can be used to make up to 215 pairs of jeans. Other crops grown for their fiber traits include hemp and flax. Some animal fibers are used to make clothing as well. Wool is the most widely produced animal fiber. **Wool** is the hair-like material that covers sheep. Wool is harvested by shearing the sheep several times a year, much like getting a haircut every few months. Other animal fibers are silk, fur, and certain animal hairs, like mohair, which comes from angora goats.

Synthetic fibers make use of important natural resources to produce clothing. Some synthetic fibers, such as rayon, are made from wood cellulose.

Agriculture's Impact on Housing

Housing protects people from harsh weather conditions and gives them a place of comfort. There are many housing options available, from apartment buildings to single-family homes or modular homes. Did you know that these housing options would not be possible without agriculture? Wood materials from trees are used in most home construction (**Figure 3.1.7**). Wood is hard, fibrous material that forms the main substance of the trunk of a tree. Wood is grown and manufactured into many different products that are

Domcobb/Shutterstock.com

Figure 3.1.6. Many products are created from cotton. Cotton is so common in our daily lives that it's easy to forget that it comes from a crop and must be harvested each and every year to meet worldwide demand. A healthy plant will flower in the fall. After blooming, the flower dies, leaving a boll. Once this boll bursts open, the cotton dries when exposed to the sun. It then looks like the cotton pictured and is ready for harvest.

Figure 3.1.7. Wood is the main construction material for building most houses. Wood is used in the interior and exterior walls, as well as the ceiling and floors. *Where does this wood come from?*

David Papazian/Shutterstock.com

useful or necessary in housing, such as lumber, plywood, and furniture. Many of the items inside your home, including beds (which are typically made of wood), curtains (which are made from cotton), and chairs (which can include bioplastics made from agricultural feedstock such as corn, soybeans, or sugarcane) (**Figure 3.1.8**) come from agriculture.

Did You Know? The US has more than 115,000 square miles of forest land, covering about a third of our land area. Some of this land is managed forest, grown specifically to be turned into products for human consumption.

The Human-Agriculture Connection

Individuals interact with the agriculture industry in one of two ways: producing or consuming. Only two percent of the US population are farmers or ranchers that produce the raw agriculture products we use every day. These people are known as ***producers***. This means that 98 percent of the US population are exclusively consumers. ***Consumers*** are people who purchase

Figure 3.1.8. There are numerous benefits of using bioplastics. Bioplastics help reduce fossil fuel use and lessen the carbon footprint of the conventional plastic products they replace. Compostable bioplastics also reduce municipal landfill waste.

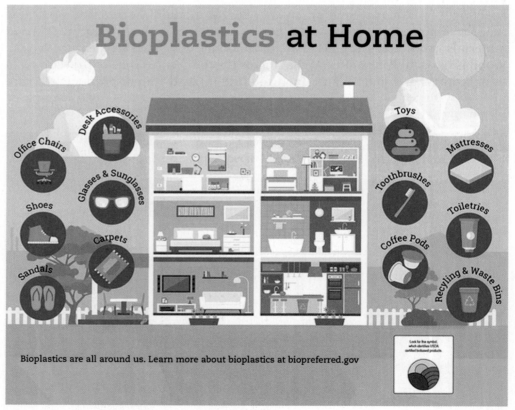

Bioplastics are all around us. Learn more about bioplastics at biopreferred.gov

USDA

and use agricultural products. Everyone is a consumer, including producers. We all purchase and use agricultural products! Every time you make a purchase at a grocery store, you are a consumer. Each year, Americans spend about 10 percent of their income on food. Recent trends show that out of that 10 percent, 46 percent is for eating food prepared at home while 54 percent is for food eaten away from home in restaurants or other food venues.

Now that you know more about the agriculture industry, could you imagine a world without agriculture?

Career Corner ▶ ROW CROP PRODUCER / FARMER

A **row crop producer or farmer**, like me, plants and harvests traditional row crops—such as corn, soybeans, wheat, cotton, potatoes, canola, sunflowers, flax, sugar beets, field peas, etc.—independently or for a large farm or company.

What Does This Job Entail? Row crop producers operate farm equipment, inspect crops, manage crop health and fertility levels, and harvest crops once they are ready.

What Do I Need to Be Good at This? You need to enjoy working outside and working with your hands. You need to be organized and have a good work ethic.

What Education Do I Need? While no formal training is needed, a high school diploma is beneficial to be a row crop farmer. An associate or bachelor's degree in a relevant or supplementary field—such as agribusiness, plant science, or agriculture—would be helpful.

How Much Money Can I Make? Row crop producers make an average of $55,000 per year.

How Can I Get Started? Start small in your own yard, porch, or patio growing fruits and vegetables for your family. Interview farmers at a local farmers' market, or job shadow a local farmer in your area.

What Did We Learn?

- Many things you use every day are made from agricultural products, including toothpaste, sports equipment, gasoline, and school supplies.
- Farmers produce the food and fiber used to provide nutrition, clothing, and shelter to consumers.
- Producers are the people who grow or raise raw agriculture products, and they make up only two percent of the US population.
- Consumers are the people who purchase and use agriculture products. Everyone is a consumer!

Let's Check and See What We Know

Answer the following questions using the information provided in Lesson 3.1.

1. *True or False?* You use products provided by agriculture every day.
2. *True or False?* Footballs are made from wool.
3. Glue is made from a by-product of ____.
   A. soybeans
   B. beef
   C. cotton
   D. corn
4. Which of the following states is one of the top five production states in terms of crop and animals?
   A. Nebraska
   B. Florida
   C. Ohio
   D. South Dakota
5. USDA stands for ____.
   A. United States Department of Agriculture
   B. Unified Department of Agriculture
   C. United States Defense of Agriculture
   D. United States Department of Animals

6. The most common fiber produced from animal hair is ____.
   A. cotton   C. flax
   B. silk   D. wool
7. *True or False?* Synthetic fibers are commonly made from the cellulose of trees.
8. Agriculture ____ make up two percent of the United States population.
   A. consumers
   B. producers
   C. governors
   D. owners
9. What percentage of the US population are exclusively consumers?
   A. 25   C. 70
   B. 48   D. 98
10. People who purchase and use agricultural products are known as ____.
    A. consumers
    B. producers
    C. farmers
    D. owners

Let's Apply What We Know

1. Think about the agricultural products that were discussed in the lesson. List 10 additional things that you have used in the last 24 hours that have ties to agriculture.
2. According to the USDA, more consumers are spending their money eating away from home rather than eating at home. What effect could this have on producers in the future?

Academic Activities

1. **Math.** Study the trends shown in **Figure 3.1.4**. Which food consumptions have changed the most over the five years shown on the table? Which do you think will change the most in the future? Why? Draw a graph that illustrates your findings.

Communicating about Agriculture

1. **Listening and Speaking.** In a group, create a presentation sharing specific examples of how agriculture interacts with a field of science. Present two or three examples of how this field of science is used in a specific sector of agriculture. Be prepared to share your examples with the class.

SAE for ALL Opportunities

1. **Research SAE.** Create a survey for customers of your local farmers' market or grocery store to determine their shopping habits, needs, or overall opinions of agricultural production (locally, nationally, or both). Share your findings with local producers, farmer market vendors, or grocery story managers.
2. **School-Based Enterprise SAE.** Start a school vegetable garden and donate the produce to your lunchroom or school lunch provider. Discuss a plan with your cafeteria manager for how to get school-grown produce into the school lunchroom. It could be one item once a week or more depending on the size of your operation and space available at the school.

SAE for ALL Check-In

- How much time have you spent on your SAE this week?
- Have you logged your SAE hours?
- What challenges are you having with your SAE?
- How can your instructor help you?
- Do you have the equipment you need?

The History of American Agriculture

chippix/Shutterstock.com

Learning Outcomes

By the end of this lesson, you should be able to:
- Describe how agriculture has changed since America's earliest days. (LO 03.02-a)
- Summarize the impact that new inventions and technology have had on agriculture. (LO 03.02-b)

Words to Know

boll weevil	plantations	Pure Food and Drug Act
Dust Bowl	precision agriculture	sharecropping

Before You Begin

How long do you think you would survive if you had to grow your own food? What if there were no grocery stores or convenience stores, or there were no farmers growing food for you? What do you think? Would you make it a day? A week? A month?

Less than 150 years ago, 90 percent of the US population did have to grow their own food, and they didn't have the advantages provided by the technology and inventions we have today.

Let's pretend that you have to grow your own food for the rest of your life. Thinking about this, take a minute to think up one thing that doesn't already exist, but should be invented, that would make this task easier.

The agriculture, food, and natural resources industry has changed a lot over the years because of other people doing exactly what you just did. Creating new inventions, having big (maybe crazy) ideas, and applying new technologies to production has helped American farmers produce more food, more efficiently. If you are curious to hear more about it, let's read on!

A Brief History of American Agriculture

Agriculture has changed greatly over the last 500 years. When the United States was a new country (and before), almost the entire population grew their own food to survive. Today, about two percent of the American workforce are farmers. Can you imagine how different your life would be if you had to produce your own food? What if you were one of the first settlers and there were no grocery stores? Do you think you would find ways to grow food more quickly and easily? That is exactly what is happening in American agriculture: Farmers are constantly adapting to change. Today's agriculture is very different than food production 300, 100, 50, or even 20 years ago.

Early Years of Settlement (1400s–1700s)

The earliest people to widely develop agriculture in North America were Native Americans. Many of the earliest European settlers to this land had little or no agricultural experience. They came to America with hopes of finding natural resources they could take back to Europe to trade or sell. They were not prepared to gather and hunt their own food. Many of the early settlers perished for lack of food, and many more would have met the same fate without the help of the Native Americans who taught them to cultivate plants for food.

As more settlers from around the world arrived, different agricultural products were introduced to the continent. Spanish settlers introduced cattle in the 16th century, and by the 17th century, most of our modern domestic livestock had been imported to America. The crops grown varied with new settlers bringing native foods from different countries. European settlers brought clover, alfalfa, and small grains to America. African settlers introduced grain and sorghum, along with okra and some melons. Potatoes, tobacco, and peanuts were introduced by settlers traveling from South America to North America. As the new world grew, so did the types of crops grown and produced (**Figure 3.2.1**).

Centers of origin of selected crops

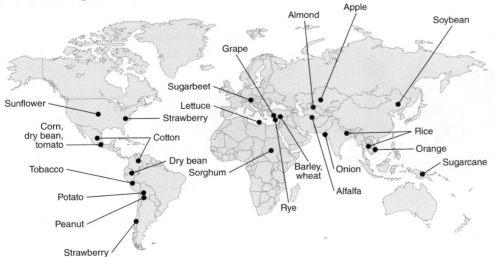

Note: The pointer locations indicate general regions where crops are believed to have first been domesticated. In some cases, the center of origin is uncertain. Other geographic regions also harbor important genetic diversity for these crops.

Source: This map was developed by the General Accounting Office using data provided by the National Plant Germplasm System's Plan Exchange Office

USDA

Figure 3.2.1. Many of the foods we consider to be American were brought here by settlers from other countries. *Do you see some of your favorite foods on the map? What countries are they from? Is this different from where you expected?*

Revolutionary Period (1764–1789)

Following the Revolutionary War, the newly formed United States of America was even more reliant on agricultural exports to European markets than the colonies had been. The nation's population was expanding quickly, and cities were growing larger.

khathar ranglak/Shutterstock.com

Figure 3.2.2. A sickle is one of the oldest harvesting tools. The short handle forces the user to harvest the crop in a squatting or bent over position. Over time, working in this position becomes uncomfortable, which makes harvesting using a sickle very slow. *How would you change the sickle's design to make it a more efficient hand tool?*

In New England, factories were built to refine raw crops into usable products for exporting. In the South, large farms expanded into plantations. ***Plantations*** were large properties on which crops were grown by resident labor, including slaves. Plantations grew a few crops that could be exported to meet the growing demand in Europe, especially cotton, tobacco, and sugar. The main export crop was cotton. Most crops were harvested by hand (**Figure 3.2.2**).

Industrial Revolution (1760s–1840s)

The Industrial Revolution greatly affected the agriculture industry. Farmers at that time made up 90 percent of the workforce in America. (It should again be noted that some large percentage of these people were enslaved, and were forced to labor in agricultural jobs.) Inventions during this time were aimed at mechanization and maintenance. Farmers were building equipment that would make farming easier and faster. They also needed machines that sped up the harvesting and processing of raw crops.

Eli Whitney's Cotton Gin

The first major mechanical advancement took place in 1793 when Eli Whitney invented the cotton gin (**Figure 3.2.3**). The cotton gin was a machine that could remove the seeds from cotton fibers.

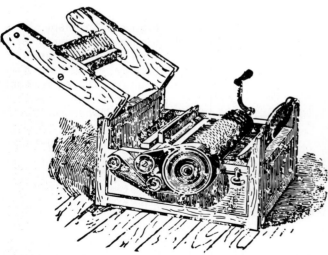

Morphart Creation/Shutterstock.com

Figure 3.2.3. Whitney's cotton gin invention worked like a strainer. Cotton was run through a wooden drum with a series of hooks inside. The hooks caught the cotton fibers and dragged them through a mesh screen. The holes of the mesh screen were too small to let the seeds pass through, but the cotton fibers could pass through easily. *What equipment replaced Whitney's cotton gin?*

Before the cotton gin, workers removed seeds from about one pound of cotton per day by hand, whereas the cotton gin cleaned 50 pounds of cotton in a single day. This invention increased production an almost unthinkable amount!

It is important to note that this mechanical advancement had a dark side effect: As cotton became more profitable to grow, slavery became a more profitable enterprise. The amount of slave labor in the US increased exponentially with the advent of the cotton gin.

Cypress McCormick's Mechanical Reaper

The Industrial Revolution also introduced the mechanical reaper in 1834. Cypress McCormick was a blacksmith, farmer, and inventor. He noticed that harvesting wheat required many workers. If they could be found, the cost of hiring them was high. McCormick's machine resembled a two-wheeled, horse-drawn chariot, with a vibrating cutting blade, a reel to bring the grain within its reach, and a platform to receive the falling grain. This was the first automated grain reaper (**Figure 3.2.4**). McCormick's machine, powerful as it was, pales in comparison to the large combines in use today. Wheat harvesting has come a long way since the Industrial Revolution!

John Deere's Steel Plow

Another major agricultural invention that came out of the Industrial Revolution is the steel plow, invented by John Deere in 1837 (**Figure 3.2.5**). Traditional plows were made of heavy cast iron, and farmers had problems with clay in the soil sticking to the surface and making the plow difficult to use. Farmers would have to stop often to clean the blades with a wooden paddle before plowing could continue. Deere, who was a blacksmith, noticed that soil would not stick to a steel sawmill blade. This inspired him to create a new plow that would decrease downtime for farmers. Like the cotton gin and mechanical reaper, this plow was a technological leap forward, making farmers' work easier while rapidly increasing output.

Did You Know?

The Land Act of 1820 encouraged purchasers (exclusively white men) to buy up public land. The law reduced the size of tracts that could be sold to as few as 80 acres, and lowered the minimum price to $1.25 an acre. This allowed many new farmers to purchase rich farmland in Ohio and Missouri.

3

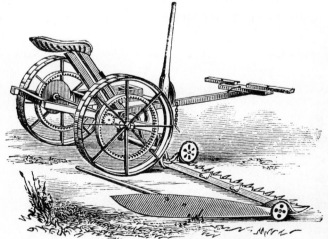

Morphart Creation/Shutterstock.com

Figure 3.2.4. The first grain reaper had some defects, one being that the clatter from the metal pieces was so loud it frightened the horses pulling the machine through the field! McCormick did not give up, though. He continued working until he perfected his invention in 1847, 13 years after his first model came out. With his improvements, McCormick sold close to 800 reapers in 1847. *What benefits did the reaper present to farmers? When was this machine replaced with another machine?*

Sham-ann/Shutterstock.com

Figure 3.2.5. In the 1830s, Americans were moving west and settling the prairie lands. The soil was rich and black and much different than the sandier soils on the East Coast. Native prairie plants had strong roots that the cast-iron plow had difficulty cutting. John Deere used his blacksmithing skills and insightful ideas to invent a steel plow and establish a company that is still on the forefront of advancing agricultural technology today. *What was John Deere's next invention?*

Reconstruction Period (1865–1877)

The years after the Civil War were a changing time for the agriculture industry. The country was working to rebuild, settlers were moving west, and slavery had been abolished. There were just over two million farms in America, with the typical farm averaging 200 acres. Farmers made up 58 percent of the workforce.

Did You Know? In 1860, the total US population was 31,443,321. The farm population was estimated to be 15,141,000, making up 58 percent of the labor force. There were 2,044,000 farms in the US that averaged 199 acres.

Sharecropping

Sharecropping was a method of agricultural labor in which the landowner leased farmland to a farmer or tenant in exchange for a share of the crop produced. Sharecropping agreements differed across the country, but typically the landowner furnished the seeds and farming supplies. Sometimes they furnished the food, clothing, and medical expenses of the tenant. In these cases, the landowner typically kept two-thirds of the profit from the crop, while the tenant workers were paid the remaining one-third. In other cases, the tenant may have furnished all the equipment and had decision-making power in the operation of the farm.

The Steam Tractor

In 1868, the steam tractor was invented as the first engine-driven farm implement (**Figure 3.2.6**). This led to a decrease in the use of draft animals (mules, horses, and oxen) for agriculture power. The machine, however, was very heavy and not well-suited for farm work. The idea was good, though, and this innovation paved the way for more engine-driven farm implements.

Figure 3.2.6. Steam tractors made farm work less reliant on human strength or animal power. Wood, coal, and straw were used to fuel the fire in the engine, which heated water into steam. Despite the large size of these steam tractors, most engines only generated 30 horsepower. The machines were heavy and required a second vehicle to carry the water to the tractor. *Why do you think this machine didn't catch on widely?*

Joseph Glidden's Barbed Wire

With cities expanding in the West, the cattle industry was also expanding in North America. The invention of barbed wire in 1874 by Joseph Glidden changed the way ranchers raised cattle (**Figure 3.2.7**). Previously, cattle had been free to graze on open range, and cattle drives were used to move cattle from grazing lands to market. Barbed wire allowed cattle ranchers to control their herd's grazing and nutrition, as well as breeding and genetics by keeping the herds separate.

Progressive Era (1890s–1920s)

By 1900, farmers made up 38 percent of the workforce and the number of farms in America was increasing. The general-purpose tractor first appeared on farms during the Progressive Era. No one is certain who invented the first general-purpose tractor, but it is widely accepted that the 1924 Farmall Tractor, manufactured by the International Harvester Company, was the first. The widespread use of tractors had a huge impact on the productivity of American agriculture.

Maciej Biedowski/Shutterstock.com

Figure 3.2.7. Before barbed wire, ranchers made fencing for their livestock with wood. Wood was expensive, as trees were scarce in some parts of the prairies and the wood had to be shipped by train. Without fencing, cattle grazed freely and wandered into other ranchers' herds, competing for grass and water. Barbed wire fences were inexpensive and kept cattle limited to smaller areas where they could be monitored for health and nutrition.

Did You Know? In 1900, the total US population was 75,994,266. The farm population was estimated to be 29,414,000, making up 38 percent of labor force. There were 5,740,000 farms in the US at that time, averaging 147 acres.

Pure Food and Drug Act

In 1906, the US Congress passed the Pure Food and Drug Act. The **Pure Food and Drug Act** was legislation that would prevent "the manufacture, sale, or transportation of misbranded or poisonous foods, drugs, or medicines." This was the first law that protected the American consumer by regulating product quality and providing product labeling. The weight or measure of food was checked for accuracy, and labels could not be false or misleading.

The Boll Weevil

The boll weevil entered the United States in the late 1800s, where they were first spotted in Texas. The **boll weevil** is a beetle that feeds on cotton buds and flowers (**Figure 3.2.8**). Thought to be native to central Mexico, it migrated into all US cotton-growing areas by the 1920s. Since it appeared in the US, the boll weevil has cost US cotton producers over $15 billion and put many cotton farmers out of business.

Carlos Rudinei A Mattoso/Shutterstock.com

Figure 3.2.8. The boll weevil feeds on cotton pollen. It damages the plant by laying eggs on cotton flower buds. The infestation stops bud development and then the buds fall off as the beetle larvae eat through them. This causes the plant to produce little or no usable cotton, making entire fields worthless. Farmers would sometimes hand pick the beetles from the fields to try and control them. There was even a song written about the boll weevil in 1941.

Everett Historical/Shutterstock.com

Figure 3.2.9. George Washington Carver was born enslaved, a year before slavery was outlawed. In 1894, Carver was the first African American to earn a Bachelor of Science degree from Iowa State Agricultural School (now Iowa State University). He went on to earn a Master of Agriculture degree. Upon his death in 1943, President Franklin Roosevelt signed legislation for Carver to receive his own national monument in Missouri. This honor had previously only been granted to presidents.

On a more positive note, this small beetle and its large appetite for cotton was strong enough to create a unique partnership between farmers, legislators, and scientists. When scientists and farmers work together with the backing of legislators, big problems can be handled. The National Boll Weevil Eradication Program is still in place today and continues to protect US cotton crops.

George Washington Carver

The first decades of the 20th century brought a great inventor to the agriculture industry. George Washington Carver was the director of agricultural research at Tuskegee Institute in Alabama (**Figure 3.2.9**). Carver is known for identifying valuable crops, such as peanuts, sweet potatoes, and soybeans, that farmers could grow in place of cotton.

His work in soil chemistry showed that after years of use, farmers had depleted the soil of nutrients, which resulted in low crop yields. He found that by rotating crops in the fields and growing crops such as peanuts, sweet potatoes, and soybeans, the soil's nutrients could be replenished. These three crops add nitrogen to the soil.

Farmers were thankful for the increased cotton yields, but did not know what to do with the surplus of peanuts produced in the off months. Carver was afraid that farmers would abandon the new planting practices if they could not make money from the peanut harvest. To ensure farmers would continue using these practices, he discovered over 300 food, commercial, and industrial uses for the peanut. His discoveries included peanut-based medicines, cooking oils, cosmetics, soaps, and wood stains.

Did You Know?

In 1930, 58 percent of all farms had cars, 34 percent had telephones, and 13 percent had electricity.

The Great Depression (1928–1936)

In October of 1929, the US economy collapsed with the stock market crash, and American agriculture took a turn along with it. On the East Coast, factories and businesses were going out of business and laying off workers. Americans were cutting expenses wherever possible. Little income was left to purchase the farmers' products. Desperate bankers began calling in farmers' bank loans, but farmers had no income and could not pay their debts. Soon, farms were being foreclosed. Close to 750,000 farms were lost between 1930 and 1935 through bankruptcy and foreclosure.

During this same period, a severe drought was occurring in the Midwest and the Southern Great Plains. Farmers could not afford to plant fields because of the poor economic times. The drought conditions left previously plowed soils bare and vulnerable to erosion. Eroding soil led to massive dust storms over the entire Midwest, which became known as the *Dust Bowl* (**Figure 3.2.10**).

Figure 3.2.10. The worst dust storm, known as Black Sunday, took place on April 14, 1935. A cloud of blowing sand and dust formed what looked like a wall in the Oklahoma Panhandle and spread east. Entire monuments, including the Statue of Liberty and the US Capitol, were completely obscured by the dust. It is estimated that 3 million tons of topsoil were blown off the Great Plains in this single day.

By 1934, an estimated 35 million acres of farmland were useless, forcing farmers to move to other parts of the country. The Dust Bowl had devastating effects on the land, the farmers, and the society in general.

Green Revolution (1940s–1960s)

As the Great Depression ended, farmers made up 18 percent of the workforce in America. Scientists began developing new technologies that would increase agricultural production without destroying the land used to grow the crops. Norman Borlaug was a scientist who studied plant genetics. His scientific research on the genetics of wheat and wheat production earned him a Noble Peace Prize and saved more than a billion people around the world from starvation. Borlaug is considered the father of the Green Revolution (**Figure 3.2.11**).

The Green Revolution was the emergence of new varieties of crops, specifically wheat and rice, that were able to double (if not triple) production. Most of the trials testing these crops began in Mexico. The varieties of wheat that Borlaug developed became a model for what could be done with other staple crops around the world. In Mexico, once the new varieties of wheat were being widely reproduced, malnutrition diminished across the country. The Green Revolution expanded around the world. In India and Pakistan, new varieties of wheat were introduced in the 1960s to help prevent future famines. The Green Revolution helped small farmers and decreased hunger in countries that had previously not been able to grow much of their own food.

Figure 3.2.11. Norman Borlaug was a leader in addressing world hunger and food security. He founded the World Food Prize in 1986, which is an annual $250,000 award honoring individuals making breakthrough advancements in improving the quality, quantity, and availability of food in the world.

Did You Know? In 1954, the number of tractors used on farms was greater than the number of horses and mules for the first time.

The Green Revolution brought many advancements in the agriculture industry including:

- Higher-yield crops
- Implementation of irrigation systems
- Improved pest and disease control
- Synthetic fertilizers

Modern Era (1980s–Present)

By the 1980s, farmers made up less than four percent of the American workforce. The number of farms was down to just over two million. As the population continued to increase and the number of farmers and farms continued to decrease, farmers had to learn how to feed more people on less land. This is why technology continues to play a major role in agriculture.

Did You Know? In 1980, the total US population was 227,020,000. The farm population was estimated to be 6,051,000, making up 3.4 percent of the labor force. There were 2,439,510 farms in the US, averaging 426 acres.

Hands-On Activity

Irrigation Systems

After the Dust Bowl, agriculture researchers began working on ways to keep crops watered, even in drought conditions. Crop irrigation was not new to farming. Early irrigation systems can be found dating back to the Egyptians. Agriculture researchers knew they had to develop systems that would both work for farmers and protect the environment.

Irrigation demonstrates the ability of humans to control natural conditions using technology. Though the technology has advanced over time, the basic principle of directing water toward agricultural land remains the same. Can you build a small-scale irrigation system using simple household materials?

Suggested materials:

- Paper or plastic cups
- Drinking straws
- Tape and scissors
- Modeling clay
- Water

Here are your directions:

1. In a small group, brainstorm ideas for moving water from one cup to another using the materials provided.
2. Lay out the steps you will take to create your irrigation system. What do you expect your irrigation system to look like and be able to do?
3. Follow your design plan and build your system.
4. Finally, test your system!
5. How did it compare with your expected outcomes?
6. How could you improve your design?

Consider This

1. Where do growers obtain the majority of crop irrigation water? Might there be conflicted interests in these water sources? Explain your answer.
2. How much irrigation water is used each day by American growers? Does this number surprise you? Has crop irrigation left areas with insufficient water? Explain your answer.
3. Who owns our water?

GPS Technology

In 1994, the first global positioning system (GPS) was used for agricultural purposes. GPS technology allowed farmers to use precision agriculture. *Precision agriculture* is the method of managing farmland with the assistance of computers or satellite information (**Figure 3.2.12**). Farmers now use GPS technology for a variety of crop-related tasks, including:

- Crop field preparation
- Crop planting
- Controlling fertilizer and pesticide application
- Monitoring soil erosion
- Collecting information related to crop yields
- Monitoring crops for pests and diseases
- Troubleshooting irrigation issues in a field
- Harvesting crops

Monopoly919/Shutterstock.com

Figure 3.2.12. Tractors equipped with GPS systems allow farmers to receive real-time data with accurate position information for their fields. GPS-based applications in precision farming are being used for farm planning, field mapping, soil sampling, tractor guidance, crop scouting, fertilizer and pesticide rate application, and crop yield mapping.

Career Corner — RESEARCH AND DEVELOPMENT MANAGER

Research and development (R&D) managers *oversee research activities and develop knowledge-based products for a company.*

What Do They Do? Research and development managers create research programs incorporating current technologies or research interests, oversee all aspects of these programs and experiments, and report to senior management on their findings.

What Education Do I Need? A master's degree in the field in which you are applying your knowledge (agriculture, food science, chemistry, biology, agronomy, etc.) is required.

How Much Money Can I Make? Research and development managers make an average of $138,000 per year.

Rawpixel.com/Shutterstock.com

How Can I Get Started? You can build an SAE around studying current needs of the agriculture industry, and brainstorming ways to address those needs. Create an agriscience project in social systems, determining people's perceptions of industry needs. Maybe you can conduct an experiment designing and testing a new product idea. You could also contact someone who currently designs new products for an agriculture business, such as a food company, and interview them about their job.

Biotechnology

Biotechnology is the use of science to modify the genetic makeup of living cells. The modified cells produce new functions and enhance the value of those organisms. Biotechnology includes genetic engineering, which is when a scientist uses genes from one living organism and inserts them into the genetic makeup of another living organism. Biotechnology has advanced many crops over the last few decades and has become a valuable tool for agricultural production. In 1994, the Flavr Savr tomato was the first genetically engineered crop introduced to consumers. Biotechnology was used to slow the ripening process, which prevented it from softening too quickly and allowed for longer shipping periods.

Making Connections # STEM

Strawberry DNA

All living creatures have DNA as part of their genetic makeup. Part of biotechnology is looking at the genetic makeup of crops to determine ways to improve our current food supply. Have you ever wondered how scientists complete these kinds of tasks? This activity will allow you to remove and see a strawberry's DNA.

The procedure for extracting DNA from a strawberry is simple and usually easy to observe without any special equipment. In this procedure, you will crush a strawberry and add detergent and salt to break down the cell walls to release the DNA within the nucleus. You will then filter the liquid from this crushed strawberry into a test tube, where it is mixed with alcohol.

Suggested materials:

- DNA extraction buffer (made by mixing 1000 mL of deionized water, 50 mL clear dish soap, and 1 teaspoon of salt)
- Fresh strawberry
- Ziploc bag
- Ethanol or 91% isopropyl alcohol
- Stirring rod
- Coffee filter
- Funnel
- Test tube and beaker

Procedure:

1. Add a strawberry to a Ziploc storage bag.
2. Add 10 mL of DNA extraction buffer and seal tightly.
3. Mash the strawberry and buffer for one minute.
4. Use a funnel and coffee filter to filter the strawberry juice into a beaker.
5. Transfer the strawberry juice into a test tube until it is halfway full. Avoid transferring any foam that may have formed.
6. Slowly pour or drip cold alcohol over the top of the strawberry liquid inside the test tube. You want a single layer on top of the strawberry liquid.
7. White strands will form in the ethanol layer; use a stirring rod to spool the strawberry DNA (white strands) together.

Kitt Peacock/Shutterstock.com

Consider This

1. What does a strawberry's DNA look like?
2. Why it is important for scientists to be able to remove DNA from cells?
3. Is there DNA in other types of food? How do you know?

What Did We Learn?

- During the Revolutionary period, cotton was the main export crop. It was grown on plantations in the Southern states.
- The Industrial Revolution brought many great inventions that helped make farming more efficient. These farm implements include the cotton gin, the mechanical grain reaper, and the steel plow. These innovations made the work of agriculture easier, but had some terrible consequences.
- After the Civil War, plantation owners and large farm operations used sharecropping to produce crops. The sharecroppers traded labor for part of the harvested crop.
- During the Reconstruction period, the steam tractor and barbed wire were invented. The steam tractor was impractical because of its weight, but inexpensive barbed wire fences transformed the livestock industry.
- Tractors became standard equipment on most American farms during the Progressive Era. The United States government also began monitoring the manufacturing and sale of food through the Pure Food and Drug Act.
- The Great Depression was a tough time for all Americans, and farmers were no exception. Many lost their farms due to foreclosure and bankruptcy. Bare lands and extreme drought conditions in the West caused the Dust Bowl. The Dust Bowl was one of the most devastating natural disasters in the United States.
- The Green Revolution of the 1940s introduced the agriculture industry to advancements in irrigation, pest and disease control, and higher-yielding crops.
- Present day farmers use advanced technology and precision agriculture to keep their farms running efficiently. Using biotechnology, they can plant crops that produce higher yields on less land.

Let's Check and See What We Know

Answer the following questions using the information provided in Lesson 3.2.

1. *True or False?* Potatoes and peanuts are native to the United States.
2. *True or False?* During the Industrial Revolution, farmers made up little of the American workforce.
3. George Washington Carver is most famous for his inventions using ____.
   A. soybeans
   B. peanuts
   C. cotton
   D. corn

4. The boll weevil is a small bug that feeds on ____ buds and flowers.
   A. wheat
   B. peanuts
   C. sugar cane
   D. cotton
5. Congress passed the ____ to prevent the manufacturing and selling of mislabeled or dangerous foods.
   A. Pure Food and Drug Act
   B. Fair Food Act
   C. Fair Deal Act
   D. Food and Drink Act

6. Severe drought conditions and poor economic times left many farms unused and soils bare, which resulted in a natural disaster known as the ____.
   A. Dirt Bowl   C. Dust Bowl
   B. Erosion Bowl   D. Drought Bowl
7. *True or False?* Norman Borlaug is known as the father of the Green Revolution.
8. The first biotechnology crop introduced in America was the ____.
   A. cotton   C. onion
   B. tomato   D. apple
9. *True or False?* Cotton was the main crop exported to Europe during the Revolutionary period.
10. One of the first agricultural tools invented was the ____, which was a handheld tool with a single curved blade.
    A. plow   C. tractor
    B. sickle   D. reaper

Let's Apply What We Know

1. Imagine you were alive and owned a small farm during the Industrial Revolution. How do you think agricultural advancements would affect your farm? What technology would you have implemented on your farm and why?
2. What is agriculture? Conduct research to gather information on people's perceptions of agriculture. Ask at least one person born in every decade going back as far as you can find participants. Write a summary on your findings. Make sure you explain how thoughts have changed over the decades.

Academic Activities

1. **Engineering.** Choose a piece of agricultural machinery and research the changes in the machinery since it originated. Note any of the mechanical advancements and how they have made the machinery more efficient. What changes do you foresee in the future for this type of equipment? Be sure to consider any unintended consequences or side effects of these inventions as well.
2. **Language Arts.** Many famous writers have written poems and songs about farming. Conduct research to find and read one or more poems or songs written about American agriculture. Note the rhythm of the words and speech patterns. You may want to read the poems aloud or have someone read it aloud to you while you listen. Using the songs and poems you read as inspiration, write your own song or poem about agriculture. Share your creation with the class.

Communicating about Agriculture

1. **Writing and Speaking.** Make an agricultural collage with images of advancements over the past 250 years. Use pictures from magazines or online resources to create your collage. If possible, place the advancements in chronological order. Include the dates and names of people involved. Share your collage in a small group setting.
2. **Writing and Speaking.** Select a major agriculture invention and create an advertisement poster for that invention. Set your ad in the era in which the invention was introduced. Research a common selling price for your invention when it first came to market. Make sure your poster shares what benefit the invention brought to the agricultural industry and addresses your target audience.

SAE *for* **ALL** *Opportunities*

1. **Foundational SAE.** Research the history of agriculture in your own state or community. Determine who the local innovators were, and how they made an impact on the agriculture industry. Create a presentation about your findings to share with your class or other local civic organizations or groups.
2. **Research SAE.** Research a few hot topics related to biotechnology and the agriculture industry. Create a consumer survey about these topics to determine public perception of agricultural issues. Write a hypothesis, conduct your survey, observe and record data, and write a report on your findings.

SAE *for* **ALL** *Check-In*

- How much time have you spent on your SAE this week?
- Have you logged your SAE hours?
- What challenges are you having with your SAE?
- How can your instructor help you?
- Do you have the equipment you need?

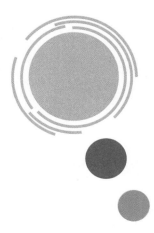

LESSON
3.3

Major US Commercial Crops

ESB Professional/Shutterstock.com

Essential Question

What do major grain, food, and fiber crops provide for us?

Learning Outcomes

By the end of this lesson, you should be able to:
- List the major grain crops grown for food. (LO 03.03-a)
- Identify major crops grown for fiber production. (LO 03.03-b)

Words to Know

cereals

feed grain

fiber crops

grains

oilseeds

Before You Begin

Now that we've talked about how much agriculture affects your everyday life, do you ever wonder about the food you eat? Where does it come from? Let's see if we can figure it out as a class!

Make a list of everything you ate yesterday. Do you know what was in each of the foods you ate? See if you can narrow down each of the foods you ate to a specific crop or animal. Next, pair up with one to three classmates and see if you can work together to determine the origin of all the foods on your lists.

Did anyone eat the same food as you? Did anyone disagree about the origin of something you ate?

I n the United States, we have an abundance of low-cost food available to us almost anywhere we go. Almost all cities and small towns have at least one grocery or convenience store that sells food to local consumers. Even our recreational activities are surrounded by food. What food do you eat when going to a movie? Did you ever eat a hot dog at a baseball game? What is your favorite fair food? Many of our gatherings with family and friends are centered around the food being served. Have you ever taken the time to think about where your favorite foods come from? What crops or animals are harvested to make that food? Let's look at some of the most common crops that are grown to create your favorite foods.

Grain Crops

When you think of grain, what do you think of? Fields of wheat? Rice, oats, corn? **Cereals** are any grass grown for the edible parts of its grain. **Grains** are cereals suitable as food for human beings. **Oilseeds** are those grains that are also valuable for the oil content they produce. Grains and oilseed crops are used in the production of most of the foods we eat every day. This includes crops such as barley, corn, flax, mustard, millet, rice, rye, sorghum, sugarcane, sunflower, and wheat.

Corn

Corn is the most widely produced feed grain in the United States, with more than 80 million acres in production annually. **Feed grain** is any grain used as the main energy ingredient in livestock feed. It is also used in a wide range of food and industrial products, including boxed breakfast cereals and sweeteners. There are many different types of corn, including dent, flint, flour, sweet, and popcorn.

According to the USDA, the average acre of corn produces around 165 bushels per year. There are more than 3,500 different uses for corn products (**Figure 3.3.1**). The average single bushel of corn can make 33 pounds of sweetener to be used in soft drinks and other sweet treats, or 2.8 gallons of ethanol, or 22.4 pounds of polyactic acid polymers, which are used in sustainable plastics.

Singha Songsak P/Shutterstock.com

Figure 3.3.1. An ear of corn is actually part of the flower, and an individual kernel is a seed. An average ear of corn has 800 kernels in 16 rows. There are over 3,500 different uses for corn products, such as the development of penicillin, biodegradable plastic, bath soaps, cosmetics, and starch for ironing clothes.

Each year in the United States, approximately:

- 48 percent of the US corn crop is used in animal feed;
- 12 percent is exported;
- 30 percent is used for ethanol fuel; and
- The remaining 10 percent is used for human consumption in high fructose corn syrup, sweeteners, starch, cereal, beverages, and other processed corn-based products.

Did You Know?

The biggest corn-producing states in the US are Iowa and Illinois.

Rice

Rice is the primary food grain for more than half of the world's population. Although rice is grown in many different places around the world, it has very specific growing requirements which limit it to only small areas of the United States (**Figure 3.3.2**). Arkansas is the leading state in rice production, growing about 57 percent of rice grown here, but rice is also grown in Mississippi, Missouri, Louisiana, Texas, and the Sacramento Valley of California. Each year, close to 18 billion pounds of rice are grown and harvested in the US. Different varieties of rice are grown, including long, medium, and short grain, as well as a few specialty varieties.

There are five different products that can be produced from rough rice: hulls, bran, brown rice, whole kernel milled rice, and second heads, which are ground into flour and pet food. The first stage of milling harvested rice produces brown rice. The second stage produces the bran layer which is mostly used in animal feeds. The next layer is the white rice that we are most familiar with. Some less common items that are also made from rice include vinegar, noodles, bread, and rice milk. Rice husks are also used for some common household items, including fertilizer, insulation material, and pillow stuffing.

RozenskiP/Shutterstock.com

Figure 3.3.2. Farmers flood rice fields with water to control weeds and increase crop yields. Rice is a semi-aquatic plant that grows well in flooded soils.

Sorghum

Sorghum is one of the five largest cereal crops grown in the world. The United States is the world's largest producer of sorghum, growing over 5.5 million acres each year. Farmers harvest an average of 70 bushels per acre. Kansas is the leading state in sorghum production, growing over 2.8 million acres. It is also grown in Texas, Colorado, Oklahoma, and South Dakota (**Figure 3.3.3**).

Most sorghum is used for livestock feed and ethanol production, but it is becoming more popular in recent years in the food industry. There are four main types of sorghum grown:

- Grain, which is typically made into flour;
- Forage, which is typically used in livestock feed;
- Biomass, which is used in the production of bioenergy; and
- Sweet, which is harvested to make syrup and other sweeteners.

Sorghum is also used in building materials, fencing, brooms, and you might even notice it popping up in floral arrangements.

Sayanjo65/Shutterstock.com

Figure 3.3.3 Grain sorghum originally came from Africa and India, and is known in other countries as milo. A single head of grain sorghum has between 750 and 1,250 seeds.

Wheat

Wheat is the third largest crop produced in the US (**Figure 3.3.4**). The yearly average wheat harvest is around two billion bushels. Most of

Annzee/Shutterstock.com

Figure 3.3.4. Did you know the average semi load of wheat can make around 42,000 loaves of bread? Think about how much money a single semi-truck full of wheat is worth. *How much does a loaf of bread cost at your local grocery store? Now multiply that cost by 42,000. Wow!*

the country's wheat is grown in Kansas, but other large producing states are North Dakota, Montana, Washington, and Oklahoma. There are five major classes of wheat:

- Hard red winter wheat
- Hard red spring wheat
- Soft red winter wheat
- White
- Durum

Making Connections #STEM

Using Satellites to Protect Farmland

The ever-growing population keeps constant pressure on American farmers to seek new and better ways to improve their crop production. In the past, the only way to increase production was to make improvements to farm equipment or change the fertilizers and chemicals used in the field. Now improvement techniques have gone digital.

From monitoring and forecasting drought conditions to mapping plant health, satellites are helping farmers produce more food on less land. Satellites can check how green fields are, allowing them to monitor plant health and other nutritional issues. They can also be used to estimate the amount of water available in fields at any given time to help with irrigation and water usage. The environmental data that is provided by satellites helps the USDA, farmers, and livestock producers across America. By collecting data, satellites enable farmers to make crop production predictions so they can make better management decisions in the future.

Andrey Armyagov/Shutterstock.com

Consider This

1. How can these satellite records benefit people other than farmers?
2. What other businesses and industries use data collected from satellites?
3. How have our lives changed since satellite data has become available?

Hard red winter wheat and spring wheat make up 60 percent of the wheat grown in the United States, and are mainly used to produce flour. Soft red winter wheat makes up 23 percent of wheat production and is used in baking cakes, crackers, and cookies. White wheat accounts for 15 percent and if you eat noodles, hard crackers, or cereal, you have almost certainly eaten it. The remaining two percent of wheat production is durum wheat used to produce pasta.

Other Major Food Crops

There are major food crops grown on farms that are not grains, but are crucially important to the food supply (and our economy). Let's consider those here.

superoke/Shutterstock.com

Figure 3.3.5. Peanuts are sometimes called "ground nuts" or "ground peas" in other countries. However, peanuts are not nuts at all, they are actually legumes. Peanuts are the edible seeds inside pods that form underground on the roots of the plant.

Peanuts

Peanuts are a multimillion-dollar industry in the Southeast United States. Ten states grow 99 percent of the US peanut crop: Georgia (which grows 42 percent of all US peanuts), followed by Texas, Alabama, Florida, North Carolina, South Carolina, Mississippi, Virginia, Oklahoma, and New Mexico. There are four different types of peanuts: runner, Virginia, Spanish, and Valencia. There are just over 1.3 million acres of peanuts planted each year. The average acre of peanuts produces roughly 3,300 pounds of peanuts annually. The United States is one the world's largest exporters of peanuts (**Figure 3.3.5**).

Did You Know? US peanut farmers produce three million tons of peanuts each year.

Did you know it takes 540 peanuts to make a single jar of peanut butter? There are also enough peanuts in a single acre to make about 35,000 peanut butter sandwiches. That is not all that is made from peanuts, though. Peanuts also make several other food products, including flour, meat substitutes, cooking oils, and cooking sauces. You might be surprised to know though that there are also peanuts in laundry soap, face and shaving cream, charcoal, and even glue.

Soybeans

Soybeans are one of the most versatile crops grown in the United States. US soybean farmers plant more than 80 million acres each year, which produce roughly 4.3 billion bushels of soybeans. This adds $40 billion to the US economy each year! The average field yields about 50 bushels of soybeans in a single acre annually. The United States is the largest producer of soybeans in the world, and exports 47 percent of the crop to other countries.

Soybean plants, like peanuts, are high in fiber. They are a type of legume related to peas, clover, and alfalfa. After harvest, soybeans are cleaned, cracked, dehulled, and rolled into flakes. This process removes the oil from the beans to be used in margarine, salad dressings, and oils (**Figure 3.3.6**).

Did You Know? One acre of soybeans can produce 82,368 crayons.

Did you know soybean oil is used to keep the chocolate and cocoa butter from separating in candy bars? The remaining flakes are made into soy protein, soybean meal, soy hulls, and soy flour. These products are used for animal feed and in food products, like edamame. Shopping in your local grocery store, you might find additional products made from soybeans: soy sauce, tofu, soy flour, soy milk, and infant formula. Did you know you can also find soybeans in carpet, candles, ink, and crayons?

Fiber Crops

Fiber crops are field crops grown for their threads which are used to make paper, cloth, or rope. Any fiber produced by plants or animals is considered natural fiber. The two main fiber crops grown in the United States are cotton and flax.

Cotton

Cotton is the most important fiber crop in the world, accounting for about 35 percent of all fiber crops produced. The United States remains a major producer of cotton for the international market, ranking third behind China and India. The United States also remains the leading cotton exporter in the world. Brazil, China, India, Pakistan, Turkey, and the United States are the major consumers of cotton. Cotton is produced in 17 states known as the "Cotton Belt," which stretches from Virginia to California (**Figure 3.3.7**). Texas consistently produces the most cotton, followed by Georgia and Mississippi.

Amarita/Shutterstock.com

Figure 3.3.6. One acre of soybeans can be made into 2,500 gallons of soymilk. Soymilk is a good source of plant-based protein and serves as an alternative for people who have an intolerance to cow milk or need a low-cholesterol diet.

The Cotton Belt
US Cotton-Producing States

Map: Goodheart-Willcox Publisher; Cotton: muratart/Shutterstock.com

Figure 3.3.7. Farmers select the variety of cottonseed based on their location. The USDA has grouped the cotton belt into four regions: the Southwest region (Kansas, Oklahoma, and Texas), the Delta region (Arkansas, Louisiana, Mississippi, and Tennessee), the Southeast region (Alabama, Florida, Georgia, North and South Carolina, and Virginia) and the Western region (Arizona, California, and New Mexico). The Western region grows some pima cotton, while most of the remaining states grow upland cotton. Upland cotton makes up about 97 percent of the cotton produced in the United States.

The United States produces more than 20 million bales of cotton per year, which equals about $7 billion in total value. There are two main types of cotton grown in the US: upland (which makes up most of the cotton grown in the US) and pima. Cotton is a versatile commodity used in many products, particularly clothing.

One bale of cotton can be processed into:

- 480 pounds of cleaned cotton; or
- 215 pairs of jeans; or
- 1,200 t-shirts.

Did You Know? Arizona alone grows enough cotton each year to make a pair of jeans for every living American.

Flax

Flax is one of the oldest cultivated crops, grown since the beginning of civilization. It is native to India and the Eastern Mediterranean. Today, Canada is the largest flax producer worldwide. Flax is mostly grown for use in linens.

Hands-On Activity

Farm Gate Value for Your State

Do you know what crops grow in your state? What about your county? The Department of Agriculture in your state keeps records on what agricultural products are grown in your state and what those products are worth. Research and make a list of the top 10 agriculture products in your state and the farm gate value of those products. Farm gate value is the market value of farm products minus the marketing and transportation costs.

Budimir Jevtic/Shutterstock.com

Items you will need:

- Device with internet access
- Paper
- Pencil or pen

Here are your directions:

1. Research and list the 10 largest agriculture products grown in your home state.
2. Research and list the farm gate value of each of these products.
3. Create a chart or graph that shows each of the agriculture products and their value.

Consider This

1. Which farm gate values surprise you? How are they different than you thought?
2. What impact does the agriculture industry have on your state?

Flax was first introduced in the United States by early colonists. When the cotton gin was introduced in the 1800s, flax production almost came to an end in America.

Flax production is beginning to make a comeback in the US because of the increased demand for linen clothing, and for the linseed oil that comes from the pressed seed (**Figure 3.3.8**). In the US, most flax production takes place in North Dakota, followed by South Dakota, Montana, and Minnesota. North Dakota alone plants more than 230,000 acres of flax each year, with each acre producing about 14 bushels. The entire US flax crop produces more than 3.5 million bushels each year.

The paper and pulp industry use the fiber in the stem of the flax plant for:

- Linen sheets
- Napkins
- Tablecloths
- Clothing
- Fine papers, such as parchment paper
- Bandages

mama_mia/Shutterstock.com

Figure 3.3.8. Linseed oil is a by-product of flax production and is used in paints, varnishes, and printing inks. It also has several health benefits, such as reducing high blood pressure. It can also be used to replace food oils and applied to your skin and hair as a moisturizer.

Career Corner   CROP ADVISOR

Crop advisors *are knowledgeable about plants and soil. They maintain close relationships with their clients and scout their fields for problems that may arise during the growing season. They make recommendations on concerns ranging from seed to fertilizer and from pest management to disease treatment.*

What Do They Do? Crop advisors scout for pests and diseases that may arise during the growing season, make recommendations to growers on actions to take when problems arise, provide training to area growers, identify potential weed problems, suggest crops and seeds to be used in the next growing season, and supervise chemical and fertilizer applications.

What Education Do I Need? You will need a bachelor's degree in agronomy, soils, plant science, or crop science.

How Much Money Can I Make? Crop advisors average about $68,000 per year.

Budimir Jevtic/Shutterstock.com

How Can I Get Started? You can build an SAE around crop production. Can you research different crops grown in your area, down to specific varieties and types? Can you research or design experiments about different crop pests and how they affect the plant? Maybe you could contact a local business or individual and do a job shadow, asking questions about what you want to learn.

Review and Assessment

What Did We Learn?

- Corn is the most widely grown grain and 48 percent is used in animal feed. Rice is the primary grain used around the world and requires a specific growing climate. Wheat is the third largest crop produced in the United States, and there are five different varieties of wheat grown here.
- Peanuts are grown in 10 Southern states and are commonly used in peanut butter, flour, meat substitutes, cooking oils, and sauces. Soybeans are used for both the oil they produce and the actual bean. Soybeans are high in protein and are used as meat substitutes, as well as soy flour and soy milk.
- Fiber crops are grown to make paper, cloth, or rope. Cotton is the most important fiber crop grown in the world, and is grown in a part of the United States known as the Cotton Belt. Flax is a fiber crop that is used to make linen and linseed oil.

Let's Check and See What We Know

Answer the following questions using the information provided in Lesson 3.3.

1. Any grain used as the main energy ingredient in livestock feed is known as _____.
   - A. cereal
   - B. oilseed
   - C. feed grain
   - D. corn

2. What percent of corn grown in the United States is exported to other countries?
   - A. 12 percent
   - B. 20 percent
   - C. 30 percent
   - D. 48 percent

3. *True or False?* The average acre of peanuts produces close to 3,300 pounds of peanuts.

4. _____ is the leading state in rice production in the United States.
   - A. Arkansas
   - B. Mississippi
   - C. Georgia
   - D. California

5. *True or False?* Sweet sorghum is commonly used in the production of flour.

6. The largest producer of soybeans in the world is _____.
   - A. China
   - B. the United States
   - C. Mexico
   - D. Italy

7. The United States harvests an average of _____ billion bushels of wheat a year.
   - A. one-half
   - B. one
   - C. two
   - D. five

8. *True or False?* Fiber crops are grown to produce paper, cloth, or rope.

9. The Cotton Belt is an area of Southern United States that stretches from Virginia to _____ and produces all the cotton grown in the country.
   - A. California
   - B. Texas
   - C. New Mexico
   - D. Arizona

10. *True or False?* South Dakota is the largest producer of flax in the United States.

Let's Apply What We Know

1. Research the origins of the various cereal grain species discussed in this lesson. Where did they originate and how did they get to the United States?
2. On a map of the United States, identify by color each of the various grains and fibers and where they are grown. Some states may have multiple colors if they grow multiple cereal grains.
3. Research and list the advantages of using plants to create gasolines and diesel fuels.
4. How important is flax on a global scale? Research the world production of flax.
5. Investigate by-products that come from grain crops. Use images from magazines or the internet and create a poster that displays your findings. Make sure to label which grain produces the by-product.

Academic Activities

1. **Math.** What are the average yields for cereal grains in your state? What are the current market prices for these grains? Based on the information you collected, which single crop would you grow if you owned a farm? What would be the total market price if you grew 100 acres of the crop of your choice?
2. **Language Arts.** Pick one of the grain or fiber crops from the chapter. Research the origin of one of those crops. What is the story behind that crop and how it came to the United States? Create a travel pamphlet to highlight the crop's journey.

Communicating about Agriculture

1. **Reading and Writing.** Create a poster highlighting the major grain crops grown in the United States. Find images to illustrate, and provide some additional information about the purposes of the crops.
2. **Speaking.** Are there any other types of grain crops grown around the world? Do some research about another country and the crops grown there. Prepare a 2–3 minute speech for your class about what you found and what it tells you about agriculture around the world.

SAE for ALL Opportunities

1. **Foundational SAE.** Interview a local grain or fiber producer. Set up an interview to ask them about the rewards and struggles that come with producing America's food supply.
2. **Ownership SAE.** Grain and fiber producers are constantly monitoring and recording cost and profits. They keep track of land, seed, chemical/fertilizer, irrigation, and other costs. They must also record the amount produced and sold. Make a list of any costs you have with your SAE program. Create a spreadsheet that will allow you to record these costs easily throughout the year.

SAE for ALL Check-In

- How much time have you spent on your SAE this week?
- Have you logged your SAE hours?
- What challenges are you having with your SAE?
- How can your instructor help you?
- Do you have the equipment you need?

Agriculture and Climate

Piyaset/Shutterstock.com

Learning Outcomes

By the end of this lesson, you should be able to:
- Describe the difference between weather and climate. (LO 03.04-a)
- Identify factors that affect climate change. (LO 03.04-b)
- Discuss the role agriculture plays in climate change. (LO 03.04-c)
- Explain the effect climate change has on the agriculture industry. (LO 03.04-d)
- List the ways that agricultural producers are trying to address climate change. (LO 03.04-e)

Words to Know

climate
climate change
emissions

Environmental Protection Agency (EPA)

greenhouse effect
weather

Before You Begin

Do you live in a state where the weather changes all the time? Is it hot one day, and then cold two days later? Or maybe the sun is shining brightly and it's a gorgeous day, and then out of nowhere it starts raining? Goodness, I can relate!

Does the weather always changing have any effect on you? Does it make your nose runny, or change your mood? If the weather can affect people, do you think it can affect the agriculture industry? (I don't mean that plants get runny noses; that's just silly!) Take a moment and think about how the weather might affect our food supply and other agricultural industries. Share your thoughts with a classmate.

Climate change is one of the most difficult and important challenges facing humanity, and its effects weigh heavily on the agriculture industry. Crops and livestock are exposed to wind, rain, hail, drought, and other harsh and changing weather and environmental conditions. We often hear about the effects these conditions have on crops and farmers. However, we don't necessarily look at the connection in reverse: How does agriculture affect our climate? What are the significant connections between different agricultural industries and Earth's climate?

What Is Climate?

Climate is the measure of the average temperature, precipitation, wind, and humidity of an area over a long period. *Weather* is the measure of the current temperature, precipitation, wind, and humidity of an area. The climate of an area is the long-term pattern of weather. For instance, a winter weather pattern in the Northern United States would include many days of snow, but the weather pattern in the Southern United States would very rarely include snow during winter months (**Figure 3.4.1**). The weather and weather patterns of an area combined over several decades make up its climate.

Did You Know? The ocean absorbs most of the heat produced by the planet.

What Is Climate Change?

Climate change refers to the effects of long-term changes in Earth's atmosphere on its climate. It is caused by the release of greenhouse gases, such as carbon dioxide, methane, and nitrous oxide, into the atmosphere. These gases build up in the atmosphere, trapping heat in and warming our planet.

Mia2you/Shutterstock.com

KWJPHOTOART/Shutterstock.com

Figure 3.4.1. *What is the climate of the area where you live? Do you enjoy warm sunny winter months where you live, or do you get lots of snow?* The United States has many different climates. *What climate do you think you might enjoy most? What states have that climate?*

Greenhouse Effect

Earth is not actually in a greenhouse. Our atmosphere traps the sun's heat, just like the glass walls of a greenhouse. A greenhouse is a building with clear walls and roof that is used to grow plants, such as tomatoes and tropical flowers. A greenhouse stays warm inside, even during the winter. In the daytime, sunlight shines into the greenhouse and warms the plants and air inside. At nighttime, it is colder outside, but the greenhouse stays warm because the walls keep the sun's heat trapped inside.

The greenhouse effect works much the same way on Earth. During the day, the sun shines through the atmosphere. Earth's surface warms up in the sunlight. At night, Earth's surface cools, releasing heat back into the air. Some of this heat is trapped by the greenhouse gases in the atmosphere. You can easily recreate the greenhouse using common household items.

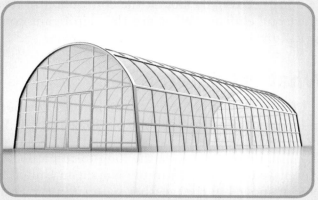

yakiniku/Shutterstock.com

6. Use the second thermometer to determine the current outside temperature.
7. At each specified time interval, take a temperature reading inside the greenhouse. Also keep a record of the outdoor temperature at the same time intervals.

Items you will need:

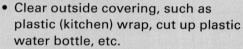

- Clear outside covering, such as plastic (kitchen) wrap, cut up plastic water bottle, etc.
- Frame materials, such as plastic straws, toothpicks, pipe cleaners, popsicle sticks
- Fastening items to secure the covering to the frame, such as tape, twist ties, wire, glue
- 2 Thermometers
- Pencil and Paper

Elapsed Time (minutes)	Inside Greenhouse Temperature (°C)	Outside Greenhouse Temperature (°C)
0 minutes		
5 minutes		
10 minutes		
15 minutes		
20 minutes		
30 minutes		

Here are your directions:

1. Create a building frame, using the framing materials. It can be a semi-circle, square, or A-frame.
2. Wrap the top, sides, and ends of the frame in a clear covering. Cover it as tightly as possible. Leave the bottom open.
3. Recreate the chart below on a separate piece of paper.
4. Take the frame outside to a sunny location.
5. Place one thermometer on the ground and set your greenhouse frame over top of the thermometer. Record the temperature inside the greenhouse.

Consider This

1. Did the greenhouse heat up faster or slower than you thought? What do you think would have happened if you left the greenhouse outside longer?
2. Compare your results with your classmates. Did the shape of the greenhouse or covering affect the temperature change?
3. How can these results be applied to climate change over the whole planet?

This is known as the greenhouse effect (**Figure 3.4.2**). The *greenhouse effect* is the trapping of the sun's solar energy in Earth's atmosphere. This works like a greenhouse used to grow plants. The sun's rays pass through the greenhouse glass or plastic where they become trapped, heating the air inside.

How Agriculture Contributes to Climate Change

Agriculture is an important industry to consider when you think about climate change. Human activities are mostly responsible for almost all the increase in greenhouse gases in the atmosphere over the last 150 years. The largest source of greenhouse gas emissions in the United States is from burning fossil fuels for electricity, heat, and transportation. *Emissions* are the gases and particles that are put into the air by various sources. According to the Environmental Protection Agency (EPA) in 2018, the agriculture industry contributes roughly 10 percent to the annual total amount of greenhouse gas emissions (**Figure 3.4.3**). The *Environmental Protection Agency (EPA)* is the United States governmental agency responsible for creating standards and laws promoting the health of individuals and the environment.

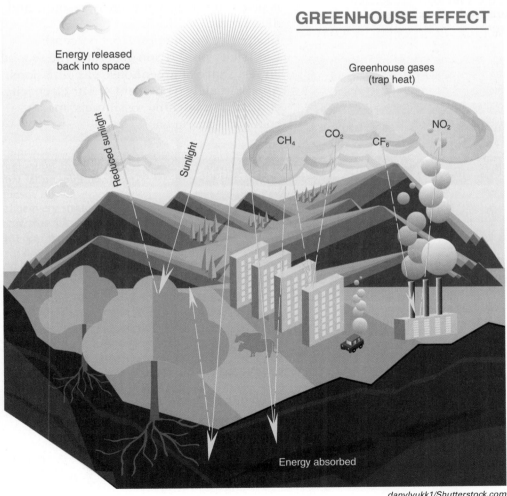

GREENHOUSE EFFECT

Energy released back into space

Reduced sunlight

Sunlight

Greenhouse gases (trap heat)

CH₄   CO₂   CF₆   NO₂

Energy absorbed

danylyukk1/Shutterstock.com

Figure 3.4.2. Although the greenhouse effect occurs naturally, the effect can be strengthened by the emission of greenhouse gases into the atmosphere as the result of human activity. From the beginning of the Industrial Revolution through the end of the 20th century, the amount of carbon dioxide in the atmosphere increased by roughly 30 percent and the amount of methane more than doubled.

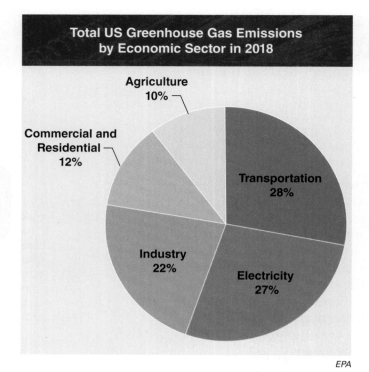

Total US Greenhouse Gas Emissions by Economic Sector in 2018

Agriculture 10%

Commercial and Residential 12%

Transportation 28%

Industry 22%

Electricity 27%

EPA

Figure 3.4.3. The Environmental Protection Agency (EPA) tracks the total US emissions each year. In 2018, the total emissions released into the atmosphere totaled 6,677 million metric tons (**Figure 3.4.4**). *Does this seem like a lot to you? Why or why not?*

In the agriculture industry, greenhouse gases come from all sorts of places: flooded rice fields, nitrogen fertilizers, improper soil management, land conversion, biomass burning, livestock production, and associated manure management. The forestry industry is referred to by the EPA as a net sink industry, which means managed forest lands in the United States have absorbed more carbon dioxide from the atmosphere than they emit.

Other sources of greenhouse gases are:

- Transportation (28.2 percent): Greenhouse gas emissions from transportation primarily come from burning fossil fuel for our cars, trucks, ships, trains, and planes. More than 90 percent of the fuel used for transportation is petroleum-based, primarily gasoline and diesel fuels.
- Electricity Production (26.9 percent): Electricity production generates the second largest share of greenhouse gas emissions. Approximately 63 percent of our electricity comes from burning fossil fuels, mostly coal and natural gas.

Figure 3.4.4. Since 1990, US greenhouse gas emissions have increased by 3.7 percent. From year to year, emissions rise and fall due to changes in the economy, the price of fuel, and other factors. *How else have trends changed since 1990?*

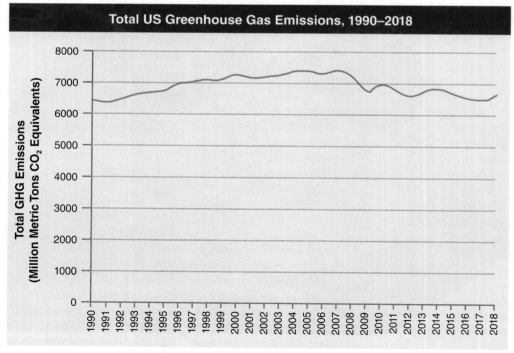

Total US Greenhouse Gas Emissions, 1990–2018

Total GHG Emissions (Million Metric Tons CO_2 Equivalents)

EPA

- Industry and Manufacturing (22 percent): Greenhouse gas emissions from industry primarily come from burning fossil fuels for energy, as well as greenhouse gas emissions from the chemical reactions necessary to produce goods from raw materials.
- Commercial and Residential (12.3 percent): Greenhouse gas emissions from businesses and homes arise primarily from fossil fuels burned for heat, the use of products that contain greenhouse gases, and the handling of waste.

In the 18th and 19th centuries, North America was covered in forest that provided food for natives and settlers and habitat for wild game, such as deer, bears, and other animals. As the population increased, people needed more food than could be provided by the forest land, and they needed open areas to build towns and settlements. These settlers began to cut down trees, using them to build homes, and converted forestland into fields for growing crops (**Figure 3.4.5**). They also burned areas to clear stumps and undergrowth. Burning trees and plant material releases carbon dioxide into the atmosphere. These actions, over such a large area, caused the atmosphere to begin to warm, and the climate in the area began to change.

Did You Know? Eighty-two percent of all the countries on this planet have agriculture in their climate change adaptation plans.

It was hundreds of years before scientists realized that clearing and burning lands could have negative effects on the environment. Today, this continued pattern of land clearing contributes about a third of the total carbon dioxide released into the atmosphere each year. In many places in Asia, it is still a common practice to burn leftover crop residue, such as rice straw. The burning of the crop residue kills insects and other pests, as well as disease-causing organisms. It also helps improve the soil quality for next year's planting. Burning is important in this area of the world because rice makes up a large portion of the crops grown and the terrain makes it difficult to work the land each year for planting.

LOUIS-MICHEL DESERT/Shutterstock.com

Figure 3.4.5. Once people began to understand the negative impacts that clearing large amounts of land can have on the environment, many countries put programs in place to protect their forest lands. Costa Rica has approximately 25 percent of their land and rainforest in protected reserves. *What percentage of land in the US is protected?*

Some of the other main agricultural methane and nitrous oxide releases come from the use of nitrogen fertilizers, prescribed burning, animal production, and animal waste. Many farmers spread nitrogen fertilizers on their fields, as nitrogen is an essential nutrient needed by plants for growth. Most of it is taken up by the plants, but some runs off into the soil and groundwater, and some enters the atmosphere. How much nitrogen is lost from the soil depends on how the fields are plowed and irrigated, and on the temperature, soil type, and weather conditions.

Making Connections  STEM

Climate Change Demonstration

The ocean is the largest solar energy (heat) collector on Earth. Not only does water cover more than 70 percent of our planet's surface, it can also absorb large amounts of heat without a large increase in temperature. This incredible ability to store and release heat over long periods gives the ocean a key role in steadying Earth's climate system.

Rising amounts of greenhouse gases are preventing heat radiated from Earth's surface from escaping into space as freely as it used to. Most of the excess atmospheric heat is passed back to the ocean. In fact, Earth's oceans are absorbing between 80 and 90 percent of the heat from climate change. Since water can withstand a lot more heat than the atmosphere, the temperature of the oceans is not changing at any significant rate.

This demonstration uses a water balloon to show how Earth's oceans are absorbing most of the heat being trapped in our warming world.

For this activity, you will need the following materials:

- Several balloons
- Lighter
- Water
- Bucket
- Safety goggles

Here are your directions:

1. Blow up a balloon and tie it off securely. This air-filled balloon represents Earth's atmosphere.
2. Using a second balloon, make a water balloon. When filling the balloon, try to remove any air bubbles. This water-filled balloon represents Earth's oceans.
3. Put on safety goggles.
4. Check your lighter to make sure it produces a small flame. The flame represents the heat from the sun.
5. Using the regular balloon filled with air, hold it out as far away from your body as possible. Using the lighter, place a flame directly under the balloon, letting the flame touch the surface of the balloon.
6. Record what happens to the air-filled balloon.
7. Using the water-filled balloon, hold it as far away from your body as possible but make sure it is directly over the bucket. Using the lighter, place a flame directly under the balloon, letting the flame touch the surface of the balloon.
8. Record what happens to the water-filled balloon.

Consider This

1. How are the results different between the air-filled balloon and the water-filled balloon?
2. Why do you think the results are different?
3. What do these results mean for the climate change occurring on Earth?

Climate Change's Effects on Agriculture

As you might guess, the changing climate is having a significant effect on the agriculture industry. Climate changes are connected to more frequent and more intense droughts in the Midwest, where many of the US grain crops are produced. Climate change affects the dairy industry: Dairy cows require cool temperatures to produce milk. Dairy farmers now need to use air-conditioned barns to raise dairy cattle. With winter months being milder, insects and other pests can live through the winter in many of the Southern United States, causing them to come back the next growing season with larger populations. How do you think this might affect crop production?

Today, farmers and ranchers experience greater extremes in temperature than any previous generation. As temperatures rise, water evaporates faster. This means that less water is available to the crops in the fields, increasing the need for irrigation. Less water and higher temperatures put animals under stress, which also affects the livestock industry. Ponds and natural water sources dry up, limiting access to water on many ranches around the country.

Rainfall amounts are also affected by climate change. Meteorologists expect that warming climates will produce more frequent and severe droughts than in previous years. They also believe that when it does rain, the rains will be more intense and cause flooding in many areas. Without consistent rain, farmers must rely on irrigation to provide water. Irrigation can be expensive to install and costs a lot to use regularly.

Did You Know?

It is predicted that by 2050, three percent of Africa's land will no longer be able to grow maize, and will transition from mixed crops and livestock to livestock-only farming systems.

Agriculture's Role in Addressing Climate Change

To help protect our agricultural future, farmers are taking steps to help reduce climate change and greenhouse gas emissions. Today's farmers use cutting-edge technology that allows them to remove the greenhouse gases that their farm produces. They can do this through climate-smart solutions such as digital farming, using satellites to know when to irrigate and fertilize, and improved plant breeding.

Farmers can reduce the need to plow fields with better weed control solutions. This makes it easier for the soil to hold nutrients and water, including carbon dioxide. Agricultural engineers are working to produce microbes that would help crops to use nutrients and water more efficiently. These microbes help crops such as corn, wheat, and rice use nitrogen found naturally in the air, reducing the need for fertilizers.

Researchers continue to develop more efficient varieties of crops. Scientists are developing solutions that help farmers grow more food on less land. This reduces the need to cut down trees for additional farmland around the world. In Vietnam, flooding can have a devastating effect on the rice fields. In recent years, scientists have developed seeds that can withstand being submerged for almost two weeks.

Climate change is a major challenge for all people, worldwide. The agriculture industry is working to develop solutions to reduce agriculture's impact on the environment, and to offset the effects of climate change on agricultural yields. They are working to become part of the solution to reduce climate change.

Climate change analysts *look at current and past research and data that has been recorded on our planet's climate. They determine how to use data and predictions to make recommendations for environmental practices.*

What Do They Do?
- Study climate conditions over a period of time;
- Predict future climate by utilizing mathematical models and data;
- Collaborate with scientists who gather climate data;
- Determine how to use data and predictions to make recommendations for environmental practices;
- Propose new policies regarding alternative fuels, transportation, and other factors related to climate change; and
- Communicate findings on climate change to the public.

What Education Do I Need? A bachelor's degree, although a master's degree is preferred, in environmental science, climatology, hydrology, meteorology or other related sciences is needed.

How Much Money Can I Make? Climate change analysts make an average of $73,000.

How Can I Get Started? You can build an SAE around studying the climate in your area. Can you research and record the climate in your area over an extended period and determine its effect on crops in your area? Create an agriscience project in social systems looking at perceptions of climate change on the agriculture or environmental industry. Can you contact a current climate change analyst and conduct an interview with them, in-person or virtually?

NicoElNino/Shutterstock.com

Review and Assessment

What Did We Learn?

- Climate is the measure of the average temperature, precipitation, wind, and humidity of an area over a long period of time. Weather is the measure of the current temperature, precipitation, wind, and humidity of an area.
- Climate change refers to the average long-term changes in Earth's atmosphere. Climate change is caused by the sun's solar energy getting trapped in the atmosphere, also known as the greenhouse effect.
- The agriculture industry contributes about 10 percent to the annual emissions of greenhouse gas. The main sources of these gases in agriculture include flooded rice fields, nitrogen fertilizers, improper soil management, land conversion, biomass burning, livestock production, and associated manure management.
- Climate changes, including warmer temperatures and decreased rainfall, have had a negative impact on the agriculture industry's ability to raise crops and livestock.
- Many in the agriculture industry are working hard to stop the lasting impacts of climate change by using climate-smart solutions, developing seeds better suited for drier, hotter conditions, and engineering microbes that help plants take up nutrients more efficiently.

Let's Check and See What We Know

Answer the following questions using the information provided in Lesson 3.4.

1. The measure of the average temperature, precipitation, wind, and humidity of an area over a long period of time is known as _____.
   A. climate
   B. weather
   C. climate change
   D. greenhouse effect

2. _____ is caused by the release of greenhouse gases into Earth's atmosphere.
   A. Climate
   B. Weather
   C. Climate change
   D. A hot spring

3. *True or False?* The largest source of greenhouse gas emissions in the United States is from burning fossil fuels for the agriculture industry.

4. The agriculture industry contributes an estimated _____ to the annual total release of greenhouse gases in the United States.
   A. 10 percent
   B. 25 percent
   C. 50 percent
   D. 60 percent

5. *True or False?* The agency responsible for regulating climate change and greenhouse gases being released into the atmosphere is the Environmental Protection Agency.

6. The crop related to the largest carbon dioxide release each year is:
   - A. wheat.
   - B. corn.
   - C. rice.
   - D. soybeans.
7. *True or False?* Climate change has a negative effect on the agriculture industry, due to warmer temperatures that allow insects and pests to live longer.
8. Farmers are battling longer drought conditions due to climate change by ____.
   - A. using different seeds
   - B. installing irrigation systems
   - C. not disturbing the soil
   - D. All of these.
9. Which sector of the agriculture industry is credited for absorbing more carbon dioxide than they produce?
   - A. Forestry
   - B. Agricultural mechanics
   - C. Row crops
   - D. None of these.
10. Which of the following are examples of climate-smart solutions?
    - A. Digital farming
    - B. Using satellites to know when to irrigate and fertilize
    - C. Improved plant breeding
    - D. All of these.

Let's Apply What We Know

1. Research and list ways farmers across the world are working to reduce climate change.
2. What are the major factors that influence the climate where you live?
3. How have new seed technologies affected agriculture's impact on the atmosphere and climate?
4. Research and list ways the agriculture industry is being impacted by climate change other than those discussed in the chapter.
5. How can you change your habits and actions to make a positive impact on climate change?

Academic Activities

1. **Technology.** Using satellite technology, examine changes in your local ecosystem. Look specifically at bodies of water, streams, fields, and forests. What changes have occurred over the last few years? Do you think those changes are related to climate change?
2. **Science.** Research weather for your hometown over a specific month. Find records as far back as possible and record the average high and the average low for the month you selected. Create a graph that shows how the high and low temperatures have changed over the years.

Communicating about Agriculture

1. **Listening and Writing.** Watch the National Geographic documentary *Six Degrees Could Change the World* (online on YouTube or Amazon Prime Video). Outline what each degree of temperature increase might mean to the world and how these changes would affect the world.
2. **Reading and Speaking.** Choose a country and research how water is used by its citizens. Develop a brief presentation that describes the country's water resources, how people obtain water, and how much water the average person uses. Expand the activity by researching other countries and using data to create a world map illustrating water usage by country. Determine water use trends based on climate and economy.

SAE for ALL Opportunities

1. **Foundational SAE.** Interview a farmer who has been farming for many decades. Ask them to discuss any changes they have noticed to the temperature or precipitation from the start of their career to now. Ask how they have adapted to any changes that have taken place.
2. **SAE.** Determine ways that climate change can affect your SAE program. Develop a plan that will reduce the impact your SAE could have on climate change.

SAE for ALL Check-In

- How much time have you spent on your SAE this week?
- Have you logged your SAE hours?
- What challenges are you having with your SAE?
- How can your instructor help you?
- Do you have the equipment you need?

SAE for ALL Profile
Where Is Ecuador?

A trip to the supermarket with his mother started Bryce Lanier on a successful career in agriculture. While in the produce section with his mother selecting bananas to purchase, he noticed a sticker on the bananas that said "Product of Ecuador."

"I didn't know what Ecuador was or where it was, but I was interested," said Bryce, "So I went home and Googled 'Ecuador,' and discovered that it was a country more than 4,000 miles away from my home in Eastern North Carolina. I wondered how those bananas made it all the way to my local grocery store."

Bryce's agricultural education classes at West Lenoir High School provided him with the basics of how the agricultural industry works. His courses in mathematics and social studies fed him with the desire to pursue a career that allowed him to work as an agricultural economist with the United States Department of Agriculture. But it took a lot of steps to reach his preferred career.

"It really started with my Supervised Agricultural Experience project. My agriculture teacher, E.S. Howard, helped me start a Foundational SAE that allowed me to learn about careers associated with international agricultural economics." Bryce's Foundational SAE positioned him to develop a Placement SAE project at Dogwood Gardens, a local plant nursery in his hometown. In 10th grade, Bryce worked after school and weekends selling plants and garden supplies to customers. The money he earned at the plant nursery, along with scholarships, paid for most of his four-year college degree in economics.

In his job at the United States Department of Agriculture, Bryce researches economic issues important to the agriculture industry, such as the effects of tariffs on the price of grain on the international market. He recommends solutions to economic problems to policy-makers and government officials and forecasts trends in agricultural markets. Being an economist is demanding, but Bryce's education and experience prepared him well.

"I earned my bachelor's degree in economics before starting my job with the USDA. But I knew I would need more education to do the job well. So, I enrolled in night classes at a university and earned a master's of business administration," said Bryce. "I love my job. It's important to me to do the best I can."

- Bryce's career started with a question and the curiosity to learn more. What are you curious about? What questions do you want to answer?
- How did Bryce's SAE get him on the path to his dream career?

Top: Rido/Shutterstock.com
Bottom: Laguna781/Shutterstock.com

FenliaQ/Shutterstock.com

Agriculture and the United States Economy

Kelvin H. Haboski/Shutterstock.com

Learning Outcomes

By the end of this lesson, you should be able to:

- Explain how agriculture influences the United States economy. (LO 04.01-a)
- Describe some of the jobs that agriculture creates in the United States economy. (LO 04.01-b)

Words to Know

economy	perishable	ton-miles
gross domestic product (GDP)	processing	wholesale
	retail	

Before You Begin

In this lesson, we are going to be looking at some very large numbers. The United States economy involves trillions of dollars! This number is so big it is probably difficult for you to imagine what it looks like, or how those dollars move through the economy.

Let's break down these large numbers into something easier to understand. Your first task is to find 100 pennies and arrange them on a table. Count 15 pennies out and separate them from the others. These 15 pennies represent the amount of money that farmers receive out of every dollar spent in the United States. Does this seem like a lot of money to you? Why or why not?

How Big Is Agriculture?

What's the big deal about agriculture? Most Americans do not live or work on farms, and many of us ride, drive, or walk to grocery stores minutes away, full of the foods we like to eat. We can hardly drive five miles down the road, wherever we are, without seeing some type of restaurant. Food is food, and we seem to have plenty of it. So what's the big deal?

Agriculture is critical to our national economy, and employs millions of people in every sector: production, sales and marketing, transportation and distribution, regulation, food processing and preparation, and international trade. The fact that food is available at a reasonable cost in most areas of the United States shows that agriculture is an efficient system, providing food, textiles, and energy to a growing population.

How Agriculture Fits into the Economy

Agriculture, as you may recall, is the production of plants and livestock for the benefit of humans. The clothes we wear, the food we eat, and sometimes the fuel used in our vehicles comes from the production of agricultural products. Exactly how important is agriculture? One way to determine the importance of agriculture is to look at how it affects the economy of the United States. What does it mean when we talk about the economy? The *economy* is the management and use of money, resources, and labor in a given place (in this case, the United States).

The Economic Value of Agriculture

How valuable is American agriculture to the United States? In order to explain this, we need to know a few things about the US economy. First, we measure the size of the economy by looking at the value of all the products and services we produce. If we were to add the value of every product created in the United States, we would come up with something called the gross domestic product. The *gross domestic product (GDP)* is the value of all goods and services that we create, measured over the course of a year. According to the World Bank, the gross domestic product of the United States is equal to approximately $21.4 trillion. Here is what this looks like written out: $21,400,000,000,000. That is a lot of zeros!

According to the United States Department of Agriculture, the value of just the crops and livestock produced by US farms and ranches is $132 billion. This only accounts for raw farm products, such as unprocessed milk, meat, and wool. The value of agricultural products increases after processing. When you add the value of processing and marketing agricultural products, we find that agriculture contributes about $1 trillion to the gross domestic product of the United States.

Processing adds value to agricultural products. *Processing* changes raw agricultural products into more usable forms. Do you see the block of cheese in **Figure 4.1.1**? This block of cheese is made from milk. It should also be noted that a 16-ounce block of cheese costs the average consumer more than a gallon (128 ounce) jug of milk.

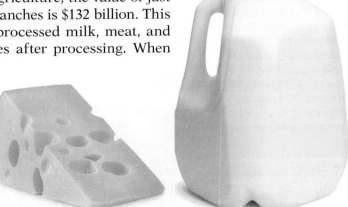

azure1/Shutterstock.com *Photo Melon/Shutterstock.com*

Figure 4.1.1. *Why does the 16-ounce block of cheese cost the average consumer more than a 128-ounce jug of milk?*

Jobs in Agriculture

More than 21.6 million jobs in the United States are related to agriculture, according to the United States Department of Labor. About one in 10 workers in America are employed in agricultural jobs. These jobs include farming, forestry, textiles, and food-service occupations. Food-service related industries make up 13.5 million jobs. There are also many jobs related to the transportation and distribution of agricultural products. It takes a lot of effort and resources to convert raw agricultural products into products useful to humans (**Figure 4.1.2**).

Employment in Agriculture, Food, and Related Industries

There are almost 22 million jobs in agriculture here in the US. The people working in some form of agriculture represent about 11 percent of all people working in the United States. Wow! So who are these people? There are almost three million farmers and ranchers in the United States. The remaining 19 million people work in jobs that do not involve farming, like forestry, fishing, manufacturing, textiles, food service, and wholesale and retail food sales. **Wholesale** foods are sold to grocery stores and restaurants for resale. **Retail** foods are sold to consumers by grocery stores and restaurants.

One of the many cool things about agriculture is that just about every job you can think of exists within the industry. You might wish to open and operate your own farm one day. If you do, you can choose from hundreds of different crops and dozens of different animals to produce. If you do not want to own a farm or ranch but like the type of work done on a farm, you can become a farm or ranch manager.

Figure 4.1.2. A. Wheat seeds are crushed into a fine flour for breadmaking. **B.** Sugar, butter, eggs, yeast, and other ingredients are added to the flour prior to baking. **C.** Bread is baked in large commercial ovens and then packaged for distribution. **D.** Bread is distributed to grocery stores and restaurants. *How much time do you think it takes to get through this process from beginning to end? How many people are involved?*

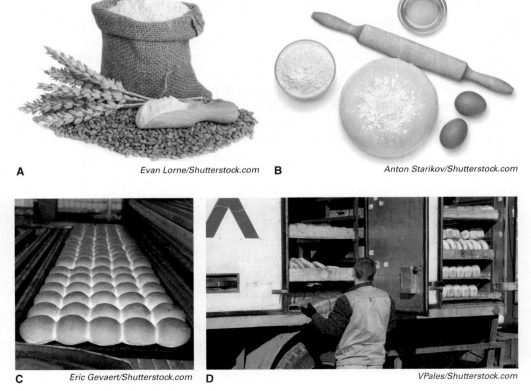

A   *Evan Lorne/Shutterstock.com*   B   *Anton Starikov/Shutterstock.com*

C   *Eric Gevaert/Shutterstock.com*   D   *VPales/Shutterstock.com*

Figure 4.1.3 shows examples of the types of jobs that fit within the scope of the industry.

There are other types of jobs in agriculture as well, ranging from agronomist to wildlife biologist. If you are unsure whether there's a place for your talents, skills, and interests in agriculture, you don't need to worry. There is definitely a place for you! If you enjoy working with people, plants, or animals, or working outside, then a career in agriculture might just be something for you to consider.

Food Manufacturing and Sales in the US

In the United States, 13 cents of every dollar spent goes toward purchasing food. That is 13 percent of your annual household income. According to the USDA, the average American household spends more for food than they do for clothing, health care, and education. A family with a household annual income of $45,000 per year spends $5,850 on groceries and restaurant meals (**Figure 4.1.4**).

Each state in the US produces agricultural products based on its climate, soil type, and the availability of good farmland and water. California produces more agricultural products than any other state. Texas, Iowa, Nebraska, and Minnesota join California in the top five agricultural production states in the United States. These five states account for almost one-third of all agricultural production in the country. This does not mean that they have a monopoly on agriculture in the US, though—every state produces unique and important products!

Bartosz/Shutterstock.com

Zivica Kerkez/Shutterstock.com

Scharfsinn/Shutterstock.com

Figure 4.1.3. These careers are not always considered to be related to agriculture, but they are. *Can you think of another career connected to agriculture?*

Figure 4.1.4. While it might not surprise you that most US income is spent on housing, food expenses (groceries and restaurants) account for 13 percent of consumer spending. *Does that surprise you? Why or why not?*

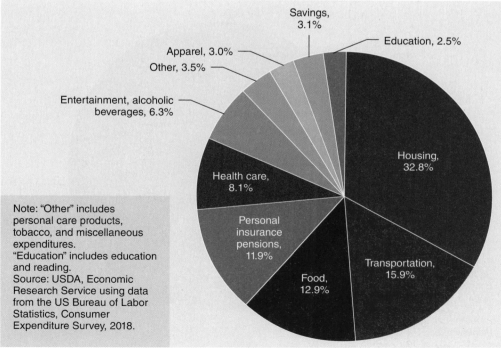

Savings, 3.1%
Education, 2.5%
Apparel, 3.0%
Other, 3.5%
Entertainment, alcoholic beverages, 6.3%
Housing, 32.8%
Health care, 8.1%
Personal insurance pensions, 11.9%
Transportation, 15.9%
Food, 12.9%

Note: "Other" includes personal care products, tobacco, and miscellaneous expenditures. "Education" includes education and reading.
Source: USDA, Economic Research Service using data from the US Bureau of Labor Statistics, Consumer Expenditure Survey, 2018.

USDA Economic Research Service, 2018 (pending new census data)

Figure 4.1.5. Food service workers make up almost 13 million jobs, by far the largest number of jobs in agriculture. *What jobs would be included in this category?*

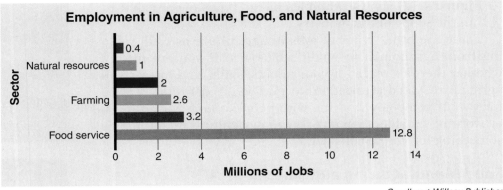

Employment in Agriculture, Food, and Natural Resources

Natural resources 0.4, 1

Farming 2, 2.6

Food service 3.2, 12.8

Millions of Jobs

Goodheart-Willcox Publisher

MBI/Shutterstock.com

Figure 4.1.6. Every school day, you come face to face with the food processing industry. Your school cafeteria is probably part of the USDA National School Lunch Program. This program provides healthy low-cost or free lunches under the National School Lunch Act of 1946.

The Food Processing Industry

The food processing industry employs 14 percent of all people working in manufacturing in the United States. It is the largest area of employment within agriculture, food, and natural resources by a wide margin (see **Figure 4.1.5**). People in food processing prepare meat, poultry, fish, fruits, grains, and vegetables for retail sale. You cross paths with people who work in food processing and foor service every day (**Figure 4.1.6**).

In 2016, the US food and beverage manufacturing sector employed more than 1.5 million people. In more than 35,000 food and beverage manufacturing plants located throughout the country, these employees transform raw agricultural materials into products for intermediate or final consumption. Meat and poultry plants employ the most workers (**Figure 4.1.7**). They are followed by commercial bakeries, and then by fruit and vegetable processing plants.

Figure 4.1.7. Food manufacturing accounts for 14 percent of all US manufacturing employees, the largest percentage of whom work in meat and poultry processing. *Do you see any categories here that surprise you?*

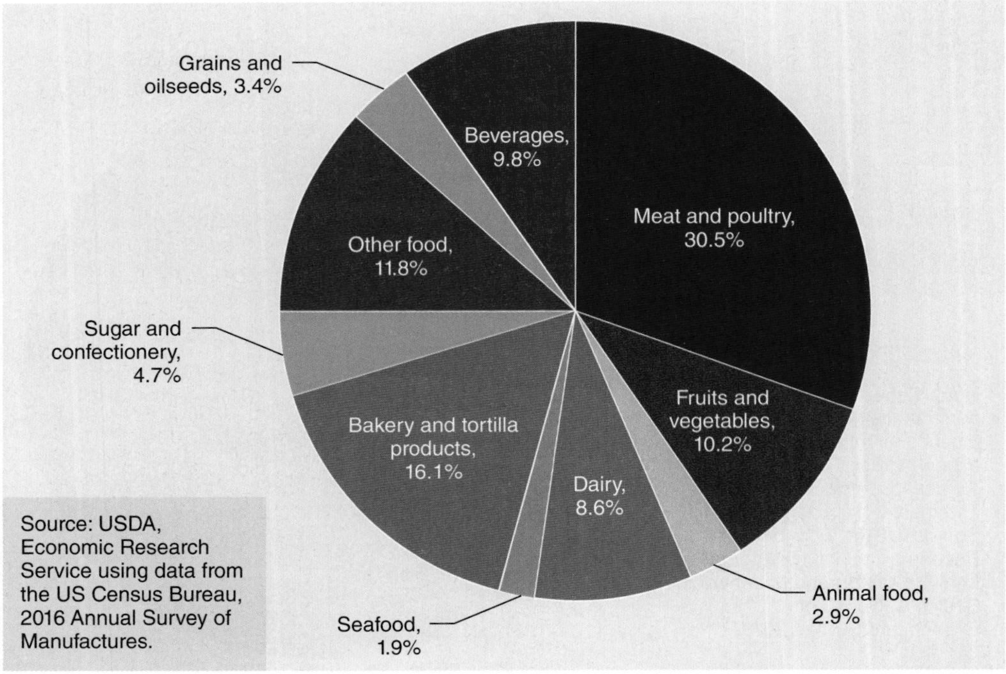

Grains and oilseeds, 3.4%

Beverages, 9.8%

Meat and poultry, 30.5%

Other food, 11.8%

Sugar and confectionery, 4.7%

Bakery and tortilla products, 16.1%

Fruits and vegetables, 10.2%

Dairy, 8.6%

Animal food, 2.9%

Seafood, 1.9%

Source: USDA, Economic Research Service using data from the US Census Bureau, 2016 Annual Survey of Manufactures.

USDA Economic Research Service, 2016

The Natural Resources Industry

The United States is the third largest nation in the world by land area, behind Russia and Canada. Our country has a rich and varied selection of natural resources that we use for recreation, work, and daily life (**Figure 4.1.8**).

Collecting and using natural resources in a sustainable way requires a network of occupations. These include people working in forestry, fish and wildlife management, geology, soil science and conservation, water quality, agricultural engineering, oil and natural gas production and management, renewable fuel production and management (**Figure 4.1.9**), climatology, meteorology, and surveying. According to the US government, the United States is one of the largest producers of metals and minerals, and manages more than 500 million acres of timber land. These resources, and the people who work to manage them responsibly and sustainably, are essential to the US economy.

Figure 4.1.8. Soil is an important natural resource. This agronomist is examining soil samples for the presence of pesticide residues. If there are no pesticides present, this field may be certified as an organic farm.

Figure 4.1.9. This dam is part of the Tennessee Valley Authority. The TVA generates electrical power through a series of dams on major rivers in six states: Virginia, Kentucky, Tennessee, Georgia, Alabama, and Mississippi. The TVA harnesses the power of water resources to control floods, make rivers navigable, and generate electricity. The TVA supports the economy by providing electricity at a reasonable cost and improving water transportation of commodities. It also prevents flood damage to farms, homes, and businesses.

Pressmaster/Shutterstock.com

Figure 4.1.10. This small business owner uses the banking industry to deposit money from sales, and to borrow money to make repairs to this greenhouse. Perhaps she needs money to build another greenhouse to expand her business.

Banking and Finance

The banking and finance sectors of the US economy are involved with saving, lending, and investing money. In the agricultural and natural resources industry, banks loan funds to farmers and ranchers to purchase equipment and produce agricultural commodities. Agribusinesses rely on banks to hold funds earned through sales and services, and to lend money for operating loans and business improvements (**Figure 4.1.10**). Banks also provide a way for agricultural businesses to save and invest money.

In agriculture, the banking industry provides agribusiness and farm credit loans. It also finances the sale of agricultural commodities in national and international markets (**Figure 4.1.11**).

Here are a few of the careers associated with agricultural banking and finance:

- Agricultural finance officer
- Farm insurance specialist
- Loan officer
- Real estate broker and appraiser
- Financial analyst
- Commodities trader

Did You Know? Did you know that the United States is the second largest exporter of goods globally? Which country is the largest exporter of goods? If you guessed China, you are correct!

Figure 4.1.11. Agricultural trade depends on the banking industry to manage the flow of dollars as goods are bought and sold. These yellow-coated traders used to buy and sell agricultural commodities at the Chicago Mercantile Exchange. *How are commodities typically traded today?*

Joseph Sohm/Shutterstock.com

Transportation of Agricultural Products

The transportation of agricultural products is a major part of the industry. Food and agricultural products move by rail, barge, and truck to markets in the United States and around the globe (**Figure 4.1.12**). Transportation is especially important to the agricultural industry because most agricultural commodities are *perishable*. This means that they will begin to spoil and rot if not stored properly and moved to market in a timely manner. The US transportation system has to move raw agricultural materials from the farm to the processing plant before going to grocery stores. There may be multiple legs of travel between stops, and multiple modes of transportation used to get from start to finish. This is a remarkably complex process! The finished product arriving at the grocery store, and then to your dinner table, depends upon an organized and efficient transportation system.

A
WvW/Shutterstock.com

B
BCFC/Shutterstock.com

C
photogal/Shutterstock.com

Figure 4.1.12. A. More than 31,000 barges move 92 million tons of raw agricultural products along US rivers each year. **B.** Railroads move large quantities of agricultural products over great distances. **C.** Trucks move agricultural products from regional distribution centers to grocery stores and restaurants.

According to the USDA, the agriculture industry is the largest user of freight transportation in the US. Freight transportation uses a funny way to measure how much freight is being moved: It is measured in ***ton-miles*** that equal the movement of 2,000 pounds (one ton) a distance of 5,280 feet (one mile). Agriculture is responsible for 31 percent of all ton-miles moved in the United States. Agricultural commodities are also shipped in large amounts to other countries. International transportation systems are complex because of the long distance, customs, different languages, and different monetary systems.

Here are just a few of the jobs associated with international agricultural trade:

- Ship captain
- Transportation dispatcher
- Forklift operator
- Government trade advisor
- US Customs agent
- Inventory control specialist

Making Connections # STEM

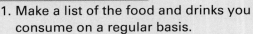

Calculating Food Miles

The purpose of this activity is to track the agricultural products you consume from raw materials to finished food products.

To complete this task, you need the following:

- Notebook
- Pen
- Device with internet access

Look at the hamburger and fries shown below. How many agricultural products were used to make the burger and fries?

ALEX S/Shutterstock.com

Here are your directions:

1. Make a list of the food and drinks you consume on a regular basis.
2. Next to each food and drink item, list from where you think each item comes from. Note: "The grocery store" is not a good answer here. Dig deeper. Where did the food items in the grocery store travel before they arrived?
3. After you have mapped your food sources, figure out which food items traveled the farthest.
4. Which items traveled the shortest distance?
5. Where did each item come from?
6. Where there any surprises to your research? If so, what are they?
7. What does your research tell you about the foods you eat?

Going Bananas

Let's go bananas for a moment. How do bananas get to our grocery store shelves and kitchen tables from where they are grown? According to the University of Florida, most of the bananas grown in the United States are grown in Hawaii and Florida, but very few bananas are grown in these two states. Bananas need a warm tropical environment to grow, and much of the United States does not have a suitable climate for banana production. So where do our bananas come from? Most banana production occurs in China, India, Ecuador, Brazil, and the Philippines. Let's follow the path of bananas from a plantation in Ecuador to your grocery shelf.

First, where exactly is Ecuador? Using web resources, find Ecuador on a world map. Also, conduct a web search to find out information about the country, its geography, and its agricultural industry. Using a reliable source, such as the United Nations Food and Agriculture Organization (www.fao.org), determine how many bananas are produced and exported by Ecuador each year.

Mikhail Ivannikov/Shutterstock.com

This photograph shows a banana flower and fruit forming. Bananas grow best in organic, well-drained soils in full sun. The weather and climate in Ecuador allows bananas to be harvested almost continuously throughout the year.

Green unripe bananas are picked in bunches called "hands" that are transported to a packing facility for processing. At the processing facility, they are washed, sprayed with a citric acid solution to prevent browning, then packaged and boxed.

barmalini/Shutterstock.com

Bananas being washed and prepared for packaging.

The bananas are then loaded onto refrigerated tractor trailers for transportation to one of Ecuador's ocean ports. From there, they are loaded into ships for transportation to the United States and other countries. Using a map online or provided by your instructor, identify a port in Ecuador where bananas could be loaded onto a ship.

Bananas arrive at US ports in California, Texas, Florida, Delaware, and Pennsylvania. Bananas traveling from a port in Ecuador to the port of Miami in Florida will travel a minimum of 2,000 miles by sea. The fruit is offloaded from ships in port and placed on tractor trailers for transport to distribution centers. From there, the fruit is transported by truck to grocery stores and markets in your neighborhood. The banana that you eat at lunch today likely traveled as many as 4,000 miles to get to your plate.

Continued

Going Bananas *Continued*

Using online mapping software, like Google Earth, plot the course and distance a ship would travel to get from an Ecuadoran port to Miami. You will notice that the shortest route will take you through the Panama Canal in Central America. Your ship's course will not be a straight line because you have to navigate around islands and land masses. How many miles did your ship travel? Most cargo ships travel at an average speed of 27 miles per hour. How long would it take for the cargo ship carrying bananas to travel from Ecuador to Miami, assuming it didn't have to stop?

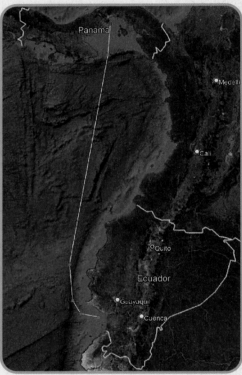

Google Earth

The yellow line on this map from Google Earth shows the possible path of a ship from a port in Ecuador to the Panama Canal.

There are other things to consider beside the logistics of getting the bananas from a plantation in South America to your dinner table. What about the price of the bananas? Who pays the growers for the fresh bananas they produce? How does the price paid to growers for fresh bananas influence the price you pay for them in the grocery store?

Most shoppers probably do not think about how the bananas they want to buy arrived in the produce section of the supermarket. They may note the price and quality of the fruit, but few people consider the complex system of transportation, pricing, marketing, quality control, and production that went into producing the fruit and getting it to where it is. There's a lot of information there!

You should also think about all of the jobs associated with moving something as simple as bananas from where they are grown to where they are consumed. What jobs can you think of?

Consider This

1. Bananas are a perishable product. How do supermarkets and food distributers prevent bananas from spoiling while in transit from the farm to your table?
2. Bananas are grown in Hawaii and Florida. Why do you think that bananas are not grown in other states?
3. Bananas bruise easily, causing them to spoil faster. How do you suppose shippers package bananas to prevent bruising and spoilage?

What Did We Learn?

- More than 22 million jobs are directly connected to agriculture in the United States, and 15 cents of every dollar spent in this country goes to the producers of agricultural products.
- Agriculture creates many jobs throughout the economy, including in food processing, natural resources, banking and finance, and transportation.

Let's Check and See What We Know

Answer the following questions using the information provided in Lesson 4.1.

1. US farms and ranches produce about _____ per year in agricultural products.
   A. $7 trillion
   B. $15 billion
   C. $100,000
   D. $132 billion

2. The agricultural industry provides more than _____ million jobs in the US economy.
   A. 12
   B. 21
   C. 45
   D. 132

3. More than _____ of all US workers are employed in the agriculture industry.
   A. 17 percent
   B. 22 percent
   C. 32 percent
   D. 100 percent

4. The average family spends about _____ percent of their annual household income on groceries and restaurant meals.
   A. 13
   B. 17
   C. 45
   D. 58

5. The state that produces the most agricultural products is _____.
   A. Alaska
   B. California
   C. Georgia
   D. Nebraska

Let's Apply What We Know

1. Research the agricultural crops and livestock grown in the five states that produce the most farm products. Which crops and livestock are produced in the largest quantities? What do you think about the information you found? Are you surprised by the results of your research?

2. Conduct research in your local community to locate agribusinesses and food processing facilities, such as farm supply stores, meat packing plants, or vegetable packing facilities. Mark the location of each facility on a map, and prepare a brief report on what you found. Were the results what you expected to find? Which of these agricultural businesses employs the most people? What does this information tell you about the agriculture industry in your community?

Academic Activities

1. **Math.** Your task is to figure out how much it costs to transport a truckload of potato chips from San Francisco to New York City. You will need to know the number of miles, the shortest route, and the cost of fuel. Assume that your tractor trailer burns fuel at a rate of 10 miles per gallon. Use an online mapping system (like Google Earth) to plot the shortest route. Based on your calculations, approximately how much does it cost to move these potato chips from San Francisco to New York City? What effect do transportation costs have on the price of the potato chips?

2. **Math.** Get shreddy! Will you save money if you shred your own cheese, versus buying it already shredded? Visit the supermarket and compare cheeses of the same type. Conduct a price comparison between shredded and block cheese. Calculate the cost per ounce of each cheese. Which is more expensive? Why?

3. **Math.** Everybody loves a good salad! Go to the produce section of your grocery store and price the ingredients you would use to make a salad. While you are pricing the items, note the origin of each ingredient. Go a step further and estimate the distance each vegetable traveled to get to your store.

Communicating about Agriculture

1. **Reading and Writing.** Research sustainable agriculture and write a two-page essay or create a visual presentation on the topic. Determine the price of foods that are grown sustainably and compare them to prices of the same foods that are not considered sustainably grown. Consider the following questions as you perform your research: Are there benefits to sustainable agriculture? What are these benefits? Are there detriments to sustainable agriculture? If so, what are they? Is it possible to make most of our agriculture production sustainable?

2. **Reading and Speaking.** Production agriculture is not without controversy. Identify and explain an example of controversy involving one or more types of agricultural production in your state or region. Write an opinion speech presenting the issue, and then arguing in favor of one position.

SAE for ALL Opportunities

1. **Foundational SAE.** Using your research into agricultural businesses in your community, dig a little deeper to see what types of jobs and careers are available at those businesses. Which jobs and careers interest you? Do any of these businesses offer part-time employment to youth?

2. **Foundational SAE.** Does anyone in your family or neighborhood work in a job that involves agriculture, food, and natural resources? Interview them to find out what they like about their job. Also ask them about the more challenging parts of the job.

SAE for ALL Check-In

- How much time have you spent on your SAE this week?
- Have you logged your SAE hours?
- What challenges are you having with your SAE?
- How can your instructor help you?
- Do you have the equipment you need?

The Global Impact of Agriculture

i_am_zews/Shutterstock.com

Learning Outcomes

By the end of this lesson, you should be able to:
- Explain why countries trade agricultural products. (LO 04.02-a)
- Describe some of the agricultural products grown in countries outside of the United States. (LO 04.02-b)
- Discuss the value of agriculture on a global scale. (LO 04.02-c)

Words to Know

malnutrition
trade

Before You Begin

Does this plant look familiar to you? Its red berries are dried, roasted, and ground up to make a drink that more than 62 percent of Americans drink every day (Figure 4.2.1). This plant, coffee, is only grown domestically in Hawaii and California, and in very small quantities. Puerto Rico grows the most coffee of any state or territory. The United States cannot produce enough coffee to meet the demand for it here. So where do we get our coffee? Team up with a classmate and make a guess about where most of this coffee is grown.

Figure 4.2.1. On most coffee plantations, the coffee beans are hand-picked to ensure only the ripe fruit is harvested. *How would this make coffee a difficult crop to grow?*

bonga1965/Shutterstock.com

Why Do We Trade Agricultural Products?

Agriculture touches the lives of every person on Earth. It supplies food, clothing, and shelter for 7.5 billion people. Food is essential for human life, and agricultural production is essential for food production. This all seems pretty straightforward, but there are challenges to feeding and clothing a global population.

Adrian Lindley/Shutterstock.com

Figure 4.2.2. According to the US Agency for International Development (USAID), the United States provided $3.7 billion in international food aid to those in need in 2018. This includes 2.5 million tons of agricultural products. According to USAID, more than four billion people have benefited from food assistance since 1954.

Feeding a Growing Population

The current food system falls short of feeding all 7.5 billion people on Earth (**Figure 4.2.2**). Each day, more than 800 million people across the world go without enough food to eat. Many more people are not getting the right types of food and are suffering from malnutrition. *Malnutrition* is a condition in which a person does not receive the necessary nutrients in their food to maintain a healthy and well-functioning body. This includes individuals who receive too much of a nutrient, such as sugar or fat. This may lead to obesity, which is also a form of malnutrition. Too much or too little of the nutrients we need may result in disease or starvation.

Agriculture is the key to feeding and clothing the people who live on this planet. Across the world, $1 out of every $3 earned goes to individuals involved in the production of food, fiber, and natural resources. Even with this large investment in agriculture, there is still a long way to go to feed our human population. It is anticipated that there will be 9.7 billion people to feed, clothe, and shelter by the year 2050. That is not very far in our future! Without enough food to feed this growing population, there exists a real possibility that human life may become unsustainable. That's a scary thought!

This is a complicated problem to address. Food production is not spread evenly across the planet. Some countries, like the United States, produce an abundance of food. We produce so much food in this country that we are able to trade with other countries, and even give away some food to help those countries that are dealing with famine. Other countries cannot produce enough food for their growing population. Moving food from an area of abundance to an area where there is great need is a difficult thing to do. As we discussed in **Lesson 4.1**, it is very expensive to ship food from one place to another. Not only is it expensive to ship food products, there are also differences in culture, language, and even the type of money used to buy and sell food. It can also be difficult to move food around within a country, from areas of abundance to areas of need. While the United States produces a surplus of food that allows for international trade, there are hungry people here, too, and many people do not have easy access to fresh, healthy, or nutritious food. Solving that problem is both difficult and critically important.

Meeting Varying Tastes and Wants

Every country on this planet has some sort of an agricultural industry, but not every country produces the same types of products. Culture, language, and tradition (along with climate and available resources) have a strong influence on the types of foods that are produced in a given country. In the next section, we'll learn a bit more about this.

Agriculture and Food Around the Planet

You might think that people eat pretty much the same things all across the planet. This is not true! As we've already said, culture, language, and tradition play a role in the foods people eat. You and your friends might go down to a local fast food restaurant to order a couple of burgers and a soft drink, and think nothing of it. What if you lived in a country where there aren't fast food restaurants? What if you were part of a family that didn't believe in purchasing fast food?

We will now take a look at the types of foods that three families from three different countries might consume in a week's time. These foods might be very different than what you are used to in your home! This virtual trip around the planet will teach us about agriculture in three countries: Cuba, Egypt, and the United States.

Did You Know?

According to the National Center for Health Statistics, more than one out of three adult Americans consume fast food on any given day.

Santiago, Cuba

Our first stop is Santiago, Cuba. Cuba is an island nation about 90 miles south of the state of Florida. Cuba has about 11 million citizens. The country produces sugarcane, tobacco, oranges and other citrus fruits, coffee, rice, potatoes, beans, and livestock (**Figure 4.2.3**).

Figure 4.2.3. Cuba produces a variety of fruits, vegetables, sugarcane, and livestock. This farmer is harvesting sugarcane for processing into granulated sugar. *Is this what you imagined sugar might look like in its raw form?*

possohh/Shutterstock.com

The average family in Cuba spends about $57.00 per week ($2,964.00 per year) on food for a family of four. The Cuban government provides a certain amount of funding to families for the purchase of food each month. In an average week, a family in Cuba will dine on a variety of foods, including:

- Bread, rice, and pasta
- Cheese and yogurt
- Chicken, pork, eggs, and fish
- Watermelon, guava, papaya, and a variety of vegetables
- Malanga, a plant that has growth characteristics similar to a sweet potato

In a week's time, the average Cuban family spends about $15.00 in a restaurant, and eats less than $10.00 worth of snacks and prepared foods.

What do you think about the food choices of a family in Cuba? Do you eat some of the same foods? What differences exist between the foods you enjoy and those of the average Cuban family? Consider these same questions at the next stop in our travels.

Cairo, Egypt

Our next stop is Cairo, Egypt. Egypt is a country in northwestern Africa, on the Mediterranean Sea (**Figure 4.2.4**). Agriculture makes up 11 percent of the economy of Egypt, in terms of revenue produced. This involves the production of rice, cotton, beans, wheat, fruits and vegetables, beef cattle, sheep, goats, and water buffalo. One out of every four citizens of Egypt works in the agriculture industry.

The average Egyptian family spends about $70 per week ($3,640.00 per year) on food. One week's grocery list includes:

- Potatoes, rice, wheat flour, bread, and pasta
- Spices
- Powdered milk
- Butter and cheese
- Fruits, such as grapes, mangoes, figs, and prickly pear
- Vegetables, such as onions, radishes, leeks, and lettuce
- Poultry, lamb, eggs, canned fish, beef, and pickled meats

Figure 4.2.4. Egypt was one of the first locations for widespread agricultural production. Farmers have been cultivating the land along the Nile River for centuries.

erichon/Shutterstock.com

Many Egyptian families rarely eat outside of the home. In one week, a family may spend about $17.00 on snacks and prepared foods, including soft drinks.

Are the foods that are enjoyed in Egypt similar to those you enjoy in the United States (**Figure 4.2.5**)? Were there any foods in their list that did not appeal to you? And what did you think about the amount of money spent on snacks and fast food? Do you spend more than $17.00 per week on these things?

Let's head back to the United States to see what foods an average family here may consume for comparison.

Atlanta, Georgia

In previous lessons, we have covered the types of agricultural commodities produced in the United States, so we will not go back over those here. Instead, compare the grocery lists and food costs of the average families in Cuba and Egypt to an average family living in Atlanta, Georgia. The typical family there spends almost $350.00 per week ($18,200.00 per year) on food. A typical grocery list in their home might include:

- Potatoes, bread, breakfast cereal, pasta, and rice
- Milk, cheese, butter, and processed cheese
- Beef, pork chops, poultry, and eggs
- Processed foods, such as bacon, ham, and canned fish
- Fruits and vegetables, like peaches, apples, and carrots (**Figure 4.2.6**)

In addition to the items on their grocery list, the average family spends about $200.00 per week on fast food, soft drinks, and snacks. How is this different from Cuba and Egypt?

What did you learn from going around the planet to see what people in other countries purchase and consume? Are the foods you purchase and enjoy similar to those we have examined in the three locations above? How are your family's habits similar or different?

Looking at how and what we eat is a good example of the ways agriculture and food systems influence us on a personal basis. On a larger scale, agriculture has a very large footprint on this planet.

The world's population is expected to increase by 2.3 billion people by 2050. In the next 30 years, we will add an average of 77 million people to the world's population each year. To meet this need (and the unmet needs we already have), we will have to increase global food production by 70 percent over that same time. To feed 9.1 billion people by 2050, by our current production standards, our worldwide agricultural systems will need to produce one billion tons of grain each year, along with 200 million tons of meat and meat products.

Elena Eryomenko/Shutterstock.com

Figure 4.2.5. Shakshuka is a traditional Egyptian breakfast dish. It is prepared with eggs, tomatoes, peppers, and other vegetables, and served with flatbread or pita bread. *Does this look like something you would enjoy eating?*

4

Darryl Brooks/Shutterstock.com

Figure 4.2.6. US consumers can find fresh, local produce at farmers markets and stores that purchase locally grown food. *Do you think that supermarkets look the same around the world?*

Exotic Fruits

Most of us are familiar with fruits and vegetables such as bananas, oranges, potatoes, and carrots. These foods are something you see every day in the grocery store. Take a look at the following photographs. These fruits are grown in other countries. Match the name of the fruit or vegetable to its picture. How did you do?

1. Guava
2. Rambutan
3. Dragon fruit (pitaya)
4. Lychee

These fruits may seem exotic, but they are probably common to the people living in the country or region where they are grown. Do you think that people who live in China, Russia, or India would consider some of the crops grown in the United States to be unique or exotic?

Consider This

1. Research each fruit to identify where it is grown and how it is used.
2. Can any of the fruits listed above be purchased at your local grocery store?
3. How far would each fruit travel from its native country to your local grocery store?

Dragon fruit
Anastasiia Skorobogatova/
Shutterstock.com

Guava
Anna Kucherova/Shutterstock.com

Lychee
Kovaleva_Ka/Shutterstock.com

Rambutan
pukao/Shutterstock.com

Global Agricultural Trade

In the last section, we visited three different countries where people have different spending habits when it comes to food. You might think that each country produces its own foods and doesn't pay much attention to what other countries are growing or producing. Not so! Every country trades agricultural products with other countries. For example, the United States is the world's largest exporter of soybeans, producing 50.1 percent of all soybeans on the world market. China, Mexico, Canada, and Japan buy more than $96 billion in agricultural products from the United States. These products play a vital role in world *trade*, or the exchange of desired goods. For centuries, countries have traded agricultural products. Different countries have different available resources, in terms of land, water, and climate. These may make a country ideal to produce certain crops, but a poor place to try and grow others. For example, US farmers grow and export rice, but Thailand produces more rice for the world market than any other country. Thai farmers produce one out of every three bushels of rice sold on the world market. The United States depends on exports from other countries for many of the things we enjoy, such as cocoa (for making chocolate), coffee, and bananas. Nations trade agricultural products with other nations so that their citizens can have all of the food and materials that they want but cannot produce.

Importing and Exporting Agricultural Products

Nearly two-thirds of agricultural products imported into the United States are related to horticultural and tropical crops. For example, the United States imports fruits and vegetables, tree nuts, sugar, tea, spices, cocoa, and rubber **(Figure 4.2.7)**. Most of our agricultural imports come from Mexico and Canada. The United States exports most of its agricultural products to Canada. China is the second largest importer of US agricultural products, followed by Mexico, Japan, and the European Union.

For some agricultural crops produced in the United States, exports are the main source of income for farmers. More than 70 percent of the volume of tree nuts (almonds, hazelnuts, pecans) are exported annually. More than 50 percent of rice and wheat production in the United States is shipped abroad. However, cotton is the most exported crop of all US agricultural products (**Figure 4.2.8**).

Agricultural trade allows the United States to acquire products from other countries. You might wonder why we need to import agricultural products, because we produce more than we need to feed our citizens. Why do you think this is? The US imports certain types of agricultural products because we cannot produce enough of them to meet our demand.

Awei/Shutterstock.com

Figure 4.2.7. China produces one-third of the world's tea. *What does tea need to grow? Why would China be a good place for it to grow?*

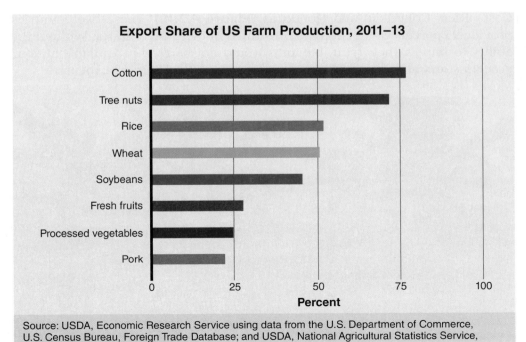

Export Share of US Farm Production, 2011–13

Cotton
Tree nuts
Rice
Wheat
Soybeans
Fresh fruits
Processed vegetables
Pork

0   25   50   75   100
Percent

Source: USDA, Economic Research Service using data from the U.S. Department of Commerce, U.S. Census Bureau, Foreign Trade Database; and USDA, National Agricultural Statistics Service, various reports.

USDA

Figure 4.2.8. Cotton leads all US agricultural export commodities, with more than 75 percent of total US production shipped to other countries. *Are there any items on this list that surprise you?*

Not every agricultural product is a source of food. We import flowers, shrubs, trees, and other horticultural products into the US. The rubber plant (*Ficus elastic*) is a species native to Asia, and we use the latex in the plant to create rubber products (**Figure 4.2.9**).

Here is another familiar example. Banana production in the United States is limited and confined to the warmest regions of the country. Most of the bananas consumed in the United States come from Guatemala, Ecuador, Costa Rica, Columbia, and Honduras (**Figure 4.2.10**). These five countries provide 95 percent of all bananas consumed in the United States. Without the ability to import their fruit, bananas would not be readily available at your grocery store, and what bananas they have would be far more expensive.

Yatra/Shutterstock.com

Figure 4.2.9. Latex is a type of fluid produced by certain plants to protect them from predators. The latex repels herbivores that attempt to eat parts of the plant. Latex should not be confused with plant sap used to carry nutrients and water throughout a plant's tissues.

Gregory James Van Raalte/Shutterstock.com

Figure 4.2.10. Most of the bananas consumed in the United States are imported.

As we learned at the beginning of this lesson, coffee is another crop that must be imported into the United States (**Figure 4.2.11**). Americans drink about 400 million cups of coffee per day; the average person drinks three cups of coffee every day. California, Hawaii, and Puerto Rico produce about 10 percent of the coffee consumed in the United States. The remaining 90 percent is produced in Brazil, Vietnam, Columbia, Indonesia, Honduras, and Ethiopia. Did any of you successfully guess one of these countries as a source of coffee beans?

There is no doubt that agricultural trade is essential for delivering food, clothing, and other agricultural products to consumers across the globe. Trade helps provide more food choices for consumers, and helps to reduce famine and food insecurity worldwide. Trade also addresses consumer desire for products that are beautiful, interesting, or useful (**Figure 4.2.12**).

According to the Organization for Economic Co-operation and Development (OECD), agricultural trade continues to grow around the globe. Improvements in agricultural production and the adoption of agricultural technologies around the world are improving our ability to feed and clothe a growing world population.

A   *Stasis Photo/Shutterstock.com*

B   *IndustryAndTravel/Shutterstock.com*

Figure 4.2.11. In the United States, coffee is grown in Hawaii, California, and Puerto Rico. **A.** Raw coffee beans with the hull removed. **B.** Coffee plantations in Kauai, Hawaii. *Why isn't more coffee produced in the US?*

Did You Know? The USDA reports that more than 300 million cut flowers are imported annually into the United States. Mexico, the Netherlands, Ecuador, and Columbia are the largest suppliers of cut flowers.

R. de Bruijn_Photography/ Shutterstock.com

Figure 4.2.12. These pink roses were produced in Ecuador. They are being packed for export to other countries around the world.

SL-Photography/Shutterstock.com

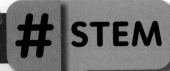
The Coconut Palm

The coconut palm is the most widely distributed tree in the world. Coconuts grow throughout most of the Southern Pacific Ocean region, south Asia, and Africa. This region encompasses millions of square miles of ocean, including thousands of islands. Coconuts were found on these islands long before the invention of ships and planes.

How did the coconut palm do it? The answer is simple—coconuts float! Coconuts produce a fruit with a hard, woody, lightweight husk that protects the seed. This husk is removed in the harvesting process, so you don't see the husk in the grocery store, only the coconut seed.

Sam Thomas A/Shutterstock.com

This husk serves as a life preserver, helping the coconut seed move through water. It works like this: A palm tree on a beach in the South Pacific produces coconuts that ripen and fall onto the beach beneath the tree. Ocean waves pick up the coconuts. Water currents carry the coconuts to other locations on the same island, or to other islands, where the waves push the coconuts onto the beach sand. The husk rots away and the seed germinates. Over thousands of years, this method of seed dispersal placed coconut palms on islands throughout the South Pacific Ocean. As the islands became inhabited, the people living on these islands placed coconuts in small boats and traded them with inhabitants of other islands.

Every country produces agricultural products, even in the harshest and most remote environments.

Consider This

1. Conduct an online search for the most remote inhabited places in the world. Determine what types of agricultural products are produced in these remote places.
2. Determine what types of foods the inhabitants of these remote places cannot produce on their own and need to import or purchase from other places.
3. If you were to visit these locations, what types of foods would you bring with you to eat?

What Did We Learn?

- Agriculture is responsible for providing food, clothing, and shelter for a rapidly growing population.
- Countries trade agricultural products to meet the demands for food, clothing, and shelter for their citizens.
- Each country produces agricultural products, but due to limitations in climate, water resources, and suitable farmland, and given different tastes and cultural desires, no country can produce every agricultural product it needs.

4

Let's Check and See What We Know

Answer the following questions using the information provided in Lesson 4.2.

1. There are approximately _____ billion people currently living on Earth.
   A. 4.2
   B. 7.5
   C. 9.5
   D. 13.5

2. More than _____ million people suffer from hunger on a daily basis.
   A. 200
   B. 400
   C. 600
   D. 800

3. The condition by which a person does not receive the necessary nutrients is called _____.
   A. attrition
   B. dysentery
   C. malnutrition
   D. deficit diet

4. The condition by which a person receives too much of certain nutrients is called _____.
   A. attrition
   B. obesity
   C. dysentery
   D. population growth

5. The United States imports _____ of the horticultural and tropical crops needed by the country.
   A. one-third
   B. two-thirds
   C. three-fourths
   D. four-fifths

6. The world population is expected to reach _____ billion people by the year 2050.
   A. 7.5
   B. 8.1
   C. 9.1
   D. 10.1

7. The country that imports the largest quantity of agricultural products from the United States is _____.
   A. Mexico
   B. Canada
   C. China
   D. the European Union

8. The largest exporter of agricultural products is _____.
   A. Brazil
   B. Russia
   C. India
   D. China

9. *True* or *False?* The European Union is the largest cut floral exporter in the world.

10. The global agricultural trade yields more than _____ per year.
    A. $1 trillion
    B. $3 billion
    C. $5 million
    D. $7 trillion

Let's Apply What We Know

1. The sweet potato is an agricultural crop grown in the Southern United States, but the crop originated in Central or South America almost 5,000 years ago. By the year 1400, sweet potatoes were being produced in the South Pacific islands. How did the sweet potato travel thousands of miles across the Pacific Ocean 500 years before airplanes and big ships?

2. In this lesson, you read about the eating habits of families in three countries. The family from Georgia tended to eat outside the home more than the families from Cuba and Egypt. Why do you think this is so? Do Americans eat outside the home (at restaurants) too much? Find some data to support your argument.

Academic Activities

1. **Technology.** More than 800 million people go without enough food to eat each day. What are some of the problems you think are causing world hunger? What are some possible solutions to the problem? What agricultural advancements in the last 50 years have helped us grow more food? What else should we do to eliminate world hunger? Use your imagination to design a tool, machine, or plant that could help eliminate world hunger.

2. **Math and Science.** Many people suffer from malnutrition. Malnutrition occurs when a person receives too few or too many of the nutrients they need. Research and determine which countries have the highest rates of hunger and malnutrition. Which countries have the lowest rates of hunger and malnutrition? Create a bar graph illustrating the differences between the countries. How do you think we should solve the problem of malnutrition?

Communicating about Agriculture

1. **Reading and Speaking.** Working with your instructor, arrange to bring an assortment of different fruits and vegetables to class. *Because some people are allergic to certain types of foods, it is best if you handle these fruits with the gloves provided by your instructor.* Examine these fruits and vegetables. How do they smell? What do they feel like? Would you purchase any of these fruits and vegetables? How do you cut or eat these fruits and vegetables? Choose one fruit or vegetable and write step-by-step instructions on how to cut and eat it. If possible, include illustrations or make a video to present to class.

2. **Reading and Writing.** Select a country in the world and research its population, total acreage, and land use. (Each student should choose a different country.) Create a chart depicting the percentage of land used for agriculture, forestry, residential, and commercial use. Compare your data with your classmates. Are there any correlations between population and land use?

SAE *for* **ALL** *Opportunities*

1. **Foundational SAE.** Research agricultural commodities to find the ones that interest you most. For instance, you may be interested in working with blueberries, beef cattle, or the dairy industry. Explore careers in two segments of the agriculture industry. Which jobs interest you the most, and which ones would you find the most difficult?

2. **Foundational SAE.** If you could travel to any country outside of the United States, which one would it be? Choose an American commodity and its sector of the industry (for example, a crop that is grown) that interests you. Research that sector and compare it to the same industry in the country of your choice. What careers are involved with the commodity? Could you pursue these careers and work in the industry in the other country? For example, you might like to work in the coffee production industry, and stay in the United States. You might find a career with a US company that imports and distributes coffee. Does your chosen career involve international travel?

SAE *for* **ALL** *Check-In*

- How much time have you spent on your SAE this week?
- Have you logged your SAE hours?
- What challenges are you having with your SAE?
- How can your instructor help you?
- Do you have the equipment you need?

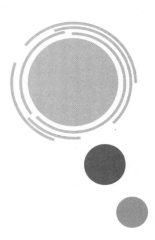

Agriculture, Economy, and the Workforce

BearFotos/Shutterstock.com

Essential Question

How does the law of supply and demand affect your life?

Learning Outcomes

By the end of this lesson, you should be able to:
- Explain how the law of supply and demand works. (LO 04.03-a)
- Explain how supply and demand influence the price of products. (LO 04.03-b)
- Describe, using examples, how agricultural products influence the economy and workforce of the United States. (LO 04.03-c)
- Explain what tariffs are, and how they work in practice. (LO 04.03-d)

Words to Know

demand

law of supply and demand

supply

tariff

Before You Begin

During the early days of the coronavirus pandemic that began in 2019, grocery store shelves were emptied of many essential food products (Figure 4.3.1). Why did the pandemic cause this food shortage? Why did people rush to purchase food at the onset of the pandemic? More importantly, why were grocery stores unable to keep up with the demand for food products?

Figure 4.3.1. *What does this mostly empty food case at a grocery store tell you about our food supply?*

Goodheart-Willcox Publisher

The transportation and distribution of food and other agricultural products in the United States is part of an international system of cooperation. Agricultural production, marketing, and trade follow the same trends and patterns that affect the national and global economies. (As a reminder, the economy is the management and use of money, resources, and labor.) If the economy is slow and the supply of money circulating is low, then people are not going to buy many products. They may only buy the things they need, and delay purchasing items that are not as important to have. Conversely, if the economy is growing rapidly and the supply of money circulating is plentiful, people are likely to spend more money, including on optional or luxury items. The demand for products goes up and down depending on the state of the economy.

If you were to track the price of a particular good, you will see prices go up and down over time. Supplies of products also go up and down, depending on the availability of and demand for those products. The constant price changes cause changes in the supply of products. Consumer demand for products can seem to be chaotic and out of control.

In fact, this is how normal economic markets work. There is even a formal economic principle that explains this: the law of supply and demand.

The Law of Supply and Demand

If the economy is healthy, most people have money to spend. People are more likely to make large purchases, like cars or houses, in this environment. In agriculture, a strong economy allows farmers to purchase new equipment and renovate (or build) farm buildings. When people have more money to spend, the demand for the things they want increases. **Demand** refers to the desire that consumers have for a particular product. For those involved in agricultural production and processing, a strong economy means a higher demand for food, fiber, and natural resources.

All of this depends on the law of supply and demand. The **law of supply and demand** says this: As the demand for something goes up, the supply of it goes down. **Supply** refers to the amount of a product available. For example, as the demand for apples goes up (people want more apples), the supply of apples available at a store decreases (more apples are purchased) (**Figure 4.3.2**). Makes sense, right? As people demand more of something, they will buy more of them, and the supply available for others to buy will go down.

That is what happened during the coronavirus pandemic. People wanted to stock up on food when they realized that restaurants would be closed, or that they didn't want to leave their homes. This increased the demand for food items, and the supply went down. This resulted in empty grocery shelves across the United States.

Price

Let's now turn our attention to another factor related to supply and demand: price. The price of something is influenced by supply and demand. As demand for something goes up, the supply goes down, and the price goes up.

Shark9208888/Shutterstock.com

Figure 4.3.2. *Have you ever stopped to look at the fruits and vegetables for sale at your local grocery store or farmers' market? What lessons can you learn about supply and demand just by looking around?*

littlenystock/Shutterstock.com

Figure 4.3.3. Pumpkins are a familiar sight every fall. Local farmers bring them into cities and sell them in open farmers' markets. *Why do you think they do this?*

Using our earlier example, let's pretend that the demand for apples suddenly goes up. This reduces the available supply. Since there are fewer apples to meet the demand, people will be willing to pay more for apples to get them.

Let's look at another example. Pumpkins are very popular in the fall (**Figure 4.3.3**). The demand is very high for pumpkins each fall, and as the supply decreases, the amount people are willing to pay increases and the price increases. In the spring and summer when the demand for pumpkins is low, the price is low. Would you pay a lot for a pumpkin in June? Probably not, right?

Supply and Demand in History

Let us apply the law of supply and demand to a real-life agricultural example from US history. During World War I (1914–1918), agricultural production soared in the United States. Farmers produced agricultural products for the war effort. The United States government provided financial incentives for farmers to increase production and to take out loans on land and equipment, and kept prices for farm commodities high.

When the war ended, the government stopped providing price incentives. As farmers were still producing crops and livestock at an increased rate, the supply of farm goods went up and the demand went down. As the law of supply and demand would predict, this decreased the prices farmers received for their crops. At the same time, farmers still had to pay for the operating loans they had secured to increase production. Farmers began going out of business because of the reduced demand for their products and the reduced income from low prices. As farms disappeared, farm production suffered on a national scale, and the supply of agricultural products diminished. This was the beginning of the Great Depression in the United States (**Figure 4.3.4**). During this period, businesses went bankrupt and many farmers lost their farms.

Everett Collection/Shutterstock.com

Everett Collection/Shutterstock.com

Figure 4.3.4. Reduced income from the sale of farm products put many farmers at risk for losing their farms during the Great Depression. (left) Many farmers and their families packed up their cars to move to cities and start over. (right) Foreclosure auctions were common as farmers were unable to pay their mortgages. *How would a collapse in agriculture affect our country today?*

How Commodity Prices Affect the Economy

Can the price of a single commodity affect the broader economy? Let's investigate. A good example of how the price of one commodity can influence other segments of the economy is ethanol. In the late 2000s, it was discovered that corn could be easily converted into ethanol. This is a type of alcohol that is added to gasoline. It makes the production of gasoline cheaper, and that brings down gasoline prices. Of course, cheaper gas means that more people buy gas, and this makes the oil and gas companies more profitable. At the same time, it made the American corn grower happy. Corn prices in the late 2000s and the early 2010s more than doubled per bushel, as oil and gas companies looked to add ethanol to their gas. A bushel of corn previously sold for a price between $3 and $3.50. When corn was diverted into ethanol production in the late 2000s, its value increased, and so did the demand for corn. The demand for corn decreased the supply of it. This took corn prices to high levels. In 2012, corn prices peaked around $7.50 per bushel. This drastically increased income for those farmers producing corn for the ethanol market, which allowed them to make improvements to farms and purchase new equipment.

As you might guess, there was also a downside to this change. Corn is an important livestock feed ingredient. When the price of corn goes up, it increases the price of feed for poultry and livestock. As feed prices go up, the prices for beef, poultry, and pork products in the supermarket will increase. The problem is that feed prices may go up more rapidly than the prices paid to farmers and ranchers for pork, chicken, and beef (**Figure 4.3.5**). A farmer's income may not be enough to cover the costs of production. Typically, farmers worry the most about the cost of production inputs (such as the cost of animal feed, fertilizer, and fuel costs) and the prices they receive for the sale of commodities.

In the following chart (**Figure 4.3.6**), you will notice that corn prices came back to their traditional levels around 2015. By then, other crops, like switchgrass, were available as alternative sources for ethanol production in the United States.

bogdanhoda/Shutterstock.com

Figure 4.3.5. These chickens are eating a corn-based feed. High corn prices drove up the cost of livestock and poultry feed in the early 2010s. *How did this ripple through the economy?*

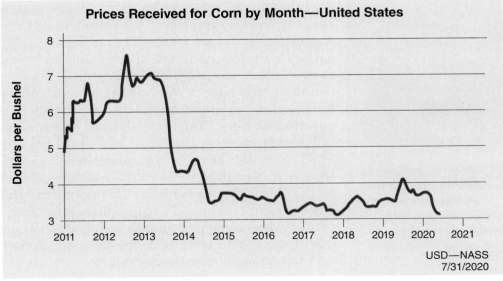

Figure 4.3.6. Corn prices climbed to nearly double their average over the course of 2011, 2012, and 2013, before plummeting quickly back to the historical average. *Why did this happen?*

Prices Received for Corn by Month—United States

USD—NASS
7/31/2020

USDA

This reduced the demand for corn and reduced the price per bushel. When the price of a commodity such as corn decreases, farmers who grow corn must find ways to reduce their operating costs. Farmers may have to delay the purchase or repair of farm machinery until prices return to a higher level.

For farmers who raise livestock that feed on corn, the news was a little better. Lower corn prices mean lower feed prices for livestock producers. The reduced corn price also brings down the cost of making plant-based fuels, which in turn reduces the price of fuel. This helps farmers and ranchers, and consumers. The reduced feed costs may also result in less expensive meats in the supermarket for consumers to purchase.

Economic Policy and Trade

Sometimes, other countries can produce products and commodities more cheaply or efficiently than we can in the United States. These countries might have more plentiful resources or less expensive labor, producing a cheaper product. Whatever the reason, this can put the American farmer or rancher at a disadvantage, because many consumers purchase the least expensive product.

To protect American producers, the United States government occasionally puts tariffs on imported goods. A *tariff* is a tax on incoming goods from another country. It is an extra amount of money added to the price of goods coming into the US, to make the product less competitive in the market.

> **Did You Know?**
>
> The US imposes tariffs on a wide range of imported goods. According to the Pew Research Center, in 2018 the highest average tariff rates were paid on imported clothing and accessories, and the lowest tariff rates were paid on petroleum and petroleum-based products.

rachel ko/Shutterstock.com

Figure 4.3.7. *If these beans are the least expensive option on the shelf, and all of the beans are similar, would you consider buying others, or would you go with this option? Why?*

Take a can of beans. Let's say that farmers in Australia can produce these beans more cheaply than farmers in the United States (**Figure 4.3.7**). Their beans will retail for $0.85. American-grown beans are priced at $1.29. That means that if the Australian beans are going to be on supermarket shelves in the United States, the US beans are going to have a hard time finding a share of the market. Not many consumers want to pay an extra $0.44 for the same product! Now, let's say that the US government puts a tariff of 50 cents on the cans of beans from Australia. Their new retail price is $1.35. Now the price of the Australian beans is slightly higher than those of the beans produced by the United States farmers. This policy has the advantage of making US commodities more competitive when compared against commodities from other nations. Why don't we do this for every product we trade on the international market?

Well, there is a disadvantage. The Australian government can also use tariffs to their advantage. They might add a tariff to any United States beans that they import—or to any other US commodity. Trade with other countries helps us get a wide variety of goods to sell to meet different demands and helps US producers find additional markets overseas. If every country throws tariffs on top of every imported good, we lose a lot of the benefits of trade.

How do you stop countries from fighting over tariffs? That's a good question! We have trade representatives in the United States government who talk to trade representatives from other countries, and they work out disagreements over tariffs. Let's go back to the example of the Australian beans. We've raised the price of Australian beans by $0.50 per can to make US-produced beans more competitive. To be fair to Australian farmers, we might allow them to put a tariff of $0.45 on cans of peas that we sell in Australia.

There are advantages and disadvantages to tariffs, but they only work if countries are willing to work together and make good decisions about the price of goods imported and exported. Tariffs are a method used to encourage fair trade between countries. It is important to note that the extra tariff expense is paid by the consumers in the country importing the product. In the US, this encourages consumers to buy products made by the US agricultural industry.

4

Career Corner ▶ INTERNATIONAL TRADE SPECIALIST

The job of an **international trade specialist** *is to help businesses develop markets for their goods and services in foreign countries.*

What Do They Do? Trade specialists help agricultural businesses by providing information about how to purchase land and buildings in a foreign country, and information about rules and regulations related to taxes and banking. Helping businesses develop international business strategies is an important part of this job. The United States Department of Commerce and the Office of the US Trade Representative hire international trade specialists, but many international trade specialists work for private companies that buy and sell products on the international market.

What Skills Do I Need? You will need good writing, speaking, and computer skills to be successful in this job.

What Education Do I Need? To qualify for these positions, you will need at least a bachelor's degree in economics or another business-related field. For higher-level positions, a master's degree may be required.

How Much Can I Make? The average annual salary for an international trade specialist is $92,000.

How Can I Get Started? In your agricultural education program in school, you should consider taking agricultural production and management courses. FFA activities that help you develop the skills needed for this job include public speaking, and the Agricultural Sales, Agricultural Communications, and Marketing Plan Career Development Events.

Hands-On Activity

Putting the Law of Supply and Demand into Practice

How does the law of supply and demand work?

For this activity, you will need the following items:

- Five small slips of paper
- Pencil or pen
- One 3 x 5 index card
- Five pieces of individually wrapped bubble gum

Before your instructor hands out the gum, please follow these rules:

1. You may not unwrap or eat any candy until your instructor gives you permission.
2. The candy is yours to keep. You may trade the gum later with other students in the class or keep it all.
3. You may not give away any candy. You can only trade with other students.

Your instructor has a bag of premium candies such as chocolate bars or similar sweets. You will have an opportunity to trade candy with your instructor. As you are asked the following questions, please write down "Yes" or "No" on a slip of paper and hand the slip to your instructor.

Question 1. Will you trade one piece of bubble gum for one piece of premium candy? Once you have voted yes or no, give your slip to your instructor. If you wrote "yes," give your instructor one piece of bubble gum in exchange for one piece of premium candy.

Question 2. Will you trade two pieces of bubble gum for one piece of premium candy? Once you have voted yes or no, give your slip to your instructor. If you wrote "yes," give your instructor two pieces of bubble gum in exchange for one piece of premium candy.

Question 3. Will you trade three pieces of bubble gum for one piece of premium candy? Once you have voted yes or no, give your slip to your instructor. If you wrote "yes," give your instructor three pieces of bubble gum in exchange for one piece of premium candy.

Question 4. Will you trade four pieces of bubble gum for one piece of premium candy? Once you have voted yes or no, give your slip to your instructor. If you wrote "yes," give your instructor four pieces of bubble gum in exchange for one piece of premium candy.

Now your instructor will record the results of the four rounds of trades on the whiteboard or chalkboard in front of the class. The following table shows what the results might look like:

Voting Results	Round 1 1 piece of gum for 1 piece of premium candy	Round 2 2 pieces of gum for 1 piece of premium candy	Round 3 3 pieces of gum for 1 piece of premium candy	Round 4 4 pieces of gum for 1 piece of premium candy
Yes	15	14	5	0
No	5	6	15	20

Now consider the following questions and discuss with your instructor and fellow students:

1. Did the demand for premium candy drop in the third and fourth rounds? If so, why?
2. As the price of premium candy went up, what happened to the demand for it?
3. As the demand for premium candy went down, what happened to the supply of it?

Consider This

1. Now, working with another classmate, write the definition of the law of supply and demand using your own words. Does this law explain what happened during the training exercise you just completed? Why or why not? Discuss your answers with the rest of the class.
2. When your instructor permits you to do so, you may enjoy the candy provided to you. As you do so, write down what you think is the most thing you learned from this exercise and give it to your instructor.

What Did We Learn?

- The law of supply and demand means that as the demand for products increases, the supply decreases. As the demand decreases, supply will increase.
- Increased demand increases the price of products. Decreased demand will lead to lower prices.
- When the demand for agricultural products is low, this generates a surplus supply. This drives down the price that farmers and agricultural processors receive.
- Tariffs are a way to encourage fair trade practices between countries.

Let's Check and See What We Know

Answer the following questions using the information provided in Lesson 4.3.

1. As the demand for apples increases, the supply of apples _____.
   A. decreases
   B. increases
   C. remains the same
   D. triples

2. As the demand for apples increases, the price of apples _____.
   A. decreases
   B. increases
   C. remains the same
   D. doubles

3. People are more likely to make large purchases when the economy is _____.
   A. improving
   B. declining
   C. nonexistent
   D. calm

4. The Great Depression, a time when the economy was in decline, occurred in the _____.
   A. 1920s and 1930s
   B. 1950s and 1960s
   C. 1970s and 1980s
   D. 1980s and 1990s

5. During the coronavirus pandemic of 2019, people rushed to the grocery store to purchase food. This created increased _____ for food and other agricultural products.
   A. demand
   B. supply
   C. economy
   D. market

6. The increase in the price of corn paid to corn producers between 2000 and 2010 was the result of the _____.
   A. corn weevil
   B. decrease in the price consumers paid for corn products
   C. weather
   D. increase in demand due to ethanol production

7. Why did agricultural production increase in the period of time between 1914 and 1918?
   A. The price of bread made from wheat increased in supermarkets.
   B. Farmers increased the price of agricultural crops.
   C. Farmers planted fewer crops and decreased livestock production.
   D. World War I created a large demand for agricultural products.

8. In order to encourage fair trade, countries will often place _____ on goods purchased from foreign countries.
   A. discounts
   B. tariffs
   C. extra income
   D. warnings

9. Switchgrass is a suitable species for the production of _____.
   A. ethanol
   B. gasoline
   C. carbon tetrachoride
   D. ammonia

10. If the price of livestock feed goes up, what is likely to happen to the price of meats in supermarkets?
    A. Prices will go down.
    B. Prices will go up.
    C. Prices will remain the same.
    D. Prices will decrease and then sharply increase.

Let's Apply What We Know

1. Visit a grocery store and find the section where sale items are located. What kinds of products are on sale for a reduced price? Why are those items on sale, and less expensive? What does this say about the supply and demand of these items?

2. Research the technology used in two agricultural careers of interest to you. Use online resources and materials provided by your school and your instructor. Explain what a person does in that agricultural career; then explain the emerging technology used in that career. For instance, truck drivers use global positioning systems and agricultural service professionals use computerized inventory systems. Next examine whether or not these new technologies have any bearing on supply and demand. If they do, how?

Academic Activities

1. **Math.** Select an agricultural commodity. This can be a crop or livestock species. Conduct an online search for news about this commodity. Based on the news reports and articles you found, predict whether the price of the commodity will be higher or lower at the end of six months. Prepare a report that includes the supply and demand information you found in news reports and articles to support your prediction.

2. **Technology.** Conduct a news search online for stories and articles related to farming and the economy. Look for stories about specific individuals, such as a farmer or rancher, who are using innovative methods to increase agricultural production or lower production costs (for example, a farmer using drones to collect data on crops). Predict how this technology may influence the economy in the future.

Communicating about Agriculture

1. **Reading and Writing.** Which sector of agriculture do you think is most important? Write a short answer with at least three reasons why you think this is the most important sector. Support your opinion with research and statistics.

2. **Reading and Writing.** Research the five top imports and exports of agricultural goods in the United States. Consider the specific products that would fall into each category. Develop theories about why the United States imports or exports these types of products, and create a presentation based on your thoughts.

SAE *for* **ALL** *Opportunities*

1. **Foundational SAE.** Prepare a list of questions that you can use to interview two or three local growers. The growers may be farmers, nursery managers, Christmas tree growers, or other horticulturists. If you cannot visit the location, set up a Facetime call or video chat on your computer. Be sure to ask about how supply and demand affect their business. Summarize your experience and create a presentation for your class.
2. **Foundational SAE.** Job shadow an agribusiness accountant. Research the position before you shadow the professional. Prepare questions that you have from your research. Log your time, as well as what you learned during your shadowing experience. Did supply and demand come up in your conversation?

SAE *for* **ALL** *Check-In*

- How much time have you spent on your SAE this week?
- Have you logged your SAE hours?
- What challenges are you having with your SAE?
- How can your instructor help you?
- Do you have the equipment you need?

UNIT 5
The Science of Agriculture

SAE for ALL Profile
Protecting Native Plants

Lee Ann Williams loves every moment of her job as a botanist and plant scientist with the United States Forest Service. She gives her middle school and high school agricultural education programs all the credit. Lee Ann never intended to take an agriculture class in middle school. A scheduling problem prevented her from taking the class she wanted, so instead she found herself in John Richardson's introductory agricultural science class at Mayfield Middle School.

"It wasn't the class I wanted," said Lee Ann, "but when I saw the plants in the greenhouse for the first time, I was hooked."

In Mr. Richardson's introductory agricultural science class, Lee Ann developed a Foundational SAE to learn more about the plant industry. On a field trip to the local arboretum, she learned the role of native wildflowers in the ecosystem. Lee Ann continued to explore her interest in native plants in high school by developing a Research SAE. Her project involved planting native milkweed species for monarch butterfly habitats and monitoring the growth and spread of these native plant species.

"Native plants are important to agriculture and the environment because they provide the habitat needed by the pollinators of the other native plants and crops. Many crops could not be produced without native insect pollinators. And these native insect pollinators need native plants, which seem to be disappearing in many places," Lee Ann noted, sadly.

Lee Ann's SAE became the foundation for her FFA agricultural science fair project in high school. Her SAE and agricultural classroom coursework, in tandem with her studies in science classes, led to her interest in botany. She enrolled at Montana State University and went to work for the United States Forest Service after graduation.

"My job with the United States Forest Service is to research the health of native plant communities, forests, and grasslands. I am now focusing on finding ways to reproduce and conserve endangered native plant species," said Lee Ann. "I have to understand the genetics behind the reproductive systems of native plants to find ways to reproduce them in the lab and eventually in the field. I love my job because I get to work outdoors and meet many interesting people. And I get to use my knowledge of botany and genetics to do something good for the environment."

As a botanist, Lee Ann conducts research in plant breeding and genetics, and studies plant growth and development. The job requires good record-keeping skills and the ability to balance indoor lab work with outdoor fieldwork.

- Lee Ann ended up in agriculture accidentally, but found a home when she kept an open mind. What appeals to you about agriculture?
- What do you know about native plants in your community, and their role in your environment?

DC Studio/Shutterstock.com

kung_tom/Shutterstock.com

SeventyFour/Shutterstock.com

Scientists' Work in Agriculture, Food, and Natural Resources

Learning Outcomes

By the end of this lesson, you should be able to:
- Describe how the scientific method works to answer questions. (LO 05.01-a)
- Assess the contributions of agricultural scientists in the improvement of agriculture. (LO 05.01-b)
- Explain how science is used to improve agriculture and natural resources. (LO 05.01-c)
- Describe how science and innovation are related to improvements in agriculture. (LO 05.01-d)

Words to Know

blacksmith	genetics	sustainable agriculture
drought	innovation	theory
engineer	robotics	

Before You Begin

Look at the photograph below (**Figure 5.1.1**). What do you think this machine is doing? Write down all the answers you can think of, then share them with the rest of the students in the class. Do not be afraid to guess! Here is a hint: This machine is doing something that certain insects do for plants. What do you think it could be? And why is it doing this? The lesson ahead will answer these questions and more, as we start to talk about how scientific principles play a role in AFNR.

Figure 5.1.1. *What is this? How do you think it is used? We'll revisit your answer in a moment.*

Love Silhouette/Shutterstock.com

The Scientific Method

Have you ever been curious about how cows make the milk we drink? How about what causes a chicken egg to hatch? Humans have a natural curiosity about how things work. We have a need to know and understand the world around us. How is it possible that some plants make their own fertilizer? Why do some plants live and thrive simply on air, without the need for soil? (See **Figure 5.1.2**.) The questions we can ask about the world around us are endless. How do we find answers?

Scientists will tell you that the key to answering any question is curiosity. Curiosity is the desire to know or learn, and we are always learning. In order to answer our questions, we can do what scientists do. Scientists use the scientific method to find answers to the things that puzzle them. The *scientific method* is the process that scientists use for thinking about and making sense of complex questions, using experiments to find answers to what we see and observe in the world around us.

The scientific method has been in use for more than 2,500 years. Aristotle, the famous Ancient Greek philosopher, popularized this research method. Many other famous thinkers and scientists that you've studied have used the same process (**Figure 5.1.3**).

The scientific method uses our abilities to reason and observe. We use our reasoning ability to make decisions about everyday things. For instance: Let's say you ate the vegetable soup in the school cafeteria the last three times it was on the lunch menu. You liked it, so you reasoned that it would be a good choice to eat again the next time it is on the menu. Or maybe you wonder which birdseed would attract the most birds to your backyard bird feeder, so you try several different brands and types of seed before choosing one that seems to attract the most birds. Perhaps you notice that an insect is eating the leaves of the bean plants in your garden. You then collect some insect specimens from the plants and identify the pest most likely causing the damage. You can then apply the appropriate method to control the pest.

Fotokostic/Shutterstock.com *domnitsky/Shutterstock.com*

Kelly Marken/Shutterstock.com

Figure 5.1.2. Soybeans are a major crop in the United States. Soybean plants have nodules on their roots that are able to convert atmospheric nitrogen into nitrogen fertilizer for use by the plant. *How does this affect a soybean plant's ability to grow in soil?*

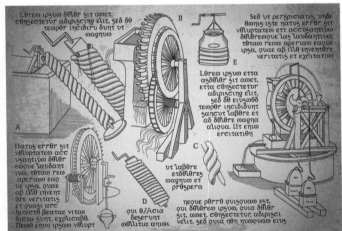

Daria Pushka/Shutterstock.com

Figure 5.1.3. Leonardo Da Vinci (1492–1519) was an Italian scientist. He is known for his scientific contributions in astronomy, botany, physics, and many other areas, as well as his art. These sketches are Da Vinci's design for pumping or moving water. *How amazing is it that someone came up with these ideas 500 years ago?*

We use scientific reasoning every day, for dozens of ordinary tasks—finding the fastest route to the supermarket, deciding when would be best time to cut the grass in the yard, and finding the healthiest pet food choices for your dog. Were you aware that you were making so many important decisions in this way?

Photodiem/Shutterstock.com

Figure 5.1.4. Scientific experiments help us understand the world in which we live. This scientist is measuring soil temperature and moisture in a field. *What questions might he be trying to answer?*

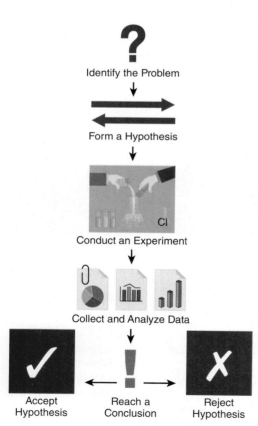

Goodheart-Willcox Publisher

Figure 5.1.5. The scientific method works like this. *What step do you think is most important?*

The Elements of Reasoning

Scientific reasoning sounds complicated, but it's actually pretty simple The reasoning process typically involves:

- **Observation.** We observe some problem or situation, noticing patterns or trends in what we see.
- **Theory.** We all have some ideas about how the world works. Sometimes when we observe something we do not understand, we try to apply a theory to what we observe. A *theory* is a set of statements that attempts to explain what we see or experience in the natural world. In other words, a theory is a type of educated guess about how things work.
- **Experimentation.** Once we have a theory about how something works, we conduct experiments to determine if our theories are correct, or to determine the meaning of what we observed (**Figure 5.1.4**).
- **Results.** After the experiment is complete, you have to look at the results and figure out what the experiment showed.
- **Conclusions and Report.** From the results, we should be able to conclude that our theory is correct, or that it has been disproved. You report what you concluded to others to share your knowledge, so that others can learn from your experiment.

The scientific method provides us with an organized way to test our theories. We have a question about how something works, or notice a problem to be solved. We do some background research to find out more about the question or problem. Then we come up with a hypothesis about the question or problem. A *hypothesis* is an informed guess as to the answer to the question or solution to the problem. It is always written as if it were true.

Here's a sample hypothesis: *Blueberry plants grow best in well-drained, sandy soils.* We then conduct an experiment to determine if our hypothesis is true. In the case of the blueberries, we might plant several blueberry varieties in sandy soil, loamy soil, and clay soil. At the end of the growing season, we look at the results of the experiment. Which blueberry plants were healthiest and had the most productive growth?

The results of your experiment help you conclude whether your hypothesis was correct. The last step of the scientific method is to report your conclusions, so others can learn from your experiment.

Use the chart in **Figure 5.1.5** to describe the scientific method.

Famous Agricultural Scientists

There are a number of famous scientists who have left their mark on agriculture, and our lives today. By following their curiosity, asking questions, and using the scientific method to conduct experiments, these people changed the world!

Rachel Carson

Rachel Carson (**Figure 5.1.6**) was a marine biologist and environmental scientist who discovered how improperly used pesticides damaged wildlife species. Carson's interest in pesticides began when a friend of hers noticed that birds were dying and believed it was because of Dichlorodiphenyltrichloroethane, or DDT. DDT is a powerful pesticide used to control insect pests (**Figure 5.1.7**). Carson conducted experiments to test her friend's hypothesis that DDT was involved. She found that DDT lasts a very long time in the environment. Birds ate insects with DDT on them; then larger birds such as hawks and eagles ate the smaller birds, and so DDT moved up the food chain.

Carson found that DDT had worked its way into groundwater, drinking wells, dairy milk, and into the bodies of humans. DDT made livestock sick, the farmers and their families sick, and caused untold sickness among American consumers. In 1962, Carson's book *Silent Spring* brought public attention to the overuse of pesticides. As a result, state and federal laws were passed to reduce the use of DDT in the environment and to protect human health from the misuse of pesticides (**Figure 5.1.8**).

US Fish and Wildlife Service

Figure 5.1.6. Rachel Carson was famous for her research into the effects of DDT. Some scientists disagreed with her findings on DDT. *Why do you think her work was controversial?*

Albert Beukhof/Shutterstock.com

Figure 5.1.7. Insects would be exposed to DDT; then birds would eat the insects, thus transferring the pesticide to the birds and their offspring. *How would this continue to move through the food chain?*

Figure 5.1.8. DDT is still used in some countries to control mosquitos. Mosquitos are responsible for the spread of malaria and other diseases. *What safety concerns do you notice in this photograph?*

hareluya/Shutterstock.com

Library of Congress

Figure 5.1.9. George Washington Carver was a prolific researcher and inventor. His work made a great difference to everyday American households. *How has Carver's work affected your life?*

George Washington Carver

George Washington Carver was a scientist and teacher at the Tuskegee Institute in Alabama in the early 1900s (**Figure 5.1.9**). He developed new uses for peanuts, sweet potatoes, soybeans, and other plants. Part of his interest in creating new uses for plants had to do with his observations of the communities around Tuskegee. He noticed that many local citizens lacked money to purchase household items they needed. Carver believed that by inventing new ways to use ordinary plants from the fields and gardens, he could help make new and less expensive household products. He invented hundreds of products, including peanut milk, paints, cosmetics, oils, and plastics. All of these came from a commonly grown peanut plant. He developed one of the first plant-based fuels for use in machinery. He also made glue, baking flour, and synthetic rubber from sweet potatoes. These discoveries were designed to help poor farm families save money on household expenses. Carver's numerous discoveries helped farmers find new markets for their crops, and improved the quality of life for rural families.

Norman Borlaug

Norman Borlaug (**Figure 5.1.10**) is known as "the father of the Green Revolution." He was a microbiologist, geneticist, and plant pathologist. In his role as a microbiologist, he studied the interactions between tiny organisms, such as viruses and bacteria. In agriculture, a microbiologist might study viruses that make plants or animals sick. Geneticists study how the characteristics of plants and animals are carried to their offspring. As a geneticist, Dr. Borlaug developed new varieties of crops that would be resistant to disease or drought. Plant pathologists study the diseases that affect plants. Dr. Borlaug studied a fungus that was destroying the wheat crops in Mexico and Central America. He used his skills to develop varieties of wheat that were not as susceptible to rust. Many of his achievements as a scientist prevented hunger and famine around the world. His research has saved countless lives.

Figure 5.1.10. Dr. Norman Borlaug developed drought-resistant wheat varieties, preventing famine in some of the world's driest regions.

Ben Zinner, USAID

USDA

Modern Leaders in Agriculture

Dr. Peggy Ozias-Akins is a scientist at the University of Georgia. She conducts experiments to find ways to improve the quality of peanuts. She works with other scientists and students in her laboratory to develop new varieties of peanuts that are resistant to disease (**Figure 5.1.11**). The most important part of her job is helping her students learn how to conduct scientific studies. She is training the next generation of scientists who will continue to improve the quality of peanuts and other crops.

Dr. George Vellidis is an engineer at the University of Georgia. He specializes in precision agriculture. He studies the design of systems that observe, measure, and respond to variations in a crop. This approach allows farmers to know exactly when and where to treat pests, apply fertilizer, or water crops. The purpose of precision agriculture is to use the exact resources needed—plant nutrients, water, pesticides—only at the appropriate time and place, in order make the process of growing more efficient. This can offer improved crop yields at a lower cost. Dr. Vellidis conducts research on the interaction between the tools that farmers use and the environment (**Figure 5.1.12**). He and the scientists in his lab find ways to help farmers farm more efficiently while conserving valuable natural resources.

University of Georgia

Figure 5.1.11. Dr. Peggy Ozias-Akins (right) works with a student researcher. *What might they looking at here?*

University of Georgia

Figure 5.1.12. Dr. George Vellidis (left) is collecting data with a student researcher. *How does Dr. Vellidis's work help the agriculture industry?*

Using Science to Improve Agriculture and Natural Resources

The agriculture industry has used scientific principles for thousands of years to improve agricultural production and to protect and conserve natural resources. Let's take a closer look at some of the ways that agriculture uses science and engineering.

Livestock and Poultry Production

Scientists find ways to breed livestock to improve productivity, animal health, and the quality of what is produced. Livestock provide meat, milk, wool, eggs, and hundreds of other products for human use, including some medicines.

Crop Production and Management

Scientists use genetics to improve the quality and vigor of plants. *Genetics* is the study of characteristics that plants and animals inherit from the parent plant or animal. Scientists can manipulate these genes to improve plants' resistance to disease, insects, and drought. *Drought* is a long period with little or no rainfall.

 **Did You Know?** Did you know that bananas are radioactive?! A small amount of the potassium in a banana emits radiation, but don't worry: The amount of radiation emitted is so small that it cannot harm you. Food scientists study the nutrients in fruits and vegetables like bananas to determine their effects on human health.

Charles Brutlag/Shutterstock.com

Figure 5.1.13. Agricultural researchers are seeking ways to make the irrigation of crops more sustainable. Using data to limit water use is one way to do this.

Agricultural Mechanization

Scientists and engineers use robotics and electronic sensors to make farm machines more efficient. *Engineers* are scientists who design and build machines and structures. Engineers design tractors, harvesters, and other machines that use computers and sensors to monitor speed, direction, and energy use. The goal is to operate agricultural machinery so that energy is saved. Efficient machines also protect valuable soil and water resources (**Figure 5.1.13**). Engineers design machines that save human labor, or make human labor more efficient. Some of these engineers may use *robotics*, which is the design and construction of machines that do the work of humans, to address these problems.

Career Corner — MECHANICAL ENGINEER

*Do you like to design and build things? Do you look at some of the machines we use in agriculture and say to yourself, "I can design something better than this"? If so, you might be a good candidate for a job as a **mechanical engineer**.*

What Do They Do? You learned from this lesson that engineers are a type of scientist who design buildings, machines, and systems to make agricultural production more efficient and effective. You will find mechanical engineers working in labs and offices, designing and testing new mechanical equipment. In the agriculture industry, mechanical engineers design sensors to determine when crops need irrigation. In the animal science industry, they design mechanical feed systems for poultry, and create devices that monitor the health of livestock. Nearly every significant technological discovery in agriculture includes the work of a mechanical engineer.

Mike_shots/Shutterstock.com

What Do I Need to Be Good at This? Mechanical engineers are curious and creative, and enjoy solving problems. You should like science and math, and have patience: It can take multiple tries to build something new correctly!

What Education Do I Need? To start working as a mechanical engineer, you will need a bachelor's degree in mechanical engineering or mechanical engineering technology.

How Much Money Can I Make? The median salary for a mechanical engineer in the US is just over $90,000.

How Can I Get Started? Look for opportunities to start building and tinkering, and develop the habit of looking for problems to solve. Create an SAE around building a new tool to help with an agricultural issue, or testing and comparing existing tools. Talk to a mechanical engineer to learn about what they're working on, and get tips for how to get started.

Sustainable Agriculture

Scientists are busy now developing **sustainable agriculture**: new ways of farming and ranching so that natural resources are not used up at a faster rate than they can be replenished. Crop scientists developed methods of planting crops that allow the soil to retain moisture and nutrients, not only for the current crop but for future crops in that field (**Figure 5.1.14**). Sustainable agriculture protects soil, reduces the use of fossil fuels, and reduces labor costs.

Innovation in Agriculture

The modern US agriculture industry was built on innovations that made the industry more productive. **Innovation** is the creation and adoption of ideas that make improvements to existing products and processes.

Gerarda Beatriz/Shutterstock.com

Figure 5.1.14. No-till farming reduces the amount of energy and resources necessary to grow crops. *How does this make the pictured soybean production more sustainable?*

When new inventions are introduced, people learn about them. They then decide whether it is worth the investment of time, effort, and resources to adopt the new technology. If a great idea is difficult to adopt because it is expensive, labor-intensive, or requires a long lead time, it may not catch on. Scientific discoveries that improve the way we work and live are only beneficial if we are able to use them.

In the 1800s, farmers plowed their fields using horses or mules and cast-iron plows. These cast-iron plows broke easily in rocky soils, and the clay soils of the Midwest stuck to the plows and made it difficult to turn the soil over. What they needed was a stronger plow that would not break easily, and that would cut through sticky clay soils.

John Deere was a blacksmith in Illinois in the early 1800s. A **blacksmith** creates metal objects out of iron and steel using tools to heat, bend, and shape the metal. Deere noticed the problems farmers were having with their iron plows, so he invented the steel moldboard plow. This plow was strong enough to withstand rocky soils. Its steel construction created a smooth surface to prevent the clay soils from sticking. This plow required less horsepower to pull through the fields, and worked well in breaking up and turning the soil for planting and cultivation.

Deere's invention, the moldboard plow (**Figure 5.1.15**), would never have gone into production unless farmers were willing to give it a try. Once they saw how well the plow worked, they purchased one for their own use, and so did their friends and colleagues. Before long, farmers across the country were using steel plows.

ilmarinfoto/Shutterstock.com

Figure 5.1.15. A modern version of John Deere's moldboard plow is now pulled through the field by tractors instead of horses and mules.

Hands-On Activity

Conduct the Radical Radish Experiment

ppl/Shutterstock.com

Hirundo/Shutterstock.com

In this experiment, we will test the ability of radish seeds to germinate under various conditions. By doing so, you will be able to describe those factors that can positively or negatively affect plant germination.

You will need the following items to conduct this experiment:

- Three large paper plates
- Paper towels
- Thirty or more radish seeds

In this experiment, you will be testing whether or not sunlight has an effect on the germination of radish seeds. How would you write this as a hypothesis? Remember to write your hypothesis as if it is true.

Here are the steps to follow to test your hypothesis.

1. Pair with another student in your class.
2. Place the three plates in front you.
3. Use a pencil or marker to label each plate with your name.
4. Number the plates 1,2, and 3, from left to right.
5. Place a folded wet paper towel in each plate.
6. Place ten seeds on the wet paper towel in each plate. Spread them out evenly.

7. Cover plate number 1 with a damp paper towel. Set plate number 1 aside, somewhere in a safe place in the classroom where it will not be disturbed. It is your experimental control. That is, this is the plate that represents how radish seeds normally germinate. Check it daily to make sure the paper towels are still damp. Add water if necessary.
8. Cover plate number 2 with a wet paper towel and place in a sunny location in the agricultural classroom, lab, or greenhouse. Check this plate every day for three days to make sure that the paper towel is still damp. Add water if necessary.
9. Cover plate number 3 with a wet paper towel and place it in a dark location. You might wish to turn an empty plate upside down and tape it in place so that no light will touch the radish seeds. Each day, lift the cover plate just long enough to make sure the paper towels are still damp. Add water if necessary.
10. After three days, gather your three plates and compare the seeds in each plate.

Consider This

1. Which plate had the most seeds germinate? Which plant had the fewest seeds germinate? What other things do you observe in your comparison of the seeds in each plate?
2. Did the seeds in plate 2, placed in the sunny location, produce more germinating seeds than the plate placed in a dark location? Compare the results of your research with the other students in class. What can you conclude from your experiment?

Allexxandar/Shutterstock.com

Remember the photograph on the first page of this lesson? That machine is a drone used to pollinate plants, shown again in **Figure 5.1.16**. It is an example of scientists using innovation to solve the problems associated with the pollination of plants. These can be used in places where native pollinators, such as bees, are not available in sufficient numbers to pollinate plants.

Innovation starts with an idea. That idea is tested using the scientific method. Once the idea proves its usefulness, it can be adopted by the agricultural industry and used to improve the work of farmers and ranchers.

Love Silhouette/Shutterstock.com

Figure 5.1.16. Drones collect field data that helps farmers make the best use of nutrients and irrigation in their fields.

LESSON 5.1 Review and Assessment

What Did We Learn?

- The scientific method of thinking about questions, using observation, reasoning, and experimentation, is an effective way to find answers. We use it in some form almost every day, to solve problems and answer questions.
- Rachel Carson's experiments and observations led to improvements in pesticide formulation and use, but her work was controversial. Science seeks to answer questions and solve problems, regardless of how unpleasant or unpopular the results may be.
- George Washington Carver's work led to more affordable and sustainable household products. Answering scientific questions can raise standards of living for all of us.
- Norman Borlaug's work to make wheat more disease- and drought-resistant is estimated to have saved more than one billion lives. Agricultural scientists can have a significant effect on human life and health.
- Scientific research in agriculture seeks ways to make the production of food, fiber, and natural resources sustainable as the demand for these things continues to grow.
- Asking questions and following the scientific method can help us to solve difficult real-world problems with new inventions. It sometimes takes time for an innovation to take hold, no matter how good the idea.

Let's Check and See What We Know

Answer the following questions using the information provided in Lesson 5.1.

1. A scientist who builds machines and structures is a(n) ____.
   A. biotechnologist
   B. engineer
   C. innovator
   D. plant pathologist

2. The protection of soil, reduction in the use of fossil fuels, and reduction of labor costs are all elements of ____.
   A. biotechnology
   B. engineering
   C. robotics
   D. sustainable agriculture

3. Which of these scientists was a marine biologist and environmental scientist who discovered how improperly used pesticides damaged wildlife species?
   A. Rachel Carson
   B. Norman Borlaug
   C. John Deere
   D. George Washington Carver

4. This scientist is known as the father of the Green Revolution.
   A. Rachel Carson
   B. Norman Borlaug
   C. John Deere
   D. George Washington Carver

5. Soybeans are able to grow in depleted soils because of their ability to absorb ____ from the soil
   A. oxygen
   B. water
   C. phosphorus
   D. nitrogen

Let's Apply What We Know

1. Let's take a moment to apply the scientific method to a problem in agriculture. Using a topic provided by your agriculture teacher, list the steps that you would follow in a scientific method to find potential solutions to this problem.

2. What is your favorite vegetable? Where did it originate, and how did it eventually find its way into the American diet? Trace the history of your favorite vegetable from discovery to modern times. Explain how the vegetable has changed over the years due to the effects of plant breeding and genetic modification.

Academic Activities

1. **STEM.** Along with a group of fellow classmates assigned by your instructor, prepare a brief skit about a scientist who conducted research in agriculture, food, and natural resources. Visit your school library and find as much as you can about this scientist. Your skit should explain what the scientist discovered, and why this discovery was important. Prepare and present your dramatic skit to the rest of the class.

Communicating about Agriculture

1. **Reading and Speaking.** Make a time line of technological advancements in agriculture in the last 20–30 years. Research advancements and the scientists who brought about these changes. Create a 10- to 15-point time line showing significant milestones that have led to new discoveries and theories. Describe your findings to the class.

2. **Reading and Speaking.** What are the purposes of genetically modified seeds in cereal gain production? Research how and why scientists have genetically modified cereal grain seeds. Prepare a 3–5-minute speech on one of these advancements in agriculture. Argue either for or against genetically modified seeds, citing evidence from your research.

SAE *for* ALL *Opportunities*

1. **Research SAE.** If you are developing your first SAE project, consider developing a research project. SAEs can involve more than just owning or working in an agricultural operation. Considering the information provided in this lesson, what type of SAE project could you develop? Even if you already have an SAE project in place that involves working for someone else in an agricultural operation, or owning your own agribusiness, consider conducting some research for your project. Make a list of three questions you would like to answer.

2. **Research SAE.** Examine your own SAE project to determine how science has made things easier for you. Make a list of the most challenging aspects of your SAE. What types of research would be necessary to find solutions to these challenges? Conduct some research in the library to determine if science has already made progress in these areas.

SAE *for* ALL *Check-In*

- How much time have you spent on your SAE this week?
- Have you logged your SAE hours?
- What challenges are you having with your SAE?
- How can your instructor help you?
- Do you have the equipment you need?

Technological Applications in Agriculture, Food, and Natural Resources

Papamoon/Shutterstock.com

Learning Outcomes

By the end of this lesson, you should be able to:
- Explain the importance of technology in agriculture. (LO 05.02-a)
- Describe some of the ways that technology is used in agriculture. (LO 05.02-b)

Words to Know

agricultural literacy
biodiversity
conservation tillage
food desert
infrared radiation

Integrated Pest
  Management (IPM)
organic farming
regenerative agriculture
Rural Electrification
  Administration (REA)

technology
tillage
urban agriculture
vertical farming

Before You Begin

Look up technology in a dictionary. Based on this definition, make a list of how technology has improved life and work for the last 20 years. How has it improved your life? Has technology made some things harder to do? Compare your list to that of another student in your agriculture class. What kinds of things were on your classmate's list that were not on your own?

Technology is the application of science, engineering, and mathematics to solve real-world, practical problems. To provide food, clothing, and shelter for an ever-increasing world population, the agriculture industry will have to develop and adopt technologies that increase production. At the same time, that production will have to be sustainable. Water shortages, climate change, and the safety of the food supply are all increasingly important concerns. Agricultural processes must be sustainable so that resources are available for future use.

Technological Applications in Agriculture

Technology helps to solve practical problems in agriculture. In the early 1900s, living and working in rural America was very difficult. There were few paved roads, no electricity, and few (if any) labor-saving devices to make farm work easier. The Great Depression of the 1920s and 1930s left many people without jobs and unable to feed their families (**Figure 5.2.1**).

As a result, the federal government began several large projects designed to put people back to work and rebuild the economy. One such program was the ***Rural Electrification Administration***, known as the ***REA***. The task of the REA was to build electrical power plants, then build transmission lines to rural areas that previously had no access to electricity. Before electricity came to rural America, people had to cook all their meals on wood- or coal-burning stoves. This took a lot of time and effort. Washing clothes involved hand washing them in a tub of soapy water, then rinsing them in a tub of clean water, wringing out the excess water and hanging the clothes on a line to dry. This was tedious and backbreaking work, taking valuable work hours. When electricity came to rural communities, it allowed for the use of electric appliances, such as clothes washers and ovens, that made hard chores faster and less tedious. Electric lamps lit homes and farms, increasing the hours available for work (**Figure 5.2.2**), and electricity powered motors for pumping well water, making clean water more accessible. Electricity allowed people to own refrigerators so they could cool and preserve foods longer.

Technology has the potential to improve the quality of life and work. What was it like in the days before modern electrical appliances? What was life like before smartphones? Ask older relatives or neighbors. Their answers might surprise you!

Everett Collection/Shutterstock.com

Figure 5.2.1. During the Great Depression, a long period of drought settled over the Midwestern and Western states. The use of unsustainable farming practices destroyed farmland and the livelihoods of thousands of farmers.

Everett Collection/Shutterstock.com

Figure 5.2.2. This rural town received electrical power for the first time in 1940. *Can you imagine what it must have been like to have electricity in your home for the first time?*

Now, think about how technology affects you. Many students have access to mobile smartphones. These telephones not only allow you to make telephone calls almost anywhere in the United States, but also to listen to music, watch movies, play games, send text messages and email, perform math calculations, browse the internet, check the news and weather, and find your way using GPS-based mapping software. In the early days of telephones, they had to be connected to hard wires that ran from your home to a network to work. Smartphones connect wirelessly, from anywhere, to satellite signals that help you map your location and find directions to the places you want to go. What a change!

Technological advances in the agriculture industry have improved the quality of life for all Americans. Technology makes foods safer to eat, less costly to produce, and more sustainably grown. Technology has helped to address some of the most pressing problems facing agriculture.

MrKumai.blogspot.com/Shutterstock.com

Figure 5.2.3. These yellow cards suspended in a greenhouse use natural hormones to attract aphids, thrips, and other insects without the use of insecticide.

Preserving Resources

It is critically important to have enough water and nutrients available to raise healthy plants and animals. Sustainability is an important issue in all areas of agriculture—production, processing, marketing, and sales and service. All areas need new technologies to improve sustainability. Remember: Sustainable agriculture is the ability to maintain production without using up natural resources.

Pest Control

The agriculture industry needs inexpensive, sustainable methods for controlling the pests that reduce yields and profits. Pests cost billions of dollars in lost production and injury to agricultural crops and animals. It is essential that the industry have safe and effective methods of pest control while preventing harm to beneficial insects and organisms (**Figure 5.2.3**). Scientists can develop new technologies to address these problems.

Labor and Robotics

Labor is an expensive part of agricultural production and processing. Some producers have a difficult time finding enough qualified labor for the most labor-intensive work. Robots are now used in many places to perform tedious and repetitive tasks that are difficult for humans to perform (**Figure 5.2.4**).

Figure 5.2.4. Robots can perform repetitive tasks such as chemical or fertilizer application, harvesting operations, and crop monitoring. *What are some specific tasks that you think a robot could do?*

MONOPOLY919/Shutterstock.com

Global Trade

Technology allows producers to package and ship products with the touch of a button (**Figure 5.2.5**). Modern communication systems allow for tracking shipments and orders, and help manage the financial transactions necessary to buy and sell agricultural products. Global positioning systems can pinpoint the exact location of products in the supply chain, and provide estimates on delivery times.

Conservation Tillage

One of the earliest technology trends in agriculture was the development of ***conservation tillage***. This tillage method involves processes which improve the quality and health of soil by reducing tillage operations. In the United States, conservation tillage is used on cropland more than any other method of tillage.

Tillage is the process of preparing soil for planting crops. It involves turning over and breaking up large chunks of soil to kill weeds and prepare a smooth seedbed for planting (**Figure 5.2.6**). This is, unsurprisingly, a lot of work!

Heavy tillage changes the structure of the soil and the way that soil particles stick together. This is an important consideration, because the "stickiness" of soil particles affects the amount of water the soil can hold. Excessive tillage reduces the amount of water the soil can hold. This leads to dry soil and poorer crop yields.

Excessive tillage also reduces the amount of useful organic matter in the soil. Farmers used to till under all residues from the previous crop to create a smooth, even-looking seedbed for the new crop. However, after it was discovered that leaving some crop residue on the soil surface helps the soil retain water and provides useful nutrients to the new crops, farmers changed their practices.

Conservation tillage also reduces soil erosion by slowing down the runoff of water from the soil surface. The crop residues and organic matter keep soil in place and slow down the speed of water moving across the soil surface. These healthy practices have also been shown to slow down wind erosion. Loose, dry soils tend to move in the wind more than those with adequate surface moisture and crop residue.

Regenerative Agriculture

Regenerative agriculture is a type of conservation tillage that is combined with other processes to increase ***biodiversity*** (the amount and variety of life on Earth), reduce soil erosion, and improve the water cycle.

David A Litman/Shutterstock.com

Figure 5.2.5. These agricultural workers are using a conveyor belt system that packages lettuce directly in the field, making it ready for shipping once it leaves the field.

LALS STOCK/Shutterstock.com

Figure 5.2.6. Traditional tillage uses methods that remove all vegetation from the soil surface, making the soil susceptible to wind and water erosion.

PHOTO FUN/Shutterstock.com

Figure 5.2.7. This Asian ladybug is hunting aphids on this plant. By removing the aphids, this insect is helping the plant grow. You don't want your pesticides targeting such a good helper!

Joseph Sohm/Shutterstock.com

Figure 5.2.8. Encountering farm equipment on public roads is a common experience in agricultural communities. *What does this mean for automobile drivers in these communities?*

Figure 5.2.9. This GPS-guided tractor is planting directly into plant residues from the previous crop. This method combines precision agriculture with conservation tillage.

Some traditional farming methods eliminate insect pests very well. However, not all insects are pests! These methods may also eliminate helpful insects, including plant pollinators, such as honeybees, or other insects that hunt the harmful pests that damage crops (**Figure 5.2.7**). Regenerative agriculture tries to address this issue by only targeting harmful insects, or using nonlethal methods to deter insect pests from causing damage. What other issues can you think of that could be addressed using regenerative agriculture?

Agricultural Technology in Communities

Agriculture is a major part of many communities across the United States. The agricultural industry employs people from all walks of life to perform the tasks that feed, clothe, and house a growing world population. Living and working in an agricultural community has distinctive features. Agricultural communities have more agricultural equipment traveling on the roads (**Figure 5.2.8**). Farms and ranches create dust, animal smells, and noise. These are the realities of farming.

Some who move to agricultural communities may have difficulty adjusting to this environment. Technology has helped to improve many agricultural operations. It has minimized the effects of dust, noise, and animal waste handling. New technology in manufacturing has produced more fuel-efficient machinery. This saves farmers and ranchers money, and reduces the smell and noise associated with their work. Conservation tillage and no-tillage practices reduce the time that farmers must be in the fields and on the roads with agricultural equipment. Conservation tillage practices reduce soil erosion by wind and water, and this reduces dust. New technologies in animal waste management have reduced the odors associated with animal production. Agricultural technology helps farmers be good neighbors in agricultural communities.

Precision Agriculture

Every element of the production of food and fiber requires good decision-making. The resources needed to produce the foods we enjoy are expensive and often limited in supply. The more data that growers have, and the more precise it is, the better their decisions become.

What kinds of decisions? Farmers must decide how much fertilizer to use on crops. If they use too much, the fertilizer can cause damage to the crop. If they use too little, the fertilizer does not help the plants grow. Using just the right amount saves money and effort, and promotes optimal growth of crop plants (**Figure 5.2.9**).

Marek Musil/Shutterstock.com

Farmers also must decide how much water is needed to irrigate cropland. Just as with fertilizer, too much or too little water can damage a crop and reduce yields.

In the food processing industry, data to make decisions is important. Imagine a conveyor belt with thousands of tomatoes whizzing by on the way to being converted into tomato sauce. How does a worker find and remove the bad tomatoes traveling at a high rate of speed? In a similar example, food safety regulations in egg-processing facilities require that every egg be examined for quality. At any given time, there are thousands of eggs moving on a conveyor system through the cleaning and packaging process. If each egg requires careful examination by a worker on the food processing line, the process will move slowly.

This is where precision agriculture enters the picture. Precision agriculture, as discussed previously, is the practice of observing conditions in the field or processing facility, then taking measurements of key factors affecting production, and making decisions based on those measurements.

Observing

Let's say that a farmer wants to know precisely where in a field he needs to apply fertilizer. He flies a drone over the whole field, measuring the amount of infrared radiation given off by plants. *Infrared radiation* is a type of light that is invisible to human eyes, but that we feel as heat radiating from objects. If you have ever walked along a sidewalk during the summer and felt the heat radiating off the sidewalk, you are experiencing infrared radiation. Everything emits infrared radiation, even you!

Scientists discovered that plants experiencing drought or stress from lack of fertilizer are a different color than normal crops in infrared photographs. Special cameras can take photographs of infrared radiation. The image on the left in **Figure 5.2.10** is a regular satellite image. The green areas are forestland. The image on the right shows the same area, photographed using infrared technology. Note that the forested area is now red in color. Variations in the red color indicate that plants may be stressed due to a lack of water, nutrients, or pest infestation. The leaves on stressed plants reflect infrared radiation in a different way than normal plants.

A
USDA-FSA

B
USDA-FSA

Figure 5.2.10.
A. Satellite image of urban and forested areas. **B.** Infrared imagery of urban and forested areas. *What differences do you see between the photos?*

Infrared photography is one method for collecting data, but there are other tools. Farmers use sensors to capture data on soil moisture, nutrient levels, and the amount of light available for crops. These sensors send data to a computer where the information is compiled for analysis.

Measuring

Sensor or infrared photography data is sent to a computer where the information is compiled for analysis. The farmer examines the data and decides what it means.

Decision-Making and Response

Using precision agriculture techniques, farmers can adjust irrigation, fertilizer applications, and pest control to improve plant health and crop yields (**Figure 5.2.11**). Data is entered into the computer systems of field implements and tractors. As the equipment travels across a field, the computer may send a signal to apply more or less fertilizer in certain areas. Some planters utilize variable-rate seeding to place crop seeds at the right location and in the right amounts for a given field. More or less seed can be planted in certain sections of a field, given the amount of soil moisture available and the specific soil type. It does all of this automatically, based on the data presented and the farmer's decisions. Isn't this amazing?

Precision agriculture is an emerging technology. Farmers and crop management specialists are using drones, sensors, and satellite photography to gather precise data about the inputs needed to grow agricultural commodities. Their information helps producers make the best decisions about the inputs needed to maximize production while protecting and conserving natural resources.

Zapp2Photo/Shutterstock.com

Figure 5.2.11. This soil sensor measures soil moisture, soil pH, and nutrient levels in greenhouse plants. It sends this information to a computer for analysis by the grower. Accurate information on growing conditions allows growers to make good decisions that increase efficiency and productivity.

Did You Know? According to the NASDAQ stock exchange, more than $12.9 billion in agricultural products are generated using precision agricultural practices.

Integrated Pest Management

Integrated Pest Management (IPM) combines biological, cultural, physical, and chemical tools to reduce the damage caused by pests to crops, human health, and the environment. This method of pest management has been in use for a long time, but new technologies are being developed.

Integrated pest management means that the farmer uses several different methods at the same time to control pests. For example, the farmer may plant disease-resistant corn, then use mechanical cultivation to remove weeds in direct competition for water and nutrients (**Figure 5.2.12**). Chemical pesticides may also be applied.

Andrii Yalanskyi/Shutterstock.com

Figure 5.2.12. Cultivation is one method for reducing the damage caused by weeds and insect pests.

Integrated pest management includes:

- Soil preparation: Rotating crops among fields and conducting nutrient testing.
- Planting strategies: Planting crops in ideal row spacing to prevent competition from weeds.
- Weather forecasting: Using weather data to predict when pests will appear.
- Trapping pests: Using traps that attract pests. Traps can be useful indicators of the numbers and types of pests present.
- Monitoring pests: Knowing which pests are in fields helps determine the best method for control (**Figure 5.2.13**).
- Cultural controls: Using plows and cultivators to destroy weeds and pest habitat in fields.
- Chemical pesticides: Using pesticides to control pests.
- Biological controls: Using natural poisons, predator insects, and other biological controls to attack crop pests.

Catherine Eckert/Shutterstock.com

Figure·5.2.13. Integrated pest management requires that you correctly identify the pest so that you can then select the method best suited to control or eliminate it.

Making Connections #STEM

Bt and the Box Tree Moth Caterpillar

The picture below shows a box tree moth caterpillar. This creature is a lean, mean, eating machine, and several of them working together can strip a plant of its leaves in a short amount of time. It is native to Asia and has spread to Europe and North America. The caterpillar eats the leaves of the boxwood plant, a common landscape plant in the United States. Many homeowners do not want to apply pesticides in their landscape plants because small children and pets may come in contact with them. It would be good to have a natural pesticide that does not harm humans or animals, yet controls this pest. Well, there is good news! We do have a naturally occurring bacteria in the soil that is toxic to the box tree moth caterpillar. It is called *Bacillus thuringiensis*, or "*Bt*" for short.

Bt produces proteins that are toxic to this caterpillar. Special pesticides are created from this bacteria and applied to plants. The caterpillar eats leaves sprayed with Bt. The bacteria breaks down the lining of the digestive system, causing infection and death. Bt is only activated as a pesticide once it enters the digestive system of the caterpillar, rendering it harmless to other creatures, including pets and humans. This is an example of how scientists have used a living organism as a pesticide to control another living organism. This is a good example of how technology improves agriculture and our quality of life.

Consider This

1. Why are homeowners concerned about spraying toxic pesticides on landscape plants?
2. How does Bt work to control the box tree moth caterpillar?

Zerbor/Shutterstock.com

Urban Farming

You might think that agricultural production in an urban area would be impossible. Where is there room to grow anything? Not only is it possible, but it is a growing and necessary part of the modern agricultural industry. **Urban agriculture** is the production and distribution of food, fiber, and natural resources in cities and towns.

In the United States, about 329 million people live and work in cities and towns, and about 46 million people live in rural communities. Major agricultural operations are typically located in rural areas, but urban agriculture is important for many reasons. Farms of less than 2.5 acres (which are small enough to find in an urban area) provide about one-third of the food supply on the planet. More than 15 percent of the world's food is grown in urban areas.

You will not see large farm fields and tractors in a city, but you will find many gardens. You may also notice agricultural operations along roadsides and small livestock grazing in open spaces. Small poultry flocks can be raised in the yards of urban homes. Some people plant community gardens where fresh produce is grown and shared. Converting a vacant lot into a garden is a good use of resources, and helps to build cooperation and a sense of community. These plots also provide recreation and wellness opportunities as people get outside and work in the garden.

Urban farmers have found ways to produce agricultural products in the city for sale in farmers' markets and for use by restaurants. Locally grown foods are considered by many restaurant operators to be a major advertising element for their businesses.

Urban agriculture also helps to alleviate the problem of food deserts. A **food desert** is an area, typically within a city or town, where the citizens have limited access to affordable and nutritious food. People in a food desert may have to travel a long distance to get the foods they need. Some city residents grow gardens to increase the quality, quantity, and variety of the foods they eat. This also reduces dependency on fast foods and other unhealthy food options, which are often more readily available in these areas. City residents can grow interesting and nutritious vegetables, supplemented by backyard-raised poultry (**Figure 5.2.14**). Every contribution to agriculture is welcome and necessary!

Figure 5.2.14. This rooftop garden uses available space to produce vegetables. Rooftop gardens also provide insulation against heat and cold, and can help reduce carbon dioxide in the atmosphere.

Alison Hancock/Shutterstock.com

One additional benefit to urban farming is that it supports the agriculture industry by increasing **agricultural literacy**: having a basic understanding of agricultural systems, how food is grown, and the effect of agriculture on the economy and environment. Urban agriculture provides an opportunity for people who live in cities and towns far away from rural agricultural communities to have a better understanding of the work involved in producing food and fiber for a growing population.

Vertical Farming

It is expected that almost 80 percent of the world's population will live in urban areas in the coming decade. This means the agriculture industry must feed and clothe a growing world population, and also distribute this food efficiently in cities and towns. One way to make this easier might be to move agriculture closer to the consumers who depend on it.

Vertical farming may be a solution to food distribution challenges in urban areas. **Vertical farming** is the production of food on vertically aligned surfaces, like stacks of planters along a wall. A tall building in a city could be converted into a vertical farm, using hydroponics and supplemental lighting (**Figure 5.2.15**).

Vertical farming can place the food production close to the intended market. This cuts down on transportation and distribution costs, and cuts down on the steps and timing required to move the food from producer to consumer. A large pool of labor to work in indoor vertical farms is also available in urban areas.

Because vertical farming does not use soil in the same way as traditional farming, growers can produce fruits and vegetables using less water. Nutrients can be injected into the hydroponics system. Pests can be monitored and controlled in indoor farms more easily than in traditional farming.

Vertical farming eliminates the expense of field tillage equipment and tractors. The controlled environment provides the right amount of light, nutrients, and water without having to worry about drought, excessive heat and rain, and other unfavorable weather conditions.

Did You Know?

You can grow many types of vegetable crops vertically. Tomatoes, squash, beans, and even small pumpkins can be grown in a small space using vertical systems.

5

Pressmaster/Shutterstock.com

Figure 5.2.15. *Did you ever imagine that a farm could look like this?*

cash1994/Shutterstock.com

Figure 5.2.16. The USDA Organic logo may be familiar to you from time spent at the grocery store or in food service. *What does it mean when you see this logo?*

Did You Know?

According to the US Department of Agriculture, organic farms in the United States annually produce $9.9 billion in agricultural crops and livestock.

Vertical farming seems like it would require too much building space to be practical. However, this method of production is very efficient regarding growing space. One acre of indoor hydroponic farming space equals four to six acres of traditional farming space, in terms of crop yields. More plants can be grown in a smaller space. A 30-story building converted to vertical farming can produce as much as a farmer grows on 2,400 acres.

Vertical farming is an innovative technology, but there are challenges. This new method of producing crops in multistory buildings requires expensive renovations to accommodate agricultural production. Several of the crops that could be grown well indoors require wind and insects for pollination. The demand for highly skilled workers will increase labor costs. The production of crops in a totally enclosed environmental system requires backup systems for power, irrigation, and lighting. A prolonged power failure in a vertical farm building would devastate growing crops.

In spite of challenges, vertical farming remains a growing alternative to traditional agricultural practices.

Organic Agriculture

Farmers who grow foods using organic methods can place this label (**Figure 5.2.16**) on their products. The United States Department of Agriculture defines *organic farming* as "the production of food that integrates cultural, biological, and mechanical practices that recycle resources, and promote ecological balance while conserving biotechnology."

Whew! That is a big definition! Let's break it down to see if we can understand it more thoroughly. Let's start with:

"the production of food that integrates cultural, biological, and mechanical practices…"

This means that organically grown food is produced without using synthetic fertilizers and traditional pesticides. This method of farming uses mechanical means to remove pests and encourage plant growth.

Next, the definition of organic farming says,

"…practices that recycle resources, provide ecological balance, and conserve biotechnology."

This means that farmers do everything they can to protect the soil and environment from damage by agricultural practices. It was a standard practice in the early days of farming to plant crops in wide rows and cultivate the soil vigorously to remove weeds. Modern practices plant crops closer together so that weeds do not have a chance to compete with crop plants. Organic farmers also use natural fertilizers such as animal manure and plant compost. These practices encourage the production of microorganisms in the soil that break down organic matter into nutrients for growing plants.

Organic farming is a different approach to traditional agriculture. Farmers using organic methods maximize the use natural resources to produce crops and livestock. This includes:

- Using animal manure to fertilize soils
- Using mechanical means to remove or control pests
- Using pesticides made from natural materials to control pests
- Eliminating synthetic growth hormones in animal production
- Spreading out farming operations to avoid the buildup of agricultural residues, such as animal manure

This maintains ecological balance by producing crops and livestock that do not harm the soil's ability break down wastes.

Hands-On Activity

The Technology of Pulleys

Technology is the use of science for practical application. Physics is a branch of science that examines the physical properties of matter and nature. The agriculture industry uses physics all the time to solve problems and make work easier.

The combine illustrated uses an auger to move soybeans from its holding tank into a trailer for transportation to market. How are the dried beans moved along a tube? The auger has an inclined corkscrew-like plane inside the tube that converts rotational energy into linear energy.

Fotokostic/Shutterstock.com

In this activity, we will make a simple machine that uses pulleys to lift an object. Pulleys make the task of lifting objects easier.

Here are the things we will need in order to conduct this activity:

- 1-Gallon jug of water
- 1 Pulley
- 1 Broom handle about 4 feet long
- 10 Inches of rope sized to fit the pulley
- A notebook and pen or pencil to record notes
- 1 Hanging scale like the one pictured here

039l/Shutterstock.com

This experiment will determine if pulleys really make lifting loads easier. Let's do this step by step:

1. The first experiment is to determine how much effort it takes to lift the gallon jug of water. Have two students hold the broom handle at shoulder height, and hook the hanging scale to the broom handle, about halfway between the two students. Then suspend the gallon jug from the hook under the scale, and record the weight in your notebook.
2. Disconnect the jug from the scale and the hanging scale from the broom handle.
3. The second experiment is to determine how much effort it takes to lift the gallon jug using a rope and pulley system. Connect the pulley to the broom handle so that it hangs on the broomstick halfway between the two students. Run the rope through the pulley, and connect one end of the rope to the handle of the water jug.
4. Connect the hanging scale to the free end of the rope and pull on the rope until the jug is completely off the ground. Record the weight reading on the scale.

Consider This

Compare the two weight readings from the first and second experiments.
1. Were the weight readings different? In what way? Was it easier to lift the gallon jug with the pulley?
2. What are some ways that pulleys would be useful in an agricultural setting?

What Did We Learn?

- Technology is the application of science, engineering, and math to solve problems.
- Technology helps solve practical problems in agriculture as we feed and clothe a growing population.
- Precision agriculture, conservation tillage, regenerative agriculture, and other new technologies have made agriculture a more sustainable endeavor.
- Urban farming and vertical farming are emerging areas of development in agriculture.

Let's Check and See What We Know

Answer the following questions using the information provided in Lesson 5.2.

1. The application of science, technology, and mathematics to solve problems is called _____.
   A. biodiversity
   B. technology
   C. sustainability
   D. regenerative agriculture

2. Repetitive tasks in agriculture can now be accomplished with _____.
   A. robots
   B. sustainable agriculture
   C. tillage
   D. conservation practices

3. Conservation tillage _____ the amount of tillage required to grow crops.
   A. increases   C. sustains
   B. decreases   D. conserves

4. The purpose of the REA was to _____.
   A. bring electrical power to rural areas
   B. reduce soil erosion on farms
   C. create research centers for urban agriculture
   D. develop new markets for agricultural products

5. The process of preparing soil for planting crops is called _____.
   A. seeding   C. tillage
   B. GPS   D. sustainability

6. Which of the following is *not* a purpose of regenerative agriculture?
   A. Increase biodiversity
   B. Reduce soil erosion
   C. Improve the water cycle
   D. Increase market prices paid for field crops

7. Which of the following is *not* the purpose of precision agriculture?
   A. Observing field conditions
   B. Measuring field conditions
   C. Decision-making
   D. Increasing fuel costs

8. More than _____ of the world's food is grown in urban areas.
   A. 5 percent   C. 15 percent
   B. 10 percent   D. 20 percent

9. An area within a town where citizens have limited access to affordable food is called a food _____.
   A. swamp   C. vacancy
   B. desert   D. market

10. *Bacillus thuringiensis* is a type of _____.
   A. plant
   B. energy source
   C. field crop fertilizer
   D. bacteria

Let's Apply What We Know

1. What are some ways that technology has the improved the quality of your life?
2. What are three ways that technology has improved agriculture?
3. In your own community, identify areas where a food desert exists. Where do people who live there have to buy their groceries?
4. As you read about the use of technology in agriculture, did you agree with its use? Are you comfortable with the use of technology in your own life?
5. Interview an older adult in your family, or someone in your community about their experiences with technology. Ask them to explain how technology has changed over their lifetime, and how it has affected them. Compare their answers to your own experiences and attitudes about technology.

Academic Activities

1. **Social Science.** What do you think about the idea of urban agriculture? Does it make sense to you? Research this topic. What do those people involved in traditional agriculture have to say about urban agriculture? What effect does a person's knowledge and skill have on their attitude toward the adoption of new technology? Do you believe that a person's personal experiences shape their beliefs about technology?
2. **Mathematics.** Center-pivot irrigation systems are deployed on farms throughout the country. Have you ever wondered how many acres one of these systems can irrigate? Using Google Earth, find a center-pivot irrigation system on a map. Using the mapping tools, measure the length of the irrigation rig. Using mathematics, determine the circumference of the circle and the area the irrigated by the rig. You may be surprised! Some irrigation systems can water hundreds of acres.

Communicating about Agriculture

1. **Reading and Writing.** Find an article about a new technology being used in agriculture. Write a description of it. Include the pros and cons of using the technology. What is your opinion about this technology?
2. **Speaking.** Prepare a three-minute presentation on the new technology you wrote about in the previous exercise. Explain how it is used, and whether or not you think it is a good idea.

SAE for ALL Opportunities

1. **Foundational SAE.** In your SAE program, look for ways that technology has made your project easier to accomplish. Which technology has made the biggest difference for you? Has it improved your project or slowed down your progress?

SAE for ALL Check-In

- How much time have you spent on your SAE this week?
- Have you logged your SAE hours?
- What challenges are you having with your SAE?
- How can your instructor help you?
- Do you have the equipment you need?

Emerging Technologies in Agriculture, Food, and Natural Resources

WitthayaP/Shutterstock.com

How has biotechnology affected the agriculture, food, and natural resources industry?

Learning Outcomes

By the end of this lesson, you should be able to:
- Identify the optimal conditions for new technologies to be adopted. (L.O. 05.03-a)
- Describe how the use of technology in agricultural contexts has changed over time. (L.O. 05.03-b)
- Describe emerging technologies in agriculture. (L.O. 05.03-c)

Words to Know

artificial intelligence
big data

Internet of Things (IoT)

radio frequency identification (RFID)

Before You Begin

There is an old saying that "necessity is the mother of invention." What do you suppose that phrase means? Pair up with another student in your class, come up with ideas about what the saying means, and share them with the rest of the class.

Accelerating Technology and Modernization

Technology is the application of science, engineering, and mathematics to solve real-world practical problems. Another thing you should know is that technology builds on itself. Inventions create the opportunity for other inventions. For example, Alexander Graham Bell received the first patent for the telephone in 1876 (**Figure 5.3.1**). Bell's telephone required that telephone calls be transmitted over wires. Bell's invention eventually led to the development of a mobile phone that allowed people to make telephone calls wirelessly from almost any location. The success of the mobile phone led to the development of the modern smartphone, with all its additional functionality. Today's phones can do so much more than Bell's telephone from 1876 that they are practically unrecognizable, even as they meet the same need we have to connect with others. In this way, technology builds on itself to greater discoveries (see **Figures 5.3.2** and **5.3.3** for another example).

Bell's First Telephone UNDERWOOD & UNDERWOOD

Everett Collection/Shutterstock.com

Figure 5.3.1. This is Alexander Graham Bell's first telephone. It's safe to say it's different than today's mobile smartphones. *What do you suppose communication was like before cell phones?*

5

Morphart Creation/Shutterstock.com

Figure 5.3.2. Eli Whitney's cotton engine, or "gin" for short, was a hugely influential invention in the 1800s. It greatly sped up the process for removing seeds from cotton fiber.

Paul R. Jones/Shutterstock.com

Figure 5.3.3. A modern cotton gin can produce thousands of pounds of cotton per minute, much faster than Whitney's first gin. This is technology at work: One innovation leads to more!

Figure 5.3.4. Early agricultural implements were made of stone and bronze. The plow probably originated in Sumeria in 3100 B.C.E.

The Agricultural Revolution: A Brief History

Humans hunted wild animals for the first 2.5 million years of human existence and gathered edible plants for food. Approximately 10,000 years ago, agriculture became humans' primary method for securing food (**Figure 5.3.4**). Scientists reason that the switch from hunting and gathering to agriculture decreased the risk of seasonal variations in the food supply. Sometimes food was available, and at other times it was in scarce supply because of a long summer drought or a cold winter. The growth and spread of plant and animal species depended on weather and growing conditions, which humans could not control.

Archeologists study history by examining ancient objects dug from the ground. By examining things left over from early generations of humans, they believe that agriculture began in several places on the planet at about the same time (**Figure 5.3.5**). For example, archeologists believe that potatoes made their first appearance as a planted crop in South America at about the same time that rice and domesticated swine appeared in China. Cultivated corn and beans appeared in Central America in the same era as pumpkins appeared in North America. This means that about 10,000 years ago, agriculture independently sprang up in several regions of the globe. That's amazing! How did people who lived thousands of miles away from each other suddenly simultaneously discover the concept of agriculture?

Agriculture increased the amount of food that humans could grow, but agriculture requires more effort and skill than hunting and gathering.

Figure 5.3.5. These ancient Egyptian hieroglyphics demonstrate farming, fishing, and the transportation of farm products.

As people got more experience with agriculture, they could produce more and more food, in line with larger populations. By this point in time, humans are experts in agricultural production. To find ways to increase productivity and yields and meet the needs of a growing population, we will rely on technological advances. For agriculture to continue to flourish, we have to use technology to find ways to reduce the effort necessary to grow food (**Figure 5.3.6**).

Satyrenko/Shutterstock.com

Figure 5.3.6. This automated system is transplanting seedlings in a greenhouse. *How much has changed in agriculture since the scenes depicted in the preceding Egyptian hieroglyphic figure?*

Introducing New Technology in Agriculture

The appearance of a problem creates the opportunity for new technologies to appear. There have been many examples in this unit; Let's look at one more. New technology was required to solve the problem of boll weevil infestations.

In the early 1900s, cotton was a significant part of the United States' economy. Cotton production traces back to the colonial period in US history, when a ready supply of cotton fiber was necessary for textile manufacturing and commercial trade (**Figure 5.3.7**). Without a strong cotton crop, the economy of the US (and the Southern US, in particular) would be dramatically weaker.

Figure 5.3.7. Cotton bales at the Port of Charleston, South Carolina in 1878. These bales were shipped to foreign and domestic markets by sea-going ships. Cotton cloth brought excellent prices in world markets, so American farmers planted as much cotton as they could.

Everett Collection/Shutterstock.com

Alf Ribeiro/Shutterstock.com

Figure 5.3.8. The boll weevil feeds on cotton bolls. *Do you see the puncture marks in the boll pictured?* These punctures damage the boll and prevent it from developing cotton fibers.

Kent Weakley/Shutterstock.com

Figure 5.3.9. The cotton boll is the "fruit" of the cotton plant. It is the protective case around the seeds and cotton fibers on cotton plants. Boll is pronounced "bowl." It contains the whitish cotton fiber used to make fabrics and textiles.

In the late 1800s, a significant threat to the US cotton crop emerged: the boll weevil (*Anthonomus grandis*). The boll weevil is a beetle that feeds on the flowers and bolls of the cotton plant. It burrows into the cotton boll, damaging it (**Figure 5.3.8**). This damage prevents the cotton plant from producing cotton fiber (**Figure 5.3.9**).

In the early 1900s, the boll weevil infested US cotton crops and caused severe damage to crop yields. In short time, the boll weevil had destroyed millions of acres of cotton. Farmers lost their cotton crops, and in some cases their farms, because of this tiny insect.

Did You Know? The boll weevil first entered the United States from Mexico in the late 1800s. By the 1920s, the boll weevil had infested all cotton-growing regions in the United States. Technological advancements in agriculture eliminated the boll weevil in 2009.

A pest management program was soon developed that pinpointed the spread of the boll weevil each year using field sensors that attracted the insects (**Figure 5.3.10**). Farmers could examine these weevil traps to determine the amount of infestation and the direction the weevils were traveling. This information was used to deploy intensive pest management practices that eventually eliminated this pest as a threat to the cotton crop.

Figure 5.3.10. This boll weevil trap uses pheromones, special chemical attractants, and bright colors to lure the insects in for capture.

a katz/Shutterstock.com

The key to the effectiveness of the boll weevil eradication program was the use of sensors to collect data on the weevils' locations. Scientists gathered data from thousands of these sensors across the American South and used that data to recommend pest management techniques to farmers. This is an example of the use of technology to improve agricultural production (**Figure 5.3.11**).

Big Data in Agriculture

When you read the term "big data," what do you think it means? ***Big data*** is a term used to describe collecting and analyzing large amounts of information. Computers now can collect and analyze millions of bits of data in mere seconds.

Here's an example: Let's suppose that a farmer wants to know how much fertilizer he will need for this year's cotton crop. In January, the farmer employs a crop consultant to sample and analyze the nutrients in the soil and prepare a report. The crop consultant goes into the field to take the soil samples. The traditional method requires that she walk across the field taking actual soil samples with a special soil probe. This was time consuming, and only provided a small number of data points for the report. Instead, now she opens a box on the tailgate of her pickup truck and removes a drone. She makes some adjustments to the infrared camera and soon the drone is flying across the field with her at the remote controls. The drone takes infrared photos of the whole area. She loads all this data into a GPS and prepares her electronic report for the farmer. The farmer reviews the data, which include thousands of measurements that pinpoint precisely where to apply fertilizer in the field. The farmer uses the report to plan for the proper amount of fertilizer needed.

In agricultural production, big data is used to make informed decisions about the use of limited resources (**Figure 5.3.12**). Farming and agricultural processing are both expensive and resource intensive. Big data is used to reduce the number of resources used to only those needed for optimal production. This saves time, money, and effort, and protects and conserves natural resources.

Andrey_Popov/Shutterstock.com

Figure 5.3.11. Citizen scientists are collecting data regarding invasive plant species in the same way that data was collected on boll weevil infestations. These observers log sightings of invasive plants—where they saw them, how many they saw—in a map on the app. Scientists use this data to track the spread and decide how to address the problem.

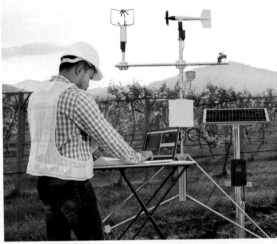

Suwin/Shutterstock.com

Figure 5.3.12. Traditionally, weather stations would gather data once or twice per day. Now, the technology exists to collect weather data thousands of times each day. This allows farm technicians and scientists to make better decisions based on precise data.

The Internet of Things

The internet is many things to us, including a main method by which many people communicate with each other. We use the internet for email and instant messages. We share our photos online and "talk" to each other through social media. Until now, the internet has been a tool for humans to interact.

Now, the internet is also a place where machines can communicate with each other. For instance, some modern household clothes washers can send messages to the manufacturer in the event of a malfunction. Modern farm equipment has technology that allows farmers to conduct onboard diagnostics whenever there is an equipment malfunction. Until recently, humans were needed to make machines "talk" to each other. Now, machines can send and receive data without human interaction. Imagine that one day in the future, the car you are driving detects a problem with the engine. The car contacts the repair shop and provides diagnostic information to the repair technicians, who recommend the necessary repairs.

Hands-On Activity

Citizen Science

The adoption of new technology depends on the work of citizen scientists. iNaturalist and Bugwood are two web-based programs that allow you to enter nature observations into a database for tracking. By entering data, citizens help scientists track the location of plants, animals, and insects. This helps them conduct scientific experiments.

Using an online citizen science program, under the direction of your agriculture teacher, locate and take photos of a plant, animal, or insect you observe outside in the wild. These can also be plants cultivated and grown in your school garden or natural area. Upload your photo into the online database and enter as much information as you are able about the plant, animal, or insect. Then search the database to see how many of the observed organisms are nearby.

This is an entry in the online citizen science database iNaturalist. Certain milkweeds are vital for the growth and development of monarch butterflies (*Danaus plexippus*).

Consider This

1. Many of the people who provide material for these online databases like iNaturalist or Bugwood are not formally trained scientists. How do these online databases make sure that the information provided by these citizen scientists is accurate?
2. What encourages people to use or participate in these online databases? What keeps people interested in contributing to them and using them?

Barry Croom

Modern farm tractors have auto-steering systems linked to global positioning systems. The tractor driver lines up the tractor, sets the GPS, and the tractor drives itself across the field, automatically adjusting speed and power settings based on field conditions. In the early days of agriculture, farmers prided themselves on being able to plow straight rows in a field. Now a modern tractor can do the same thing with the press of a few buttons.

This new reality, where machines communicate with each other with minimal human interaction, is called the **Internet of Things (IoT)**. Machines have been able to send and receive messages for some time now, but the large-scale use of the technology has only recently begun.

Consider these practical examples of how the IoT might work:

- In a greenhouse, a soil moisture meter determines that plants need water. It sends a signal to the irrigation system. The irrigation system switches on and overhead sprinklers water the plants. The soil moisture sensor also collects data on the amount of water used and reports to an app on the greenhouse manager's smartphone (**Figure 5.3.13**).
- A sensor detects a loss of power affecting the cooling system in a feeder pig house on a swine farm. The sensor sends a message to the ventilation system and the side curtains on the house open to provide natural ventilation. It then sends a warning message to the farm animal technicians responsible.

IoT technology is useful for managing power use in farm and agribusiness operations, monitoring crop and animal health, and analyzing data in real-time. Onboard computers in a modern combine now record the amount of grain handled, engine system functions, and threshing operations. All this information is analyzed to determine the most efficient strategies for equipment operation (**Figure 5.3.14**).

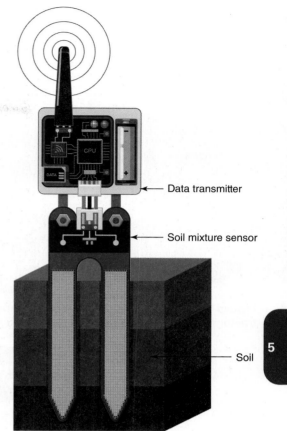

Data transmitter

Soil mixture sensor

Soil

piscari/Shutterstock.com

Figure 5.3.13. This soil moisture meter transmits commands to the irrigation system and directs the application of irrigation water at precise times during the day.

Figure 5.3.14. As this combine moves through the field, sensors record its speed, and the number and duration of stops in the field.

Aleksandr Rybalko/Shutterstock.com

Artificial Intelligence

The term *artificial intelligence* refers to the ability of computers to "think" and make decisions without human input. It is the performance of computational tasks that humans once solely accomplished. Almost everything we do today involves the use of computers. Our mobile smartphones are a type of computer that allows us to perform various tasks. Computers manage the traffic lights at intersections, your school's heating and cooling systems, and millions of other tasks every day. In agriculture, computers help maintain business records and track the shipping and transportation of agricultural products (**Figure 5.3.15**). Artificial intelligence performs many tasks, including some humans prefer not to do.

Scharfsinn/Shutterstock.com

Figure 5.3.15. The experimental autonomous-driving tractor frees up labor to perform other tasks.

Did You Know? According to Statista, companies around the world spend more than $281 billion each year on artificial intelligence software to operate computerized equipment and systems.

Stephen Barnes/Shutterstock.com

Figure 5.3.16. This Holstein dairy cow is wearing an RFID cattle tag to help the farmer determine her location and how much feed she consumes.

RFID

Radio Frequency Identification, or **RFID**, uses a tiny radio transponder to send messages and data to a computer. Let's take a look at how it works!

The technicians on a dairy farm need to keep up with every cow to make sure that each one is milked on a precise schedule. Technicians also need to keep up with the amount of feed each cow eats and track their movements in the barns and pastures to ensure they are healthy. Each dairy cow wears an RFID "necklace" around its neck (**Figure 5.3.16**). The RFID transponder pinpoints the location of each cow on the farm, and how much time they have spent at any particular location. By analyzing the eating patterns of each cow (given by time spent at the feeder), technicians can easily determine which ones may be ill. Sick cows do not eat as much as healthy cows.

In an agribusiness, RFID tags can be placed on inventory items. As these items are purchased, the RFID sends a message to a computer that manages the inventory. This allows the agribusiness to track the inventory of products they sell in real-time.

Career Corner SOFTWARE DEVELOPER

The agriculture industry needs skilled software developers to design the programs to operate equipment and machinery. Modern tractors utilize computer systems to operate efficiently, and many on-farm systems such as automated livestock feed systems and irrigation systems require computer programs to operate properly. If this sounds interesting to you, then maybe you should consider being a **software developer***!*

What Do They Do? Software developers write code, a type of computer language, that allows the computer to function properly. Most software developers work in offices and labs, but may go into the field to solve problems with computer systems in agricultural operations.

What Do I Need to Be Good at This? To be a software developer, you need to be curious and attentive to detail. You'll also need computer skills!

What Education Do I Need? Most jobs for software developers require a bachelor's degree in computer science or a related field.

How Much Money Can I Make? The average salary for this occupation is $89,000 per year.

How Can I Get Started? Take a computer class or elective if it's available to you. Many schools or communities have courses for students on computers and coding. Design an SAE around coding simple computer programs to help manage an agricultural task. Talk to a software developer about the work they do, and why they like it.

goodluz/Shutterstock.com

What Did We Learn?

- New technologies are created and adopted during the process of solving new problems in agriculture, food, and natural resources.
- Agriculturalists need technology to improve the ways we grow food.
- The collection of "big data" allows scientists to make better decisions.
- The Internet of Things (IoT) is where machines communicate with each other with minimal human interaction.
- Artificial intelligence is the ability of computers to "think" and make decisions without human input.
- Radio frequency identification (RFID) can be used to track animals and agricultural commodities and to identify problems or issues quickly.

Let's Check and See What We Know

Answer the following questions using the information provided in Lesson 5.3.

1. What would a rancher do to pinpoint the location of a particular cow, or how many times they eat at a feed trough?
   A. Use RFID to track each animal
   B. Use a drone to fly over and count the animals
   C. Send a farmhand out to watch the animals
   D. Move the animals closer

2. When computers communicate with each other over the internet, with minimal human intervention, the computers are using:
   A. the Internet of Things.
   B. artificial intelligence.
   C. big data.
   D. RFIDs.

3. Collecting and using large amounts of data to make precise decisions is using:
   A. the Internet of Things.
   B. big data.
   C. RFIDs.
   D. artificial intelligence.

4. What term is used to describe when computers utilize stored data to make simple decisions?
   A. Artificial intelligence
   B. Big data
   C. RFIDs
   D. The Internet of Things

5. Which of the following is used to separate seeds from cotton fiber in an efficient manner?
   A. RFID
   B. Big data
   C. Internet
   D. Cotton gin

Let's Apply What We Know

1. In this and previous lessons, you learned that technology is the application of science, engineering, and mathematics to solve real-world practical problems. What are some ways that technology has solved a problem for you?
2. Big data is used to make better decisions about the use of resources in agricultural production. How does big data affect you in an average day? Where do you see big data at work in your life?
3. During a normal day, do you see machines communicating with each other? Where, and when?
4. In what way do you think artificial intelligence will be most useful to humans?
5. RFIDs are used in agriculture. What devices do people use that do the same things (e.g., tracking our location, for instance)?

Academic Activities

1. **Writing.** Write an essay that explores what you think the agriculture industry will look like in 20 years. How will the work on farms be completed? How will agricultural products be transported, marketed, and sold? Use the information from this lesson to make an educated guess about the future of agriculture.

Communicating about Agriculture

1. **Reading.** Read the local newspaper, or another credible news source online, to find information about the current state of the agriculture industry. How does the agriculture industry change the way it operates based on technological breakthroughs?
2. **Writing.** Based on your reading, make a list of the most significant changes currently in process in the agricultural industry. These changes can be positive or negative; make a note of what you think about these changes.

SAE for ALL Opportunities

1. **Foundational SAE.** Examine your own SAE project to determine what types of technology you might use to improve it. For instance, you might use an inexpensive timer to turn on the plant growth lamps used in a greenhouse.
2. **Foundational SAE.** How are you using technology to keep the records associated with your SAE project? Are you using a computer to record your SAE hours and share them with your agriculture teacher? Do you use a smartphone to take pictures of your SAE as part of the record keeping process? Are there other programs you could use that might be helpful? Do a little research, or talk to other students with an SAE. Did you find anything new?

SAE for ALL Check-In

- How much time have you spent on your SAE this week?
- Have you logged your SAE hours?
- What challenges are you having with your SAE?
- How can your instructor help you?
- Do you have the equipment you need?

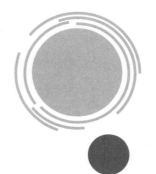

nattanan726/Shutterstock.com

Genetic Engineering in Agriculture, Food, and Natural Resources

Essential Question

In what ways does genetic engineering affect the agriculture, food, and natural resources industry?

Learning Outcomes

By the end of this lesson, you should be able to:

- Describe how biotechnology has influenced the agriculture, food, and natural resources industry. (LO 05.04-a)
- Explain how genetically modified organisms (GMOs) have benefited agricultural production. (LO 05.04-b)
- Identify the genetically modified crops approved for use in the United States. (LO 05.04-c)
- Evaluate the pros and cons of the use of genetically modified organisms in agriculture. (LO 05.04-d)

Words to Know

chromosomes
DNA

genes
genetic engineering

genetically modified organism (GMO)

Before You Begin

*Look at the picture of teosinte below (**Figure 5.4.1**). Compare it to a picture of corn growing in a field. Teosinte is the ancestor of corn. How do you think corn developed from teosinte? How long do you think the process took?*

vainillaychile/Shutterstock.com

Figure 5.4.1. Teosinte (Zea mexicana) is the ancestor of corn and originated in South America. *Can you see the resemblance to modern corn plants?*

Biotechnology's History in Agriculture

Recorded history tells us that people have been domesticating plants and animals for more than 10,000 years. Plant domestication usually involved selecting individual plants that were of interest, and then breeding those plants until the desired characteristics were achieved. For as long as anyone can remember, we have been domesticating organisms for higher yields, reduced toxicity, improved flavor, and ease of harvest.

Wheat, rice, maize, potatoes, and tomatoes were among the earliest plants domesticated. Corn is perhaps the best example. Ancient Americans domesticated teosinte through a selective breeding process. Teosinte is a grass species that has lots of lateral branches and small seed pods. Through human selection, the plant was modified into a single stalk, and one or two seed pods called *cobs*. These cobs held hundreds of large seeds, and still do.

Tomatoes, carrots, and other crops followed a similar process. Selection altered the fruit size and shape, as well as seed size. Improved taste was also a positive result. Wild tomatoes are not very large, nor are they as tasty as the domesticated versions. Genetically modified organisms in plant breeding can be traced back to Darwin's theory of evolution, and to Mendel's discovery of the basic principles of heredity.

5

Making Connections #STEM

Tying Darwin and Mendel to Genetic Engineering

Charles Darwin was a famous biologist and naturalist in the 19th century. You probably know his name because it is associated with the theory of evolution, one of the most radical and important discoveries in biology. Evolution suggests that populations of organisms change over time in order to adapt to their environments. They do this by a process called natural selection: Those individual organisms that are better adapted to the environment are more likely to thrive, survive, and reproduce, passing on their genes.

Gregor Mendel was a Hungarian scientist living in the same century as Darwin. You may have heard of him in your science class as well, as his experiments with growing peas were critical to understanding heredity, or how genes are passed down from a parent organism to its offspring. Because of Mendel, we know about dominant and recessive genes and traits and how genes are expressed (or shown).

These discoveries provided modern scientists a peek into how genes work, and the role they play in changing populations. The idea to manipulate the genes of organisms in order to grow or produce an organism with desired characteristics would not be possible without these discoveries.

Consider This

1. How did Darwin's work lead to genetic engineering?
2. How did Mendel's work lead to genetic engineering?
3. Both Darwin and Mendel were prolific thinkers and scientists, researching and learning about many other things. Take a few minutes to find a reliable source and read about one of these scientists. Can you describe another of their discoveries or areas of research?

Naturally occurring differences in plants provide an opportunity for plant breeders to identify and isolate preferred characteristics of organisms, and conduct breeding programs that incorporate those preferred characteristics in future generations. Natural mutation may take thousands of years to make changes to a plant's dominant characteristics. Controlled plant breeding takes considerably less time to get to the same results. Plant breeding is a sophisticated business. Genetic engineering allows for the development of plants that provide resistance to insect damage, to herbicides, and to drought and other environmental stressors.

What Is Biotechnology?

You have learned in your science classes that biology is the study of living things. *Biotechnology* is a field within biology that is concerned with finding ways to enhance the growth and value of plants, animals, and microorganisms through genetic modification. To be specific, biotechnology is the practice of modifying genes in an organism to create the intended output.

Agriculture offers many good examples of the use of biotechnology. Do you remember the discussion of Norman Borlaug in a previous lesson? He spent years developing varieties of wheat that were resistant to plant diseases. Today, scientists would attempt to make the plant resistant to the ill effects of plant diseases by manipulating only certain genes within a plant (**Figure 5.4.2**).

Genetic Engineering

Have you ever heard the term *GMO*? Let's take a moment to talk about what we mean by **genetically modified organisms (GMOs)**. A GMO is an organism that has had its DNA modified using genetic engineering techniques. The purpose of this modification is to develop a new trait in the organism that does not appear naturally in the species. The process of genetic engineering, or making GMOs, involves manipulating genes to produce certain beneficial features in plants or animals.

Figure 5.4.2. Some varieties of soybeans have been genetically modified to resist the application of broadleaf herbicides. This means that pesticides would eliminate the weeds but not harm the soybean plants.

oticki/Shutterstock.com

DNA

Our bodies are made from trillions of individual cells. Inside each of these cells is a nucleus. The nucleus is the nerve center for the cell. Inside each nucleus, we find strands of deoxyribonucleic acid, or **DNA** for short. DNA is like a blueprint for the construction of a living organism. DNA molecules contain all of an organism's genetic information. DNA is comprised of nucleotides, which are segments of a nitrogen base attached to a "backbone" made of sugars and phosphates (**Figure 5.4.3**).

Chromosomes

Chromosomes are long threads of DNA. They carry the genetic information of a cell. Chromosomes tell the cell what to do. If the cell needs to make proteins, for instance, the chromosomes provide the instructions on how to do so.

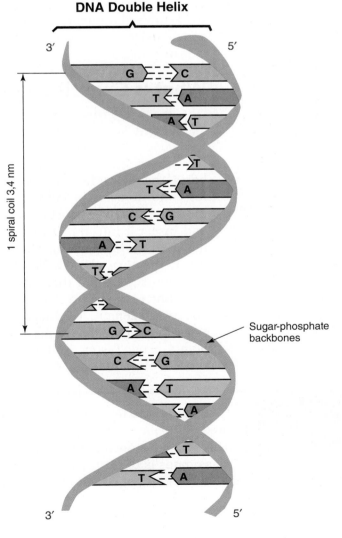

DNA Double Helix

1 spiral coil 3,4 nm

Sugar-phosphate backbones

 Adenine (A)    Thymine (T)    Guanine (G)   Cytosine (C)

Zvitaliy/Shutterstock.com

Figure 5.4.3. DNA, in the chromosome. *Can you believe that genetic information is coded in such an interesting structure?*

J. Andrew Patronik/Shutterstock.com

Figure 5.4.4. The black hair color on Black Angus cattle is the result of a specific sequence of the nucleotides in a strand of DNA.

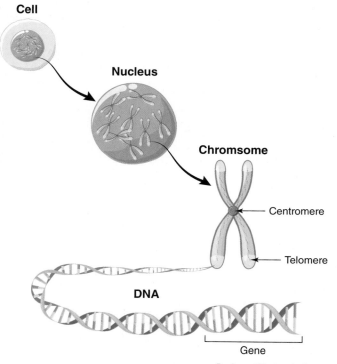

Designua/Shutterstock.com

Figure 5.4.5. The relationship between genes, DNA, and chromosomes is shown here. *How do these parts relate to each other?*

Genes

Genes are segments of chromosomes that contain specific genetic information about an organism. These segments are made of nucleotides arranged in a specific manner. For instance, the gene that determines your hair color is a strand of DNA with the nucleotides arranged in a particular order. Each hair cell in your body has the same gene, and that is why each hair is the same color (**Figure 5.4.4**).

The Relationship between DNA, Chromosomes, and Genes

You may be wondering, "How do all three of these—DNA, chromosomes, and genes—relate to each other?" The nucleus of a cell contains chromosomes. The segments of DNA that make up the chromosomes have a specific purpose in the living organism, such as giving a person red hair. Individual segments of DNA that work toward a purpose are called genes. These genes exist in a specific order within DNA (**Figure 5.4.5**).

What Is Genetic Engineering?

Genetic engineering is the process of cutting apart DNA strands and removing or inserting new genes. For example, let's say that scientists have discovered a gene in a cactus plant that helps the plant live in very dry environmental conditions. In a laboratory, scientists take the gene out of a portion of the cactus DNA and insert it into the DNA of a corn plant. The hope is that by inserting the gene into the corn plant, the corn plant will reproduce DNA with the cactus gene and that the plant will be more resistant to drought. The agriculture industry has used genetic engineering to make some crop plants resistant to pesticide damage, make plants grow faster, and increase the nutrient content of plants.

Did You Know?

According to Texas A&M University, humans are 99.9% genetically identical to each other. The only difference between you and your classmates is 0.1% of your genetic makeup. Can you believe that such a small amount of DNA can make such a large difference?

The Genes We Wear

Genetic traits are those observable characteristics that are passed down to you from your parents. While you are a unique individual, you do share some traits with others in the class. These traits are based upon your genetic makeup—the arrangement of genes in your DNA.

Look at the following chart and recreate it on your paper. Check each of the traits that you possess. Each of the traits that you checked are the result of the arrangement of genes on DNA in the cells of your body.

With the help of your teacher, get into groups of four students each. Now determine how many people in your group have each of the traits in the chart below.

Take a look at your findings. Which of the traits in the chart below occur with the most frequency in your group? That is, which traits seem to be the most common?

It is important to remember that just because some traits are more common than others, it doesn't mean that those traits are better than others. Some genetic traits simply show up more frequently than others. That does not mean they are good or bad. Don't worry if you share a lot of genetic traits with others in your class. You are still a unique individual—your DNA says so!

Trait	Check the box if you have this trait	How many people in your group have this trait, counting you?
Red hair	☐	☐
Left-handed	☐	☐
Naturally curly hair	☐	☐
Over 5 feet tall	☐	☐
Have allergies	☐	☐
Brown eyes	☐	☐
Freckles	☐	☐
Can see the colors red and green	☐	☐

Adoption of Biotechnology on the Farm

Biotechnology has improved the agricultural industry by helping to produce plants and animals that are more productive and more resistant to drought and heat. GMO plants and animals may also be less susceptible to injury by pests. Traditional plant and animal breeding programs result in more productive plants and animals that have some resistance to pests and environmental conditions. Genetic engineering has led to significant and rapid improvements as well.

The following list of plants and animals represents the genetically modified crops and livestock approved for production or consumption in the United States.

Genetically Modified Organisms

As of February 2020, the following food crops have been approved as GMOs:

nnattalli/Shutterstock.com

Figure 5.4.6. The United States is the world leader in soybean production.

Andrei Dubadzel/Shutterstock.com

Figure 5.4.7. There are about two million acres of canola grown in the United States each year, according to the USDA.

Corn

GMO corn varieties are planted on most of all corn farmland in the US. Corn is the most commonly grown crop in the United States, according to the US Food and Drug Administration. GMO corn is treated with *Bacillus thuringiensis*, a bacterium naturally occurring in soil. Bacillus thuringiensis, or Bt for short, is poisonous to certain insect species. Bt does not harm the corn plant, but it does kill the insects that attempt to eat corn plants. Bt corn is used in animal feeds, and corn syrup is used as a sweetener in certain human foods.

Soybeans

Soybean oil is used extensively in cooking oils, baked goods, and salad dressings. This textbook was printed using ink containing soybean oil and other vegetable oils. Most soybeans grown in the United States are genetically modified organisms (**Figure 5.4.6**).

Rapeseed/Canola

Canola oil, made from the rapeseed plant, is the second most widely consumed vegetable oil in the US after soybean oil (**Figure 5.4.7**). It is used as a cooking oil, and as a form of biodiesel fuel for automobiles and farm equipment. It is also used in the production of livestock feed.

Potatoes

Potatoes are important in the human diet because they are a major source of carbohydrates and the mineral potassium. GMO potatoes were developed to combat insect pests and diseases.

Some varieties have been engineered for longer shelf life, and resist browning and bruising during harvesting, processing, and storage (**Figure 5.4.8**).

Papayas

GMO papayas are grown in Hawaii. They are a good source of vitamins and help maintain healthy digestion. GMO papaya is resistant to ringspot virus that has been known to wipe out whole papaya crops in a season.

Zucchini and Squash

GMO zucchini and squash were approved for human consumption in 1995.

Beets and Sugar Beets

Sugar beets are used to make granulated sugar. They are the source of 20 percent of the world's production of processed sugar (**Figure 5.4.9**). More than half of the processed sugar on grocery store shelves comes from GMO sugar beets. GMO sugar beets are resistant to herbicides.

Alfalfa

GMO alfalfa is commonly used to feed dairy cattle, and is resistant to herbicides (**Figure 5.4.10**). This allows farmers to use pesticides to destroy weeds that affect the alfalfa crop.

Flax

GMO flax is both a food crop and a fiber crop. Flax seeds are a nutritional supplement. Flaxseed oil, or linseed oil, is used in paints. The fiber from flax is used to make clothing and linen.

nednapa/Shutterstock.com

Figure 5.4.8. Potato crops have been vulnerable in the past to blights and pest infestations, to large (and terrible) effect. GMO potatoes are designed to resist these issues.

5

DedovStock/Shutterstock.com

Figure 5.4.9. The United States is one of the world's top producers of processed sugar, mainly derived from sugar beets (pictured) and sugarcane.

B Brown/Shutterstock.com

Figure 5.4.10. The United States harvests approximately 18 million acres of alfalfa annually, according to the USDA.

bergamont/Shutterstock.com

Figure 5.4.11. The Arctic golden delicious apple was genetically modified to prevent browning.

Apples

Apples, when sliced, tend to turn a brownish color caused by exposure to oxygen in the air. This browning effect reduces the quality and appearance of the apple. The Arctic apple was genetically modified to prevent this from happening (**Figure 5.4.11**).

Plums

The GMO plum is resistant to the plum pox virus that reduces yields in non-GMO plum orchards.

Salmon

GMO salmon is a cross between salmon and a type of sea eel called the ocean pout (**Figure 5.4.12**). This makes the salmon grow year round and reach a marketable weight more quickly.

Pineapples

GMO pineapple is pink on the inside. This makes it a more attractive and marketable fruit.

Rice

GMO golden rice is not approved in the US for cultivation, but it can be imported into the United States for human consumption.

Sugarcane

GMO sugarcane is not approved in the US for cultivation, but it can be imported into the United States for human consumption.

Aristokrates/Shutterstock.com

RLS Photo/Shutterstock.com

Figure 5.4.12. GMO salmon (left) are the result of adding genetic material from an eel called the ocean pout (right) for the purposes of increasing growth rates in the salmon.

The Debate over GMOs

Some people believe that genetically modified organisms are not safe for human consumption, and cause harm to the environment. Here are some of the arguments against GMOs:

1. Some GMOs are herbicide resistant, leading to the overuse of pesticides. If crops are not harmed by herbicides, there is the risk that the grower may use too much of these chemicals to control weeds. This could lead to runoff into lakes and rivers that serve as our water supply, and the chemical residues may show up in other foods.
2. Some GMO crops are designed for increased production under optimum growing conditions. This pushes out plants created under traditional breeding programs. The traditional plants may have better ability to thrive in harsh climates.
3. The process of genetic engineering may accidently alter organisms in such a way as to be harmful to humans.
4. GMOs indirectly modify the food web. This means that GMO plants could eliminate some pests that other living organisms need in order to survive.
5. There is a concern that GMOs may make humans resistant to antibiotics. This may make it more difficult to control human diseases.

What do you think about these arguments?

Hands-On Activity

Debating GMOs

The use of genetically modified organisms is a hot topic in the United States. Many people have strong beliefs, one way or another, regarding GMO production. Let's see if you have what it takes to develop a sound argument for or against GMOs: Let's have a debate!

Your instructor will divide you into equal teams and provide you with one of the following topics for argument:
- Genetically modified organisms are safe for human consumption.
- GMOs should be banned in the United States.
- Genetic engineering is better than traditional breeding methods for crops and livestock.

Your instructor will assign you to respond to one of the statements above. Meet with your assigned team members and begin developing a plan for presenting your argument. Your task will be to support or refute the statement with facts.

1. Visit the library or conduct a web search to find information that will support your argument. Look for information supported by scientific research.
2. When directed by your instructor, your group will make its presentation to the entire class. Make posters and presentation materials that support your argument.

Consider This

1. Of the facts you assembled in preparation for your debate, which ones surprised you the most? Why?
2. People often exhibit fear as an emotional response to GMOs. What do people fear most about GMOs? How about you? Do you have concerns about the production of GMO crops? What are they?

What Did We Learn?

- Biotechnology helps to produce plants and animals that are more productive and more resistant to drought and heat.
- Major GMO crops that have been approved for use in the United States include corn, soybeans, canola, and potatoes.
- Though GMO crops have been approved for human consumption, some people do not believe that GMO crops are safe to produce. There are arguments for and against growing these crops.

Let's Check and See What We Know

Answer the following questions using the information provided in Lesson 5.4.

1. Agriculture, or the domestication of plants and animals, began about ____ years ago.
   A. 5,000
   B. 10,000
   C. 20,000
   D. 100,000
2. All of the genetic information for a particular living organism is found in its ____.
   A. DNA
   B. biotechnology
   C. genetically engineered nucleus
   D. variety

3. GMO apples were created to prevent ____.
   A. rotting
   B. injury
   C. browning
   D. tartness
4. *True or False?* GMOs have been proven to be a risk to humans.
5. Biotechnology is a field of science within ____.
   A. chemistry
   B. botany
   C. biology
   D. medicine

Let's Apply What We Know

1. What are the strongest arguments in favor of GMOs?
2. What do all the GMO crops listed in this lesson have in common?
3. Look for GMO products at the grocery store. How do you know that they are GMO products?
4. Go online and take a look at the menus of a number (4–5) of restaurants in your community, or in the general area. Do they use GMO products? Is there any way to tell which foods on the menu have GMO ingredients?
5. Do a web search to determine which countries produce the most GMOs. Why do you suppose those countries are pro-GMO?

Academic Activities

1. **Social Studies.** Conduct a web search for articles on GMOs. Can you determine how other countries view the consumption of GMOs? What does the information you find from the US tell you about the discussion in this country about GMOs?
2. **Language Arts.** Create a poster to advertise for or against GMOs. Be careful to only use information from fact-based research.

Communicating about Agriculture

1. **Writing.** Prepare a short letter to the editor of your local newspaper, in response to an article you read in the news media about a GMO issue. In your letter, explain the reasons why you would or would not consume GMO foods.
2. **Speaking.** Conduct a brief survey of students in one of your classes. Explain what GMO foods are and present the arguments for and against them. Ask them to tell you their preference: GMO or non-GMO foods. Record their responses and report your findings back to your agriculture teacher. Were you surprised by the preferences of your fellow students? Why do you think they felt the way they did?

SAE for ALL Opportunities

1. **Foundational SAE.** Conduct online research to determine the types of career options available to you in genetic engineering. What are the education requirements for these careers?

SAE for ALL Check-In

- How much time have you spent on your SAE this week?
- Have you logged your SAE hours?
- What challenges are you having with your SAE?
- How can your instructor help you?
- Do you have the equipment you need?

Skills in Food Science

SAE for ALL Profile
Local Food from School-Based Enterprise SAE

Meet members of the Spanish Fork FFA Chapter in Spanish Fork, Utah. Market Manager Kacie Jones, Ethan VonHoene, Josi Woodhouse, Brooklynn Giles, Katelyn Beckstead, Kalley Stubbs, and Brylee Ferre used their school garden plot to add to the local food system through a weekly farmers' market. The garden enabled students to complete an SAE using school resources.

Brandon Gardner, one of the advisors, realized that students who could not have a hands-on SAE at home could have one at the school if they could use the overgrown quarter-acre garden at the school for their SAE project. With help from the Tractor Supply Company Grants for Growing, the garden was prepared for planting. They laid plastic down, installed a drip system, and cultivated the plants. Students came through the summer to tend the garden. Students sold the harvested produce at the local farmers' market and took some home in exchange for their work. The money earned helped fund FFA activities and trips and paid for the next season's garden expenses. Two students paid for their National FFA trip with the earnings and used the garden as their School-Based Enterprise SAE.

The students reflected on their experience and shared what they learned. Kacie Jones said, "Taking what I learned in the classroom and being able to apply it outside of the classroom in selling products at the farmers' market taught me how to deal with others. As I want to go into the floral and plant world, these skills will be helpful in my future." Ethan VonHoene noted, "This experience taught me how important it is to work with others around you to help everyone meet a common goal." Kalley Stubbs said, "I learned that sometimes in life we need to be willing to negotiate and be flexible to help our business move and grow." Katelyn Beckstead shared the importance of the experience by saying, "I wouldn't have been able to afford attending the National FFA Convention if it wasn't for the School-Based SAE."

- What opportunities exist in your school to maintain, start, or continue a School-Based Enterprise SAE?
- How could these members take advantage of the local farmers' market to help educate the public about agriculture?
- What skills related to marketing and communications could these members learn?
- What teamwork skills do you think students working in the school garden could have developed?

Top: Spanish Fork FFA Chapter
Bottom: ampol sonthong/Shutterstock.com

Introduction to Food Science

279photo Studio/Shutterstock.com

Learning Outcomes

By the end of this lesson, you should be able to:
- Define food science. (LO 06.01-a)
- Describe the role food science has played in changing the US food system since the middle of the 20th century. (LO 06.01-b)
- Explain how food science relates to food waste, food security, and our ability to feed a growing population. (LO 06.01-c)
- Describe how and why the work of food scientists is important. (LO 06.01-d)

Words to Know

Food and Drug Administration (FDA)

food insecurity

food loss

food safety

food security

food science

food system

food waste

Before You Begin

Hi, there and welcome to the delicious world of food... food SCIENCE, that is! Before we get started, take a few minutes to think about the foods available in the cafeteria at school. What's the first food that comes to mind? I know my brain immediately goes to a bottle of milk and some apple slices and peanut butter. Sounds like a great afternoon snack, right? We're going to talk about snacks, for sure, but in this unit, we will also learn about our nation's food supply and the scientists who apply a broad range of science and technology to help us feed a growing population while minimizing waste. Let's jump in!

Our supply of safe and nutritious food doesn't magically appear each day. Our *food system*, or the combination of people, resources, and processes involved in the production, processing, transportation, and sale of food before it reaches your plate, took centuries to become what it is today. Just as society has changed over time, food and food science has changed with it. *Food science* is the study of the physical, biological, and chemical makeup of food, the causes of food deterioration, and the nature of food processing. Food scientists are responsible for significant innovation to our food supply (**Figure 6.1.1**). Let's take a stroll back in time to better understand how these developments occurred.

LEDOMSTOCK/Shutterstock.com

Figure 6.1.1. Thousands of new products are developed in a laboratory by food scientists each year, but not every product is a success. The discovery of new products and processes in the food industry requires commitment to constant experimentation using the scientific method.

Did You Know?

Food scientists are responsible for many items you eat every day, including microwaveable meals, frozen foods, packaged snacks, canned goods, and milk that stays fresh for an extended amount of time.

Hands-On Activity

Identifying the Work of Food Scientists

In this unit, we'll be learning about some of the principles of food science and the work that food scientists do. While their work affects your life every day, you may not be aware of it. This activity will help to get you more familiar with their work!

For this activity, you will need:

- Paper
- Pen
- A variety of packaged food items, or pictures of food items

Procedure:

1. Your teacher will divide the class into groups of 2 or 3 students. Each group will be provided with a handful of food items or pictures of food items (4 or 5).

2. As a group, look at the items. What evidence do you see that a food scientist worked on or with it? Consider the ingredients and their form and origin, the packaging, how the item is stored, and how long it stays fresh. Make a note of your observations.

3. Share your observations and thoughts with your class.

Consider This

1. Were you able to find evidence of food scientists' work in all your products? Why or why not?

2. What does this tell you about the role that food scientists play in our food system?

3. Did any of the tasks your class identified sound interesting to you? What were they?

The Development of Our Modern Food System

In the years around World War I, food science and methods of preservation were central to feeding our troops and meant the difference between losing and winning the war. Our nation's food system innovated in response to three major needs: the preservation of foods beyond their season, the use of toxic substances or fake material in foods, and building a food system that could support the needs of a growing United States and other countries as well. Innovations included processes such as vacuum packaging, freezing, and hydrogenation that are designed to increase the shelf life of foods. Demand and modernization of our food system during this period gave rise to many of our country's most successful food companies. PepsiCo, Land O'Lakes, the Hershey Company, the Campbell Soup Company, Mount Olive Pickles, Sunsweet, Planters, and Blue Bell Ice Cream are just a few of these companies (**Figure 6.1.2**). Are you familiar with them or their products?

In the years after World War I, through the Great Depression and World War II, food companies responded to the demands of American consumers for more meat, sweet treats, and dairy products. Despite interruptions in shipping goods such as bananas, starches, tapioca, and cocoa and a shortage of fat and sugar during the war, the mass production of food with the use of new, automated technology sped up. Food shortages resulted in innovations, such as reduced-fat foods and the use of applesauce or mayonnaise in many baked goods. The demand to support troops overseas with safe, nutritious food was

Figure 6.1.2. Some of America's favorite food companies got their start during turbulent times in our country. Since then, these products have evolved in numerous ways, including the appearance of the brand through packaging. *Can you find the original packaging of these products in a quick online search?*

monticello/Shutterstock.com

Keith Homan/Shutterstock.com

Grossinger/Shutterstock.com

Keith Homan/Shutterstock.com

Giuliano Del Moretto/Shutterstock.com

responsible for major innovations in food science, such as flour fortified with vitamins and minerals, and sterile processing and packaging. Additionally, increased focus on food waste continued to drive public opinion after World War I and into the decades after (**Figure 6.1.3**).

Consumer Concerns and Shift in Demand

In the years that followed World War II, American consumers began to express concerns over the nutritional value of their food while also expecting new and different options and products of convenience. Competition grew between companies as new products were developed, including one-dish meals and casseroles. Many products also failed during this time, but food scientists were in great demand. While innovation was in warp speed, the ultimate success of products was based less on nutritional value and more on price, convenience, and availability.

Everett Collection/Shutterstock.com

Figure 6.1.3. With challenges facing the US food supply in wartime, reducing waste was considered a patriotic duty. *Do you think the US could adopt a mindset like this again? Why or why not?*

In the 1960s and 1970s, Americans formed concerned consumer groups and began calling for more government supervision of our food supply. The United States Department of Agriculture (USDA) is the government agency responsible for overseeing federal laws related to farming, forestry, rural development, and food. It dates to 1862, when Abraham Lincoln signed legislation creating the department. They provide leadership on rural development, agriculture policy, and science, among other things (**Figure 6.1.4**). The *Food and Drug Administration (FDA)* is a government agency (part

Figure 6.1.4. The United States Department of Agriculture (USDA) is located in Washington, D.C. but has more than 4,500 offices across the country. Nearly 100,000 Americans work for the USDA. *Can you name three areas or issues that USDA employees work to address?*

melissamn/Shutterstock.com

of the Department of Health and Human Services) that regulates food quality and nutrition. It is involved in making changes to product standards, regulating nutritional labelling, and food quality. The FDA is responsible for protecting and promoting public health through the control and supervision of food safety, vaccines, cosmetics, tobacco products, and pharmaceuticals. These agencies existed prior to this increased consumer involvement, but they changed their focus and work in response to the new consumer demand for food that was nutritious and safe, in addition to being convenient.

In the years leading up to 2000, an increasing number of consumers showed interest in fresh fruits and vegetables year-round. This required more imported fruits and vegetables from other countries, where they can be grown in warmer weather. The focus on fresh international foods brought rise to concerns over ***food safety***, or the conditions and practices used to preserve the quality of food to prevent contamination and food-borne illnesses. In response, food scientists developed methods of testing to identify pests, pathogens, or contamination, and developed production and processing methods to ensure American consumers had safer and fresher foods to eat. High-pressure food processing was used by food companies to kill any microorganisms in freshly packaged foods without changing the appearance, flavor, or nutritional value of the food.

The food science industry is also using genetic engineering techniques to solve problems. The late-ripening tomato that was introduced into the market in 1994 addressed concerns about short shelf life, and about the use of chemicals to ripen most tomatoes. The Arctic apple was developed to address the issue of apples browning and becoming unappealing after they are cut. Golden rice was engineered by food scientists to provide Vitamin A to vitamin-deficient populations. Innovation doesn't stop; food scientists are working on problems and ideas today that we'll be writing about in 20, 50, and 100 years!

Making Connections #STEM

Keeping Food Safe from Contamination

Food scientists who are concerned with food contamination use all sorts of different scientific tools to detect issues and keep food safe. They may take samples from food to check for the levels of bacterial growth and test for harmful bacteria. They may look at food temperature logs, particularly for prepared food, or take measurements themselves. Some may be more concerned with chemical contamination, using test strips to look for the presence of chemicals that shouldn't be in food. There are other, more complicated methods that food scientists use in labs to detect contamination.

1. How do you see food science overlapping with other types of science, like chemistry and biology?
2. If you do a quick online search on this topic, you're likely to come across a few scientific and technical terms. Choose one of these terms and learn what it means.
3. What sorts of tools and skills might a food scientist need to do this work?

> If this is interesting to you, then maybe being a **microbiologist** *(like me!)* is a career for you to explore!

What Do They Do? Microbiologists in the dairy industry use their knowledge to grow the cultures and organisms required to make tasty products like yogurt or cheese, to test for potential causes of illness, to preserve products from pathogens, to improve texture and taste, or to improve the health benefits to humans.

What Do I Need to Be Good at This? Microbiologists need to love science! You should have curiosity, ask thoughtful questions, and pay attention to detail. Microbiologists spend plenty of time in the lab, so you should enjoy that, and you need to have strong communication and organization skills.

What Education Do I Need? To get an entry-level job, you'll need a bachelor's degree in microbiology or a related field. To get higher level jobs doing research, you will need a higher-level degree.

How Much Money Can I Make? The average salary for a microbiologist in the US is just over $61,000.

How Can I Get Started? If you are interested in this, it's time to get started in science! Pay attention in your science classes and learn all you can. Try to find a microbiologist to talk to or job shadow for an SAE. Start a research project about the issues that microbiologists in a particular industry or at a particular company are likely to be working on.

Gorodenkoff/Shutterstock

Food Loss and Food Waste

Look around your school lunchroom. How much of the food that is served in a single day do you think will end up in the garbage? The USDA reports that between 30 and 40 percent of our nation's food supply, much of which is nutritious and edible, is lost or wasted every year. There are two main categories of wasted food: food loss and food waste. The larger category of *food loss* includes any edible food that goes uneaten during any stage in the supply chain. This may be crops destroyed in the field because of pests or bad weather, food spoiled during transportation, or overproduction and lack of demand for a certain product. ***Food waste*** refers to all high-quality food that is intended for human consumption but instead is discarded by supermarkets or consumers. Food is wasted for a variety of reasons including overbuying, oversized or damaged packaging, and poor consumer planning. The impact of food waste comes with a significant price tag, costing Americans nearly $218 billion every year while also wasting valuable resources like water and farmland. While the USDA is committed to understanding and addressing food loss at the production level, there are steps we can take each day to reduce food waste at home and school.

Tips to Reduce Food Waste

- Teach other students about food waste and challenge your classmates to reduce waste in the lunchroom.
- Talk with your parents and teachers about creating separate bins at home and school to donate uneaten food or use discarded items for compost.
- Focus on preparing correctly sized portions based on what you are likely to eat.
- Use composted food waste in your garden at home or school.

Vic Hinterlang/Shutterstock.com

Figure 6.1.5. The COVID-19 pandemic left millions of Americans jobless and unable to purchase a steady supply of food. Many food banks, including this one in Austin, Texas, experienced a boom in demand, with cars lined up for miles.

Food Insecurity

It's important to consider how the reduction of food waste could benefit American families. According to the USDA, a family is considered to have ***food security*** when all members of a household always have access to enough food to provide for an active and healthy lifestyle. ***Food insecurity*** is a term that describes the opposite scenario, where a household is uncertain of having or unable to acquire enough food to meet the needs of every family member. In 2019, 90 percent of American families were considered food secure, with 10 percent, or nearly 14 million people, lacking access to sufficient food. Job loss and supply chain issues during the 2020 COVID-19 pandemic had a devastating impact on food security in the United States, with approximately 42 million people experiencing food insecurity (**Figure 6.1.5**). Most of these people had never lacked access to food before.

The Future of Food

Innovation and technology in all aspects of food science, from farming techniques and equipment to processing, packaging, and marketing, are set to transform the future of food. Several factors are driving this progress. Changing weather patterns, combined with limited space, have led many producers indoors to grow crops year-round. Greenhouse and vertical farming structures are providing food to urban communities where growing space is limited while also meeting the increasing food demands of a booming population.

Greenhouses and vertical farming techniques allow farmers to control nearly every aspect of their operation while saving valuable resources like water (**Figure 6.1.6**).

Increasing pressures to reduce the global ecological impact of food packaging may also revolutionize agriculture through science and innovation. Approximately 10 *million* tons of trash end up in our planet's waterways each year, much of which is plastic, glass, or paper. The challenge to protect marine environments has sparked the development of other, more environmentally friendly options for packaging such as "natural plastics" made from starch, sugarcane, and cellulose. These innovations are just the beginning of the work ahead of food scientists in addressing this challenging and important problem.

Lano Lan/Shutterstock.com

Figure 6.1.6. Vertical farming has revolutionized how and where producers can grow food for the world's population. These indoor farms significantly reduce the amount of space needed to grow fruits and vegetables. Because producers can completely control the conditions inside, they can be built almost anywhere!

6

LESSON 6.1 Review and Assessment

What Did We Learn?

- Food science is the study of the physical, biological, and chemical makeup of food, the causes of food deterioration, and the nature of food processing.
- Our food system in the United States has evolved over time, improving the quantity and quality of our food supply through production, processing, transportation, and sales. Innovation in food science has been driven by consumer wants, needs, and concerns throughout history.
- Food science can reduce food loss and food waste and may be one answer to reducing food insecurity in the United States and around the globe.
- Innovations in food science are changing the future of agriculture and meeting the challenges of producing a bountiful food supply, while also reducing the environmental impact of food packaging.

Let's Check and See What We Know

Answer the following questions using the information provided in Lesson 6.1.

1. *True or False?* The United States Food and Drug Administration is responsible for protecting and promoting public health through the control and supervision of food safety.

2. Innovations by food scientists include which of the following items?
   A. Canned vegetables
   B. Packaged brownies
   C. Frozen pizza
   D. All of these.

3. What term refers to the conditions and practices used to preserve the quality of food to prevent contamination and food-borne illnesses?
   A. Biotechnology   C. Food safety
   B. Nutrition   D. Consumer

4. *True or False?* The United States Department of the Interior is primarily responsible for food safety at the federal level.

5. *True or False?* Vitamin-rich grains, like golden rice, were developed by food scientists using biotechnology.

6. Which of the following terms describes good quality food that is uneaten and discarded each year?
   A. Processing   C. Food security
   B. Food waste   D. Producer

7. Vertical farming provides farmers the opportunity to do which of the following?
   A. Change the nutritional value of produce
   B. Grow more food with less land
   C. Control all elements of water and temperature
   D. Both grow more with less land and control all elements of water and temperature

8. The combination of people, resources, and processes involved in the production, processing, transportation, and sale of food before it reaches your plate is also referred to as our:
   A. food process.   C. food science.
   B. food system.   D. economy.

9. Every day in the United States, ____ percent of the population deals with food insecurity.
   A. 2   C. 10
   B. 5   D. 30

10. _____ organized in the 1960s and 1970s to put pressure on the US government to improve food safety and nutrition.
    A. Kids
    B. Pediatricians
    C. Grocery store owners
    D. Consumers

Let's Apply What We Know

1. Why is food science important?
2. Describe how the work of food scientists has changed our food supply in the last 75 years.
3. Create a time line showing the evolution of our modern food system. Start with the dates mentioned in this lesson. Then, ask a question about our food system (e.g., When was the first commercially canned good sold? When were our modern guidelines for healthy eating established? When was (food company) founded?) and go online to find an answer. Add this to your time line. What other questions can you think of to ask?
4. Explain the importance of food science to food security.
5. List areas of innovation in the future of food.

Academic Activities

1. **Statistics.** We're often unaware of the large amount of food that goes to waste due to spoilage, cooking incorrectly, or poor planning. Design a week-long study to really understand food waste in your lunchroom. How would you keep a record of the foods purchased, eaten, and discarded by you and your classmates? Who could help you gather this information? Share your ideas with members of your class and make a plan to conduct your study. What statistics can you find to share with your classmates about food waste at your school? How can you address these issues?

Communicating about Agriculture

1. **Social Media.** With the rise of social media, young people have increased their awareness and influence on a variety of different topics, including food. If you had the opportunity to discuss food with your friends, how might you start the conversation? Consider the best ways to use online platforms for initiating discussions around our food system and create short topical videos to get the conversation started.

SAE for ALL Opportunities

1. **Foundational SAE.** Consumers are critical to the food science industry. Think back to the information in this lesson about consumers organizing to advocate for healthier, more nutritious food. How do consumers wield power now? Imagine your own feed about food safety (or other consumer wants) on social media. How might you share new products with other consumers? How might you raise awareness about food safety issues, or other food-related issues important to other consumers? Design a campaign using social media to advocate for food-related causes important to you and your classmates.

2. **Research SAE.** The demands that consumers have of their food are constantly changing. Organize a research study to find out what issues people in your community are concerned about when it comes to food. Write and administer a survey to a large group of your neighbors. Review and summarize the findings for your class, presenting the data with visuals to help other students understand these issues.

SAE for ALL Check-In

- How much time have you spent on your SAE this week?
- Have you logged your SAE hours?
- What challenges are you having with your SAE?
- How can your instructor help you?
- Do you have the equipment you need?

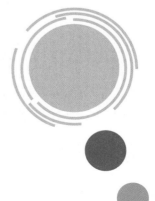

Toidi/Shutterstock.com

Food Safety

Learning Outcomes

By the end of this lesson, you should be able to:
- Describe the risks that unsafe food handling practices cause to our food supply. (LO 06.02-a)
- Summarize the conditions for harmful bacterial growth. (LO 06.02-b)
- Explain how the Four Cs of Food Safety control bacterial growth. (LO 06.02-c)
- Describe food safety practices that support a safe food supply on the farm, in transit, and in retail establishments. (LO 06.02-d)
- List strategies for safely handling food at home. (LO 06.02-e)

Words to Know

antibiotics	foodborne illness	United States Centers for Disease Control (CDC)
cross-contamination	food safety	
E. coli	salmonella	

Before You Begin

Did you know that each year, about one-sixth of our nation's population experiences mild to severe illnesses caused by contaminated food — and more than 3,000 people die from them? Look around your classroom. How many students would be affected if one-sixth of the class got sick? Maybe you or someone in your family experienced this yourself as something we commonly call food poisoning. It isn't fun, is it? Can you think of ways we might prevent this from happening at home or school? Take a moment and write down three ways that you can think of to prevent getting sick this way. Food safety is everyone's job, so let's look into the ways we can keep our food safe at every stop on its journey from the farm to our fork.

We are fortunate to live in a country with one of the safest food supplies in the world, but we can always look for ways to improve practices and processes to prevent foodborne illnesses. A *foodborne illness* is any illness that is caused by consuming a food that was contaminated with a disease-causing agent such as bacteria, fungi, or another pathogen. Food science and food safety work together to control these disease-causing agents and keep our nation's food supply safe at every point in the process (**Figure 6.2.1**). Specifically, the term *food safety* refers to the conditions and practices used to preserve the quality of food to prevent contamination and foodborne illnesses. In this lesson, we will discuss steps we can take to prevent the spread of foodborne illnesses by understanding common ways this occurs at the source, during transportation, at retail stores and restaurants, and even in our homes.

Festa/Shutterstock.com

Figure 6.2.1. While not all bacteria are harmful, some are. Bacteria can quickly multiply and spread when food safety precautions are not taken when handling food, and this can make people really sick!

The Risks to Our Food Supply

Every day, members of the agriculture industry in every sector work to develop, implement, and improve food safety plans and procedures to ensure safe food products make it into the mouths of nearly 330 million Americans and others around the globe. As consumers, we also hold a great deal of responsibility in this process and can implement safe practices by understanding the fundamental concepts of food preparation and safety measures when storing, preparing, and eating food at home. Foodborne illnesses and outbreaks are likely when an error is made at some point in the chain, but many can be prevented with strict safety plans and attention to detail.

Although the food industry and government agencies such as the *United States Centers for Disease Control (CDC)* work around the clock to prevent outbreaks, many risks make food safety an issue of growing importance. The global nature of our food supply, a decrease in the number of meals we eat at home, and a rise in bacteria that are resistant to antibiotics are all likely to lead to an increased need to be aware of good food safety practices.

Meals Prepared Away from Home

Today, nearly 50 percent of the money we spend on food goes toward buying food that others prepare, such as takeout and restaurant meals. Plus, a growing number of Americans eat meals prepared and served in hospitals, schools, nursing homes, concerts and sporting events, and daycare and senior centers. Have you eaten meals out recently? Where at?

Eating more meals away from our homes means that we as consumers are trusting others to keep that food safe. If we handle it at home, we know that we can follow safe practices. We may not know this about the senior center where your grandparents live, or the movie theater where you grabbed a snack. Unsafe procedures or errors in food handling and preparation in public places can lead to large outbreaks of illness.

Did You Know?

In 2021, there were multiple outbreaks of foodborne illnesses across the United States. Outbreaks of listeria, salmonella, nonviral hepatitis, and E. coli were linked, at various points, to bagged salads, raw onions, soft cheese, unpasteurized yogurt, bottled water, ground turkey, frozen chicken, frozen shrimp, and uncured salami. Did you hear about any of these outbreaks?

6

Food from around the Globe

Food in your local grocery store comes from all over the world, which may introduce microorganisms that are uncommon in the US. While many of these may be harmless, there may be others that can cause illness. This presents a whole new set of modern food safety challenges. In **Lesson 6.1**, we talked about how food scientists test food coming into the country for potential pathogens, and raise alarms when contamination is detected. If a shipment of something contaminated made it to your grocery store, it probably made it to other stores also, and the risk of a larger outbreak is very real.

Resistant Bacteria

In 1950, scientists knew of five foodborne pathogens. By 2000, there were at least 25 known foodborne pathogens, including 20 newly discovered ones. While scientific advances are undoubtedly part of this growing knowledge, it is also true that bacteria are evolving in ways that can pose a threat to human health and the food supply.

Bacterial infections are treated with **antibiotics**, or drugs that are designed to kill a specific strain of bacteria. Antibiotics have been crucial to controlling many harmful diseases and painful childhood maladies, from ear infections to pneumonia. However, as antibiotics have become more commonly used to address bacterial infections in people, animals, or food, we have seen the development of bacteria that show resistance to these drugs, making them less effective as treatments.

From the standpoint of food safety, disease-causing bacteria that are resistant to antibiotic treatment are a real concern. If a pathogen can't be contained with the tools we have in place, it could cause a major outbreak of a foodborne illness (**Figure 6.2.2**).

Giovanni Cancemi/Shutterstock.com

Figure 6.2.2. Keeping our food safe is a team effort that involves detailed safety plans from the farms where foods are produced, proper transportation and handling at grocery stores and restaurants, and taking necessary precautions once the food arrives in your home.

Conditions for Harmful Bacterial Growth

As we start to talk about food safety, it is also important to understand how bacteria grow. When do they thrive? What can we do to control or stop their growth in our food?

Time and Temperature

Under certain conditions, some bacteria can double their numbers within minutes and form toxins that cause illness within hours. To minimize bacterial growth in food, it is important to keep food temperatures below 40° F (4° C) or above 140° F (60° C). The level in between this temperature range is known as the *Danger Zone*. It is important to minimize the amount of time that food spends in the Danger Zone.

Nutrients

Bacteria need many of the same nutrients as humans to thrive (glucose, amino acids, and some vitamins and minerals). This means that bacteria grow rapidly in high-protein foods such as meat, eggs, dairy, and seafood.

pH

Most microorganisms thrive in a relatively neutral pH range. Acidic foods like vinegar and citrus juices are not favorable foods for pathogenic bacteria to grow because the environment is too harsh for them. It is important to note, though, that even though they may not grow quickly in these environments, they may survive.

Moisture

Most bacteria thrive in moist environments; they don't grow on dry foods. Dry foods like cereals or grains can safely sit out at room temperature.

Hands-On Activity

How Clean Are Your Surfaces?

It's important to understand when, how, and where bacteria grow as we work to keep our food supply safe. You may be aware that bacteria grow on all kinds of surfaces, but do you have any sense of just how many bacteria are in the world around you? What surfaces are relatively clean or dirty? This activity will help to answer those questions!

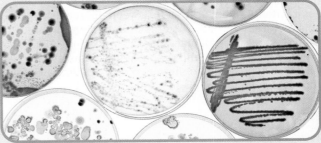

luchschenF/Shutterstock.com

For this activity, you will need:

- Petri dishes (6–8 per group)
- Cotton swabs (6–8 per group)
- Agar
- Microwave-safe bowl (and access to a microwave)
- Masking tape or painter's tape
- Marker or pen
- Ziploc bags

Procedure:

1. Your teacher will divide the class into groups of 3–4 students. Each group will be provided with petri dishes and cotton swabs.
2. The agar (growth environment for bacteria) will need to be prepared. Work with your teacher to microwave the agar to the proper temperature, and then place warm agar in each petri dish until it is about half full. Let these cool for an hour.
3. With your group, decide what locations you want to test for bacteria. Take a clean cotton swab and run it along the surface you want to test to collect a good sample.

4. Carefully, take the used swab and draw a small pattern in the agar of a petri dish. Throw the swab away and cover the dish. You don't want to get more than one sample in each dish!
5. Using a piece of tape, label the location of the bacteria on the petri dish.
6. Repeat steps 3, 4, and 5 until all petri dishes have samples.
7. Place the petri dishes in Ziploc bags, seal them shut, and set them somewhere warm and dark to grow.
8. After some time, check on the dishes. What do you see? Be sure not to open the plastic bags – we want to keep all bacteria inside for proper disposal.

Consider This

1. Did bacteria grow in your dishes? How did the bacteria across dishes look different from each other?
2. What does this tell you about the how bacteria grow, and how clean the surfaces are in your school?
3. Think about your house. What areas do you think have the most bacteria? What could you do about that?

How the Four Cs of Food Safety Control Bacteria

Bacteria are everywhere and can spread between surfaces, people, and food. Not all bacteria are harmful; in fact, most bacteria are beneficial to us. We pay attention to the spread of bacteria mostly to control the spread of those bacteria that can be harmful. This spread can be controlled by practicing the *Four Cs of Food Safety*.

- **Cleaning** removes bacteria from hands and surfaces. To prevent the spread of harmful bacteria, proper cleaning of both surfaces and hands is especially important.
- **Cooking** kills bacteria by breaking down their cell walls and destroying enzymes, which they need to survive.
- **Chilling** slows down the bacteria's metabolism, thus slowing their growth.
- **Combating *cross-contamination*** (separating foods) prevents bacteria from spreading from one food item to another, or between foods and hands, surfaces, or utensils (**Figure 6.2.3**).

kathrinerajalingam/Shutterstock.com

Figure 6.2.3. By taking simple precautions to prevent cross-contamination, we can help keep ourselves safe. In this picture, you'll see the cross-contamination of raw chicken and vegetables. This is a major no! Always be sure to clean, chop, and prepare raw meats and vegetables separately on their own clean cutting boards, using a clean knife and clean hands.

Keeping the Food Supply Safe, from Farm to Fork

Now, let's look at what people can do at each stage of the food system to prevent foodborne illness.

Harmful Bacteria at the Source

There are many places on a farm that can be contaminated by harmful bacteria, so farmers must make sure the areas where food is handled are kept clean and at the right temperature. Many innovations on the farm help prevent the growth of bacteria, like special areas for washing vegetables, refrigerated storage areas for milk and eggs, and portable sanitation in fields.

Salmonella is a harmful foodborne bacterium sometimes found in the intestines of chickens. It can be passed on in chicken meat, and also inside the chicken's eggs. The best way to reduce the risk of foodborne illness from eating contaminated chicken is to prevent salmonella from living in the animal in the first place. Scientists have developed mixtures of beneficial bacteria to prevent bad bacteria like salmonella from infecting the chickens.

Another way farmers prevent the spread of harmful bacteria is through composting. Compost is made up of the decomposed parts of all the residuals that come from the farm operation: the waste from the animals, leftover food the animals didn't eat, hay/straw, etc. It gets mixed together and piled up so that bacteria can eat it, break it down, and create compost, which farmers use to fertilize their crops. The bacteria get a workout from eating all the organic materials. As they digest the wastes in the compost, the temperature of the compost rises. The heat plays an important role in bacteria management because many harmful bacteria can't survive in temperatures above 131° F (55° C).

Compost isn't a perfect cure-all, though. **E. coli**, a harmful bacterium commonly found in the intestines of humans and other animals that can cause severe illness, may be found in the manure that is used in compost. If the compost doesn't get hot enough, some of these bacteria may sneak through. Farmers need to be careful about cross-contamination when compost is used on crops, particularly low-growing crops, such as lettuce and strawberries. Scientists are working to develop ways for farmers to ensure that their compost reaches temperatures high enough to kill pathogens and make the compost safe for their crops.

Food Safety During Transportation

The Four Cs of Food Safety play a very important role during the transportation of foods from the farm. Keeping food safe and in good condition as it's shipped across the country or around the world is critical. There are many steps to shipping food safely, and science behind each step. Cold temperatures must be maintained throughout the loading process, in transit, and during receiving and unloading. The food is cleaned and precooled as it comes from the field or plant. The cooling extends product life by reducing field heat, rate of ripening, loss of moisture, rate of respiration, and the spread of decay. As you probably recall, cooler temperatures also slow the rate of bacterial growth.

Each product is placed in packaging that will prevent damage as it travels, whether that is a box, a crate with protective layers, or plastic wrap. The shipping container that the food will travel in is cleaned and properly loaded, making sure that the boxes are stacked tightly to lock in the cold during transit. Proper temperature control can be tracked by satellites. Refrigerated containers usually have equipment that automatically records refrigeration system functions and the air temperature inside the container. This information provides a detailed record of refrigeration system performance through the trip (**Figure 6.2.4**). Food is then properly stored and cooled at the warehouse. Food may move through multiple trucks or warehouses on its way to a retail location.

Figure 6.2.4. This truck isn't just any truck: the trailer is specially insulated and temperature-controlled to get the perishable food safely to its destination.

pfluegler-photo/Shutterstock.com

Figure 6.2.5. Proper safety precautions taken at restaurants help prevent the spread of harmful bacteria. If you're visiting a new restaurant, pay special attention to the food safety rating posted by the door and that all workers are following food safety procedures.

Food Safety in Retail

Receiving areas or unloading docks of super-markets or grocery stores are maintained at cold temperatures of 41° F (5° C) or below to maintain the cold chain that started all the way back in the field. Storage areas and display cases are kept clean and cool, with temperature monitoring in place to prevent bacterial growth. Food preparation areas within grocery stores are kept clean and are set up to avoid cross-contamination. Red meats, fish, and poultry will never be mixed together, or mixed with fruits or vegetables.

In any restaurant or food service establishment, the Four Cs are critical. Humans are one of the biggest sources of food contamination in restaurants. Proper handwashing is critical to keep food safe. Contamination can occur when someone doesn't wash his or her hands and then prepares or serves food (**Figure 6.2.5**). Sometimes, the Four Cs can be assisted by technology. For example, in fast food restaurants, employees use an engineer-developed two-sided "clam shell" grill that has a temperature sensor to prevent human error. It cooks burgers on both sides simultaneously, using a sensor to ensure that all the burgers reach a safe internal temperature.

The Four Cs at Home

Even with great technology, food can still become contaminated, so it's important for you to always practice the Four Cs of Food Safety at home. Once you purchase food and take it home, the responsibility for food safety is literally in your hands! Here are a few best practices to protect yourself, and your family, from foodborne illness:

Clean: Wash your hands and disinfect surfaces often. When you wash your hands, wash with warm, soapy water for 20 seconds at a time (you can sing the ABCs to time it out). Wash cutting boards, dishes, utensils, and surfaces with hot, soapy water before and after food preparation. Make sure that the water is hot enough to kill bacteria! Also, be sure to wash your raw fruits and vegetables thoroughly before consuming.

Cook: Cook foods to proper temperatures. If your food doesn't get hot enough, you can't know that you've eliminated the risk from bacteria. Use a food thermometer to ensure that foods reach a safe internal temperature. This is most applicable to cooking meat, where bacteria can grow quickly and are more likely to have a foothold, but it is also important to pay special attention to egg dishes and leftovers. **Figure 6.2.6** shows the proper internal temperature for common foods to reach to prevent foodborne illness.

Chill: Refrigerate or freeze your food promptly after you purchase it. It isn't a good idea to go to the grocery store to buy your food, and then drive around to run a bunch of other errands or go to soccer practice, even if you don't have ice cream or something else that will melt. Refrigerate or freeze foods quickly; cold temperatures keep harmful bacteria from growing and multiplying. Follow the Two-Hour Rule: Refrigerate or freeze perishables, prepared foods, and leftovers within two hours. It is also important to make sure your refrigerator and freezer are cold enough to prevent bacterial growth.

Safe Minimum Internal Temperatures

Food	Type	Internal Temperature (°F)
Ground meat and meat mixtures	Beef, pork, veal, lamb	160
	Turkey, chicken	165
Fresh beef, veal, lamb	Steaks, roasts, chops **Rest time: 3 minutes**	145
Poultry	All Poultry (breasts, whole bird, legs, thighs, wings, ground poultry, giblets, and stuffing)	165
Pork and ham	Fresh pork, including fresh ham **Rest time: 3 minutes**	145
	Precooked ham (to reheat) **Note:** Reheat cooked hams packaged in USDA-inspected plants to 140°F	165
Eggs and egg dishes	Eggs	Cook until yolk and white are firm
	Egg dishes (such as frittata, quiche)	160
Leftovers and casseroles	Leftovers and casseroles	165
Seafood	Fish with fins	145 or cook until flesh is opaque and separates easily with a fork
	Shrimp, lobster, crab, and scallops	Cook until flesh is pearly or white, and opaque
	Clams, oysters, mussels	Cook until shells open during cooking

Foodsafety.gov

Figure 6.2.6. Cooking food properly can help prevent some pretty nasty illnesses from coming your way. *Does anything on this chart surprise you?*

Your refrigerator should have an internal temperature of 40 degrees F or less, in order to stay out of the Danger Zone. Your freezer should have a temperature of 0 degrees.

Combat Cross-Contamination (Separate Foods): Keep raw meats, poultry, eggs, and seafood — and any of their juices— away from other foods in your shopping cart, shopping bag, on kitchen counters, and in your refrigerator. It is often a good idea to store these foods lower in your refrigerator, where they can't drip onto other foods. When you are preparing foods, keep raw meats, poultry, eggs, and seafood separate from other foods. Cut them on different cutting boards, use different knives, store them in different bowls. Wash your hands thoroughly and often when handling raw meat, and between handling meat or eggs and other food or utensils. Your goal is to keep any bacteria related to your meat or eggs with the meat or eggs only.

The safe handling of food in your home is your responsibility! Make sure you take it seriously, and your home can be a safe and pleasant place to eat (**Figure 6.2.7**).

Prostock-studio/Shutterstock.com

Figure 6.2.7. Although we have the safest food supply in the world, it is our responsibility to do everything we can to ensure our food is safe once we get home. Make the Four Cs of Food Safety a common practice in your home and help keep your family safe!

What Did We Learn?

- A foodborne illness is any illness that develops from consuming a food that was contaminated with a disease-causing bacteria, fungus, or other harmful microorganism.
- Conditions that foster the growth of harmful bacteria include extended time in warm temperatures, nutrient-rich foods, neutral pH, and plenty of moisture.
- The Four Cs of Food Safety (cleaning, cooking, chilling, and combating cross-contamination) are critical at every stage in the food chain from the farm, to transit, to retail of food products, and at home.
- Food science and food safety work hand in hand to control these disease-causing agents and keep our nation's food supply safe at every point in the process.
- Healthy food safety practices include controlling temperature throughout the process, looking to eliminate bacteria in animal production with helpful bacteria or compost, and keeping different types of food separate.

Let's Check and See What We Know

Answer the following questions using the information provided in Lesson 6.2.

1. *True or False?* Some strains of E. coli are associated with foodborne illness.
2. What is the best way to reduce the risk of foodborne illness from chicken?
   A. Prevent salmonella from living in the animal
   B. Move animals between housing structures
   C. Improve animal nutrition
   D. Compost
3. Perishables, prepared foods, and leftovers should be refrigerated or frozen within how many hours?
   A. 7   C. 24
   B. 12   D. 2
4. *True or False?* The United States Centers for Disease Control is responsible for tracking outbreaks of foodborne illnesses.
5. *True or False?* Foods kept between the temperatures of 40° F and 140° F are outside the Danger Zone.

6. The Four Cs of Food Safety include all of the following except:
   A. Cooking   C. Covering
   B. Cleaning   D. Chilling
7. The most important thing you can do at home to protect yourself from foodborne illness is to:
   A. Wash your hands and clean your work surfaces.
   B. Leave the refrigerator open to cool the room down.
   C. Heat all foods in the oven.
   D. Keep the bathroom door closed.
8. All of these items have been linked to foodborne illness outbreaks in the US recently, except for:
   A. Bagged salads
   B. Raw onions
   C. Steaks
   D. Soft cheese

9. Which factors promote bacterial growth?
   A. Cold temperatures, acidic pH
   B. Extended time in the Danger Zone, moisture-rich environments
   C. Warm temperatures, dry conditions
   D. Plenty of nutrition, acidic pH

10. Chicken and other poultry need to be cooked to a temperature of _____ degrees F to be considered safe to eat.
   A. 180   C. 150
   B. 120   D. 165

Let's Apply What We Know

1. Why is it important to safely handle food products?
2. What is a foodborne illness and how does it spread?
3. Identify bacteria that commonly cause foodborne illnesses.
4. Describe the importance of food safety to our nation's food supply.

Academic Activities

1. **Science.** Keeping our food safe from contamination can be simple if we create a plan based on the foods we consume. Take an inventory of the lunches you eat each week, listing each individual item. Conduct research on each of the items and common foodborne illnesses associated with them. Once you have a good idea of the common issues, think about how the lunchroom at your school needs to follow the Four Cs of Food Safety. Create a plan to handle, prepare, and store each food safely. Once you've created a clear plan, see if you can arrange for a tour of your school kitchen to see how many of your recommendations are already in place!

Communicating about Agriculture

1. **Communicating Information.** Improper handling and sanitation can lead to the spread of foodborne illnesses. Create a public service announcement (PSA) for your school's announcements that discusses the importance of food safety. The PSA should address common food mishaps and include specific examples such as improper methods of thawing meats or the length of time foods can be left out of the refrigerator at room temperature.

SAE for ALL Opportunities

1. **Foundational SAE.** Community gardens are popular in rural, urban, and suburban communities and produce fresh herbs, fruits, and vegetables for people to share during warmer months. Helping others understand the importance of properly handling and storing these foods is important. See if you have a community garden near your home, and create a flyer that describes the Four Cs and other food safety procedures to help educate the garden's users.

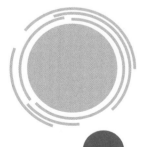

SAE for ALL Check-In

- How much time have you spent on your SAE this week?
- Have you logged your SAE hours?
- What challenges are you having with your SAE?
- How can your instructor help you?
- Do you have the equipment you need?

casanisa/Shutterstock.com

Principles of Food Preservation

Learning Outcomes

By the end of this lesson, you should be able to:
- Describe how preserving food products protects our food supply. (LO 06.03-a)
- Identify factors that contribute to the deterioration of food. (LO 06.03-b)
- Summarize basic food preservation methods and techniques. (LO 06.03-c)

Words to Know

additives	fermentation	freeze drying
canning	food deterioration	shelf life
dehydration	food preservation	sterilization

Before You Begin

*Welcome back! Today's lesson will build on our understanding of food science and technology by expanding our knowledge past food production and safety to processing and preserving food. The abundant food supply in the United States exists despite the fact that fresh foods like fruits, vegetables, and meat require special methods to extend their freshness beyond a few days (**Figure 6.3.1**). You may have helped your family with one method of food preservation before when freezing foods to eat later. Can you think of any other methods we might use to help extend the life of our food?*

Extarz/Shutterstock.com

Figure 6.3.1. A trip to the grocery store is a convenient weekly task for most American families, but access to large quantities of food is not a reality for thousands of people around the world. *How can we help extend access to a safe and abundant food supply to other countries?*

H ave you ever considered the large number of food items available for purchase at your local grocery store? If not, try to count the number of potato chip varieties available next time you round the corner of Aisle Three looking for a salty snack. The quantity and variety of food options available in the United States exist in part because of advancements in food science and technology.

A quick look at food packaging provides insight into one of the most notable innovations in food science, with ***shelf life***, or the amount of time food stays good, getting longer for many food items. ***Food deterioration*** is the process of food breaking down to the point where it is no longer healthy or safe to eat. Food scientists have pinpointed the causes of this deterioration and can now ensure food products last longer before going bad. Why might that be important? Why would we want to extend the length of time an item is good to eat? And how would we go about it?

Why Do We Preserve Foods?

We are lucky to have a fairly predictable food supply in the United States, with most foods available throughout the year. While our consumption of food is constant, we have to consider the time needed for plants and animals to grow and be processed through the food system in maintaining a steady supply. Many foods are produced and harvested at specific times of the year, based on location and climate in that region of the country. For example, oranges are grown in several states, including Florida, where the heat and humidity help to cultivate ripe and juicy fruit for us to enjoy. Navel oranges are usually harvested between November and June, making it difficult for us to enjoy freshly squeezed navel orange juice in October, unless that fruit is imported from a warmer part of the world. In fact, it is because of innovations like the use of preservatives that we can easily find a gallon of orange juice on our grocery store shelves most every day (**Figure 6.3.2**)! Preserving foods helps to keep the supply consistent with consumer demand, during all seasons, and in all conditions.

Niloo/Shutterstock.com

Figure 6.3.2. Do you ever get overwhelmed by the number of options available for a single item? One of the many advantages of a large food supply is variety, but what should you consider before making a purchase? For many people, the decision comes down to cost, but you might also take into account the nutritional information provided on the label. *What other factors influence our purchasing decisions?*

Food Deterioration and Preservation

It is also important to consider factors that contribute to the deterioration or spoilage of food. Food goes bad and becomes inedible when interactions occur that change the chemical, biological, or physical makeup of the food's structure. Chemical interactions may include enzymes that break down carbohydrates, proteins, and fats. Biological changes involve the interaction of bacteria, yeast, and mold with a food item. These changes quickly change the taste, smell, texture, and appearance of foods and make them less desirable to eat. As you may remember from the previous lesson, the appearance of bacteria or mold in foods can do more than make foods less desirable: It can make them dangerous to eat. Harmful microorganisms can thrive in spoiled food!

A wide variety of bacteria cause food to deteriorate, but all have similar requirements for survival. Bacteria thrive in environments that provide water, food, oxygen, and warmth, making many foods perfect candidates for growth. Techniques used to preserve food limit the ability of bacteria to grow by altering elements of the environment such as temperature, access to oxygen, and pH.

Food preservation, or the science of maintaining the form, health, and nutrition of our edible food, is as old as humankind, with some of the earliest evidence of its occurrence found in Middle Eastern and Asian cultures in approximately 12,000 B.C., where early civilizations learned to sun dry certain foods to be eaten later in the year. History is also peppered with stories of epic battles between nations that were won or lost based on the ability to feed armies of soldiers. Unfortunately, the downfall of many can be found in the inability to preserve foods for long periods of time. The discoveries of salt, sugar, and spices significantly advanced food preservation techniques used around the world. In addition to salting and canning, American colonists and pioneers began to use the cold and cool temperatures of winter and early spring to preserve meats, butter, milk, and eggs.

Modern conveniences like refrigerators were not widely available in our homes until the early 20th century, but refrigerated train cars revolutionized the food industry with widespread transportation beginning in the 1880s (**Figure 6.3.3**). In the years since refrigeration technology changed our ability to extend the shelf life of perishable foods, science and innovation have revolutionized the food industry. Now, new advancements will be necessary to feed a growing population in the decades ahead.

KAMONRAT/Shutterstock.com

Figure 6.3.3. Refrigerated train cars and containers completely transformed our nation's food supply by creating a method of transporting perishable foods like milk and meat to every corner of the United States. Today, approximately 70 percent of food is transported by truck, 17 percent by train, 8 percent by ship, and 5 percent by air. Here, an engineer checks refrigerated containers before they are moved to truck or train.

Methods of Food Preservation

Multiple approaches can be used to preserve fruit, meat, and vegetables, with different factors to consider before a method is chosen. You might look at: 1) the availability of preservation equipment needed for a particular method; 2) the best method to increase the shelf life of a specific food product; and 3) the best method to preserve and/or enhance the nutritional value, flavor, and texture of the food. Additionally, it's important to consider methods that may destroy any harmful microorganisms or bacteria. Let's outline the specifics of popular techniques below.

Heat

P A/Shutterstock.com

Figure 6.3.4. We don't often think of sterilization when opening a can of corn or jar of spaghetti sauce, but the process of heating and storing our food is essential to ensuring food is safe and long-lasting.

Because most harmful bacteria and microbes are destroyed at temperatures greater than 180 degrees, heat is commonly used to ensure food is safe through sterilization and canning. The process of *sterilization* involves heating food to a temperature of approximately 240 degrees for at least 15 minutes. *Canning* uses heat and pressure to raise the temperature of foods above 212 degrees and provide an airtight seal around the lids of cans or glass jars. This method was introduced in the early 1800s and continues to be used in the food industry to preserve canned goods, but individuals also use this method to preserve many fruits and vegetables for consumption year-round (**Figure 6.3.4**).

Cold

On the opposite end of the spectrum, cold temperatures are used to slow the growth of microbes and bacteria through refrigeration and freezing. By reducing the temperature of foods to between 33- and 40-degrees Fahrenheit (0 to 4 degrees Celsius), we can significantly extend the shelf life of those items. Freezing requires the temperature be lowered to zero degrees or below to stop any harmful growth (**Figure 6.3.5**).

New Africa/Shutterstock.com

Figure 6.3.5. Freezing our food is one easy way to ensure we have quick access to fresh meals. *What steps should you take to ensure frozen foods stay fresh and flavorful for an extended period of time?*

Did You Know? In a peer-reviewed study published in 2017, researchers found that frozen foods are six times less likely to go to waste, versus fresh food. Why might this be? What does this tell you about the effects of food preservation on food waste?

Making Connections #STEM

Food Preservation in History

You may remember reading in history class about European settlers sailing around the world with rations of salt pork, tongue, hardtack, oatmeal, and pickled vegetables. If you are growing up in the West, you may be familiar with pemmican, a mix of dried meat and fruit that Native Americans ate. If you enjoy stories of explorers and pirates in centuries past, you may recall hearing about a strange disease called scurvy, which resulted from a lack of Vitamin C.

Consider This

1. Using your knowledge of history or reading you have done (for school or fun), can you name other examples of preserved foods, or issues related to the inability to preserve food?
2. Looking at these examples, what role did food preservation play in the success of global explorers? Would their journeys, innovations, and discoveries have been possible without food preservation?

jgorzynik/Shutterstock.com

Drying

We can also use drying techniques to preserve foods by removing moisture in two ways: dehydration and freeze-drying. While each of these techniques limits the growth of harmful microbes and bacteria, **dehydration** uses hot air to preserve food, while **freeze drying** involves first freezing food and then removing the moisture under pressure. Both processes help preserve foods, but freeze drying tends to be better at maintaining flavor (**Figure 6.3.6**).

Fermentation

Sometimes referred to as pickling, the process of **fermentation** preserves food and slows the growth of harmful bacteria using helpful microorganisms or bacteria. This process creates unique flavors through the conversion of carbohydrates to alcohol or acids (**Figure 6.3.7**).

Additives

Additives, or chemicals added to foods to inhibit the growth of microbes, are also used to preserve foods. Salts and sugars create high levels of pressure when added to food, which limits microbial growth. Other additives, such as smoke or spices, slow the growth of harmful microbes. Additives may also be included to enhance flavors. Some popular foods that use additives to preserve food include bacon, jam, wine, and packaged cereal.

Vacuum Packaging

Some foods are placed in plastic packages, and then a machine pulls out the air in the packaging to create a seal. Think about a package of beef jerky or dried fruit, or a Thanksgiving turkey. Vacuum packaging works as a method of preservation for some types of food because bacteria and molds that might work to spoil your food need oxygen to live and multiply. Without air in the package, bacterial growth is significantly slowed.

PremiumVector/Shutterstock.com

Figure 6.3.6. NASA astronauts have relied on dehydration techniques for preserving space-ready food for years. Innovations in dehydration mean that astronaut favorites like shrimp cocktail, ice cream sandwiches, and butterscotch pudding are now available to enjoy in space. *What food would you want if you were on a long mission to space? Is it available?*

Marian Weyo/Shutterstock.com

Figure 6.3.7. The fermentation process is widely used in food preservation and responsible for the development of many new flavors. From cucumbers to cabbage, okra, and carrots, almost any fruit and vegetable can be pickled and preserved.

Hands-On Activity

Does Preservation Work?

As you learned previously, it's important to understand when, how, and where bacteria grow as we work to keep our food supply safe. Food preservation works to keep food from spoiling or growing harmful microorganisms by addressing the conditions in which bacteria grow. Does it work, though? This activity will help you see the answer for yourself!

For this activity, you will need:

- Banana (two per group)
- Three Ziploc bags
- Jar
- Vinegar
- Salt
- Marker or pen
- Masking tape or painter's tape
- Food dehydrator (to be shared in the classroom, if one is available)

Procedure:

1. Your teacher will divide the class into groups of 3 or 4 students. Each group will be provided with supplies.
2. Cut the bananas into five total sections. Make notes of any differences between sections—if one section is already particularly brown or bruised, for instance, it may throw off your results! If it is possible to remove and not use spoiled or bruised sections, that would be ideal.
3. Label all of your Ziploc bags with your group's name(s). Using tape, label your jar also.

4. With your group, you will be trying to show how different methods of preservation work. One piece of banana will be placed in an unsealed Ziploc bag and placed on a counter or desk at room temperature. This will be your control, or the basis against which you can compare the other bananas in your experiment.
5. One piece of banana will be placed in cold. Put another piece of banana in a second Ziploc bag, and then put it in a refrigerator or freezer.
6. One piece of banana will be dried, if your classroom has access to a food dehydrator. If you do, run a slice or two of your banana through the dehydrator with your instructor's help. If this isn't available, skip this step.
7. One piece of banana will be placed in a jar full of vinegar. Make sure there is enough vinegar to cover the banana.
8. One piece of banana will be rolled in salt, and then packed into a third Ziploc bag.
9. When you get to class over the next day or two, check on the bananas. What do you see?

Consider This

1. Did you see differences between the bananas over two or three days?
2. What banana was in the best shape? Why do you think this was?
3. Would you want to eat these bananas? Why or why not? What does this tell you about food preservation methods?

6

What Did We Learn?

- Food preservation is important to extending the shelf life of meats, fruits, and vegetables, and to providing a steady supply of food.
- Over time, chemical, physical, and biological changes to food break it down and make it inedible. Food preservation works to prevent this from happening, or slow down the rates of deterioration, by addressing the temperature, available oxygen, or pH of the food.
- Preservation techniques include the use of heat, cold, drying, fermentation, additives, and vacuum packaging.

Let's Check and See What We Know

Answer the following questions using the information provided in Lesson 6.3.

1. *True or False?* Food preservation was originally developed in the 20th century following with the invention of the refrigerated train car.
2. Removing all the oxygen from a food package is called:
   A. Sterilization
   B. Freeze drying
   C. Heating
   D. Vacuum packaging
3. *True or False?* Most harmful bacteria and microbes are destroyed at temperatures greater than 180 degrees.

4. *True or False?* Additives such as smoke and spices are used to speed up the growth of microbes.
5. Which process removes all moisture from a food product?
   A. Fermentation
   B. Irradiation
   C. Freezing
   D. Dehydration

Let's Apply What We Know

1. Why are foods preserved?
2. Describe the methods used to preserve food products.
3. How is heat used to aid in food preservation?
4. List specific causes of food deterioration.

Academic Activities

1. **Science.** Have you ever noticed that fruits and vegetables seem to spoil faster than you can eat them? Set up an experiment in your classroom with two apples, placing one in a refrigerator and leaving the other on a counter. Which do you believe will spoil first? Record the physical changes you see occurring to the apple each day and consider other alterations to the experiment such as using different fruits and vegetables, storage containers, and changing temperatures.

Communicating about Agriculture

1. **Marketing.** Have you ever seen spoiled or rotten images in an advertisement or commercial selling food? Of course not. Food companies and restaurants spend millions of dollars each year in communications and marketing to visually entice us to buy their products. Consider elements of marketing next time to see a grocery flyer in the newspaper or a restaurant commercial on television. Be sure to pay attention to how the food is presented, whether as a fresh, cooked, or packaged item. Make a list of best practices that a marketing campaign might use in presenting food.

SAE for ALL Opportunities

1. **Entrepreneurship SAE.** Canning is widely used around the United States to put up vegetables after harvest each year. This might include tomatoes, green beans, peas, and cucumbers, among others. Consider asking an adult to help you can fresh fruits and vegetables from the farmers' market, grocery store, or your own garden. You may be able to give these canned items as gifts or start selling your canned food products.

SAE for ALL Check-In

- How much time have you spent on your SAE this week?
- Have you logged your SAE hours?
- What challenges are you having with your SAE?
- How can your instructor help you?
- Do you have the equipment you need?

Decoding Food Labels

Song_about_summer/Shutterstock.com

Learning Outcomes

By the end of this lesson, you should be able to:
- Describe why nutrition labels are important. (LO 06.04-a)
- Name trusted sources for nutrition information. (LO 06.04-b)
- Explain nutrition information located on a food label. (LO 06.04-c)
- Interpret other labels found on food packaging. (LO 06.04-d)

Words to Know

caloric value
food nutrition label

ingredients
percent daily value

serving size

Before You Begin

Have you ever heard the expression "You are what you eat"? My grandfather would always say that to me, and then reference a silly animal that I would never even consider eating! The message behind that expression, though, is true. The choices that you make about what to eat affect your health and energy. According to leading health experts, your ability to make informed decisions about the types and amounts of food you eat is extremely important to your overall health and wellness now and for years to come. Think back to when you ate lunch yesterday. Did you think about the nutritional contents of what you were eating? Why or why not? What usually affects your decision-making when eating or purchasing food?

Why Do Nutrition Labels Matter?

Food nutrition labels provide basic information about the contents and healthfulness of the food contained in the attached package. Food nutrition labels in the United States have only been around since the 1970s, but have gone through major revisions to ensure valuable information is presented to consumers in a clear and informative way. Although it is commonplace now, this was not always the case. It wasn't until the Nutrition Labeling and Education Act (NLEA) was passed in 1990 that food manufacturers were required to provide a detailed Nutrition Facts label with standard information on every food item produced. The NLEA was signed into law to help Americans make healthier choices while also holding food companies accountable for producing more nutritious foods.

Today, the FDA recommends that middle school students should consume between 1,400 and 2,600 calories per day. Middle school is a crucial age for learning healthy eating habits that will continue into adulthood, and it's important to make sure that the calories you consume are coming from the best, most nutritious sources. Today, most middle schoolers do not eat enough fruits, vegetables, or dairy products, and almost all consume too much added sugar, salt, and saturated fat. It is important to make healthy choices to keep your body functioning at its best. Food labels provide you with information that allows you to make smart, informed choices (**Figure 6.4.1**). Let's learn more!

Stokkete/Shutterstock.com

Figure 6.4.1. *How often do you look at the label on your favorite foods? Are you most concerned with the number of calories per serving? Or maybe the number of ingredients listed?* Eating healthfully can be difficult but understanding how to read a Nutrition Facts label can help us make healthy food choices.

Trusted Information

With such a large amount of information available on the importance of healthy living and nutritious food choices, it's sometimes difficult to know who to trust. It is important to seek out reliable information from reputable sources. In the United States, the three most trusted sources of information on guidelines for a healthy diet are:

1. **The Dietary Guidelines for Americans.** These dietary guidelines are used to help people over the age of two eat a nutritious diet.
2. **MyPlate.** MyPlate is a visual food guide that uses five main food groups (fruits, vegetables, protein, grains, and dairy) to show us how much to eat each day to be healthy (**Figure 6.4.2**). It's important to eat a variety of foods from each of the food groups to ensure we get necessary nutrients.
3. **The Nutrition Facts label.** The Nutrition Facts label is found on the packaging of all manufactured food products in the United States. We'll learn more about this next!

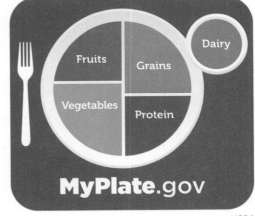

USDA

Figure 6.4.2. *What's on your plate?* The United States Department of Agriculture (USDA) provides interactive and personalized resources for healthy eating on the MyPlate website: https://www.myplate.gov. Take a few minutes to explore your food choices with the help of the USDA.

What's On a Food Label?

Let's break down the information found on a Nutrition Facts label (**Figure 6.4.3**) by starting at the top and working our way down.

Serving Size and Servings per Container. At the top of every food label, you will find information on the number of servings you can expect to find in the box or package, based on a predetermined serving size. A *serving size* is a standard amount a person should eat at one time. According to the label provided in **Figure 6.4.3**, a serving size for this product is equal to one cup, with approximately six servings in this container.

Calories per Serving. Below, you will find information describing the food's *caloric value*, or the energy produced by the food when burned or metabolized by the body. To help consumers stay attentive to the number of calories eaten in one serving, calories are listed in bold. In **Figure 6.4.3** you will find this product contains 358 calories per serving. With what we've learned about serving sizes, how many calories would be consumed if you ate the entire package of food?

Nutrients. The next section of a food label is dedicated to nutrient information required by the United States Food and Drug Administration (FDA). Because there are several nutrients we should limit when making healthy choices, the FDA recommends daily amounts on each label. It does this using the Percent Daily Value. *Percent Daily Value* is a reference that tells us how much of each nutrient can be found in one serving of the labeled food, based on a 2,000 calorie diet. The Percent Daily Value provides key information about whether there is a lot or a little of a nutrient in a specific food item.

Figure 6.4.3. The FDA requires nutrition labels to be on all processed foods. *When was the last time you stopped to look at one? Do you know what any of these items mean?*

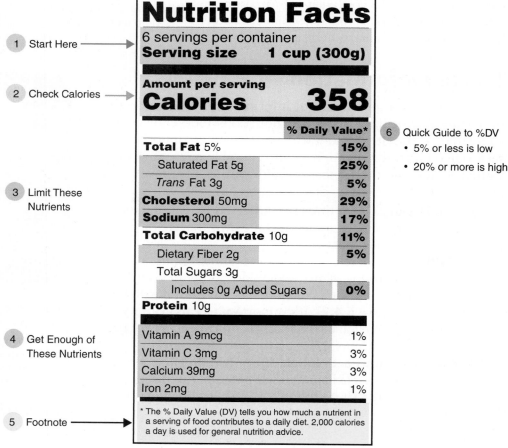

robuart/Shutterstock.com; FDA

For example, 50 milligrams (50mg) of cholesterol and 300 milligrams (300mg) of sodium are listed on the label in **Figure 6.4.3**, representing 29 percent of the cholesterol and 11 percent of the sodium recommended per day for someone eating 2,000 calories.

Did You Know?

Why is Percent Daily Value based on 2,000 calories? This is the recommended total of calories for most adults. If you are growing and you eat more than 2,000 calories per day (some middle school students do, and appropriately so), what can you take away from those numbers? What if you eat fewer calories?

Research suggests we should limit five nutrients in our diet: saturated fat, trans fat, cholesterol, sodium, and added sugars. Too much of any of these nutrients can hurt your health; they may affect your blood pressure, weight, and heart health. Added sugars is a relatively new element of the Nutrition Facts label and includes all types of sugars, including syrups and granulated sugar added during processing. Information for each of these nutrients is provided per serving, according to Percent Daily Value. How much trans fat does this product contain?

Did You Know?

The Dietary Guidelines for Americans recommend that added sugars should be limited to 10 percent of total daily calories. Consuming too much added sugar can make it difficult to meet our body's nutrient needs each day while not eating too much.

Because many Americans do not get enough daily nutrients, the Nutrition Facts label also includes five nutrients we should eat more of: dietary fiber, Vitamin D, Calcium, Iron, and Potassium. You may often find additional nutrients listed that are not required of the manufacturer, such as Vitamin A or Vitamin C, because Americans are less likely to be deficient in these nutrients today.

At the very bottom of the label, you will find a final footnote defining the Percent Daily Value. Why is it so important to understand this when reading a label?

Other Labels Found on Food Packaging

Ingredients List

If you have a food allergy, you might be most interested in the ingredients list, found just below the Nutrition Facts label. Packaged foods are required to list the *ingredients*, or individual foods combined to make a food product or recipe, of any product with more than one ingredient (**Figure 6.4.4**). These ingredients are listed on the package in order by weight, from most to least. If you have food allergies, the ingredient list can help you identify foods that you may need to avoid.

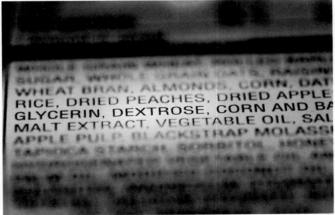

Benoit Daoust/Shutterstock.com

Figure 6.4.4. *Have you ever found yourself wondering what all those words mean in the ingredients list on the back of the package?* Next time you're three handfuls deep into your favorite salty snack, take a moment to read the ingredients and look them up online. If we are what we eat, it's important to know exactly what those words mean!

Label Language

You may come across marketing labels on your packaged food products using words like *free, low, reduced, high, less, more, light, good source of, lean,* or *extra lean*. The United States Food and Drug Administration (FDA) currently allows the use of this language to describe a food's nutritional value. This language brings attention to the information found on the Nutrition Facts label as a way to help guide consumers to these options.

Hands-On Activity

You Are What You Eat

Learning how to make healthy choices is an important part of growing up. Nutrition labels can help you do this! In this activity, you will use nutrition labels and facts to show your classmates the difference between healthy choices and unhealthy choices.

For this activity, you will need:

- Paper
- Magazines or other printed-out images of food
- Glue or glue stick
- Pen or marker
- Calculator
- Device for internet access

Procedure:

1. On two sheets of paper, create two similar outlines of a person.
2. Going through magazines, cut out a variety of food pictures. Sort them into "healthy" and "unhealthy" piles. Note: This can be hard! There are some foods that are obviously healthy, and some that are clearly not, and there are others that exist in the middle. Use your best judgment and be prepared to explain your choices!
3. Taking your piles of food, glue the "healthy" foods onto one figure. Glue the "unhealthy" foods onto the other.

milsamil/Shutterstock.com

4. Next, go online and try to find some nutritional information for the foods that you glued onto each figure. Make a note of anything that surprises you.
5. In groups, share your collages with others, and ask for feedback.

Consider This

1. Did anything that you found surprise you? Did your classmates share anything you hadn't thought about?
2. What types of food were difficult to classify? Why?
3. Do you think there is a place in a healthy lifestyle for unhealthy foods? Why or why not?

The FDA regulates all labels and language that may be used on food. While some of this language is useful in bringing the health benefits of a product to your attention, it is also important to understand that some of these terms do not mean a lot, or do not mean what you think they do (**Figure 6.4.5**). If you see the word "light" on a salad dressing bottle, what do you assume about it? Do you assume that it is a healthy choice? Labeling something as "light" or "lean," or even "free," may not mean what you think it does. Always look at the nutrition label to understand what you are getting.

Vlad Klok/Shutterstock.com

Figure 6.4.5. There are a number of labels on this image of a steak. *Are any of these labels meaningful? Why or why not? What information might you look for on the Nutrition Facts label to verify these claims?*

LESSON 6.4 Review and Assessment

What Did We Learn?

- A food nutrition label describes nutritional content of a food product. Information is based on a 2,000 calorie diet, including servings and serving sizes, calories per serving, and nutrients found in a food product.
- Trusted sources of information such as the Dietary Guidelines for Americans, MyPlate, and food nutrition labels are helpful when looking for resources to maintain a healthy diet.
- Additional information provided on food packaging includes a list of ingredients and language that describes the nutritional content. These claims should always be verified by checking the food nutrition label.

Let's Check and See What We Know

Answer the following questions using the information provided in Lesson 6.4.

1. *True or False?* All packaged foods with more than one ingredient are required to list the ingredients on the package label.
2. The Nutrition Facts label includes five nutrients that most Americans don't eat enough of. Which of these is included?
   A. Water   C. Vitamin D
   B. Nitrogen   D. Saturated fat
3. All information listed on a Nutrition Facts label is based on a recommended daily diet consisting of how many calories?
   A. 2,000   C. 4,100
   B. 2,800   D. 800
4. *True or False?* The Nutrition Labeling and Education Act (NLEA) requires all food manufacturers to place a Nutrition Facts label on every food product produced.
5. *True or False?* The Dietary Guidelines for Americans recommends added sugars should be limited to 25 percent of our total daily calories.

6. All of the following are reliable sources for nutritional information *except*:
   A. MyPlate
   B. Health and Wellness Facebook Group
   C. Nutrition Facts label
   D. The Dietary Guidelines for Americans
7. *True or False?* The FDA regulates what language may be used on food packaging.
8. Most middle school students should consume between _____ calories per day.
   A. 850 and 1,200   C. 2,000 and 2,600
   B. 1,400 and 2,600   D. 1,400 and 1,600
9. Most middle school students consume too much:
   A. Added sugar   C. Saturated fat
   B. Sodium   D. All of these.
10. Most middle school students do not consume enough:
    A. Grains   C. Protein
    B. Vegetables   D. All of these.

Let's Apply What We Know

1. Why is it important to have manufacturers and producers provide accurate food labels?
2. How do food labels help consumers make good decisions?
3. Where can you go to get helpful, accurate information about food and dietary guidelines?
4. Take a look at this nutrition label, and answer the questions below.
   A. How large is a serving? How many servings are in the container?
   B. How many calories would you consume if you had two servings?
   C. How much trans fat is in a serving?
   D. Look at the difference between the column for one serving, and the column for three servings. Why do you think it's important to pay attention to serving sizes when looking at this information?

Nutrition Facts

3 servings per container

Serving size		1 cup (180g)

Calories	Per serving 245	Per container 735

		%DV*		%DV*
Total Fat	12g	14%	36g	43%
Saturated Fat	2g	10%	6g	30%
Trans Fat	0g		0g	
Cholesterol	8mg	3%	24mg	8%
Sodium	210mg	9%	630mg	27%
Total Carb.	34g	12%	102g	36%
Dietary Fiber	7g	25%	21g	75%
Total Sugars	5g		15g	
Incl. Added Sugars	4g	8%	12g	24%
Protein	11g		33g	
Vitamin D	4mcg	20%	12mcg	60%
Calcium	210mg	16%	630mg	48%
Iron	3mg	15%	9mg	45%
Potassium	380mg	8%	1140mg	24%

* The % Daily Value (DV) tells you how much a nutrient in a serving of food contributes to daily diet. 2,000 calories a day is used for general nutrition advice.

maradaisy/Shutterstock.com

Academic Activities

1. **Science and Financial Literacy.** Sticking to a balanced diet of 2,000 calories per day while also making sure you get an appropriate amount of servings from all five food groups can be a difficult task, especially when you're on a budget. In this activity we will attempt to plan a full day's worth of meals based on the recommendations of MyPlate (see **Figure 6.4.3**) with only $10. Visit a local grocery store, look at a grocery ad from the newspaper, or look up product information online and make a list of items you would purchase (and their prices) to meet those daily requirements for a healthy, well-balanced diet. Don't forget to check the Nutrition Facts label for help!

Communicating about Agriculture

1. **Interpreting Information.** Have you ever noticed food labels with claims that simply did not make sense or seemed misleading to the consumer? These types of food labels (seperate from those regulated by the FDA) have become more common over the past decade and may make unnecessary or misleading claims about food products. Do some research about these claims—what they look like, what they claim, and why they have started popping up. How might this be detrimental to the agriculture industry? How can agriculturalists communicate the importance of science-based research in the proper labeling and packaging of food products in the United States?

SAE for ALL Opportunities

1. **Service Learning SAE.** A healthy diet is important no matter how old you are, but it is especially vital to the healthy development of children. Our local food banks and pantries are hard at work trying to provide healthy and nutritious foods to families at no cost to them. Have you ever considered volunteering to sort or deliver food items at your local food bank? Even a couple of hours a month could make a huge difference in the lives of people in your community.

SAE for ALL Check-In

- How much time have you spent on your SAE this week?
- Have you logged your SAE hours?
- What challenges are you having with your SAE?
- How can your instructor help you?
- Do you have the equipment you need?

UNIT **7**
Soil Science Exploration

Scott Hutto loves every moment of his job as a district soil conservationist with the Natural Resources Conservation Service (NRCS). As a high school student, Scott volunteered at Lake Pontchartrain and Fontainebleau State Park in Louisiana. This sparked his passion for working outside and giving back to nature.

While in high school, Scott took an agricultural mechanics class. One of the class projects was building bird houses, duck boxes, and bat houses for the lake and state park. Scott always enjoyed woodworking and was excited to get to work with his hands. He thought he wanted to work with his hands as a career.

Those plans changed when Scott went to the park to hang all the boxes and houses his class had built. He fell in love with the area and found himself taking friends and family there to visit. This developed into a Service Learning SAE where Scott gave his time to complete needed maintenance projects at the state park. His project involved cleaning nature trails and making repairs to handrails, ramps, and bridges.

During his SAE, Scott built relationships with park rangers by being dependable. They allowed him to volunteer on days when the park rangers were working on erosion control methods around the park and the lake with the local NRCS office. He became fascinated by the way hurricanes and ocean currents changed the land and waterways.

"What started as something to do on a single Saturday turned into two years' worth of volunteer work and making connections with people I would later be working with," said Scott. "Now I get paid to do what I love: conserving the environment and its natural resources so people will have places like Fontainebleau State Park to enjoy for years to come."

As a district soil conservationist, Scott designs and explains resource conservation plans to landowners and farmers. Soil conservationists share practical knowledge of the methods and techniques of soil, water, and environmental conservation as they relate to agricultural operations and land use. They also help landowners with issues with soil, water, air, plants, and animal resource concerns.

- Scott's interests over the course of his SAE changed and grew with his experience. What can you learn from that?
- Are you aware of local or state natural areas near you that might need help or care? Who could you contact about that?

Top: Chatchawal Phumkaew/Shutterstock.com
Bottom: Pamela Wertz/Shutterstock.com

Volodymyr_Shtun/Shutterstock.com

Components of Soil

Kichigin/Shutterstock.com

Essential Question

How do soils differ from each other?

Learning Outcomes

By the end of this lesson, you should be able to:

- Explain why healthy soil is important to people, plants, and animals. (LO 07.01-a)
- List the living organisms that play critical roles in soil health. (LO 07.01-b)
- Describe the components of soil. (LO 07.01-c)

Words to Know

algae	minerals	silt
bacteria	mole	soil
clay	nematode	soil horizon
humus	protozoa	soil profile
loam	sand	

Before You Begin

Have you ever thought about soil before? I would guess probably not, but soil plays an important role in our lives. Soil gives our food supply nutrients to grow, supports our homes and schools, and serves as the main ingredient in mud pies like I made as a kid. Next time you go outside, pinch some soil and rub it between your fingers. Does it feel gritty or smooth? Is the soil sticky or dry? Is it soft, or hard and difficult to break apart? What color is it? Does it have bits of rock or organic matter?

Soil comes in many forms, but all soils provide a natural environment to support the growth and development of plants.

Why Does Soil Matter?

Soil is a combination of minerals, gases, liquids, living organisms, and organic matter that supports the growth and development of plants and animals. At first glance, soil may not look like much more than something we walk or build on and in which we grow plants. On closer observation, we see that soil is full of living organisms.

In the soil, all things are connected. Soil is the home for a variety of organisms that have found a way to work and live together. This series of complex relationships provides nutrients and food for all the species present, the plants, and for people and animals who eat the plants (**Figure 7.1.1**).

Did You Know?

There are more living organisms in a handful of soil than there are people on Earth!

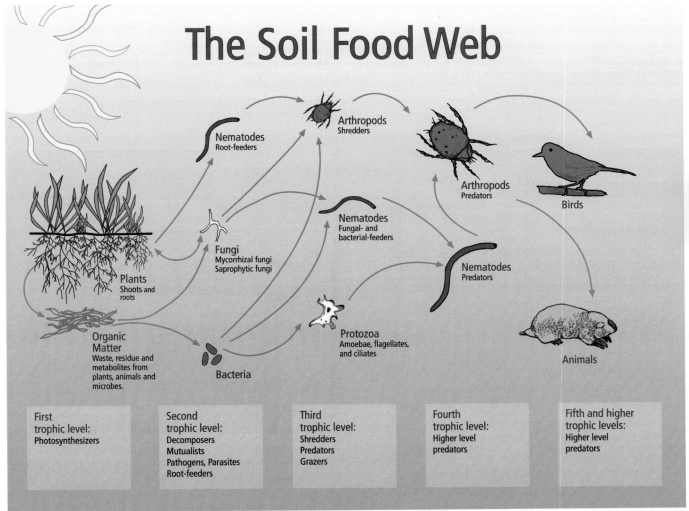

The Soil Food Web

Nematodes
Root-feeders

Arthropods
Shredders

Arthropods
Predators

Birds

Fungi
Mycorrhizal fungi
Saprophytic fungi

Nematodes
Fungal- and
bacterial-feeders

Nematodes
Predators

Plants
Shoots and
roots

Protozoa
Amoebae, flagellates,
and ciliates

Organic
Matter
Waste, residue and
metabolites from
plants, animals and
microbes.

Bacteria

Animals

First
trophic level:
Photosynthesizers

Second
trophic level:
Decomposers
Mutualists
Pathogens, Parasites
Root-feeders

Third
trophic level:
Shredders
Predators
Grazers

Fourth
trophic level:
Higher level
predators

Fifth and higher
trophic levels:
Higher level
predators

USDA

Figure 7.1.1. The soil food web is the community of organisms living all or part of their lives in the soil. As organisms decompose material found in the soil, or consume other organisms, nutrients are converted into usable forms, and are made available to plants and other soil organisms.

Who Lives in the Soil?

Soil is filled with bacteria, fungi, algae, and protozoa (**Figure 7.1.2**). It is also a home to larger organisms including mites, nematodes, earthworms, and ants. These organisms spend most or all of their lives underground. The largest organisms that contribute to soil health include burrowing rodents.

Bacteria are tiny organisms found in great abundance in the soil. Bacteria are among the simplest living things; they contain cell walls but lack a nucleus and organelles. One teaspoon of soil can contain as many as one billion bacteria.

Protozoa are single-celled organisms that feed on bacteria. They release important nutrients for plants back into the soil. ***Nematodes*** are multicellular organisms that resemble worms and feed on bacteria. Nematodes also release nutrients into the soil, but some types of nematodes harm plants by feeding on their roots.

Another common soil organism is ***algae***. Algae need sunlight and moisture to grow. Algae plays an important

drical/Shutterstock.com

Figure 7.1.2. Soil is filled with many small organisms that can only be viewed under a microscope. Soil scientists use microscopes to help them determine the amount of organic matter or organisms present in the soil.

Making Connections # | Technology

Soil Mapping

Soil mapping is beneficial to the agriculture industry. Farmers can use tablets or computers while tilling the soil to conduct deep soil scans. Soil mapping is a geographical representation which shows the different types of soils and soil properties such as pH, texture, organic matter, depth of horizons, and other helpful information. Soil maps also provide the location of particular soil types, properties of those soils, and potential uses. Soil maps are commonly used for land evaluation, agriculture extension, environmental protection, and development projects. Soil mapping originated before 1900, and up until 2005 soil maps were printed and available for viewing through county, state, and federal Natural Resources Conservation Service offices. Until the digital era, these printed soil survey reports were the main source of soil information for farmers and other individuals. Today, digital soil mapping provides farmers with current information about the soil.

MONOPOLY919/Shutterstock.com

Consider This

1. How could soil mapping benefit people outside of agriculture?
2. Why would knowing the soil type, characteristics, and other information help building contractors before they start building a large mall or apartment complex?

role in producing organic matter needed in healthy soils. Organic matter holds nutrients and moisture for plants.

Soil is home to other creatures that are easier to see than bacteria and protozoa, like earthworms. Earthworms are a source of food for birds, reptiles, and mammals. Earthworms burrow through the soil, consuming and digesting organic matter like leaves and roots. Earthworms leave waste that provides nutrients to plants. Earthworms also improve soil health by transporting nutrients as they move. Their tunnels allow air to enter the soil (**Figure 7.1.3**).

Did You Know? Earthworms eat up to one-third of their body weight in one day!

Soil also serves as a home and hiding place for larger creatures. **Moles** burrow through the soil, eating roots and uprooting plants as they dig.

Figure 7.1.3. Earthworms increase the amount of air and water in the soil and break down organic matter for plants.

Career Corner SOIL SCIENTIST

If this is interesting to you, you might want to be a **soil scientist***!*

What Do They Do? Soil scientists study soil characteristics, map soil types, and investigate responses of soils under certain conditions. They provide advice on how to use land, analyze soil contents, and conduct research experiments to see what plants thrive in different soils.

What Education Do I Need? A bachelor's degree in chemistry, crop science, soil science, biology, or a related field, such as horticulture, plant physiology, or environmental science is required to become a soil scientist.

How Much Money Can I Make? Soil scientists make an average of $70,000 per year.

How Can I Get Started? You can build an SAE studying the soils in your area. Can you gather details about the soil in your area over an extended period and determine its effect on crop yields or plant growth? Create an agriscience project determining major factors affecting soil erosion in your area. Contact a soil scientist and conduct a personal interview with them.

tchara/Shutterstock.com

Figure 7.1.4. Moles live underground and feed on insects, but they also damage gardens and lawns by tunneling and damaging plant roots.

Moles consume earthworms and other organisms living in the soil. Even though moles look like mice, they are not rodents. They are small mammals that feed almost exclusively on insects. While hunting underground for these insects, they dig tunnels that damage lawns and gardens. However, these tunnels and holes are good for the soil (**Figure 7.1.4**).

What Is Soil?

Minerals are inorganic particles that make up the largest component of soil. These minerals include clay, sand, gravel, stones, and silt, which make up about 45 percent of soil. Soil also contains water and air. About 25 percent of a handful of soil is comprised of water and another 25 percent of air in the spaces between the particles (**Figure 7.1.5**).

Humus makes up the remaining five percent of soil composition. *Humus* is organic matter formed by the decomposition of plant material in the soil. The exact percentage of mineral matter, organic matter, air, and water in a soil sample may differ, depending on location and other environmental factors.

Soil Particles

Most soils contain a mixture of sand, silt, and clay. *Sand* is the largest particle, and individual sand grains are easy to recognize. *Silt* is the next largest particle. Silt is difficult to see without a microscope. The smallest particle is *clay*. This particle is so small that it cannot be seen without a microscope.

Soil particle size is important to soil health. The large sand particles allow air to enter the soil, which causes sandy soils to dry out quickly. Soils with silt and clay hold more water than sandy soil, but may not allow air to enter. Plant roots need both water and air to function properly (**Figure 7.1.6**).

Loam is soil made with sand, silt, and clay. Loam should consist of equal parts of each soil type. This combination creates an ideal composition for plant growth.

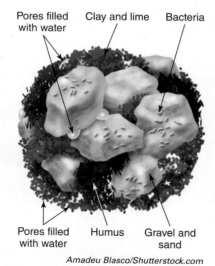

Amadeu Blasco/Shutterstock.com

Figure 7.1.5. The spaces between the mineral particles are filled with water molecules and air.

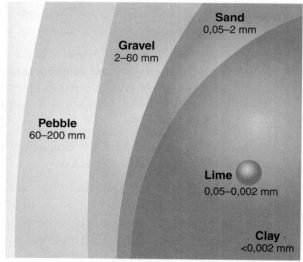

Amadeu Blasco/Shutterstock.com

Figure 7.1.6. The relative sizes of gravel, sand, silt, and clay. Note that the clay particle in the figure is about the same size as the period at end of this sentence.

Soil Profile

If you were to remove soil from an area next to your house and compare it to a sample from an open field, do you think they would be different? No two soils are the same. Even soils that are geographically very close to each other will be unique.

If you look at a soil pit or in a roadside cut (**Figure 7.1.7**), you will see different layers of soil. These layers are called the *soil horizon*. A soil horizon is a layer parallel to the soil surface whose physical, chemical, and biological characteristics differ from the layers above and beneath. The arrangement of these horizons in the soil is called the *soil profile*. Soil horizons differ in many ways such as color, texture, structure, and thickness. Other properties are less visible, such as chemical or mineral content. These properties are used to define types of soil horizons.

Tatyana20/Shutterstock.com

Figure 7.1.7. Soil profiles help scientists determine how the land should be used and if it can support large buildings or other structures.

Soil scientists use the capital letters **O**, **A**, **B**, **C**, and **R** to identify the major sections in soil horizons, and lowercase letters for smaller distinctions. Most soils have three major horizons: the surface horizon (**A**), the subsoil (**B**), and the substratum (**C**). The surface horizon is also commonly referred to as *topsoil*. Some soils have an organic horizon (**O**) on the surface. Hard bedrock, which is not soil, uses the letter **R** (**Figure 7.1.8**).

Figure 7.1.8. It can take thousands of years to create an inch of topsoil. Soil is mostly derived from rock. Rock must be broken into small pieces by physical weathering.

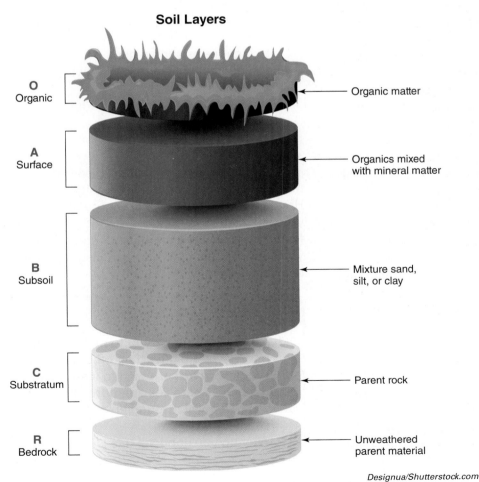

Soil Layers

O — Organic — Organic matter

A — Surface — Organics mixed with mineral matter

B — Subsoil — Mixture sand, silt, or clay

C — Substratum — Parent rock

R — Bedrock — Unweathered parent material

Designua/Shutterstock.com

Hands-On Activity

The Soil Shake Out

In this activity, we will see if we can "shake out" a sample of soil and identify the different components.

You will need the following materials:

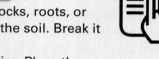

- About 1 cup of soil (from home or collected at your school)
- 1 clear pint jar with a lid
- 6-inch ruler
- 1 teaspoon of alum (aluminum sulfate, a chemical compound that helps the soil settle faster)

Here are your directions:

1. Remove any large rocks, roots, or plant material from the soil. Break it up with your hands.
2. Place the soil in the jar. Place the ruler alongside the jar and measure the height of the soil. Make sure you have at least 2 inches of soil in the bottom of the jar. Add or remove soil until the level is sufficient.
3. Next, add one teaspoon of alum under the direction of your teacher.
4. Add enough water to fill the jar to within 1 inch of the lid. Dry any water left on the jar.
5. Screw the lid on tightly and make sure it does not leak.
6. Shake the jar vigorously for about two minutes, or until your instructor tells you to stop. Make sure the water and soil are well mixed. Keep a good grip on the jar so you do not drop it.
7. Set the jar down on a flat table surface and watch the soil particles as they settle to the bottom.
8. After about two minutes, place the ruler against the glass jar, without disturbing the soil and water mixture. Measure the depth of the soil at the bottom of the jar. This is the fraction of sand in the soil mixture. Record the measurement in your notes.
9. Place the jar in a safe place where it will not be disturbed.
10. In the next class period, observe the layers of soil in the jar, without disturbing the soil and water mixture. You should see a second layer of soil that formed on top of the layer you measured in the last class period. Measure the depth of this second soil layer and record the measurement in your notes. The second layer is the silt layer.
11. Set the jar aside in a space where it will not be disturbed and leave it for one week.
12. The soil particles that remain suspended in water may take a long time to settle. These particles are so small that they may weigh only slightly more than the water they would have to displace to settle.
13. To speed up your measurement of the amount of clay particles, you can do a neat math trick. Do you remember that you placed 2 inches of soil in the jar on the first day of the experiment? All you need to do is subtract the sand measurement and the silt measurement from the 2 inches. The remaining difference represents the amount of clay in the soil sample. Look at the following example:

silt (second layer)	= ½ inch
total soil	= 2 inch
sand (first layer)	= 1 inch
clay (total soil − sand + silt)	= ½ inches

14. Now convert the measurements into percentages as shown here:
Sand percentage (sand/total soil × 100) = (1 ÷ 2) × 100 = 50%
Silt percentage (silt/total soil × 100) = (½ ÷ 2) × 100 = 25%
Clay percentage (clay/total soil × 100) = (½ ÷ 2) × 100 = 25%

Compare the results of your experiment with other students. How did your results differ? How were they similar? Did you get the results you expected? You may notice that some matter floats on or near the top of the water. What do you suppose this is? Which particles were the heaviest—sand, silt, or clay?

Consider This

1. In the experiment, you found that some particles are bigger than other particles. Would you be surprised to learn that the bigger particles are more likely to be washed away through erosion than the smaller particles?
2. Why do you suppose this is?

What Did We Learn?

- Soil is a combination of minerals, gases, liquids, living organisms, and organic matter that supports the growth and development of plants and animals.
- Soils contain bacteria and other organisms that break down organic matter and provide nutrients for growing plants. Soil needs many kinds of organisms to remain healthy, from microscopic bacteria to animals as large as moles.
- The four basic components of a soil are sand, silt, clay, and humus. Sand is the largest soil particle, then silt. Clay is the smallest.
- Soil has layers known as the soil horizons. Each horizon is comprised of soils that are physically and chemically different than the layers above and below.
- The vertical section of soils showcasing the different horizons is known as the soil profile.

Let's Check and See What We Know

Answer the following questions using the information provided in Lesson 7.1.

1. One spoonful of soil can contain as many as one ____ bacteria.
   A. hundred   C. million
   B. thousand   D. billion

2. Which organism can be harmful to plants by feeding on plant roots?
   A. Actinomyces   C. Nematodes
   B. Protozoa   D. Algae

3. Which organism increases the organic matter in the soil?
   A. Actinomyces   C. Nematodes
   B. Protozoa   D. Algae

4. *True or False?* Organic matter is important to soil because it holds nutrients and provides moisture to plants.

5. *True or False?* Earthworms are harmful to the health of the soil.

6. *True or False?* Moles are rodents.

7. Soil composition contains which of the following?
   A. Minerals   C. Air
   B. Water   D. All of these.

8. Which of the following has the largest particle size?
   A. Sand   C. Clay
   B. Silt   D. Humus

9. Which of the following has the smallest particle size?
   A. Sand   C. Clay
   B. Silt   D. Humus

10. Which soil layer is also commonly referred to as topsoil?
    A. O   C. B
    B. A   D. C

Let's Apply What We Know

1. Collect soil samples from the school grounds or an outdoor area near your home. Compare each sample. How are they alike and how are they different? What do you suppose causes the differences?
2. Notice the plants rooted in the soil where you took your soil samples. Do the plants look healthy or unhealthy? Why do you suppose they look as they do?
3. Healthy soil has a balance of certain types of soil organisms. From your study of this lesson, which soil organisms seem to be the most useful to plant growth and development?
4. What is the connection between the air level in soil and its ability to hold water?
5. Think about the area that you live in, compared to the same area 50 years ago. Do you think the rate of soil erosion is increasing or decreasing? Why?

Academic Activities

1. **Engineering.** Demonstrate the effects of tillage and crop growth on soil erosion on agricultural lands in your area. Compare tillage techniques and other factors that affect erosion.
2. **Science.** Test soil compaction around your school. Where water soaks into the soil quickly, the soil is less compact; where water pools or slowly soaks into the soil, there is high compaction. Identify five areas around your school where you think there are different compaction levels.
3. **Science.** Create your own soil horizon using a small piece of card stock and 1-inch double sided tape. Use different colored soils and rocks to represent each layer, making sure to label each horizon correctly.

Communicating about Agriculture

1. **Reading and Speaking.** Determine the type of soil that you have at school through the Natural Resources Conservation Service Web Soil Survey. Write a report in which you describe how that specific type of soil can be used. Share your findings with your peers.

SAE for ALL Opportunities

1. **Research SAE.** Contact your local or state Natural Resources Conservation Service and obtain a soil map for the area you are conducting your SAE. Determine if the type of soil has or could have any effect on your SAE.
2. **Research SAE.** Conduct an experiment about earthworms. Determine if different diets affect the earthworms and the soil they enrich.

SAE for ALL Check-In

- How much time have you spent on your SAE this week?
- Have you logged your SAE hours?
- What challenges are you having with your SAE?
- How can your instructor help you?
- Do you have the equipment you need?

Soil Conservation

Fotema/Shutterstock.com

*How can
we protect
our soils
for future
generations?*

Learning Outcomes

By the end of this lesson, you should be able to:
- Describe the reasons that soil loss is a problem. (LO 07.02-a)
- Explain how soil erosion occurs. (LO 07.02-b)
- Name ways to prevent soil erosion and protect soil. (LO 07.02-c)

Words to Know

erosion
soil biodiversity

soil conservation
practice
soil deposition

soil detachment
soil transport

Before You Begin

*Is it possible to "lose" soil? Think about some of your
favorite natural places. Have they changed over the
years? Think about hillsides, sandbars, vegetation, and
development. What role did soil play in
how those changes occurred? Soil loss
can be a real reason for concern. Talk with
a classmate about your experience with
seeing areas change over time. Can you
come up with a theory about what caused
those changes, and why?*

Figure 7.2.1. *Have you ever seen the effects of soil loss? What factors do you think caused the soil loss shown here?*

alsamua/Shutterstock.com

Meryll/Shutterstock.com

Figure 7.2.2. Soil erosion removes valuable topsoil, which is the most important part of the soil profile for agriculture. When topsoil is gone, erosion can cause rills or gullies like this.

Our soil is valuable. It is important to our everyday lives and how we meet our essential needs. Soils provide nutrients for plants, animals, and billions of microorganisms. It is important that we protect our soils and maintain healthy ecosystems where plants, trees, and animals can thrive (**Figure 7.2.1**). It is also important to keep our soils productive to ensure healthy agricultural yields.

Soil Loss

When soil is lost, that soil has degraded or decreased its productivity. Most soil loss is caused by a type of erosion. ***Erosion*** is the physical wearing away of Earth's surface. Erosion can move soil with ice, gravity, human activity, water, or wind. Erosion is a concern because it removes the topsoil, which is vital for holding water, minerals, and organisms that are essential for plant growth (**Figure 7.2.2**). The two most common causes of soil erosion are wind (**Figure 7.2.3**) and water.

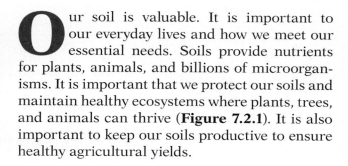

Did You Know? It is estimated that 26 billion tons of soil are lost each year.

Figure 7.2.3. Wind erosion causes damage by drying out the land and reducing the amount of nutrients in the soil. It can also cause air pollution. *What other problems might a dust storm of this size cause?*

Caleb Holder/Shutterstock.com

Soil Erosion

There are three steps to erosion. The first step is ***soil detachment***, when soil is physically removed by a type of force. For example, soil detachment occurs every time wind blows across the beach and picks up sand, or landslides move large sections of Earth. Step two is ***soil transport***, when soil particles are moved from one location to another. This might happen when heavy rains wash soil into water, turning it muddy. ***Soil deposition*** is the third step and occurs when soil particles settle in a new location. For example, when a flooded river recedes back to normal levels, the riverbank may become wider from soil deposited during flooding (**Figure 7.2.4**).

There are many reasons erosion can be dangerous. Typically, erosion affects the most fertile layer of soil, reducing organic matter. This can make the land less productive for growing crops or other plants. Erosion decreases the amounts of air, water, and nutrients available in the soil.

Did You Know?

Scientists estimate that the Amazon River contributes 20 percent of the soil deposition that enters oceans around the world.

Did You Know?

According to the US Army Corps of Engineers, the Maumee River in Ohio deposits 1.4 million tons of sediment each year into Lake Erie.

CameraLensPro/Shutterstock.com

Figure 7.2.4. Erosion can affect our water supply. Rainwater washes soil particles into this water, turning it brown. *What would it take to make this muddy water into drinking water?*

Hands-On Activity

Soil Splash Zone

In this activity, we will see what effect rain has on soil erosion.

You will need the following materials:

- Bull's-eye drawn or printed out on white paper
- Teaspoon of dry soil
- Eyedropper
- Small container of water

Here are your directions:

1. Place the soil sample in the center of the bull's-eye. Label the remaining areas of the bull's-eye "zone 1", "zone 2", "zone 3", etc. starting at the center and working outward.

2. Fill the eyedropper with water.
3. Hold the eyedropper about 18 inches above the soil sample.
4. Drop 5 drops of water directly onto the soil sample. If a drop misses the soil, continue until 5 drops hit the soil.
5. Record the number of water splashes/spatters that contain soil particles in each zone.

Consider This

1. What did you observe? How did the soil particles move from the center of the target?
2. What do you think would happen if the drops of water were larger?
3. What conservation methods could be put in place to prevent splash erosion?

Matthew J Thomas/Shutterstock.com

Figure 7.2.5. Without erosion control measures, the resulting damage can be dangerous and costly. The cost to repair this stretch of road could be more than half a million dollars.

Erosion can also cause problems for:

- Wildlife. Increased soil sediment in water can be dangerous for animals to consume and for fish and other aquatic species.
- Water quality. Erosion can deposit runoff from fields and other lands into streams, lakes, and rivers, increasing the chemical level and changing the pH of the water.
- Roadways. Roads can be damaged or washed out (**Figure 7.2.5**).
- Buildings and other structures. Foundations can be exposed, making them unstable or more likely to collapse.

Making Connections # Science

Soil Erosion

Soil erosion is easy to see; it does not take microscopes or advanced skills to witness its effects. Conduct an experiment to show how different conservation methods can affect soil erosion.

You will need the following materials:

- Three aluminum casserole pans
- Enough soil to fill all three pans
- Plant residue (such as grass clippings, leaves, or organic debris)
- Sheet of sod the size of the casserole pan (grass seed may be planted in place of this step, but must be done several weeks in advance to ensure seeds sprout and cover the area of the pan)
- Scissors
- 12 cups of water
- Watering can
- 3 clear plastic containers
- Timer

Here are your directions:

1. Fill each of the casserole pans with soil.
2. In one pan, cover the soil with plant residue. In a second pan, cover the soil with the sheet of sod. Leave the third pan with bare soil.
3. Cut a V-shaped notch in the end of each of the pans.

Soil erosion

Arpit Deepchand Kahar/Shutterstock.com

4. Tilt the pans so they sit at a slight angle.
5. Just below the V-shaped notch, place a clear container under each pan.
6. Fill the watering can with 4 cups of water.
7. Using the watering can, pour the water evenly over pan 1. Make sure to cover the entire pan and pour the water as evenly as possible, not focusing in any certain area.
8. Repeat steps 6 and 7 for the other two pans.
9. Let the pans drain into the plastic containers for 5 minutes.
10. Record your results.

Consider This

1. After examining the plastic containers, which pan produced the cleanest water? Why do you think the water was cleanest from this pan?
2. How do your results relate to water runoff and soil erosion?

Soil Preservation

It is important that we take care of our soils. Soil scientists have determined that it takes at least 100 years to form a single inch of topsoil and it could be longer, depending on climate, vegetation, and other factors.

Methods of preventing soil erosion include:

- Contour plowing
- Protecting ground surface with crop residue
- Planting winter cover crops
- Planting windbreaks (**Figure 7.2.6**)
- Maintaining a protective cover on the soil all year
- No-till cropping
- Shortening the length and steepness of slopes and/or planting grass to stabilize the soil
- Strip cropping
- Terracing farm fields to prevent runoff (**Figure 7.2.7**)

John T Callery/Shutterstock.com

Figure 7.2.6. Windbreaks are a row of trees or shrubs that reduce the force of the wind blowing over a field. This can reduce soil erosion, increase crop yields, and protect livestock from heat and cold.

Did You Know? After the Dust Bowl, a project called the Shelterbelt Project was started to reduce wind erosion. This project planted 220 million trees over a 100-mile area stretching from the Canadian border to South Texas.

It is also important to maintain soil biodiversity to preserve our soils for future generations. **Soil biodiversity** is the healthy variety of organisms that exist within the soil, including bacteria, fungi, and earthworms. These different organisms and organic matter play an important role in the development and health of the soil. Through years of research, the agriculture industry now understands ways to better maintain soil diversity through rotating crops, limiting cultivation of the soil, monitoring salt buildup from irrigation, and limiting the use of pesticides.

Top Photo Corporation/Shutterstock.com

Figure 7.2.7. Terracing farm fields is a common practice in Asia, where farming is often done on hilly or mountainous terrain. Terraced fields decrease both soil erosion and surface runoff.

Conservation Efforts

When extreme weather events, such as the Dust Bowl, occur, scientists do research and put preventive measures into place to stop these types of events from happening again. They also help build organizations and advocate for laws to help should something similar happen again.

In 1933, in the middle of the Dust Bowl, President Franklin D. Roosevelt established the Soil Erosion Service. Its purpose was to educate farmers throughout the nation about soil conservation practices. A *soil conservation practice* is a method of farming that prevents the erosion of soil. Today, this same service has been renamed the Natural Resources Conservation Service (NRCS). Agriculturists now rely on the NRCS to provide programs, educational workshops, and research about conserving the Earth's natural resources, including soil, water, forests, wetlands, and grasslands.

Career Corner > CONSERVATION OFFICER

If this is interesting to you, maybe you should consider being a **conservation officer**! *Conservation officers work to secure and protect natural resources.*

What Do They Do? Conservation officers assist with research projects about management practices, construct and maintain watershed and soil-erosion structures, patrol backcountry areas and other secure sites, provide emergency services, manage wildlife/human interactions on the roadways or in campgrounds, provide public education for visitors to conservation sites, and offer advice regarding water management, forage production methods, and brush control.

Sheila Fitzgerald/Shutterstock.com

What Education Do I Need? A bachelor's degree in natural resources, crop science, soil science, biology, or a related field, such as horticulture, plant physiology, or environmental science is required. An associate degree or training in law enforcement is an additional benefit for this job.

How Much Money Can I Make? Conservation officers make an average of $40,000 per year.

How Can I Get Started? You can build an SAE around natural resources conservation in your area. You can research what conservation laws are currently in effect for your community and your state and create an informational pamphlet or brochure to spread awareness. You can visit a local, state, or natural park and job shadow a conservation officer. You can contact a current conservation officer and conduct a personal interview.

What Did We Learn?

- Most soil loss is caused by erosion. Wind and water are the two most commonly occurring types of erosion.
- Soil erosion is a three-step process: soil detachment, transport, and deposition.
- Soil erosion can cause problems for wildlife, water quality, farmlands, roadways, buildings, and other structures.
- Soil biodiversity is important to preserve our soils for future generations. The agriculture industry has many methods to maintain or increase soil diversity.
- US soil conservation practices began in the 1930s and are still being practiced today through the Natural Resources Conservation Service (NRCS).

Let's Check and See What We Know

Answer the following questions using the information provided in Lesson 7.2.

1. How many billion tons of soil are lost each year to erosion?
   - A. 5
   - B. 50
   - C. 31
   - D. 26

2. Erosion is a concern because it typically removes the _____, which is vital for plant growth.
   - A. topsoil
   - B. bedrock
   - C. minerals
   - D. fertilizers

3. The two most commonly occurring forms of erosion are water and _____.
   - A. ice
   - B. wind
   - C. gravity
   - D. human activities

4. The first step in soil erosion is soil _____.
   - A. hardening
   - B. detachment
   - C. transport
   - D. deposition

5. *True or False?* Erosion can be damaging to roadways and buildings.

6. Scientists have determined it can take _____ or more years to form one inch of topsoil.
   - A. 5
   - B. 40
   - C. 100
   - D. 500

7. *True or False?* The agriculture industry uses methods such as contour plowing, no-till cropping, and strip cropping to prevent soil erosion.

8. *True or False?* Soil biodiversity has *no* effect on preserving soils for future use.

9. US soil conservation practices began in the _____.
   - A. 1930s
   - B. 1950s
   - C. 1970s
   - D. 1990s

10. *True or False?* Today, farmers rely on the Natural Resources Conservation Service to provide programs, educational workshops, and research about the conservation of soil and other natural resources.

Let's Apply What We Know

1. How have farmers attempted to reduce soil erosion in their fields?
2. What are the two main types of erosion? Explain how each of these methods transfer soil.
3. Describe the three processes that occur during soil erosion using real-life examples or situations.
4. Research educational programs offered by the NRCS in your home state. Present your findings to the class. Make sure to include the value of the program to the agriculture industry.
5. Using what you know about the negative effects of water erosion on soil, research the effects of water erosion on other parts of the ecosystem.

Academic Activities

1. **Technology.** Research criminal cases in which soil evidence was used to solve crimes. This evidence might include soil particles found under nails that were specific to a certain area or muddy foot prints left behind at the scene. What do you think about using this evidence?
2. **Science.** Test the effects of organic matter and/or cover crops on soil erosion. Choose two soils with different organic matter contents, such as a garden or a lawn. Try to get intact samples of the soils with roots and vegetation. Cut a piece of hardware cloth so it will support a soil clod at the top of a jar filled with water. Place each soil clod in a separate jar so it rests on the hardware cloth just under the surface of the water. Observe the difference between the soil clod with high organic matter versus the one low in organic matter.

Communicating about Agriculture

1. **Speaking.** Working with your teacher, find aerial photographs of your community over a period of years. Make a note of differences that you see between the maps. Present your findings to the class, talking about how erosion may have played a part in the changes you identified.
2. **Social Media.** How can you use your knowledge of soil conservation and health to help your friends and neighbors take better care of their yards and gardens? Put together a social media post with pictures and text encouraging others to adopt better conservation practices at home.

SAE for **ALL** Opportunities

1. **Foundational SAE.** Contact your local or state Natural Resources Conservation Service office and set up a day to shadow a field agent.
2. **Research SAE.** Conduct an experiment to determine whether no-till or plow-based farming practices are better at retaining surface moisture and preventing surface runoff.

SAE for **ALL** Check-In

- How much time have you spent on your SAE this week?
- Have you logged your SAE hours?
- What challenges are you having with your SAE?
- How can your instructor help you?
- Do you have the equipment you need?

Soils and Fertilizers

Simon Kadula/Shutterstock.com

Learning Outcomes

By the end of this lesson, you should be able to:
- List the nutrients required for plants to grow. (LO 07.03-a)
- Explain the difference between macro and micronutrients. (LO 07.03-b)
- Discuss the importance of soil pH. (LO 07.03-c)
- Describe the role fertilizers play in maintaining healthy soil. (LO 07.03-d)

Words to Know

complete fertilizer	macronutrients	soil test
fertilizer	micronutrients	
incomplete fertilizer	soil pH	

Before You Begin

Let's play a game: How many nutritious foods can you name in 30 seconds?

I had 15—did you beat me? That was pretty easy. What makes foods nutritious? They are full of vitamins and nutrients. Did you know soil can be healthy too? Soil is also full of nutrients. In fact, scientists test soils to find their nutrient levels. With a partner, discuss what scientists might be looking for when they do this.

Figure 7.3.1. Carbon, oxygen, and hydrogen are non-mineral nutrients. They are absorbed by plants in the form of gases (from the air or dissolved in water) and in the form of water itself.

KPG-Payless/Shutterstock.com

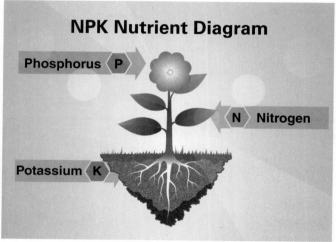

NPK Nutrient Diagram

Phosphorus P

N Nitrogen

Potassium K

AOTTORIO/Shutterstock.com

Figure 7.3.2. Nitrogen, phosphorus, and potassium are the three most important nutrients to a plant.

Pencil case/Shutterstock.com

Figure 7.3.3. Plants absorb macronutrients from the soil. When soil dries out or lacks the needed nutrients, plants start to discolor and drop their leaves.

Soil's physical, chemical, and biological properties affect how plants grow. Chemical properties are especially important, as they regulate the nutrients available to plants. Without nutrients from the soil or applied as fertilizer, plant growth would slow or stop.

Plant Nutrition

To thrive and reproduce, plants must be able to meet their nutritional needs. Scientists have determined there are 17 elements that are essential for healthy plant growth and development. Carbon (C), hydrogen (H), and oxygen (O) are primarily received from water and carbon dioxide in the air (**Figure 7.3.1**). The remaining 14 elements are classified as either macronutrients or micronutrients and should be found in the soil.

Primary Macronutrients

Macronutrients are nutrients that plants need in large quantities. Macronutrients help create plant cells, leading to growth and reproduction. A shortage of these elements can lead to stunted growth, discoloration, poor crop yields, or death. The primary macronutrients are the major building blocks of all fertilizers and make up the bulk of the fertilizer produced.

The primary macronutrients are nitrogen (N), phosphorus (P), and potassium (K) (**Figure 7.3.2**). Nitrogen helps foliage growth and color. Phosphorous assists with the growth of roots and flowers. It also helps plants survive harsh climates and environmental stressors. Potassium contributes to early growth and root development. It also helps plants resist diseases and insects.

Secondary Macronutrients

Calcium (C), magnesium (Mg), and sulfur (S) are absorbed into plants in smaller quantities, making them secondary macronutrients. Plants usually receive calcium and magnesium by mineral weathering or limestone powder. Sulfur may be added to the soil through the air or in fertilizers. Magnesium makes plants green (**Figure 7.3.3**). Sulfur helps plants resist disease and form seeds. It aids in the production of amino acids, proteins, enzymes, and vitamins. Calcium builds cell walls, which helps plants resist disease.

Micronutrients

Micronutrients are nutrients needed in small or limited amounts that are essential for plant growth and development. These nutrients support all aspects of plant growth, including structural integrity, vitamin production, and increasing crop yields.

The eight micronutrients are:

- **Iron (Fe):** makes chlorophyll in plants.
- **Manganese (Mn):** helps iron make chlorophyll, and activates enzymes in the growth process.
- **Zinc (Zn):** is essential in root and plant growth.
- **Boron (B):** regulates the metabolism of carbohydrates in plants. This is critical for new growth and assists in pollination and fertilization.
- **Copper (Cu):** activates enzymes in plants.
- **Chlorine (Cl):** is required for photosynthesis and root growth.
- **Molybdenum (Mo):** processes nitrogen.
- **Nickel (Ni):** creates viable seed.

Did You Know?

Healthy soil has amazing water-retention capacity. Every 1 percent increase in organic matter results in as much as 25,000 gallons of available soil water per acre.

Soil pH

Soil pH is a measurement of the hydrogen activity in soil. Do you remember how plants get hydrogen? Hydrogen mostly comes from water (**Figure 7.3.4**). Any time you water a plant, the hydrogen in the water enters the soil and changes the pH (or hydrogen activity) of the soil. You can see this effect on the

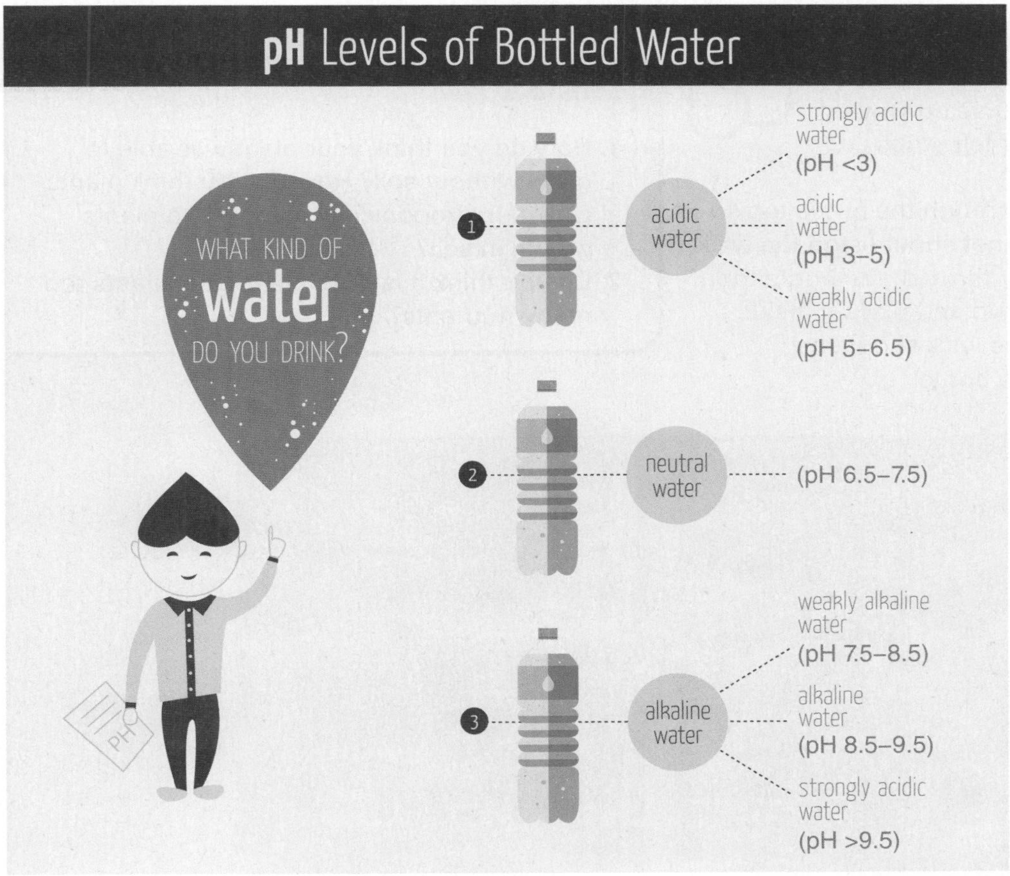

pH Levels of Bottled Water

WHAT KIND OF **water** DO YOU DRINK?

1 — acidic water
- strongly acidic water (pH <3)
- acidic water (pH 3–5)
- weakly acidic water (pH 5–6.5)

2 — neutral water — (pH 6.5–7.5)

3 — alkaline water
- weakly alkaline water (pH 7.5–8.5)
- alkaline water (pH 8.5–9.5)
- strongly acidic water (pH >9.5)

ShelleyFox/Shutterstock.com

Figure 7.3.4. *Have you ever noticed the notation* pH *on a bottle of water? What does* pH *mean?*

7

Hydroponics

Soil gives most plants the nutrients they need to grow. Plants can also grow without soil, though, provided they get the necessary nutrients. Growing plants without soil is called *hydroponics*.

Collect the following items:

- Two-liter plastic bottle
- Scissors or utility knife
- Rubber gloves
- Safety glasses/ goggles
- Hydroponic clay pebbles or coconut coir
- Green leafy plant, such as spinach, lettuce, oregano, or basil
- Filtered, purified, or bottled water
- 2 cotton or felt strips
- Hydroponic growing solution/complete liquid fertilizer
- Measuring spoons and cup
- Large container with lid

Here are your directions:

1. Cut the top off the two-liter bottle, just where it starts to curve downward.
2. Knot the two cotton or felt strips together at one end.
3. Push the loose ends through the bottle top so they hang down; the knot should stop the wick from going all the way through the bottle top.
4. Flip the top upside down and place it inside the two-liter bottle. The wick should be hanging down into the bottle.

5. Fill the top of the bottle with the clay pebbles or coconut coir.
6. Put on your gloves and safety goggles.
7. Measure 1 quart of filtered, purified, or bottled water into your large container.
8. Read the label of the liquid nutrient bottle or hydroponic solution to determine how much solution is needed for 1 quart of water. Mix the determined amount into your container of water. You may not use all of it at first, so put on a lid to store the remainder for later use.
9. Place your plant into the clay pebbles or coconut coir, placing the roots directly on top of the wick.
10. Pour the water solution over the plant. Keep pouring into the reservoir until about half of the wick is underwater.
11. You may have to water your plant with your leftover water solution for a few days until the roots get established. Then, the plant will be able to pull water and nutrients up through the wick.

Consider This

1. How do you think your plant was able to grow without soil? How do you think plants grown hydroponically compare to plants grown in soil?
2. Do you think it is possible to give plants too many nutrients?

Brostock/Shutterstock.com

The pH Scale

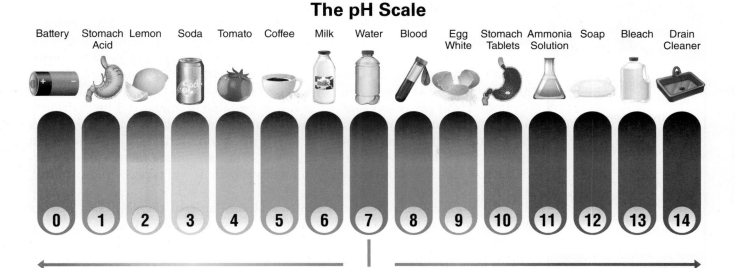

Battery | Stomach Acid | Lemon | Soda | Tomato | Coffee | Milk | Water | Blood | Egg White | Stomach Tablets | Ammonia Solution | Soap | Bleach | Drain Cleaner

0 1 2 3 4 5 6 7 8 9 10 11 12 13 14

Acidic   **Neutral**   **Alkaline**

BlueRingMedia/Shutterstock.com

Figure 7.3.5. Think of the pH scale like a thermometer. Lower numbers indicate an acidic substance (shown in red) and higher numbers indicate an alkaline substance (shown in blue). *Do the pH levels of any of the items pictured surprise you?*

pH measurement scale (**Figure 7.3.5**). The middle number, 7, shows a neutral pH. Numbers below 7 mean the soil has higher levels of free hydrogen in the soil which makes the soil acidic. Soils that have fewer hydrogen ions in the soil have pH measurements higher than 7, which means the soil is alkaline.

Water is not the only factor that affects soil pH. Soil pH can also be changed by chemicals. These chemicals can be added as fertilizers, come from pollution (**Figure 7.3.6**), or naturally exist in the surrounding environment.

Hydrogen ion concentration as pH of precipitation, 2002

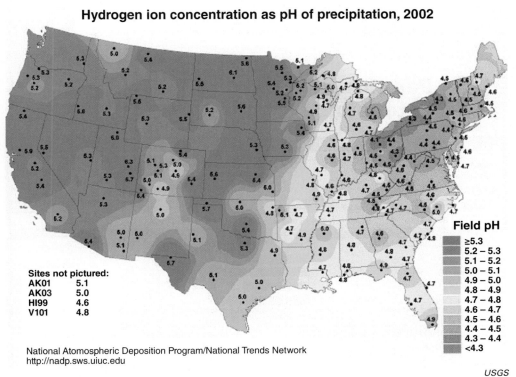

Sites not pictured:
AK01   5.1
AK03   5.0
HI99   4.6
V101   4.8

Field pH
≥5.3
5.2 – 5.3
5.1 – 5.2
5.0 – 5.1
4.9 – 5.0
4.8 – 4.9
4.7 – 4.8
4.6 – 4.7
4.5 – 4.6
4.4 – 4.5
4.3 – 4.4
<4.3

National Atmospheric Deposition Program/National Trends Network
http://nadp.sws.uiuc.edu

USGS

Figure 7.3.6. Rainfall can affect the pH of the soil. Locate your home state on the map. *What is the typical pH of the rainfall in your area?*

Why does pH matter? When the pH of soil is extremely low or extremely high, plants cannot take in needed nutrients. In strongly acidic soils, manganese and aluminum reach toxic levels. At these levels, calcium, phosphorus, magnesium, and molybdenum are less likely to be absorbed by a plant's roots. When soil pH value is neutral or high, phosphorus, iron, copper, zinc, boron, and manganese are less likely to be absorbed.

If your soil pH is too high or low, it is possible to adjust it. Finely ground agricultural limestone (called *lime*) is commonly used to decrease the acidity of the soil and increase the soil pH. If soil pH is too high, sulfur can be added to decrease alkalinity and lower the pH. There are many factors that determine how much treatment needs to be used, such as soil type, organic matter, and soil use.

Soil tests calculate the nutrient levels available in the soil. Test results are interpreted by a laboratory and fertilizer suggestions are given. The standard soil test checks for pH, nitrogen, phosphorus, potassium, calcium, magnesium, sodium, sulfide, and salinity (salt). Most soil tests can be done through your state university extension service for a small fee (**Figure 7.3.7**).

Did You Know?

If you grow crops, you should have your soil tested every two or three years. It is important to test your soil at the same time each year so test results can be fairly compared.

Fertilizer Composition

A **fertilizer** is a substance that is spread on the ground or mixed in the soil to provide nutrients for plant growth and health. Fertilizer bags are labeled with at least three numbers, listing the amount of nitrogen (N), phosphorus (P), and potassium (K) (commonly referred to as N-P-K).

Figure 7.3.7. *Can you locate the soil pH and major plant nutrients on these soil test results? What do these results show?*

PennState Extension

Agricultural Analytical Services Laboratory
The Pennsylvania State University
111 Ag Analytical Svcs Lab
University Park, PA 16802

(814) 863-0841 aaslab@psu.edu www.aasl.psu.edu

SOIL TEST REPORT FOR:			ADDITIONAL COPY TO:				
Your name and contact information			You can add your local extension educator's contact information to get a copy of your report				

DATE	LAB #	SERIAL #	COUNTY	ACRES	ASCS ID	FIELD ID	SOIL
00/00/2019	S00-LABID	00000	County field in	10		Field name	

SOIL NUTRIENT LEVELS			Below Optimum	Optimum	Above Optimum
¹Soil pH	5.9				
²Phosphorus (P)	31	ppm			
³Potassium (K)	82	ppm			
⁴Magnesium (Mg)	74	ppm			

RECOMMENDATIONS: (See back messages for important information)

Limestone*: 5000 lb/A for a target pH of 7.0.   **Magnesium (Mg):**   NONE

*Calcium Carbonate equivalent

Extension.PennState.Edu

The numbers represent the amount of each element in percentage that can be found in 100 pounds of fertilizer (**Figure 7.3.8**). For example, a 100-pound bag of 5-10-15 fertilizer would contain five pounds of nitrogen, 10 pounds of phosphorus, and 15 pounds of potassium. The remaining 70 pounds is filler which helps to spread the fertilizer evenly and avoid burning the plant.

A **complete fertilizer** contains the three primary macronutrients: nitrogen, phosphorus, and potassium. An **incomplete fertilizer** is missing one or more of the primary macronutrients. For example, an incomplete fertilizer bag might be labeled 0-45-0 or 18-46-0.

Figure 7.3.8. A complete fertilizer has all three of the primary macronutrients: nitrogen, phosphorus, and potassium.

Hands-On Activity

Soil Lab Analysis

Analyzing soil test results requires complex equipment and knowledge. This activity will give you a sense of how to read the results of a soil test.

You will need the following materials:
- Fun-size bag of M&Ms
- Colored pencils
- Paper

Here are your directions:

1. Open your bag of M&Ms (do not eat any!) and separate each color into piles.
2. On your paper, list how many M&Ms you have of each color. Each color represents a different element in the soil.
   A. Red = Nitrogen (N)
   B. Orange = Phosphorus (P)
   C. Yellow = Potassium (K)
   D. Green = Calcium (Ca)
   E. Dark Brown = Organic Matter (OM)
   F. Blue = pH
3. Create a bar graph to show the analysis of your soil. Draw a chart with 6 columns. Draw 8 rows across your columns, creating a grid.
4. Using your red colored pencil, color in the number of red M&Ms you have, one square per M&M. Start at the bottom of the chart; then color up for the number of M&Ms. If you do not have any red M&Ms, leave this column empty.

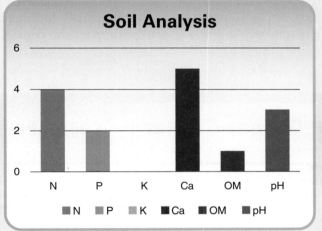

5. Complete the rest of your chart with the remaining colors—one square per M&M starting at the bottom and working your way to the top.
6. Now label your chart. Put the correct letter representing each element underneath the corresponding column.

Consider This

1. Looking over your chart, what can you tell about your soil?
2. What recommendations would you make to improve your soil?
3. Compare your results to some of your classmates'. Which had the best soil? What made you come to this conclusion?

Fertilizer Application

The timing and method of fertilizer application are affected by the soil. Sandy soils require more frequent fertilizer applications and lower amounts of nitrogen than clay soils. Other factors affecting application include the crop being grown, frequency and amount of watering, and the type of fertilizer. For example, sweet corn requires high levels of nitrogen and may require fertilization two or more times during the growing season. However, if tomatoes are fertilized with too much nitrogen, they might not bear any fruit.

There are different methods of applying fertilizer. These include:

- Broadcast. A recommended amount of fertilizer is spread evenly over the area and can be mechanically incorporated into the soil.
- Band. Narrow bands of fertilizer are applied in furrows 2 to 3 inches to the side of the planting area and 1 to 2 inches deeper than the seeds or plants that are to be planted.
- Side-dress. Fertilizer may be applied to the side of the row 6–8 inches away from the plants that are already growing.
- Foliar Feed. Plants are fed by applying liquid fertilizer directly to the leaves. This method is used when insufficient fertilizer was applied before planting, a quick response of plant growth is needed, or when the soil is too cold for a plant to take up nutrients.
- Starter Solutions. Small amounts of fertilizer are applied near the seed to meet the demands of seedlings until a root system develops.

Career Corner   SALES REPRESENTATIVE

A **sales representative** *is responsible for an assigned territory and a product (such as fertilizers) that they sell to agricultural suppliers, dealers, or producers.*

What Do They Do? Sales reps are busy! They maintain and grow their business accounts, introduce new products, recommend programs for brand building and line expansion of existing brands, and provide educational training to accounts through seminars and trade events.

What Education Do I Need? A bachelor's degree in agricultural business or an agricultural major that relates to the product you are representing (i.e., agronomy, animals, poultry, horticulture) is required.

How Much Money Can I Make? Sales representatives make an average of $61,000 per year.

How Can I Get Started? Learn more about the Agricultural Sales CDE offered through the National FFA Organization and train a team to compete. Research and write a report on different sales tactics that have been proven successful in the agriculture industry.

Budimir Jevtic/Shutterstock.com

What Did We Learn?

- There are 17 essential elements needed for plant growth and development.
- Nitrogen, phosphorus, and potassium are the three most important elements for plant growth. These are known as the primary macronutrients.
- Micronutrients are nutrients needed by plants in smaller or limited amounts.
- Soil pH affects how well nutrients are absorbed by plants.
- Soil tests should be conducted before planting to determine the current pH and nutrients in the soil.
- A complete fertilizer contains all three primary macronutrients, while an incomplete fertilizer is missing one or more of the primary macronutrients.

Let's Check and See What We Know

Answer the following questions using the information provided in Lesson 7.3.

1. Scientists have determined there are _____ elements essential for plant growth and development.

   A. 5
   B. 7
   C. 17
   D. 20

2. Which three elements are primarily received from the water and carbon dioxide in the air?

   A. hydrogen, oxygen, and carbon
   B. nitrogen, phosphorus, and potassium
   C. hydrogen, oxygen, and nitrogen
   D. nitrogen, carbon, and oxygen

3. Which elements are the primary macronutrients needed for plant growth?

   A. hydrogen, oxygen, and carbon
   B. nitrogen, phosphorus, and potassium
   C. hydrogen, oxygen, and nitrogen
   D. nitrogen, carbon, and oxygen

4. Which element is responsible for helping grow healthy roots and flowers?

   A. Nitrogen
   B. Sulfur
   C. Phosphorus
   D. Potassium

5. *True or False?* Primary and secondary macronutrients are needed by plants in equal amounts.

6. Soil _____ is an indication of the hydrogen levels in the soil.

   A. elements
   B. testing
   C. pH
   D. limestone

7. *True or False?* A soil pH of 7 indicates neutral soil.

8. *True or False?* If soil pH is too high, sulfur can be added to lower the pH of the soil.

9. A 100-pound bag of 5-10-15 fertilizer would contain _____ pounds of potassium.

   A. 5
   B. 10
   C. 15
   D. 100

10. A(n) _____ fertilizer is missing one or more of the primary macronutrients.

    A. complete
    B. incomplete
    C. developed
    D. undeveloped

Let's Apply What We Know

1. Create a chart listing each of the macro and micronutrients and what each element does to help with plant growth and development.
2. In a complete fertilizer, what do each of the three numbers in the analysis mean? If a 100-pound bag of fertilizer reads 6-22-9, how many pounds of nitrogen are in the bag?
3. How does pH affect the nutrient balance of the soil and plant growth?
4. Find pictures of plants showing signs of missing each of the primary and secondary macronutrients. Create a collage of your findings, labeling each nutrient deficiency.
5. Collect soil samples from your school or home. Turn them in to your local extension office for testing. Share your results with your classmates.

Academic Activities

1. **Math.** Research the recommended rate of 10-10-10 fertilizer to be used on a 100-square-foot, 500-square-foot, and 1,000-square-foot vegetable garden.
2. **Science.** Plants need nitrogen to grow healthy stems and leaves. Compare plants grown without nitrogen fertilizer to plants grown with nitrogen fertilizer. What differences do you see?

Communicating about Agriculture

1. **Visual Aids.** Create a poster explaining to your classmates what soil pH means and why it matters.

SAE for ALL Opportunities

1. **Foundational SAE.** Contact a fertilizer sales professional and ask them to come speak to your class about information that is important for them understand in their line of work.
2. **Research SAE.** Conduct an experiment on the range of soil pH and how it affects nutrients, minerals, and plant growth.

SAE for ALL Check-In

- How much time have you spent on your SAE this week?
- Have you logged your SAE hours?
- What challenges are you having with your SAE?
- How can your instructor help you?
- Do you have the equipment you need?

Fertilizer Sources

Essential Question

Why do farmers choose to use synthetic fertilizers instead of organic, and vice versa?

Learning Outcomes

By the end of this lesson, you should be able to:
- Describe the sources of nutrients needed for plant growth. (LO 07.04-a)
- Compare organic and inorganic fertilizers. (LO 07.04-b)

Words to Know

by-product	inorganic fertilizer	organic fertilizer
compost	manures	

Before You Begin

What does organic *mean? Organic materials are derived from living matter or the waste or by-product of living matter. With a partner, make a list of items that might be used to make organic fertilizer.*
No peeking ahead for an answer!

7

Simon Kadula/Shutterstock.com

Figure 7.4.1. Selecting the right fertilizer is important for plant growth and health.

Farmers, gardeners, landscapers, golf course managers, and many others rely on fertilizers to keep their crops, plants, and turf growing and healthy. Fertilizers provide nutrients to soil and plants in synthetic (man-made) or natural sources. There are several types of fertilizer that can be used (**Figure 7.4.1**).

Career Corner ▸ AERIAL APPLICATOR / AGRICULTURAL PILOT

If this interests you (and you have a sense of adventure!), perhaps being an **Aerial Applicator** *is for you! Agricultural pilots fly small planes at low altitudes to apply pesticides, fertilizers, or fungicides on fields.*

What Do They Do? Agricultural pilots apply essential pesticides, fertilizer, or fungicides to crops; assist firefighters in the containment and extinguishing of forest fires; and mix and add chemicals to be applied.

What Education Do I Need? You must obtain a commercial pilot license through the Federal Aviation Administration (FAA) with the required flying hours and medical examinations. You must also attend agricultural pilot training to pursue a career as an agricultural pilot.

How Much Money Can I Make? Agricultural pilots make an average of $80,000 per year.

Grindstone Media Group/Shutterstock.com

How Can I Get Started? You can build an SAE around aerial application in your area. Can you research what laws are currently in effect for your community and state and create a class presentation about applying chemicals by plane? Can you interview landowners or farmers who currently use aerial pilots? Can you contact the forestry service in your area and discuss ways agricultural pilots help with wildfires?

Sources of Fertilizer

Many different substances are used to provide the essential nutrients needed to make fertilizer. These compounds can be mined from the earth or gathered from naturally occurring sources. Some examples include sodium nitrate, seaweed, potash, and phosphate rock (**Figure 7.4.2**). Fertilizer compounds can also be created from chemical raw materials like ammonia, nitric acid, or ammonium phosphate.

Organic Materials

An *organic fertilizer* is a naturally occurring nutrient material that originates from plants or animals. Organic fertilizers release nutrients into the soil as they decompose. The rate of decomposition determines how much fertilizer is needed. Organic fertilizers improve the texture, aeration, and drainage of the soil. Home gardens often use kitchen and yard wastes as organic fertilizer.

Large commercial growers do not normally use organic fertilizers. One reason is that large amounts of organic fertilizer are needed to get improved results. Another reason is that some types of organic fertilizers may contain unwanted seeds or undesirable odors (**Figure 7.4.3**).

Yarlander/Shutterstock.com

Figure 7.4.2. Seaweed shares no diseases with land plants, which makes it a good choice for an organic fertilizer.

Did You Know?

Native Americans used fish as a fertilizer, planting fish skeletons in the rows with their corn crops.

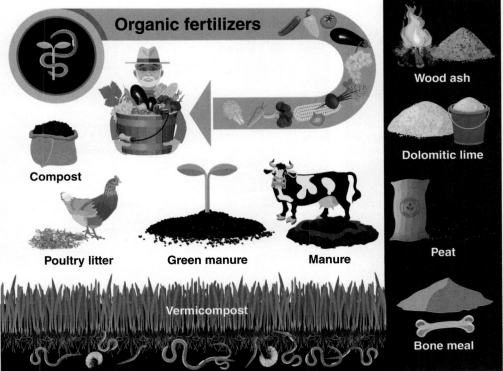

Organic fertilizers

Compost

Poultry litter   Green manure   Manure

Vermicompost

Wood ash

Dolomitic lime

Peat

Bone meal

Hennadii H/Shutterstock.com

Figure 7.4.3. Organic fertilizers come from many different sources. You might be curious about the worms: *vermiculture* is the product of using worms to create a mixture of decomposing food waste, bedding materials, and vermicast (worm feces). *Can you think of any other forms of organic fertilizer?*

David Tadevosian/Shutterstock.com

Figure 7.4.4. Chicken feathers are collected, dried, and ground into meal to be used as an organic fertilizer. Feather meal can have a nitrogen content of 12–14 percent.

Manures and Animal By-products

Manures are a combination of animal feces and bedding material, such as straw or wood shavings. Manure's nutrients can help build and maintain soil fertility. Animal manures vary greatly in the nutrients they provide, depending on the livestock species, diet, and the storage and age of the manure. Manure that has been properly composted is typically an excellent source of nitrogen, phosphorus, potassium, and other micronutrients.

Animal by-products are also an excellent nutrient source. A *by-product* is a secondary product made in the manufacturing of something else. In animal production, these include items like feather meal from a poultry house or bone meal from slaughterhouse waste (**Figure 7.4.4**). Animal by-products are often high in nitrogen.

Compost

Compost is decomposed organic matter that can be added to the soil to help plants grow. The process of composting uses naturally occurring microorganisms to break down organic matter into useful soil nutrients. Adding composted material to soil improves its physical structure, which helps with drainage and water-retention.

Successful composting (**Figure 7.4.5**) requires:
- Green plant material, such as grass clippings
- Dried material, such as fallen leaves or sawdust
- Microorganisms naturally found in soil
- Aeration (without air, decomposition cannot take place)
- Moisture

Did You Know?

It would take 7 billion cattle to produce enough manure to replace synthetic fertilizers.

JurateBuiviene/Shutterstock.com

Martin Christopher Parker/Shutterstock.com

Figure 7.4.5. Compost bins provide a place for organic material to decompose. Bins can be purchased commercially or made from wood or other material. Whether store bought or homemade, it is important that your compost has plenty of ventilation. *Does your school have a composting system for lunchroom waste?*

Hands-On Activity

Incomplete Fertilizers: Can Plants Survive on One Nutrient?

You have learned a lot about plant nutrition needs, but do you think plants can survive on only one nutrient? What do you think will happen if plants do not get all the nutrients they need? Complete the hands-on activity and draw your own conclusions.

You will need the following materials:

- 5 clear plastic cups
- Potting soil
- Spoon or popsicle stick
- A high-nitrogen fertilizer (such as blood meal)
- A high-phosphorus fertilizer (such as bone meal)
- A high-potassium fertilizer (such as kelp meal or potash)
- A balanced/complete fertilizer (such as 10-10-10)
- Permanent marker
- 1 package of bean seeds – 15 seeds total
- Water
- Notebook
- Pencil
- Ruler

Here are your directions:

1. Using a pencil, poke a small hole at the bottom of each cup for drainage.
2. Fill each cup about 2/3 full of potting soil.
3. Label the cups with the marker: Cup 1 – N, Cup 2 – P, Cup 3 – K, Cup 4 – All, Cup 5 – None. Add a teaspoon of the correct fertilizer to Cups 1 through 4 and stir well. Keep Cup 5 without fertilizer for comparison.
4. Press three bean seeds into each cup, making sure to keep them against the side of the cup where you can see them sprout. Place the beans evenly around the cup, and plant them about an inch deep.

uparmar34617/Shutterstock.com

5. Water slowly until water begins to drain out the bottom.
6. Place the cups in a sunny location and wait for the beans to sprout. Sprouting should take 2–5 days.
7. Once sprouting occurs, keep good records of the dates of each, noting differences for each fertilizer cup.
8. As roots, stems, and leaves begin to form, measure daily and note the growth for each cup.

Consider This

1. How did each nutrient and fertilizer type affect the beans?
2. What were the differences in plant growth and appearance for each nutrient?
3. Is there a "best" nutrient for plant growth?

saoirse2013/Shutterstock.com

Figure 7.4.6. Inorganic fertilizers are created in large factories. The process starts by taking the raw product and converting it into the desired granulated fertilizer form, before it is blended with other minerals and filler. The end product is bagged or sold in bulk.

Inorganic Fertilizers

Inorganic fertilizers (also known as *synthetic fertilizers*) are nutrient compounds that are derived from materials other than plants and animals, such as mineral salts. During World War II, the United States government built factories to transform large amounts of nitrogen collected from the atmosphere into nitrogen compounds for explosives. After the war, these factories were converted to produce fertilizers (**Figure 7.4.6**). By 1985, American farmers were using approximately 11 million tons of nitrogen fertilizers each year.

The most widely used nitrogen fertilizer is pure ammonia, kept in its liquid form under pressure in steel tanks.

Making Connections #|STEM

Organic versus Inorganic

There has long been a debate whether organic fertilizers work as well or better than inorganic fertilizers. Conduct an experiment, record your results, and settle the debate for yourself.

Collect the following items:

- 18 bean seeds
- 6 containers
- Soil
- Water
- Labels
- Marker
- Organic complete fertilizer
- Inorganic (synthetic) complete fertilizer

Here are your directions:

1. Fill 6 containers with the same soil.
2. Plant 3 bean seeds in each container.
3. Mix organic fertilizer according to the directions provided and water/treat two of the containers with organic fertilizer. Label both containers "organic."
4. Mix the inorganic fertilizer according to the directions provided and water/treat two of the containers with inorganic fertilizer. Label both containers "inorganic."
5. Leave the last two containers and water with untreated water. Label both containers "no treatment."
6. For one month, monitor the containers and continue to water as needed with the correct treatment for each container.
7. Record how many days before your seeds sprouted.
8. Record plant height of each container for one month.
9. Graph your results.

Consider This

1. According to your results, which treatment option worked best?
2. Did these results surprise you? Why? Or why not?
3. What recommendations would you make to someone looking to purchase fertilizer for their home garden? Provide facts to back up your recommendation.

Ammonium sulfate, ammonium nitrate, and ammonium phosphate are solid nitrogen fertilizers. Urea is another solid nitrogen form that releases its nitrogen over a long time. The phosphorus compound in fertilizer is made using sulfur, coal, and phosphate rock. The potassium comes from potassium chloride, a primary component of potash. Calcium is obtained from limestone. The magnesium in fertilizers comes from dolomite (**Figure 7.4.7**). Sulfur is mined and added to fertilizers.

Olga Maksimava/Shutterstock.com

Figure 7.4.7. Potassium and magnesium are mined using large excavators.

Did You Know?

Ammonia was discovered in 1908 by German chemists Fritz Haber and Carl Bosch. This was a breakthrough in creating modern day fertilizers on an industrial scale.

Many organic fertilizers release nutrients over time as they decompose, while inorganic fertilizers are produced in a form that is readily absorbed in full. There are two common types of inorganic fertilizers: quick release and slow release. Quick release fertilizers, or water-soluble fertilizers, are not coated, and the nutrients are released when the fertilizer is exposed to moisture in the air or when it encounters plant material or soil. Slow release fertilizers have a thin resin-based coating which gradually wears away and releases nutrients over an extended period. These are also known as water-insoluble fertilizers. See **Figure 7.4.8** to see a breakdown of the differences.

Figure 7.4.8. There are advantages and disadvantages to all types of fertilizers. Consumers must determine which option is best for them and their plant needs.

Type of Fertilizer	Advantages	Disadvantages
Slow Release Fertilizer	• Fewer applications • Low plant burn potential • Slower nutrient release rate	• Cost is high • Release rate is governed by factors other than plant needs
Quick Release Fertilizer	• Fast acting • Low cost • Easily found in stores/garden centers	• Greater potential to burn plants • Solidifies in the bag when exposed to moisture
Organic Fertilizer (such as manures)	• Low plant burn potential • Relatively slow release • Contains micronutrients • Conditions soil	• Can be bulky or difficult to handle • Higher in salt and can be inconsistent • Odor • Expensive per pound of actual nutrient • May contain unwanted seeds

M. Riley

What Did We Learn?

- Organic fertilizer is a naturally occurring nutrient material that comes from plants, animals, or their by-products.
- Organic fertilizers slowly release their nutrients as they decompose. Manures and compost are the two most-used organic fertilizers.
- Inorganic fertilizers are derived from materials other than plants and animals and are often mined from the earth.
- Inorganic fertilizers provide nutrients that are readily available for plant use.
- Inorganic fertilizers are produced in two forms: slow release and quick release.

Let's Check and See What We Know

Answer the following questions using the information provided in Lesson 7.4.

1. Organic fertilizers release their nutrients slowly as they _____.
   A. decompose
   B. form
   C. improve
   D. develop
2. *True or False?* Large commercial greenhouse growers typically use organic fertilizers.
3. A fertilizer composed of animal waste and bedding material is _____.
   A. compost
   B. waste
   C. manure
   D. inorganic
4. Feather meal from a poultry house is an animal _____ that can be used as a fertilizer.
   A. by-product
   B. waste
   C. manure
   D. compost
5. Decomposed organic matter that can be added to the soil to help plants grow is _____.
   A. by-product
   B. waste
   C. manure
   D. compost
6. *True or False?* An important part of a successful composting bin is aeration.
7. *True or False?* Many inorganic nutrient compounds are mined from the earth.
8. _____ release fertilizers have a thin resin-based coating that wears away over time, providing nutrients to plants.
   A. Quick
   B. Slow
   C. Control
   D. Timed

Let's Apply What We Know

1. Create a chart listing the pros and cons of organic and inorganic fertilizers.
2. Compare the different types of organic fertilizers and how they are used.
3. What are the key components for successful composting? Explain why each is necessary.
4. Research inorganic fertilizers and how they are made.

Academic Activities

1. **Language Arts.** Using pictures from magazines or free online resources, create a collage that showcases the difference between organic and inorganic fertilizers. Show and discuss your collage in a group.
2. **Science.** Most plant species and crops have different nutrient requirements. Research a plant or crop that you find interesting. Write a report that summarizes your findings on their nutrient needs and fertilizing requirements.

Communicating about Agriculture

1. **Persuasive speech.** Pretend that you are a farmer trying to decide what type of fertilizer to use on your field. Put together a 2–3 minute speech arguing in favor of one type of fertilizer over another. Present it to your class, and then take a poll to see if they found your argument persuasive.

SAE for ALL Opportunities

1. **Entrepreneurship SAE.** Create and sell your own homemade fertilizers.
2. **School-Based Enterprise SAE.** Build a composting bin at your school. Collect waste from the school cafeteria to use in the compost. Create compost that can be used in your school greenhouse or gardens, or sold to local gardeners.

SAE for ALL Check-In

- How much time have you spent on your SAE this week?
- Have you logged your SAE hours?
- What challenges are you having with your SAE?
- How can your instructor help you?
- Do you have the equipment you need?

SAE for **ALL** Profile
It All Started with Flowers

Theresa Fletcher can't remember a time in her life that didn't involve flowers. Theresa often spent summers and holidays at her grandmother's house while her parents worked. Her grandmother had a flower garden that took up almost her entire yard. There were perennial flower beds and vegetable gardens and everything in between. Theresa loved spending all day outside with her grandmother transplanting, trimming, watering, or whatever needed to be done.

When she reached middle school, she was excited to take agriculture class and spend time in the school greenhouse. Those were her favorite days! When her teacher started talking about needing an SAE project, Theresa already had something in mind. She started a roadside flower market at her grandmother's house. She divided the ever-spreading plants in the many gardens and potted them for sale. She cut flowers and learned from her grandmother how to create little bouquets. She split a small portion of her earnings with her grandmother and used some of the money to put back into the garden, planting new varieties of plants.

By high school, Theresa had over 50 different varieties of perennials and cut flowers in her market. Her SAE grew and once she was able to drive, she moved her flower market to the local farmers' market more centrally located in town. Theresa took every horticulture class offered in high school and was always eager to learn more about plants. Her teacher helped feed her passion for plants by assigning her individual tasks in the school greenhouse, which also led to her role as the school greenhouse manager.

After graduating high school, Theresa had no doubts she was going to college to major in Horticulture at North Carolina State University. While in college she learned about a summer internship program through Ball Horticulture company. Though she was nervous about traveling to Chicago for the summer, she was excited about the things she could learn working in the industry. That internship solidified Theresa's interest in entering horticulture as a career, and she returned to Ball after college. As a Product Development Manager, Theresa works directly with the Ball FloraPlant Research and Development Team evaluating crops, visiting trial sites, taking notes, making reports, communicating findings, and getting pictures of various trials (Advanced Grower Trials, Culture Trials, Production Trials, and Breeder Trials). She loves what she does – and it all started with growing summer flowers with her grandmother!

- Do you participate in any activities now that could be a springboard to learning something new?
- Have you ever spent time with beautiful plants? How did it make you feel to be around them?

Top: michaeljung/Shutterstock.com
Bottom: Elena Rostunova/Shutterstock.com

Maria Sbytova/Shutterstock.com

What Is Horticulture?

Essential Question

What role does horticulture play in your daily life?

Learning Outcomes

By the end of this lesson, you should be able to:
- Define horticulture. (LO 08.01-a)
- Describe the segments of horticulture. (LO 08.01-b)

Words to Know

floriculture
greenhouse
hardscapes
horticulture
horticulturist

hydroponics
interiorscaping
landscapers
nursery
olericulture

ornamental horticulture
pomology
turf grass
viticulture

Before You Begin

Hi, everyone! My name is Luisa, and you may remember that I'm a greenhouse manager. Plant jokes are one of my favorite things. I try to share at least one on each episode of my podcast! Want to hear a few? What did one flower say to another? Hey, bud, how's it growing? Ok, I can do better than that. What kind of flower grows on your face? Tu-lips! I crack myself up! We have time for one more, right? Why do flowers drive so fast? Because they put their petals to the metal! Do you get it? It's corny, but it's funny!

You might be wondering why I love flower and plant jokes so much. As a greenhouse manager, it is my job to know a lot about plants. Over this next unit, you will learn all kinds of cool things about plants and how they grow. Can you come up with a plant joke of your own?

Figure 8.1.1. Plants are enjoyed by people all over the world. Even people who live in skyscrapers can enjoy outdoor garden spaces through rooftop gardens and patio plantings.

IndustryAndTravel/Shutterstock.com

The word *horticulture* comes from the Latin *hortus*, meaning "garden," and *culture*, meaning "cultivation." There is more to the horticulture industry, though, than gardening. Horticulture is a highly technical industry that is valued at nearly $50 billion annually in the United States. Horticulture is found everywhere, from rooftop gardens in big cities (**Figure 8.1.1**) to small garden centers, to large farms producing fruits and vegetables. Horticulture is a huge industry that supports the demand for diverse plant products. This lesson will take a closer look into what horticulture means to our everyday life.

What Is Horticulture?

Horticulture is the production, processing, and sale of plants for food, comfort, and beauty. This might seem like it covers everything, but it does not include traditional row crops like grain and fiber. It also does not include forestry products, which focus on timber production. A *horticulturist* is a person who studies and specializes in raising horticultural crops. Horticulturists may grow plants for food or to landscape a new home (**Figure 8.1.2**). There are two main divisions of the horticulture industry: ornamental horticulture and food crop production.

Ornamental Horticulture

Ornamental horticulture is the production of plants for their beauty. Flowers, leaves, or other parts of plants are appealing for their blooms, colors, or smell. These plants are kept indoors in homes or businesses or are used to enhance the beauty of outdoor spaces. The ornamental horticulture industry can be divided into four areas: floriculture, landscaping and nursery production, interiorscaping, and turf grass.

BearFotos/Shutterstock.com

Figure 8.1.2. Horticulturalists typically specialize in either ornamental horticulture or food crop production. Specialized programs are offered at community colleges, universities, in fieldwork, and in global study abroad programs.

Floriculture

Floriculture is the study, cultivation, and marketing of flowers and ornamental plants. Floriculture is where you find the creative and decorative aspects of the horticulture industry. Plants grown for the floriculture industry include bedding plants, houseplants, potted or container plants, cut flowers, and foliage.

UfaBizPhoto/Shutterstock.com

Figure 8.1.3. Florists are professional floral designers who arrange, cut, and dry flowers to create corsages, centerpieces, bouquets, elaborate arrangements, and wreaths for gifts, holidays, weddings, and funerals. They also treat and care for the greenery and flowers in the shop.

Most of this plant production takes place in greenhouses. A *greenhouse* is a structure made of plastic or glass that provides the ideal growing environment for plants to thrive. Many plants in this industry are grown for seasonal demand, such as roses for Valentine's Day or poinsettias for Christmas. Cut flowers are sold to florists who arrange them and sell them to the public (**Figure 8.1.3**). Flowering potted plants are sold in the same container they are grown in and usually live for one growing season. Foliage plants are grown for their leaf colors and are used as houseplants. Annual flowers are used as bedding plants for gardens.

Did You Know? The Netherlands has some of the largest greenhouses in the world. Greenhouses cover more than 26,000 acres in this small country.

Did You Know? Landscaping can add as much as 14 percent to the resale value of your home.

Landscaping and Nursery Production

The landscaping and nursery production industry produces and uses plants to make our outdoor spaces more appealing. A *nursery* is a place where shrubs, ornamental trees, and plants such as perennials, ground covers, and vines are grown for transplanting into landscape areas. When these plants reach maturity, they are planted around homes, businesses, and public areas (like parks) by landscapers.

Landscapers design plans for plant layout, install plant material, and maintain plants in the environment. Landscapers can also design and install irrigation systems, water features such as ponds or fountains, and *hardscapes* (constructed areas within a landscape such as walkways, patios, arbors, and retaining walls). Landscape work is done outside in all climates and types of weather (**Figure 8.1.4**).

Figure 8.1.4. Landscape designs set the theme in every area of Walt Disney's theme parks. Every day, workers weed gardens, remove dead flower heads, and replace withering flowers with new ones to ensure the parks are in pristine condition for guests.

Valerija Polakovska/Shutterstock.com

Interiorscaping

Interiorscaping is the design, installation, and maintenance of foliage plants inside buildings. Most offices and businesses use plants to create an attractive indoor environment. If you have visited an indoor shopping mall, you might have noticed tall plants reaching up to the second story or lots of green plants near the food court (**Figure 8.1.5**). Interiorscapes help filter indoor air, produce oxygen, improve employee productivity, and make the environment beautiful. These plants may be changed out seasonally or to celebrate holidays.

Elnur/Shutterstock.com

Figure 8.1.5. Many businesses provide interiorscaped areas as a place to enjoy lunch, have casual conversations with coworkers, and take breaks.

Turf Grass

Turf grass, or sod, is a collection of green plants that form a ground cover. This includes the cultivation of lawn grasses for homes or commercial sites. It also includes sod grown for athletic fields and golf courses (**Figure 8.1.6**). Sod, like most crops, requires fertilization, maintenance, and a pest management plan to grow a successful crop that can take heavy foot traffic. Sod growers must cut the grass during production to maintain proper height and promote spreading the grass.

Food Crop Production

Many horticulture crops are grown for food. Do you like pizza? Many of the ingredients on your pizza are horticulture food crops. There are the tomatoes that make the sauce, and the olives, peppers, mushrooms, and onions (among others!) that serve as toppings. Horticulture food production can be broken down into three major areas: olericulture, pomology, and viticulture.

Did You Know?

Just 50 square feet of turf grass can release enough oxygen to support a family of four.

karamysh/Shutterstock.com

Figure 8.1.6. Golf course turf grasses require extensive maintenance. Golf course grasses must meet specific standards that include everything from their color and texture to the way a golf ball rolls when it lands on the green.

nicepix/Shutterstock.com

Figure 8.1.7. Hydroponics is the growing of plants with their roots in a nutrient solution, instead of soil. This allows growers to produce many vegetables year-round, in highly controlled environments.

Olericulture

Olericulture is the science, cultivation, processing, storage, and marketing of herbs and vegetables. A vegetable can be the roots, stems, leaves, flowers, or fruit of a plant.

- Vegetables harvested from a plant's roots: sweet potatoes, radishes, carrots
- Vegetables harvested from a plant's stem: celery, asparagus
- Vegetables harvested from a plant's leaves: spinach, kale, cabbage, lettuce
- Vegetables harvested from a plant's flower: broccoli, cauliflower
- Vegetables harvested from the fruit of a plant: beans, peas, tomatoes, cucumbers, squash, peppers

Vegetables are produced for fresh consumption and for processing. Olericulture is found throughout North America in fields, greenhouses, or using *hydroponics* (**Figure 8.1.7**).

Did You Know? California is the leading vegetable producer in the United States. Almost 50 percent of the fresh vegetables purchased in supermarkets across America are grown in California.

Making Connections  # Economics

Tomato: Fruit or Vegetable?

Many people debate whether a tomato is a fruit or a vegetable. In 1893, a case (*Nix v. Hedden*) went to the Supreme Court of the United States to make a final determination on this question. In 1883, President Chester A. Arthur signed the Tariff Act of March 3, which placed a 10 percent tariff on vegetables shipped to the United States from other countries. This law was meant to protect American vegetable growers against growing competition from overseas. The tax was not placed on fruit.

Mr. John Nix was a large wholesale produce magnate in Manhattan, NY. In 1886, he imported tomatoes from the Caribbean. The collector of the Port of New York, Edward L. Hedden, taxed Mr. Nix on the arrival of the tomatoes, which he considered a vegetable. Mr. Nix disputed the tax and sued Mr. Hedden on the grounds that tomatoes were botanically a fruit and therefore should not be taxed.

The case went all the way to the Supreme Court, and the justices decided that while tomatoes do botanically fit the definition of a fruit, consumers do not eat or prepare tomatoes like other fruits. Because consumers relate to the tomato as a vegetable, that is how they should be legally defined.

Conduct your own research on *Nix v. Hedden*. After compiling your research, make your own decision on whether a tomato should be considered a fruit or a vegetable. Find classmates that agree and disagree with your decision. Organize and conduct a class debate and let your class come to a final decision.

Consider This

1. Do you think tomatoes are a fruit or a vegetable?
2. Do you think the results of the court case were fair? Why or why not?
3. What other types of fruits or vegetables do you think might not have a clear classification?

Pomology

Pomology is the cultivation, processing, storing, and marketing of fruits and nuts. Oranges, apples, peaches, cherries, berries, pecans, walnuts, almonds, and pistachios are examples of popular American pomology crops. California is known for almond production, Georgia and South Carolina are known for peach production, Florida is known for citrus production, Washington is known for apple production, and Hawaii is known for macadamia nuts. Fruits and nuts can grow in all parts of the United States, but not every crop can be grown in every climate (**Figure 8.1.8**).

Steven Beck/Shutterstock.com

Figure 8.1.8. Berries such as blueberries, blackberries, and raspberries are often grown under protective shelters called hoop houses. This allows a grower to have a better control of the temperature and give the fruit a longer shelf life.

Hands-On Activity

Propagate Your Own Apple Tree

Apples are a common fruit and a favorite of many people. Have you ever thought about where apples come from or how they grow? Did you know you can grow your own apple tree?

You will need the following materials:

- An apple
- Paper towel
- Sealable plastic bag
- Knife
- Water
- Cool, damp storage location
- Marker

Here are your directions:

1. Carefully cut down the middle of an apple and remove the seeds. You may need a few apples because not all seeds will sprout.
2. Fold your paper towel in half and wet it thoroughly.
3. Place your seeds in the paper towel and fold it in half again, making a square.
4. Place your paper towel with the seeds into your plastic bag. Make sure that the seeds are all still completely covered.
5. Use the marker to write your name and the date on the outside of the bag.
6. Seal your bag completely and store in a cool, damp location such as a refrigerator.
7. Check your bag periodically and make sure the paper towel stays damp. Add more water if needed. It can take up to 30 days for your seeds to sprout.

Consider This

1. Did you find the growing conditions odd? Why do you think the seeds needed cool or cold temperatures to grow?
2. What other types of fruits do you think can be grown from collected seeds?

Lukasz Szwaj/Shutterstock.com

Figure 8.1.9. Grapevines need training to grow on an arbor, wire trellis, or fence. This cuts back on the risk of disease and allows harvesters to have easy access to the fruit.

Viticulture

Viticulture is the cultivation of grapes to be eaten fresh or used in making juice, raisins, jams, jellies, and wines. Today, grapes grow all over the world and in every state of the United States. Ninety percent of these grapes are grown in California, but Michigan, New York, Virginia, and Pennsylvania also produce a lot of grapes. Grape production takes a lot of work. Workers must pay special attention to pruning and training the grape vines to grow properly on the support system (**Figure 8.1.9**). Grapes also require properly fertilized soils.

Career Corner HORTICULTURIST

If this is interesting to you, you might want to come work with me! **Horticulturists** *oversee research programs for a variety of ornamental and vegetative crops.*

What Do They Do? Horticulturists manage crop scheduling for appropriate planting and harvesting; perform propagation, irrigation, and pest management; control plant growth by monitoring fertilizers; and are responsible for greenhouse environment management and maintenance.

What Education Do I Need? Postsecondary education is beneficial but not required. Consider obtaining an associate degree or equivalent experience and/or training. Certifications such as a Pesticide Applicator's License are often preferred.

How Much Money Can I Make? Horticulturists make an average of $55,000 per year.

How Can I Get Started? You can build an SAE around horticulture. You can grow your own vegetables and start a roadside stand or booth at a farmers' market. You can design and conduct experiments testing factors such as fertilizer or light and the effect they have on plants. You can volunteer at a community garden or garden center.

BearFotos/Shutterstock.com

What Did We Learn?

- Horticulture is the production, processing, and sale of plants for food, comfort, and beauty.
- There are two main divisions to the horticulture industry: ornamental horticulture and food crop production.
- Ornamental horticulture is the production of plants for their beauty, and consists of four divisions: floriculture (growing and marketing flowers and ornamental plants), landscaping and nursery production (producing and installing outdoor plants), interiorscaping (installing and maintaining decorative indoor plants), and turf grass (growing and maintaining sod and turf for lawns and specialized surfaces).
- Food crop production can be broken down into three divisions: olericulture (herbs and vegetables), pomology (fruits and nuts), and viticulture (grapes). Food crop production takes place across the world, including every state in the United States.

Let's Check and See What We Know

Answer the following questions using the information provided in Lesson 8.1.

1. The production, processing, and sale of plants for food, comfort, and beauty is known as _____.
   A. horticulture
   B. ornamental horticulture
   C. landscape and nursery production
   D. floriculture

2. *True or False?* There are two main divisions of the horticulture industry: ornamental horticulture and food crop production.

3. _____ is the production of plants for their beauty.
   A. Horticulture
   B. Ornamental horticulture
   C. Landscape and nursery production
   D. Floriculture

4. A _____ is a structure made of plastic or glass that provides the ideal growing conditions for plants to thrive.
   A. nursery
   B. garden
   C. hydroponics
   D. greenhouse

5. *True or False?* Foliage plants are grown for their leaf colors and are usually used as house plants.

6. A place where shrubs, ornamental trees, and plants are grown for transplanting into landscape areas is called a _____.
   A. nursery
   B. garden
   C. hydroponics
   D. greenhouse

7. *True or False?* Hardscapes are constructed areas within a landscape such as walkways and patios.

8. *True or False?* Interiorscapes are areas designed using foliage plants on the outside of a building.

9. _____ is the science, cultivation, processing, storage, and marketing of herbs and vegetables.
   A. Floriculture
   B. Olericulture
   C. Pomology
   D. Viticulture

10. _____ is the science of cultivating, processing, storing, and marketing fruits and nuts.
    A. Floriculture
    B. Olericulture
    C. Pomology
    D. Viticulture

8

Let's Apply What We Know

1. Create a chart explaining the different divisions of the horticulture industry and what each of them represents.
2. Compare and contrast the floriculture, landscape and nursery production, and interiorscaping industries.
3. Compare and contrast each of the different areas of food production horticulture.
4. Select an area of horticulture and investigate its role in your community.
5. List the differences between a greenhouse and a nursery.

Academic Activities

1. **Language Arts.** Create a classification chart for each of the areas of the horticulture industry. Use pictures from magazines or free online resources to show examples of each of the areas (so, a picture of fruit to show pomology). Show and discuss your chart in a group.

Communicating about Agriculture

1. **Speaking.** Create a 3–5 minute presentation about one of the segments of horticulture covered in this lesson. Talk about which states play the largest role in that area, what major crops are produced in different parts of the US and the world, jobs in that area, and what economic impact this segment has on your local community and the US as a whole.

SAE for ALL Opportunities

1. **Foundational SAE.** Job shadow a florist in your community.
2. **Research SAE.** Harvest or purchase several of the same pieces of produce (heads of lettuce, apples, grapes, cucumbers). Design an experiment about different postharvest treatments for your crops. You might look at the effects of washing (or not), different storage temperatures or humidities, or exposure to light. What do your results tell you?

SAE for ALL Check-In

- How much time have you spent on your SAE this week?
- Have you logged your SAE hours?
- What challenges are you having with your SAE?
- How can your instructor help you?
- Do you have the equipment you need?

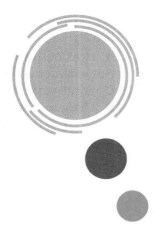

Classifying Plants

Wirestock Creators/Shutterstock.com

Learning Outcomes

By the end of this lesson, you should be able to:
- Explain what scientific names are and why they are important. (LO 08.02-a)
- Define characteristics that scientists use to classify plants. (LO 08.02-b)
- Describe the different life cycles of plants. (LO 08.02-c)

Words to Know

angiosperm	dicot	herbaceous plant
annual	dormancy	monocot
biennial	evergreen plant	perennial
cultivar	gymnosperm	scientific name
deciduous plant	hardiness	woody plant

Before You Begin

When you were small, did you play games where you had to group things together? Do you remember some of the ways you grouped similar items? Maybe you sorted items by shape, or by color, or by size.

In a similar way, scientists group plants into categories. Some groupings are straightforward and simple, like trees, shrubs, or flowers. Others are more technical, such as the shape of their leaves or types of roots. What other ways might plants be grouped?

8

Have you ever taken a walk around your community or hiked a nature trail? Did you notice the different types of plants? There might be small flowers, large shrubs, or trailing ground covers. There are many different types of plants, just as there are many different types of animals. Plants differ in size, shape, color, growth season, or leaf shape (**Figure 8.2.1**). In this lesson, we will take a closer look at how plants are classified and how they can be told apart.

How Are Plants Named?

Imagine someone yelling out the name John in a crowded area. How many people do you think would turn around? Would it help if someone's full name were yelled out instead of just their first name?

Did you know that plants have full names, like people, to help us know what they are? The famous Swedish botanist Carl Linnaeus simplified the naming of plants by giving all plants a **scientific name**: a Latin, two-part name consisting of the genus and species. This naming system is called the *binomial system* for naming plants. Scientific names are unique to each plant; no two plants have the same scientific name. The first name is the *genus*. All plants belonging to the same genus share similar characteristics and are more closely related to each other than plants of any other genus. The second name is the *species*. Plants of the same species will consistently produce plants of the same type. From here, plants can be divided into different varieties, or cultivars. **Cultivars** are a group of plants from the same species grown for desirable, reproduceable traits.

Figure 8.2.1. *Did you know there are more than 1,000 different tree species in North America?* Look at these leaves. *Do any of these look familiar to you?*

Birch · Oak · Viburnum · Chestnut

Linden · Alder · Walnut · Maple

Apen · Rowan · Poplar · Sycamore

adecvatman/Shutterstock.com

Making Connections # Technology

Virtual Herbarium

An *herbarium* is a storage area for collected plant material. Plants can be pressed and mounted to suitable paper and stored in cabinets that can preserve them. Many herbaria are now creating digital collections of their specimens. This gives scientists and researchers across the globe access to a wide variety of plant material. For this activity, create your own virtual herbarium.

Collect 5–10 samples of plant material around your school or home. You will need to collect leaves, bark, or stems, as well as any flowers or fruit. Collect each sample in its own bag labeled with the location and date of collection. Take photos of each of the plants you collected. You will need photos of the leaves and other plant parts you collected for each sample.

Upload your pictures to a computer file or online storage source. For each of your photos, record any plant classification traits you recognize from this lesson such as evergreen or deciduous. Have a teacher, family member, or friend help you identify the plant material, labeling it with the correct common and scientific name.

Consider This

1. What are some of the benefits of creating a virtual plant collection?
2. Do you think there could be benefits to adding pictures to your herbarium during each of the four seasons?

The scientific name helps scientists and horticulturists identify plants regardless of the language they speak or where they live because scientific names are always Latin (**Figure 8.2.2**). Conversely, common names for plants and animals are used by local people, in local languages. These can be totally different from one country to another, one state to another, and even from one county to another. One plant may have many common names, or two different plants can share the same common name. This can be very confusing, as in the case of *Calendula officinalis* and *Tagetes patula*, both commonly referred to as a marigold (**Figure 8.2.3**).

DavidTB/Shutterstock.com

Figure 8.2.2. Scientific names are a worldwide system of naming plants and stay the same no matter where you live. This tree is the *Acer rubrum*, commonly known as the red maple. *Acer* is the scientific name for the genus of all maple trees. In Latin, *rubrum*, the species name, means red. The species name often gives important information about the plant, such as the color.

Elena Koromyslova/Shutterstock.com

mizy/Shutterstock.com

Figure 8.2.3. Both the *Calendula officinalis* (left) and the *Tagetes patula* (right) are referred to as marigolds. *Which one do you think of when you think of marigolds? Why might this be confusing?*

Classifying Plants

Scientists use plant similarities to sort them into groups and help people identify them. The similarities can be based on types of stems, sizes, types of fruit, life cycles, or even leaf shapes. All plants are from the kingdom *Plantae*, but from there they vary greatly.

Reproduction

There are four classifications of plants; angiosperm and gymnosperm are the most common in the horticulture industry, as they both reproduce using seeds. A **gymnosperm** is a plant that carries its seed in a cone, such as a pine tree. An **angiosperm** is a seed-bearing plant that protects its seed in an ovary, such as an apple tree. Within the angiosperm class are two important sub-classes, *Monocotyledonae* (monocots) and *Dicotyledonae* (dicots). **Monocots** are plants characterized by one embryonic leaf in their seedling stage, parallel-veined leaves, and flower parts arranged in multiples of three. Examples of monocots include orchids, bamboo, and grasses. **Dicots** are plants character-ized by two embryonic leaves in their seedling stage, leaf vines that resemble veins or lines crossing, and flower parts in multiples of four or five. Examples of dicots include oak trees, rhododendrons, and herbs. There are more than 175,000 known species of dicots and more than 50,000 species of monocots.

Hands-On Activity

Comparing Monocot and Dicot Seeds

Angiosperms can be separated into two distinct categories: monocots and dicots. What makes the two types different and why is it important to understand which is which? The differences start from the very beginning of the plant's life cycle with the seed. The plant's embryo is in the seed. Monocots have one vein and dicots have two. This small difference at the start of the plant's life cycle leads to each plant developing very differently. Try for yourself and see if you can tell the difference!

You will need the following materials:

- 2 or 3 corn seeds
- 2 or 3 bean seeds
- One sealable plastic bag
- Water
- Paper towel
- Tape
- Marker
- Sunny location, such as a window

Here are your directions:

1. Carefully wet your paper towel until it is soaked through.
2. Fold your paper towel in half and place it in the sealable plastic bag.
3. Place your seeds against the wet paper towel, making sure to group the corn seeds together on one side and the bean seeds together on the other side.
4. Use the marker to write your name and the date on the outside of the bag.
5. Seal your bag completely and tape it to a warm sunny location such as a window. Keep your seeds visible so you can watch them grow.
6. Check your bag periodically and make sure the paper towel stays damp. Add more water if needed.

Consider This

1. Did you notice any distinct differences in the corn and bean seeds? What did you notice?
2. Make a list and categorize other plants you are familiar with as either monocot or dicot.

Stem Type

In the horticulture industry, plants are often classified by their stem type, foliage retention, or life cycle. **_Herbaceous plants_** have stems that are soft and die back to the ground each year. Herbs, vegetables, and ferns are some examples of herbaceous plants. **_Woody plants_** have hard stems and buds that survive through the winter months above ground. The ability of a plant to withstand cold temperatures typical during winter months is known as **_hardiness_** (**Figure 8.2.4**). Shrubs, trees, and some vines are all examples of woody plants (**Figure 8.2.5**).

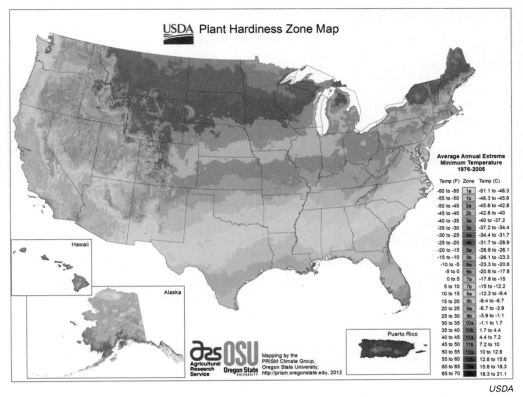

USDA

Figure 8.2.4. The USDA Plant Hardiness Zone Map tells gardeners and homeowners which plants will grow best in their location. The map is based on the average winter temperatures in an area. All plants are assigned numbers to inform consumers if the plant will survive in the hardiness zone they are located.

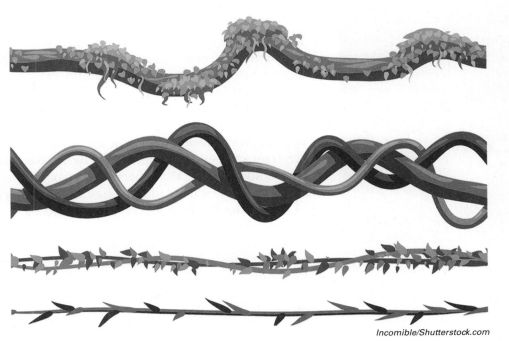

Incomible/Shutterstock.com

Figure 8.2.5. Woody vines and stems are usually brown in color, thick, and hard to break between your fingers. Herbaceous vines and stems are softer and usually green in color. They require very little effort to break. *Which of the pictured stems are herbaceous? How can you tell?*

8

Mario Krpan/Shutterstock.com

Figure 8.2.6. Evergreen trees, like the ones on the left, often have needle-like leaves or flattened leaf blades and keep their leaves all year. Deciduous trees, like the ones on the right, lose their leaves for a portion of the year. *Are the trees around your school or home evergreen, deciduous, or do you have some of both?*

Foliage Retention

When classifying plants, *foliage retention* refers to whether plants lose their leaves in the fall or keep them all year. **Deciduous plants** are leafless during a portion of the year. Deciduous plants typically lose their leaves in the fall and grow them back in the spring. **Evergreen plants** are plants that stay green and keep their leaves year-round (**Figure 8.2.6**). Pines and hollies are examples of evergreen plants.

Did You Know? It can be tricky to tell whether some plants are deciduous or evergreen. Let's look at the live oak as an example. It is a deciduous tree, but it loses leaves year-round and old leaves are not pushed off the stem until new leaves form. This makes it appear evergreen!

Plant Life Cycles

Life cycles refer to the length of a plant's life. Some live only one season, while others live for many years. Plants typically fall into three groups: annuals, biennials, and perennials. An **annual** is a plant that completes its life cycle in one growing season or a single year. Annual plants typically begin their life in either the spring or the fall. They grow, flower, produce seed or fruit and then die when it becomes too cold or too hot for them to grow. Pansies, vinca, and petunias are all examples of annual plants.

A **biennial** is a plant that requires two growing seasons or two years to complete its life cycle. The first year of the plant's life is dedicated to growth and the development of stems and leaves. The second year is more focused on fruit or seed development. These plants typically die after the second year. Hollyhocks, garlic, and celery are examples of biennial plants.

Perennials are plants that have an unspecified life span, meaning they can live for many years. Perennial plants do not die after producing fruit or seed. Perennials can be herbaceous or woody plants. Most herbaceous perennials die back in the winter months before they return in the spring (**Figure 8.2.7**). This period of slow or inactive growth is known as **dormancy**. Dormancy can occur in seeds as well as mature plants. Strawberries, asparagus, and chrysanthemums are examples of herbaceous perennials and holly, ivy, hosta, and junipers are examples of woody perennials.

Elena Elisseeva/Shutterstock.com

Figure 8.2.7. Perennials are commonly planted in home landscaping since many of them are low maintenance and can continue growing for years. Perennials can be flowering or non-flowering, deciduous or evergreen.

PLANT BIOLOGIST

A **plant biologist** *specializes in topics such as plant breeding or genetics. They conduct and support research of plant production.*

What Do They Do? Plant biologists conduct field research programs and perform tasks required for each research trial; study the effects of pollution on plants; produce plants from tissue cultures; identify new species of plants; and monitor, diagnose and treat insect, disease, physiological, or cultural conditions.

What Education Do I Need? Most plant biologists obtain an undergraduate or graduate degree in botany, horticulture, plant pathology, weed science, or agronomy. A pesticide applicator's license is also helpful.

How Much Money Can I Make? Plant biologists make an average of $70,000 per year.

How Can I Get Started? You can build an SAE around horticulture in your area. You can conduct experiments with cross pollination of plants. Can you volunteer at a botanical garden or university in your state? You can interview a plant biologist about their typical day and everyday duties.

Chokniti Khongchum/Shutterstock.com

What Did We Learn?

- Scientific names are used to identify plants worldwide and consist of two Latin names, the genus and the species.
- Botanists use several different methods to identify and name plants, including stem type, foliage retention, and life cycle.
- Gymnosperms carry their seeds in a cone, where angiosperm carry their seeds in the ovary of a plant.
- Angiosperms can be divided into two different categories: monocot or dicot. Monocots have one embryonic leaf in their seedling stage, parallel-veined leaves, and flower parts arranged in multiples of three. Dicots have two embryonic leaves in their seedling stage, leaf vines that resemble veins or lines crossing, and flower parts in multiples of four or five.
- Herbaceous plants have soft stems and die back to the ground each year. Woody plants have hard stems and buds that survive the winter months.
- Deciduous plants lose their leaves for a portion of the year and evergreen plants keep their leaves all year.
- An annual plant completes its life cycle in one year or season, biennial plants require two growing seasons, and perennials have an unspecified life cycle and can live for many years.

Let's Check and See What We Know

Answer the following questions using the information provided in Lesson 8.2.

1. Scientific names are always written in _____.
   A. French   C. English
   B. Latin   D. Greek
2. *True or False?* The scientific name is comprised of the genus and the cultivar.
3. A(n) _____ is a plant that carries its seed inside a cone.
   A. angiosperm   C. monocot
   B. gymnosperm   D. dicot
4. Dicots are characterized by _____ embryonic leaves in their seedling stage.
   A. one   C. three
   B. two   D. four
5. *True or False?* Herbaceous plants usually die back to the ground each winter.

6. _____ is a plant's ability to withstand cold temperatures.
   A. Hardiness   C. Evergreen
   B. Deciduous   D. Dormancy
7. *True or False?* Evergreen plants keep their leaves year-round.
8. *True or False?* An annual plant dies back to the ground in the winter and returns in the spring.
9. Plants that can live for many years through multiple life cycles are _____.
   A. annuals   C. perennials
   B. biennials   D. herbaceous
10. _____ is when plant growth slows or becomes inactive.
    A. Hardiness   C. Evergreen
    B. Deciduous   D. Dormancy

Let's Apply What We Know

1. Explain the difference between the scientific name and the common name of a plant. What problems can arise from using common names?
2. List characteristics botanists and scientists use to classify plants.
3. Chart the differences between monocot and dicot plants. Give examples of each.
4. Sketch a deciduous tree compared to an evergreen tree and a woody stem compared to an herbaceous stem.
5. Compare the differences in annual, biennial, and perennial plants.

Academic Activities

1. **Language Arts.** Create a horticulture haiku using a term from this lesson. A haiku is a three-line poem consisting of 17 syllables. The standard format is the first line has five syllables, the second line has seven syllables, and the third line has five syllables. Example: Pretty maple tree, dropping its leaves in the fall. It's deciduous.
2. **Technology.** Use the USDA Plant Hardiness Zone map to determine what hardiness zone you live in. Research and make a list of annual, biennial, and perennial plants that can survive in your area.

Communicating about Agriculture

1. **Writing and Speaking.** Look at the plants outside your school. Work in a group to identify as many as you can, and describe the characteristics of each as you learned in this lesson. Present your findings to the class, using your new vocabulary, photos, and visual aids.

SAE for ALL Opportunities

1. **Foundational SAE.** Create an herbarium specimen for identification of plants students must know when competing in an FFA Career Development Event. Create a classroom collection of specimens to help CDE teams learn about plant material.
2. **Foundational SAE.** Interview a landscape designer in your community. Determine what plant characteristics they use to help them select plants for their design projects such as plant life cycles and leaf retention. Write a report of your interview and present it to your class.

SAE for ALL Check-In

- How much time have you spent on your SAE this week?
- Have you logged your SAE hours?
- What challenges are you having with your SAE?
- How can your instructor help you?
- Do you have the equipment you need?

8

Parts of a Plant

Learning Outcomes

By the end of this lesson, you should be able to:
- Name the six main parts of an angiosperm. (LO 08.03-a)
- Describe the function of each of the plant parts. (LO 08.03-b)

Words to Know

adventitious roots	fruits	seed
bud	internode	seed coat
embryo	leaves	stem
endosperm	node	taproot
fibrous roots	phloem	terminal bud
flowers	root	xylem

Before You Begin

Did you know that plants have six main parts? Can you name any of them?

If I had to pick a favorite plant part, I think mine would be seeds. I love to snack on ranch-flavored sunflower seeds. But I really like celery too, with peanut butter. Yum! Can I change my favorite plant part to stems? Or cherry tomatoes, I love to snack on those too. Any guesses which plant part a cherry tomato might be? Tell a classmate what your favorite plant part is and why.

You may have noticed that some plants have large leaves and some have small. Some have showy flowers, and others have none. Why is this? What role do individual plant parts play and why are they essential to the survival of the plant?

Parts of a Plant and Their Functions

There are six main parts to plants: roots, stems, leaves, flowers, fruits, and seeds. While not all plants have all six parts, most angiosperms do (**Figure 8.3.1**). Roots, stems, and leaves are the vegetative parts of a plant. Flowers, fruits, and seeds are the plant's reproductive parts.

Roots

The roots of a plant may not be regularly visible, but they are one of the most important parts of the plant. A ***root*** is a structure that anchors the plant in the ground and takes up water and nutrients.

Roots have three main functions for plant growth. They anchor and support the plant, take water and nutrients up from the soil, and store food and nutrients to be used by the plant.

The first structure to emerge from a seed is the primary root of the plant. The root immediately starts absorbing water to keep the seed alive and help it grow. Secondary roots begin to emerge from the primary root to anchor the plant and protect it from winds and heavy rains. The roots will develop root hairs. As the plant matures, most water and nutrient absorption occurs through the root hairs. Roots move through the soil with root caps leading the way (**Figure 8.3.2**).

Did You Know?

The roots of a South African wild fig tree can grow 390 feet deep!

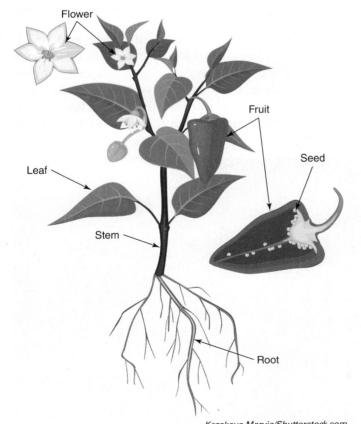

Kazakova Maryia/Shutterstock.com

Figure 8.3.1. Most plants have six main parts to serve their needs in their growing environment and reproduce.

Root Structure

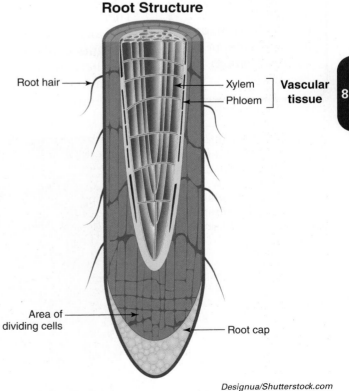

Designua/Shutterstock.com

Figure 8.3.2. When a root cap runs into an obstacle, it will grow around it. *What purpose do you think root hairs serve?*

Tap root system

Fibrous root system

Kazakova Maryia/Shutterstock.com

Figure 8.3.3. Taproots and fibrous roots are the most common root systems.

Figure 8.3.4. Mangrove trees have aerial roots because they grow directly in water or on sandy beaches. These aerial roots lift the plant out of the water, allowing the roots to get oxygen.

There are several types of root systems.

- A *taproot* is one large root that grows directly downward and has a few small roots that branch off the main root. Examples of taproots are carrots, beets, radishes, and dandelions. The length of taproots depends on the soil and availability of water. Taproots occur mostly in dicots.
- *Fibrous roots* are shallow, with many roots spreading through the soil. Turf grasses, wheat, rice, and corn all have fibrous root systems. Fibrous roots occur mostly in monocots (**Figure 8.3.3**).
- *Adventitious roots* are roots that grow from the stem of the plant. These can be found on many climbing vines. You can also find adventitious roots on tomato plants that keep the plant upright while it grows. Adventitious roots are used to propagate new plants.
- Aerial (air) roots are roots that grow above the soil surface. Since roots require oxygen to survive, many plants need aerial roots to grow in wet places. Mangroves and other swamp trees have aerial roots (**Figure 8.3.4**).

Stems

The *stem* is the part of the plant that moves nutrients and food throughout the plant. Stems vary in appearance and thickness from plant to plant. The stem also holds leaves, flowers, and fruit, provides support for the plant, and makes food (green stems only).

There are three main external parts to a stem: node, internode, and bud. A *node* is an area where leaves develop. The *internode* is the space between each node. *Buds* contain undeveloped leaves or flowers and are a stem's growing point. There are three different kinds of buds: vegetative buds, flower buds, and terminal buds. Vegetative buds contain immature leaves. Flower buds contain immature flowers. The large bud at the tip of the stem is the terminal

Sakarin Sawasdinaka/Shutterstock.com

bud. The ***terminal bud*** is the primary growth point of the stem and usually forms the main trunk of the plant (**Figure 8.3.5**).

Water, minerals, and food are transported through tissue inside the stem. The two tissues that transport these substances are the phloem and the xylem. The ***phloem*** is the tissue in the plant stem that transports the food made in the leaves to the rest of the plant. The ***xylem*** is the tissue in a plant stem that transports water and nutrients from the roots to all other parts of the plant. The xylem moves one-way, up from the roots. The phloem transports substances up and down the plant (**Figure 8.3.6**). The roots would die if the phloem could not move food made in the leaves down to the roots.

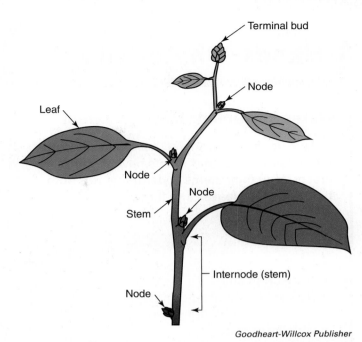

Goodheart-Willcox Publisher

Figure 8.3.5. *Did you realize that the structure of a stem was so complicated?*

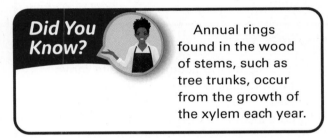

Did You Know? Annual rings found in the wood of stems, such as tree trunks, occur from the growth of the xylem each year.

Figure 8.3.6. The xylem and the phloem are important parts of the plant. They serve as the transportation system of the plant, moving water and nutrients where they are needed.

8

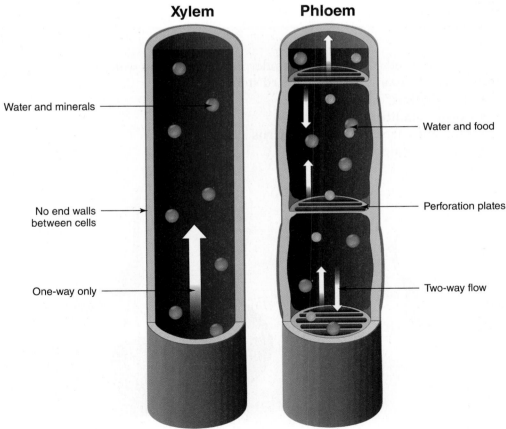

Designua/Shutterstock.com

Hands-On Activity

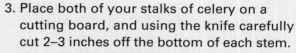

Water Movement through Xylem

Have you ever wondered how plants absorb nutrients from the soil? Plants have tube-like structures in their stems (xylem) that run from the roots to the leaves and flowers. How do these work? Let's find out.

You will need the following materials:

- Water
- Measuring cup
- Red or blue food coloring
- 2 one-pint glass jars
- 2 stalks of celery
- Knife
- Cutting board
- Magnifying lens

Here are your directions:

1. Pour one cup of water into each glass jar.
2. Drop 20 drops of food coloring into the water in one of the glasses.
3. Place both of your stalks of celery on a cutting board, and using the knife carefully cut 2–3 inches off the bottom of each stem.
4. Immediately stand one celery stem in the colored water and the other in the jar with plain water.
5. After 24 hours, observe and record the appearance of the stalks and leaves. Lift each stalk and observe the bottom cut.
6. Using a knife, dissect each celery stalk by cutting down the center to see the inside of the stem.

Consider This

1. How was the food coloring able to move from the glass to the top of the stalk and the leaves?
2. Create a colored drawing of what you observed. Make sure to label the parts of the stem in your drawing.

Some modified stems serve as underground food and water storage organs (**Figure 8.3.7**). Examples of modified stems are:
- Bulbs: tulips, onions, lilies
- Corms: gladiolus, crocus
- Rhizomes: irises, asparagus, ferns
- Tubers: potatoes, yams

Figure 8.3.7. Bulbs, corms, rhizomes, and tubers are examples of modified stems. *Can you think of any vegetables we eat that are modified stems?*

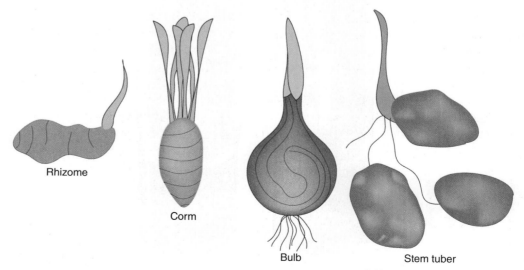

K.K.T Madhusanka/Shutterstock.com

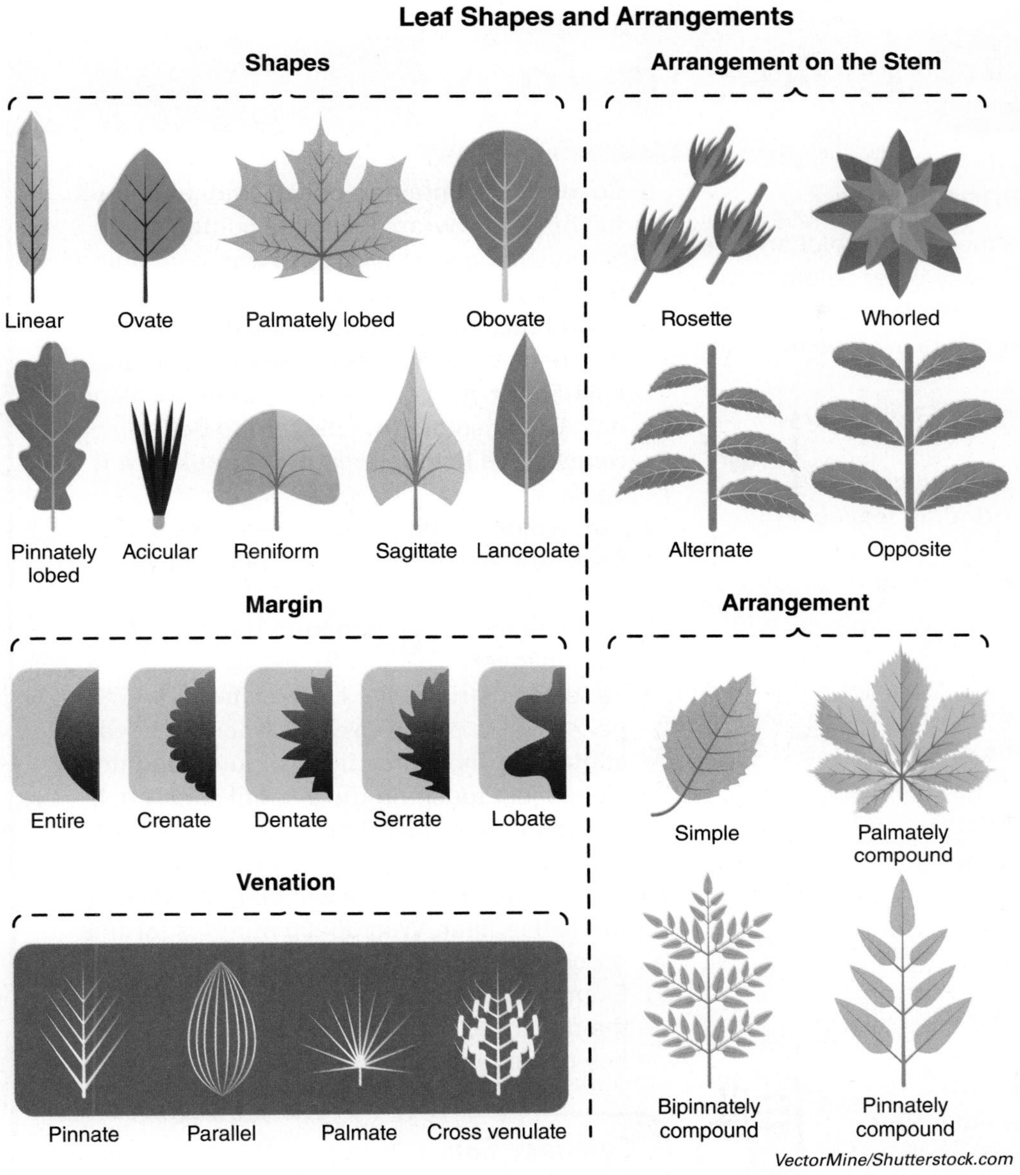

Leaf Shapes and Arrangements

Shapes

Linear

Ovate

Palmately lobed

Obovate

Pinnately lobed

Acicular

Reniform

Sagittate

Lanceolate

Margin

Entire

Crenate

Dentate

Serrate

Lobate

Venation

Pinnate

Parallel

Palmate

Cross venulate

Arrangement on the Stem

Rosette

Whorled

Alternate

Opposite

Arrangement

Simple

Palmately compound

Bipinnately compound

Pinnately compound

VectorMine/Shutterstock.com

Figure 8.3.8. Leaf shape and arrangement are factors to note when identifying plants. Hollies, for example, always have an alternate leaf arrangement, whereas *Holly Osmanthus* (False Holly), an evergreen shrub that looks remarkably like holly, has opposite leaves.

Leaves

Leaves are a part of the plant attached to the stem that capture sunlight so the plant can carry out photosynthesis. *Photosynthesis* (which we will talk more about in **Lesson 8.5**) is the process by which plants make food. Leaves are used to help identify the plant. Horticulturists note the leaf arrangement (**Figure 8.3.8**) to help with identification. There are also several different leaf shapes that aid in identification.

Leaves are made of different parts (**Figure 8.3.9**). The major parts of a leaf are:

- Petiole: the stem that attaches the leaf to the plant stem
- Blade: the flat surface of the leaf that captures the most sunlight
- Midrib: the large vein in the center of the leaf
- Veins: the pathways that move water, minerals, and nutrients in and out of the leaf blade
- Tip: the end of the blade

Structure of a Leaf

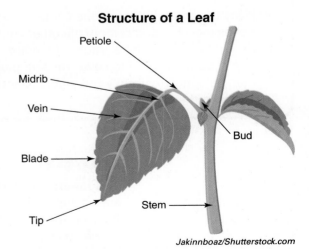

Petiole

Midrib

Vein

Blade

Bud

Stem

Tip

Jakinnboaz/Shutterstock.com

Figure 8.3.9. The leaf's main function is to create food for the plant. The blade of the leaf is designed to collect sunlight for photosynthesis.

Why Do Leaves Change Colors?

Chlorophyll gives leaves their green color and the color is so dominant it hides any other colors in the leaf. In the fall, chlorophyll in the leaves begins to break down, allowing other colors to come through.

You will need the following materials:

- 2–3 green leaves
- 2–3 leaves that have changed color
- Mortar and pestle for grinding leaves (optional)
- 2 - 8–10 oz glass or plastic containers
- Marker
- Labels
- Rubbing alcohol
- Plastic spoon
- Aluminum foil or plastic wrap
- Shallow tray or pan
- Hot water from a faucet
- White coffee filter
- Scissors
- 2 pencils
- Tape
- Ruler
- Hand lens

Here are your directions:

1. Tear or cut up the different leaves into small pieces. Leaves can be ground with a mortar and pestle if available.
2. Place the green leaves in one container and the leaves that changed color in the other container.
3. Using the marker, label your containers.
4. Add rubbing alcohol to cover the leaf pieces. Using a spoon, carefully but vigorously stir the leaves in the alcohol.
5. Cover each container loosely with aluminum foil or plastic wrap. Place the containers carefully in a shallow tray or pan containing one inch of hot water.
6. Swirl the jars around carefully every five minutes. Keep the jars in the water for at least a half hour, longer if needed, until the alcohol has turned colors (the darker the better). Replace the hot water in the water tray if it cools off.
7. Cut two thin, long strips of the coffee filter. Each strip needs to be 2 cm wide and 15 cm long.
8. Remove the containers from the water tray and uncover.
9. Tape each strip of the coffee filter to a pencil. Lay a pencil over the opening of each container. The coffee filter should hang down so it is just touching the alcohol solution. If it is too long or too short, roll your pencil to make the filter piece longer or shorter.
10. The alcohol and colors will begin to travel up the coffee filter. Wait 30–90 minutes for this process to be complete.
11. Examine the strips with a hand lens. Measure the distance the colors traveled and record any observations.

Consider This

1. What differences did you see between the strips?
2. What can be determined by the color pigments represented on the strip?
3. Why do you think some colors traveled higher than others on the strip?

Flowers

Flowers are grown for their beauty and fragrance. Flowers vary widely in their size, color, and scent. Each unique flower is designed to attract a specific insect, bird, or animal. Some flowers attract honey bees or hummingbirds, for instance. *Flowers* are the reproductive part of a plant. Flowers allow mature plants to produce seeds that will later become new plants. Some flowers attract insects to help with reproduction, while others are designed to use wind or rain for that purpose. We will cover more about the reproductive parts of the plant in **Lesson 8.6**.

Fruits and Seeds

Fruits act as protective tissue around seeds and aid in seed dispersal. Some fruits are fleshy and serve as edible horticulture crops, like peaches, apples, and plums. Other fruits are hard and dry and less desirable to eat, such as acorns.

Seeds are the reproductive unit of plants. Seeds range in size from a few millimeters to several centimeters. They can also be different shapes, including flat or rounded.

The seed has three main parts: the embryo, endosperm, and the seed coat (**Figure 8.3.10**).

- The seed *embryo* is the immature plant. The embryo has a root, stem, and one or two seed leaves. It draws on food reserves in the endosperm to develop.
- The *endosperm* is specialized tissue that stores food to aid in growth and development.
- The *seed coat* is a protective covering for the developing seed. The seed coat prevents embryo growth until conditions are right.

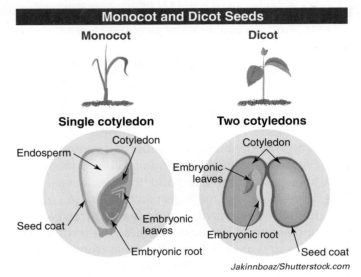

Monocot and Dicot Seeds

Monocot — Single cotyledon: Endosperm, Cotyledon, Seed coat, Embryonic leaves, Embryonic root

Dicot — Two cotyledons: Cotyledon, Embryonic leaves, Embryonic root, Seed coat

Jakinnboaz/Shutterstock.com

Figure 8.3.10. The seed has three main parts: the embryo, endosperm, and the seed coat.

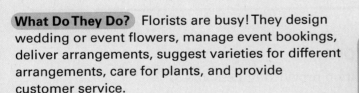

Career Corner — FLORIST

Florists *design and create arrangements of flowers for us to enjoy.*

What Do They Do? Florists are busy! They design wedding or event flowers, manage event bookings, deliver arrangements, suggest varieties for different arrangements, care for plants, and provide customer service.

What Education Do I Need? A high school diploma or an associate degree in business management, floral design, or plant landscaping is needed.

How Much Money Can I Make? Florists make an average of $50,000 per year.

How Can I Get Started? You can build an SAE around growing your own flowers and using them in cut flower arrangements, participate in the FFA Floriculture CDE, volunteer at a local flower shop in your area, or work with your FFA advisor to host a floral design workshop for local community members.

BearFotos/Shutterstock.com

8

What Did We Learn?

- There are six main parts to a plant: roots, stems, leaves, flowers, fruits, and seeds.
- Roots anchor the plant to the ground and absorb water and nutrients from the soil. The root is the first thing to emerge from a seed.
- There are two main types of roots: tap and fibrous. Tap roots have one central root with several smaller roots coming from the main root. Fibrous roots are shallow and spread out in the ground.
- Stems move nutrients and food throughout the plant. They also help support the plant and hold the leaves, flowers, and fruits.
- There are three main parts to a stem: the node, internode, and buds. The nodes are where leaves attach to the stem. Buds contain undeveloped leaves or flowers. The terminal bud is the primary growth point and usually forms the main trunk of the plant.
- The two main plant tissues located inside the stem that move water and nutrients through the plant are the xylem and phloem. The phloem transports food that is made in the leaves throughout the plant. The xylem moves water and nutrients from the roots to the rest of the plant.
- Leaves are the part of the plant attached to the stem that captures sunlight to carry out photosynthesis.
- Flowers are the reproductive part of the plant.
- Fruits serve as the protective tissue around a seed and aid in seed dispersal.
- Seeds are the reproductive unit of the plant. The seed has three main parts: the embryo, endosperm, and the seed coat.

Let's Check and See What We Know

Answer the following questions using the information provided in Lesson 8.3.

1. Which of the following plant parts anchors the plant and absorbs water and nutrients from the soil?
   A. Roots
   B. Leaves
   C. Stems
   D. Seeds

2. *True or False?* The root is the first thing to emerge from a developing seed.

3. ____ roots grow above the soil surface to absorb oxygen.
   A. Tap
   B. Fibrous
   C. Adventitious
   D. Aerial

4. Which of the following plant parts move nutrients and food throughout the plant?
   A. Roots
   B. Leaves
   C. Stems
   D. Seeds

5. *True or False?* The bud at the tip of the stem that serves as the primary growth point for the plant is called the vegetative bud.

6. The ____ is tissue in the plant stem that transports food made in the leaves to the remaining parts of the plant.
   A. cambium
   B. xylem
   C. phloem
   D. modified stem

7. Which of the following plant parts absorb sunlight so the plant can carry out photosynthesis?
   A. Roots
   B. Leaves
   C. Stems
   D. Seeds
8. *True or False?* The large vein in the center of the leaf is called the midrib.

9. The reproductive part of the plant is the ____.
   A. flower
   B. seed
   C. fruit
   D. stem
10. Which of the following plant parts is the reproductive unit of the plant?
   A. Roots
   B. Leaves
   C. Stems
   D. Seeds

Let's Apply What We Know

1. Collect a plant, making sure to keep the roots intact. Tape the plant to a piece of paper and label the individual components of the roots, stem, leaves, and flowers, if available.
2. Describe the four different types of root systems, and what each does for the plant.
3. List the six main parts of the plant and explain the function of each part.
4. Sketch how the phloem and the xylem work in plant stems.
5. Collect a large seed, such as a bean seed, and carefully split it in half. Label the three main parts of the seed.

Academic Activities

1. **Science.** Using Play-Doh or modeling clay, create a 3-D plant model showcasing all six major plant parts.
2. **Technology.** Research a dichotomous key for plant identification. Collect leaves from two or three plants. Make additional notes about the plant such as stem type, leaf arrangement, and flowers. See if you can correctly identify the plant using a dichotomous key.

Communicating about Agriculture

1. **Visual Communication.** Create a poster labeling the different parts of a plant. Include all the information you are able to, explaining the purpose of each part in detail. Display your poster in your classroom.

8

SAE for ALL Opportunities

1. **Entrepreneurship SAE.** Start a small business growing and selling modified stems such as bulbs, tubers, and corms. Research how to properly divide each stem system so you can grow new plants each year to keep your business growing.
2. **Research SAE.** Create a plant growth study. Record how different plant parts react to different stimuli in the environment, such as different amounts of light or water. Make a hypothesis, create your experiment, observe and record data, and write a report on your findings.

SAE for ALL Check-In

- How much time have you spent on your SAE this week?
- Have you logged your SAE hours?
- What challenges are you having with your SAE?
- How can your instructor help you?
- Do you have the equipment you need?

Environmental Conditions for Plant Growth

sickmoose/Shutterstock.com

Learning Outcomes

By the end of this lesson, you should be able to:
- Discuss how light affects plant growth. (LO 08.04-a)
- Describe plants' responses to temperature. (LO 08.04-b)
- Explain why oxygen is important to plant growth. (LO 08.04-c)
- Explain how water aids plant growth and development. (LO 08.04-d)

Words to Know

freeze injury	light quality	phototropism
humidity	light quantity	
irrigation	photoperiodism	

Before You Begin

Humans grow. Think about how big you were when you were born compared to how big you are now. What things were needed for you to grow? Jot down a quick list of the things you can think of.

How would your list be different if you were talking about what plants need to grow? Put a check mark next to the items on your list that you think plants might also need.

Plants have basic needs for survival, just as humans do. Plants need nutrients, just as we need food to meet our nutritional needs. Plants also need water, just like people. Let's explore what plants need for healthy growth.

Environmental Conditions for Plant Growth

To be successful in horticulture, it is important to understand how plants grow and what they need to survive. When a plant grows in nature, its growing environment provides what the plant needs. When plants are cultivated in fields, greenhouses, and other structures, it is up to the grower to provide the nutrients a plant would otherwise get from its natural environment (**Figure 8.4.1**). There are many environmental factors that a grower can control, such as light, air, temperature, and water. When growers understand how to control these things and how they affect the plant, they can grow a healthy crop.

Light

All plants require light to grow and develop into mature plants. Light is critical for the process of photosynthesis (we will cover this more in **Lesson 8.5**). Sunlight is the best light for plants, but sometimes artificial light is necessary.

Light is necessary for many reasons, including:
- Color formation in the leaves
- Leaf drop and fall coloring, and new growth
- Seed germination
- Promoting the flowering and fruiting of the plant

Light Quality and Quantity

The intensity and amount of light is important for plant growth (**Figure 8.4.2**). *Light quality* refers to the intensity, or how bright the light is. Every plant species has a preferred light intensity. Seeds even require specific amounts of light to grow. Some seed varieties need to be left on top of the soil to grow. Other seeds require certain intensities of light to bring them out of dormancy in the spring.

Did You Know?

If grown in proper conditions, bamboo can grow three feet in a single day, making it the fastest growing plant on the planet!

Did You Know?

Fluorescent lights provide the best artificial light for interior plants, and some are designed to produce specific colors or approximate natural sunlight.

8

Den Edryshov/Shutterstock.com

Figure 8.4.1. When plants are grown inside greenhouses, it is up to the grower to control the water, light, and temperature to maximize growth.

JPBC/Shutterstock.com

Figure 8.4.2. Light quality refers to how intense or bright the light is that a plant receives. Some seeds require very intense light to sprout and grow.

R R/Shutterstock.com

Figure 8.4.3. Poinsettias are a photoperiodic plant that require shortened periods of light to turn their familiar red color.

Anest/Shutterstock.com

Figure 8.4.4. When plants do not receive enough light, they will bend toward a light source. *Why do you think this happens?*

Manhattan001/Shutterstock.com

Figure 8.4.5. When temperatures get too low or high, plants experience severe stress. This vegetable plant is stressed from freezing temperatures damaging plant tissue. This can result in the plant's death.

Light quantity is the amount and duration of light emitted by a light source. Sunlight decreases in the fall and increases in the spring, naturally increasing and decreasing the light quantity. Light can cause plants to flower at different times of the year. Plants that need a lot of sunlight to flower are more likely to do so in the summer months. Growers can increase the light quantity by using artificial light and can decrease light quantity by covering the greenhouse with a cloth. Plants that do not receive enough light tend to have poor growth patterns, such as leggy stems that cannot support the weight of the plant.

Plants have growth responses based on the number of hours of light they receive each day. This response is known as *photoperiodism*. The length of light a plant receives is important, but it is equally important they receive uninterrupted periods of dark. For example, for poinsettias to turn red, they need as few as seven hours of light in a single day (**Figure 8.4.3**).

Another growth response related to light is *phototropism*. This is the bending of a plant toward or away from a light source. Have you ever seen a plant bending or leaning toward a light? This is phototropism at work (**Figure 8.4.4**). The exact opposite can also happen: Plants that require low amounts of light will bend away from a light source.

Temperature

From the arctic reaches of Alaska to the deserts of New Mexico and the warm, tropical climate of Florida, temperature plays an important role in what plants grow in different places. Most horticultural crops thrive in temperatures between 50- and 85-degrees F. The temperature requirements of the crop help growers better plan when to produce their desired crops.

Temperature controls plant responses, such as when plants go dormant for the winter. When temperatures are too low or too high, plants experience stress. Examples of these plant stresses include frost or freeze damage, sunscald, and heat stress. *Freeze injury* is a condition in which plants are damaged when low temperatures freeze the water in plant tissue (**Figure 8.4.5**).

There are several things that growers can do to help manage temperatures for better plant

Hands-On Activity

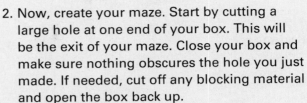

Creating a Plant Maze

Have you ever noticed how plants grow toward a light source? You can see phototropism in action by building a simple plant maze and watching your plant wind its way through the maze to reach the light.

You will need the following materials:

- Bean seeds
- Soil
- Small pot
- Small box – shoebox size
- Black construction paper or card stock
- Knife
- Tape or glue
- Pencil
- Ruler

Here are your directions:

1. Plant the bean seeds in a pot (about one inch under the surface) and add water so that the soil is moist. You may want to plant several seeds ahead of time to make sure the beans sprout. Place the pot near a windowsill and wait a few days until a small plant sprouts. This will ensure you are starting the experiment with a good plant.

2. Now, create your maze. Start by cutting a large hole at one end of your box. This will be the exit of your maze. Close your box and make sure nothing obscures the hole you just made. If needed, cut off any blocking material and open the box back up.

3. Measure the inside of the box to create a divider or wall in the center of your box. Mark your measurement on your black paper but add about a ½-inch border to three sides. Those will act as a glue flap in the later steps.

4. Draw and cut out a circle in the divider/wall. You will need to place this hole on the opposite side of the exit hole at the end of your box.

5. Place your plant inside the box. Fold the ½-inch border on your black paper and glue or tape the folds inside your box above your plant leaving room for the plant to grow.

6. Close the box up and place it near a window. Make sure to keep your plant well hydrated. Check your box and keep records to determine when the plant travels through the light maze.

Consider This

1. How did the plant grow toward the light?
2. Do you think the brightness of the light going into the box makes a difference in how quickly the plant is able to make it through the maze?

Aldona Griskeviciene/Shutterstock.com

Figure 8.4.6. Plants grown under tunnels are planted directly in the ground and covered with plastic material. This layer of plastic provides warmer air and soil temperatures for crops to grow, making an earlier spring and a longer growing season.

growth. Greenhouses give growers the ability to control temperatures through the year, whether by adding heat in the winter or cool air in the summer (**Figure 8.4.6**). Growers can also manage temperatures by:

- Growing plants under high or low tunnels
- Implementing overhead irrigation to provide protection from frost
- Installing wind machines

Air

Plants take in oxygen from the air around them. Having enough oxygen is vital for the growth of plants, as they need it for photosynthesis. ***Humidity*** is the amount of water vapor (containing oxygen) in the air. Plants typically grow best in places with high humidity (**Figure 8.4.7**). However, too much humidity can lead to plant diseases.

In controlled growing environments, it is important for growers to monitor the air quality. Carbon dioxide levels in closed greenhouses can get too low without monitoring. Vents are important to keep air circulating through the building.

Water

Plants cannot survive without adequate water. In fact, water makes up between 80 and 90 percent of horticultural plants. When water is low, plants wilt and stop growing (**Figure 8.4.8**). If plants are given too much water, they may exhibit stunted or slowed growth, leaf discoloration, and increased disease. Regular watering is needed to maintain proper plant growth and soil texture. The amount of water needed varies by the type of plant.

Figure 8.4.7. Many plants, especially tropical plants, thrive in high moisture environments. *Why do you think this is?*

Figure 8.4.8. Plant wilt occurs when plants do not receive enough water. *Will plant wilt typically kill a plant? What do you think?*

Light Quality

Plants respond to the quality of light available. Plants primarily absorb light in the visible wavelength range of 400-700 nanometers. The greatest impact on plant growth is at peaks in red light and blue light. If only blue light is applied to plants, you might notice shortened or dark plants. If only red light is applied to plants, you might notice long, leggy stems. Red and blue light together promote flowering in plants. Conduct your own experiment and observe how plants respond to different colors of visible light.

nikkytok/Shutterstock.com

You will need the following materials:

- Yellow, green, red, blue, and white (regular) light
- Five identical plants
- Five pots
- Potting soil
- Watering can
- Ruler
- Marker

Here are your directions:

1. Plant each plant in the pot with potting soil. Label each pot with the color of light that plant will be exposed to during the experiment.
2. Place each plant under the assigned color light bulb. Make sure the lightbulbs are set at least two feet apart.
3. Turn each light bulb on for 12 hours, then leave the plant in darkness for the other 12 hours.
4. Make sure your plant has enough water to grow.
5. Measure and record the height of the plants each day. Record any observations about the plants' appearances, such as leaf color.

Consider This

1. What effect did different colors have on plant growth?
2. What do you think would happen if you grew a plant under two light colors, such as red and yellow, instead of only one?
3. How do you think plants receive all needed light colors when growing outdoors?

8

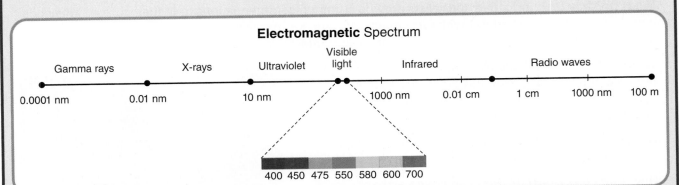

Electromagnetic Spectrum

Gamma rays   X-rays   Ultraviolet   Visible light   Infrared   Radio waves

0.0001 nm   0.01 nm   10 nm   1000 nm   0.01 cm   1 cm   1000 nm   100 m

400 450 475 550 580 600 700

AlexVector/Shutterstock.com

Fahroni/Shutterstock.com

Figure 8.4.9. Some lawns are watered with an automatic sprinkler system. This type of system can be controlled by an electrical box which guarantees even watering to the entire area.

Irrigation

When plants grow in nature, they get their water from the soil. If there is not enough precipitation or when growing indoors, growers use irrigation to water plants. ***Irrigation*** is the process of mechanically adding water to soil. Irrigation can take place in many ways, including:

- Surface irrigation. Water is distributed over and across land by gravity.
- Localized irrigation. Water is distributed under low pressure, through a piped network and applied to each plant.
- Drip irrigation. A type of localized irrigation in which drops of water are delivered at or near the root of plants. In this type of irrigation, evaporation and runoff are minimized.
- Sprinkler irrigation. Water is distributed by overhead high-pressure sprinklers or guns from a central location in the field or from sprinklers on moving platforms (**Figure 8.4.9**).
- Center-pivot irrigation. Water is distributed by a system of sprinklers that move on wheeled towers in a circular pattern. This system is common in flat areas of the United States.
- Lateral move irrigation. Water is distributed through a series of pipes, each with a wheel and a set of sprinklers. The sprinklers move a certain distance across the field and then need to have the water hose reconnected for the next distance. This system tends to be less expensive but requires more labor than others.
- Sub-irrigation. Water is distributed across land by raising the level of water underground, through a system of pumping stations, canals, gates, and ditches.
- Manual irrigation. Water is distributed across land through manual labor and watering cans. This system is very labor intensive.

Career Corner   IRRIGATION SPECIALIST

Irrigation specialists *are responsible for the sales, programming, operating, and maintenance of irrigation systems to help producers.*

What Do They Do? Irrigation specialists provide system recommendations to fit their customers' needs; coordinate watering schedules with farm managers and agronomists; prepare ground for installation and install pipes, wiring, and valves; determine water rates; and design and read blueprints.

What Education Do I Need? A high school diploma is required. A higher degree in turf grass management, agricultural engineering, or landscape architecture could be beneficial, depending on the type of job you are pursuing.

How Much Money Can I Make? Irrigation specialists make an average of $50,000 per year.

How Can I Get Started? You can create an experiment on plant watering requirements for an SAE. You could volunteer with a local irrigation installation or landscape company. Maybe you could create a blueprint for an irrigation system for your own yard or the school.

Virrage Images/Shutterstock.com

What Did We Learn?

- Environmental conditions are elements that are typically controlled by nature, such as air, water, temperature, and light, but can be modified by a grower to improve plant growth.
- Light is necessary for color in plant leaves, seed germination, and promoting flowering and fruiting of the plant.
- Light quality refers to the intensity of the light, whereas light quantity refers to the amount of light emitted by a light source.
- Plants require different temperatures to grow, but most plants grow between 50- and 85-degrees F.
- Hot or cold temperatures can cause stress to a plant, potentially leading to plant death.
- Humidity is the amount of water vapor in the air. Plants typically grow best in areas with high humidity.
- Water is essential for plant growth and survival. When plants do not receive enough water, they wilt or die.
- Growers use irrigation to mechanically add water to the soil.

Let's Check and See What We Know

Answer the following questions using the information provided in Lesson 8.4.

1. Which of the following is *not* an environmental condition that affects plant growth?
   A. Temperature   C. Light
   B. Water   D. Fertilizer

2. *True or False?* Light affects when plants drop their leaves in the fall and grow new leaves in the spring.

3. _____ refers to light intensity.
   A. Light quality
   B. Light quantity
   C. Photoperiodism
   D. Phototropism

4. _____ is the bending of a plant toward or away from a light source.
   A. Light quality
   B. Light quantity
   C. Photoperiodism
   D. Phototropism

5. *True or False?* Plants grow best in temperatures between 30 and 50 degrees F.

6. _____ is a condition in which low temperatures freeze the water inside the plant, usually resulting in the death of the plant.
   A. Heat stress   C. Sunscald
   B. Freeze injury   D. Light quantity

7. The amount of water vapor in the air is _____.
   A. freeze injury   C. irrigation
   B. humidity   D. dew point

8. *True or False?* Growers use vents and fans to circulate air through the greenhouse.

9. Which of the following is *not* a sign of overwatering?
   A. Slowed or stunted growth
   B. Discoloration of the leaves
   C. Freeze injury
   D. Disease

10. _____ is the process of mechanically adding water to the soil.
    A. Freeze injury   C. Irrigation
    B. Humidity   D. Dew point

Let's Apply What We Know

1. Discuss each environmental factor and its effect on plant growth.
2. Explain the difference between photoperiodism and phototropism.
3. List ways growers can manage temperature.
4. Explain the stresses plants undergo when exposed to extreme temperatures.
5. Briefly describe three different types of irrigation and how they are used.

Academic Activities

1. **Technology.** There are many different types of light bulbs you can buy: fluorescent, LED, incandescent. Create an experiment for growing plants under different types of light. What light quality and quantity do they provide to the plant? How does each light source affect the growth of your plants?
2. **Language Arts.** Plants have different growing requirements. Select a plant that you find interesting. Research the environmental conditions that plant needs to grow. Outline the growing needs of the plant (light, temperature, water) so any gardener could grow your selected plant.

Communicating about Agriculture

1. **Writing.** What information do you need to know when you buy a plant for your garden or home? Create a sample label that provides the necessary information in an effective way. Share it with your class.

SAE for ALL Opportunities

1. **Foundational SAE.** Visit a vegetable grower and observe how the field is planted. Take notes on how the rows are laid out, spacing, irrigation, and any other environmental conditions. Interview the grower and ask about how they manage the needs of the plants, such as water and temperature.
2. **Research SAE.** Create an environmental conditions study. Record how plants react to different temperatures and different humidity levels. Form a hypothesis, observe, record data, and write a report on your findings.

SAE for ALL Check-In

- How much time have you spent on your SAE this week?
- Have you logged your SAE hours?
- What challenges are you having with your SAE?
- How can your instructor help you?
- Do you have the equipment you need?

Photosynthesis, Respiration, and Transpiration

New Africa/Shutterstock.com

Learning Outcomes

By the end of this lesson, you should be able to:
- Identify the roles photosynthesis, respiration, and transpiration play in plant development. (LO 08.05-a)
- Describe the processes of photosynthesis and respiration. (LO 08.05-b)
- Explain how plants benefit from transpiration. (LO 08.05-c)

Words to Know

chlorophyll	photosynthesis	stomata
chloroplasts	respiration	transpiration

Before You Begin

Did you know that humans aren't the only living creatures with sweat glands? We are, however, one of the few species that produces large amounts of perspiration to cool off. It's a little gross to think about! Are there any other species that you can think of that produce perspiration?

Did plants make your list? While we might not think of it as sweat, plants release large amounts of water to cool off. How crazy is that?! The next time you are outside on a hot day, you should check around for plant sweat! You will learn about this process—and lots of other cool processes—in this lesson.

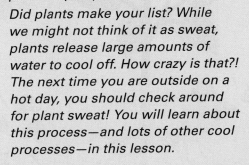

8

Plants are living organisms, just as people are. Complex chemical processes help them grow and develop. The human body engages in complex process like digestion, respiration, and nutrition. Plants engage in some of these same processes, which include photosynthesis, respiration, and transpiration. In both humans and plants, these processes are regulated by hormones and enzymes. This lesson will investigate each of these major processes and what roles they play in plant development.

Photosynthesis

All green plants are able to produce their own food. Plants manufacture food through a process called photosynthesis. *Photosynthesis* is a series of chemical reactions in which plants take carbon dioxide from the atmosphere, add water, and use the energy from sunlight to produce sugar (**Figure 8.5.1**).

Photosynthesis is one of the most important reactions on this Earth. Photosynthesis is responsible for nearly all the energy we need to survive, as well as the oxygen we breathe.

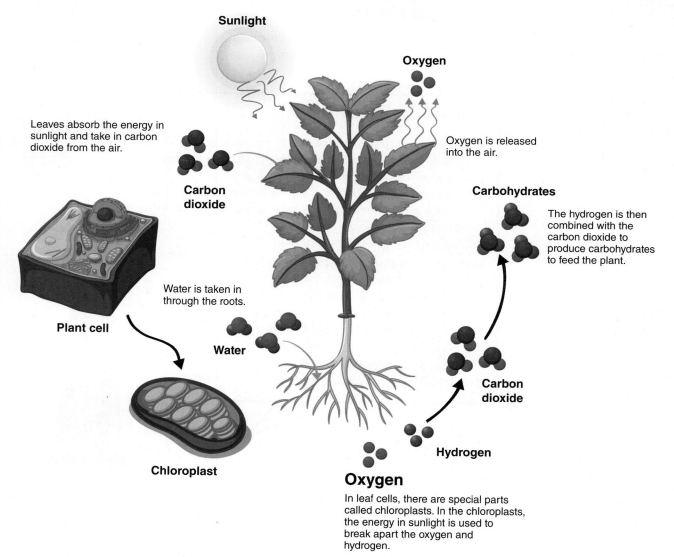

Process of Photosynthesis

Sunlight

Oxygen

Oxygen is released into the air.

Leaves absorb the energy in sunlight and take in carbon dioxide from the air.

Carbon dioxide

Carbohydrates

The hydrogen is then combined with the carbon dioxide to produce carbohydrates to feed the plant.

Plant cell

Water is taken in through the roots.

Water

Carbon dioxide

Hydrogen

Chloroplast

Oxygen

In leaf cells, there are special parts called chloroplasts. In the chloroplasts, the energy in sunlight is used to break apart the oxygen and hydrogen.

GraphicsRF.com/Shutterstock.com

Figure 8.5.1. Photosynthesis is complex, but not hard to understand when you take the time to look!

Oxygen is produced as a by-product of photosynthesis. The simplified chemical equation showing photosynthesis can be seen in **Figure 8.5.2**.

Photosynthesis takes place in the chloroplasts inside plant cells. **Chloroplasts** are specialized structures within individual leaf cells. Chloroplasts contain **chlorophyll**, a pigment that captures light energy. The light energy is converted to chemical energy through the steps of photosynthesis.

The reactions of photosynthesis can be divided into two major types: light-dependent reactions and light-independent reactions. The light-dependent reactions convert energy from the sun into a form that the chloroplasts can use to make sugar from carbon dioxide, in the process producing oxygen as a waste product. The simple sugars that are produced fuel plant growth and development. The light-independent reactions use energy to make glucose (a more complex sugar) from carbon dioxide and water.

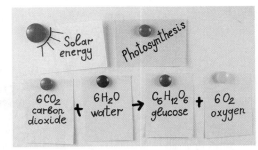

Botyev Volodymyr/Shutterstock.com

Figure 8.5.2. In photosynthesis, carbon dioxide and water are combined in the presence of light to form glucose (sugar) and oxygen. *Do you think that solar energy is the only light that will work for this process?*

Making Connections #Science

Exploring the Rate of Photosynthesis

There are many variables in photosynthesis, but light is the primary variable. The amount of light has a direct connection to the rate of photosynthesis. There are various set-ups that can be used to measure the rate of photosynthesis, but each relies on measuring the oxygen produced during the reaction.

You will need the following materials:

- 2 test tubes
- Elodea cuttings (aquatic plant)
- Baking soda (sodium bicarbonate)
- 2 beakers with water
- Lamp or flashlight

Here are your directions:

1. Add a pinch of baking soda to the bottom of both test tubes and then fill three-quarters of the way with water.
2. Cut two elodea stems at an angle. Place the end of the stem between your fingers and crush the stem at the end where you made the cut. Each stem should fit into a test tube and should be completely submerged in water.
3. Fill each of the beakers halfway with water.
4. Place test tube 1 inside the beaker with the opening of the test tube facing upward. Place the beaker and test tube in front of the light. As the water warms up, count the bubbles as they escape from the test tube to measure the rate of reaction.

QuinxGhoul/Shutterstock.com

5. Place test tube 2 inside the beaker with the opening of the test tube facing down/inverted. Place the beaker and test tube in front of the light. As the water warms up, the air bubble that forms at the end of the test tube can be measured to determine how much oxygen was collected.
6. If you do not see bubbles right away, re-cut and crush the stems, or experiment with moving the apparatus closer to the light. Your goal is to find a way to accurately measure the rate of photosynthesis using either of these designs.

Consider This

1. Which test tube trial worked better for your experiment? Why?
2. Brainstorm variables which may affect the rates of photosynthesis.

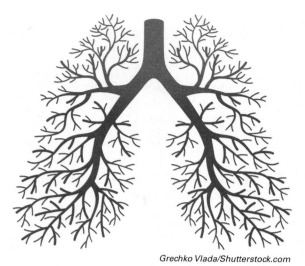

Grechko Vlada/Shutterstock.com

Figure 8.5.3. The roots are much like the lungs of the plant. They absorb oxygen from the soil for respiration to take place. This can be a problem for plants grown in waterlogged or overwatered conditions.

Respiration

Stop and take a deep breath, filling your lungs with oxygen (**Figure 8.5.3**). What important thing does oxygen do for your body? As with humans, oxygen is important to plants and is a key component of respiration. Respiration is the opposite of photosynthesis. In ***respiration***, sugars made in photosynthesis combine with oxygen to produce energy in a form that can be used by plants. The energy that is produced is applied toward the growth and development of the plant, as well as flower and fruit formation. The chemical equation for respiration is:

$C_6H_{12}O_6$ (glucose or sugar) + $6O_2$ (oxygen) $\rightarrow$ $6CO_2$ (carbon dioxide) + $6H_2O$ (water) + energy

Plant growth that is fueled by respiration takes place mostly at night (**Figure 8.5.4**). The rate of respiration controls how quickly a plant grows. Temperature is directly related to the rate of respiration. Rising temperatures can increase growth rates and lower temperatures can decrease growth rates. Managing temperature is important for growers because it is directly connected to the rate at which plants grow.

Figure 8.5.4. Photosynthesis occurs in the presence of light and uses carbon dioxide and water to produce sugar and oxygen. Respiration is the opposite. It occurs at night, and uses sugar and oxygen to produce carbon dioxide, water, and energy. Think of it as inhaling and exhaling: You cannot have one without the other.

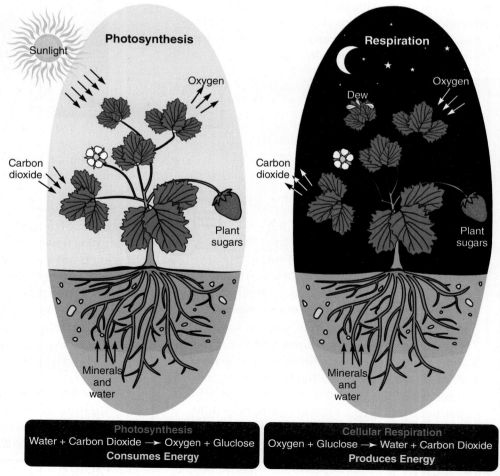

udaix/Shutterstock.com

Transpiration

Have you ever exhaled on a cold day and seen your breath? This is because when you exhale, you are releasing water. Plants also "breathe" through their leaves. Leaves have tiny openings on their surface called *stomata* (**Figure 8.5.5**). When stomata are open, water vapor is released. Most stomata are found on the underside of the leaf.

Roots absorb water from the ground in liquid form. The water is drawn upward through the xylem to the leaves and other plant parts. Water is released from the leaves as water vapor through the process of *transpiration*.

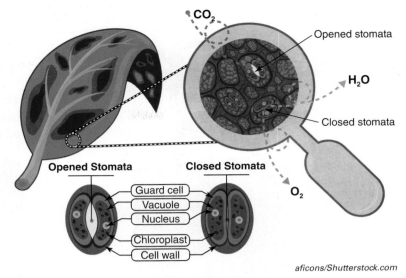

aficons/Shutterstock.com

Figure 8.5.5. Water is released from the plant as water vapor through openings in the leaves called stomata.

Hands-On Activity

Leaf Transpiration Lab

Plant transpiration is an invisible process. As the water is evaporating from the leaf surfaces, you cannot see the leaves "breathing." However, just because you can't see the water escaping doesn't mean it is not being released into the air. One way to see transpiration is to put a plastic bag around a group of plant leaves and collect the released water vapor. Shall we get started?

Atjanan Charoensiri/Shutterstock.com

You will need the following materials:

- Sealable plastic bag
- Rubber band or twist tie
- Plant
- Measuring cup, beaker, or graduated cylinder

Here are your directions:

1. Select a plant limb with several leaves on the end.
2. Place as many leaves as you can inside the sealable bag without removing the leaves from the plant.
3. Close the seal on the bag as tightly as possible with the leaves inside with either a rubber band or a twist tie. Avoid damaging the plant.
4. Leave the bag sealed for 24 hours.

5. After 24 hours, carefully remove the bag and avoid letting any of the water escape. It may be helpful to shake the bag gently to get the water to fall to the bottom of the bag.
6. Using the graduated cylinder, measure and record the amount of water collected in the bag.

Consider This

1. How much water did you collect from your bag? Find six other students and record the amount of water they each collected. How are their amounts different from yours?
2. What environmental factors do you think would affect transpiration?

Water evaporates from the leaves

Veins carry water into the leaves

Water is drawn up the stem to the leaves

Roots take up water from the soil

BlueRingMedia/Shutterstock.com

Figure 8.5.6. Transpiration helps to regulate the temperature of the plant while allowing carbon dioxide and nutrients to enter from the atmosphere.

As much as 99 percent of the water taken up by the roots is lost through transpiration. When the stomata are open and releasing water vapor, carbon dioxide can enter the plant for photosynthesis (**Figure 8.5.6**). Transpiration is also important for absorbing nutrients from the atmosphere, as well as evaporative cooling on hot days.

Did You Know? Since cacti don't have leaves, they only have a few stomata in their green stems. They transpire very little. How might this benefit them?

Career Corner   HYDROPONIC PRODUCER

Hydroponic producers *grow crops without soil using a water solution that is rich in nutrients, typically in greenhouses.*

What Do They Do? Hydroponic producers choose plants to be grown, develop a system and schedule for planting, develop business relationships with local chefs to supply locally grown produce, determine what nutrients the plants need, and find ways to make hydroponic systems more efficient. They also test and maintain water quality for growing.

What Education Do I Need? A high school diploma and farming or gardening experience is needed, but no degree is required.

How Much Money Can I Make? Hydroponic producers make an average of $30,000 per year for small operations. Working in larger operations, producers can make closer to $75,000.

How Can I Get Started? You can start an SAE building your own hydroponic system. You can create an experiment comparing plants grown hydroponically to plants grown traditionally, or conduct research on different hydroponic systems and present it to your class. Interview a hydroponic grower or set up a virtual field trip for your class.

Lano Lan/Shutterstock.com

Review and Assessment

What Did We Learn?

- All plants make their own food through photosynthesis.
- In photosynthesis, plants take carbon dioxide from the air, add water, and use the energy from the sun to produce sugar for plant food. Oxygen for people to breathe is another important by-product of photosynthesis.
- Photosynthesis takes place in the chloroplasts, which are filled with chlorophyll. Chlorophyll captures the sun's energy.
- Respiration is the opposite process to photosynthesis. It takes the sugars made in photosynthesis and combines them with oxygen to produce energy. Respiration takes place mostly at night and is affected by temperature. The rate at which respiration takes place controls how quickly a plant grows.
- Leaves have tiny openings on their surface called stomata which release water vapor. This process is called transpiration. Up to 99 percent of the water absorbed through the roots is lost through the leaves during transpiration. Transpiration helps cool the plant and allows the plant to absorb nutrients from the atmosphere.

Let's Check and See What We Know

Answer the following questions using the information provided in Lesson 8.5.

1. The chemical reaction that makes food for plants is known as ____.
   A. respiration
   B. photosynthesis
   C. transpiration
   D. miosis

2. During photosynthesis, ____ is absorbed from the air.
   A. carbon dioxide
   B. sugar
   C. water
   D. oxygen

3. *True or False?* Photosynthesis takes place inside the chlorophyll.

4. ____ is a pigment that is important in capturing energy from light.
   A. Chloroplast
   B. Stomata
   C. Chlorophyll
   D. Transpiration

5. *True or False?* Respiration is the opposite of photosynthesis.

6. ____ is the chemical reaction that produces energy in a form that can be used by plants.
   A. Respiration
   B. Photosynthesis
   C. Transpiration
   D. Miosis

7. *True or False?* Temperature affects the rate of respiration.

8. The tiny openings in the leaves of plants that release water vapor are called ____.
   A. chloroplasts
   B. stomata
   C. chlorophyll
   D. transpiration

9. The process of releasing water from the plant is known as ____.
   A. respiration
   B. photosynthesis
   C. transpiration
   D. miosis

10. *True or False?* Transpiration is important to help the plant absorb nutrients from the atmosphere.

Let's Apply What We Know

1. What is photosynthesis and why is it important to all living things on Earth?
2. Explain the process of respiration.
3. Where does respiration occur within the cell?
4. How is transpiration beneficial for a plant?
5. How does raising and lowering temperatures affect transpiration?

Academic Activities

1. **Science.** Transpiration can be measured by weighing a potted plant at several points. Take a small potted plant and water it thoroughly. After the water has drained, weigh it and record the weight. Cover the soil with plastic wrap to prevent any water from evaporating from the surface of the soil. Observe the plant over several days and record the weight of the plant each day. Do not add any additional water to the soil while you're observing. Make a chart of the different weights you recorded.
2. **Language Arts.** The primary purpose of descriptive writing is to describe a person, place, or thing in such a way that a picture is formed in the reader's mind. Write a summary of photosynthesis and respiration using descriptive writing techniques.

Communicating about Agriculture

1. **Visual Communication.** Presenting information visually can be a good way to help others understand complex ideas. Create a poster explaining photosynthesis, respiration, and transpiration. Use informative (but brief) labels to explain key concepts. Share your poster with your class.

SAE for ALL Opportunities

1. **Research SAE.** Conduct an experiment to determine if different types of light have any effect on photosynthesis. You can model your experiment after the one outlined in this lesson, setting up your experiment using LED light, fluorescent light, black light, and any other light source you find interesting. Formulate a hypothesis, create your experiment, observe, and record data and write a report on your findings.
2. **Research SAE.** Conduct an experiment on the effect of temperature on the rate of transpiration. You can model your experiment after the one outlined in this chapter, setting up your experiment at different temperatures. Formulate a hypothesis, create your experiment, observe, and record data and write a report on your findings.

SAE for ALL Check-In

- How much time have you spent on your SAE this week?
- Have you logged your SAE hours?
- What challenges are you having with your SAE?
- How can your instructor help you?
- Do you have the equipment you need?

Reproductive Parts of a Plant

Anton Nikitinskiy/Shutterstock.com

Learning Outcomes

By the end of this lesson, you should be able to:
- Identify and describe the reproductive parts of a plant. (LO 08.06-a)
- Explain the process of pollination. (LO 08.06-b)
- Describe the differences between complete, incomplete, perfect, and imperfect flowers. (LO 08.06-c)

Words to Know

complete flower	perfect flower	pollination
cross pollination	petals	sepals
imperfect flower	pistil	stamens
incomplete flower	pollen	

Before You Begin

AAA-CHOO! My goodness, my allergies have been awful lately. AAA-CHOO! Excuse me, I have been sneezing all day. It is springtime where I am and there is a lot of pollen in the air. While I do love all the pretty flowers, pollen sure does make me sneeze. Is there a lot of pollen where you live?

Why do you think plants make pollen? Do you think it has a purpose, other than making people sneeze? Take a moment and discuss your thoughts with a classmate. You might even need to pass them a tissue if it's pollen season.

We learned in a previous lesson that flowers are the reproductive part of a plant. Flowers have different shapes, sizes, and colors, but most of them are made up of the same parts. Their unique features are designed to help them attract particular creatures, like bees, butterflies, or birds. These animals are important in helping some plants reproduce. Other flowers take advantage of natural elements such as wind or rain to help them reproduce. Each separate part of the flower plays a specific role in the reproduction process.

Flower Parts

There are four main parts to a flower: sepals, petals, stamens, and pistil. Each of these main parts is made of smaller parts that play unique and important roles.

- **Sepals** are the green leaf-like structures beneath the petals that serve as a protective covering of the flower before it opens. Some sepals have spines or spikes that protect the flower from insects and other animals. Some sepals produce a chemical making them undesirable to eat, driving away insects and animals. The calyx is made up of all the sepals on one flower.
- **Petals** are brightly colored modified leaves that surround the male and female organs and serve to attract pollinators. Attracting pollinators is essential for most plants to be able to reproduce.
- The **stamens** are the male organs of the flower (**Figure 8.6.1**). The stamen consists of a stalk called a filament and an anther. The anther produces and holds the pollen. **Pollen** is the male sex cells of a plant.
- The **pistil** is the female organ of the flower. The pistil is in the center of the flower and is surrounded by the stamens. The pistil consists of three parts: the ovary, the stigma, and the style. The stigma can be found at the end of the pistil and has a round sticky surface to catch pollen. The neck or center tube is the style. The style connects the stigma to the third part, the ovary. Egg cells are developed in the ovary (**Figure 8.6.2**). Once fertilized by pollen, the ovary grows to become the seed or the fruit.

gsrsirji/Shutterstock.com

Figure 8.6.1. Most flowers have several stamens but only one pistil.

Did You Know? The largest flower in the world can be up to ten feet tall and three feet wide, and it can weigh up to 24 pounds! It is called the Titan Arum and it has a distinctive smell of rotting flesh, which is why it is also known as the "corpse flower."

Figure 8.6.2. The pistil is the female reproductive part of a flower and is made up of the stigma, the style, and the ovary. The stamen is the male reproductive part of a flower and is made up of the anther and filament.

Common Flower Parts

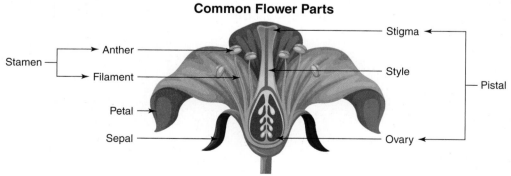

Stamen — Anther
Filament

Petal
Sepal

Stigma
Style
Pistal
Ovary

BlueRingMedia/Shutterstock.com

Hands-On Activity

Flower Dissection

Some flowers have flower parts that are visible to the naked eye. You can even see the individual parts of both the male and female flower parts. In this activity, dissect a flower and label its individual parts.

You will need the following materials:

- Flower (such as an azalea or lily)
- Knife or scissors
- White paper
- Tape
- Marker or pen

Here are your directions:

1. With your knife or scissors, carefully remove the flower from the stem at the base of the sepals.
2. Carefully remove the sepals from around the petals. Tape them to your paper and label them correctly.
3. Remove the petals from the flower, being careful to not disturb the center plant parts. Tape them to your paper and label them correctly.
4. Carefully pull the stamens from around the pistil. Tape them to the paper and label the stamens and their individual parts.
5. Attach the pistil to your paper and label it and its individual parts.

Consider This

1. Were you able to see the pollen grains on the style? What would have been required to transfer those pollen grains to your piece of paper?
2. Go an extra step and cut the ovary in two. What do you see?

Pollination

Pollination occurs when pollen grains from the anther transfer to the sticky surface of the stigma. In nature, this process happens naturally when birds, insects, and other animals are attracted to the colorful scented flower petals. They pick up pollen from the anther on their legs, wings, and hairs (**Figure 8.6.3**). When they visit another flower, they deposit the pollen onto the stigma. Other plants rely on wind to transfer pollen from one plant to another.

Figure 8.6.3. Bees are excellent pollinators. Many large-scale greenhouse growers put small beehives inside their greenhouses to help with pollination in an indoor environment. *Why would pollination be an issue inside?*

MERCURY studio/Shutterstock.com

Once pollen enters the stigma, the fertilization process begins. The pollen that lands on the stigma sprouts like seeds and sends long stalks down the style to the ovary and egg cells. The pollen sperm cell fertilizes the egg cells and seeds begin to develop.

When the pollen of a plant pollinates the flower on the same plant, it is known as self-pollination. Self-pollination is only possible for a small number of plants. When the pollen of one plant pollinates the flower of another plant, it is known as *cross pollination* (**Figure 8.6.4**). Cross pollination occurs between plants that are closely related. Plant breeders are known to control cross pollination to help them develop new plant varieties.

Did You Know? Vanilla flowers are very delicate; they only open for a couple of hours at a time and need to be pollinated by hand to produce a vanilla bean.

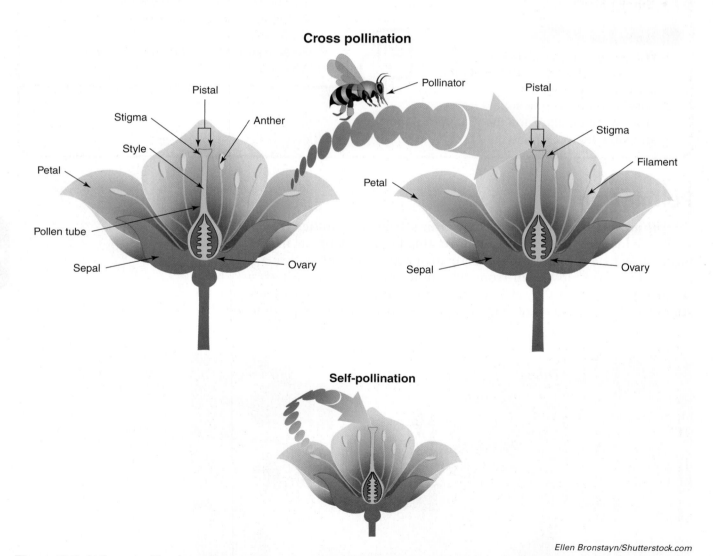

Ellen Bronstayn/Shutterstock.com

Figure 8.6.4. Cross pollination takes place when pollen from one plant is transferred to another plant, usually by an insect or wind. Self-pollination is when insects or wind move pollen from the anther to the stigma inside the same flower. *Which is more common? Why?*

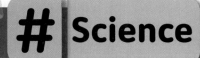

Making Connections # Science

Cross Pollination

Some flowers cannot pollinate themselves, even though they have both male and female parts. They need the pollen from one plant to mix with the pistil of another plant. This is known as cross pollination. Typically, cross pollination is the result of insects or wind transferring the pollen. However, you can play the role of an insect and transfer the pollen yourself.

You will need the following materials:

- 1 white flowering plant
- 1 plant of the same species with a different-colored flower
- Small paint brush

Here are your directions:

1. Select your colored flowering plant and locate the stamen. Lightly swipe your fingernail over the yellow pollen. If the pollen sticks to your nail, the flower is ready to pollinate other flowers.
2. Select your white flowering plant and locate the stigma. Lightly touch it with your finger. If it is shiny and sticky to the touch, it is ready for pollination.

Elena Masiutkina/Shutterstock.com

3. With your paint brush, collect the pollen from the anther of the colored flower.
4. Immediately brush the pollen across the stigma of the white flower.

Consider This

1. What do you think will happen when the white plant flowers again?
2. How difficult was transferring the pollen? What characteristics do insects have that make them well-equipped to do this task?

Types of Flowers

Flowers can either be complete or incomplete. A flower that has all four major flower parts is called a ***complete flower***. Roses, tulips, and vegetables such as pea plants all have complete flowers (**Figure 8.6.5**). A flower that lacks one or more of the major parts is an ***incomplete flower***. For example, calla lilies do not have sepals, and are incomplete flowers (**Figure 8.6.6**).

Flowers can also either be perfect or imperfect. A flower that contains both the pistil and stamen is known as a ***perfect flower***. Most flowers are perfect flowers, but tomatoes, morning glories, petunias, and irises are a few examples.

Kicky_princess/Shutterstock.com

Figure 8.6.5. Complete flowers contain all four major plant parts: petals, sepal, pistil, and stamen. Lilies are a great example of complete flowers. *Can you identify each part?*

A flower that lacks either the female or male part of the flower is an ***imperfect flower***. Grass, sweet corn, and squash plants are all examples of imperfect flowers (**Figure 8.6.7**). Complete flowers cannot be imperfect, and perfect flowers cannot be incomplete.

Donna Ellen Coleman/Shutterstock.com

Figure 8.6.6. Calla lilies are an example of an incomplete flower. They do not have sepals that surround their flower.

nam_nueng/Shutterstock.com

Figure 8.6.7. Corn is an example of a plant with imperfect flowers. Corn flowers are known as tassels and rely on wind to transfer their pollen to other corn plants.

Career Corner > PLANT PATHOLOGIST

Plant pathologists *study the health of plants. They work to identify diseases, pests, and other health problems a plant may experience.*

What Do They Do? Plant pathologists sample plant tissues, provide mechanical immunization of pathogens in growing areas, perform research with insects or vectors, perform data collection, and design and implement new screening methods for diseases.

What Education Do I Need? Plant pathologists typically obtain an undergraduate or graduate degree in botany, horticulture, plant pathology, or biology. Having a doctorate degree is recommended.

How Much Money Can I Make? Plant pathologists make an average of $75,000 per year.

How Can I Get Started? You can build an SAE comparing different natural treatments of plant pests, conduct research on plant viruses and present it to your class, interview a plant pathologist, or set up a virtual field trip to a lab for your class.

Chutima Chaochaiya/Shutterstock.com

What Did We Learn?

- There are four main parts to a flower: sepals, petals, stamens, and pistil.
- The pistil is the female reproductive part of the plant and is made up of the stigma, style, and ovary.
- The stamen is the male reproductive part of the plant and is made up of the anther and filament.
- Pollination occurs when the pollen grains from the anther transfer to the sticky surface of the stigma.
- Self-pollination occurs when a pollen is transferred from the anther to the stigma on the same plant. Cross pollination occurs when the pollen is transferred to a different plant.
- Complete flowers have all four flower parts and incomplete flowers are missing one or more of these parts.
- Perfect flowers contain both the male and female parts within the same flower. Imperfect flowers have either the male or female part, but not both.

Let's Check and See What We Know

Answer the following questions using the information provided in Lesson 8.6.

1. The green leaf-like structures that protect the petals as they are developing are the _____.
   A. sepals
   B. stigma
   C. style
   D. stamen

2. The male reproductive organ of the flower is the _____.
   A. sepal
   B. pistil
   C. stamen
   D. anther

3. *True or False?* The pistil is in the center of the flower and is surrounded by the stamen.

4. The process of transferring pollen grains from the anther to the stigma is known as _____.
   A. pollination
   B. reproduction
   C. pistil
   D. stamen

5. *True or False?* Insects are solely responsible for the pollination of plants.

6. _____ occurs when pollen is transferred from the anther of one plant to the stigma of another plant.
   A. Reproduction
   B. Stamen
   C. Self-pollination
   D. Cross pollination

7. *True or False?* All flowers are complete flowers.

8. A flower that has all four major plant parts is a(n) _____.
   A. complete flower
   B. incomplete flower
   C. perfect flower
   D. imperfect flower

9. A flower that contains both male and female parts within a single flower is a(n) _____.
   A. complete flower
   B. incomplete flower
   C. perfect flower
   D. imperfect flower

10. *True or False?* Complete flowers cannot be imperfect flowers.

Let's Apply What We Know

1. In a notebook, draw and label a complete flower.
2. How is pollination accomplished?
3. Explain the difference between self- and cross pollination.
4. How are perfect and imperfect flowers different?
5. Go on a nature walk around your campus or neighborhood. Collect several different kinds of flowers. Determine whether each flower is complete or incomplete, perfect or imperfect.

Academic Activities

1. **Reading and Speaking.** Using your textbook, library resources, and the internet, research ornamental crop breeding. Determine what is necessary to create a new plant and sell it to consumers. Make a poster illustrating your findings and present it to your peers.
2. **Technology.** Using a handheld fan and sticky piece of paper, demonstrate how wind can move pollen from one plant to another. Tape your sticky paper (which you can cover in glue or tape) to the wall. Hold a flower with large pollen grains like a lily up in front of the paper. Use your handheld fan to try and move the pollen. Experiment with different distances between the fan and the flower. What did you find?

Communicating about Agriculture

1. **Speaking.** What types of flowers (perfect, imperfect, complete, incomplete) are the most common? Why do you think so? Put together a 2–3 minute persuasive speech for your class, explaining why you think your argument is correct.

SAE for ALL Opportunities

1. **Research SAE.** Conduct an experiment using cross pollination techniques on different plant material. Make a hypothesis, create your experiment, observe, record data, and write a report about your findings.

SAE for ALL Check-In

- How much time have you spent on your SAE this week?
- Have you logged your SAE hours?
- What challenges are you having with your SAE?
- How can your instructor help you?
- Do you have the equipment you need?

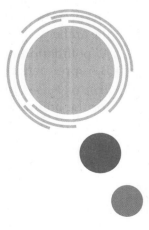

Sexual Propagation

Essential Question

What challenges do plants face when reproducing from seed?

Learning Outcomes

By the end of this lesson, you should be able to:
- Define propagation. (LO 08.07-a)
- Describe the process of sexual propagation. (LO 08.07-b)
- Explain the differences between direct and indirect seeding. (LO 08.07-c)

Words to Know

direct seeding
germinate
indirect seeding

plant propagation
scarification
seedlings

sexual propagation
stratification

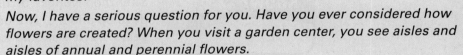

Before You Begin

Do you know where flowers go to recharge after a long day? A power plant! I bet you have been missing my jokes. I can do better than that, though. What makes some plants better at math than others? Square roots! Ha, that is one of my favorites!

Now, I have a serious question for you. Have you ever considered how flowers are created? When you visit a garden center, you see aisles and aisles of annual and perennial flowers. At the grocery store you might notice the large display of cut flowers ready to go home. Have you ever stopped and thought about what it takes to get that flower to you? In this lesson, we will be learning about how these beautiful flowers are produced for consumers, season after season and year after year.

Have you ever snacked on sunflower seeds at a ballpark or enjoyed pumpkin seeds at a fall festival? People have been eating, collecting, and planting seeds for thousands of years. When European settlers first came to America, the Native Americans provided them with seeds to grow food. Around the world, seeds have been sold, traded, and passed down through generations. Why do you think people take the time to collect, store, and sell seeds? Did you know that seeds have a particularly important role when it comes to producing new plants?

Plant Propagation

All living organisms, including plants, reproduce. **Plant propagation** is the process of producing a new plant. Everyone benefits from the production of new plants. Without new plants, we would not have many of the vegetables we eat, turf grasses for sports, and flowers for bouquets.

Most plants naturally reproduce with seeds. For many years, creating new plants from seed was the only propagation method used by horticulturists. Today, other methods are used to produce new plants, even though seeds are still common. Advanced technology allows us to produce new plants from plant tissue. Biotechnology allows us to create disease-resistant plants and plants with other unique, beneficial features (**Figure 8.7.1**).

Propagation methods can be divided into two categories: sexual and asexual. **Sexual propagation** is the reproduction of plants from seed. Asexual propagation is the reproduction of plants from other vegetative parts, such as the leaves. The best method of propagation depends on the type of plant. While some plants, such as sunflowers, produce lots of seeds for reproduction (**Figure 8.7.2**), others, such as pothos plants, are better reproduced by using the vegetative parts of a plant. We'll talk more about asexual propagation in **Lesson 8.8**.

Did You Know?

The most expensive flower ever sold was a Shenzhen Nongke orchid. It took eight years to develop and it only blooms once every four to five years, but it was sold for $200,000 at auction!

Sergey Bezgodov/Shutterstock.com

Figure 8.7.1. Limelight hydrangeas are the result of advanced technology in the horticulture industry, creating a new plant variety that is much more sun tolerant than traditional hydrangeas.

Jesada Sabai/Shutterstock.com

Figure 8.7.2. Sunflowers are easily reproduced by seed. A single flower can produce hundreds of seeds.

Propagation from Seeds

Since most plants naturally reproduce from seed, sexual reproduction is often the easiest and least expensive method for plant reproduction. For successful growth, seeds must have the correct environmental and cultural conditions, including temperature, moisture, light, and soil medium. Some seeds need light to *germinate* (or sprout from dormancy) while other seeds, like delphinium, need darkness. Some seeds need to be completely buried into the soil medium for germination, while others, such as impatiens, only require a fine sprinkling of soil for planting. It is important for the grower to research and understand the specific germination requirements for that seed type (**Figure 8.7.3**).

Outside of environmental factors, seeds sometimes have additional conditions that must be met for germination. Some seeds have a hard seed coat that must be soaked or scratched before they will sprout. This process is known as *scarification*. The red bud tree is an example of a seed that requires the thick seed coat to be weakened before germination.

Other seeds require a moist, cold period of rest (dormant stage) at temperatures between 32- and 50-degrees F for an extended period to germinate. This process is known as *stratification*. The temperature and amount of time depends on the seed type. Stratification happens naturally for plants in outdoor environments (**Figure 8.7.4**). Greenhouse growers collecting and planting seeds that require stratification sometimes use artificial cold, such as refrigeration, to help seeds go through the process. It is important for growers to read the label on the seed packet and understand the growing requirements.

Bogdan Wankowicz/Shutterstock.com

Figure 8.7.3. There are many conditions that must be met for seeds to germinate, including the correct planting depth, moisture, and light.

Craig A Walker/Shutterstock.com

Figure 8.7.4. Some seeds require an extended period of cold before they will germinate. This happens naturally in the winter months for plants grown outdoors.

Scarification

Some plants produce seeds with a hard seed coat. What environmental factors would favor a hard seed coat over a seed with a soft outer shell? Sometimes, another action must take place, such as cracking or softening the seed coat, before the seed can sprout.

You will need the following materials:

- Morning glory seeds
- 4 small pots with holes for drainage
- 2 plastic dishes
- Soil
- Water
- Marker
- Sandpaper
- Hot water
- Lemon juice

Here are your directions:

1. Pour hot water into a small plastic dish, enough to cover the seeds. Drop 2–3 seeds into the hot water and let them soak for 30 minutes.
2. Pour lemon juice into a small plastic dish, enough to cover the seeds. Drop 2–3 seeds into the lemon juice and let them soak for 30 minutes.
3. Collect 2–3 seeds and carefully scratch the surface of the seed with sandpaper. You want to lightly remove the seed coat from the seeds.

Thosapon.s/Shutterstock.com

4. Label 4 pots with "no treatment," "hot water," "lemon juice," and "sandpaper."
5. Fill each of your pots with soil.
6. Plant 2–3 seeds directly into the pot labeled "no treatment." This will be your control.
7. Plant the other treated seeds into the correctly labeled pot.
8. Water each of the pots and put them in a warm, well-lit area to germinate.

Consider This

1. What was the purpose of the lemon juice in the experiment? What effect do you think it had on the seed coat?
2. Which treatment method was the most successful in your experiment? What made it the most successful?

Planting Seeds

Direct Seeding

Many vegetables and grass seeds are planted directly into the soil outdoors. This is known as *direct seeding*. Growers typically use direct seeding with plants that germinate easily in the soil and with very small seeds that are difficult to handle. Direct seeding can be done both by hand and with machinery (**Figure 8.7.5**).

There are several factors one should consider before direct seeding:

- Site selection. Soil should have good drainage and plenty of light.
- Soil type. Soil should not be too compact or hard. Soil should be loose and finely textured to allow for absorption of water. Soil should also be free of weeds, which will compete with the seeds for water and light as they grow.
- Planting date. Seeds have specific temperatures in which they like to grow. Planting during the correct season is important to hitting the soil temperature needed for successful germination.
- Planting depth and spacing. Each seed has its own unique depth in the soil for planting. Planting a seed too deep or not deep enough could affect your germination rate (**Figure 8.7.6**). The plant size at full maturity should be taken into consideration when spacing your seeds.
- Moisture. All seeds need moisture to germinate. The soil should not be allowed to dry out from the time the seed is planted until it is established.

Indirect Seeding

Indirect seeding occurs when seeds are planted in a different place from where the plants will grow to full maturity. Young plants are called *seedlings*. The seedlings will be transplanted one or more times before reaching their permanent growing location. By starting seeds in a greenhouse, growers can extend the length of the growing season for the plant (**Figure 8.7.7**).

photowind/Shutterstock.com

Figure 8.7.5. Grass seeds used in lawns are often spread using a seed spreader. The seed is poured into the machine bine, and then the plates inside are set to the correct spacing. As the spreader is rolled across the lawn, seeds are released from multiple locations, dropping directly onto the soil below.

amenic181/Shutterstock.com

Figure 8.7.6. Proper planting depth and spacing is important for successfully planting seeds.

Figure 8.7.7. If you have ever purchased plants or flowers from a garden center, you have purchased plants that were indirectly seeded. These plants were started in a greenhouse to later be purchased by a homeowner or landscaper and planted in a new location.

wavebreakmedia/Shutterstock.com

hamhaful/Shutterstock.com

Figure 8.7.8. Seed trays are often used to start seeds indirectly. They have individual cells for each seed and hold only the small amount of soil needed for germination.

Seeds are planted into trays or flats (**Figure 8.7.8**). It is important for growers to label these seed trays with the name, variety, and planting date. This helps the grower keep track of what is planted as the seeds are developing and will be used later when transplanting. Once the seedlings have produced true leaves and a strong root system, they are transplanted into larger containers where they will finish growing until they are ready to be sold or moved to their permanent location.

Hands-On Activity

Reading a Seed Packet

Looking at a seed packet, see if you can determine the answers to the following questions.

You will need the following materials:

- Various seed packets
- Pencil and paper

Answer the following questions from your seed packet:

1. What are you growing?
2. How many days will it take to germinate?
3. Should you start your seeds inside? Are there any other items noted for seed germination?
4. What is the planting depth?
5. Does it require sun, shade, or both?
6. Is the plant an annual or a perennial?
7. How many days to maturity? What does this mean?
8. Is there an expiration date on the seed packet?

HelloSSTK/Shutterstock.com

Consider This

1. How does this information benefit the grower? What information was not included that you think you may need?
2. Why do you think seeds have an expiration date?
3. Design and create your own seed packet.

Career Corner   PLANT BREEDER

*If this is interesting to you, then maybe being a **plant breeder** is a career for you to explore!*

What Do They Do? Plant breeders study seed characteristics and work to improve those characteristics that are most desirable for a plant, such as yield, size, quality, maturity, and resistance to frost, drought, disease, and insect pests.

What Education Do I Need? A bachelor's degree in crop science, plant genetics, or agronomy is required. In many cases a Ph.D. or master's degree in plant science is required.

How Much Money Can I Make? Plant breeders make an average of $75,000 per year.

How Can I Get Started? You can build an SAE around starting and identifying a seed collection. You can create an experiment comparing scarification and stratification, conduct research on different seed planting methods and present it to your class, or interview a plant breeder. Maybe you can set up a virtual or in person field trip to a greenhouse for your class.

SeventyFour/Shutterstock.com

LESSON 8.7   Review and Assessment

What Did We Learn?

- Plant propagation is the process of producing a new plant. Sexual propagation is producing a new plant from seed, and asexual propagation is producing a new plant from vegetative plant parts.
- Sexual propagation is the most common method of plant propagation because it is easiest and least expensive.
- Successful seed germination requires the grower to know the individual needs of each seed regarding environmental factors such as light, temperature, and water.
- Seeds with a hard seed coat require scarification, which is the wearing down of the seed coat allowing the seed to germinate. Other seeds require extended periods of cold known as stratification.
- Direct seeding is planting seeds directly into the soil outdoors. This is done both by hand and with machinery.
- Indirect seeding is the process of planting seeds in a place that is different from where the plants will grow to maturity.

Let's Check and See What We Know

Answer the following questions using the information provided in Lesson 8.7.

1. The process of producing a new plant is _____.
   A. sexual propagation
   B. asexual propagation
   C. plant propagation
   D. germination
2. Producing plants from seed is known as _____.
   A. sexual propagation
   B. asexual propagation
   C. plant propagation
   D. germination
3. *True or False?* Asexual propagation is the easiest and cheapest method of plant propagation.
4. Soaking or scratching a hard seed coat to help with germination is known as _____.
   A. scarification   C. stratification
   B. propagation   D. germination
5. Exposing seeds to an extended period of cold to help with germination is known as _____.
   A. scarification   C. stratification
   B. propagation   D. germination

6. *True or False?* Greenhouse growers use artificial cold, such as refrigeration, to promote stratification.
7. *True or False?* Direct seeding can only be completed by hand planting.
8. Which of the following is *not* a consideration when direct seeding?
   A. Soil type
   B. Planting depth or space
   C. Planting date
   D. Annual or perennial
9. Starting seeds in a location that is different than their final planting location is known as _____.
   A. direct seeding
   B. indirect seeding
   C. plant propagation
   D. sexual propagation
10. *True or False?* Plants grown from seed are called seedlings.

Let's Apply What We Know

1. Explain plant propagation and how it has changed over the years.
2. What conditions must be met for a seed to germinate?
3. Explain the difference between stratification and scarification.
4. What factors should be considered when selecting vegetable seeds for an outdoor garden that you plan to directly seed?
5. When is it beneficial to use indirect seeding?

Academic Activities

1. **Math.** As a small greenhouse grower, you have a limited budget, $500, to purchase seeds, and want to grow five different varieties of plants. You have five greenhouse tables, each 50 feet long and 4 feet wide. Research the growing requirements and spacing needed for your desired crops, and create a plan for your purchase. What plants will you grow? How many can you grow? How will you use your budget?
2. **Language Arts.** Imagine you are the marketing director for a seed company that sells vegetable seeds. One of your job responsibilities is to write articles for the grower's information center. Write an article about how to choose the best variety of vegetable seeds for your garden. Limit your article to three different varieties of vegetable seeds.

Communicating about Agriculture

1. **Social Media.** Find two social media accounts that provide useful information about growing seeds, selecting seeds, or creating a planning calendar. What do you think about how the information is presented? Using resources at the library, quickly fact check three facts you see presented. Why is it important to make sure that information you get on social media is reliable?

SAE for ALL *Opportunities*

1. **Foundational SAE.** Create a germination calendar for a summer vegetable garden. What seeds do you need to begin indoors? Which can be planted directly? Include planting time, depth, and spacing for each plant. Outline the importance of having a planting calendar.
2. **Placement SAE.** Research a garden center, governmental agency, or plant conservation society near you that collects seeds. Volunteer to harvest, extract, and clean seeds.
3. **Research SAE.** Conduct a germination experiment. Test different environmental conditions needed for successful seed germination. Make a hypothesis, create your experiment, observe and record data, and write a report on your findings.

SAE for ALL *Check-In*

- How much time have you spent on your SAE this week?
- Have you logged your SAE hours?
- What challenges are you having with your SAE?
- How can your instructor help you?
- Do you have the equipment you need?

Beekeepx/Shutterstock.com

Asexual Propagation

Essential Question

How can plants reproduce without seeds?

Learning Outcomes

By the end of this lesson, you should be able to:
- Explain why asexual propagation methods are important, and when they are used. (LO 08.08-a)
- Describe different methods of asexual propagation. (LO 08.08-b)

Words to Know

asexual propagation	grafting	stem cuttings
cuttings	leaf cuttings	
division	separation	

Before You Begin

I don't mean to alarm anyone, but did you know there are plants out there that don't produce seeds? Ferns are one of the most common plants that do not produce seeds. This may be a little confusing, because I see ferns everywhere and I even have a few ferns hanging on my porch. If they don't have seeds, how do they keep growing and spreading?

In Lesson 8.7, we learned about producing plants from seed. If ferns don't make seeds, how do they reproduce? Do you have any guesses? Jot down some ways you think a plant might reproduce without seeds.

Did you know that not all plants have flowers? Consider aloe plants, English ivy, and snake plants: Each of these plants rarely, if ever, flower. How do you get new plants if there are no flowers to make seeds? Not all plants reproduce sexually through seed production. You can also reproduce plants from the leaves, stems, or roots of a parent plant, known as *asexual propagation* (**Figure 8.8.1**). Asexual propagation is required for plants that do not produce seeds or are difficult to grow from seed.

Figure 8.8.1. Not all plants reproduce from seed. This plant is producing new plants from the leaf.

Asexual Propagation

Asexual propagation allows growers to produce plants quickly, especially in situations where plants produce few or no seeds or are hard to grow from seed. One of the benefits of asexual propagation is that the plants produced are genetically identical to the parent plant. This means they look the same as the plant they came from and have the same traits as far as height, leaf shape, and color (**Figure 8.8.2**). There are several methods to reproduce a plant asexually, including cuttings, separation, division, and grafting.

Cuttings

While there are several asexual propagation methods, taking cuttings is the most common method in the horticulture industry. *Cuttings* are detached parts of a parent plant, such as the leaves, stems, or roots, that grow into complete new plants. In this method, the detached portion of the parent plant is placed in potting soil, or in water, and provided with the nutrients it needs to grow and develop into a new plant. This new plant looks identical to the parent plant and can be reproduced much faster than growing the same plant from seed.

Figure 8.8.2. When plants are reproduced asexually, the new plants share the exact same traits as the parent plant.

8

Amanda Feltz/Shutterstock.com

Figure 8.8.3. Pothos is a plant commonly reproduced through stem cuttings. Each one of the stem sections above will start to grow its own root system and develop into new plants.

Pegasene/Shutterstock.com

Figure 8.8.4. African violets are commonly propagated through leaf cuttings. The petiole (stem) of each leaf is planted in the soil. Eventually these leaves and petioles will start to form roots and grow into new plants.

The age of the plant when the cutting is taken, growing season, and the type of stem that the cutting is taken from all need to be considered when taking cuttings. Cuttings typically require similar conditions as seeds to grow, needing the right temperature, correct amount of light, moisture, and oxygen. The two most common method of cuttings taken are stem cuttings and leaf cuttings.

Stem cuttings use 3–6 inch portions of a stem that contains a plant bud. These stem cuttings are placed in potting soil and new roots form at the bottom of the stem where the cut was made (**Figure 8.8.3**). Stem cuttings can be classified into two main categories: softwood and hardwood cuttings. Softwood cuttings are stems cut while they are soft and green, usually during periods of new growth. Hardwood cuttings are fully mature stems cut while the plant is dormant and have lost their leaves for the winter.

Leaf cuttings are taken by removing a leaf blade from the parent plant. Some leaf cuttings require you to keep the stem (petiole) attached to the leaf for propagation, while other methods only use the leaf. Houseplants and foliage plants are commonly reproduced from leaf cuttings (**Figure 8.8.4**). It is important that once the leaves are cut, they are planted immediately or placed in a plastic bag under refrigeration to prevent them from wilting. Leaf cuttings root best if placed under mist irrigation with a high level of humidity.

Separation and Division

Some parent plants can easily be reproduced by separation or division of vegetative parts of the plant, which often occurs naturally in nature. Some plants create new plants through vegetative pieces that are attached to the parent plant. **Separation** occurs when these vegetative parts are completely removed from the parent plant and planted to grow on their own. The separated plant already has everything it needs to survive, provided it is given adequate water, light, air, and temperature. These parts are usually removed during the plant's dormant season. **Division** occurs when parts of the parent plant are physically cut into sections that will grow into new plants (**Figure 8.8.5**). Division does not hurt the parent plant; often, it helps by giving the parent plant more room to grow, rejuvenating growth.

Figure 8.8.5. Division is a common method of propagation that has been used for many centuries. Because this method of asexual propagation keeps the roots attached, the new plants sometimes grow faster than those reproduced by cuttings.

VectorMine/Shutterstock.com

Hands-On Activity

Taking a Leaf Cutting

Try your hand at asexual propagation by taking a leaf cutting from a snake plant (*sansevieria*). Be sure to follow the directions very carefully.

You will need the following materials:

- Snake plant
- 6-inch pot with drain holes
- Soil
- Water
- Scissors or knife

Complete the following steps:

1. Fill the pot completely with damp soil.
2. Select a healthy stem from your plant. This stalk should not be wilted or have edges that appear to be browning or burnt.
3. Remove the entire stem from the plant, cutting as close to the soil as possible. Make sure to track which end of your stem is up.
4. Starting from the bottom, measure 3–4 inches up and cut straight across the stem, removing the bottom portion of the plant. Again, remember to keep track of which end is up. You can place an up arrow with a marker if necessary before you make your cut to help you remember.
5. Plant the bottom piece that was removed from the plant into your pot. You need to bury at least an inch of the stem and firm the soil around it so it stays upright in the soil.
6. Repeat the last two steps until you have used up your entire stem.
7. Your cuttings should be spread out evenly around the pot. You want plenty of room around each cutting for your plant to grow.
8. Water your cuttings, keeping the soil damp at all times.
9. Avoid the temptation of pulling your new cuttings out of the soil to check for roots! This could damage the tiny root hairs and take longer for the plant to establish a root system. After 2–3 weeks, carefully check your pot for roots by turning the pot over on its side and leaning it forward. This will allow the plant to fall forward as well.

Pixelbender36/Shutterstock.com

10. Being careful not to disturb too much of the root system, check to see if your new plant has a nice supply of healthy white roots. If it does, you can separate the healthy cuttings into their own pots.

Consider This

1. What other plants do you think could be started from leaf cuttings?
2. What environmental factors could affect the success of propagating plants from leaf cuttings?

dani daniar/Shutterstock.com

Figure 8.8.6. Shallots are a type of onion. They form small bulblets, as seen in this image, that can be separated from the parent bulb. Once planted, these bulblets will form a new plant.

There are many plants that can be propagated by division and separation. The most common types of plants that are reproduced this way are those grown from bulbs, corms, rhizomes, and tubers. Bulbs, such as tulips and daffodils, as well as corms, such as crocus and gladiolus, are commonly reproduced through separation. Bulbs produce smaller bulbs, called bulblets, on the side of the parent bulb that can be separated into new plants (**Figure 8.8.6**). Corms are similar in that they produce small cormels that can be separated. Division, on the other hand, is common with plants that produce rhizomes or tubers. The underground structure of a rhizome or tuber is taken out of the soil and cut into several pieces that will each produce into a new plant. Examples of plants that reproduce through division of rhizomes are irises, canna lilies, and calla lilies.

Career Corner   GREENHOUSE MANAGER

If you like this, explore doing what I do! **Greenhouse managers** *are responsible for the daily operations involved with running a greenhouse or nursery.*

What Do They Do? Greenhouse managers oversee inventory of plants, plan planting calendars, supervise greenhouse workers, know the stages and diseases of plant life, harvest plants or their fruits, make recommendations on fertilizer and potting mediums, manage greenhouse equipment and computer controls, explore ways to improve processes for more efficient growth and harvest yields, and maintain breeding records.

What Education Do I Need? A bachelor's degree in horticulture is preferred, but an associate degree is sometimes accepted.

myboys.me/Shutterstock.com

How Much Money Can I Make? Greenhouse managers make an average of $65,000 per year.

How Can I Get Started? You can build an SAE working as a student manager in your school greenhouse, create a production schedule that could be used to prepare for a spring plant sale, conduct research on plant growing requirements and present it to your class, interview a greenhouse manager, or set up a virtual field trip for your class.

Grafting

Grafting is the process of merging two different plants or plant parts together so that they will connect and continue to grow as one new plant. For grafting to occur, you need a strong healthy plant that has a stem section with roots, known as *rootstock*, and a portion including a stem with at least one bud, known as the *scion*, that you would like to merge with the rootstock (**Figure 8.8.7**). These two pieces are cut at specific angles and mechanically merged with the help of tape or rubber bands. Over time, the rootstock and the scion will grow together, forming one plant. Grafting is a complicated process that takes a lot of practice and does not work with every plant. For grafting to be successful, the two plants must be closely related.

Did You Know? The practice of grafting can be traced back 4,000 years to ancient China and Mesopotamia.

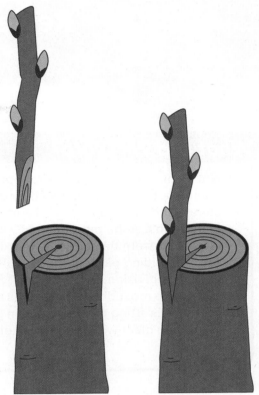

Kazakova Maryia/Shutterstock.com

Figure 8.8.7. The grafter must be sure everything is lined up correctly and the scion and stock are a tight fit to ensure the two pieces join.

Making Connections # Science

Grafting

We just mentioned that plants had to be closely related for a graft to be successful. Do you think oranges and grapefruit are closely related enough to grow on the same tree? What about limes and lemons? Oranges and tangerines? Research grafting more than one fruit tree together and write up a summary of what you discover.

Consider This

1. Do you think non-fruit tree plants could be grafted together? Why or why not?
2. What would be the benefit of grafting multiple plants together?

What Did We Learn?

- Not all plants require seeds to reproduce. There are many plants that can be reproduced by using pieces of their leaves, stems, and even roots, called cuttings.
- Plants can also be reproduced by removing vegetative structures from the parent plant, known as separation, or by simply dividing parts of the parent plant into smaller pieces, known as division.
- One of the most difficult methods of reproducing plants is grafting. Grafting occurs when an individual tries to merge the top of one plant with the bottom of a different (but closely related) plant.

Let's Check and See What We Know

Answer the following questions using the information provided in Lesson 8.8.

1. Producing plants from the vegetative parts of other plants is known as _____.
   A. sexual propagation
   B. asexual propagation
   C. plant propagation
   D. germination

2. *True or False?* Asexual propagation allows growers to produce plants faster than other methods of propagation.

3. The asexual propagation method that uses detached leaves, stems, or roots to create new plants is known as _____.
   A. cuttings   C. grafting
   B. division   D. separation

4. *True or False?* The growing season of the plant has *no* effect on the success of propagating a plant from cuttings.

5. Cuttings made while the stems are still soft and green are known as _____.
   A. grafting
   B. hardwood cuttings
   C. leaf cuttings
   D. softwood cuttings

6. Propagating a new plant using the leaf of a parent plant is known as _____.
   A. grafting
   B. hardwood cuttings
   C. leaf cuttings
   D. softwood cuttings

7. *True or False?* House plants are commonly propagated through leaf cuttings.

8. When the parent plant is physically cut into sections that will grow on their own, this is propagation through _____.
   A. cuttings   C. grafting
   B. division   D. separation

9. Small bulbs that form on the side of parent bulbs that can be separated into new plants are known as _____.
   A. bulblets   C. cuttings
   B. cormels   D. separation

10. *True or False?* The top portion of a graft that is attached to the existing stem section is known as the rootstock.

Let's Apply What We Know

1. Define asexual propagation and explain why it is important.
2. List factors that should be considered before taking cuttings.
3. Discuss the differences between leaf and stem cuttings.
4. Describe the differences in plant division and plant separation.
5. Explain the process of grafting.

Academic Activities

1. **Science.** Plants are not the only living things that multiply through division. The human body is made up of cells that also go through a process called cell division to reproduce. Research to discover if there are any similarities between plant division and cell division.
2. **Language Arts.** Create an informational poster about plants that can be propagated through cuttings. Include the different type of cutting techniques and give examples of plants propagated by each cutting method.

Communicating about Agriculture

1. **Speaking.** With a partner, take turns explaining to each other how you would reproduce the following plants: snake plant, iris, onion, and two different (but related) plants. Take the time to give each other feedback about the information shared. What did you miss in your explanations? What was difficult to explain? Go back and reread any sections of the lesson that will help to clarify. What does this tell you about your understanding?

SAE for ALL Opportunities

1. **Foundational SAE.** Create a propagation calendar for three of your favorite plants that are produced from cuttings. When should you take cuttings? Include things such as time of year to take cuttings, length of cuttings, and other important details for each plant.
2. **Entrepreneurship SAE.** Start a small business with plants that are produced through division or separation such as bulbs or daylilies. Dedicate an area of your yard to your parent plants and sell the new plants you propagate each year. Make sure to keep records on planting dates, money spent and earned, and other important aspects of your new business.

SAE for ALL Check-In

- How much time have you spent on your SAE this week?
- Have you logged your SAE hours?
- What challenges are you having with your SAE?
- How can your instructor help you?
- Do you have the equipment you need?

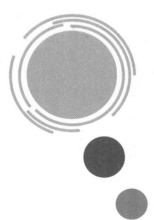

Agricultural Pests

Tomasz Klejdysz/Shutterstock.com

Essential Question

What effects do pests have on plants?

Learning Outcomes

By the end of this lesson, you should be able to:
- Identify major types of pests. (LO 08.09-a)
- Explain why integrated pest management (IPM) is required to manage pests effectively. (LO 08.09-b)
- Discuss the four main pest control options. (LO 08.09-c)

Words to Know

biological pest management

chemical pest management

cultural pest management

fungi

insect

mechanical pest management

pathogen

pest

pesticide

plant disease

virus

weed

Before You Begin

The best way to wrap up this unit is with a few more jokes! What does a caterpillar do on New Year's Day? Turns over a new leaf! What did one firefly say to the other? Got to glow now! I just love that one. Ok, one more: What do you get if you cross a centipede with a parrot? A walkie talkie! Ha!

As a greenhouse manager, I have more than plant jokes. We know a lot about insects, too. Some insects can cause problems for plants. How would a greenhouse manager check for insect damage? Make a list of ideas and review it at the end of the lesson. You can also write down your best plant or insect joke to share later with a friend!

Types of Pests

All plants are subject to pests. A ***pest*** is anything that can cause loss or damage to a plant. Pests can damage plants by causing them to produce fewer flowers or less fruit. Pests can damage the appearance of flowers or fruit, making them less desirable. Pests can affect plant reproduction, or even kill a plant. Pests can be insects, weeds, diseases, and other animals. Let's take a closer look!

Insects

An ***insect*** is a small animal that has six legs, three distinct body parts, and one or two pairs of wings (**Figure 8.9.1**). Insects have an *exoskeleton*, which is a hard, shell-like covering over the outside of their body. Insects usually have a pair of antennae on top of their head and multipart eyes. Mites, spiders, centipedes, and slugs are not technically insects by this formal definition, but some can also be considered pests.

There are more than 800,000 species of insects in the world, and more than 100,000 exist in the United States. Only about 100 of these species cause damage to horticultural plants and turf grass. Insects damage plants by feeding on plant parts. Some insects feed on multiple plant parts, while others feed on specific parts. Some insects feed on young seedlings, and others feed on mature crops.

The insects' feeding behavior determines what damage they will cause to the plant (**Figure 8.9.2**). Their feeding behavior also determines whether they will affect only the appearance of the plant or if they will affect yields or kill the plant.

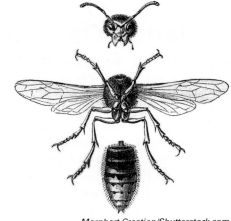

Morphart Creation/Shutterstock.com

Figure 8.9.1: All insects have six legs, three distinct body parts, and one or two pairs of wings.

Did You Know?

Insects have existed for about 350 million years, while humans have only been around for 300,000 years.

Figure 8.9.2. How an insect (or related animal) feeds is determined by the structure of its mouth. *What differences do you notice about the animals shown here?*

8

OlegD/Shutterstock.com

khlungcenter/Shutterstock.com

Peeter Teedla/Shutterstock.com

Wollertz/Shutterstock.com

Inventing Insects

Insects are characterized by having an exoskeleton, a segmented body divided into three parts, one pair of antennae, three pairs of legs, and (usually) wings. If you were designing your own insect, what kind of legs would it have? Would it have one pair of wings or two? For this activity, let the dice help you decide.

Working with a partner or by yourself, follow the steps below to build a unique insect. Roll the dice only once for each body part. The number you roll each turn will determine the type of body parts your insect will have.

You will need the following materials:

- A single die
- Paper
- Pencil

Here are your directions:

1. Roll the die to determine each body part below.

Body Part	Number on die = body type
Legs	1 = Digging legs 2 = Predatory (Grasping) legs 3 = Jumping legs 4 = Swimming legs 5 = Running/Walking legs 6 = Predatory (Clasping) legs
Antennae	1 = Flabellate (fan-like) 2 = Geniculate (hinged or bent like an elbow) 3 = Plumose (brush- or feather-like) 4 = Serrate (saw-toothed) 5 = Stylate (ending in a long, slender point) 6 = Clavate (gradually clubbed at the end)

Body Part	Number on die = body type
Mouthparts	1 = Biting and chewing (grasshopper) 2 = Sipping (nectar) and chewing (pollen) (bees) 3 = Probing and sipping (butterflies) 4 = Sponging and lapping (houseflies) 5 = Piercing and sucking (aphids) 6 = Boring (beetle)
Wings	1 = No wings 2 = Elytra (wings are hardened, beetle) 3 = Lepidoptera (scaled wings, butterfly) 4 = Tegmina (wings like parchment, grasshopper) 5 = Membranous (thin and transparent, dragonfly) 6 = Hemelytra (thick and leathery, cotton bug)
Compound Eyes	Odd Number – Eyes on side of the head Even Number – Eyes on front of head

2. Draw your insect. Make sure you attach each of your selected body parts to the correct part of the insect. Take the time to research any terms or parts that are not familiar.
3. Name your insect!

Consider This

1. Based on your insect's mouthparts, what might your insect eat?
2. What predators might eat your insect?

Feeding behaviors of insects and related pests include:
- Biting and chewing. Insects bite, crush, or chew the plant parts.
- Boring. Insects drill or bore into plant tissue or parts.
- Piercing and sucking. Insects use their mouthparts to extract fluid out of plant materials.
- Rasping. Insects scrape the surface of the leaf to feed on the fluid from the top layer of cells.
- Gall-producing. Insects deposit their eggs into plant tissue, causing the plant to swell.

Weeds

A **weed** is an unwanted plant or a plant growing out of place. Weeds grow and spread in landscaped areas or lawns. Weeds distract from the goal of the landscaped area, which is to be visually appealing in color, texture, and appearance. Weeds compete with landscape plants and turf grass for light, water, nutrients, and growing space. Weeds can also be a host for insects and diseases that will feed on or damage other plants.

Figure 8.9.3 shows some common weeds found in landscapes.

Fotofermer/Shutterstock.com

Nataliia Iliuk/Shutterstock.com

haireena/Shutterstock.com

Madeleine Steinbach/Shutterstock.com

Kathy Clark/Shutterstock.com

Viesturs Ozolins/Shutterstock.com

panotthorn phuhual/Shutterstock.com

simona pavan/Shutterstock.com

Orest Iyzhechka/Shutterstock.com

Figure 8.9.3. These weeds thrive in a variety of climates and conditions. Some of them are even very pretty. *Why would something beautiful be considered a weed?*

Plant Disease Triangle

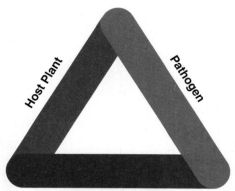

Goodheart-Willcox Publisher

Figure 8.9.4. For an abiotic plant disease to occur, there must be a host plant, a pathogen, and the necessary environmental conditions.

Did You Know?

Agricultural pests eliminate up to half of the food we produce during crop growth, transport, and storage.

Diseases

Plant diseases are irregular conditions that affect plant appearance, growth, and fruit or flower production. Some plant diseases affect the entire plant, while others target specific plant parts, such as the fruit or leaves. There are two main types of diseases: *abiotic* (parasitic) and *biotic* (environmental). Biotic diseases are noninfectious; they do not spread from plant to plant. Biotic diseases are caused by the plant's environment. These may include pollution injuries from a factory, chemical injuries, or damage from machinery.

Abiotic diseases are caused by **pathogens**. Pathogens are microorganisms or viruses that can cause disease. For an abiotic disease, there must be three things: a susceptible host plant, a pathogen, and the necessary environmental conditions (**Figure 8.9.4**).

Fungi, bacteria, and viruses cause most abiotic diseases. Fungi cause more plant diseases than any other pathogen.

Fungal Diseases

Fungi are small, single-celled spore-bearing organisms that form parasitic or symbiotic relationships with plants or animals. Some fungi can survive cold weather in soil or plant debris. Fungal diseases can be spread by water, animals, insects, or even people. Plant diseases caused by fungus include mildew, some blights, black knot, some rots, and rust (**Figure 8.9.5**).

Bacterial Disease

Bacteria are single-celled microscopic organisms with cell walls that reproduce when one cell splits into two. Bacteria can overwinter in the soil and in or on plant material that does not decompose. Bacteria is usually introduced to a plant through an opening or wound in the plant. Bacterial diseases include cankers, crown gall, bacterial wilt, some blights, some rots, and scab (**Figure 8.9.6**).

Julie Vader/Shutterstock.com

Figure 8.9.5. This fungal disease is known as powdery mildew. It will cause this squash plant to lose its leaves, reduce the amount of fruit produced, and decrease fruit quality.

Nataliia Reshetnikova/Shutterstock.com

Figure 8.9.6. Crown gall is a bacterial disease that causes rough, woody, tumor-like galls to form on roots, trunks, and branches of trees and shrubs. Galls can stop the flow of nutrients and water through the plant and reduce overall growth.

Viral Diseases

Viruses are particles that infect other living organisms and use host cells to replicate. Plant viruses are commonly spread through insects that introduce the virus to the plant while feeding. Viruses can also be transmitted through nematodes, fungi, vegetative propagation, and pruning using contaminated equipment. Plant viruses include diseases such as mosaic, psorosis, curly top, and spotted wilt (**Figure 8.9.7**).

Kazakov Maksim/Shutterstock.com

Figure 8.9.7. The unique coloring of these tulips is caused by tulip breaking virus. The virus is transmitted by aphids and causes striped colors in the flowers. Tulips with this stripe pattern were once sold for extraordinarily high prices.

Managing Pest Issues

Many agricultural pest problems can be successfully managed, and some can be prevented. The best method depends on the plant and the pest. All growers should have an *integrated pest management* plan (IPM), which is a pest management strategy that combines best management practices to reduce pest damage with the least disruption to the environment. IPM uses the best control method possible and takes into consideration the economic, social, and environmental conditions. It is also important to consider that there may be specific laws in your state that regulate how pests can be managed.

Research has shown that no single control measure works consistently over long periods of time. Growers must understand which techniques work best in each situation. There are four main methods that are best for controlling pest populations: biological control, chemical control, cultural control, and mechanical control.

Making Connections # Science

Start an Insect Collection

One of the best ways to learn about insects is to collect them, either through live specimens or photographs. You can learn so much about insects by observing, collecting, and handling them. Collectors learn about relationships between insects and their environment, the importance of habitat, keys to species survival, and the relationships between species groups such as hosts, predators, and parasites.

You may decide to collect and release the insects after you catch them, or you may want to make a collection of preserved specimens. Insects live almost everywhere, so it takes only a little time and patience to find them. If you want to collect a variety of insects, try looking in the following places: under rocks and boards; on plant leaves, flowers, or fruit; or near a light source at night. Insects prefer warm weather and damp environments. Check a variety of habitats, at different times of day, and during different times of the year.

ju_see/Shutterstock.com

You will need to research the best and safest way to catch or trap your insects as well as preserve and pin them (should you choose to do so). You may also choose to document your collection through photographs.

Consider This

1. What is the difference between what most people think of as a bug and a true insect?

Biological Pest Management

Figure 8.9.8. Lady beetles eat aphids as part of their natural diet.

Jolanda Aalbers/Shutterstock.com

Biological pest management uses living organisms to reduce pest populations. All living organisms have natural predators in the environment; frogs are a natural enemy to many insects, as an example. Less than one percent of insect species feed on plants in a harmful way. There are many more beneficial insects than harmful ones. Beneficial insects feed on the pest insects and naturally help control the pest population. For example, lady beetles (lady bugs) help control aphid populations (**Figure 8.9.8**). Other beneficial insects include praying mantises, assassin bugs, lacewings, and parasitic wasps.

Biological pest management may also include bacteria and fungi. Certain species of fungi infect insects and some bacteria attack insects, both leading to the death of harmful pests.

Chemical Pest Management

Chemical pest management uses pesticides to control harmful pests. A **pesticide** is any chemical used to control pests. Pesticides can be classified by how they interact with the pest. Pesticides require less time and labor to apply than other methods. They can also be selected to handle specific weeds, diseases, or insects (**Figure 8.9.9**). While pesticides are effective, they can be costly to purchase, and may have other side effects.

Pesticides may require specialized equipment for application. It is important to read the pesticide label and apply the correct amount. Some pesticides require a specialized application license that is granted by the Department of Agriculture in each state. It is also true that many pesticides can have consequences to the health of the surrounding environment, or the people who apply it. Before applying pesticides, you should consider these questions:

- Do you need certification?
- Did you read the pesticide label?
- Did you select the proper protective equipment? (**Figure 8.9.10**)
- Do you know the correct amount to use?
- Did you mix and apply it correctly?
- Did you plan to keep people and animals away from the spray until it is safe to enter the area?

Pesticide	Pest Controlled
Insecticide	Insects
Fungicide	Fungus
Bactericide	Bacteria
Herbicide	Weeds
Miticide	Mites
Molluscicide	Snails and Slugs
Nematicide	Nematodes

Goodheart-Willcox Publisher

Figure 8.9.9. Chemical pest management uses different chemicals to target specific pests. These can be further specialized, like purchasing an herbicide that specifically attacks broadleaf weeds or an insecticide that specifically attacks whiteflies.

David Moreno Hernandez/Shutterstock.com

Figure 8.9.10. Following proper safety procedures is important when working with pesticides. *Why would this be?*

Cultural Pest Management

Cultural pest management uses environmental management techniques to control pests. Cultural pest control includes proper maintenance programs, sanitation of facilities and equipment, and selecting plants that have been bred to be resistant to different insects or diseases. Some additional cultural practices include:

- Providing appropriate amounts of water, light, nutrients, and space for growing plants.
- Planting cover crops.
- Rotating crops each year.
- Watering plants at the proper time.
- Removing dead and diseased plant foliage.

Floki/Shutterstock.com

Figure 8.9.11. Heat and flames are used to burn and kill weeds through thermal weed control.

Mechanical Pest Management

Mechanical pest management uses tools to control pests. Mechanical pest management can be costly, since it requires labor, time, equipment, and fuel to run the equipment. Mechanical methods for control include:

- Tillage/plowing
- Mowing
- Hand pulling/removing
- Weeding tools
- Thermal weed control/flaming (**Figure 8.9.11**)
- Solarization
- Mulching

Career Corner — ENTOMOLOGIST

An **entomologist** *is a type of scientist who focuses specifically on the study of insects. They examine their growth, behavior, nutrition, and how they interact with plants.*

What Do They Do? Entomologists monitor insect feeding behavior and biology, collect samples, apply experimental and commercial insecticides, and design and implement research plans to support the selection of new insecticide products.

What Education Do I Need? A doctoral degree in entomology, biology, or zoology is required to become an entomologist.

How Much Money Can I Make? Entomologists make an average of $75,000 per year.

How Can I Get Started? You can build an SAE around scouting for insects in your school greenhouse. You can create an integrated pest management system for use in the greenhouse, conduct an experiment comparing natural and chemical insect control methods, start an insect collection, or conduct research on common greenhouse pests and present it to your class. You might interview an entomologist or set up a virtual field trip for your class.

Simia Attentive/Shutterstock.com

What Did We Learn?

- A pest is anything that can cause loss or damage to a plant such as an insect, weed, or a disease.
- Insects damage plants by feeding on plant parts. This can affect growth, production, and appearance of the plant.
- A weed is any plant that is unwanted or growing out of place. Weeds compete with the desired plants for light, water, nutrients, and growing space. Weeds can also serve as a host for damaging insects.
- Plant diseases are irregular conditions that affect plants' appearance, growth, and production. Abiotic diseases are typically caused by viruses, fungi, or bacteria.
- Pest issues can be managed by having an integrated pest management plan. There are four main ways to manage pests: biological, chemical, cultural, and mechanical pest management.

Let's Check and See What We Know

Answer the following questions using the information provided in Lesson 8.9.

1. Anything that can cause loss or damage to a plant is known as a(n) _____.
   A. insect
   B. disease
   C. virus
   D. pest

2. Insects that drill into plant tissue or plant parts have _____ mouthparts.
   A. biting
   B. boring
   C. rasping
   D. gall-producing

3. *True or False?* Weeds compete with desired plants for light, water, nutrients, and growing space.

4. Any microorganism or virus that can cause a disease is a(n) _____.
   A. pathogen
   B. plant
   C. insect
   D. weed

5. Viruses are most commonly spread by _____.
   A. insects
   B. water
   C. people
   D. wind

6. *True or False?* Agricultural pest problems cannot be prevented.

7. *True or False?* Each state has their own laws and regulations for managing pests.

8. _____ pest management uses living organisms to reduce the population of pests.
   A. Biological
   B. Chemical
   C. Cultural
   D. Mechanical

9. Which type of pesticide is best for treating insects?
   A. Herbicide
   B. Fungicide
   C. Insecticide
   D. Miticide

10. *True or False?* Wearing the correct protective equipment is important when applying pesticides.

Let's Apply What We Know

1. Define pests and give examples of things that qualify as pests.
2. Describe how different insect mouthparts affect how they feed on a plant.
3. Discuss the benefits of integrated pest management (IPM).
4. Explain the differences in the four types of pest management controls.

Academic Activities

1. **Science.** Research dichotomous keys and find a key that will help you identify plant diseases. Visit your school greenhouse, garden, or outdoor classroom. Use your plant key to help you correctly identify plant diseases you find.
2. **Technology.** Investigate technology that allows growers to digitally map insect infestations in crops. Create a presentation that describes the process and how it is useful to the grower.

Communicating about Agriculture

1. **Visual Communication.** Research common plant diseases in your area. Make a flyer describing the causes, symptoms, and treatments for 3–4 common issues. Distribute the flyer in your classroom.

SAE for ALL Opportunities

1. **Foundational SAE.** Find an area of your school or community that has a weed or insect problem. Research and correctly identify the pest. Determine the best treatment options. Write a summary about your findings, including how this pest is affecting desired plants.
2. **Placement SAE.** Volunteer at a local garden center or greenhouse business. Focus on learning about their integrated pest management plan. Write a report about that plan, highlighting practices that can be implemented in your school's growing areas.
3. **Research SAE.** Conduct an experiment testing the effectiveness of organic herbicides on different weed species. Make a hypothesis, create your experiment, observe, record data, and write a report on your findings.

SAE for ALL Check-In

- How much time have you spent on your SAE this week?
- Have you logged your SAE hours?
- What challenges are you having with your SAE?
- How can your instructor help you?
- Do you have the equipment you need?

UNIT 9
Animal Science Exploration

SAE for ALL Profile
All the Things That Slither

Dustin Johnston founded the Johnston House of Reptiles in Chicago, Illinois 11 years ago. As a child, Dustin has always loved animals. Frogs and other reptiles were some of his favorites. As a small child, much to his parents' displeasure, he often caught new "friends" outside his apartment complex and brought them inside. By the time he was seven, he had nine terrariums in his bedroom filled with different kinds of snakes, frogs, lizards, and even a tarantula.

As a middle school student, Dustin took his first agriculture class. He wasn't very happy about it at first. He, after all, lived in an apartment building in a big city and knew nothing about farming. He tried to talk his mom and dad into letting him get out of the class, but they encouraged him to stick it out. Those nine weeks changed Dustin's life.

His agriculture teacher, Mrs. Linder, told everyone they needed to have an SAE project for class, and they got to choose their own topic. Dustin had been having a difficult time finding mice, crickets, and mealworms to feed to all his pets. It was difficult and costly to have those items shipped and there was not a convenient store that sold them nearby. His parents had even talked with him about finding his pets new homes because they were concerned with the health of the animals.

Dustin started raising his own rodents and other insects to feed reptiles. After doing a lot of research on the most nutritious insects and proper way to raise and harvest rodents, he convinced his parents to let him start his Entrepreneurship SAE project. Within its first year, Dustin grew his business from one customer inside his own apartment complex to five customers. As word of mouth began to spread about his small business, so did the demand. By the end of his high school agriculture career, Dustin had more than 150 customers and three employees.

Dustin says "I've learned a lot along the way. I credit my middle school ag teacher, Mrs. Linder, for encouraging my crazy idea and my high school ag teacher, Mr. Williams, for all the animal science lessons. It's hard to believe that a class project turned into owning my dream business." At Johnston House of Reptiles, Dustin has expanded the business beyond selling feed for various types of reptiles to selling the reptiles themselves. He even takes his reptiles into schools for educational programs. He hopes to show young students that it is ok to dream big and think outside the box.

Top: AJR_photo/Shutterstock.com
Bottom: Yatra4289/Shutterstock.com

Baronb/Shutterstock.com

Fundamentals of Animal Science

Essential Question

How have animals enriched human lives and standards of living?

Learning Outcomes

By the end of this lesson, you should be able to:
- Define animal science and describe the work of animal scientists. (LO 09.01-a)
- Describe the history and importance of livestock production. (LO 09.01-b)
- Summarize the factors that make some animals more suitable for domestication than others. (LO 09.01-c)
- Discuss factors to consider when raising livestock. (LO 09.01-d)

Words to Know

carnivore	herbivore	traits
feed conversion	nomad	
fodder	temperament	

Before You Begin

Hi, everyone! I am Annalise. I work as a veterinary technician, which means I work at an animal hospital assisting our veterinarian with animal surgeries and other tasks around the office. Together, we are going to learn all about animals in this unit and I am so excited to share lots of cool things with you!

I have always loved animals and even as a small child, I knew I wanted to work with animals. My favorite pet growing up was a guinea pig named Snickerdoodle. What has been or would be your favorite pet? Tell a classmate about a pet you have owned or hope to own one day.

What Is Animal Science?

Animal science, or the study of the biology and health of animals, is a very diverse field of study. Earlier in this book, we touched briefly on livestock, or animals raised for food or products. Think about the wide variety of animals that exist in the livestock industry—it's a lot! Then, consider companion animals, or animals bred and raised to serve as pets. Can you imagine being a veterinarian and having to know the anatomy of a cow and a horse when you make a visit to a farm, or diagnosing problems with cats and dogs on the same day (**Figure 9.1.1**)? Medical doctors only deal with a single species (humans), who all have a common anatomy and health concerns. It requires a tremendous amount of knowledge of different animals, different systems, and different conditions to become a veterinarian. It's amazing!

Many years of research and study in animal science have given us a better understanding of animals and what they need to thrive, leading to great gains in nutrition, genetics, and environmental management. In this lesson, we'll be looking at livestock. Companion animals will be covered in more detail in **Lesson 9.11**.

Globally, livestock production is an invaluable part of agriculture. Efficient production on this massive scale is the result of farmers' decades of experience and animal scientists' research. As our knowledge of animals has gotten better, so have our abilities to raise and care for them. From a production standpoint, our food supply has become more efficient, supporting an increase in human population and quality of life.

Syda Productions/Shutterstock.com

Figure 9.1.1. A veterinarian called to this location might have to work on a cow, a horse, a dog, or all three.

A Brief History of Livestock

Any discussion of animal science should acknowledge that our relationships to animals have changed over time. From the earliest days of humanity, people hunted wild animals for food or, as space and conditions allowed, gathered them to raise for food. Animals would graze wild plants, and as the amount of forage dwindled, they would move (or be moved) to an area with more abundant food supply. This meant that the caretakers of livestock lived as ***nomads***, moving from place to place (**Figure 9.1.2**). The population and health of the livestock were subject to the availability of food, which could be affected by environment or by climate. For example, a season of drought could reduce forage plant growth, which in turn would reduce the livestock population. This would have a direct impact on the health and well-being of their caretakers.

As people began to grow crops, they settled into a less nomadic lifestyle. With more plentiful food, they could stay in one place, and they could use some of their crops to help raise animals. The livestock were provided ***fodder***, or food given to domestic animals like cattle, sheep, or horses instead of food they forage for themselves. Providing feed to animals meant less moving around, and having stable herds meant that breeding programs were possible. Remaining in one location brought challenges also, like maintaining animal health and waste management. Animal scientists still work on those issues today!

Did You Know?

Worldwide, more than 68 percent of all protein that humans consume comes from animal sources.

canyalcin/Shutterstock.com

Figure 9.1.2. *Did you know that there are people in the world today who live nomadic lives with their livestock? What are the benefits and challenges of this lifestyle?*

9

Animal Science and Domestication

Why did some animals, like horses, cows, and dogs, develop close relationships with humans, while others, like tigers or bison, didn't? Some species of animals are more easily domesticated, or trained to live in close association with humans. Many animals display common characteristics that allow them to be successfully bred and raised near people.

Temperament

Domesticated livestock live comfortably in groups or herds that have a community structure easily controlled by humans. They also tend to be calmer than non-domesticated or wild animals. This calm nature is easier to manage than an animal that panics at imagined threats. An animal's demeanor or natural behavioral and emotional state is referred to as its ***temperament***.

While domestic animals can display aggression, they are not characterized by it. Animals that are aggressive are dangerous to themselves, other animals, and humans, making them unsuitable for agriculture. Livestock tend to be docile and comfortable with humans. This favorable relationship makes it easier and safer to raise an animal with less risk of human or animal injury.

Hands-On Activity

Animal Science Is a Puzzling Industry

The animal science industry addresses the needs of a lot of different animals. It can be puzzling to figure out where they all fit!

For this activity, you will need the following:

- Small animal-related puzzles (50 pieces or fewer) removed from boxes and placed in brown paper bags
- Paper
- Pencil or pen

Here are your directions:

1. Divide up into groups of 2 or 3 students.
2. Each group should get a bagged puzzle.
3. Work as a team to complete the puzzle without the box or other photos.
4. Write a short summary explaining how this puzzle relates to the animal science industry.

BestPhotoPlus/Shutterstock.com

Consider This

1. Was this activity challenging without a picture? What strategies did your group use to complete the task?
2. What are some careers in the animal science industry that might relate to your puzzle?
3. What industry/sector of agriculture does your puzzle relate to? Research a fact about your industry/sector that can be shared with the class. The fact should show the impact this industry has on people.

Efficient Eating and Growth

Large herds of animals can be expensive to feed. To make good livestock, animals need to gain weight efficiently, eating feed that is readily available and inexpensive. Most livestock, such as cattle and sheep, are **herbivores**, which eat only plant-based material, and their feedstuffs are generally readily available. They can be fed grass on a pasture, hay, or surplus grains from other farming activities. Why do you think it would cost more to raise **carnivores**, or animals that eat meat?

Livestock today have been bred to be efficient at converting their food into mass, allowing them to reach maturity faster. An animal that grows quickly can be sold and harvested more quickly, making it a less expensive investment for producers (**Figure 9.1.3**). If an animal is being raised for consumption, this is an important characteristic.

Livestock Species	Age at Harvest
Dairy Cow	5.5 years
Breeding Bull	3 years
Beef Cattle	18–24 months
Calf (Raised for Meat))	8 months
Sow Pig	< 3 years
Fattening Pigs	6–7 months
Boar Pig	2 years
Laying Hen	20 months
Broiler	40 days
Male Chicks	1 day
Milk Sheep	5 years
Breeding Sheep/Rams	3 years
Lamb (Raised for Meat)	3–4 months
Wool Sheep	2 years

Goodheart-Willcox Publisher

Figure 9.1.3. This table shows how long an average animal will live in animal production. *What are the advantages of short turnaround times to the producer?*

Breeding in Captivity

If you ever visit a zoo, you will see that one of the challenges of raising captive non-domesticated animals is breeding. Some animals will not breed in captivity. The ability to breed animals in captivity is necessary for the efficient production of large numbers of animals, especially as individual animals grow quickly and need to be replaced. Breeding in captivity also allows the producer to build herds with specific and desirable **traits**, or genetically determined characteristics.

The Work of Animal Scientists

All of these traits that are critical to domestication—docile temperament, efficient **feed conversion** (the ability to efficiently convert feed rations into a desired output, like beef or milk), ability to reproduce in captivity—have been studied extensively by animal scientists. Animal scientists help breed herds of cows with the right temperament to live on a dairy farm, using available data to select the right cows to give birth to replacement calves in that operation. Animal scientists have used data and farmers' observations to identify particular breeds of pigs that gain weight more quickly and efficiently than others, or to design feed rations that maximize weight gain for minimal cost. Animal scientists have studied breeding, designing equipment that keeps baby piglets safe from being crushed by their mothers, artificial insemination processes that maximize the likelihood of pregnancy, and production schedules that chart the development of replacement animals in a calendar year exactly as they're needed. Animal scientists do amazing things!

Considerations in Raising Livestock

Have you ever had a pet? If so, then you know that pets need to be properly fed and cared for. They need somewhere to sleep, somewhere to use the bathroom, somewhere to eat. They need attention, grooming, and occasional medical care. Someone has to pick up their messes, too!

Mai.Chayakorn/Shutterstock.com

Figure 9.1.4. *Do chickens need the same amount of space to grow and thrive as swine or cows? Why or why not?*

What's true of your pet is true of livestock, too, on a larger scale. Livestock need appropriate room to live. Space requirements vary depending on the size of the animal, the purpose of the animal, how it is fed, how much waste it produces, and the environment where the operation is located. Beef cattle require more space to grow than chickens. It can take as much as two acres of quality pastureland to feed one cow for 12 months compared to a chicken farm, which may contain as many as 20,000 chickens (**Figure 9.1.4**) in a single 16,000-square-foot facility.

Livestock need to be kept safe from predators or other hazards. Consideration must be paid to the type of livestock operation and the type of housing. How much land will be needed? What type of fencing must be used to keep the animals contained? What preventive measures are necessary to ensure herd growth and health? These might include vaccinations, regular pest control sweeps, or setting traps for predators.

Making Connections #STEM

Calculate the Number of Cattle a Pasture Can Hold

Ranchers commonly have to calculate the number of cattle that can graze on a pasture over time. They use this information to determine how many head of cattle to place on a determined acreage.

To make this calculation we need the following information:

- Length of time on the pasture in days
- Average weight of one animal in pounds
- Total size of pasture in acres
- The average yield of pasture grass in pounds per acre per year
- The daily utilization rate for cattle (This number is a constant 4 percent (0.04) of the animal's weight.)

Use the following formula and numbers to make your calculations:

Total Number of Animals = Total Acreage × Average Yield (in pounds) Per Acre Per Year/Utilization Rate (0.04) × Average Animal Weight × Number of Grazing Days

- You have a 20-acre pasture available.
- Your average yield of Bermuda grass is 10,000 pounds per acre per year.
- The average weight of your cows is 1,000 pounds.
- You plan to use the pasture all year (365 days).

1. Determine the maximum number of cattle your pasture can hold for one year. Remember to round your answer down to the nearest whole number. You can't put a partial cow in the pasture!

Large animal operations produce large amounts of animal waste that must be properly handled and processed. If it isn't, a smelly nuisance can become a real health concern to larger communities! Swine farmers and dairy farmers in particular are required by law to deal with large volumes of waste products. Much of this waste is now placed in tanks with bacteria that break it down into less noxious components, before it is used as fertilizer. In this way, animal scientists have worked to make something that can pose problems to animal and human health into something that benefits agriculture.

Livestock need the right food to eat. How do you know that your animals are staying healthy and getting the right nutrition to grow? If you guessed that the answer was animal science, you'd be right! Animal scientists have studied what nutrients are required to raise healthy animals, and in what amounts. They can design healthy and cost-effective feed rations for any livestock animal.

Some animal scientists, circling back to where we started, specialize in veterinary medicine, making sure that animals are healthy. They can identify animals with nutritional or other medical issues, and solve their problems. They keep animals healthy and happy, and the agricultural production system running smoothly. Hooray for animal scientists!

Career Corner — ANIMAL WELFARE SPECIALIST

An **animal welfare specialist** *typically performs animal handling/welfare inspections and food safety audits on farms, animal production facilities, and animal processing facilities.*

What Do They Do? Animal welfare specialists work for animal processing companies and research labs to ensure company and legal standards are maintained in terms of animal handling/welfare, euthanasia, and food safety; develop and implement corrective actions identified in audits and provide follow-up to ensure completion; properly report and record animal welfare misconduct; and provide animal welfare practice training as needed.

What Education Do I Need? A bachelor's or master's degree in animal science, animal welfare, or a related field is required.

How Much Money Can I Make? Animal welfare specialists make an average of $50,000 per year.

How Can I Get Started? You can build an SAE researching legal standards for animal facilities, welfare, and food safety; build a survey around consumer knowledge of animal welfare standards; set up a tour at a farm or animal production facility and ask them about their animal welfare policies; or interview an animal welfare specialist.

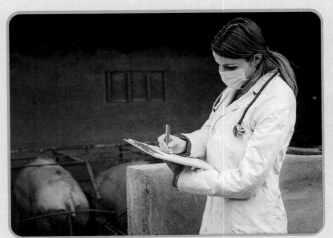

Dusan Petkovic/Shutterstock.com

9

What Did We Learn?

- Animal science is the study of the biology and health of animals.
- Animal scientists work to care for livestock and companion animals, research how to raise healthy animals, improve production practices, and collect data to improve our understanding of animals.
- Throughout history, people have raised animals for food and goods.
- Domesticated animals possess common characteristics, like a calm demeanor, ability to gain weight efficiently, and ability to reproduce in captivity, that allow them to live in close association with humans.
- When raising livestock, it is important to consider how to take care of them. All animals need space to live, food to eat, proper medical care and grooming, and protection from predators. They also produce waste that needs to be managed!

Let's Check and See What We Know

Answer the following questions using the information provided in Lesson 9.1.

1. The study of the biology and health of animals is known as:
   A. Animal science
   B. Animal nutrition
   C. Animal biology
   D. Animal genetics

2. People who move around following the animals that they depend on for food are called:
   A. Farmers
   B. Ranchers
   C. Shepherds
   D. Nomads

3. Feed given to domestic animals is called:
   A. Forage
   B. Grain
   C. Fodder
   D. Grazing

4. *True or False?* Providing feed to animals meant that animals and their caretakers could stay in one place.

5. Animals trained to live in close relationship with humans are considered ____.
   A. wild
   B. domesticated
   C. feral
   D. non-domesticated

6. Livestock animals are herbivores, which means they can be fed:
   A. Grass
   B. Hay
   C. Grains
   D. All of these.

7. *True or False?* Breeding animals in captivity prevents the grower from being able to select desirable traits to improve animal quality.

8. Working with animal genetics and breeding animals for specific and desirable traits is the work of ____.
   A. animal scientists
   B. farmers
   C. ranchers
   D. veterinarians

9. *True or False?* Space requirements for livestock animals vary depending on the size of the animal, the purpose of the animal, how it is fed, how much waste it produces, and the environment where the operation is located.

10. *True or False?* Swine farmers and dairy farmers are *not* required by law to deal with animal waste products created on their farms.

Let's Apply What We Know

1. Imagine you were a nomad living 200 years ago. What processes do you think you would have to take to domesticate a wild animal? What kinds of things would you need to provide for it once it was domesticated?
2. What is temperament? Why is temperament important when raising livestock?
3. Animal scientists research many important things to keep domesticated animals happy, productive, and healthy. Write a short paragraph describing some of the ways animal scientists help animals.

Academic Activities

1. **Engineering.** Find a piece of machinery created by animal scientists and research the changes in the machinery since it originated. Note how this new technology helped the animal industry. What changes do you foresee in the future for this type of equipment?
2. **Language Arts.** Many popular nursery rhymes relate to animals: Hey Diddle Diddle; The Three Little Pigs; and Baa, Baa, Black Sheep. Listen to several nursery rhymes that involve animals. Using these as inspiration, write your own nursery rhyme about a domesticated animal. Include important details about your animal species. Share your creation with the class.

Communicating about Agriculture

1. **Reading and Writing.** Who are the animal scientists in your community? Using your creativity if necessary, come up with a list of 3–5 animal science occupations in your community. Go online and read what you can about one of these occupations. What might their day look like? What animals do they work with? What training do they have?

SAE for ALL Opportunities

1. **Foundational SAE.** Research the history of livestock animals in your state or community. Determine who the local ranchers were, and how they made an impact on your local animal science industry. Create a presentation about your findings to share with your class or other local civic organizations or groups.
2. **Foundational SAE.** Research different species of livestock animals and determine their food, housing, and general care needs. Create a poster or presentation comparing three or four different animals and share your findings with your class.

SAE for ALL Check-In

- How much time have you spent on your SAE this week?
- Have you logged your SAE hours?
- What challenges are you having with your SAE?
- How can your instructor help you?
- Do you have the equipment you need?

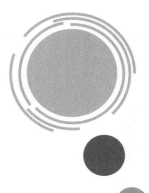

The Importance of the Livestock Industry

Essential Question

How does the livestock industry affect your daily life?

Learning Outcomes

By the end of this lesson, you should be able to:

- Explain the difference between livestock and companion animals. (LO 09.02-a)
- Identify the major animals that make up the livestock industry in the United States and the main products and by-products that they produce. (LO 09.02-b)
- Define the value of the livestock industry to the US economy. (LO 09.02-c)

Words to Know

Bos indicus	finishing
Bos taurus	fowl

Before You Begin

Take a moment to think about your ride to school each morning. Make a list of any animals you might see along the way. Compare your list with a classmate. Add any animals to your list that you may have left off your original draft.

Think about the word domestic *from the previous lesson. Put a check next to any animal on your list that you think would be a domestic animal. Divide your combined list into two groups, one column for large animals and another for small animals. Can you find any similarities between the animals you put in each column?*

A nimals are found all over the planet. We have birds in the air, fish in the sea, and animals that roam the land. Some are large and others are small. Some are wild, and some are domesticated. Domesticated animals fall into two major categories: livestock and companion animals. Livestock are animals used for work, food, or food products, while companion animals provide companionship. Livestock raised for food provide valuable nutrients that contribute to a balanced, healthy diet. Animal products and by-products comprise a valuable industry, providing billions of dollars to our economy.

Livestock Production

Livestock are domesticated animals raised to provide labor; food products like meat, milk, and eggs; or by-products such as wool, fur, and leather. These animals are typically raised on a farm. Beef cattle, dairy cattle, and horses are classified as large livestock. Swine, sheep, goats, poultry, and other niche animal production (rabbits, for instance) are classified as small livestock.

Livestock are raised in large numbers to meet demand for the products and by-products they produce. To successfully raise these animals in quantity, the animal production industry must consider many factors, including nutrition requirements, health concerns, and general welfare for each type of animal. Farmers, ranchers, veterinarians, research scientists and others work together to address these issues. Healthy, well-fed animals are efficient producers, and that is important to consider with our increasing world population and the growing need for animal products.

Beef Cattle

Modern cattle are descendants of two major species. **Bos taurus** are domestic cattle that originally came from Europe, Northeast Asia, and parts of Africa. *Bos taurus* cattle are better adapted to cooler climates and include breeds such as Angus and Hereford. **Bos indicus** are domestic cattle that originate from South Asia. *Bos indicus* cattle are better suited to withstand hotter climates and include breeds such as Brahma and Santa Gertrudis. Cattle are not native to the United States and the first to arrive were brought by Christopher Columbus on his second voyage in 1493. Other settlers brought cattle after that, with the English being the first to bring large numbers in 1611, when they founded the Jamestown colony.

The cattle population of the United States grew as the human population grew and spread west. The greatest growth happened as cattle were moved into the Great Plains area of the United States because of the vast grazing area available (**Figure 9.2.1**). Today, beef cattle provide the largest proportion of total United States farm income from livestock; the estimated total economic impact in 2014 was $165 billion. Beef cattle produce many popular meat products, like hamburger, roasts, and steaks.

Ken Schulze/Shutterstock.com

Figure 9.2.1. This herd of beef cattle is grazing on grassland in Nebraska. *Why did the cattle population explode as European settlers moved into the Midwest and Western United States?*

Dairy Cattle

Dairy cattle produce larger volumes of milk for a longer period of time than other cows, and are now bred and raised for milk production. Dairy cattle are typically of the species *Bos taurus*. Historically, cattle produced meat and milk. Over time, the cattle industry has become more specialized, with some species of cattle raised for beef production (usually those that put on disproportionate body mass) and others raised for milk production. One of the most recognizable dairy breeds is the Holstein, because of their large population and their easily identifiable black and white markings. Dairy production is huge business in the United States, with annual sales of more than $600 billion.

In 1624, ships transported the first dairy cattle from England to the Plymouth Colony in Massachusetts. For many years, dairy cattle were used to provide milk to individual families. As the US population grew and people moved into cities, dairy cattle began to be kept in herds (**Figure 9.2.2**) to produce milk for people who did not have the land or space to raise a cow. Today, more milk is produced with fewer cattle as a result of selective breeding methods.

Horses

Horses are primarily used in the US for work and recreation (**Figure 9.2.3**). Although there is some debate about the early history of horses, most scientists think that horses were first domesticated in Asia between 4,200 and 6,000 years ago. Domesticated horses spread across Asia and Europe for agricultural work, transportation, and warfare. The Bible mentions the Egyptian use of the horse for chariot warfare about 4,000 years ago. The Arabian breed (one of the oldest modern horse breeds) is one of the most popular breeds and originated in the Arabian Peninsula.

Horses are raised worldwide today and make up an important part of the animal industry. Some countries use horses for food in addition to work and recreation, though that practice is illegal in the United States. In the US, horses are used as pets, and for riding, rodeos, ranching, competition showing, and other recreational activities. The sale of horses and horse-related equipment, medicine, and feeds account for more than $50 billion in direct economic impact to the US economy.

tarczas/Shutterstock.com

Figure 9.2.2. The average dairy herd size in the US is now more than 300 cows. *Why do you think dairy cows are kept in herds?*

Diane Garcia/Shutterstock.com

Figure 9.2.3. Barrel racing is a popular recreational activity for horse owners. *What other recreational activities can you name that involve horses?*

Swine

Swine are produced worldwide for human consumption. The East Indian Pig (*Sus vittatus*) and the European Wild Boar (*Sus scrofa*) are the ancestors of American swine. The first record of swine being tamed was written around 5,000 years ago in modern-day China. Recent DNA analysis shows that swine may have been domesticated in the Middle East between 8,000 and 8,500 years ago. As with beef cattle, swine were brought to the Americas by Christopher Columbus on his second voyage in 1493. Spanish and English settlers brought additional swine herds and they grew rapidly. The rapid growth of the early swine population allowed pork and pork products to be a major export from the US from its early days. Now, the industry is a multi-billion-dollar industry, and the US remains one of the largest pork producers in the world.

Did You Know? Pigs say more than just oink! Pigs communicate with one another through distinct grunts and squeals. More than 20 vocalizations have been identified, each one conveying a different message.

Swine are raised all around the United States, with large populations found in the Midwest states because of the availability of corn feed for finishing (**Figure 9.2.4**). *Finishing* livestock involves feeding livestock an energy-dense diet to encourage rapid muscle growth and even distribution of fat in preparation for slaughter. Pork consumption in the US today is about 50 pounds per person annually and accounts for more than 23 percent of the meat consumed.

Sheep

Sheep were among the first animals domesticated for food. Early evidence of sheep raised for human use can be found in the earliest passages of the Bible, in Egyptian script, and in cave drawings. Evidence of the use of sheep, such as bones and wool fabric, has been found in archaeological sites throughout Europe and Asia. The ancestry of sheep is not as well known as other animals. More than 200 breeds of sheep can be found throughout the world.

The only sheep native to North America are wild: Bighorn (also called the Rocky Mountain sheep), Dall, and Stone sheep (**Figure 9.2.5**). These can be found in the wild in the Northwestern areas of the United States and Canada. The domesticated sheep found throughout the US today were brought by European explorers and settlers. Today, sheep are raised throughout the United States, with large populations in the Northeast and West. Sheep are not as popular for food in the US as beef, chicken, or pork, but are also raised for wool.

Mai.Chayakorn/Shutterstock.com

Figure 9.2.4. These baby pigs are nursing in a swine facility. Pigs are quickly raised and slaughtered, so producing healthy litters is important to efficient production.

Natalia Puschina/Shutterstock.com

Figure 9.2.5. Bighorn sheep live in the Rocky Mountains. Most breeds of domesticated sheep are not close relatives of these wild sheep.

Goats

Goats were some of the first animals to be domesticated. Goats share many characteristics with sheep (**Figure 9.2.6**). Goats are believed to have descended from several species of Middle Eastern ibexes, a type of wild goat. The Bible mentions the goat more than 150 times, with specific references to them being used for meat, milk, and hair for clothing.

Goats were imported into Virginia in the 17th century by Captain John Smith. Goats are now raised all over the United States for milk, meat, and hair production. Most goat farms in the United States are small, with a few larger herds being found in the West. Like sheep, goat meat is not as popular in the United States as beef, chicken, or pork.

Did You Know? There is a stereotype about goats being willing to eat anything, but that is not true. Goats are picky eaters. They have very sensitive lips, which they use to "mouth" things in search of tasty food. They may refuse to eat hay that has been walked on or lying around loose for a day.

Poultry

The poultry industry includes chicken, turkeys, ducks, geese, and other *fowl*, or birds. Chickens were being raised in Asia as many as 10,000 years ago. The ancient Egyptians used chickens and chicken products like eggs and feathers. Turkeys, ducks, and geese were all domesticated thousands of years ago, with records of each in Europe, Africa, and parts of Asia. The small amount of time required to grow poultry to maturity make them efficient for individual families to raise, as well as for commercial production.

Chickens are the most popular fowl raised for food and today, in the United States, more chicken is consumed per person than any other meat product. Americans eat more than 95 pounds of chicken per person annually. The poultry industry, compared to other animal industries, has only recently (in the past 50 years) grown into a major commercial business. In the past, poultry raised for human use was mostly by individual families for their own personal use (**Figure 9.2.7**).

Celiafoto/Shutterstock.com

Figure 9.2.6. Sheep and goats have several characteristics that are similar. *What might those be?* Make a list of the similarities and differences that you can see.

only_kim/Shutterstock.com

Figure 9.2.7. Many people raise chickens for fresh eggs. *Do you know anyone who raises chickens in your area?*

Hands-On Activity

Mapping Livestock

Livestock are raised all over the United States. Do you know about the animal production industry in your area? It's time to take a closer look.

For this activity, you will need the following:

- Map of your state, showing cities and counties
- Pencil or pen
- Colored pencils or markers

Here are your directions:

1. Assign each livestock category a different colored pencil or marker. Make a key at the bottom of your state map. For example, cattle may be blue, and horses may be yellow.
2. Using a state agriculture census, research livestock populations in each county of your state.
3. Color each county for the livestock that can be found there. Be mindful that more than one livestock species can be found in one place. You may need to leave room for a county to have more than one color.

Consider This

1. Did you notice any trends in your state? Were the swine farms located in a particular region of the state, for example?
2. How do livestock populations in your state compare to human populations in your state? What does this tell you?
3. Create a pie or bar graph showcasing the livestock in your home county.

Bardocz Peter/Shutterstock.com

Companion Animals

Companion animals include dogs, cats, and many other species of animals raised for the purpose of living with humans for pleasure, companionship, and sometimes work (think of service dogs, military dogs, or comfort animals). Evidence of people having companion animals can be found in some of the earliest written documents. The first domesticated animal was probably used for companionship rather than for livestock purposes.

Cast of Thousands/Shutterstock.com

Figure 9.2.8. Pets bring joy to the lives of many people. *Do you own a pet? Why or why not?*

Today, these animals are usually referred to as pets. Pets bring value to their owners by providing joy and relationships (**Figure 9.2.8**). Pet owners spend large amounts of money each year on food and care for their animals. It costs an average of more than $1,000 per year to own a dog, the most popular pet in the US. Some veterinarian offices in the United States only take care of pets because of the demand for pet care.

Popular companion animals include dogs, cats, rabbits, hamsters, reptiles (like turtles and snakes), small birds, and fish. The production of and care of pets has a tremendous impact on the US economy; the companion animal industry provides more than $220 billion in economic impact to the US each year! We'll talk more about companion animals and how they fit into the world of agriculture in Lesson 9.11.

Career Corner — FOOD/LARGE ANIMAL VETERINARIAN

A **food, or large animal veterinarian,** *prevents, diagnoses, treats and researches illnesses in food production animals such as cattle, swine, poultry, and fish.*

What Do They Do? Food veterinarians provide health care for livestock and laboratory animals, examine animals to diagnose illness or injury, provide necessary medical treatment, monitor animal performance and provide feedback, provide technical expertise in health management, research new diseases, medicines, and procedures, and test for, vaccinate, prevent, and control contagious animal diseases.

Hedgehog94/shutterstock.com

What Education Do I Need? A doctorate of veterinary medicine is required.

How Much Money Can I Make? Food veterinarians make an average of $95,000 per year.

How Can I Get Started? You can build an SAE conducting research on the need for large animal veterinarians in the US. You can set up a tour to a food veterinarian facility, interview a food veterinarian, or job shadow them for the day.

What Did We Learn?

- Livestock are domesticated animals raised to provide labor, food products, or by-products.
- Domesticated animals can be classified as livestock or companion animals.
- Beef cattle, dairy cattle, horses, swine, sheep, goats, and poultry are some of the most common types of animals raised as livestock.
- Each type of animal produces different products and by-products, edible and otherwise.
- Companion animals are raised for relationship and companionship.
- Livestock production and companion animal production contribute billions of dollars to the US economy each year.

Let's Check and See What We Know

Answer the following questions using the information provided in Lesson 9.2.

1. _____ are domesticated animals raised to provide labor; food products like meat, milk, and eggs; or by-products such as wool, fur, and leather.
   A. Livestock
   B. Companion animals
   C. Exotic pets
   D. Pests

2. *True or False?* Modern beef cattle are descended from five major species that originated in Europe.

3. *True or False?* The Holstein is the most recognized dairy breed in the world.

4. _____ livestock involves feeding an energy-dense diet to encourage rapid muscle growth and even distribution of fat in preparation for slaughter.
   A. Completing   C. Finishing
   B. Racing   D. Plucking

5. Which of the following animals are *not* used for meat production in the United States?
   A. Beef cattle   C. Horses
   B. Pigs   D. Chickens

6. What animal was the first to be domesticated for food?
   A. Cattle   C. Chickens
   B. Swine   D. Sheep

7. *True or False?* Wool is a fiber produced from cattle.

8. Which of these animals would be considered fowl?
   A. Cat   C. Horse
   B. Pig   D. Turkey

9. *True or False?* Americans consume more beef per person than any other meat.

10. Which of these would be considered a companion animal?
    A. Cat   C. Hamster
    B. Dog   D. All of these.

Let's Apply What We Know

1. Compare and contrast livestock animals to companion animals. Create a list of their similarities and differences.
2. How has the livestock industry changed since it first began in the United States? Select one species of livestock covered in this lesson and research the development of that industry in the United States. Find at least five major events that affected the course of the industry.
3. Work with a partner and select one species of livestock covered in this lesson. Research different breeds and make a chart explaining the benefits and challenges of the different breeds.

Academic Activities

1. **Math.** Milk collected from a dairy cow is often reported in pounds, but is sold in gallons and consumed in glasses (8 oz). Research how many pounds of milk a dairy cow produces in a single day; then, covert that amount to gallons and glasses.
2. **Language Arts.** There are many ethical issues surrounding the livestock industry. Research an issue related to raising animals in a production setting. Write a persuasive argument for your position. Present it to the class, and look for counterarguments.

Communicating about Agriculture

1. **Visual Communication.** Make a poster or flyer explaining how some aspect of domesticated animal production affects the US economy. Do some additional research to add whatever statistics or information you are curious about. Use visuals (images, infographics, graphs, photos, drawings) to convey information. Share your product with the class.

SAE for ALL Opportunities

1. **Entrepreneurship SAE.** Research what is needed to show your own livestock animal. Talk with your FFA advisor and purchase your own livestock animal to exhibit in youth livestock shows.
2. **Entrepreneurship SAE.** Research what is needed to raise chickens for eggs. Start a backyard chicken flock and sell your eggs to local community members.

SAE for ALL Check-In

- How much time have you spent on your SAE this week?
- Have you logged your SAE hours?
- What challenges are you having with your SAE?
- How can your instructor help you?
- Do you have the equipment you need?

Major Breeds of Livestock

solomonphotos/Shutterstock.com

Learning Outcomes

By the end of this lesson, you should be able to:
- Explain why it is important for different breeds to exist, and what factors might lead farmers to choose to raise one breed over another. (LO 09.03-a)
- Identify common breeds of livestock. (LO 09.03-b)

Words to Know

breed	maternal	selection
drake	mohair	switch
hand	mutton	terminal
marbled	polled	withers

Before You Begin

Let's play a quick game of 20 questions. Find a partner! One person will be asking the questions and one will be answering. If you're the partner answering questions, look around the room and decide on one object that you want the person to guess. Now, your partner gets to ask up to 20 yes-or-no questions to figure out what object you picked. If they guess the correct object before the 20 questions are up, they win! If they don't get the object within 20 questions, you win! Ready, set, go!

How easy was that game? Were you able to identify things based on their similarities and differences? You may not even think about it while you are doing it, but we use similarities and differences every day to identify things. We also look at an item's characteristics to classify or group items. Think about this with animals. If you don't know what a certain animal is, you might use its characteristics to compare how similar or different it is to other animals to help you figure out what you are observing.

9

Animal Breeds

Animals are grouped in similar ways. Each animal has characteristics like size, color, or shape that allow them to be sorted into specific groups. These groups are called breeds. An animal **breed** has a distinctive appearance that is typically developed by intentional **selection**, or breeding with the hope of maintaining desired characteristics. There are obvious differences between a cow and a horse, but it is more difficult to identify differences between two cows. Animal breeds each have specific characteristics that differ from other breeds in the same species.

The development of breeds started when farmers chose specific animals with desirable characteristics to reproduce. For example, if a farmer lived in an area with limited water, she would choose to breed animals that could survive better in dry conditions. Over time, this method of selective breeding resulted in animals that did well in dry conditions. Many other characteristics like color, size, and temperament were used to develop animal breeds that were desirable for specific conditions. We see examples of this in all domesticated animal species.

Why do farmers choose to raise one breed over another? Different breeds have different characteristics. In the example above, the farmer chose to raise animals that did well in dry conditions. Other farmers might look for animals that produce large litters, handle heat or cold well, resist parasites or disease, live long lives, produce lots of milk, put on weight quickly, or are docile and easily handled. The best animal for a producer depends on their situation. The ability of different breeds to thrive in different circumstances, and for different purposes, makes it possible for all kinds of people to raise livestock, to meet all kinds of needs. As human populations continue to grow, this diversity helps increase production, and protects the health and long-term viability of the industry.

Elkin Restrepo/Shutterstock.com

Figure 9.3.1. Black Angus cow and calf

Wollertz/Shutterstock.com

Figure 9.3.2. Brahman bull

Beef Cattle Breeds

Modern beef cattle breeds have been developed from two main species of cattle: *Bos taurus* and *Bos indicus*. There are more than 250 beef cattle breeds in the world, and over half of them are found in the United States. Beef producers may select a specific breed to raise or they may choose to crossbreed cattle to make a new combination.

Angus

Angus cattle (**Figure 9.3.1**) originated in Scotland and were imported to the US in 1873. Angus cattle are the most popular breed of cattle in the US, with an estimated population that is greater than all other cattle breeds combined. Angus cattle are black and **polled**, or without horns. They produce a high-quality carcass that is well **marbled**, with a desirable mixture of fat and lean meat.

Brahman

Brahman cattle (**Figure 9.3.2**) were developed in the United States from Indian Zebu cattle (*Bos indicus*). Several types of Zebu cattle were crossbred with British cattle breeds with the intention of developing a breed suited to the climate of the southwestern United States.

The resulting cattle are mostly light-to-medium gray with a noticeable hump, loose skin, large drooping ears, and a remarkably high heat tolerance. Brahmin cattle are also noticeably resistant to insects and disease, with the ability to survive on poor-quality rangeland.

Coatesy/Shutterstock.com

Figure 9.3.3. Charolais bull

Charolais

Charolais (pronounced **sheh**-rah-lay) (**Figure 9.3.3**) is one of the oldest breeds of cattle. As their name might indicate, they come from France. Charolais are a large, heavily muscled breed that are typically white with pink skin. Most Charolais are naturally horned. They are known for their high feed efficiency and for adapting well to many areas. Charolais cattle are commonly used in crossbreeding programs.

Hereford

Hereford cattle (**Figure 9.3.4**) originated in England, with early breeders selecting for a high beef yield and economical production. They are known for their red bodies with white on their faces, bellies, legs, and *switch* (tip) of the tail. Herefords are docile and easily handled. They are well adapted to the climate of the western United States, with superior ability to produce calves under difficult conditions.

Simmental

Simmental cattle (**Figure 9.3.5**) originated in Switzerland and have been around for centuries. In the early 1800s, Simmental were introduced as a dual-purpose breed, producing milk and meat. Simmental were brought to the United States from Canada in 1969 and remain a popular breed. Original Simmental colors included a white-to-tan faces with red-to-dark red-spotted bodies. Today, many Simmental cattle are black or black with white faces. Simmental cattle produce a thick, muscled carcass without excess fat which, along with their adaptability to a variety of range types, makes them a popular breed for cattle producers.

Other popular breeds of beef cattle in the US include Gelbvieh, Limousin, and Red Angus.

dcwcreations/Shutterstock.com

Figure 9.3.4. Hereford cattle in a pasture

Alf Ribeiro/Shutterstock.com

Figure 9.3.5. Simmental bull

Dairy Cattle Breeds

Dairy cattle are selected and raised for their ability to produce milk in large amounts. The average dairy cow produces around seven gallons of milk per day for a total of more than 2,500 gallons per year. Dairy cows are milked at regular intervals, two to three times per day.

inavanhateren/Shutterstock.com

Figure 9.3.6 Brown Swiss cow

Brown Swiss

Brown Swiss (**Figure 9.3.6**) dairy cattle originated in Switzerland and are believed to be one of the oldest dairy breeds. They are solid brown, varying from light to dark, with a black nose and tongue. They are large cattle, averaging around 1,400 pounds when mature. Their milk is known for its high protein content, making it an excellent choice for cheese production.

Holstein

Holstein cattle (**Figure 9.3.7**) are the most popular and well-recognized of the dairy breeds. They are also large, with mature cows averaging 1,400 pounds or more. They are most frequently black and white, though some are red and white. Holsteins rank first in average milk production per cow.

Did You Know?

Holstein cattle rose in popularity when a well known fast-food chain used this breed of cows and the message 'Eat mor Chikin' on a billboard in 1995. It's been so successful that Chick-fil-A has stuck with the campaign ever since.

Jersey

Jersey cattle (**Figure 9.3.8**) are typically a cream-to-light brown color. They have a black muzzle and either a black or a white switch. They are known as the smallest of the dairy breeds, with mature cows averaging 1,000 pounds or less. Because of their lighter weight, they produce less milk per cow than other breeds, but they rank first among all dairy breeds in the protein and butterfat content of their milk. This makes their milk an excellent choice for products like ice cream.

Other popular dairy breeds include the Guernsey, Ayrshire, and Milking Shorthorn.

HannieVerhoeven/Shutterstock.com

Figure 9.3.7. Holstein cows

Cameron Watson/Shutterstock.com

Figure 9.3.8. Jersey cow

Horse Breeds

Horses in the United States are primarily owned and used for recreational purposes or work. They vary in size and type from small ponies to large draft horses. Horse height measurement is different from other animals, as the unit used is a **hand**. A hand is a unit of measurement equal to four inches. The horse family also includes donkeys and zebras.

Thoroughbred

Thoroughbreds (**Figure 9.3.9**) originated in England and were bred to carry weight at a sustained speed over long distances. These horses average between 15 and 17 hands tall and are a solid dark color. They have a well-defined head with a long neck, deep chest, lean body, and long legs. They are widely used in racing, showing, and casual riding.

Did You Know? Thoroughbred horses are the most common breed in horse racing. Most Thoroughbreds begin their racing career at two years old.

PJ photography/Shutterstock.com

Figure 9.3.9. Thoroughbred horses

American Quarter Horse

There are more registered American Quarter Horses (**Figure 9.3.10**) in the United States than any other breed, making it the most popular breed in the nation. Its name comes from its ability to quickly sprint a quarter mile. Quarter Horses range in height from 14 to 16 hands tall. These horses are known for having an attractive head and neck with a long shoulder, strong back, and uniform muscularity throughout their body. They are versatile and are commonly used in ranch work, racing, showing, and casual riding.

Jaco Wiid/Shutterstock.com

Figure 9.3.10. American Quarter Horse

Arabian

Arabian horses (**Figure 9.3.11**) are descendants of ancient Middle Eastern horses. Arabians range in height from 14.1 to 15.1 hands tall. They have large, wide-set eyes; a broad forehead; small, curved ears; and large nostrils. They also have a long, arched neck with a short back, and a high tail set. Arabians are commonly used for showing, endurance racing, and casual riding.

Other popular horse breeds include Appaloosas, Morgans, Warmbloods, Painters, Tennessee Walkers, and Andalusians. Draft horse breeds include Shires, Clydesdales, and Belgian Drafts.

Olga_i/Shutterstock.com

Figure 9.3.11. Arabian horse

Hands-On Activity

Measuring the Height of a Horse

Measuring the height of a horse can be confusing because the standard unit for measuring a horse's height is not feet or inches, but hands. A hand is equal to four inches. A horse's height is taken by measuring from the ground up to the highest point on the **withers** (the ridge between the horse's shoulders). If a horse measures 60 inches, the horse would be 15 hands high (60 ÷ 4 = 15). What if the measurement is not evenly divided by four? This is where horse measurement is even stranger. In hands, the decimal point represents a ¼ fraction of a hand. For example, if the horse measures 61 inches at the withers, the height would be recorded as 15.1 hands, 62 inches would be 15.2 hands, and 63 inches would be 15.3 hands. Once we reach 64 inches, the horse would measure 16 hands high.

For this activity, you will need the following:

- Tape measure
- A horse (if a horse isn't readily available, you can look up the average height of a specific breed and mark it along a wall to see what it looks like in real life)
- A partner

Optional items:

- Measuring tape (specific to horses in hand units)
- String or twine (to mark and measure later with a ruler)

Here are your directions:

1. Make sure the horse is standing on flat, level ground.
2. Make sure the horse's front feet are as close to even as possible.
3. Measure from the ground right behind the front foot up to the highest point of the withers in a vertical line.
4. Record your measurement and convert to hands using the method described above.
5. If you do not have a horse, look up a breed or two and measure them along a wall.
6. Measure your partner using this procedure and convert the number to hands.

Consider This

1. Did the height of horses surprise you?
2. Are there horse breeds that are similar in size to you and your partner?
3. Why do you think that horses are measured in hands?

Swine Breeds

There are more than 70 swine breeds worldwide. Swine breeds are commonly divided into two major types: terminal and maternal. Many colored breeds are **terminal** breeds and are known for their muscling and carcass quality. The solid white breeds are commonly called **maternal** breeds and are known for their mothering ability, litter size, and longevity.

acceptphoto/Shutterstock.com

Figure 9.3.12. Duroc hogs

Duroc

Duroc hogs (**Figure 9.3.12**) originated in the United States. Red Durocs from New York and Jersey Reds from New Jersey were crossbred to produce the modern Duroc. They are usually dark red but can range from almost yellow to a mahogany color. Their ears droop forward and they are known for fast growth and quality muscling.

Yorkshire

Yorkshire hogs (**Figure 9.3.13**) originated in York County in England and are also known as English Large Whites. They are the most recorded breed in the United States and were first imported to Ohio in 1830. They are solid white in color with large, erect ears. Yorkshires are known as the "mother breed" with a reputation for excellent mothering ability, litter size, and length of body.

Hampshire

Hampshire pigs (**Figure 9.3.14**) originated in Southern Scotland and Northern England. They are black with a white belt encircling their body, including both front legs and feet. Hampshires are known for their muscling, quality carcass, rapid growth, and durability.

Other popular breeds of swine include Berkshire, Chester White, Landrace, and Spotted Poland.

Sheep Breeds

Sheep are produced worldwide and were brought to the Americas by Christopher Columbus. Today, they are produced in every state in the United States for meat and wool. There are more than 200 breeds of sheep in existence, with multiple ways to categorize them. The most common method of grouping sheep is by the type of wool or hair that they grow. Hair sheep are becoming more common in the southern US, where they are better able to withstand the warm, humid climate than wool-bearing sheep.

Dorset

Dorset sheep (**Figure 9.3.15**) originated in England and were imported to the US in the 1800s. They are a medium-sized sheep, solid white in color, and may be horned or polled. They are heavily muscled and are considered a medium-wool sheep. They are a dual-purpose sheep, producing both *mutton* (sheep meat) and wool.

Hampshire

Hampshire sheep (**Figure 9.3.16**) also originated in England and were imported into the US in the 1840s. They are a large breed known to produce fast-growing market lambs. They are solid white with a black head and legs. They have a wool cap on top of their head and wool going down their legs. They are often crossed with other breeds to increase size and promote rapid growth.

Figure 9.3.13. Yorkshire hogs

Figure 9.3.14. Hampshire pig

Figure 9.3.15. Dorset ewes

Figure 9.3.16. Hampshire sheep

9

Jesse Seniunas/Shutterstock.com

Figure 9.3.17. Suffolk sheep

critterbiz/Shutterstock.com

Figure 9.3.18. Katahdin sheep

Suffolk

Suffolk sheep (**Figure 9.3.17**) were developed in Southern England from Southdown rams and Norfolk Horned Ewes. They were imported into the US in 1888. Suffolk sheep are white with a black head and legs and no wool extensions. They are a large breed known for growth and muscling. They are also used in crossbreeding programs to produce market lambs that grow rapidly.

Katahdin

Katahdin sheep (**Figure 9.3.18**) are a hair breed. They were bred in the United States (Maine) and are primarily used for meat production. They are easy to care for and breed and come without the maintenance that comes with shearing sheep.

Other popular sheep breeds include Rambouillet, Merino, Dorper, and Southdown.

Goat Breeds

Goats, like sheep, are recognized as one of the earliest domesticated animals. Goats are produced for their hair and hides, meat, and milk, making them a multi-use animal with value throughout their lifespan. They are classified based on their primary breed product.

Boer

Boer goats (**Figure 9.3.19**) are a meat goat breed that originated in South Africa in the early 1900s. They were introduced in the US in 1993 and have quickly gained popularity. Boer goats are white with a red head and white blaze in the center of their face. They cross well with other breeds and are efficient growers.

Nubian

Nubian goats (**Figure 9.3.20**) are a dairy goat breed that originated in Africa. It is the most common dairy goat breed in the United States. They can be a variety of colors and are medium to large in size with short hair. They have the highest milk fat content of any dairy goat breed. Nubian goats have distinctive long ears.

THAICOWBOY MAGAZINE/Shutterstock.com

Figure 9.3.19. Boer goats

Liliya Kulianionak/Shutterstock.com

Figure 9.3.20. Nubian goat

Angora

Angora goats (**Figure 9.3.21**) are known for their production of long silky hair called *mohair*. Mohair is the primary product of this breed. Angora goats are an ancient breed that originated in Turkey and were imported into the United States in the mid-1800s. They do well in mild, dry climates, with most of the Angora goat population in the US living in Texas.

Other popular goat breeds include Pygmy, Nigerian, Saanan, Alpine, Kiko, and La Mancha.

EcoPrint/Shutterstock.com

Figure 9.3.21. Angora goat

Poultry Breeds

Poultry production includes chickens, ducks, geese, turkeys, and other species of birds. Chickens make up most poultry production in the United States. Poultry are grown mainly for their meat and eggs, with some secondary products like feathers being collected also. They are fast-growing compared to other livestock and are widely raised throughout the world.

Chickens

Most poultry raised for food are chickens. Modern chickens are descended from the red jungle fowl of Southeast Asia and were selected for fast growth rates and high meat production.

Cornish chickens (**Figure 9.3.22**) were developed in England. They are known for their robust meat production, with adult chickens weighing up to 10 pounds. They are very popular as part of meat production (broiler) operations. They may be white, white with red, or dark in color with some variations. Cornish chickens are unique in that the male and female, unlike most chickens, look very much alike.

Rhode Island Red chickens (**Figure 9.3.23**) were developed in Rhode Island and Massachusetts in the late 1800s. They were developed to be disease resistant, good foragers, good egg layers, and weigh 7 to 8 pounds at maturity. They lay extra-large brown eggs and have plumage that varies in color from rust brown to darker maroon and sometimes nearly black.

White Leghorn chickens (**Figure 9.3.24**) were imported into the United States in 1852. They are known for being excellent egg layers and active foragers with a small appetite, making them very efficient for egg production.

Julia Kaysa/Shutterstock.com

Figure 9.3.22. Cornish chicken

Ariene Studio/Shutterstock.com

Figure 9.3.23. Rhode Island Red chicken

slowmotiongli/Shutterstock.com

Figure 9.3.24. White Leghorn rooster

Andi111/Shutterstock.com

Figure 9.3.25. Pekin duck

Tom Curtis/Shutterstock.com

Figure 9.3.26. Aylesbury duck

Florian Teodor/Shutterstock.com

Figure 9.3.27. Khaki Campbell duck

Ducks

Ducks are raised for their meat, eggs, and feathers. They are used as pets and for showing. Ducks are native to the United States and are found in the wild, with most domesticated ducks being descendants of the mallard.

Pekin ducks (**Figure 9.3.25**) are the primary breed used for meat production in the United States. They were imported from China to New York in 1873. They are a large breed that grow and mature quickly to between 8 and 11 pounds. Pekin ducks are also excellent layers and can produce an average of 200 eggs per year. They are primarily white with orange bills and legs.

Aylesbury ducks (**Figure 9.3.26**) originated in England. They are white in color and similar in size to the Pekin. They can be distinguished from the Pekin by their lighter, tanner bills and light orange legs and feet.

Khaki Campbell ducks (**Figure 9.3.27**) are known for their exceptional egg laying ability. They average an egg a day! The *drake*, or male duck, is khaki-colored with bronze lower back, tail, head, and neck feathers. Adults average about 4 ½ pounds at maturity.

Did You Know? Most ducks have a lifespan of 8 to 12 years.

Geese

Goose production in the United States is less than that of other poultry, with most goose products being exported to other countries. Geese have a longer life span than most other poultry and are generally disease free with a rapid growth rate.

Embden geese (**Figure 9.3.28**) were imported into the US in 1820. They are fast growing, large birds that mature at weights up to 20 pounds. They are good layers and produce large eggs. Embden geese are white with a yellow-orange bill, legs, and feet.

Toulouse geese (**Figure 9.3.29**) are used for meat and egg production. They are large, with mature birds weighing 18 to 20 pounds and producing up to 40 eggs annually. They are commonly grey but can be buff-colored. They have orange bills, legs, and feet.

Edward Westmacott/Shutterstock.com

Figure 9.3.28. Embden geese

slowmotiongli /Shutterstock.com

Figure 9.3.29. Toulouse goose

Create a Breed Social Media Profile

If a cow or a pig had a social media profile, what do you think that would look like? Each member of the class should be assigned a different breed to create a social media profile.

Category	Description
Profile Picture	Include a picture of your animal
Overview	Include a general description of your animal
Places Lived	Research what areas of the world your animal currently lives
Hometown	What is your animal's native country?
Five Interesting Facts	Include five unique facts about your animal

Buff geese (**Figure 9.3.30**) were developed in North America and are descended from the Greylag goose found in Europe and Northern Asia. The American Buff is an apricot-fawn color with buff-colored feathers on its back and sides that are edged with creamy white. Their feet and legs are orange. The bill is orange with a pale pink tip. They have brown eyes and mature at a weight of 16 to 18 pounds.

Turkeys

All domestic turkeys are descended from wild turkeys from North and South America. Spanish explorers brought turkeys back to Europe. Turkeys are primarily grown for meat production in the US. There are not as many breeds of turkeys as there are of other poultry.

Standard Bronze turkeys (**Figure 9.3.31**) were developed by crossing American wild turkeys with Black European turkeys. This cross produced a turkey that was larger and more vigorous than the European birds and tamer than the wild turkeys. Mature adults often weigh as much as 25 pounds. Their plumage has a coppery-bronze colored metallic sheen which gives the variety its name.

Broad Breasted White turkeys (**Figure 9.3.32**) were developed in the United States and are the most widely used domesticated turkey for meat production. They are large and have been bred to produce a carcass with more breast meat than any other turkey breed. They are completely white with pink legs, black beards, and a red head.

Diane Kuhl/Shutterstock.com
Figure 9.3.30. Buff goose

Ed Phillips/Shutterstock.com
Figure 9.3.31. Standard Bronze turkeys

Jennifer Yakey-Ault/Shutterstock.com
Figure 9.3.32. Broad Breasted White turkey

Career Corner   POULTRY HATCHERY MANAGER

Hatchery managers *oversee all aspects of a poultry hatchery, which includes the management of personnel, eggs, and equipment.*

What Do They Do? Hatchery managers coordinate hatchery activities; monitor facility equipment and repairs; oversee sanitation; keep up with industry trends; manage animal welfare programs; manage chick care, maintenance, packing, and transfer activities; and manage and maintain weekly hatch schedules.

What Education Do I Need? To be a hatchery manager, a bachelor's degree in animal or poultry science is required.

How Much Money Can I Make? Hatchery managers make an average of $75,000 per year.

How Can I Get Started? You can build an SAE conducting research on hatchery schedules and present your findings to the class, raise your own backyard chickens, set up a tour to a poultry hatchery facility, interview a hatchery manager, or job shadow them for the day.

branislavpudar/Shutterstock.com

What Did We Learn?

- Animals have characteristics, including size, color, and shape, that allow them to be identified in specific groups. These groups are based on specific character traits and are called breeds.
- While some differences in breeds occur naturally, other differences have been deliberately cultivated by farmers to meet specific needs.
- Crossbreeding animals of different breeds is one way to create new breeds or try and create the desired mix of traits.
- There are more than 250 breeds of beef cattle in the world. The most popular breeds of beef cattle in the US include Angus, Brahman, Charolais, Hereford, and Simmental.
- Dairy cattle are selected and raised for their ability to produce milk in large amounts. The most popular breeds of dairy cattle in the US include Brown Swiss, Holstein, and Jersey.
- Horses in the United States are primarily owned and used for recreational purposes or work. They vary in size and type from the small pony to the large draft horse. The most popular breeds of horses in the US include the Thoroughbred, American Quarter Horse, and Arabian.
- Swine breeds are split into terminal breeds, raised for meat and carcass quality, and maternal breeds, raised for their litter size and mothering ability. Popular swine breeds for production in the US include Duroc, Yorkshire, and Hampshire.
- Sheep are one of the oldest domesticated animals in the US. They are grown for meat and wool, and are often classified by the type of wool they grow. Popular breeds include Dorset, Hampshire, and Suffolk.
- Goats are another of the longest-domesticated animals in the US. They are grown for hair, meat, and milk, and breeds are classified according to their primary product. Popular breeds in the US include Boer, Nubian, and Angora.
- Poultry production includes chickens, ducks, geese, and turkeys.
- Chickens make up most poultry production in the United States. Chickens are raised mainly for their meat and eggs. Popular breeds of chickens include Cornish, Rhode Island Red, and White Leghorn.
- Ducks are raised to produce meat and eggs. Popular breeds of ducks include Pekin, Aylesbury, and Khaki Campbell.
- Geese are raised for meat and eggs, though these products are not as popular in the US as other poultry. The most common breeds of geese include Embden, Toulouse, and Buff.
- Turkeys are native to the US and are bred for meat. There are fewer breeds of turkeys than other poultry. The most popular breeds for meat production include Standard Bronze and Broad Breasted White.

Let's Check and See What We Know

Answer the following questions using the information provided in Lesson 9.3.

1. Which of the following cattle breeds was developed from the Zebu cattle of India?
   - A. Angus
   - B. Simmental
   - C. Hereford
   - D. Brahman

2. Identify the cattle breed below that is not considered to be a dairy breed.
   - A. Simmental
   - B. Holstein
   - C. Jersey
   - D. Brown Swiss

3. Most Thoroughbred horses mature at a height of at least 15.2 hands. How tall is that measurement in inches?
   - A. 152 inches
   - B. 62 inches
   - C. 60.2 inches
   - D. 15.2 inches

4. Which of the following horse breeds was developed in the Middle East?
   - A. Quarter Horse
   - B. Arabian
   - C. Thoroughbred
   - D. Appaloosa

5. Pigs can also be referred to as:
   - A. Bovine
   - B. Equine
   - C. Swine
   - D. Avian

6. Which of the following is *least* likely a product or benefit of sheep?
   - A. Wool
   - B. Mutton
   - C. Meat
   - D. Work

7. Identify the breed of goat from the following list that was most recently brought to the United States.
   - A. Nubian
   - B. Angora
   - C. Boer
   - D. Rambouillet

8. Which of the following animals would not be classified as poultry?
   - A. Chickens
   - B. Sheep
   - C. Ducks
   - D. Turkeys

9. Identify the animal from the following list that is native to North America.
   - A. Sheep
   - B. Horse
   - C. Pig
   - D. Turkey

10. Most domesticated ducks are descendants of the:
    - A. Goose
    - B. Turkey
    - C. Mallard
    - D. Pekin

Let's Apply What We Know

1. Create an informational pamphlet or brochure about the different breeds of a specific livestock animal. Include traits that make each breed recognizable as well as the benefits they bring to the livestock industry.
2. Using a map, label which livestock breeds are native to the United States.
3. Why might polled animal breeds be preferred to horned breeds? Provide reasoning for your answer.

Academic Activities

1. **Engineering.** Cattle-handling facilities are very diverse. In small groups of two or three students, research the common facilities and components of a cattle handling area. Create a 3D model of a livestock handling facility using what you learned from your research. Be sure to include the individual components such as a squeeze chute, holding pen, and other needed pens.
2. **Language Arts.** There are many books that have been written about livestock animals including *Black Beauty* by Anna Sewell, *Charlotte's Web* by E. B. White, and *A Horse Called Wonder* by Joanna Campbell. Choose a book to read that is about a type of livestock animal and write a detailed book report.

Communicating about Agriculture

1. **Writing.** Using the information in this lesson and any additional research you wish to conduct, make a list of five breeds of livestock animals that you believe would thrive in your area. Write up a proposal to a local farm, recommending that they purchase these animals. Make a persuasive argument and present as much information as you can about why you think these animals would be a good investment.

SAE *for* ALL *Opportunities*

1. **School-Based Enterprise SAE.** Create an elementary agriculture lesson about the different livestock breeds. Include an interactive activity with your lesson. Present your lesson to an elementary class or after-school program.
2. **Entrepreneurship SAE.** Research what is needed to hatch and sell different types of poultry. Hatch different types of poultry eggs and sell the offspring.

SAE *for* ALL *Check-In*

- How much time have you spent on your SAE this week?
- Have you logged your SAE hours?
- What challenges are you having with your SAE?
- How can your instructor help you?
- Do you have the equipment you need?

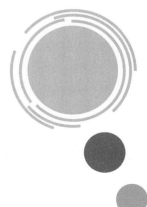

Introduction to Livestock Selection

Mysikrysa/Shutterstock.com

Learning Outcomes

By the end of this lesson, you should be able to:
- Explain how an animal's end use is related to the characteristics that are valued in selection. (LO 09.04-a)
- Describe the process of selection, including the four characteristics that are considered for all animals. (LO 09.04-b)
- Discuss the factors that are considered in evaluating all major species of livestock. (LO 09.04-c)
- Summarize how performance records can be used to help producers make decisions for their herd. (LO 09.04-d)

Words to Know

breeding animal	dam	sire
broilers	layers	udder
conformation	market animal	
cull	performance records	

Before You Begin

Get in a group of 3 or 4 students. As a group, find items that all of you have with you at your desk, like pencils, erasers, notebooks, binders, or backpacks. Compare these items, making notes about their similarities and differences. After making some notes, rank the items from best to worst. Your rankings do not have to match anyone else's. However, you should be prepared to defend your answer!

Livestock producers are constantly faced with the responsibility of evaluating their animals. They must make decisions about which animals to sell and which animals to buy. Having experience and animal knowledge helps them make decisions that will improve the quality of their operation.

Livestock animals are used for many purposes. Many are grown for their meat or milk production. Some are grown for their hair or wool. All livestock must be evaluated for their ability to reproduce if you want to continue producing animals. Healthy animals live a longer time and produce more products like milk, eggs, or hair. How can you tell if an animal is healthy? There are different selection criteria for different types of animals. Each animal's characteristics must be considered when making choices about which ones to keep and which ones to *cull*, or remove, from the herd (**Figure 9.4.1**).

Selection for Specific Purposes

What will the animal be used for? Will it be used to produce a product like meat or milk, or will it be used to produce other animals that will then be raised to produce those products? Once you know what you are using an animal for, you can evaluate livestock to choose the right animal.

Breeding

Livestock raised for reproduction are called *breeding animals*. Breeding animals are selected for desirable traits or physical characteristics that make them more suited to reproduction. Animals that are fertile, structurally correct, and fast-growing are desirable for reproduction. Breeding animals should also display the characteristics that you want from their offspring. Dairy cows and goats chosen for reproduction should be good milk producers. Beef cattle and swine should produce a high-quality meat carcass, and sheep should produce good quality wool or quality carcass traits. Female animals chosen for reproduction should have strong mothering traits, capable of raising young animals quickly and efficiently.

Market

Market animals are livestock raised for food consumption (**Figure 9.4.2**). You might recall from a previous lesson that this includes beef cattle, swine, sheep, goats, and poultry. These animals are chosen for quality muscle shape and structure, or *conformation*. Market animals should be fast growers that are able to convert feed into muscle efficiently, allowing them to reach a mature weight and be ready for market quickly. They should have good bone structure and be able to move about freely to eat, drink, and exercise to keep them healthy and growing.

visuall2/Shutterstock.com

smereka/Shutterstock.com

Figure 9.4.1. These cattle will not be used for the same purpose: one will produce beef, and one will produce milk. *What different characteristics might be important in selecting the right animals for each purpose? Would you expect them to be the same?*

Cirkovic Milos/Shutterstock.com

Figure 9.4.2. Market animals are raised for their meat. *What makes an animal good for meat production?*

2xSamara.com/Shutterstock.com

Figure 9.4.3. These goats are part of a milk producing operation. *What characteristics make these goats well-suited to this?*

Milk Production

Dairy animals (cows, goats) raised to produce milk and milk products tend to be leaner in their conformation than market animals (**Figure 9.4.3**). They are selected and bred to turn feed consumed into milk, rather than meat. Dairy animals are selected for high milk production, fertility (dairy animals must become pregnant to produce milk), disease resistance, and longevity. Less emphasis is placed on muscling. There is more emphasis placed on the proper structure of the animal's frame, legs, udder, and teats. An animal that will be producing milk needs to have a structurally sound and firmly attached udder, teats that are not prone to infection, and legs that are properly built to handle the weight of pregnancy and milk production.

Did You Know? The dairy industry processes raw milk into an array of products including butter, cheese, cream, yogurt, condensed milk, dried milk, ice cream, and produces various by-products including buttermilk, whey, and skim milk.

Animal Fiber Production

Livestock raised to produce animal fibers include sheep, goats, certain types of cattle, llamas, alpaca, rabbits, and geese (**Figure 9.4.4**). These animals are selected and bred for their outer coats which produce useful and beautiful fibers. Goose down can be found in mattresses and pillows. Other fibers produce socks, coats, and sweaters. These animals are selected for their ability to produce quality fibers quickly.

Figure 9.4.4. Alpacas, cashmere goats, and angora rabbits are raised for their soft wool or hair. *What products are made from these fibers?*

Wasim Muklashy/Shutterstock.com

Oliver Hoffmann/Shutterstock.com

Evita Trankale/Shutterstock.com

Animal Selection Considerations

Each species of livestock is evaluated based on unique characteristics. Within each species, selection considerations can be customized based in the animal's intended use. Basic characteristics that are evaluated in all species include:

- Frame—size of the skeleton
- Design/Structure—balance, proportion, and movement
- Muscling—how much muscle is on the animal's frame, how well it is formed, and how quickly it gains muscle
- Capacity/Volume—the animal's ability to take in and convert feed

These characteristics are prioritized differently when evaluating animals for different purposes. For example, in selecting market animals, muscling would be your top concern, but in dairy animals, muscling would likely be the least valuable of these characteristics. Livestock grown for their hair or wool would also have more emphasis placed on frame, structure, and body capacity.

Livestock evaluation happens by examining the animal from various viewpoints. A front view will allow an examination of chest width, foot placement, front shoulder structure, and head/neck structure. A rear view will reveal hip width, rear leg placement, muscling, and dairy characteristics. A side view gives a good indication of capacity, volume, frame size, and overall balance when moving. Smaller market animals can be viewed from overhead to evaluate loin width. It is important to view an animal from all sides and observe their movements when selecting or evaluating them.

Did You Know? Livestock judging events such as FFA and 4H Livestock Evaluation, Horse Evaluation, Dairy Evaluation, and Poultry Evaluation are highly effective at giving students practice and training for observing and evaluating different animals. Students that do well as high school students can also earn scholarships and compete on collegiate evaluation teams across the nation.

Cattle Selection

Cattle are divided into two main production categories: beef and dairy. Beef cattle are grown to produce market animals for meat consumption. Dairy cattle are grown to produce milk. These cattle types have differing requirements from one another because of their end purpose.

The evaluation of beef cattle emphasizes muscling, then design/structure, then capacity/volume, and finally, frame. Beef cattle used for breeding would be evaluated slightly differently: capacity/volume, design/structure, frame, and muscling. Dairy cattle are evaluated like breeding cattle with an extra emphasis on feminine characteristics and conformation of the *udder*, or milk-producing organ of cattle (**Figure 9.4.5**).

Cattle herds are improved by selecting animals with desirable traits to pass on to their offspring. Purebred cattle have specific requirements and can only be selected and bred with animals from the same breed. Commercial cattle are made up of a mixture of two or more breeds and may result in a mixture of traits related more to conformation than breed characteristics.

Eric Isselee/Shutterstock.com

Figure 9.4.5. *What characteristics would be used to evaluate the dairy cow in this picture?*

Figure 9.4.6. Sheep come in a variety of sizes. *What differences might you find between large and small sheep?*

Figure 9.4.7. Angora goats are produced for their beautiful fleece. *What differences do you see between these and other goats? What characteristics would be most valued here?*

Figure 9.4.8. Most swine produced in the US are valued for being heavily muscled resulting in larger amounts of retail meat cuts.

Sheep Selection

Sheep are produced in the United States for wool and meat. When selecting sheep (breed or individual) for an operation, producers must consider how well the sheep grow and how they will do in the area they are to live in, whether breeding stock is available, what market exists for their products, and whether the sheep are able to reproduce easily.

Body conformation and frame size are important considerations, as larger-framed sheep produce heavier fleece, and more muscle for meat production (**Figure 9.4.6**).

Goat Selection

Goats produce meat, dairy products, and fiber. The primary product of a goat herd will determine the specific characteristics selected when building that herd. Market goats must display large, high-quality muscling to meet consumer demand. Dairy goats are selected for being good milk producers. Goats raised for fiber production, like the Angora goat, are judged and selected based on the quantity and quality of their fiber production and body conformation (**Figure 9.4.7**). There must be a balance between the two because too much emphasis on fleece production can reduce the average goat size over time, reducing the volume of product produced.

Swine Selection

Swine are raised for meat consumption or for breeding stock. Consumer demand for leaner pork in recent years has led producers to select more heavily muscled, leaner animals than previously. Many market hogs today are the result of crossbreeding programs, as producers have realized that crossing certain breeds will result in lean, high-quality meat that will bring a premium price at market.

Selection of swine for market emphasizes muscularity and lean growth (**Figure 9.4.8**). Structure, dimension, and balance are secondary market considerations. When selecting breeding animals, the emphasis is on structure and movement. Growth, balance, and muscling are also important breeding swine characteristics.

Horse Selection

Horses in the United States are used primarily for recreational or pleasure purposes, though they are considered livestock. Breeding, showing, working, and sport make up the other common uses (**Figure 9.4.9**). Horse selection will be based on these uses and the following factors:

- Age—older, well-trained horses are better suited for younger or less experienced riders
- Gender—stallions, or male horses, are much more difficult to control than a gelding (castrated male horse) or a mare
- Breed—some horse breeds are better suited to particular activities than other breeds
- Conformation—depending on intended use, foot and leg structure are two of the most important features to evaluate when selecting a horse
- Color—based on personal preference, the basic horse colors are bay, black, brown, chestnut, and gray (**Figure 9.4.10**)

Nicole Ciscato/Shutterstock.com

Figure 9.4.9. Many horses are bred for showing at competitions. *What characteristics might benefit horses in this context?*

Black Bay Chestnut Flaxen chestnut

Seal brown Grulla Buckskin Palomino

Cremello Gray Skewbald Piebald

ProfiTrollka/Shutterstock.com

Figure 9.4.10. Horses come in different colors. *Why might you want a particular color?*

Hands-On Activity

Evaluate a Class of Animals

Livestock producers continually evaluate the condition of their animals. To improve their herds, they must evaluate and determine which animals to keep and which ones to cull. Let's look at a group of animals, make observations and notes about their physical characteristics, and rank them in order from best to worst based on your observations.

hadot 760/Shutterstock.com

For this activity, you will need the following:

- A set of four animals:
  - Cattle, swine, goats, sheep, etc.
  - You can use a smaller number for comparison, but four animals are used in most events
  - They need to be of the same type (market, breeding, etc.)
  - You may substitute photographs or videos for live animals
  - If using photos, provide front, rear, side and top views of each animal
- Paper or note cards for writing down descriptive observations
- A list of livestock terminology (for reference) specific to the animals chosen
- Pencils/pens
- Clipboard

Here are your directions:

1. Select and safely contain animals to be judged.
2. Identify each animal with a number.

3. Have students observe animals and make notes about the physical appearance of the animals. If need be, this activity can be done as a full class to help everyone get the hang of it.
4. Compare each animal and rank them based on comparison of physical appearance from best to worst.
5. Discuss the outcome and allow students the opportunity to compare their results.

Consider This

1. Did everyone come to the same conclusions in this exercise? Why or why not?
2. What were the easiest characteristics to judge? Why? What were the most difficult?
3. Did you enjoy doing this? Why or why not?

Poultry Selection

Chicken production involves two major commercial enterprises: egg production and broiler production. **Layers** are chickens raised to produce eggs. **Broilers** are chickens raised for meat production. Commercial egg and broiler production involve huge numbers of animals, making individual evaluation and selection time-consuming and difficult. Commercial hatcheries provide chicks to producers, and they manage the selection process there. Chickens have a short life cycle from egg to adult (seven weeks, for broilers), making it much easier to select for desired traits and make changes in structure and conformation to a group in a brief time.

In small-scale operations, growers may raise small populations of specific breeds (**Figure 9.4.11**). Selection in this situation allows more time for individual examination and evaluation. As with other animals, chickens are evaluated for quality structure and conformation. They should move well and be healthy with no visible defects in their form. Similar evaluation methods are used for other poultry, like turkeys and ducks.

Figure 9.4.11. Small-scale operations may raise more different and unique breeds of chickens. *Have you seen any of these before?*

Polish

Plymouth rock

Minorca

Faveroles

Vorwerk

White leghorn

Faenkova Elena/Shutterstock.com

9

Clara Bastian/Shutterstock.com

Figure 9.4.12. This tag helps to keep track of animals. *Why might it be important in production to know the location of animals on a farm or ranch?*

Performance Records

One of the best methods for selecting quality beef and dairy cattle is by using performance records. ***Performance records*** are a method of collecting production data on livestock to select the most productive animals. Accurate records are kept on each animal for:

- Animal identification—taken from an ID tag or marking (**Figure 9.4.12**)
- Yearling weight—taken between 320 and 440 days of age and adjusted to 365 days
- Birth weight of calves—taken within 24 hours of birth
- Weaning weight of calves—taken between 120 and 280 days old (average at 205 days)
- ***Sire*** (male parent) records
- ***Dam*** (female parent) records

Performance records provide valuable data for producers and allow them to make good decisions about which animals to keep or cull, selection of replacement animals, and setting herd improvement goals. Some systems even track animals via their ID tags and offer producers additional information, like how much they eat, how much milk they produce, when they are fertile for breeding, how much weight they gain, and so on.

Making Connections # STEM

Livestock Identification Tagging

Many livestock have electronic (EID or RFID) tags. The EID tag or device contains a microchip that can be read electronically in a fraction of a second by a suitable reader. EID in animals is based on low-frequency radio waves or Radio Frequency Identification (RFID). A reader sends out a radio signal which is picked up by the microchip. The chip sends back the unique identity number. The reader transfers the unique number to a computer which uses it to store whatever information is necessary about the animal being identified.

Electronically reading animal identities can deliver many benefits for everyone handling animals. Many farmers, particularly on pig and dairy units, have been using EID for many years. Farm management practices can be greatly improved by using EID systems, saving both time and labor in the yards.

EID tags help livestock owners:
- Manage less paperwork
- Decrease the risk of injuries in handling animals

Clara Bastian/Shutterstock.com

- Increase herd efficiency
- Keep records
- Know how many livestock are on the farm at any given time
- Keep up with medicines, vaccines, and any treatments the animals have received
- Track pregnancies

Review and Assessment

What Did We Learn?

- The markers that a producer may look for in an ideal animal will vary from species to species, depending on what a producer needs and wants. Animals are selected for breeding, market, milk production, and fiber production, depending on the animal and purpose of the operation.
- Producers selecting animals for breeding operations place emphasis on their fertility, desirable traits for offspring, and structural correctness.
- Producers selecting animals for market operations place emphasis on muscling and capacity/volume.
- Producers selecting animals for milk production will be looking for milk production, fertility, disease resistance, longevity, and conformation.
- Producers selecting animals for fiber production will be looking for the quality and quantity of their fur/feathers/fiber and their ability to grow it quickly.
- All livestock are evaluated based on frame, design/structure, muscling, and capacity/volume. The way that these characteristics are ranked and prioritized depends upon the purpose of the animal.
- Production records are kept for many animals. The data from those records help producers to make the best choices about how to manage the animals.

Let's Check and See What We Know

Answer the following questions using the information provided in Lesson 9.4.

1. Which of the following factors would be *least* important to a livestock producer when planning for selection of animals to add to their herd?
   A. Muscling
   B. Structure
   C. Health
   D. Previous owner

2. Which of these factors would be of greatest value to a dairy producer?
   A. Milk production
   B. Muscling
   C. Eye color
   D. Size

3. _____ are grown primarily for the high quality of their hair coat.
   A. Beef cattle   C. Ducks
   B. Dairy goats   D. Angora goats

4. _____ is a characteristic related to an animal's ability to convert feed into muscle.
   A. Volume/capacity
   B. Frame
   C. Muscling
   D. Balance

5. _____ help use data to select the most economically productive animals.
   A. Judging cards
   B. Performance records
   C. Weaning weights
   D. Health records

6. A _____ is the male parent of a livestock animal.
   A. dam
   B. mare
   C. pullet
   D. sire

7. Bay, black, brown, chestnut, and gray are the basic colors of _____.
   A. chickens
   B. geese
   C. cattle
   D. horses

8. Which of the following is *not* a primary use of horses in the United States?
   A. Pleasure riding
   B. Work
   C. Meat consumption
   D. Showing

9. _____ are chickens raised to primarily produce eggs.
   A. Broilers
   B. Layers
   C. Sires
   D. Angora

10. *True or False?* Layers are chickens that are raised for their meat.

Let's Apply What We Know

1. Explain the difference between breeding animals and market animals, and why they are evaluated differently.
2. List and explain the basic characteristics that are used to evaluate all livestock species.
3. What items should be included in animal performance records and how are performance records beneficial?

Academic Activities

1. **Science.** Typically, poultry eggs are evaluated for stains, shape, shell texture, and shell thickness, among other things. Research what goes into USDA requirements for egg grading. Secure fresh eggs and work with a partner to grade the eggs.
2. **Math.** Research at what weight market hogs are sent to harvest. Determine weeks from birth to harvest and how many pounds hogs should gain each week to reach the suggested market weight.

Communicating about Agriculture

1. **Communicating Information.** Choose a livestock species. Create a flyer describing what characteristics are important to consider in selecting animals of that species. Use illustrations to help make your flyer informative.
2. **Social Media.** Create a post advertising for an upcoming livestock judging event. What do students need to do to prepare? What factors will be evaluated? How can you get younger students interested in showing up to watch or help?

SAE for **ALL** *Opportunities*

1. **Foundational SAE.** Interview a local livestock producer. Discuss with them how they select their livestock for breeding and decide which animals to cull. Discuss how they keep records on their herd and how this benefits their operation. Create a presentation on your experience and present it to your class.
2. **Placement SAE.** Volunteer at a local animal farming operation. Discover the daily responsibilities of running a livestock operation.

SAE for **ALL** *Check-In*

- How much time have you spent on your SAE this week?
- Have you logged your SAE hours?
- What challenges are you having with your SAE?
- How can your instructor help you?
- Do you have the equipment you need?

Livestock Products and By-Products

CEPTAP/Shutterstock.com

Learning Outcomes

By the end of this lesson, you should be able to:

- Identify common livestock products and their importance in daily life. (LO 09.05-a)
- Recognize the use and value of livestock by-products in daily life. (LO 09.05-b)
- Explain how by-products reduce waste, and why that is important. (LO 09.05-c)

Words to Know

abattoir	edible	pasteurization
by-products	fertilized egg	unfertilized egg
collagen	offal	variety meats

Before You Begin

Take a moment to think about what you have eaten lately. Make a list of the five foods that you eat most often. How many of them come from livestock? Maybe you eat eggs or bacon for breakfast, or turkey on your sandwich at lunch each day. Most of us consume at least some meat and livestock products.

*What to know something crazy? The average US consumer eats more than 200 pounds of meat each year (**Figure 9.5.1**)! That's more than a half-pound per day, per person. This means the average person eats 0.18 pounds of meat at every single meal! Think about how much meat you consume each week. Do you eat more or less than that average?*

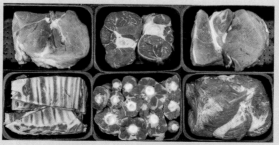

Figure 9.5.1. Meat products make up some part of most people's diets. *What role do they play in your diet?*

Sandro Pavlov/Shutterstock.com

Livestock Products

Most livestock are produced to provide food for humans. Human food from livestock comes primarily from meat. Livestock are raised to produce the most valuable livestock products. Examples of these primary products include meat from cattle or chickens, milk from goats or cattle, eggs from chickens or ducks, and wool from sheep (**Figure 9.5.2**). In recent years, US consumers have started eating more protein, contributing to the increased production of meat products.

Meat makes up about half of a market animal's weight. The remaining parts of the carcass, called *offal*, are used to make secondary animal products, or *by-products*. Because so much of the animal remains when the meat has been removed, it is beneficial to find uses for the offal to reduce waste and provide additional income to farmers. Today, almost all of an animal is used to make products and by-products, leaving very little waste.

margouillat photo/Shutterstock.com

Figure 9.5.2. *Which of the foods pictured don't come, at least in part, from animals?* There aren't many, and you will need keen eyes to spot them!

Meat

Meat from animals is an important source of nutrition and a primary source of protein for many people around the world. Market livestock are raised to produce meat products which come from the large muscles of the animal (**Figure 9.5.3**). Meat production in the world has quadrupled over the last 50 years. Globally, more than 661.5 billion pounds of meat is produced annually for human consumption!

Worldwide, pork and chicken are the most consumed meats, followed by beef, mutton, and goat meat products. In the US, chicken is the most consumed meat, with beef, pork, and lamb coming after. Poultry consumption is increasing at a faster rate than other livestock and is expected to become the most consumed meat globally if current trends continue.

Did You Know?

The Meats Evaluation and Technology Career Development Event (CDE) is an event for high school FFA members who want to learn about the meat industry. During this team event, students evaluate beef carcasses for quality and yield grade, identify various meat cuts and place carcasses, and identify wholesale and retail cuts.

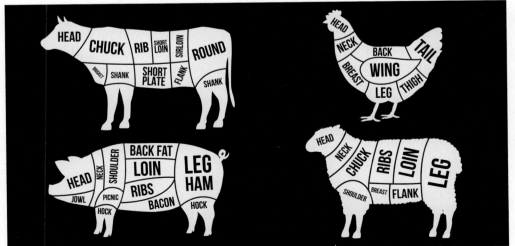

DenysHolovatiuk/Shutterstock.com

Figure 9.5.3. Every meat-producing animal produces a variety of meat products, as shown here. *Do you see any unfamiliar cuts?*

Kzenon/Shutterstock.com

Figure 9.5.4. Butchers at a processing facility carve animal carcasses into the cuts of meat you see in packaging at the grocery store.

In the US, people can raise and process their own animals, but most consumers purchase meat that is raised commercially. Commercial growers raise animals according to strict government requirements for feeding and health. Animals are sent to a processing facility (**Figure 9.5.4**) once they reach market size and age. The processing facility is also required to meet strict government requirements for health and safety. The animals are slaughtered, packaged, and sent to retail stores where people can purchase the meat products.

Did You Know? Americans eat nearly 50 billion hamburgers each year, with the average person eating around 150.

Dairy

Dairy products include milk, cheese, butter, cream, and yogurt (**Figure 9.5.5**). Most dairy products come from cows' milk. Milk consumption has steadily increased over the past 50 years. Current global cow milk production is estimated at more than 500 million metric tons, which works out to 1.1 trillion pounds per year. Wow!

India has the largest population of milk cows and produces more than a fifth of the world's milk. The United States and China also produce large amounts of cow milk. France, Germany, Ireland, and other countries in the European Union are well-known for their quality cheese and butter, with more than one-third of the world's dairy exports coming from Europe. Germany alone exports more than $4.5 billion worth of cheese annually. Compare that to US exports of all dairy products, totaling around $7 billion, and you can see that Germany produces a lot of (delicious) cheese!

margouillat photo/Shutterstock.com

Figure 9.5.5. Dairy products are a rapidly growing segment of food production. *Do you like any dairy products?*

Decades ago, families usually collected raw milk for consumption from a single dairy cow or small family herd. Any extra milk could be sold to neighbors who had no cows. As consumption of milk and demand for milk products increased, dairy herds have grown to meet that need. Today, commercial dairy herds produce large quantities of milk that is collected and sent to processing facilities for ***pasteurization***, or heat sterilization to make it safer for consumption. It is then packaged and sent to retail stores or on to other commercial facilities to make other dairy products like butter, cheese, or yogurt.

Fiber

People use many types of animal fibers. The most common is sheep's wool. Australia is the top wool-producing country in the world, currently responsible for about 25 percent of all wool produced each year. Other countries that are top producers include China, the United States, and New Zealand. The US is responsible for producing about 17 percent of the world's wool, which amounts to around 17 million pounds each year.

Sheep produce most wool (**Figure 9.5.6**). The hair of some breeds of goats, llamas, alpacas, camels, musk oxen, and rabbits also produce fibers called wool. The Angora goat produces a long, shaggy coat called mohair. Cashmere, another high-quality fiber, comes from the hair of the cashmere goat. Angora rabbits are used to produce wool and each rabbit can produce about three pounds of wool per year.

Wool is classified based on several characteristics including fiber diameter, strength, length, uniformity, color, and number of contaminants in the fiber. Wool is placed into one of three categories: fine, medium, and coarse. Fine wool is used to make clothing because of its softness, while coarser wools are used to make yarns or rugs.

Juice Flair/Shutterstock.com

Figure 9.5.6. Many sheep are raised to produce wool, which will be used to make all kinds of consumer goods.

Eggs

Egg production worldwide has more than doubled in the past 30 years. The US produces more than 113 billion chicken eggs each year, and they aren't the largest producer in the world (**Figure 9.5.7**)! Eggs are a valuable source of protein and are one of the most popular protein sources in the United States. China is the largest egg producer in the world, by a wide margin. According to Statista, India, Indonesia, the US, and Brazil combine with China to produce more than 991 *billion* eggs each year.

All poultry produce eggs, but chicken eggs are, by far, the most consumed in the US. Chickens have such market domination in the United States that the USDA does not collect data on sales of any other kind of egg. Duck eggs produced in the US have a small market compared to chickens, but in China and Japan there is a large market for their eggs. Other poultry, like geese and turkeys, are also produced commercially in small operations, but that egg production focuses on fertilized eggs for hatching young.

Two types of chicken eggs, *fertilized* and *unfertilized*, are produced by growers. A hen, or female chicken, can lay eggs without the presence of a rooster, or male chicken, resulting in **unfertilized eggs**. Unfertilized eggs are grown and sold for human consumption. When a hen mates with a rooster and lays an egg, that is a **fertilized egg**. Fertilized eggs are used in hatcheries to raise chickens to replenish layer or broiler operations.

Janon Stock/Shutterstock.com

Figure 9.5.7. Eggs produced at commercial facilities are processed in preparation for going to US markets. *How many eggs does your family go through each year?*

Livestock By-Products and Waste Reduction

Livestock by-products are the secondary products that result from the production of an animal. Livestock by-products can be classified into two basic types: *edible*, able to be eaten; and inedible. After slaughter, the by-products are separated into edible and inedible parts. Edible by-products account for a little more than half of the offal. Edible parts include beef tongue, pig hearts, and chicken feet that are not necessarily common on US grocery store shelves (**Figure 9.5.8**). By producing livestock by-products, growers reduce the environmental impact of carcass disposal.

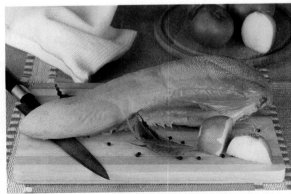

Vladimir Shulenin/Shutterstock.com

Figure 9.5.8. Beef tongue is a relatively common edible by-product. *Have you or a friend or family member ever eaten it?*

NUM LPPHOTO/Shutterstock.com

Figure 9.5.9. Organ meats are often ground up into variety meats that can be used in other products. Using by-products like this makes it possible to reduce the amount of waste produced in livestock production.

Animal by-products are used to make many products that contribute to our quality of life. The economic benefit of using most of the animal adds value to the livestock industry. Profits from animal processing by-products result in about ten percent of the value of the live animal. It is important to research and develop new ways to use offal to reduce waste and to sustain our increasing population.

Edible By-Products

Many livestock by-products are used to create items that can be eaten by humans or other animals. Butcher shops sell whole organs like the liver, or *variety meats*, intestines and internal organs ground up and blended (**Figure 9.5.9**).

Hands-On Activity 👍

Where Do I Come From?

Many everyday items are made from animal by-products. You may or may not be aware that common items in your pocket, purse, or classroom are manufactured from livestock by-products.

For this activity, you will need the following:

- Sheet of paper
- Marker or pen
- Magazine or computer for images
- Glue
- Ruler

Here are your directions:

1. Using the entire piece of paper, create the following chart. Use a ruler to make sure the lines of your chart are neatly spaced.

2. Find at least three by-products from each of the species on the chart.
3. Find a picture of each by-product from a magazine or an online source.
4. Glue each picture into the appropriate square on your paper.
5. Write under each picture what part of the animal it came from (e.g., bones, fat, etc.).

Consider This

1. Did any of the by-products you researched surprise you? Which ones and why?
2. Do you think consumers would change the items they purchase if they were more aware of how they were made?

Species	By-Product	By-Product	By-Product
Beef/Dairy			
Sheep			
Swine			
Goats			
Poultry			

Some edible products may also include meat by-products from multiple species blended together. Gelatin (**Figure 9.5.10**) is an example of an edible by-product made from livestock tissue high in **collagen**, a structural protein found in connective tissues, like bone and skin.

In the US, edible by-products must be inspected and observed to be without infection or abnormal growth before being approved for human consumption. They are further processed by washing, heat treating, or other means to remove or reduce any health risk to consumers. Certain by-products are rich in nutrients, beneficial to humans, and in some instances are considered a delicacy.

cigdem/Shutterstock.com

Figure 9.5.10. Gelatin is what makes your Jell-O bouncy —and gives it its name. *Did you know that gelatin is an animal by-product?*

Making Connections #STEM

Making Jell-O

Many of the food products we eat every day contain animal by-products. You may not see or taste these items, but they are a vital part of the recipe.

For this activity, you will need the following:

- 4 cups of 100% fruit juice (do not use pineapple)
- 2 tablespoons unflavored beef gelatin
- 2 tablespoons honey
- Single burner
- Medium saucepan
- Bowl
- Whisk
- Baking dish or individual cups
- Refrigerator

Here are your directions:

1. Add ¾ cup of fruit juice to a bowl and sprinkle with gelatin powder.
2. Whisk together until combined and let sit for 3–5 minutes to "bloom." The granules will plump, and the mixture will look like thick applesauce or take on a lumpy appearance.
3. Pour the remaining fruit juice into the saucepan and heat over medium heat until it starts to boil.

Ildi Papp/Shutterstock.com

4. Remove from heat and whisk in honey and bloomed gelatin mixture. Stir until dissolved.
5. Pour into a baking dish or into individual cups.
6. Refrigerate for four hours or until set.

Consider This

1. What did you find interesting about the "blooming" process when you mixed the gelatin into the fruit juice?
2. Look at the list of ingredients here. Does it surprise you that Jell-O contains animal by-products?

There are edible by-products that might not be fit for human consumption but can still be used in animal feeds. The high nutrient and protein content provide a good food source or food additive for certain livestock and pet feeds. There are stricter requirements that must be met if using the by-product to feed an animal that will ultimately be used for human consumption.

Inedible By-Products

Other by-products are not edible but are useful in other ways. Medicine, cosmetics, soap, and many other items contain livestock by-products (**Figure 9.5.11**). Skin and hides are used to make leather products and clothing. Bone, hoof, and horn are used to make items like buttons. Organs and glands are used to gather enzymes and hormones for medicinal purposes. Intestines are used as strings for musical instruments. Blood may be used in fertilizers.

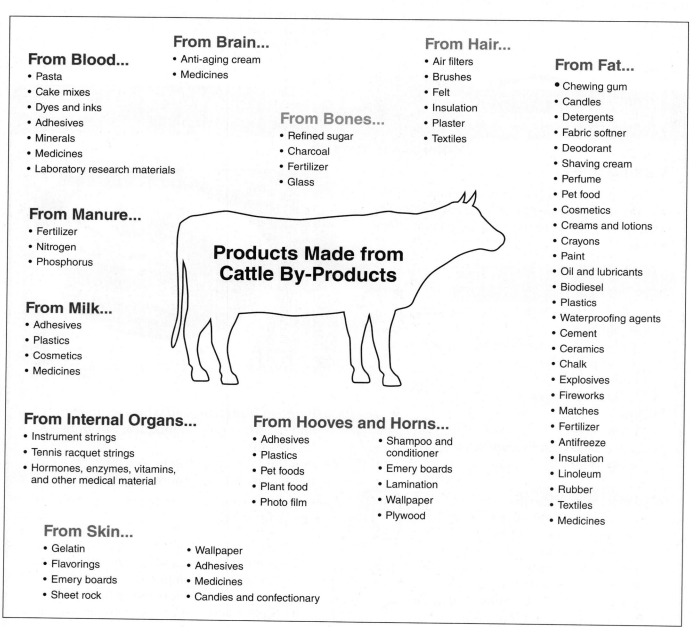

From Blood...
• Pasta
• Cake mixes
• Dyes and inks
• Adhesives
• Minerals
• Medicines
• Laboratory research materials

From Brain...
• Anti-aging cream
• Medicines

From Bones...
• Refined sugar
• Charcoal
• Fertilizer
• Glass

From Hair...
• Air filters
• Brushes
• Felt
• Insulation
• Plaster
• Textiles

From Fat...
• Chewing gum
• Candles
• Detergents
• Fabric softner
• Deodorant
• Shaving cream
• Perfume
• Pet food
• Cosmetics
• Creams and lotions
• Crayons
• Paint
• Oil and lubricants
• Biodiesel
• Plastics
• Waterproofing agents
• Cement
• Ceramics
• Chalk
• Explosives
• Fireworks
• Matches
• Fertilizer
• Antifreeze
• Insulation
• Linoleum
• Rubber
• Textiles
• Medicines

From Manure...
• Fertilizer
• Nitrogen
• Phosphorus

Products Made from Cattle By-Products

From Milk...
• Adhesives
• Plastics
• Cosmetics
• Medicines

From Internal Organs...
• Instrument strings
• Tennis racquet strings
• Hormones, enzymes, vitamins, and other medical material

From Hooves and Horns...
• Adhesives
• Plastics
• Pet foods
• Plant food
• Photo film
• Shampoo and conditioner
• Emery boards
• Lamination
• Wallpaper
• Plywood

From Skin...
• Gelatin
• Flavorings
• Emery boards
• Sheet rock
• Wallpaper
• Adhesives
• Medicines
• Candies and confectionary

Goodheart-Willcox

Figure 9.5.11. Look at this list of products and by-products that come from producing beef cattle. *What here is most surprising to you? Are there any parts that haven't been used?*

Abattoirs, or animal processing facilities, produce a large amount of waste. Careful management must be taken to ensure that these wastes are properly handled. There are researchers working on ways to use this waste as a biofuel component. Certain waste components can be composted and used as fertilizers.

Did You Know? The medical industry uses pig organs and skin for many medical procedures. In severe burn cases, doctors graft pig skin into the human body, as it will join with and perform like human skin. In January 2022, doctors successfully completed the first heart transplant from a genetically modified pig to a man in Baltimore.

Career Corner — SLAUGHTER PROCESSOR GENERAL OPERATOR

A **general operator** *processes the slaughter of beef, sheep, goat, fish, pork, and poultry to meet site compliance standards and targets.*

What Do They Do? Slaughter processor general operators know animal anatomy and cuts of meat, follow safety procedures and policies, prepare carcasses for cutting by cleaning and removing parts, cut carcasses for further processing, separate meat and by-products, inspect meat products for defects and blemishes, process prime parts into cuts that are ready for retail use, pack final products, and check products to ensure they are up to USDA standards.

What Do I Need to Be Good at This? A high school diploma is required for this career.

How Much Money Can I Make? Slaughter processors make an average of $30,000 per year.

How Can I Get Started? You can build an SAE researching different cuts of meat and where they come from in the animal. Join an FFA meats evaluation team and volunteer to help train younger students. You can conduct a survey of consumer preference of different cuts of meat or interview a butcher or slaughterhouse manager.

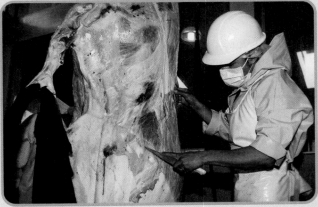

Ammit Jack/Shutterstock.com

9

What Did We Learn?

- Most livestock are raised to make products for human consumption.
- In modern agriculture, almost the entirety of the animal is used to make various products and by-products, meant to reduce waste.
- Most livestock are raised to produce meat. Meat is made from the large muscles of animals and provides protein to people around the world.
- Milk, primarily from cattle, is another important livestock product. Milk is used to make other dairy products.
- Wool is the most common animal fiber harvested or grown for human use, with different breeds of animals producing different qualities and textures of hair.
- Egg production has grown in recent years. Chicken eggs dominate the US egg industry.
- By-products are made from offal and can be classified as edible and inedible.
- Using every part of an animal's carcass that can be used in some form helps to reduce waste. Animal by-products are found in all kinds of items in our daily lives, from adhesives to shampoos to some medicines.

Let's Check and See What We Know

Answer the following questions using the information provided in Lesson 9.5.

1. _____ is the fastest-growing meat consumed in the world.
   A. Beef
   B. Pork
   C. Chicken
   D. Mutton

2. _____ would be an example of a livestock by-product.
   A. Steak
   B. Ham
   C. Chicken legs
   D. Beef liver

3. Which of the following animals produce dairy products?
   A. Chickens
   B. Ducks
   C. Turkeys
   D. Cattle

4. Which of the following livestock products would *not* be a dairy product?
   A. Eggs
   B. Cheese
   C. Butter
   D. Yogurt

5. _____ is the most common animal fiber used by humans.
   A. Goat hair
   B. Rabbit fur
   C. Sheep's wool
   D. Camel hair

6. If the total value of an animal and all its parts were $100, what would be the estimated value of its by-products?
   A. $10
   B. $25
   C. $50
   D. $75

7. *True or False?* Gelatin is an example of a by-product made from the collagen of a livestock animal.

8. *True or False?* Edible livestock by-products like liver and other organ meats provide a good source of protein and other valuable nutrients when consumed by humans.

9. *True or False?* Livestock can never be fed the by-product of other livestock because of the poor nutrient value and high cost of processing.

10. Which of the following examples would be a good use of the waste product that remains after the slaughter and processing of livestock?
    A. Biofuel
    B. Human food additives
    C. Livestock feed
    D. Edible by-products

Let's Apply What We Know

1. Create a collage of everyday items that are derived from animal by-products.
2. Select a species of livestock animal and label the cuts of meat that are obtained from that animal.
3. Explain the difference between fertilized and unfertilized eggs and how they are used in the livestock industry.

Academic Activities

1. **Social Science.** There is a lot of controversy about the terms used to label meat products, such as "antibiotic free," "no hormones added," or "free range." Research common terms used in labeling meat products and what they mean. Work with a partner to create an informational brochure.
2. **Engineering.** Eggs must be candled before they can be sold for human consumption. Research, design, and build a candler from everyday household items. Test your candler, keeping in mind the correct timing and temperature.

Communicating about Agriculture

1. **Persuasive Speaking.** Animal by-products are used in all sorts of unexpected products. How do you feel about this? Is this a good thing, or a bad thing? Research two livestock animals and the by-products that are made from their production. Put together a three-minute persuasive speech, citing examples, to argue your position to the class.

SAE for ALL Opportunities

1. **Entrepreneurship SAE.** Raise your own livestock animals for market or show. Once they have reached full maturity, process and sell the meat.
2. **Placement SAE.** Volunteer at a butcher shop. Learn the different cuts of meat and what part of the animal they are derived from.

SAE for ALL Check-In

- How much time have you spent on your SAE this week?
- Have you logged your SAE hours?
- What challenges are you having with your SAE?
- How can your instructor help you?
- Do you have the equipment you need?

Pressmaster/Shutterstock.com

Livestock Physiology

Learning Outcomes

By the end of this lesson, you should be able to:

- Identify the major body processes of livestock and explain the importance of each. (LO 09.06-a)
- Identify the major parts of livestock skeletal, muscular, and digestive systems. (LO 09.06-b)
- Explain the difference between monogastric, ruminant, and avian digestive systems. (LO 09.06-c)

Words to Know

cardiac muscle
cartilage
circulatory system
digestion
ligament
marrow

monogastric
muscular system
nervous system
physiology
regurgitate
respiratory system

ruminant
skeletal muscles
skeletal system
smooth muscles
tendon

Before You Begin

Do you ever daydream? Yesterday, I started asking myself how my body moves. Why can you move the fingers on one hand while everything else is still? Why don't you have to think about breathing? Where does the energy to run all these processes come from? Have you ever thought about anything like that?

There are medical doctors who spend their entire careers focused on one specific part of the human body, like the heart. Now, imagine being a veterinarian who has the responsibility of treating several different species of animals with entirely different systems! Each animal that arrives in our office during the day can have an entirely different size or body type. To add another layer, our patients are not able to talk about how they feel or explain what the problem might be. That's a lot to manage!

Physiology is the branch of biology that deals with the normal functions of living organisms and their parts. In this lesson, we will look at some of the major body systems of livestock, their importance, and the basics of how they work. Producers need to understand how the bodies of their livestock function. This knowledge allows them to make better decisions for nutrition, housing, and care that increase productivity and profits. There are several body systems of importance in every animal. We will focus on the skeletal, muscular, and digestive systems as the primary systems to understand from a production standpoint.

Skeletal System

The main functions of an animal's *skeletal system* (**Figure 9.6.1**) are to provide structure, protection, and support for the body. The skeletal system is composed of bones, cartilage, joints, tendons, and ligaments. There are two major types of livestock skeletal systems: mammalian and avian. The mammalian skeleton is found in mammals. Mammals are warm-blooded animals from the class Mammalia that have fur and produce milk such as the cow, pig, and goat. The avian skeleton is found in birds and poultry. Avian animals come from the class Aves which have feathers, wings, and lay eggs, like the chicken, duck, and turkey.

Young animals have lots of cartilage in their skeleton that is replaced by bone as they mature. *Cartilage* is a tough and flexible connective tissue that provides support between the bones at joints of both skeletal types. Cartilage also forms structures such as the nose and ears, and other internal structures, such as the trachea.

Bones are made of calcium, calcium compounds, and small amounts of minerals. Bones are rigid, allowing the animal to maintain its shape and form. Bones contain blood vessels, lymph vessels, and nerve fibers. Bones store minerals and produce blood cells in the *marrow,* or soft center of the bone. Muscles are attached to the bone structure by *tendons* and bones connect to other bones with *ligaments*. Tendons and ligaments work with the bones to provide movement and support.

The mammalian and avian skeletal systems have some similarities, but also major differences. Avian bones are thinner, harder, and more brittle than mammals' (**Figure 9.6.2**). Avian bones contain more cartilage. The avian rib cage is more stationary than the mammalian, which stretches in the breathing process. The avian skull contains no teeth but does have a beak.

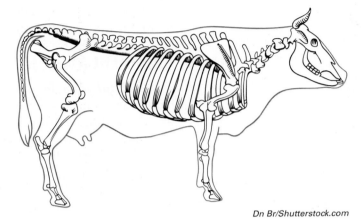

Dn Br/Shutterstock.com

Figure 9.6.1. The skeleton of a mammal includes bones, cartilage, joints, tendons, and ligaments. *Can you locate examples of those parts on your own body?*

miha de/Shutterstock.com

Figure 9.6.2. A chicken's skeleton looks very different than the cattle skeleton shown previously, but there are functional differences as well. *What are some of the differences between avian and mammalian skeletons?*

Did You Know?

Cows have 207 bones in their bodies and horses have 205. Chickens have 120 bones, even though they are about one-tenth the size of a horse or a cow.

Muscular System

The ***muscular system*** controls a body's movement. Muscles are body tissue composed of cells or fibers that contract to produce movement. The animal body contains three types of muscles: skeletal, smooth, and cardiac (**Figure 9.6.3**). These muscle types are differentiated based on the type of work they perform, their structure, and location in the body. All three muscle types are able to contract. Muscle contractions are *voluntary* or *involuntary*. Voluntary muscle contractions are controlled by thought and are used for movement. Involuntary muscle contractions are automatic and happen naturally, like heartbeat, breathing, and other organ functions.

Muscles work with the skeletal system to provide form, support, and bodily movement. The chemical process of muscle contraction also produces heat to help maintain body temperature. Muscles even aid in digestion and breathing.

Skeletal Muscle

Skeletal muscles attach to bones or skin to provide movement. They are voluntary muscles, controlled by thought, and make up much of the flesh of livestock that we think of as meat. Skeletal muscles are long (nearly 12 inches in some locations) and cylindrical in appearance.

Skeletal muscles are attached to bone by tendons, which are made up of strong, fibrous collagen tissue. The muscle is attached to two bones on either side of a joint. The contraction of the skeletal muscle causes the bones to move along the joint. Energy for muscle activity at the cellular level comes from the food that animals consume.

Smooth Muscle

Smooth muscles occur in the walls of hollow organs like the stomach or bladder and in blood vessels and the respiratory tract. Smooth muscles operate involuntarily. They have a shape that is tapered at both ends.

Cardiac Muscle

Cardiac muscle tissue is found only in the heart. It is an involuntary muscle which contracts to pump blood and maintain blood pressure throughout the body. It is branched and rectangular in shape. Cardiac muscle contraction is regulated by the nervous system, which can speed up or slow down the process as needed by the body.

Types of Muscle Cells

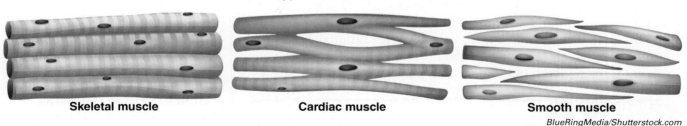

Skeletal muscle   Cardiac muscle   Smooth muscle

BlueRingMedia/Shutterstock.com

Figure 9.6.3. Muscle cells come in three types: skeletal, cardiac, and smooth. *What different functions do these cells and structures serve?*

Pig Heart Valve Replacement

For more than 30 years, pig heart valves have been successfully transplanted in humans. A pig's heart is similar in size, weight, and structure to a human heart. Contrary to what some believe, pigs are not grown specifically for the harvesting of their hearts. The pigs that are used for medical purposes are grown for human consumption.

Under sterile conditions, the heart is removed from the pig's body. The excess tissue and myocardium are then removed from the heart, leaving the valves. The valves are "sized," so they fit appropriately when implanted into a human. Heart valves come in different sizes, like different-size feet.

The pig valve is typically mounted to a stent (frame) that can be reinforced with cloth and sutures. After mounting, the valve is checked under a microscope and cultures are taken. To preserve the tissue, the valve is often placed in chemicals to reduce possible failure or rejection by the recipient. Porcine valves last an average of ten years, but some have lasted up to 17 years! Typically, pig valve replacements wear out faster in younger, more active individuals.

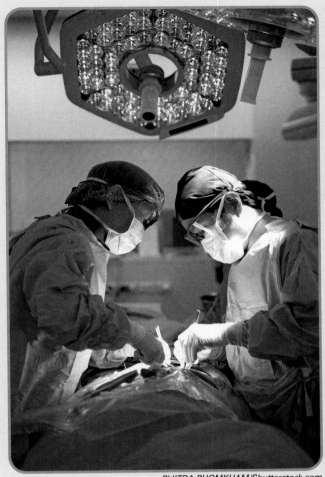

PIJITRA PHOMKHAM/Shutterstock.com

9

Digestive System

A basic understanding of the major livestock digestive systems is helpful in feeding livestock for proper production, growth, and development. **Digestion** is the process of breaking down feed into simple substances that can be absorbed by the body for energy and nutrient value.

The main livestock digestive system types are *monogastric*, *ruminant*, and *avian*. These digestive system types differ in various ways, but the main difference is the design of the stomach of each system and how feed is broken down in that organ. These differences between digestive systems are why it is necessary for a producer to have a basic understanding of the digestive process of their livestock.

Part	Function
Mouth	Mechanical action of chewing, adds saliva which contains enzymes
Esophagus	Tube connecting the mouth to the stomach
Stomach	Provides some limited mechanical action, adds hydrochloric acid and other enzymes, ruminant stomach has four parts
Small Intestine	Most absorption of nutrients takes place here
Large Intestine	Undigested material from the small intestine ends up here where mucus is added to allow easier passage from the body
Rectum/Anus	Final passage of undigested body waste leaves through here

Goodheart-Willcox Publisher

Figure 9.6.4. The parts of a simple digestive system should sound familiar to you, as you have all of them! *Do you think your digestive system differs from an animal's in any way?*

Several parts make up the digestive system. Starting at the mouth, food enters the animal's digestive system and travels through the esophagus to the stomach. From the stomach, food enters the small intestine and then the large intestine. The large intestine ends with the rectum, where undigested material is expelled from the body through the anus. The function of these parts is explained in **Figure 9.6.4**.

Monogastric

Monogastric livestock have simple stomachs with a single compartment for digestion. When food enters the stomach of a monogastric animal, gastric juices from the stomach wall containing acid and enzymes start to chemically break down food. The muscles in the stomach wall churn and squeeze the food, pushing liquids into the small intestine and allowing gastric juices to continue breaking down the solids that remain. This process continues until the partly digested food leaves the stomach and begins again as more food enters from the mouth.

People have monogastric digestive systems, if all of this sounds familiar. Swine and rabbits also have monogastric systems. Horses also have a single stomach, but have the special ability to digest larger amounts of roughage than typical monogastric animals because they have an organ called the *cecum*. The cecum in a horse's digestive system is found at the beginning of the large intestine, where it joins the small intestine (**Figure 9.6.5**). The cecum uses bacteria to digest roughage more effectively.

Figure 9.6.5. The horse's digestive system is unique, as it contains an extra organ called the cecum. *What difference does the cecum make?*

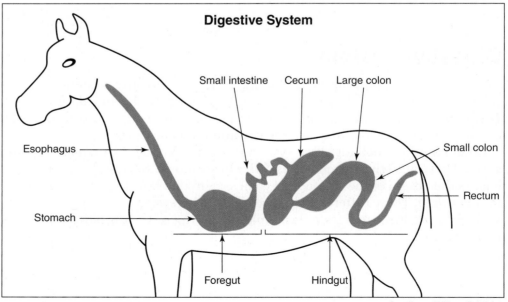

Goodheart-Willcox Publisher

Hands-On Activity

Digestion Process

Digestion is the process of breaking down food in the digestive system of the body. In this activity, we will make a simple stomach model and break down a common food item using processes like those in the digestive system of animals.

For this activity, you will need the following:

- Resealable plastic bags (thicker freezer bags work better)—one per student or group
- Sliced bread (inexpensive sandwich bread works well)—one slice per student or group
- Acidic beverage like cola—enough for each student or group to have 3–4 ounces
- Paper towels and trash bag—for cleanup (potential for a mess is above average)
- Clock, stopwatch, or other time-keeping means

Here are your directions:

1. As a group, make sure you have plastic bags, bread, cola, and paper towels.
2. Place one slice of bread into the plastic bag along with the premeasured amount of cola and seal it.
3. Observe and record what begins to happen to the bread when the cola is added.
4. On your teacher's signal, (very gently) squeeze and massage the contents of the bag for about two minutes. You need to be careful not to spill the contents!
5. Observe and record what has happened to the bread after the mechanical "digestion."

Consider This

1. What does the bag in this activity represent? What about the cola? The bread? The movement from your hands?
2. Discuss the similarities and differences between the activity and an actual digestion process.

Ruminant

Ruminant animals are even-toed mammals that digest food using a multipart stomach. Cattle, sheep, and goats are all ruminant livestock. The ruminant stomach is comprised of four parts: the *rumen, reticulum, omasum,* and *abomasum* (**Figure 9.6.6**). Ruminant livestock can consume different kinds of food (roughage) far more effectively than nonruminant animals because of the structure of their stomach. A list of ruminant and nonruminant livestock is found in **Figure 9.6.7**.

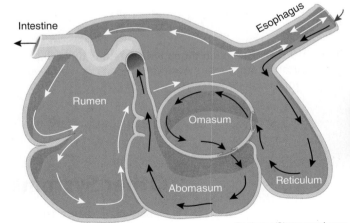

Designua/Shutterstock.com

Figure 9.6.6. A ruminant stomach has four parts. *How do these additional parts help these animals to digest roughage?*

Ruminant Livestock	Monogastric Livestock
Cattle	Swine
Sheep	Horses*
Goats	

*modified

Figure 9.6.7. The major livestock species are classified as ruminant and nonruminant. *Where would other species of animals fit?*

Ruminants do not chew much of their feed before they swallow it. They consume large amounts of feed rapidly, with the solid part going into the rumen and any especially dense or metallic parts going into the reticulum. When the rumen is full, the animal lies down and **regurgitates**, or forces the consumed feed back into their mouth. This is where rumination occurs, or "chewing the cud." Cattle are known to chew their cud for about eight hours per day. Muscles in the rumen and reticulum (along with microorganisms like bacteria and protozoa) break the food into smaller particles. These bacteria change the low-quality protein of the roughage into the compounds that the animal needs.

The remaining 15 percent of the ruminant stomach is made up of the omasum and the abomasum. The omasum mechanically churns food with its strong muscles and with fermentation. The abomasum, also known as the true stomach, carries on digestion the same fashion as in the non-ruminant stomach.

Avian

The avian digestive system (**Figure 9.6.8**) contains organs not found in other species of livestock. Feed consumed by chickens and other poultry travels to the *crop*. Feed is stored in the crop where it is softened by saliva and other secretions from the crop wall. The softened feed moves from the crop into the proventriculus. The *proventriculus* is like the true stomach in ruminants and the monogastric stomach in non-ruminants in that it uses hydrochloric acid and enzymes to aid in digestion. Food then travels into the muscular stomach, called the *gizzard*, where it grinds the softened feed. The gizzard is unique to the avian digestive system. Poultry often consume small pebbles or stones that are passed into and used by the gizzard to aid in the grinding process, as birds have no teeth. The small intestine receives ground material from the gizzard and is where most nutrient absorption occurs. The large intestine is where the last of the water reabsorption occurs before passing into the cloaca. The *cloaca* is where digestive wastes mix with wastes from the urinary system and pass out of the vent. Unlike other animals, poultry do not have separate urinary and fecal waste exits.

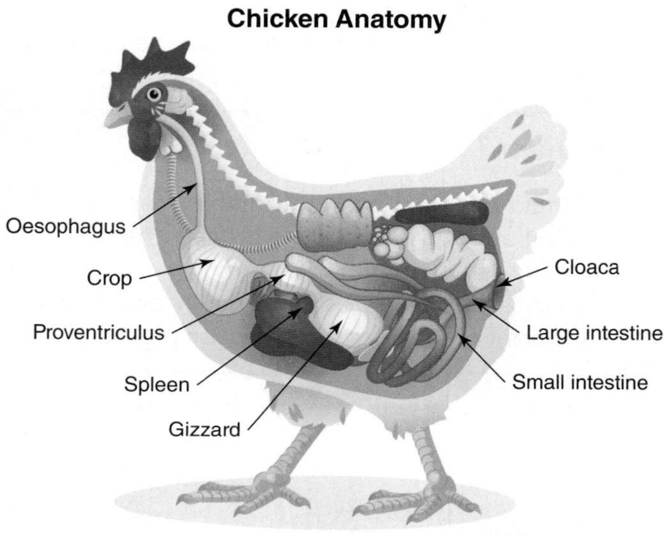

Chicken Anatomy

Oesophagus
Crop
Proventriculus
Spleen
Gizzard
Cloaca
Large intestine
Small intestine

VectorMine/Shutterstock.com

Figure 9.6.8. The avian digestive system is different from mammal stomachs. *What differences do you see? What similarities do you see?*

Other Systems

Several other important body systems exist in livestock animals. A brief description of respiratory, circulatory, and nervous systems below will help with understanding the basic functions of each.

Respiratory

Like people, animals need oxygen for their bodies to complete many basic processes. Breathing is the process of taking air into and expelling air from the lungs. The **respiratory system** manages oxygen intake and absorption. Muscles enlarge the chest cavity, forcing air in, and muscles reduce the chest cavity, forcing air out. Air enters the body through the animal's mouth or nostrils passing through to the pharynx, then the trachea, and finally into the lungs.

In addition to taking in oxygen and expelling carbon dioxide, the respiratory system's constant air intake also helps regulate the animal's body temperature.

Air coming into the lungs has higher oxygen content than air leaving the lungs. Oxygen is absorbed by the lungs when an animal breathes in. The oxygen passes into the air sacs in the lungs (*alveoli*) which contain very thin blood vessels called capillaries. The *capillaries* carry oxygenated blood throughout the body and return it with an excess of carbon dioxide. The exchange of oxygen and carbon dioxide in the lungs is a process called *diffusion* where gases go from an area of higher concentration to an area of lower concentration. Carbon dioxide is flushed out of the lungs when the animal exhales.

Not all animal respiratory systems are the same (**Figure 9.6.9**). The avian respiratory system differs from the mammalian respiratory system in several ways.

Mammals	Birds
Nostrils	Nares
Larynx – produces vocal sounds and found at the upper end of the trachea	Syrinx – produces sounds and found at the lower end of the trachea
Alveoli – where the exchange of oxygen/carbons dioxide occurs	Do not have alveoli – they have small air capillaries in the lung tissue
Diaphragm – contracts to breathe in and exhale air	No diaphragm – breathing is accomplished by abdominal contraction and wing movement

Goodheart-Willcox Publisher

Figure 9.6.9. Mammalian and avian respiratory systems are similar, but not quite the same. *What differences do you see?*

Circulatory

The ***circulatory system*** moves blood (and nutrients) throughout the body. The major parts of the circulatory system include the heart and blood vessels (**Figure 9.6.10**). The heart is the muscular organ that pumps blood throughout the body. Blood vessels include *arteries*, *veins*, and *capillaries*. Blood leaves the heart through the *aorta*, which is the largest artery in the body. The aorta branches into smaller arteries, which branch into capillaries to reach tissues in all parts of the body. From the capillaries, deoxygenated blood enters the veins and returns to the heart.

Capillaries are the smallest and most numerous blood vessels and they form the connection between the vessels that carry blood away from the heart (arteries) and the vessels that return blood to the heart (veins). The primary capillary function is the exchange of materials between the blood and tissue cells. Nutrients and oxygen go into the cells while water, carbon dioxide, and waste products enter the blood from the cells.

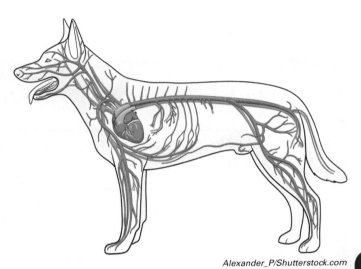

Alexander_P/Shutterstock.com

Figure 9.6.10. The circulatory system of a dog is similar to other animals and consists of the heart and blood vessels.

Blood is the liquid material that flows throughout the circulatory system providing the basic functions listed below:

- Transports nutrients from the digestive system to body cells and tissues
- Transports oxygen from lungs to cells (primary function of red blood cells)
- Transports carbon dioxide from cells to lungs
- Transports hormones as needed in the body
- Transports waste products from cells to be eliminated from the body
- Regulates body temperature by absorbing and diluting heat
- Clots to reduce blood loss at injury sites
- Fights diseases (primary function of white blood cells)

Nervous

The **nervous system** is responsible for transferring information throughout the body (**Figure 9.6.11**). It receives information about the environment and generates responses to that information. The nervous system can be divided into two main parts: the *central nervous system* and the *peripheral nervous system*. The central nervous system is composed of the brain and the spinal cord. The peripheral nervous system is made up of the nerves that reach throughout the body.

SciePro/Shutterstock.com

Figure 9.6.11. The nervous system carries information around the body.

Did You Know? Sponges are the only multicellular animals without a nervous system.

Career Corner — ANIMAL PHYSICAL THERAPIST

An **animal physical therapist** (*also called a rehabilitation veterinarian*) *rehabilitates animals that have had surgery or an amputation, been injured, or suffer from chronic pain.*

What Do They Do? Animal physical therapists observe animal patients and perform exams; determine proper rehabilitation or recovery plans; communicate with clients and veterinarians to agree on a rehabilitation plan and determine medical treatment; utilize laser therapy, ultrasound, shockwave, and radiography equipment; assist animal patients in strengthening and exercise routines; perform stretches and massages on animal patients; and organize check-ups with clients.

What Education Do I Need? Most animal physical therapists are licensed as human physical therapists and treat animals; a minimum of a master's degree in physical therapy is required for this path.

msgrafixx/Shutterstock.com

How Much Money Can I Make? Rehabilitation veterinarians make an average of nearly $100,000 per year.

How Can I Get Started? You can build an SAE conducting research on different animal physical therapy options and report your findings to your class. Maybe you can conduct an experiment on the benefits of animal physical therapy. You can certainly try to interview an animal physical therapist or job shadow them for the day.

What Did We Learn?

- Producers need to understand how the bodies of livestock work to make informed production decisions and provide good care.
- The skeletal system provides structure, protection, and support for the body. It includes bones, cartilage, tendons, ligaments, and joints.
- The muscular system controls movement. Skeletal muscles are attached to bones and skin and are responsible for movement. These muscles are consciously controlled. Smooth muscles make up organs, and cardiac muscles are found in the heart. These muscles are involuntary.
- Digestion is the process of breaking down feed into substances that can be absorbed into the body for energy and nutrient value. Mammal stomachs can be classified as monogastric, or simple, stomachs, and ruminant, or multiple-compartment, stomachs. Birds have a completely different digestive system, containing several additional specialized organs.
- Respiratory (breathing), circulatory (blood flow), and nervous (information processing) systems are also important to animal health. These tend to be more similar across species.

Let's Check and See What We Know

Answer the following questions using the information provided in Lesson 9.6.

1. Which of the following animal systems is responsible for structure, support, and protection?
   A. Circulatory system
   B. Skeletal system
   C. Digestive system
   D. Nervous system

2. _____ is a tough and flexible connective tissue that provides support between the bones at joints of the skeletal system.
   A. Cartilage   C. Blood
   B. Marrow   D. Calcium

3. Which of the following is *not* a type of muscle found in the animal body?
   A. Skeletal   C. Cardiac
   B. Smooth   D. Respiratory

4. Which of the following digestive system types has a stomach with four compartments?
   A. Monogastric   C. Ruminant
   B. Avian   D. Nonruminant

5. Which stomach compartment of cattle is more like the "true stomach" of a pig?
   A. Rumen   C. Omasum
   B. Reticulum   D. Abomasum

6. Which of these digestive system parts is unique to poultry?
   A. Esophagus   C. Small intestine
   B. Crop   D. Cecum

7. *True or False?* There are no muscles in the stomach of an animal.

8. Horses can consume large amounts of roughage in their diet because of which digestive part?
   A. Cecum
   B. Stomach
   C. Small intestine
   D. Crop

9. *True or False?* The respiratory system is responsible for moving blood through the body of an animal.

10. The brain of an animal is a part of which of the following systems?
   A. Circulatory   C. Skeletal
   B. Digestive   D. Nervous

Let's Apply What We Know

1. Explain the different types of muscles in the muscular systems and how they affect livestock.
2. Explain how a mammal's respiratory system is different from a bird's.
3. Describe a ruminant digestive system, and how it is different from yours.

Academic Activities

1. **Science.** Pick a livestock animal. Draw and compare the internal systems of that animal to yours. How do those differences benefit each species?

Communicating about Agriculture

1. **Visual Communication.** Pick one of the systems and a livestock animal covered in this chapter and create an informational diagram on a poster. Do some additional research, if necessary. Your poster should provide information about what the system does, what the parts are, and why it is important for livestock producers to understand how it works. Present your poster to your class, and gather feedback for how it might be improved.

SAE for ALL Opportunities

1. **Placement SAE.** Line up an internship at a local veterinarian's office. Talk with the staff and watch them with animals to learn about why it is important to understand animal systems. What animals do they see? How many different types of systems do they need to understand? Present this information to your class.

SAE for ALL Check-In

- How much time have you spent on your SAE this week?
- Have you logged your SAE hours?
- What challenges are you having with your SAE?
- How can your instructor help you?
- Do you have the equipment you need?

Livestock Feeding and Nutrition

Oaisu/Shutterstock.com

Learning Outcomes

By the end of this lesson, you should be able to:
- Name the major nutrients needed by livestock and describe why they are necessary to healthy animal development and growth. (LO 09.07-a)
- Explain the difference between roughages and concentrates. (LO 09.07-b)
- Describe the purpose of feed additives and how they are regulated. (LO 09.07-c)
- Identify the components that should be present on a feed label. (LO 09.07-d)
- Explain the characteristics of a balanced livestock ration. (LO 09.07-e)

Words to Know

amino acids	feed additives	palatable
balanced ration	fiber	proteins
carbohydrates	generally recognized as safe (GRAS)	ration
Center for Veterinary Medicine (CVM)		roughages
	legumes	simple carbohydrates
complex carbohydrates	lipids	vitamin deficiency
concentrates	nutrients	

Before You Begin

When I was a kid, I loved to eat a cheeseburger whenever I was hungry. Once, I bought more cheeseburgers than I could eat, and I decided to offer one to one of my family's cows. To my surprise, she ate the entire cheeseburger in one bite, swallowing it whole! I called her "Cannibal" after that.

While feeding a cow one cheeseburger, one time, might not cause any problems, it is not the proper food for cows. The specific dietary needs of cattle cannot be met with human food, just like you would not be very satisfied or healthy if you ate hay. In this lesson, we will look at animal nutrient requirements and learn how to feed them to meet their needs, no cheeseburgers necessary!

Livestock Nutrition and Health

Livestock nutrition has improved after decades of research, and it continues to be an area of study as we seek to feed animals more efficiently. As the world population grows, the need for efficient production becomes more evident. In modern agricultural production, the nutritional requirements of livestock are well known and may be supplied by natural forage and fodder alone, or supplemented with concentrated nutrients in manufactured feeds (**Figure 9.7.1**).

Nutrients are substances that provide nourishment for growth and the maintenance of life. Nutrients are not made in the body and must be provided in the diet or feed of livestock. Nutrients may be *organic* or *inorganic*. Organic nutrients contain carbon, while inorganic nutrients do not. Each species must have the right kinds of nutrients in the proper amounts and balance, and species' needs differ. Not enough or too much of a nutrient may cause health issues. Producers who follow the research-based recommendations of animal nutritionists will have healthier livestock. The six major nutrients that all animals need are carbohydrates, fats and oils, proteins, vitamins, minerals, and water.

Evgeniy Kazantsev/Shutterstock.com

Figure 9.7.1. These manufactured feeds contain additional nutrients for livestock. *What do you think some of these nutrients might be?*

Water

Let's start with the most critical nutrient for life: water. Water makes up 60 to 70 percent of the body of most livestock (**Figure 9.7.2**). Fresh, clean, quality water is critical to healthy livestock; without water, livestock would perish within days. Livestock need a quality water supply to help them absorb and use the other major nutrients from feed. Water transports nutrients to cells and waste from cells. Water also assists in digestion and regulates body temperature.

Did You Know?

You should never give very cold water to a sweaty, overheated horse. It can shock their system, making them ill or contribute to a colic episode, which is the leading cause of death in horses. Yikes!

Figure 9.7.2. These cows are drinking what looks like some nice, clean water. *Why is drinking water so important to cows? Are there lessons there for people, too?*

FloKai/Shutterstock.com

Carbohydrates

Carbohydrates are the main energy nutrient found in livestock rations. They are made up of sugars, starches, cellulose, and lignin. The basic makeup of a carbohydrate includes the elements carbon, hydrogen, and oxygen. Sugars and starches are called **simple carbohydrates** because they are easily digested. Cellulose and lignin are called **complex carbohydrates** or **fiber** and they are more difficult to digest than simple carbohydrates. The energy produced by carbohydrates fuels muscular movement and produces heat to maintain correct body temperature. Extra carbohydrates are stored as fat.

Sources of simple carbohydrates include corn, wheat, oats, and other cereal grains (**Figure 9.7.3**). Fiber comes from pasture grasses and hay such as alfalfa, Bermuda grass, and fescue. Ruminant animals like cows, sheep, and goats can digest larger amounts of fiber than other livestock, which means that their feed source can contain more of this material.

Figure 9.7.3. Grains like these contain many nutrients, but the main nutrient they contain is simple carbohydrates. What livestock might you feed these types of grains to?

Did You Know? The largest market for US corn is animal feed, as it provides a good source of energy. Nearly 40 percent of the corn grown in 2021 was used as animal feed.

Fats and Oils

Fats and oils are also called **lipids**. They provide energy to produce body heat like carbohydrates, and are also made up of carbon, hydrogen, and oxygen. Fats contain more carbon and hydrogen than carbohydrates, giving them an energy value that is 2.25 times higher. Fats are solid at body temperature, while oils are liquid at body temperature. Fats come from both animal and vegetable sources.

Proteins

Proteins are compounds made up of amino acids. **Amino acids** are made up of carbon, hydrogen, oxygen, and nitrogen. Proteins serve as the structural components of body tissues like muscle, hair, and ligaments. If livestock consume too much protein, the nitrogen is separated and passed in urine. The remaining material is used to produce energy or is stored as body fat.

Protein sources can come from plants or animals. Animal sources of protein come from bone meal, fish meal, and other sources. Animal protein sources are considered high quality because they contain high levels of the amino acids needed by livestock. Plant sources of protein come from legumes. Legumes are plants from the pea family which contain higher levels of protein than other plants (we'll talk about them more in a moment). Soybeans, a popular legume, are a great source of protein!

Amino acids for livestock are categorized as *essential* or *non-essential*. Essential amino acids must be sourced in the food they eat. Non-essential amino acids can be produced from other (essential) amino acids, which means they do not have to be provided in the feed ration.

Vitamins

Vitamins are organic compounds that livestock need in very small amounts. All vitamins contain carbon, but beyond that are chemically quite different. They are divided into two groups: *fat-soluble* and *water-soluble* (**Figure 9.7.4**). Fat-soluble vitamins can be dissolved in and carried by fat. Water-soluble vitamins can be dissolved in and carried by water. Vitamins are important for good health, disease resistance, bone development, muscle development, and growth (**Figure 9.7.5**). They are also important for reproductive health. Commercial feeds often include necessary vitamins. A **vitamin deficiency**, or lack of vitamins, leads to many growth and disease problems that could be easily prevented.

Vitamins for Livestock

Fat-Soluble	Purpose
A	Healthy eyes, conception, disease resistance
D	Bone development, mineral balance in the blood
E	Reproduction, muscle development, strengthen immune system
K	Clotting of blood

Water-Soluble	Purpose
C	Teeth and bone formation, prevent infection
B_1 (thiamine)	Increase growth, improve appetite and reproductive function
Riboflavin	Increase growth, improve appetite and reproductive function
Niacin	Increase growth, improve appetite and reproductive function
Pyridoxine	Increase growth, improve appetite and reproductive function
Pantothenic Acid	Increase growth, improve appetite and reproductive function
Biotin	Pain relief, increase growth, improve appetite, sleep, reproductive function and energy
Folic Acid	Increase growth, improve appetite and reproductive function
Benzoic Acid	Increase growth, improve appetite and reproductive function
Choline	Detoxification, increase growth, improve appetite and reproductive function
B_{12}	Increase growth, improve appetite and reproductive function

Goodheart-Willcox Publisher

Figure 9.7.4. These are the vitamins necessary for raising healthy livestock. Vitamins are necessary for human nutrition and well being. *How do you benefit from these vitamins?*

Figure 9.7.5. These sheep are stripping bark from tree branches. *Why might they do this?*

Photosebia/Shutterstock.com

Minerals

Minerals are inorganic materials that livestock need in small amounts. They contain no carbon. They provide material for bone and tissue growth. Minerals also aid muscular activities, reproduction, digestion, tissue repair and formation, and energy release for body heat. A mineral deficiency may cause basic body functions to falter or lead to various diseases.

There are two categories of minerals: *macrominerals* and *microminerals*. Macrominerals are needed in larger amounts than microminerals (**Figure 9.7.6**). There are geographic locations known for mineral deficiencies in livestock forages which must be supplemented in the feed to keep livestock healthy. Other locations may have excess minerals in plants, where consuming them could be toxic to livestock. Producers must be aware of the situation in their area to prevent problems associated with excess minerals or deficiencies.

Minerals

Macrominerals (needed in larger amounts)	Microminerals (smaller amounts needed)
Sodium	Iron
Chloride	Iodine
Calcium	Copper
Phosphorus	Cobalt
Magnesium	Zinc
Potassium	Manganese
Sulfur	Boron
	Molybdenum
	Fluorine
	Selenium

Goodheart-Willcox Publisher

Figure 9.7.6. These minerals are needed to raise healthy livestock. *Are you familiar with any of these? Why?*

Feed Classifications

Livestock feeds can be classified as *roughages* or *concentrates*. This classification is based on the amount of fiber in the feed. **Roughages** contain more than 18 percent crude fiber and are high in carbohydrates, while **concentrates** contain less than 18 percent crude fiber. Livestock need a proper balance of each type of feed based on their digestive system, age, and end purpose. Ruminants (and horses) can consume large amounts of roughages, while nonruminants cannot. Animals like horses or oxen that are used for work and need lots of energy are typically fed higher amounts of concentrates to meet those demands.

Roughages

Roughages are grouped into two main categories: legume and non-legume (**Figure 9.7.7**). **Legumes** are a type of plant with nodules (small swellings) on the roots, containing bacteria. These bacteria obtain nitrogen for the plant. This higher concentration of nitrogen results in greater protein levels. Non-legume plants do not have the ability to absorb nitrogen and are lower in protein content.

Stockonya/Shutterstock.com

Figure 9.7.7. Corn silage contains different levels of nutrients than the corn grain. *Why?*

RGtimeline/Shutterstock.com

Figure 9.7.8. Corn is an energy feed. *Why might you choose to feed your livestock corn, instead of corn silage?*

Concentrates

The two types of concentrates are *protein supplements* and *energy feeds* (**Figure 9.7.8**). Protein supplements are livestock feeds containing more than 20 percent protein. Livestock feeds with less than 20 percent protein are called energy feeds. Most grains are energy feeds, with corn being the most widely used energy feed for livestock.

Feed Additives

Manufactured feeds allow the addition of additives to improve health and better balance the animal diet. **Feed additives** improve feed efficiency, promote faster growth, and treat diseases and parasites. These materials are not usually considered nutrients and are used in small amounts. Additives are added to basic feed mix, and require careful attention be paid to the amount and the intended animal. For example, market animals cannot be fed additives that would leave residue present in their carcasses.

A safe food supply helps keep animals and people healthy. In the United States, the **Center for Veterinary Medicine (CVM)** approves safe food additives and manages the Food and Drug Administration's medicated feed programs. They also monitor for animal feed contaminants. Food product use is governed by the Federal Food, Drug, and Cosmetic Act. This law requires that any substance added to or expected to become a component of animal food must be used in accordance with food additive regulations unless it is **generally recognized as safe (GRAS)** for that intended use. Forages, grains, most minerals, and most vitamins fall under the GRAS category. All other substances are considered additives and when added to feeds must be regulated.

GW Feed Version 1
To be fed to Beef Calves on Pasture

GUARANTEED ANALYSIS

Crude Protein	min	14.00%	Salt	min	0.30%
Crude Fat	min	3.50%	Salt	max	0.80%
Crude Fiber	max	14.00%	Potassium	min	0.80%
Calcium	min	1.00%	Copper	min	30 ppm
Calcium	max	1.50%	Selenium	min	0.45 ppm
Phosphorus	min	0.70%	Vitamin A	min	5.000 IU/lb

List of Ingredients

Processed Grain By-Products, Roughage Products, Grain Products, Plant Protein Products, Calcium Carbonate, Forage Products, Molasses Products, Sodium Chloride, Lignin Sulfonate, Monocalcium Phosphate, Potassium Chloride, Hemicellulose Extract, Hydrated sodium calcium aluminosilicate, Magnesium-Mica, Manganous Oxide, Diatomaceous Earth (flow agent), Copper Sulfate, Zinc Sulfate, Zinc Oxide, Manganese Sulfate, Sodium Selenite, Cobalt Carbonate, Lactic Acid, Origanum Oil, Thyme White Oil, Cinnamaldehyde, Vitamin E Supplement, Minderal Oil, Vitamin A Suppliment, Vitamin D3 Suppliment, Fenugreek Flavor Extract, Ethylenediamine Dihydriodide, Dried Seaweed Meal (Fucaceae, Bangiaceae, Ulvaceae), Chicory Root, Red pepper, Cloves, Anise Oil, Saccharin Sodium, Natural Flavor.

Feeding Directions

To be fed to beef calves as a self-feed on pasture. An adequate supply of good quality water should be available at all times.

CAUTION: Follow label directions. The addition to feed of higher levels of this product containing selenium is not permitted.

WARNING: This product, which contains added copper, should not be fed to sheep or related species that have a low tolerance to copper.

Manufactured by:
GWP, Inc.
PO Box 123, Tinley Park, IL 18451
1800-323-0440.

NET WEIGHT: 50 lb. (22.7 kg) OR BULK

Goodheart-Willcox Publisher

Figure 9.7.9. Food labels are required for use at the state and federal levels. *Can you identify all of the requirements listed above, using this label as a sample?*

Feed Labels

Feed labels (**Figure 9.7.9**) contain information describing the product and instructions for safe and effective use. Animal food labels must include the following information:
- Brand name (if any)
- Product name
- Purpose statement
- Guaranteed analysis
- List of ingredients
- Directions for use
- Warning or caution statements
- Name and address of manufacturer
- Quantity statement

Create Your Own Feed Bag

Use the information in the Feed Label section to create your own feed bag. Your feed bag can be created electronically or drawn out on paper. Research other feed companies and labels to help you gain ideas as well as to collect information.

Your feed bag should include:

- Name of your feed
- What animal it is for – make sure to be specific (adult, breeding, etc.)
- At least one image
- Purpose statement (what is your feed intended for or how will it help the animal)
- List of ingredients
- Directions for use
- Warning or caution statements
- Name and address of manufacturer
- Quantity statement

Pixel-Shot/Shutterstock.com

There are federal labeling requirements, and individual states may have additional requirements. Individual state labeling requirement information may be obtained from state government agencies.

Did You Know? The three largest animal feed manufacturers in the US in 2020 were Cargill, Land-O-Lakes, and Tyson Foods.

Rations

Livestock must receive the proper balance and amounts of nutrients to produce eggs, meat, milk, wool, or to complete work efficiently. A *ration* is the amount of feed given to livestock over a twenty-four-hour period to meet its needs. A *balanced ration* has all the nutrients needed by the animal in the correct proportions and amounts. Cost is an important factor to consider in livestock rations. Most of the cost of raising livestock is providing feed. To feed livestock profitably, it is important to develop rations that are cost-effective.

Rations must be *palatable*, or taste good to the animal, or they will not be eaten. Feed must be consumed to provide any value. The species of animal and the type of digestive system must be considered when determining the ratio of concentrates and roughages. The purpose for which the animal is being fed is another important consideration. The functions of a proper ration are maintenance, growth, fattening, production, reproduction, and work.

Balancing Rations

The livestock ration must meet the nutritional needs of the animal. The following steps explain the basics of balancing livestock rations:

1. The kind of animal, its age, weight, and purpose must be identified when determining the content of a ration.
2. Determine the nutrient requirements of the animal.
3. Select the feeds to be used in the ration and determine their nutrient content.
4. Calculate the amounts of each feed to use in the ration.
5. Double check the ration formulation against the needs of the animal. Adjust if necessary.

Hands-On Activity

Yummy Diet Lab

Now that we understand what animals need to stay healthy and reproduce, let's try our hand at creating a livestock feed. Balancing a ration is not always an easy task, but it will get easier with practice.

For this activity, you will need the following:

- Strips of paper labeled *Cat, Dog, Dairy Cow, Beef Cow, Pig, Rabbit, Horse*
- 2 paper plates
- Marker or pen
- 1 scoop of each:
  - Cheerios = Fiber
  - Raisins = Protein
  - Candy Corn = Carbs
  - Mini-Marshmallows = Fat
  - Chocolate Chips = Minerals and Vitamins

Here are your instructions:

1. Your teacher will walk around the room, letting students draw a slip of paper. This will be your animal!

2. On the bottom of one of the plates, write your animal name.
3. On the other paper plate, dump out your "feed options."
4. Over two minutes, take feed options from one plate and place them on the other to create a proper diet for your assigned animal.
5. Share with the class what you thought would work, and why.
6. When you are done with one animal, pass your animal plate to a neighbor and try again with a new animal.

Consider This

1. What feed source do you think the animals need most in their feed ration? Why?
2. What would a feed ration look like if you were creating it for a human? How would a teenager's ideal diet look different than someone's in their 50s?
3. Go online and find a real diet ration for your assigned animal. How close were you to being right? Did you miss anything?

A **sales representative** for feed and animal healthcare products sells and services products produced by their company for the animal industry.

What Do They Do? Animal feed sales representatives share technical information with veterinarians; analyze, develop, and implement sales goals and plans; maintain and grow the market share within a specific sales territory; attend regional, annual, and vendor meetings; work with supplier representatives to install programs or products; and attend training programs and education workshops to ensure customer satisfaction.

What Education Do I Need? A sales representative needs an associate's degree in animal science, agricultural business, or related major.

How Much Money Can I Make? Feed and animal health sales representatives make an average of $65,000 per year.

How Can I Get Started? You can build an SAE conducting research on animal feeds and report your findings to your class. You could also conduct an experiment on the benefits of animal feed additives, research tips for being successful in the sales industry, interview an agricultural sales representative, or job shadow them for the day.

Pressmaster/Shutterstock.com

9

What Did We Learn?

- Livestock nutrition is critically important to raising healthy and productive livestock.
- As world populations grow and the demand for agricultural products increases, it becomes more important to feed animals efficiently.
- Nutrients provide nourishment for growth and natural body processes. They are not made in the body, so they must come from feed or supplements. The six main nutrients are carbohydrates, fats and oils, proteins, vitamins, minerals, and water.
- Water is the most critical nutrient to livestock health.
- Carbohydrates are the main nutrient found in livestock feed. Simple carbohydrates are found in feeds like corn, wheat, oats, and grain, and are easy to break down. Nonruminant animals will have a diet high in simple carbohydrates. Complex carbohydrates, also called fiber, are found in roughage, and are tougher for the stomach to break down. Ruminant animals will have a diet higher in complex carbohydrates.
- Fats and oils help the body to produce energy.
- Proteins are made up of amino acids and can be made from animal or plant sources. Animal protein contains more essential amino acids.
- Vitamins and minerals are needed in small doses in livestock feed to keep animals healthy and prevent disease.
- Livestock feeds are grouped into two categories: roughages and concentrates.
- Feed additives can be added to manufactured feed to improve quality of the final product or meet industry demands. The Federal Food, Drug, and Cosmetic Act regulates the use of food additives in the industry.
- Food labels are required to provide information about what goes into animal feed.
- A ration is the portion of food required to feed an animal, meeting all of its nutritional needs and requirements, for a 24-hour period.

Let's Check and See What We Know

Answer the following questions using the information provided in Lesson 9.7.

1. Organic nutrients contain _____, while inorganic nutrients do not.
   A. carbon
   B. hydrogen
   C. oxygen
   D. sulfur

2. Which of the following items would *not* be considered one of the six major nutrients?
   A. Carbohydrates
   B. Fats
   C. Protein
   D. Silage

3. *True or False?* Fats have an energy value that is 2.25 times higher than that of carbohydrates.

4. What happens when the livestock diet contains more carbohydrates than the animal needs?
   A. The extra carbohydrates are passed in the urine.
   B. The extra carbohydrates are stored as fat.
   C. Extra energy is given to the animal.
   D. The animal may experience weight loss.

5. Which of these animals can digest large amounts of fiber?
   A. Swine
   B. Chickens
   C. Turkeys
   D. Cattle

6. ____ are also known as the building blocks of protein.
   - A. Carbohydrates
   - B. Amino acids
   - C. Fats
   - D. Minerals

7. A ____ is a lack of proper vitamins, which can lead to growth and disease problems that could otherwise be easily prevented.
   - A. nutrient
   - B. protein
   - C. deficiency
   - D. ration

8. ____ is the most important nutrient of all.
   - A. Water
   - B. Fat
   - C. Vitamins
   - D. Minerals

9. Feeds classified as roughages contain more than 18 percent crude ____.
   - A. oil
   - B. fat
   - C. protein
   - D. fiber

10. A(n) ____ is the amount of feed given to an animal in a 24-hour period to meet its basic nutrient needs.
   - A. diet
   - B. ration
   - C. additive
   - D. supplement

Let's Apply What We Know

1. List and explain the different components needed for a healthy animal feed and how each of these benefit the animal.
2. Discuss the vitamins needed by livestock animals and what they do for the body.
3. What is the meaning of a ration and how does it relate to livestock feed? How do rations ensure that animals receive the correct amount of different nutrients?

Academic Activities

1. **Math.** Research and create a balanced ration for a livestock animal of your choice. The diet should include all the needed vitamins, proteins, and other ingredients. Determine how much of each ingredient is needed to make a balanced diet.

Communicating about Agriculture

1. **Speaking.** Balancing a ration can be a complicated concept to understand. Craft a 2–3 minute speech that describes in clear and straightforward language what a balanced ration is, how it is created, and how it may differ across different species.

SAE for ALL Opportunities

1. **School-Based Enterprise SAE.** Create an elementary agricultural lesson on the importance of animal nutrition and health. Discuss why animals need a balanced diet and include a hands-on activity. Present your lesson at a local elementary school or after-school program.
2. **Placement SAE.** Volunteer at a livestock production facility. Learn about their feeding program and how they ensure the animals receive a healthy, balanced diet.

SAE for ALL Check-In

- How much time have you spent on your SAE this week?
- Have you logged your SAE hours?
- What challenges are you having with your SAE?
- How can your instructor help you?
- Do you have the equipment you need?

Dusan Petkovic/Shutterstock.com

Keeping Livestock Healthy

Essential Question

Why is important for producers to pay attention to animal health?

Learning Outcomes

By the end of this lesson, you should be able to:
- Discuss why it important to maintain healthy livestock. (LO 09.08-a)
- Describe what goes into a livestock health management plan. (LO 09.08-b)
- Explain what a disease is, and describe general symptoms that would alert you to a problem. (LO 09.08-c)
- List the three main types of diseases in livestock, and describe how they can be addressed. (LO 09.08-d)

Words to Know

contagious	parasites	vaccination
disease	prevention	withdrawal time
host	quarantine	zoonotic diseases
infectious	symptoms	

Before You Begin

Do you remember how you felt the last time you were sick? Did you run a fever? Were you tired? Did your head or body ache? When I am sick, I try my best to stay away from other people to make sure that I don't spread my germs to others.

Did you know that when animals are sick, they feel the same way? They have no interest in eating or activity. Their sickness may spread to other animals. Identifying the symptoms early and providing proper treatment will help them feel better. Paying close attention to your animals will allow you to help them heal—or prevent their illness in the first place. Let's learn more!

Why Are Healthy Animals Important?

Healthy animals are more productive than unhealthy animals. They have additional energy for eating, growing, and reproduction. They grow faster and reproduce, which increases profits. All livestock programs benefit from keeping their animals healthy.

Health Management

Livestock producers need a plan to manage their animals' health. Successful plans focus on preventing health problems. The same basic principles of prevention and management can be applied to all livestock species.

Good livestock health management involves:

- Working with a veterinarian (**Figure 9.8.1**) to develop a program for preventing, diagnosing, and treating illnesses
- Keeping accurate (identification, feeding, medical) records (**Figure 9.8.2**)
- Understanding the nutritional requirements of your animals and feeding them to meet those needs
- Providing fresh, clean water
- Vaccinating against diseases
- Controlling parasites
- Maintaining a clean, healthy housing environment
- Observing livestock for signs of problems
- Isolating and observing new animals until you are certain they are disease-free
- Properly (and quickly) disposing of dead animals

Prevention

The best method of keeping animals healthy is *prevention*, or stopping disease from occurring in the first place. The three types of prevention are *primary*, *secondary*, and *tertiary*. Primary prevention means keeping healthy animals from getting sick by stopping the sickness before it happens. An effective means of primary prevention is vaccination. *Vaccination* (**Figure 9.8.3**) involves the use of a vaccine to stimulate an animal's immune system so that it recognizes a disease and protects the animal from future infection. Vaccination works with known diseases and can provide protection for either a short period of time or for the life of the animal. Other methods of primary prevention involve providing proper nutrition, clean drinking water, and a sanitary living environment. Livestock that are fed properly have lower odds of getting sick because their bodies are stronger.

Julia Zavalishina/Shutterstock.com

Figure 9.8.1. Working with a veterinarian is a critical part of maintaining healthy animal stocks.

Studio Romantic/Shutterstock.com

Figure 9.8.2. Keeping accurate records is another important part of maintaining healthy animal production facilities. *Why do you think this might be the case?*

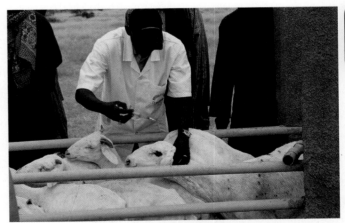

BOULENGER Xavier/Shutterstock.com

Figure 9.8.3. A veterinarian vaccinates these goats to protect them from disease. *What type of prevention would this be: primary, secondary, or tertiary?*

As a result, their immune system does a better job of protecting the animal from attacking diseases or parasites.

Secondary prevention involves identifying sickness early. Early detection can make it possible to prevent the disease from worsening or spreading. This can reduce complications and limit the disease spread. Regular observation, screening for symptoms, and isolating symptomatic or potentially diseased animals are all methods of secondary prevention.

Tertiary prevention is effectively managing diagnosed sickness. It is important to accurately diagnose any illness. Then, appropriate treatment methods can be used to manage and cure the disease before it gets worse or spreads. The use of medication and isolation are methods of tertiary prevention. When using medication to treat illnesses in livestock, be sure to follow the label instructions and observe the withdrawal times before sale or slaughter. A *withdrawal time* is the time from injection until the medicine is no longer detected in the animal's body. Withdrawal times ensure that medicines are no longer effective or active in the animal or its meat when it is harvested.

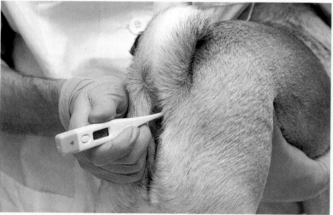

Yekatseryna Netuk/Shutterstock.com

Figure 9.8.4. Animals that are behaving strangely or showing a lack of energy should be tested with a thermometer to see if they're running a fever. *Do you remember a time you had a fever? Was the fever a symptom of an illness like the flu or a stomach virus?*

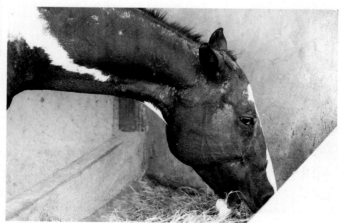

AgriTech/Shutterstock.com

Figure 9.8.5. *What symptoms do you see on this horse that indicate something might be wrong?*

Symptom Recognition

One of the best methods of prevention is observation. Being familiar with your animals and their normal habits and behavior will help you quickly recognize when animals are not acting normally. Sick animals display **symptoms,** or signs of their sickness. Animals may display general symptoms of sickness like fever, depression, or low energy (**Figure 9.8.4**). Other general symptoms of animal illness include watery eyes, dull hair or coat, and a lack of appetite. When an animal isn't acting normally, it is important to determine the cause quickly and find a remedy.

Animals also may display symptoms like respiratory, digestive, behavioral, or skin abnormalities that indicate sickness. Respiratory symptoms include coughing, wheezing, a runny nose, or trouble breathing. Digestive symptoms include diarrhea, vomiting, swollen stomach, colic, or weight loss. Physical symptoms include blisters, rashes, ulcers, warts, blotchy skin, itchiness, hair loss, and physical discoloration (**Figure 9.8.5**). Behavioral symptoms include restlessness, agitation, lack of coordination, stumbling, trembling, convulsions, seizures, leaning, inability to stand, or head hanging to the side. Those don't sound like healthy animal behaviors, do they?

Any of these symptoms can indicate disease or illness. A veterinarian will be helpful in diagnosing and treating disease. Early detection and treatment will often lead to a better response to medications and quicker healing time than for animals that continue fighting sickness without any help.

Diseases

A **disease** is a condition that negatively affects the structure or function of a living creature, and is detected by the presence of consistent, specific signs and symptoms. Some diseases are **contagious,** which means they can be easily transmitted from one animal to another. **Infectious** diseases can infect or cause harm to another creature. All contagious diseases are infectious, but not all infectious diseases are contagious. Some diseases are specific to a particular species, while others will infect many species. Diseases can be caused by bacteria, viruses, parasites, fungi, nutritional deficiencies, toxic chemicals/substances, and other factors. These factors can be grouped into three major disease types: infectious diseases, diseases caused by parasites, and diseases caused by improper nutrition.

Infectious Diseases

Infectious diseases are the greatest threat to livestock health. When disease-carrying viruses or bacteria enter an animal's body, they multiply and interrupt normal body function, causing disease. The disturbance of normal function is caused by the production of harmful chemical substances that are by-products or waste products of the invading organism. Organisms that cause infectious diseases include bacteria, viruses, and fungi (**Figure 9.8.6**). A veterinarian will help diagnose and properly treat the cause of infectious disease. Some infectious diseases target specific species of animals, while some target multiple species of animals, and others can be transferred to humans.

Diseases that can be transferred between humans and animals are called **zoonotic diseases**. Many emerging diseases are zoonotic diseases, including coronaviruses, West Nile virus, flu, rabies, and Lyme disease. You can protect human health by being careful around animals, monitoring animal health and behavior, and taking necessary safety precautions.

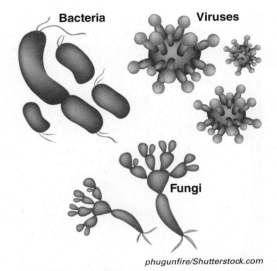

Bacteria

Viruses

Fungi

phugunfire/Shutterstock.com

Figure 9.8.6. Infectious diseases are caused by a variety of agents, and must be treated by a variety of different techniques and strategies.

Did You Know? The Centers for Disease Control (CDC) estimates that more than six out of every ten known infectious diseases in people come from animals, and three out of every four new or emerging infectious diseases in people come from animals.

Parasites

Parasites are organisms that live in, with, or on another organism (**host**) and derive nutrients at the other's expense. Parasites do not normally kill the host organism unless the parasite population is especially large or they introduce a disease. It benefits the parasite for the host to live; if the host organism dies, the parasite will no longer have a food source.

Parasites fall into two main categories: *external* and *internal*. Both internal and external parasites may weaken the livestock immune system, allowing other diseases to attack.

Did You Know?

Foot and mouth disease (a viral infection) alone costs farmers $21 billion each year.

Figure 9.8.7. *Can you imagine what it would be like to have these embedded in your skin and hair?* Good animal care addresses external parasites like these.

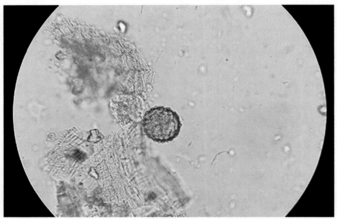

Figure 9.8.8. Roundworms, like those that hatch from the egg shown here, are an internal parasite that can be devastating to livestock.

Figure 9.8.9. Alfalfa beetles can be toxic to any number of livestock animals, particularly horses. *How might you develop a plan to keep them away from your animals?*

External parasites (**Figure 9.8.7**) often annoy their hosts by biting or embedding into the skin. They include fleas, flies, lice, mange, mites, mosquitos, and ticks. External pests can cause damage to an animal's hide, fur, or wool. These parasites can also introduce more serious diseases. The transfer of disease from one animal to another by external parasites can lead to economic loss for a producer.

Internal parasites live in the blood or tissue inside an animal's body (**Figure 9.8.8**). They enter the animal through contaminated food or water. Sometimes they burrow through the skin to reach the bloodstream. Once in the blood, these parasites will travel to a desired location within the body to mature and reproduce. These parasite populations can interfere with digestion, causing poor growth. Injuries and damage from internal parasites can be temporary or permanent and can result in the death of livestock if not treated.

Improper Nutrition

Nutrient deficiencies, imbalances, and excesses are potential hazards to animal health. The animal body can adapt to short-term deficiencies or problems, but long-term problems cause cell and tissue damage. Many potential problems arise from poor quality or inadequate feed supply. Good feed management programs help prevent nutritional problems. The nutritional requirements of livestock vary by species and producers must be familiar with the requirements of their animals. Producers must also be aware of deficiency symptoms to quickly identify and correct any problems with the feed ration.

Producers must also be aware of what livestock are eating. Livestock must be kept away from poisonous plants. Red maple, tansy ragwort, and milkweed are all common plants that are not toxic to people, but are toxic to horses. There are insects like the alfalfa (blister) beetle (**Figure 9.8.9**) that are toxic to sheep, cattle, and horses. Just a few grams of the toxin produced by the beetle can be deadly to a horse. Care must be taken when working around feed storage, mixing, and handling areas so that wire, nails, screws, or other sharp metal objects are not accidentally dropped in the feed. Cattle and other ruminants swallow large amounts of feed without chewing and can easily consume these objects. Swallowing metals like this may lead to disease-like symptoms from digestive complications.

Hands-On Activity

Infectious Disease Spread

Infectious agents can be spread between animals easily when they are housed together in large areas with a shared feeding and water location.

For this activity, all students in the class are healthy cattle or other livestock animals, except one. One of the students has an infectious disease. It is feeding time and all the animals are coming to one location to eat, causing several of them to come in contact with the infected animal.

P Maxwell Photography/Shutterstock.com

For this activity, you will need:

- Saturated baking soda solution - Fill a cup half-full of distilled water. Add baking soda to the water, then stir or shake vigorously. Repeat until no more baking soda will dissolve into the solution. Let this saturated solution stand until it is clear.
- Distilled water
- Numbered clear plastic cup for each student
- Contamination indicator - Phenolphthalein indicator solution **or** water from boiled red (purple) cabbage. To make the cabbage solution: Boil a half of a head of red (purple) cabbage in a large pot of distilled water (using tap water will affect the pH of the solution and the results of the activity). Strain out the cabbage and keep the water. The water should turn purple. Refrigerate any portion that you do not use immediately. This can be refrigerated for a day or two.

Here are your instructions:

1. Your teacher will pass cups out to students. Most students will have a cup full of distilled water, but one student (you won't know who) will have a cup full of the clear baking soda solution.

2. You will have three minutes to move around the room and meet up with five other students. During each visit, one student will pour all their liquid into the other student's cup. Then they will pour half of the liquid from the full cup back into the first (empty) cup. Each student should end the visit with their cup half full. Repeat this a total of five times with five different students.

3. At the end of three minutes, everyone should get in a line. Your teacher will go down the line and put a few drops of phenolphthalein (or cabbage water) into each cup. If the student is "infected," the contents of his or her cup will turn pink. All "infected" students will move to a designated part of the room and the "noninfected" students move to a different part of the room.

4. As a class, determine who had the original infected cup.

Consider This

1. How does this model reflect many real-world situations?
2. If you get a cold, does every member of your family also get a cold every time? Why? If not, why do you think this is?

9

Disease Control

Disease control methods vary based on the source of the problem. Quality health management and prevention are the best methods of control. When a disease does occur, other methods of treatment are necessary. This is where careful observation and veterinarian involvement will help get potentially dangerous situations under control. Keep in mind that even animals that look healthy can still spread germs to other animals. *Quarantine*, or isolate, sick animals and keep accurate records of treatments. Provide vaccinations for known disease problems.

Medical control can be established by using drugs that are known to treat specific illnesses. Antibiotics are used to treat bacterial infections. Certain viral infections are not treatable with medicine. In these cases, isolation and checking nutrition requirements are the best forms of treatment. There are chemicals that are used to treat internal parasites, and pesticides can be effective against external parasites. Each of these treatment methods should be given by or at the advice of a veterinarian.

Career Corner — VETERINARY PATHOLOGIST

Veterinary pathologists *are doctors of veterinary medicine who diagnose diseases by examining animal tissue and body fluids.*

What Do They Do? Veterinary pathologists complete postmortem exams on small and large animals, prepare tissue for testing, test blood and tissue samples, analyze the results of laboratory tests and studies, determine an animal's cause of death, diagnose disease, develop drugs and animal health products, and conduct research studies.

Gorodenkoff/Shutterstock.com

What Education Do I Need? Beyond earning a doctorate of veterinary medicine, veterinary pathologists must complete an extensive anatomical or clinical pathology residency at a veterinary teaching hospital. To become board certified, individuals must complete a minimum of three years of clinical training and pass the board certification exams.

How Much Money Can I Make? Veterinary pathologists make an average of $150,000 per year.

How Can I Get Started? You can build an SAE conducting research on animal diseases and report your findings to your class. You might conduct an experiment on using home remedies to treat animal diseases, conduct an agriscience project in animal systems over an animal disease topic, or interview a veterinary pathologist, or job shadow them for the day.

What Did We Learn?

- Healthy animals are more productive, so it is important for all animal producers to have a plan in place to keep their animals in good condition.
- A good livestock management plan addresses medical care, recordkeeping, good nutrition, sanitary housing, parasite control, and livestock observation.
- Preventing diseases or health problems is the best method for maintaining healthy animal stocks. Preventing illness includes taking measures like vaccinating animals, observing symptoms, and treating sick animals promptly.
- Diseases interfere with the healthy and normal functioning of physical systems. They can be infectious, parasitic, or caused by poor nutrition.

Let's Check and See What We Know

Answer the following questions using the information provided in Lesson 9.8.

1. Healthy animals:
   A. Have slower growth rates.
   B. Are less profitable.
   C. Reproduce at slower rates.
   D. Are more productive.

2. Which of the following would *not* be a part of a quality livestock health management plan?
   A. Observation of animals for symptoms
   B. Consultation with a veterinarian
   C. Placing new animals into the herd as quickly as possible
   D. Providing sanitary conditions and fresh, clean water

3. The best method of keeping animals healthy is:
   A. Prevention
   B. Medication
   C. Isolation from other animals
   D. Antibiotics

4. _____ are used to stimulate an animal's immune system so that it recognizes a disease and protects the animal from future infection.
   A. Antibiotics
   B. Medicated feeds
   C. Vaccines
   D. Parasites

5. *True or False?* Livestock that are fed properly have lower odds of getting sick because their bodies are stronger.

6. When an animal departs from its normal function and displays signs of sickness, these signs are called:
   A. Symptoms
   B. Parasites
   C. Vaccines
   D. Evidence

7. Difficulty breathing would be evidence of a _____ disease.
   A. digestive
   B. respiratory
   C. behavioral
   D. neurological

8. *True or False?* All contagious diseases are infectious, but not all infectious diseases are contagious.

9. _____ are external parasites.
   A. Roundworms
   B. Bacteria
   C. Ticks
   D. Viruses

10. Antibiotics are used to treat certain types of _____ infections.
   A. viral
   B. bacterial
   C. fungal
   D. parasitic

Let's Apply What We Know

1. Create a list of things that should be included in a good livestock health management plan.
2. Explain the three types of prevention related to animal health.
3. Describe what makes a disease infectious.

Academic Activities

1. **Science.** Research how to prepare an agar solution in a petri dish for growing bacteria. Swab several surfaces animals regularly touch, like bird feeders, water dishes, food bowls, or toys. Rub the swab over the agar solution. Label and seal the petri dishes and observe for several days. Keep records of what develops.
2. **Language Arts.** Many pet owners have a difficult time recognizing health issues with their pet and getting that information to a veterinarian. Create an informational chart of symptoms animals display when they might not be healthy. Categorize the symptoms as general or specific to a disease or condition. Share your informational chart with the class.
3. **Science.** Research an emerging zoonotic disease. Where did it come from? What animals carry it? How is it passed to humans?

Communicating about Agriculture

1. **Speech.** Pretend that you are a consultant, working to help livestock producers get a good livestock management plan in place. Create a presentation designed to inform producers about what all a livestock management plan should include. Create a PowerPoint presentation to accompany your presentation, and add your own research. Make your speech to your class.

SAE for ALL Opportunities

1. **Research SAE.** Research and conduct an agriscience experiment on how disease is spread in the livestock industry or in animal shelters or kennels.
2. **Foundational SAE.** Research common health concerns for all major livestock breeds. Find the major causes of these concerns and how to prevent them. Create a presentation on your findings and report back to your class.

SAE for ALL Check-In

- How much time have you spent on your SAE this week?
- Have you logged your SAE hours?
- What challenges are you having with your SAE?
- How can your instructor help you?
- Do you have the equipment you need?

Safety around Livestock

Why is it important to follow safe practices around livestock?

Learning Outcomes

By the end of this lesson, you should be able to:

- Describe why it is important to follow approved safety practices. (LO 09.09-a)
- List the human factors and attributes that affect safety around livestock. (LO 09.09-b)
- Explain what hazards livestock may pose to human workers, and how those hazards can be mitigated. (LO 09.09-c)
- Describe the personal protective equipment that should be used around livestock. (LO 09.09-d)
- Identify practices associated with using chemicals safely. (LO 09.09-e)
- Identify and describe the safety hazards associated with livestock facilities, and how the risks they pose can be reduced. (LO 09.09-f)

Words to Know

demeanor
experience
flight zone

halter
livestock confinement housing

personal protective equipment (PPE)

Before You Begin

Have you ever heard a saying that just sticks with you? One of my favorites is "An ounce of prevention is worth a pound of cure." It's much easier to stop something from happening in the first place than to repair the damage after it has already happened. The same advice applies to working around livestock. A cautious and prepared approach to working with animals will result in a safer environment for humans and animals alike. Can you think of three things you should do to keep yourself safe around animals? Share your ideas with a classmate or have a class discussion.

Why Is Safety Important?

Following safe practices protects us from danger, risk, or injury. What are safe practices? Safe practices are acting with care, thought, foresight, and preparedness when entering a potentially risky situation. Accidents are more likely to happen when we don't make safe choices. Accidents can lead to damage or injury, to you, a colleague, or an animal you are working with. Injuries can lead to lost income, unexpected expenses, or even death. It is important to take a cautious and safe approach to working with animals and making sure that your environment is a safe as possible. Acting in a safe manner comes with experience and proper training.

There are multiple factors that affect safety around livestock. Some of these factors are more easily controlled than others. Animals are involved in thousands of farm injuries and many deaths each year. Recent studies show that animals and animal-related accidents are a factor in more than ten percent of injuries reported on farms. In fact, animal-related injuries are second only to machinery-related injuries.

Human Factors Affecting Safety

Humans can act unpredictably and often make mistakes. These mistakes and their frequency are affected by training, experience, age, physical ability, mental ability, and attitude. Each of these factors has a positive or negative effect on your likelihood of making a mistake. That effect can be compounded when multiple factors are combined.

Initial safety training is standard for most companies. It is often ignored in smaller companies or on family farms. There are numerous resources available to help with safety training and these resources should be used to educate all workers. Self-training is available for many activities and should be used with anything that is new to the operation or to a worker.

Experience is practical contact with and observation of facts or events. To gain experience in risky situations, you need to observe or be involved in activities that are best suited to experienced individuals. This is why proper training is important. It is also valuable to gain experience in a controlled environment with limited exposure to unsafe conditions until enough experience makes you more aware of and better able to anticipate and react to potentially unsafe actions.

Human physical abilities differ. Younger people are faster and more agile. They react quickly, and they can handle more hard physical labor than older adults. However, the experience of older people allows them to make decisions that make up for slower, weaker physical bodies. Age is a factor to consider when making safety decisions in and around livestock facilities. Care must be taken to keep young children away from potential dangers beyond their experience and ability (**Figure 9.9.1**).

Similarly, human physical ability requires careful consideration around livestock. Livestock operations require physical activity that can be strenuous and difficult. Knowledge of one's ability to complete necessary tasks is important.

FamVeld/Shutterstock.com

Figure 9.9.1. Young children should be taught to interact with livestock in a safe and controlled environment. *What is positive about the interaction pictured here?*

Another important human factor to consider in handling animals is your ***demeanor***, or the way you present your emotions in your actions. Working in a calm, methodical, and predictable manner is important when you are around animals. When working with unpredictable creatures, it is of extra importance that you control your own actions. By behaving calmly and deliberately, you can help to minimize the risk of accidents—for you and the animals you work with.

With animals, you always need to pay attention! Working with animals is not a good fit for people who want to spend the day daydreaming or regularly zone out. Animal behavior is not something you can control, so you need to show up ready to focus on the situation at hand, take in clues about how your animals are feeling from their behavior, and react quickly to any unexpected situations.

Safe workplaces are prepared. Safe workplaces may keep checklists or other information on-site to help people stay safe. They have emergency plans in place in case of an incident, a fire, or an illness. When you are working somewhere, ask if there are emergency contacts posted or plans in place for an incident dealing with livestock. If there isn't a plan in place, work to help create one. Find backup care for animals, create written instructions for important daily tasks, and have extra supplies on hand.

No dangers can be fully eliminated from a workplace. Maintaining a safe work environment is everyone's concern (**Figure 9.9.2**). Individuals who engage in unnecessarily risky behavior should be identified and encouraged to act in a more responsible way. Attitude affects action, meaning that a positive attitude toward safety is more likely to result in safe actions. Correcting negative attitudes and refusing to tolerate irresponsible behavior will help to develop a safe working culture.

skipper_sr/Shutterstock.com

Figure 9.9.2. Safe workplaces are everyone's job. Know the rules, get proper training, and look out for others on your team. Unsafe behavior is not okay!

Animal Factors Affecting Safety

Each species of livestock has its own unique set of safety considerations. Larger animals like cattle or horses can overwhelm you with their size. Pigs, while not as large as cattle or horses, still weigh more than the average human and can bite with enough force to cause serious injury. While these animals rarely attack, some are easily startled and can move suddenly and quickly, causing harm by trampling, butting, kicking, or biting (**Figure 9.9.3**). Males, females in heat, and females with young tend to be more aggressive than other animals, and it is wise to exercise extra caution around them. Castrated males and females without young tend to be calmer.

Kwadrat/Shutterstock.com

Figure 9.9.3. *What clues do you see in this horse's posture that you should take care in approaching?*

Smaller animals may not be obvious threats to inflict major damage, but they can cause harm. Common accidents with sheep and goats involve being butted. Goat horns may not be large, but they can cause serious injuries! This is especially dangerous for small children and elderly adults. Goats usually keep their horns, but some species of cattle and sheep are dehorned to minimize these risks. Poultry are relatively harmless, but they can peck or pinch using their beaks or bills. Chickens grow spurs on their legs and use them for defense and can cause minor injury. Pets are also capable of causing injuries and must be watched around small children.

It is important when handling animals of all sizes to understand their temperament and behavior. Some animals are prone to getting skittish and excited, and behaving in unpredictable manners. Animals like this should not be handled without experienced help. Some animals are wary of people and may be hesitant to interact. Other animals are quite docile and easily managed once you are familiar with their needs and patterns of behavior.

Animals may behave differently in the company of other animals than they would alone. It is important to understand these dynamics, and to always be appropriately prepared for the situation you are entering (**Figure 9.9.4**).

Animals also have physical limitations that their handlers should understand. Many animals have blind spots in their vision. It is important to approach them from where they can see you coming – no one wants to surprise a bull by sneaking up behind! Some animals don't hear terribly well, so it's especially important for them to see you.

Many herd animals have *flight zones*, or areas of personal buffer space that they need to feel calm and safe. If you get into an animal's flight zone, they may panic and behave in an unpredictable or dangerous manner. If you are working with animals, make sure you understand their need for personal space. When working with unfamiliar animals, approach them slowly, and from where they can see you. When they begin to move away from you, you will have reached the edge of their flight zone. Familiar animals or animals that regularly interact with people will have smaller flight zones than animals who rarely see people.

It is critical to understand how to use the proper equipment in handling animals. Well-designed equipment, used properly, can help to minimize the risk of injury in working with animals. Working with animals requires training, supervision, and care, even when working in an operation as small as a family farm. All people working with livestock should know how to properly use a *halter*. A halter is a type of collar that fits behind the ears of an animal and around the animal's muzzle, and it usually has a lead rope attached to allow someone to lead the animal. Halters are used for all mammalian livestock with the exception of pigs and can be very useful in managing and handling animals safely.

Animal diseases that are species-specific are not usually a cause for concern when thinking of handling animals safely. However, there are several kinds of diseases that spread from animals to humans. It is important to be aware of animal health to protect yourself from illness or infection. Keep facilities clean and animals properly vaccinated against known problem diseases to avoid human exposure and safety risks. If there is an outbreak, take extra precautions to be safe, like wearing appropriate personal protective equipment.

OUTDOOR_MEDIA/Shutterstock.com

Figure 9.9.4. Pigs are highly social animals with complicated herd dynamics. *What does this producer need to understand to stay safe?*

Hands-On Activity

Putting on a Halter

Placing an animal in a halter is an important skill in managing and interacting with livestock safely. In this activity, you'll practice how to do this.

For this activity, you will need the following:

- Large stuffed animal with a protruding nose and ears (Great if it's a horse or cow, but any animal with the proper positioning would be acceptable)
- Halter
- Masking tape
- Marker or pen

Steps:

1. Get into groups of 2–3 students.
2. Looking at the halter and the diagram here, identify the nose band (1), free end (2), loop (3), adjustable chin loop (4), and large adjustable loop (5). Using the masking tape, label each part with its number.
3. Holding your stuffed animal, place the large adjustable band behind its ears.
4. Next, place the nose band over the nose, and the chin band under the chin.
5. Tighten the straps to make sure the halter fits the animal properly. Have your teammates and teacher check your work.
6. Pass to another teammate to practice.

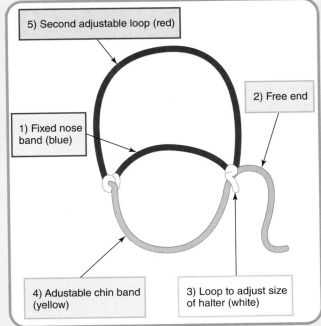

5) Second adjustable loop (red)

2) Free end

1) Fixed nose band (blue)

4) Adustable chin band (yellow)

3) Loop to adjust size of halter (white)

Goodheart-Willcox Publisher

Hannah_Muhlbeier/Shutterstock.com

Consider This

1. How does confident and capable haltering affect safety at an animal facility?

9

Personal Protective Equipment

Personal protective equipment, or **PPE**, is equipment worn to reduce exposure to hazards that can cause serious injuries and illnesses. These hazards may result from contact with chemicals, physical dangers, or other potentially harmful situations. When working with livestock, proper PPE may include:

- Eye and face protection (when dealing with chemicals)
  - Safety glasses
  - Safety goggles
- Foot and leg protection
  - Close-toed shoes or boots
  - Long pants
- Hand and arm protection
  - Gloves – various types for different applications
- Hearing protection
  - Earplugs (single use or reusable)
- Respiratory protection (for dealing with chemicals, poor ventilation, or sick animals)
  - Mask
  - Respirator

Did You Know?

Many states have hazardous waste disposal days where residents can bring products labeled as toxic, poisonous, corrosive, flammable, or combustible to a central location for free professional disposal.

Nattawit Khomsanit/Shutterstock.com

Figure 9.9.5. It is important to keep materials on hand to clean up any spills.

Safe Chemical Use

Livestock production may require the use of chemicals. Insecticides may be used to reduce flying insect populations in a cattle feedlot. Fungicides and other chemicals or medicines may be used to treat animal illnesses. Storing animal feeds can attract pests that require chemical control. Sometimes chemicals can be used to control the growth of undesirable weeds in and around livestock facilities. Fuels, solvents, cleaners, fertilizers, and veterinary chemicals or medicines also present potential safety concerns. Situations where chemicals are used require careful practices to protect yourself and others.

Some suggestions for using agricultural chemicals safely include:

- Always read and follow the instructions on the label for proper use and storage.
- Keep chemicals in their original containers, with their original labels, and only store fuel in approved containers.
- Regularly inspect containers for damage or corrosion and replace if necessary.
- Make sure chemicals are properly labeled and never remove labels.
- Store chemicals in a locked, well-ventilated cabinet or shed with floors that will contain spills.
- Store personal protective equipment separately from chemicals to prevent contamination and exposure.
- Store liquid chemicals beneath solids to prevent accidental mixing in the case of spills.
- Keep different classes of chemicals separate to prevent reactions.
- Store animal feeds, seeds, and fertilizers separately from chemicals.
- Have mop-up materials on hand, such as sand, soil, or commercial absorbent. (**Figure 9.9.5**)
- Keep ignition sources away from chemicals.
- Keep a record of the chemicals you buy, store, and use.

Using Livestock Facilities Safely

Livestock facilities vary based on the type of livestock they house. Most livestock facilities are used to house and handle animals or animal feed. These facilities can be as small and simple as a chicken coop in the backyard or as complex as a facility that houses 3,000 feeder pigs complete with watering, feeding, and climate control systems. In ***livestock confinement housing***, where large populations of animals are kept in small areas, there are several factors that pose a threat to human safety (**Figure 9.9.6**). Dust from animals, animal feed, and animal waste and gases from animal waste can rise to harmful levels. Swine and poultry confinement housing can be particularly bad in this regard.

It is impossible to eliminate the formation of dusts and gases, which means that removing these contaminants is particularly important. Delivering ground feed through an enclosed system to covered feeders helps reduce the amount of airborne dust. Professionally designed ventilation systems that remove dust-filled air and replace it with clean air help improve air quality. They also help with potentially toxic fumes. People working in confinement housing should wear dust masks to reduce exposure. Waste management should be designed to keep people from being exposed to high gas concentrations. Manure stored for use on the farm should be kept in safe, well-ventilated areas to prevent illness. Manure pits (in dairy and swine operations) should be constructed outside with proper ventilation and all workers should be trained to be aware of and avoid potential dangers.

Safe agricultural workplaces also include good sanitation practices. Clean facilities are less likely to harbor disease or pests. It is easier to find necessary equipment and follow protocols in clean facilities. Animals stay healthier in clean facilities, with fewer diseases and injuries. All parts of the animal facility need to be regularly cleaned and disinfected. Equipment must be kept in good repair. Old feed and bedding should be discarded and replaced with new. Ventilation and air-conditioning systems should be checked regularly to make sure they are in good working order.

rafapress/Shutterstock.com

Figure 9.9.6. Confinement can pose risks to human and animal safety, even as it has benefits. *What are some factors you should be aware of if you are working in or around one of these facilities?*

Grain Storage

Grain storage bins and grain handling equipment can pose serious hazards to workers (**Figure 9.9.7**). Grain bins store large quantities of feed for livestock. In certain situations, workers are required to enter the bins. Flowing grain can quickly trap and suffocate someone if proper precautions are not taken. Never enter a grain bin or storage container while it is in operation. Storage bins are usually very tall and are loaded from the top. Care must be taken not to fall when climbing on the bins to load or unload grain. Workers must also be aware that an accumulation of dust in a grain bin can be explosive. Accidental sparks can ignite the dust and cause a fire or explosion.

Max Lindenthaler/Shutterstock.com

Figure 9.9.7. Working in and around grain silos can pose many potential hazards to workers. *Why wouldn't this be a good place to go and play?*

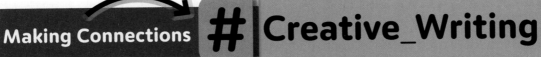

Making Connections # Creative_Writing

Safety Word Cloud

What is the first image that comes to mind when someone says the word *safety*? Is it a hard hat, pair of gloves, or maybe an orange safety cone or vest? What words do you associate with safety? For this activity, you will create a safety word cloud.

For this activity, you will need:

- Plain white or colored paper
- Markers, crayons, or colored pencils

Steps:

1. Draw an image on your paper that you associate with safety.
2. Fill your safety image with words that you also associate with safety.
3. The most important words should be the largest inside your safety image. Use different colors for different words.

kenary820/Shutterstock.com

ricochet64/Shutterstock.com

Consider This

1. Which of your safety words apply to all industries and not just the agricultural industry?
2. Look at a classmate's word cloud. Do you see any words you missed or would like to include?

Fire Safety

Fire can present a serious hazard around livestock facilities. Most fires start from electrical equipment. Dust from feed and other sources can accumulate on and around electrical equipment. The use of water and chemicals around agricultural facilities can compromise electrical systems, resulting in a spark that ignites the dust. People should know what to do in case a fire starts. Firefighting equipment should be available, and a plan in place for handling fires. Some suggestions for preventing fires include:

- Store fuels and combustible chemicals in safe locations and proper containers.
- Keep areas clean and free from debris and dust buildup.
- Make sure electrical wiring meets code requirements for agricultural buildings.
- Be careful using heaters in livestock buildings.
- Use fire-resistant materials in new construction.
- Make sure fire extinguishers are available.
- Provide instruction in methods of fire prevention and what to do in case of fire to all workers.

Livestock Equipment

Mechanical equipment around a livestock facility should be kept in the best possible condition. Working equipment and capable operators equal a safer work environment. Machinery is one of the top causes of agricultural injuries. Functional equipment is very important, but most accidents occur due to distracted, overtired, or unprepared equipment operators. There are several machinery-related safety tips to help avoid accidents:

- Read, understand, and follow the instructions in equipment manuals.
- Make sure you know, understand, and follow any federal, state, and local laws that apply to your activity.
- Dress appropriately. Proper attire can reduce the risk of injuries.
- Make sure you are well-rested and aware of your surroundings and activities.
- Only allow well-trained operators to operate machinery.
- Avoid alcohol or medications that impair judgment or otherwise affect your ability to operate machinery or equipment.
- Keep machinery and equipment properly maintained and adjusted. Make sure all guards, shields, and other protective parts are in place.
- Keep children and animals away from working areas and dangerous equipment.
- Be familiar with the methods of safe operation for any equipment that you use.

Career Corner | SANITATION COORDINATOR

Sanitation coordinators *are responsible for maintaining the cleanliness of their assigned production areas.*

What Do They Do? Sanitation coordinators direct all daily sanitation activities within the processing plant, investigate any issues and recommend corrective actions, manage spills and leaks policy, provide training on proper sanitation practices, perform weekly inspections, inspect the entire facility once per month, and monitor maintenance of sanitation equipment.

What Education Do I Need? A high school diploma is required.

How Much Money Can I Make? Sanitation coordinators make an average of $30,000 per year.

SimpleBen.CNX/Shutterstock.com

How Can I Get Started? You can build an SAE creating a safety plan for your school's outdoor and indoor lab areas, research common safety concerns in the livestock industry and create an informational poster or chart for your class, conduct an agriscience project to gather information about people's safety concerns with student livestock projects, interview a sanitation coordinator, or job shadow them for the day.

What Did We Learn?

- Safe practices protect us from danger, risk, or injury. While not all risks can be eliminated from any situation with animals, most risks can be reduced.
- Human factors that affect safety include experience, training, age, physical ability, mental ability, demeanor, preparedness, and attitude.
- Animal factors that affect safety include the animals' size, age, sex, temperament, physical abilities, herd dynamics, and health.
- Proper personal protective equipment for working around livestock includes close-toed shoes, long pants, eye protection, gloves, and respiratory protection (as necessary).
- Livestock production can require chemical use, which can pose health risks. Learning to handle chemicals properly will keep you safe.
- Livestock facilities can be full of potential risks to health and safety, including odors and unclean air, poor sanitation, grain storage bins, fire risks, and machinery. Proper training and working systems will help you to stay safe.

Let's Check and See What We Know

Answer the following questions using the information provided in Lesson 9.9.

1. ____ is practical contact with or observation of facts and events.
   A. Experience   C. First aid
   B. Safety   D. Sanitation

2. Which of the following is *not* a human factor that can affect safety?
   A. Age
   B. Experience
   C. Environment
   D. Physical ability

3. *True or False?* Older adults have a faster reaction time and are less likely to have an accident than younger adults.

4. ____ is a way of thinking or feeling that is typically reflected in a person's behavior.
   A. Training   C. Experience
   B. Attitude   D. Common sense

5. *True or False?* Females of the larger animal species (cows, horses) tend to be more aggressive than males.

6. Keep animals ____ to avoid catching or spreading known diseases.
   A. trained   C. vaccinated
   B. contained   D. clean

7. Which of the following would be used to control a fly problem affecting cattle in a feedlot?
   A. Herbicide   C. Fungicide
   B. Insecticide   D. Vaccine

8. *True or False?* Dust from feeds and other materials in large concentrations can be explosive.

9. The cause of most farm injuries is machinery, followed by:
   A. Contact with animals.
   B. Chemical spills.
   C. Falling asleep.
   D. Pests.

10. Which of the large animals below poses the lowest risk to handlers?
    A. A bull
    B. A cow with calf
    C. A young female horse
    D. A stallion

Let's Apply What We Know

1. Compare and contrast human factors and animal factors that affect safety.
2. Make a list of different types of personal protective equipment that should be used when working with livestock, and the body part they protect. Why is it important to wear this equipment?

Academic Activities

1. **Social Science.** Think of three different situations when middle school students should use personal protective equipment. Interview students in your school and find out how many students wear PPE in each of the situations. Create a chart showing your findings.
2. **Language Arts.** Create safety posters for different areas in your agricultural classroom, lab areas, livestock facilities, or school greenhouse.

Communicating about Agriculture

1. **Reading.** Temple Grandin is a famous name in agriculture. Go to your school library and do some research about her life and work. Read a full book about her, if you can! What does her work have to do with this lesson? Share your findings with the class.

SAE for ALL Opportunities

1. **Service Learning SAE.** Create a Personal Protective Equipment and Safety on the Farm presentation with several members from your FFA chapter. Speak to different civic organizations in your local community about safety on the farm and PPE they should be wearing.
2. **Placement SAE.** Intern for the day with a job or crew foreman for an agricultural business. Interview them about how they keep their staff (and animals, if they have them) safe.

SAE for ALL Check-In

- How much time have you spent on your SAE this week?
- Have you logged your SAE hours?
- What challenges are you having with your SAE?
- How can your instructor help you?
- Do you have the equipment you need?

Specialty Livestock Production

Allexxandar/Shutterstock.com

Learning Outcomes

By the end of this lesson, you should be able to:
- Identify and explain what it means to be classified as a specialty animal. (LO 09.10-a)
- Explain what role specialty animal production plays in agriculture. (LO 09.10-b)
- Identify examples of specialty animals and their products. (LO 09.10-c)

Words to Know

apiaries	camelid	pseudo-ruminant
apiculture	heliciculture	specialty animals
aquaculture	hypoallergenic	
bison	marine aquaculture	

Before You Begin

Some of my favorite days at work in the veterinary office are when really unique animals come in or when we make farm visits to some of their homes. Don't get me wrong, I love dogs and cats and cows, but my favorite day is when we visit a nearby alpaca farm. Have you ever seen an alpaca? Did you know that alpacas, like ruminants, do not have top front teeth? That's why they always look like they have an adorable underbite! But they don't need those teeth; they are vegetarians and only eat two pounds of food a day. I think just one of my meals weighs more than two pounds!

There are lots of interesting animals that are raised on small farms or in specialized facilities for a variety of purposes. Can you imagine what some of these might be? Make a list of as many types of animals as you can think of that might be raised for commercial purposes. See how many from your list we cover in this lesson.

What Is a Specialty Animal?

Specialty animals are livestock that are less common than the typical animals found on large production farms such as cattle, pigs, sheep, and chickens. Many specialty animals are domesticated from wild species. Game animals, exotic livestock, fish farms, and niche breeds are all part of specialty livestock production. Specialty animals may raised for meat, eggs, fiber, or other livestock products. The demand for specialty livestock products may be low or concentrated in a small area, or the cost to produce them may be high, which would explain why they are not as widely produced. Many animals could fall into the category of specialty animal, but for the purpose of this unit we will limit our discussion to some of the most common types.

Types of Specialty Animal Operations

Apiculture

Apiculture is the care and management of honey bees to produce honey and wax (**Figure 9.10.1**). Bees are typically raised for honey. Although honey is an ingredient used in many food products, beeswax is also important for its use in cosmetics and medical products. It is used to make candles, coat cheeses, as a food additive, and make products like furniture polish. Bees are also valuable for the role they play in the pollination of plants and crops. Honey bees are effective pollinators and work to increase crop yields.

Honey bees are raised and bred commercially in apiaries. *Apiaries* are areas where large numbers of beehives are placed, and the bees can be taken care of and managed to produce honey and beeswax. Honey bee colonies contain thousands of bees made up of a queen bee that lays eggs, female worker bees that collect pollen and nectar, and male drone bees that fertilize the eggs laid by the queen. Honey bees can be found around the world and are raised in hives made from wood, pottery, straw, or hollow trees. Honey bee populations can hurt native bee populations in an area, so it is important to manage them carefully.

StudioSmart/Shutterstock.com

Figure 9.10.1. Honey bees are a critical part of ecosystems, in addition to producing honey and wax. *Have you ever visited a beekeeper?*

Aquaculture

Aquaculture is the production or culture of aquatic organisms under controlled conditions (**Figure 9.10.2**). Animals grown in aquaculture include fish, shrimp, crawfish, and other crustaceans. Aquacultural practices take place in the open ocean, in natural or man-made ponds, and other man-made bodies of water. About half of all seafood consumed comes from some type of aquaculture production. Aquaculture is a fast-growing method of producing food and is expected to continue to grow rapidly in the coming years as production methods are refined and improved. The two major types of aquaculture are *marine* and *freshwater*.

leo w kowal/Shutterstock.com

Figure 9.10.2. *Did you ever imagine that fish can be farmed in the ocean?* It happens every day!

Why Are Honey Bees Dying?

Honey bees are important to our environment and food production. We could survive without them, but our food choices would look quite different. Without honey bees, we would not have plentiful coffee, avocados, lemons, limes, and oranges. Most of the fruits and vegetables we enjoy are pollinated by honey bees. While our favorite veggies and fruits wouldn't disappear entirely, they would be rarer and more expensive.

Honey bees are worth more than $18 billion to the US economy each year, primarily because of the work they do to pollinate our plants. Beekeepers across the United States lost 45.5 percent of their managed honey bee colonies from April 2020 to April 2021, according to a nationwide survey conducted by the nonprofit Bee Informed Partnership. Entomologists are currently studying the reasons behind this enormous bee die-off. According to the USDA, most recent evidence points to four Ps being the culprit: parasites, pathogens, poor nutrition, and pesticides.

Parasites

Current scientific research indicates that parasites, and the diseases they carry, are the main threat to the lives of honey bees. The most dangerous parasite threatening beehives currently is a mite with a descriptive name: Varroa destructor. Commonly known as Varroa mites, these parasites infect bees before they reach adulthood. Their parasitic relationship with bees is similar to that of ticks and mammals; the main issue lies in the diseases the mites carry, such as Deformed Wing Virus. When a hive is already weakened, a Varroa mite infestation can wipe it out. There are other parasites attacking colonies as well, including the small hive beetle.

Pathogens

Weakened immune systems leave hives susceptible to bacterial and viral diseases as well. Two of the most well known diseases to infect bees are American Foulbrood and Deformed Wing Virus.

American Foulbrood affects larvae less than a day old, preventing them from surviving until adulthood, while Deformed Wing Virus prevents bees from being able to fly.

Poor Nutrition

When only one crop is grown on a piece of land, it limits the bees' diet to one type of pollen for extended periods of time. Think of it as if a human were limited to eating only one type of food for every meal. That is not very healthy! These malnourished bees are more susceptible to chemical pesticides, parasites, and pathogens, as their immune systems are weaker.

Pesticides

Pesticides are also a contributing factor to honey bee decline. Scientists are continuing to research the effects different types of pesticides have on bees' health. The most studied chemical culprit is a class of agricultural pesticides called *neonicotinoids*. These chemicals are systemic, meaning the plant takes them in and spreads them to all tissues. Farmers like them because they are efficient, cost-effective, and less prone to runoff, and thus less dangerous to humans, birds, and livestock. However, studies have found trace amounts of neonicotinoids in pollen grains. One pollen grain with trace chemicals wouldn't be a large problem, but scientists have found that the chemicals build to dangerous levels as they accumulate in the beeswax. Exposure over a long period of time has been shown to change bees' foraging behavior, their communication, and their larval development. Many companies have stopped using neonicotinoids, but the EPA does still allow some to be used.

Consider This

1. What are steps that you can take in your home and community to help the bee population?
2. What can the agriculture industry do to help with the decline of the bee population?

Marine

Marine aquaculture refers to the growth and harvesting of saltwater species (**Figure 9.10.3**). In the United States, marine aquaculture mostly produces clams, mussels, oysters, shrimp, salmon, and other marine fish. As demand for seafood continues to grow, wild-caught sources cannot support the numbers necessary to meet that demand. Safe, nutritious, and sustainable methods of marine aquaculture provide a steady source of seafood for consumers around the world.

CreativeZone/Shutterstock.com

Figure 9.10.3. This ocean hatchery is made of a series of containers placed in the ocean for fish to be raised in.

Freshwater

Freshwater aquaculture (**Figure 9.10.4**) is specifically devoted to raising freshwater organisms in a controlled environment. Most freshwater species are raised in ponds, but systems can include cages and other types of recirculating systems. Freshwater products include catfish, tilapia, freshwater shrimp, crawfish, gamefish, baitfish, and ornamental fish for your aquarium.

Did You Know? Ornamental fish production is Florida's most profitable aquaculture sector, with annual revenues of almost $29 million. More than 100 species are produced, including guppies, gouramis, cichlids, and koi (**Figure 9.10.5**).

tanakornsar/Shutterstock.com

Figure 9.10.4. These red tilapia are raised in an indoor freshwater pond. *What might be the benefits of raising fish in an indoor environment?*

Vital Safo/Shutterstock.com

Figure 9.10.5. *Did you ever think about where fish for ornamental ponds or your aquarium come from?* Aquaculture is the answer!

Fish Model Taxidermy Project

When you hear the word *farming*, fish farming is probably not the first thing to come to mind. With the demand for fresh fish from restaurants increasing each year, aquaculture is booming. Fish farms have become a billion-dollar industry, and more than 30 percent of all fish consumed each year are now farmed. The United Nations Food and Agriculture Organization reports that the aquaculture industry is growing three times faster than the land animal industry. Markets for fish farms are expanding rapidly because fish are highly nutritious, low in fat and calories, and may contribute to a reduced incidence of heart disease.

For this project, you will be selecting a fish to learn more about and building a 3D paper model replica of that fish.

For this activity, you will need the following:

- 3–4 sheets of clean white paper
- Marker, colored pencils, or crayons
- Scissors
- Stapler or tape

Here are your directions:

1. Select which fish you want to recreate and research the internal and external parts of the fish. Your choices are catfish, tilapia, salmon, trout, carp, and cod – the most farmed species in the US, in that order.
2. One sheet of paper will be colored to look like your fish with the outer parts labeled. The outer parts should include: gills, dorsal fin, caudal fin, pectoral fin, pelvic fin, lateral line, anal fin, scales, eyes, and mouth.
3. The other sheet of paper will show the internal parts of the fish. The internal parts should include: swim bladder, vent, intestines, stomach, liver, heart, gills, spinal cord, kidney, and reproductive system.
4. Once you are done with both sides, staple or tape them together, leaving a small opening.
5. Cut or rip your extra unused paper into small pieces.
6. Stuff the shredded pieces of paper into the opening of your fish to give it a 3D look. Don't forget to tape or staple the opening when you are finished.

Consider This

1. Why would it be important for a fish farmer to know the internal and external parts of a fish? How would that help their farm be more successful?

Ricardo Reitmeyer/Shutterstock.com

Figure 9.10.6. Bison are classified as a specialty animal for livestock production, but have historical significance to the United States. *What role have they played in our history?*

Bison

Bison (also called American buffalo) are mammals similar to cattle. They are native to North America, where they lived in large numbers (more than 50 million across the Central and Western states) up until the late 1800s. They provided food, hides for clothing and shelter, and bone, sinew, and horn for tools to the Native American populations of the Great Plains (**Figure 9.10.6**). During westward expansion by European settlers, bison were nearly hunted to extinction, with the population being reduced to fewer than 1,000 animals. Now, bison can be found in all 50 states on public and private lands, and populations have rebounded well.

Bison are large and powerful animals weighing up to 2,000 pounds. Despite their size, bison are known to be quick and agile. They are mainly sold for meat, by-products, and breeding stock in much the same way as cattle. Bison meat is a popular alternative to other livestock meat because it is full of nutrients but lower in fat, calories, and cholesterol than beef. The current bison livestock population in the US is estimated at 400,000, with about 50,000 slaughtered under state and federal inspection each year. In comparison, about 125,000 beef cattle are slaughtered each day in the United States.

Did You Know?

Bison are the national mammals of the United States.

Did You Know?

Bison can be bred with cattle to produce offspring with quality meat and a more docile nature. These are called beefalo!

Camelids

The three main groups of camelids found in the United States are alpacas, llamas, and camels. *Camelids* are even-toed mammals with a three-compartment stomach. Camelids are *pseudo-ruminant*, chewing their cud like cattle and sheep, but lacking the fourth stomach compartment (rumen). They are herbivores feeding primarily but not exclusively on grasses. They are generally found in arid or semi-arid climates, and some camelids (especially the camel) have an impressive ability to conserve and go long periods without drinking water. They are grown and raised for use as pack animals or for their hair, which is used to make many products.

Alpacas

Alpacas are domestic livestock raised and bred specifically for fiber (**Figure 9.10.7**). A single animal can produce up to ten pounds of fleece each time it is sheared. The number of times an alpaca can be sheared varies depending on factors like climate, but they are usually sheared yearly. Alpaca fleece is recognized for being soft, fine, durable, lustrous, and warm. Huacaya and Suri are the two breeds of alpacas, with Huacaya representing about 90 percent of all alpacas. Alpacas can weigh up to 200 pounds and live up to 20 years.

Canna Obscura/Shutterstock.com

Figure 9.10.7. Alpacas are raised primarily for their fleece, or hair. *What do you notice about their fleece as you look at this picture?*

Llamas

Llamas are also raised for their fiber (**Figure 9.10.8**). The fibers produced by llamas are extremely warm, coarse, strong, and durable. The fiber does not contain lanolin, which makes it *hypoallergenic*, or relatively unlikely to cause an allergic reaction. Adult llamas can weigh up to 500 pounds and live up to 25 years. Llamas make excellent guards for herds of smaller animals like sheep. The most noticeable difference between llamas and alpacas is their head shape and size.

Did You Know?

Llamas can shoot green spit up to ten feet away to ward off predators.

Noe Besso/Shutterstock.com

Figure 9.10.8. Llamas are also raised for their fleece. *What is a difference between llama and alpaca fleece?*

Ali A Suliman/Shutterstock.com

Figure 9.10.9. Camels are often associated with desert living. *Why would they be well-suited to living in this kind of landscape?*

Camels

Camels (**Figure 9.10.9**) are large ruminant animals known for their ability to go for long periods without drinking, making them useful as pack animals in dry desert climates around the world. In addition to being used as pack animals, camels are raised for milk, meat, wool, and hides. There are two major types of camel which are easily distinguished by their humps. The Arabian camel has a single hump, while the Bactrian camel has two humps. Camels are generally docile but are known to bite or kick when annoyed. When excited, they huff so sharply that saliva is expelled, giving them the reputation of spitting. Camels can weigh up to 1,400 pounds and live up to 40 years.

The current camel population in the United States is around 3,000, with most living in zoos, petting zoos, or camel ride operations. There are a few camel milk producers in the United States. The limited supply makes it expensive, which keeps the demand low.

Poultry

There are many types of specialty poultry. Small, unique populations of specialty poultry are raised to meet the needs or demands of a niche market. Certain breeds of pigeons, ducks, geese, and chickens can be considered specialty animals, though we will not cover them here.

Milan Zygmunt/Shutterstock.com

Figure 9.10.10. *What types of products might ostrich feathers be used in?*

Ostriches

An ostrich is a large, flightless, fast-running bird native to Africa with long legs and a long neck (**Figure 9.10.10**). They are the largest birds in the world, weighing up to 250 pounds and measuring up to nine feet tall. An ostrich can run faster than 40 miles per hour and defend themselves with powerful legs and claws. Ostriches are raised on farms around the world, mostly for their feathers. Ostrich skin can be used for leather products and its meat is considered a delicacy in many parts of the world.

Did You Know? An ostrich has the largest eyes of any land animal. Their eyes are two inches across!

colacat/Shutterstock.com

Figure 9.10.11. *What types of products contain emu oil?*

Emus

Emus (**Figure 9.10.11**) are large, flightless birds that are native to Australia. They are the second largest birds in the world. They have long, powerful legs built for running. Emus have three-toed feet, while the ostrich has two-toed feet. Emus are raised for their meat, skin, oil, and feathers. An emu's body contains about two gallons of oil. This oil is used in lotions, soaps, shampoos, and many other healthcare products.

Rheas

The rhea (**Figure 9.10.12**) is a large bird native to South America. They are smaller than ostriches and emus, but very similar. They are also flightless with long necks and long, powerful legs built for running. They have larger wings than ostriches and emus but use them for balance rather than flight. Adult rheas are raised for their meat, feathers, oil, and leather. The feathers of a rhea tend to be grey-brown in color, with adults sometimes weighing over 85 pounds.

Quails

Quails (**Figure 9.10.13**) are small birds from different genus and species. They are native to most continents in some form. Some, such as the king quail, are sold as pets, while many others are farm-raised for meat or eggs. Some are popular as game fowl and are hunted in the wild. Quails can be raised to be hunted on game farms and may be released to supplement the wild population. Quails are ground-nesting birds and although they can fly short distances, they spend most of their time on the ground. The most common domesticated quail is the Japanese quail. Quail eggs are known to be rich in iron, protein, and many vitamins.

Rabbits

Rabbits (**Figure 9.10.14**) are small mammals that weigh, on average, between two and four pounds. Many species of rabbits raised for meat are larger, and can grow up to 12 pounds. In the wild, rabbits tend to only live a few years, while domesticated rabbits can live ten or more years. Rabbits have particularly good eyesight and can see almost 360 degrees. This ability is extremely helpful in avoiding predators. Rabbits are mainly raised for their meat and fur. Rabbit meat is known for its high protein content, which is about 18 percent higher than the protein content of chicken meat. More than half of the world's rabbit population is found in North America.

Ondrej Prosicky/Shutterstock.com

Figure 9.10.12. This rhea is similar in many ways to an emu. *How are they the same? How are they different?*

Wirestock Creators/Shutterstock.com

Figure 9.10.13. Quails are small birds bred for a variety of livestock purposes. *Have you ever eaten quail, or seen it on a menu?*

Cora Mueller/Shutterstock.com

Figure 9.10.14. Rabbits are raised for meat and fur. *Do wild rabbits live near your home?*

Obraz/Shutterstock.com

Figure 9.10.15. *Did you know that snails are raised in conditions like these? What might be fun or interesting about raising snails?*

Other Animals

Many other animals can be raised as specialty animals. Those animals include species of goats, elk, partridges, pheasants, deer, reindeer, and snails. Snails, for example, are also grown for food (**Figure 9.10.15**). Snail farming is called *__heliciculture__* and involves raising edible land snails for human consumption and cosmetic production.

Career Corner — ANIMAL GENETICIST

Animal geneticists *analyze the genetic makeup of animals to discover which genes cause them to behave certain ways. Geneticists may also study animal health to determine what causes animals to be immune to specific diseases or fail to thrive in certain environments.*

What Do They Do? Animal geneticists research and evaluate genetic trends, improve selection decisions, investigate disease resistance, and crossbreed animals to obtain new combinations of desirable characteristics.

What Education Do I Need? A bachelor's degree in genetics or a related field such as animal science, biology, poultry science, or dairy science, followed by a master's degree, is required. Some roles may require a doctorate-level degree.

tilialucida/Shutterstock.com

How Much Money Can I Make? Animal geneticists make an average of $115,000 per year.

How Can I Get Started? You can build an SAE volunteering on a horse or cattle farm to learn more about animal selection for breeding. You can research desirable genetic traits for different livestock species and share your findings with your class, interview an animal geneticist, job shadow them for the day, or ask them to be a class speaker.

What Did We Learn?

- Specialty animals are breeds of livestock raised for production, but on a smaller scale or for more specialized purposes than in large-scale operations. They produce meat, eggs, fiber, and other products for smaller markets.
- Honey bees are raised in apiaries for honey and beeswax.
- Aquaculture is the practice of raising fish and crustaceans in controlled environments. Both marine and freshwater species can be cultivated in this way.
- Bison are raised for meat, breeding, and by-products.
- Camelids are raised for meat, milk, or fiber, as well as their ability as pack animals. These animals include llamas, alpacas, and camels.
- There are many types of birds raised as specialty poultry, including ostriches, emus, rheas, and quails.
- Rabbits are raised on farms for meat or fur.

Let's Check and See What We Know

Answer the following questions using the information provided in Lesson 9.10.

1. Which of the following would be considered a specialty animal?
   A. Cattle
   B. Swine
   C. Sheep
   D. Quail
2. _____ is the care and management of honey bees to produce honey and wax.
   A. Apiculture
   B. Horticulture
   C. Silviculture
   D. Aquaculture
3. Which of the following would *not* be grown in an aquacultural application?
   A. Shrimp
   B. Catfish
   C. Tilapia
   D. Snails
4. *True or False?* Marine aquaculture refers to the culture and growth of saltwater species of fish or other aquatic life.
5. *True or False?* Shrimp, depending on the variety, can be grown in freshwater or saltwater (marine) aquaculture applications.

6. *True or False?* Bison *cannot* be bred to cattle to produce offspring.
7. _____ are plant eaters that feed primarily but not exclusively on grasses.
   A. Herbivores
   B. Carnivores
   C. Omnivores
   D. Reptiles
8. _____ are known to chew their cud like cattle or sheep, but have a three-compartment stomach which lacks a rumen.
   A. Ruminants
   B. Pseudo-ruminants
   C. Omnivores
   D. Monogastric animals
9. _____ are known to be the largest flightless birds in the world.
   A. Quails
   B. Rheas
   C. Emus
   D. Ostriches
10. *True or False?* Rabbit meat has a very low protein content compared to other animals.

9

Let's Apply What We Know

1. Explain how raising fresh and marine species for human consumption differ from one another. In what ways are they similar?
2. Research and explain how specialty animal production has changed in the United States over the last 10, 25, and 50 years.
3. Working with a partner, select one of the specialty animals covered in this lesson. Research where in the United States specialty farms are located for this species. Why do you think they are located there?

Academic Activities

1. **Science.** Aquaculture systems must be tested and monitored closely for oxygen, pH, and nutrients in the water. Collect different water samples from ponds, streams, lakes and other water sources in your community. Test the water to determine the amount of oxygen, dissolved nitrogen, and the pH of the water.
2. **Engineering.** Beef cattle and bison operations are similar in many ways. However, bison are much larger and stronger than cattle and the equipment required to work them must be must stronger than what is used in the cattle industry. Draw and explain how you would adapt standard beef cattle equipment to work for the bison industry.

Communicating about Agriculture

1. **Writing.** Pick a sector of specialty animal production discussed in this lesson. Do some research and then write an informative essay (two or three paragraphs) about how large the industry is in the US, where it is located, what it produces, and how much it produces and sells each year. Share what you learned with your class.

SAE for ALL Opportunities

1. **Entrepreneurship SAE.** Research what is needed to raise and exhibit show rabbits. Start your own rabbit breeding program and provide other students with high-quality show rabbits.
2. **Placement SAE.** Research specialty animal farms in your area or state. Volunteer at that farm for the day and learn more about the specialty animals they raise.

SAE for ALL Check-In

- How much time have you spent on your SAE this week?
- Have you logged your SAE hours?
- What challenges are you having with your SAE?
- How can your instructor help you?
- Do you have the equipment you need?

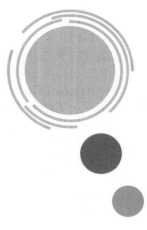

Companion Animals

Learning Outcomes

By the end of this lesson, you should be able to:
- Explain what role the companion animal industry plays in agriculture. (LO 09.11-a)
- Define companion animal. (LO 09.11-b)
- Identify popular companion animals in the US. (LO 09.11-c)
- Describe what factors should be considered in choosing an appropriate companion animal. (LO 09.11-d)

Word to Know

service animals

Before You Begin

Do you have a best friend? I do! We do just about everything together. My best friend is my dog, Roxie. She is a four-year-old black labradoodle, and I've had her since she was six weeks old. She loves to go for car rides and hikes. She will even ride in my kayak with me when we go to the river. Her favorite thing is when I take her to the dog park, though. She just loves to get out and run with the other dogs!

The industry that produces and sells pets is a large one in the US, but it's one that is invisible to many people. Often, people don't give much thought to where these animals that we love come from. Have you ever thought about how dogs, cats, birds, and other animals get to a pet store? Does your community have an animal shelter? How do you think those animals got there? Discuss your thoughts with a classmate.

9

Companion Animals and Agriculture

You might be surprised to find a lesson all about companion animals in your agriculture text, but breeding, raising, and providing medical care to companion animals are all a part of the animal science pathway, and an important part of the agricultural industry (**Figure 9.11.1**). Companion animals are a $5 billion industry in the US, with most of that revenue produced in terms of providing medical care to our pets.

Breeding and raising pets for people happens on a smaller scale than mass livestock production, but many of the principles are the same. Good breeders look for healthy animals, select for desirable characteristics, and help their offspring start their lives healthy before sending them off to new and loving homes. There is demand for animal health care that comes along with this, and other businesses spring up around providing services (walking, boarding, grooming, training) or products (food, toys, beds) for your pets.

Companion Animals

Companion animals include dogs, cats, and other species of animals raised for the purpose of living with humans for pleasure, companionship, and help. Evidence of people having companion animals can be found in some of the earliest written documents. According to the archaeological record, the first domesticated animal was the wolf, the ancestor of all modern-day dogs. Wolves were not domesticated for livestock purposes, but rather as companions.

Today, these animals are usually referred to as pets. Seventy percent of Americans own pets (**Figure 9.11.2**). Dogs are the most popular pets in the US by a wide margin, followed by cats, fish, and reptiles. Pet owners spend large amounts of money each year on food, gear, and care for their animals. Some veterinarian offices only take care of pets because of the large pet population.

Types of Companion Animals

Dogs

Dogs are thought to be the earliest animal tamed for companionship and human use. Descended from the wolf, modern dogs have been selectively bred for specific uses and vary greatly in size and type (**Figure 9.11.3**). The smallest breeds grow to about two pounds and the largest to a size of more than 200 pounds.

Figure 9.11.1. Many people buy their puppies from a local breeder, or someone who runs a small business producing puppies of a particular breed or type. *Do you know of any of these businesses in your area? Would you have thought of this as agriculture?*

Figure 9.11.2. Pets bring joy and companionship to our lives. *Have you ever had a pet?*

Figure 9.11.3. Dogs come in a wide variety of temperaments and sizes. *What kind of qualities would best match your family's lifestyle and space?*

Dogs are used for hunting, herding, work, entertainment, and companionship. They have wide-ranging temperaments. Breed characteristics must be considered when choosing a dog. You would not want a highly active large breed to live in a small apartment. In the same way, a small toy breed would not be useful for herding cattle on a large ranch.

When cared for properly, some dog breeds can live for 15 years. Owning a dog requires a long-term commitment and careful thought about the responsibility to provide proper care. Carefully research the characteristics of dog breeds before making a choice so that you can be more knowledgeable of your requirements as an owner. Dogs make wonderful companions for many people.

Hands-On Activity

In-Class Dog Show

Have you ever watched a dog show on TV? To the untrained eye, a dog show can seem like a beauty pageant. However, dog breeders would be quick to tell you there's more to it. A dog show involves the presentation of purebred dogs to be judged for conformity to their respective breed standards. The decisions judges make in the ring affect the decisions breeders make when planning the next generations.

Canden Scales/Shutterstock.com

For this activity, you will need:

- Poster board
- Pencil or pen
- Markers/crayons/colored pencils
- Scissors
- Device for internet research

Here are your instructions:

1. Select a dog breed from the American Kennel Club (AKC) list of registered dog breeds. Make sure to pick a backup breed in case your first choice is already taken!
2. Recreate a life-size version of your dog (or as close as your poster will allow). Use the internet to research accurate height and length measurements. The dog should be drawn in side profile. The coat color should also be an acceptable color by the breed association.
3. Write a biography about your dog to be read by the announcer during the in-class dog show.

Your dog's biography should include their name, gender, age, and note at least two special characteristics that represent your dog breed.
4. Research how to properly show your breed of dog. It is very helpful to watch AKC dog shows that display your breed being shown in the ring.
5. Prepare your dog for the in-class dog show. Your dog must have a collar and be on a leash. These can be real or handmade.
6. On the day of the show, a course will be laid out in a circle or square. You and your dog should line up and follow along the edge of the course. You should "trot" with your dog to accurately reflect a real dog show.

Consider This

1. Do you think having a real dog would have been easier or harder for this activity? Why?

Etienne Outram/Shutterstock.com

Figure 9.11.4. It is believed that cats were first domesticated, at least in part, to manage pest control around food and grain storage sites. Cats are still useful for this purpose today!

Cats

Cats were domesticated thousands of years ago. Cats were probably originally desired for their ability to control rodent populations. This was especially helpful around grain and food storage sites. Even today, cats are helpful around farms at controlling unwanted mice and other rodents (**Figure 9.11.4**).

Cats can be quickly and easily trained to use a litter box from a young age. Cats are quiet animals and are known for their attention to personal cleaning and care (**Figure 9.11.5**). Cats make great pets for people with small homes who do not have much outdoor space. There are many breeds of cats, with two main classifications based on the length of their hair. *Longhair* breeds have longer hair and require more grooming and care than the *shorthair* breeds.

Xseon/Shutterstock.com

Figure 9.11.5. Cats are known to be clean animals and can be a good choice for small spaces.

Did You Know?

Have you ever heard the saying a cat has nine lives? The origin of the saying can be traced back to Ancient Egypt, as they believed that Atum-Ra, a god in feline form, gave birth to eight other gods.

Other Companion Animals

Many other animals are classified as companion animals, and the list seems to keep growing. Potential companion animal types may include:

- Rabbits
- Hamsters, gerbils, guinea pigs, or other small rodents (**Figure 9.11.6**)
- Reptiles like snakes, turtles, and lizards
- Birds, such as parakeets, cockatiels, parrots, and finches (**Figure 9.11.7**)
- Fish (saltwater or freshwater)
- Exotic pets like tarantula spiders

Miroslav Hlavko/Shutterstock.com

Figure 9.11.6. Guinea pigs remain a popular pet. *Why do you think guinea pigs are popular companion animals?*

Kateryna_Moroz/Shutterstock.com

Figure 9.11.7. Finches, like this rainbow finch, can be kept as pets. *What qualities might make a bird a good pet?*

Special care must be taken when choosing a nontypical pet. Some species of animals require special care that can be time consuming or very costly. The work or expense in caring for an unfamiliar animal may not be evident upfront, and you want to be prepared to make a wise decision (**Figure 9.11.8**). It is also unlawful to keep certain species of animals. Make sure you consider all the potential challenges before choosing an animal for a companion. It can be difficult to overcome the problems that come with a poor choice.

Service Animals

Some companion animals serve functions beyond providing companionship and love to humans. **Service animals** serve as aides and helpers to their owners, from seeing-eye dogs to military service animals (**Figure 9.11.9**). Service animals are almost exclusively dogs, and service animals receive special training to assist people with disabilities. People who suffer from severe anxiety or psychiatric issues may have emotional support animals that help them manage their symptoms. Emotional support animals may include any type of animal, including some you would find surprising: chickens, parrots, and miniature horses are popular choices!

Did You Know? Magawa, a giant African pouched rat, was trained to detect landmines. Over his career in Cambodia, he detected dozens of unexploded mines, and is largely credited with saving many human lives!

Wild Animals for Pets

Wild animals should not be kept as pets and in some locations, it is illegal to do so. Wild animals are not domesticated and even if raised from a young age they can behave in unpredictable ways that can be unhealthy for them and risky or dangerous to you and others (**Figure 9.11.10**). Wild animals have specific needs from their natural habitat that differ from what can be provided in a home environment. They may also carry diseases that spread to other pets or even to humans. It is always best to appreciate and enjoy wild animals in their natural habitat for their safety and for yours.

Figure 9.11.8. Burmese pythons have become a problem in the Florida Everglades, where the large numbers of house pets released into the wild have done great damage to the ecosystem. It is never the answer to release a pet you can't care for into the wild!

Figure 9.11.9. Service dogs are specially trained to provide help to people with all types of special needs. They are more than just pets!

Figure 9.11.10. This raccoon may look like it would be fun to play with, but owning one is likely to be anything but a party. *What problems might arise bringing this animal into your home?*

9

Choosing a Companion Animal

How do I choose a pet or companion animal that is right for me? This question requires a lot of thought. When choosing a pet, you are deciding to take care of another living creature for its entire life. Pets can live for 20 years or longer, making it valuable to make an informed decision about the type and kind of companion animal you choose. It may be helpful to consider the following factors:

- **Cost** – The cost of caring for a pet varies. Some species or breeds of animals have special needs that can be expensive. Food, housing, grooming, veterinary care, and other items must be provided.
- **Time** – Some pets are more dependent on humans than others. Some animals are needy and require lots of attention, while others are very independent and require little care or attention. Consider how long you are away from home each day and what to do for extended stays away from home like vacations. Some pets need frequent exercise and feeding and might not be the best choice if they need to be left home alone for lengthy times. It is also important to consider who will take care of your pet during extended absences.
- **Space** – Living in a small apartment with no access to outside exercise or play areas requires a different type of pet than locations with large yards or outside areas where pets can exercise and play. If you live in a rental property, check with the property owner first to make sure they allow pets.
- **Activity** – Some pets, like dogs, are regularly active, while others, like reptiles, live their entire lives in a small area. Consider the activity level of the pet you plan to have and make sure that you can accommodate their specific needs (**Figure 9.11.11**). Some animals are more inclined to limited activity and make excellent lap pets, while others are highly active and require lots of exercise and room to run or play.
- **Size** – Most all pets start small. Some remain small, while others grow to a considerable size and require lots of room. If you have a specific pet size in mind, it is better not to guess how large an animal will grow but to do your research and make an informed decision about potential size.

Figure 9.11.11. *Would this dog be happy lying around the house?* Keeping your pet happy is an important consideration when deciding to introduce one into your home.

Shevs/Shutterstock.com

- **Age** – Younger animals require lots of attention, care, training, and effort (**Figure 9.11.12**). Older animals are usually more predictable, requiring less time and attention. Consider the age of a pet that you choose and the special requirements (if any) of the animal as it grows to maturity. If a mature adult animal is what you ultimately choose, you can adopt a pet from a shelter.
- **Other Pets** – Animals usually respond well to other animals and even like having another pet around as a friend. However, this is not always the case, and consideration must be given to potential challenges with introducing new pets to existing pets.
- **Advice** – A veterinarian is a good source of advice about your pet. Be sure to talk to your local veterinarian while planning and before deciding to get a new companion animal (**Figure 9.11.13**). Veterinarians can help you understand the needs and requirements of specific animals and how they might fit with the lifestyle and resources of your family.

Responsible Pet Ownership

Owning a pet can be full of joy and rewards, but there is also responsibility that comes with the job. Owning a pet or companion animal is a privilege that should not be taken lightly. The animal must be fed, kept healthy and clean, and cared for. Pet ownership is a lifelong commitment to that animal that requires an investment in time and money. Selection of a pet for a companion animal needs to be an informed decision that is carefully made. We have already talked about how to choose the right pet, but to be a responsible owner, you should also think about the following factors:

- Where will you get your pet? There are plenty of places you can find one, from a shelter or rescue to a breeder or a pet store, but not all businesses are equal in the way that they produce and care for animals. Take the time to understand how the operation works and who is involved, and then make an informed decision about what type of organization you wish to support (**Figure 9.11.14**).
- Choose an animal that you can care for appropriately (food, shelter, health care, companionship).

Anna Hoychuk/Shutterstock.com

Figure 9.11.12. Puppies are so cute! You may not feel that way, though, if you have to spend the time to train it and pick up after it.

FamVeld/Shutterstock.com

Figure 9.11.13. Before introducing a new animal into your home, it may be a good idea to consult your veterinarian and get any questions answered. *What questions might you want to ask?*

marcinm111/Shutterstock.com

Figure 9.11.14. Adopting a dog from a shelter can be a choice that makes everyone feel good. If you adopt a dog from a shelter, though, make sure you understand any special needs the animal may have.

Syda Productions/Shutterstock.com

Figure 9.11.15. Getting your pet out for exercise makes for a happier animal, depending on what type of pet you have. *Can you imagine taking your hamster for a stroll?*

- Consider the number of animals that you can provide appropriate care for and do not exceed that number. Too many animals in one location can be unsafe and stressful for you and the animals.
- Animals that spend lots of time outdoors or that require lots of outdoor time should be confined in such a way to provide appropriate care, minimize stress, and shelter them from extreme weather conditions.
- Establish a relationship with a veterinarian to maintain quality health and prevention of parasites or diseases.
- Make sure you are aware of local laws and requirements for having a pet, including any license requirements, leash requirements, and waste disposal, noise control, and containment ordinances.
- Provide plenty of exercise for your pet and properly train them (when appropriate) for their well-being and for the benefit of other animals and people (**Figure 9.11.15**).

Career Corner ▶ SMALL ANIMAL VETERINARIAN

A **small animal veterinarian** *is responsible for the diagnosis and treatment of sickness, disease and injury in companion animals like dogs, cats, rabbits, caged birds, and other small pets.*

What Do They Do? Small animal veterinarians provide clinical and surgical care and disease surveillance to small animals, prescribe and administer medications, set fractures, administer vaccinations and draw blood for testing, euthanize terminally ill or fatally wounded animals, perform reproductive surgeries, give technical advice on products for small animals or pets, operate medical equipment like X-ray machines and ultrasound technology, and advise on measures to prevent the occurrence or spread of diseases.

Ermolaev Alexander/Shutterstock.com

What Education Do I Need? A doctorate of veterinary medicine (DVM) is required to become a small animal veterinarian.

How Much Money Can I Make? Small animal veterinarians make an average of $80,000 per year.

How Can I Get Started? You can build an SAE creating a care plan for common pets, research common vaccines dogs and cats require and create an informational poster or chart for your class, conduct an agriscience project to determine what companion animals are most popular as pets in your area, interview a small animal veterinarian, or job shadow them for the day.

What Does a Pet Cost?

The cost of owning a pet is more than just the cost of food. As an example, let us think about a dog. The list below contains potential annual dog ownership expenses:

- Cost of dog—$0 to $1,000+
- Food and Treats—$200 to $800
- Toys—$5 to $50
- Bedding—$25 to $100
- Collar/Leash—$10 to $50
- Grooming/Grooming Products—$25 to $500
- Routine Veterinary Care—$500 to $1,000
- Training—$0 to $500
- Exercise—$0 to $5,000
- Boarding—$25 to $500

The cost of the items above can vary greatly. For example, you may receive a free dog as a gift, or your dog may have an allergic reaction to food items which requires you to purchase a more expensive food. Veterinary costs can add up quickly if your dog develops a health condition that must be professionally treated. The size of your dog will determine how much food she eats, along with how much exercise she needs.

Using the math above, a dog can cost between $790 and $9,500 per year! Follow the instructions below to gather information to estimate and add the costs of owning a dog breed of your choice. Compare your results with others in your class.

1. Select a breed and research the cost to purchase an animal. Remember, even shelter pets come with costs for vaccinations and other medical expenses that are listed as rescue fees.
2. Research dog food brands and estimate the amount of food per year based on the label recommendations. For example, a dog weighing 60 pounds would need about three cups of food per day. If four cups of food equals one pound, then a 50-pound bag of dog food would last about two months. It would take six bags of food to feed your dog for one year. If each bag cost $40, the annual cost would be $240.
3. Determine what toys you want to buy and how much they cost.
4. Find the price of an appropriate dog bed or shelter.
5. Estimate the size of collar you will need for your dog when it is a puppy, when it is half grown, and when it is full grown. You will probably need at least three collars over the lifespan of your dog along with at least one leash. Add those costs to your total.
6. Research dog grooming costs for your breed and add those costs.
7. Contact a local veterinarian and get an estimate of routine health care costs for your chosen dog breed.
8. Will you train your dog? If so, will you do it yourself or pay for classes? Add those costs to your total. Any books or materials you purchase should be included in the cost.
9. Depending on the breed, your dog will require regular exercise that you will need to provide or pay for a dog walking service. Be sure and add those costs to your total.
10. You can't always take your dog with you when you go on vacation or for other extended stays away from home. Add any potential boarding costs to your yearly fees.

Consider This

1. What did you come up with for a total cost? Did it surprise you?
2. Compare your costs with other classmates to see what they come up with for cost of ownership. How are they different?
3. Comparing with your class, what breeds are the most expensive? What are the least expensive? What factors contribute the most to that difference?

9

What Did We Learn?

- Companion animal production and care are major industries in the US, contributing $5 billion to the economy annually.
- Companion animals are raised for the purpose of living with humans for pleasure, companionship, or service.
- Dogs are thought to have been the first animals domesticated. A wide variety of species of dogs exist now, bred for all kinds of different needs and purposes.
- Cats were domesticated in Ancient Egypt, with the main purpose of controlling pests around food storage sites.
- The types of animals raised for companionship continue to grow as peoples' tastes and desires change.
- Service animals provide care and aid to people with special needs, and are carefully trained to help.
- Wild animals are not appropriate pets.
- There are many types of considerations to keep in mind when looking to obtain a companion animal, including cost, time, size, age, space, and activity level.
- Responsible pet ownership requires time, care, and thought, and it is not optional; owning a pet is a privilege, and taking care of their needs becomes your job. Responsible owners consider where to buy their pets, how to shelter them, how to get care for them, provide them exercise, and follow local ordinances.

Let's Check and See What We Know

Answer the following questions using the information provided in Lesson 9.11.

1. Which of the following animals should *not* be a companion animal?
   A. Dog
   B. Cat
   C. Parakeet
   D. Cow

2. Which of these animals is most likely to become a service animal?
   A. Miniature horse
   B. Rat
   C. Cat
   D. Dog

3. Which of the following are thought to be the earliest animals tamed for human use and companionship?
   A. Horses
   B. Cats
   C. Dogs
   D. Cattle

4. Identify the pet below that is known for its independent behavior and ability to care for itself.
   A. Dog
   B. Cat
   C. Finch
   D. Guinea Pig

5. Select the animal from the list below that would be the *least* likely to make a good pet.
   A. Dog
   B. Parakeet
   C. Guinea pig
   D. Squirrel

6. What should you do if you can no longer provide appropriate care to your companion animal?
   A. Release the animal into the wild
   B. Euthanize the animal
   C. Drop it off at the veterinarian's office
   D. Find someone else who can meet the animal's specific needs and give it to them

7. Which of the following items need to be considered when selecting a companion animal?
    A. Size of the animal
    B. Age of the animal
    C. Cost of raising the animal
    D. All the these should be considered.

8. Which of the following people would probably be the best source of advice for selecting a companion animal as a pet?
    A. Lawyer   C. Car salesman
    B. Doctor   D. Veterinarian

9. *True or False?* Properly training a pet or companion animal is *not* an important factor to consider when raising them.

10. *True or False?* Wild animals make excellent pets and are often raised with great success as companion animals.

Let's Apply What We Know

1. What are some important things that should be considered when deciding on a companion animal?
2. Describe some of the responsibilities of a good companion animal owner.
3. Research and create a top ten list of the most common companion animals in the United States.

Academic Activities

1. **Social Science.** Conduct a survey of people in your school or community asking if they consider themselves a cat or dog person. Include data such as their age, gender, and if they own any companion animals currently. Use this data to create some fun charts and infographics for your class.
2. **Language Arts.** There are many poems written about pets. Authors often find a deep connection with their companion animals and want to put that connection into words. Identify five poems about animals and pets. Use your research to start creating your own animal or pet poem.

Communicating about Agriculture

1. **Visual Communication.** Create a poster encouraging responsible pet ownership. Use photos, diagrams, drawings, and infographics to make the information you present clear to viewers. Hang the poster in your classroom and share with your classmates.

SAE for ALL Opportunities

1. **Placement SAE.** Volunteer for the day, week, or month with a local veterinarian. Job shadow them to learn about their daily responsibilities.
2. **Service Learning SAE.** Volunteer at a local animal shelter. Learn proper ways to care for the animals at the shelter and help feed, exercise, and take care of their daily needs.

SAE for ALL Check-In

- How much time have you spent on your SAE this week?
- Have you logged your SAE hours?
- What challenges are you having with your SAE?
- How can your instructor help you?
- Do you have the equipment you need?

Wildlife and Natural Resources Management

SAE for ALL Profile
Going Where the Game Is

Jerry Davis grew up on a wheat farm in Oregon's Willamette Valley. Whenever he wasn't working on the farm or taking care of school work and FFA responsibilities, you could find Jerry climbing the hillsides, searching for wild game. "During my high school years, I'll bet I spent more time outside than inside," said Jerry. "I was at home exploring the hills and valleys around the farm."

Jerry enjoyed hunting and fishing with family and friends and learned the values of wildlife conservation through his studies and natural resources classes in the high school agriculture program. In his junior year, his agriculture teacher invited a wildlife law enforcement officer to speak to the class about his work in enforcing game regulations and wildlife conservation. "I soaked up every word the speaker said," said Jerry, "and by the end of this presentation in our class, I knew I wanted to be a game warden."

Jerry's SAE was about small grain production, based on his work on the family farm in high school. However, he added a supplemental project in wildlife conservation in his junior year. "I must have built 100 bird nesting boxes and duck nesting boxes in high school," said Jerry. "During my senior year, my SAE became the basis of a community service project in our FFA chapter. The agricultural mechanics classes help me build and place waterfowl nesting boxes around a local lake. The newspaper published a feature on the project, and it brought some good publicity to our FFA chapter."

Jerry graduated from high school and attended the local community college, where he earned an associate degree in wildlife management and law enforcement training and certification. Today, Jerry Davis is a senior trooper in the Oregon State Police, Fish and Wildlife division. He enforces fish and game laws and mentors new officers who enter the service.

"There is a lot to do, but there is never a dull moment in this job. One day I may be investigating poachers on Oregon's public lands, and the next day be working the coastal fisheries to make sure that commercial fishers are following wildlife regulations. A fellow trooper and I recently pulled a fisherman from the Willamette River after his boat capsized. I have the best job in the world," said Jerry.

- Did you know that taking care of wildlife falls within the AFNR umbrella?
- What wildlife species live in your area? Have you seen game officers working with them?

Top: Bill Morson/Shutterstock.com
Bottom: Bob Pool/Shutterstock.com

NOOR RADYA BINTI MD RADZI/Shutterstock.com

Wildlife and Natural Resources Careers

Preechar Bowonkitwanchai/Shutterstock.com

Learning Outcomes

By the end of this lesson, you should be able to:

- Describe how demand for wildlife, forestry, and natural resources career professionals developed in the US. (LO 10.01-a)
- Identify the pathways for WFNR careers, and careers within each path. (LO 10.01-b)
- Describe the knowledge, skills, and disposition needed for careers in wildlife and natural resources. (LO 10.01-c)

Words To Know

cartography
dendrology
ecological
  management

environmental
  management
forestry
geology

hydrology
meteorology
wildlife refuge
wilderness area

Before You Begin

Take a look at the photo in **Figure 10.1.1**. Is this the type of "office" you would like to work in every day? Do you enjoy working outdoors in nature? This lesson provides information on potential careers in wildlife, forestry, and natural resources management. What kind of work do you enjoy doing?

Francisco Blanco/Shutterstock.com

Figure 10.1.1. *Is this what you imagine when you think about a workplace? Why or why not?*

Many people have a fascination with the outdoors. We like to get outside and enjoy outdoor sports and recreation. While many outdoor activities are purely fun, there are also plenty of job and career opportunities. You may have a part-time job or SAE that takes you outside to work. There are also career opportunities in wildlife, forestry, and natural resources (WFNR) that allow you to work outside or in a laboratory or office. Regardless of where you work, there are many career pathways to choose from in WFNR. Most of these careers interact with nature – plants, animals, mountains, rivers, rocks, dirt, trees. Does that sound like fun?

WFNR in the US

People have always shared the world with wildlife and natural resources. Native Americans who lived in North America long before European settlers depended on fish and game for food. When European settlers arrived in the 1500s, they hunted and fished. The earliest of these settlers wrote about the abundance of fish and game and the forests thick with trees and wildlife. They harvested trees to build homes, farms, and towns. The abundance of wildlife led them to believe that it wasn't necessary to manage wildlife resources. As settlements grew and more settlers came, wildlife populations diminished. In time, the public became concerned and took steps to manage wildlife and natural resources sustainably.

As public interest in protecting wild resources increased, so did the market for careers in that area. Game wardens were needed to enforce state and federal laws. Biologists were needed to develop habitat restoration plans to increase the quality and diversity of wildlife. The science of wildlife management created new jobs and brought many intelligent and energetic people into the field.

And WFNR is a large field! In the United States, 568 national wildlife refuges encompass 95 million land acres and 760 million acres of water. All 50 states and five US territories have wildlife refuges and wilderness areas. A **wildlife refuge** is a protected place for wildlife (**Figure 10.1.2**). A **wilderness area** is an area where the impact of human activities is minimal or nonexistent (**Figure 10.1.3**).

Marisa Estivill/Shutterstock.com

Figure 10.1.2. The Kilauea Point National Wildlife Refuge is on the island of Kauai in Hawaii. *What kinds of jobs do you think are available in this refuge?*

Bob Grabowski/Shutterstock.com

Figure 10.1.3. The High Peaks Wilderness Area in New York is protected from human interference. *Why is it important to have these areas?*

10

LMspencer/Shutterstock.com

Figure 10.1.4. Park rangers at Grand Canyon National Park provide educational programs for all age groups. *What would you need to be good at this job?*

Thomas Barrat/Shutterstock.com

Figure 10.1.5. Visiting Fort Jefferson in the Dry Tortugas National Park requires a boat or seaplane trip from the mainland. *Where is this national park located? What kinds of jobs might be involved in just getting visitors to and from this national park?*

The US Department of the Interior handles the task of managing the nation's wild areas. Managing wildlife species requires applying many different fields: biology, genetics, plant physiology, plant and animal reproduction, and forestry.

Other careers in wildlife management enforce laws and regulations. Game wardens, park rangers, wildlife refuge managers, and park administrators share this responsibility.

The National Park Service needs wildlife interpreters who can explain to park visitors the history and characteristics of the wildlife they view there (**Figure 10.1.4**). Interpreters must design and deliver educational programs in settings from classrooms to forests. More than 318 million visitors enter a national park in the United States each year. There are 419 national parks, monuments, and historical areas in the United States, encompassing 85 million acres in all 50 states (**Figure 10.1.5**).

These are just the natural resources under the supervision of the federal government. Each state has its own state-managed and private parks, natural areas, memorials, and monuments. These parks need highly qualified people to manage park resources.

Did You Know? The need for zoologists and wildlife biologists is growing each year in the US. If you like studying wildlife in their native habitat, you may wish to explore these career options!

Careers in WFNR May Not Be What You Think

There are hundreds of different jobs and careers in wildlife, forestry, and natural resources. There are so many jobs available that it is impossible to name them all in this lesson! When most people think of WFNR jobs, they probably think of forest rangers, game wardens, and zookeepers. These jobs are good examples of career fields in WFNR, but what about these?

- IT specialist
- Chemist
- Mathematician
- Writer
- Graphic designer
- Architect
- Firefighter
- Diesel mechanic
- Meteorologist
- Airplane pilot

All of these careers exist in WFNR. Let's take the position of IT specialist. We think of IT specialists as people who install and maintain computer, video, and communications systems. IT specialists play an essential role in WFNR by designing and maintaining equipment for tracking wildlife.

They also create and manage systems that measure climatic conditions and weather patterns. These systems help meteorologists predict future weather patterns that affect wildlife. Meteorologists examine current weather patterns and predict future weather patterns, and they notify forest rangers when dry and windy weather conditions make forest fires a severe threat. In this way, meteorologists play an essential role in preventing forest fires (**Figure 10.1.6**).

If a forest fire breaks out, forest firefighters who have specialized training show up to handle the blaze (**Figure 10.1.7**). Bulldozer operators plow fire lines that prevent fire from spreading. It takes a skilled technician to drive a bulldozer and good diesel mechanics to keep it and other heavy equipment in good working order.

Airplane pilots can help put out forest fires by dropping fire-retardant chemicals and water into the fire to prevent further spreading (**Figure 10.1.8**). Chemists develop the fire-retardant chemicals used in forest firefighting to extinguish the fire without damaging the environment.

Chemists also analyze the water quality in lakes and streams to determine the types of pollutants present. They measure the amount of natural fertilizer in soils to determine if enough is present to sustain plant life.

What about the other jobs on the list above? Mathematicians? Really? Yes! Mathematicians design programs that help wildlife biologists analyze wildlife species data. We know, for example, that 68 percent of wildlife species have decreased over the last 30 years. How do we know that? Well, mathematicians and wildlife biologists gathered the data and built the models to tell us.

Figure 10.1.6. Meteorologists analyze weather patterns and predict when conditions indicate a high probability of fire danger.

Figure 10.1.7. Forest firefighters are accustomed to working in highly hazardous outdoor conditions.

Figure 10.1.8. Airplane pilots drop water and fire-retardant chemicals on forest fires.

10

Studying Bee Behavior

Bees pollinate approximately 130 agricultural crops around the world. Without bees, it would be hard to meet the food and nutrition needs of humans. Apiaries raise bees for crop pollination and honey production. In order to raise bees, it is important to understand how they communicate. Some scientists have actually studied this!

How do bees tell others where the best pollen is? They use a special dance during flight that gives clues to the other bees about the flowers in a certain area. When they see the dance, they go investigate the area on their own.

How did we discover that bees dance? Wildlife biologists conducted experiments with bees and observed their behavior.

MERCURY studio/Shutterstock.com

Consider This

1. What are some things you can learn about wildlife just by observation?
2. How do you suppose other animal species communicate?
3. If you were a wildlife biologsist, how would you design an experiment to study wildlife communication?

What about being a writer or graphic designer in WFNR? Writers prepare books and other documents that help people understand important information about WFNR. Writers and graphic designers prepare brochures and maps for local, state, and national parks. These help the public understand what they are seeing and experiencing (**Figure 10.1.9**). Cartographers design maps, and graphic designers prepare them for public service.

Architects work with engineers to design structures to help us interact with and protect wildlife. Architects design public park facilities and engineers build roads and other structures that benefit wildlife (**Figure 10.1.10**).

Joseph Sohm/Shutterstock.com

Figure 10.1.9. Writers and graphic designers create documents and maps that help park visitors understand what they're seeing.

steve estvanik/Shutterstock.com

Figure 10.1.10. Fish ladders aid salmon on their upstream migration at the Bonneville Dam on the Columbia River in Oregon. Engineers designed this structure to help salmon populations return to their homes and avoid dangerous boat channels.

Pathways in Wildlife, Forestry, and Natural Resources Careers

Ecosystem Management

Ecosystem management includes *ecological* and *environmental management*. People who work in **ecological management** look for ways to conserve and protect natural ecosystems. They do this while being mindful of the needs of humans. ***Environmental management*** brings nature and people closer together, into a relationship where both can benefit. Examples include city planners, code enforcement officers, city park rangers, construction and engineering jobs, horticulturists, urban forestry specialists, and landscape designers.

Here are some examples of ecosystem management at work:

- Natural resources conservation engineers work with a golf course to establish a buffer zone between the course and a wetland area. This buffer zone prevents fertilizers and pesticides applied on the golf course from entering the wetlands (**Figure 10.1.12**).
- City storm drains may carry liquid waste and oil from roadways into freshwater streams. An environmental management technician develops buffers to keep contaminated stormwater out of those streams.
- To provide more opportunities for outdoor recreation, city planners enlist environmental management specialists to develop nature parks, urban gardens, and hiking trails (**Figure 10.1.13**).

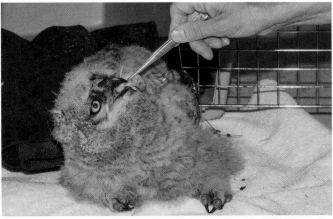

Jay Ondreicka/Shutterstock.com

Figure 10.1.11. Wildlife rehabilitators are feeding this injured infant great horned owl. *Do you notice any differences between an infant and an adult great horned owl?*

Alizada Studios/Shutterstock.com

Figure 10.1.12. Well-maintained golf courses are beautiful and pleasant to play, but streams like this are vulnerable to pesticide and fertilizer runoff. These chemicals injure wildlife and may enter municipal water systems.

Ingus Kruklitis/Shutterstock.com

Figure 10.1.13. New York's Central Park is one of five parks that provide outdoor recreation opportunities for the more than 8 million people who live in the city. Park rangers, educators, dendrologists, and wildlife biologists work within this urban park to make nature available for New Yorkers.

10

Bandersnatch/Shutterstock.com

Figure 10.1.14. The US Forest Service manages federal forests. There are almost 189 million acres of national forests in the United States.

pikselstock/Shutterstock.com

Figure 10.1.15. Working in the lumber industry requires knowledge of tree species and lumber grades.

Forestry and Forest Management

Forestry as a career is thousands of years old—it's almost as old as agriculture itself. *Forestry* involves managing, conserving, and creating forests (**Figure 10.1.14**). Modern forestry involves a variety of different career opportunities. Forestry consists of the production of lumber and wood products (**Figure 10.1.15**). Timber is often harvested in or near fragile ecosystems. A forester might engage in watershed management, erosion control, and wildlife habitat restoration.

Natural Resources Engineering

Natural resources engineers look for ways to improve the sustainability of natural resources for human use. One of the most significant responsibilities of natural resources engineers is controlling and removing non-native species from the environment (**Figure 10.1.16**).

Natural resources engineers also study how water moves through soil and develop methods for reducing erosion. They monitor the use of natural resources by humans for commercial purposes. The role of natural resources engineers has become more important as natural areas make way for human settlement, and conservation of resources is ever more important.

Fish and Wildlife Management

Fish and wildlife management is probably the best known of all WFNR career pathways. Fish and wildlife management balances the needs of wildlife with the requirements of humans (**Figure 10.1.17**).

Yingna Cai/Shutterstock.com

Figure 10.1.16. A natural resources engineer might design and develop a special harvester for removing invasive plants from a pond or lake.

mikeledray/Shutterstock.com

Figure 10.1.17. This ornithologist is placing a tracking band on a tree swallow (*Tachycineta bicolor*) to better understand the species' migration patterns. Ornithologists study the physical characteristics of birds and bird behavior.

Figure 10.1.18. The passenger pigeon went extinct in the late 1800s. These slow-flying birds were an essential source of food for humans. Unrestricted hunting decimated bird populations.

Wildlife management as a career came about as a need to protect game animals from being overharvested through hunting and fishing (**Figure 10.1.18**). Wildlife management likely began in Great Britain. Gamekeepers there managed large estates to maintain wildlife, prevent poaching, and control invasive plant and animal species. In the United States, professional guides and hunting preserve managers perform similar tasks today.

Minerals and Geology

Minerals and geology are at the edges of WFNR. *Geology* is the study of the rocks that make up Earth. It also looks at how mountains, ridges, valleys, and other natural formations come into being. The wildlife, forestry, and natural resources industries depend on Earth's ability to support wildlife, plants, animals, and humans. A study of the planet's geology helps us understand how we might sustainably utilize natural resources. There are a wide variety of careers available in this pathway, including:

- *Meteorology* - the study of Earth's weather patterns
- *Hydrology* - the study of the movement of water on Earth's surface
- *Cartography* - the creation of maps using global positioning systems

Outdoor Recreation

Outdoor recreation covers many careers that may be interesting to active people who love the outdoors. There are opportunities to serve as a local or state park ranger, a forestry technician, or a wildlife and fisheries biologist within local, state, and federal government agencies. All of these individuals work to ensure that natural resources are available for people to enjoy.

There is an educational component to many outdoor recreation careers. There are opportunities to serve as a park naturalist and provide educational programs that explain natural features to park visitors. Across the United States, there are camps and facilities where people can get away from work responsibilities and enjoy nature. Camping program directors, challenge course facilitators, and educators work in camps to provide outdoor recreation programs for visitors.

Some basic information about a number of jobs in WFNR can be found in **Figure 10.1.19**.

Jobs in WFNR

Career	Job Duties	Entry-Level Education Required	Average Salary
Fish and game officer	Enforce fish and game laws	High school diploma plus law enforcement training	$58,000
Wildlife biologist	Study animals and other wildlife	Bachelor's degree	$66,000
Forester	Manage land and forest resources	Bachelor's degree	$64,000
Forest firefighter	Control and extinguish forest fires	High school diploma plus additional certification	$52,500
Environmental engineer	Use engineering principles to solve problems in the environment	Bachelor's degree	$92,000
Logger	Harvest forest trees	High school diploma	$42,500
Veterinarian	Care for and protect animal health	Doctoral or professional degree	$99,200
Extension agent	Assist the public with natural resources issues	Bachelor's degree	$45,000

US Department of Labor

Figure 10.1.19. These careers are only a few of those available in wildlife, forestry, and natural resources. *Does anything here surprise you?*

Villiers Steyn/Shutterstock.com

Figure 10.1.20. Wildlife biologists use radio telemetry to measure the travel patterns of wildlife that are tagged with a tracking device. Taking field measurements is the hands-on application of biological sciences and mathematics.

Robert Kneschke/Shutterstock.com

Figure 10.1.21. These foresters are measuring the diameter of this tree. Foresters use special instruments to measure tree diameter and height. This procedure provides an estimate of how much usable lumber is in the tree before harvest.

Education Requirements for WFNR

If a career in WFNR interests you, start preparing for it by doing your best in all of your academic subjects. Nearly every job in the WFNR sector requires good reading, math, and science skills. They all require at least a high school diploma, so make sure that you keep your grades up. Many jobs require additional education beyond high school. For some occupations, such as a veterinarian, a doctoral degree is required. For others, such as firefighters, a certificate of specialized training in firefighting may be all that is needed.

Each career in WFNR requires specific knowledge and skills. Here is a brief list of the different types of skills and knowledge required in different WFNR careers:

- Wildlife telemetry: using radio-tracking equipment to locate and track the movements of animals in nature (**Figure 10.1.20**)
- Tree mensuration: the estimation of the volume of usable wood in a tree (**Figure 10.1.21**)
- Writing reports: preparing reports to share with government officials and the public
- Small engine maintenance and repair: chain saw repair, all-terrain vehicle (ATV) maintenance
- Heavy construction equipment operation: operating bulldozers, cranes, and heavy trucks
- Horticulture: germinating and planting tree seedlings
- Animal science: understanding the life cycles and unique characteristics of wildlife animals

Hands-On Activity

Dendrology Scavenger Hunt

Dendrology is the study of the botany of woody plants. Dendrologists study trees to determine the best methods for planting and reproduction. In this activity, you will conduct a dendrology scavenger hunt by collecting leaf samples from live tree and shrub specimens under the supervision of your instructor.

Your instructor will arrange for you to be in teams of three or four students. Your task is to visit a nature area near the school to find examples of leaves that meet the criteria determined by your instructor.

For this activity, you will need:

- A plant identification list that describes the plants from which you will collect leaves. Your instructor will provide this list.
- A printed copy of the botanical terms chart, also provided by your instructor.
- Pruning shears. *Caution:* Shears are sharp and dangerous. Your instructor will explain how to properly and safely use them. Each team will need one set of shears.

Once your instructor has given you directions about where and how to collect leaf specimens, you are to find and collect these for further review in the classroom.

Your instructor may allow you to use smartphones and the iNaturalist app to take photographs of the plants from which you remove leaf specimens. Use this app to confirm that you have selected leaves from the intended plant species.

After you have collected your specimens and returned to the area designated by your instructor, place the samples on a table or flat surface. Identify the type of leaf margin, leaf venation, and leaf shape using the botanical terms chart.

Botanical Terms Chart

Leaf Tips: Acute, Acuminate, Cuspidate, Obtuse, Emarginate

Leaf Attachment: Stalked, Sessile, Perfoliate, Rosette

Leaf Shapes: Needle, Linear, Oblong, Elliptic, Ovate, Obovate, Lanceolate, Oblanceolate, Spatulate, Orbicular, Rhomboidal, Deltoid, Reniform

Leaf Arrangement: Alternate, Opposite, Whorlet

Leaf Bases: Attenuate, Acute, Obtuse, Truncate, Oblique, Auriculate, Coradate, Sagittate, Hastate, Transversely oblong-peltate, Orbicular-peltate, Sagittate-peltate

Leaf Venation: Parallel, Palmate, Pinnate

Leaf Margins: Entire, Undulate, Crenate, Dentate, Serrate, Lobed, Pinnatifid, Pinnate-trifoliate, Pinnate, Bipinnate, Palmately lobed, Pedately lobed, Palmate-trifoliate, Palmately compound, Pelate-palmate, Tendrils, Stipulate

prodepran/Shutterstock.com

Next, read about plant species from which you collected leaves using the iNaturalist app or other resource materials.

Elect a team leader to report your findings to the rest of the class when directed by your instructor.

Consider This

1. Was it difficult to tell any species apart?
2. What stood out to you about the differences between leaves as you collected samples?
3. Did you enjoy being out in nature? Why or why not?

10

Being Career Ready

Working in wildlife, forestry, and natural resources occupations requires other career-ready skills. Although you will be spending most of your time in a natural setting, you still need to deal with people. For example, game wardens check hunters' licenses during hunting season. Practically everyone that they meet out in a remote area is going to have some type of firearm. It's essential to have practical communication skills to manage this and other situations! Responding to people calmly and authoritatively is necessary for wildlife, forestry, and natural resources careers.

LESSON 10.1 Review and Assessment

What Did We Learn?

- People have always coexisted with wildlife and natural resources. Over time, we have learned that we need to manage our relationships with wildlife and natural resources to protect and preserve them for the future.
- There are hundreds of careers in wildlife, forestry, and natural resources. These careers range from biology to law enforcement, and from geology to zoology. Not all WFNR jobs look like you might assume!
- Pathways in WFNR careers include ecosystem management, forestry and forest management, natural resources engineering, fish and wildlife management, minerals and geology, and outdoor recreation.
- Ecosystem management careers involve conserving and protecting natural ecosystems. They also find ways to do this while balancing humans' needs and wants.
- Forestry involves managing, conserving, and creating forests.
- Natural resources engineers look for ways to improve the sustainability of natural resources for human use.
- Wildlife management protects game animals from being overharvested through hunting and fishing.
- Geology helps us understand how we might sustainably utilize natural resources.
- If a career in WFNR interests you, start preparing for it by doing your best in school. All careers in WFNR require a high school diploma, and others require more education. Almost all WFNR careers require additional skills and training, and more general skills like knowing how to deal with people.

Let's Check and See What We Know

Answer the following questions using the information provided in Lesson 10.1.

1. *True or False?* Wildlife management as a career came from a need to protect game animals from being overharvested.

2. If you are planning a hiking trip for your FFA chapter and want to check the weather forecast for the dates of the event, you would consult a forecast prepared by a(n):
   A. Hydrologist.
   B. Meteorologist.
   C. Weather statistician.
   D. Agriculture teacher or extension agent.

3. A forest firefighter, in addition to having a high school diploma, must have:
   A. Certification to fight fires.
   B. A college degree.
   C. A pilot's license.
   D. Mathematical modeling expertise.

4. Using radio-tracking equipment to find and track the movements of wildlife is called:
   A. Global positioning.
   B. Global information systems.
   C. Wildlife telemetry.
   D. Wildlife sustainability.

5. Which of the following career paths requires the most formal education?
   A. Agriculture teacher
   B. Architect
   C. Botanist
   D. Veterinarian

Let's Apply What We Know

1. What is an invasive species? Are there any of these species in your community? If so, what are they?
2. If you were to work in a national forest for the US Forest Service, what are some of the tasks you might find yourself doing?
3. How did wildlife management develop into a career path?
4. What are some career options in your community related to outdoor recreation?
5. Identify the parks, natural areas, and forestlands nearest your home. Visit a local, state, or federal park near you. See if you can identify some of the careers described in this lesson.

Academic Activities

1. **Social Studies.** Pretend that you are a biologist and conduct research on invasive plant or animal species in your community. Locate the origin of the invasive species and determine whether or not this plant or animal is an invasive species in its native habitat. Does this plant or animal have any commercial value? Is it used as a food source for humans or livestock? Share your results with the class.

2. **Language Arts.** Create a poster that describes a specific career in wildlife, forestry, and natural resources. Once you have completed this task, describe the job portrayed on your poster to your classmates. Display your poster in your agriculture classroom.

10

Communicating about Agriculture

1. **Writing.** Start keeping an outdoor journal. Include a description of the parks you visit and the activities you do. Describe any hiking or camping trips you have taken. Write about your experiences.
2. **Reading.** Go to your school or community library and check out books related to nature and other outdoor topics. Pick something that you enjoy reading. Read for fun on topics that interest you. Keep a list of the books you read.

SAE for ALL Opportunities

1. **Foundational SAE.** Conduct a job shadowing experience with someone who works in one of the careers discussed in this lesson. Your agriculture teacher can help you arrange this.
2. **Foundational SAE.** Visit the US government jobs website at www.usages.gov. Use the search function to look up jobs that interest you. Note the qualifications for the position, education requirements, the location of the job, and the salary. Then write a plan that lists the steps needed to become qualified for that job. Of course, you will need a high school diploma, and some college education may be required. What other tasks would you need to do to prepare yourself for your preferred career?

SAE for ALL Check-In

- How much time have you spent on your SAE this week?
- Have you logged your SAE hours?
- What challenges are you having with your SAE?
- How can your instructor help you?
- Do you have the equipment you need?

Understanding and Researching WFNR Issues

DisobeyArt/Shutterstock.com

Essential Question

Why is it important to find accurate and reliable information about wildlife and natural resources issues?

Learning Outcomes

By the end of this lesson, you should be able to:
- Describe why it is important to understand wildlife, resources, and natural resources issues. (LO 10.02-a)
- Describe good research skills. (LO 10.02-b)
- Describe how facts, opinions, and propaganda differ. (LO 10.02-c)

Words to Know

bias	issue	propaganda
facts	opinions	social media

Before You Begin

Let's say you come home from school, and there is a letter waiting for you. Who could be writing you a letter? You look at the front of the envelope and it is from a local conservation group. You tear open the envelope and read this:

The monarch butterfly population has been declining steadily for decades now. This is due to habitat loss, climate change, and other factors. We urgently need your help to prevent the monarch butterfly from being completely wiped out (**Figure 10.2.1**).

The letter goes on to ask you to make a financial contribution to the work of the conservation group. You put down the letter, and you think, how do I know that this is a real issue? How do I know that this group is not trying to rip me off? And how in the world did they get my name and address!?

*These are all excellent questions. How do you know that the news you're reading and the information you're receiving online or in the mail are accurate? How can you make good decisions about spending your money, time, and other resources without knowing if the information you're receiving is correct? (**Figure 10.2.2**)*

Sean Xu/Shutterstock.com

Figure 10.2.1. The monarch butterfly is a species of concern. Conservation efforts are underway to protect the species.

Figure 10.2.2. *How can you confirm that this information is correct? Where would you go?*

Supriya07/Shutterstock.com

Issues in WFNR

An ***issue*** is a problem that typically involves a dispute between two or more parties. Issues address topics of concern to a broad range of people. For instance, climate change is a significant issue in the United States. *Climate change* is a term describing how Earth is becoming gradually warmer due to human activity. Some people believe that we are not doing enough to reduce its effects. Others believe that climate change is not a significant problem, and still others do not believe it is real.

In wildlife, forestry, and natural resources, there are plenty of issues to discuss and problems to solve. There are issues related to climate and energy. There is considerable interest in moving toward clean energy sources, for example. There is also wide interest in protecting and conserving wilderness areas and wildlife habitats. At the same time, there are people who would like to allow private industry access to wilderness areas to extract fossil fuels. Clean drinking water is an issue in some cities and towns. There is concern there that some drinking water sources are contaminated with chemicals that cause cancer in humans and injure wildlife (**Figure 10.2.3**). Is the government doing enough to ensure that everyone can have clean drinking water? Everyone is entitled to have an opinion about these issues, but how do we talk about them with each other? How do we know what the facts are? How can we make progress in addressing these problems when not everyone agrees? Building our communication on good information is a start.

harnchoke punya/Shutterstock.com

Figure 10.2.3. The availability of clean and safe drinking water is an important issue across the United States.

Did You Know?

Body language is essential to the communication process. In a face-to-face conversation with someone, your body language, hand gestures, and facial expressions determine how the information you are conveying is interpreted.

Conflict in WFNR Case Study: The Northern Spotted Owl

People differ in their attitudes and opinions about wildlife and resources management issues. People may disagree about the facts, and these disagreements may end up in the news or legislature for resolution. One example of how conflicts can form over wildlife issues involves the case of the northern spotted owl (**Figure 10.2.4**).

In 1990, the US Fish and Wildlife Service granted the northern spotted owl protection under the Endangered Species Act. The act provides a framework for the conservation and protection of wildlife species in danger of extinction. Sometimes, though, the conservation efforts on behalf of endangered species get in the way of human activity. In the 1980s, the northern spotted owl's habitat was being destroyed by logging in old-growth forests in the Northwestern United States. *Old-growth forests* are forests where the trees are ancient and where logging activity has rarely occurred. They provide a habitat for a wide variety of wildlife species (**Figure 10.2.5**). The protection of the spotted owl prevented loggers from cutting down these trees, reducing their income and that of the communities that supported logging.

These protection efforts led to public protests and clashes between environmental groups and the logging industry. In 1991, a federal court ordered that the US Forest Service halt logging operations in forests where spotted owls live. Even with the court decision, there remains opposition. The logging industry and environmental groups continue to clash over policies related to the Endangered Species Act and the protection of the spotted owl.

C.M.Corcoran/Shutterstock.com

Peter K. Ziminski/Shutterstock.com

Figure 10.2.4. The northern spotted owl (pictured top) doesn't look like a troublemaker, but an honest desire by some people to protect it rubbed other people the wrong way (pictured below). *Are you surprised that this unassuming bird was at the center of such a big controversy?*

Jim Schwabel/Shutterstock.com

Figure 10.2.5. A nature trail cuts through the old-growth forest in the Olympic National Park in Washington. *Why would animals live in this habitat?*

10

Each group involved in the issue – environmental groups, the logging industry, and government agencies – is working to get their message out to the public. Who would you side with? Before you decide, let's talk about research skills.

Research Skills

Getting involved in an issue takes a willingness to hunt down the facts. You need good research skills if you're going to get the best information about a subject. If you support your opinion with solid research from credible sources, you have a better chance of getting your point across to others. But how can you be sure that the information you find during the research process is accurate?

Finding fact-based information starts with looking at the types of sources you use in your research. The internet is an excellent source of entertainment, and it also has valuable sources of information. You have to be careful and selective in what you use from the internet, though. Some information found online is misleading and inaccurate. Through practice, you'll be able to determine which websites have helpful information and which ones do not.

Pay attention to detail as you conduct research. One of the ways that you can do this is by taking good notes. You can take notes in a variety of ways, whether on a laptop, notes apps, or with paper and pen. A pen and notebook may seem like an old-fashioned approach to collecting information, but it does have benefits (**Figure 10.2.6**). If you choose to use a paper notebook and pen, find one you like and keep it with you. A notebook and pen doesn't do you any good if it's not with you! Record information in a manner that helps you retrieve it later.

Make sure to use your notes to capture the thoughts and information necessary for the research you're doing. Good researchers have practical time management skills (**Figure 10.2.7**). In other words, you have to break down a research project into manageable parts and then make detail-oriented plans to complete each section.

Witthawat/Shutterstock.com

Figure 10.2.6. Taking notes on paper helps you remember important facts and information.

fizkes/Shutterstock.com

Figure 10.2.7. Conducting research requires patience.

#School_Success

Good Study Habits

One of the hard parts of being a student is remembering the facts and information that your teacher shares with you in class. Keeping up with this information can be hard to do, and studying for a test can be hard.

There is one good trick you can use to remember information accurately and for a longer period of time. It's pretty simple, too. All you have to do is write down the key facts and information in a paper notebook. That's right: a paper notebook! Research studies indicate that people remember facts longer and can recall information more easily if they keep handwritten notes instead of notes on a computer or smartphone.

Three good reasons for writing things down on paper:
1. Many people can write faster on paper than on a computer keyboard.
2. The notes you take by hand tend to be more accurate and detailed.
3. Handwriting triggers your brain to remember the things you're writing.

Consider This

1. What are some things you might do to take better notes of important facts or information?
2. What kind of strategies work best for you for organizing your notes?

Social Media

We live in a connected world. According to the Pew Research Center, more than 97 percent of Americans over 18 own a mobile phone. The largest group of phone owners are 18 to 29 years old. Not only do almost all adults own a mobile phone, but seven in ten adults use social media for entertainment and news. Once again, 18- to 29-year-olds make up the largest group of social media users. About half of all Americans get their news from social media. What is social media? **Social media** is defined as electronic communication between members of an online social network created to share information, ideas, messages, and other content (**Figure 10.2.8**).

Social media can be great for many things, but there is a problem. Social media is not a reliable news source, in general. The Pew Research Center found that the people who get their news from social media are less likely to receive accurate information. When asked about important issues facing the country, 85 percent of respondents said they could not agree on the basic facts! Social media has many positive uses, but it is not the best place to find facts and accurate information about issues.

Wachiwit/Shutterstock.com

chrisdorney/Shutterstock.com

Worawee Meepian/Shutterstock.com

Figure 10.2.8. Popular social media sites include Facebook, Reddit, Instagram, WhatsApp?, TikTok, and Twitter.

Identify Fact-Based Sources of Information

You can do many things to ensure that the sources you're using for information are as factual as possible. Find out as much as you can about the author or organization that prepared the information you're looking at. Are they qualified to provide information on the subject? Have they written other publications on similar topics? When looking at a source of information, see if you can determine who the intended audience is. Who is the author trying to reach? Would almost any person interested in the subject matter find that the information is written in a fair and balanced manner (**Figure 10.2.9**)?

The next step is probably the hardest, and that is to determine whether the information is *fact*, *opinion*, or *propaganda*. **Facts** are objective. That means that they are widely known to be accurate. **Opinions** are often based on emotions and feelings. **Propaganda** is communication that attempts to use your emotions and feelings to lead you to think a certain way about something. Let's look at them in action!

Fact: Theodore Roosevelt was President of the United States from 1901 to 1909.

Opinion: President Theodore Roosevelt did the best job of all presidents in the area of wildlife conservation (**Figure 10.2.10**).

Propaganda: The only people who think President Roosevelt did an excellent job at conservation are lazy people who don't want to work.

Pay close attention to the language used in the information. Does the language appear to be emotional, or does it appear to be logical? Is the language designed to make you angry or cause you to feel an emotion? Read an informative article and make a note of the facts listed in it. Then cross-check that information with other credible sources to verify it. Lastly, check to see how old the source of information is. Facts that are more than five or ten years old may be out of date. The world changes rapidly, and information has to keep up with the rapid changes in society. What may have been true five years ago may not be true today.

Maxx-Studio/Shutterstock.com

Figure 10.2.9. National, state, and local news are more easily accessed today than they ever were previously. *How can you verify the accuracy of news sources?*

Everett Collection/Shutterstock.com

Figure 10.2.10. President Theodore Roosevelt at Glacier Point overlooking Yosemite Valley, May 17, 1903. Roosevelt was a famous early conservationist.

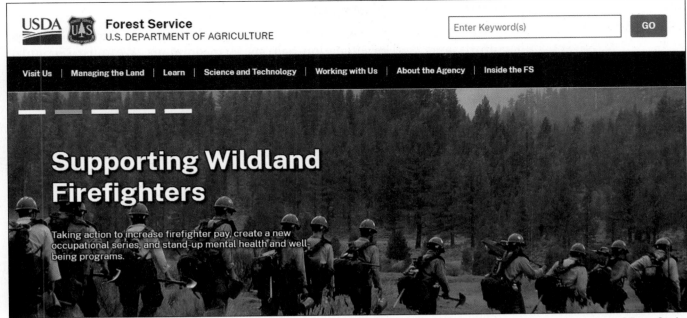

US Forest Service

Figure 10.2.11. The US Forest Service website is a reliable source of information.

Finding Quality Resources on the Internet

Finding quality resources can be a problem if you don't know where to look. Local, state, and federal websites are typically good sources of information because taxpayer dollars fund them. Government sources have a responsibility to be accurate and truthful (**Figure 10.2.11**).

For commercial websites and non-government organizations, see if there is an "About" tab on the webpage that tells you more about the organization. If you cannot find out who created the online resource, it may not be an accurate source of information.

Online news media are good resources for revealing the issues around a topic. However, do some digging to find out the source of the news reporting. If the online news article refers to a report or research document, see if you can find the same document on your own.

Using the Library

Let's face it: going to a library may not sound like that much fun, but going to the library can be exciting if you know how to use its resources to your advantage. Using library resources to prepare a report can save you a lot of time and effort. The library staff is knowledgeable in all kinds of resources. The reference desk in a library can direct you to the right places to find the information you are looking for and even help you find that information. Not only do libraries have the resources necessary for you to research and prepare your report, but they can also provide you with computing resources to type your paper and print it.

If you do not live near a library, you may still be able to use library resources. Many libraries have online resources available (**Figure 10.2.12**). You can check out books and read them online, or search for items in an online database using the library's website.

Monkey Business Images/Shutterstock.com

Figure 10.2.12. If you cannot visit the library in person, bring the library to you through online resources.

Dealing with Personal Bias

We all see the world from our own unique perspectives. Our experiences, culture, and level of education help us form opinions. We must recognize that we all have biases that affect how we view certain subjects. A **bias** is an unreasoned outlook or opinion about something. Let's say that you are out with friends, and someone says, "I don't know what it is about that restaurant, but I just don't like eating there." This is an *unreasoned statement*, which means that it was given without any thought. Bias is evident when you like or dislike something without a reasoned basis for doing so.

To guard against bias while working on a research project, remember to check for it. You can do this by:

- Using only high-quality information sources.
- Having someone else read your report. Have them point out the flaws and biases in your argument.
- Looking for information on both sides of an issue. Use that information in your message.

Hands-On Activity

Protecting the Red-Cockaded Woodpecker

The red-cockaded woodpecker, or RCW for short, used to be found in great numbers in Southern pine forests. The RCW population exceeded 3 million birds in the early 1900s. Now, fewer than 15,000 birds are living in the wild, a 99.5 percent population decrease! The US Fish and Wildlife Service placed the red-cockaded woodpecker on the endangered species list in 1970, and it remains there today.

feathercollector/Shutterstock.com

Your task is to prepare a report on the red-cockaded woodpecker. Do some detective work to determine why RCW numbers declined. Then develop a list of possible solutions that would help to increase the RCW population.

1. First, review news and media sources online to find relevant information. Check reputable online sources such as the US Fish and Wildlife Service, the Cornell Ornithology Lab, and the Nature Conservancy.
2. Outline your research report.
3. Using the outline, prepare a report that explains the issue. Include how the subject came about, and describe possible solutions for solving the problem or concern. Include examples of where these potential solutions have worked.
4. Include all of the references you used to write the report. Check your spelling and grammar.
5. Ask another student to read your report and offer suggestions for improvement.
6. Before submitting your report to your teacher for grading, give a brief oral report to the rest of the class on your findings.

What Did We Learn?

- An issue is a problem that typically involves a dispute between two or more parties. People have different opinions and feelings about many issues. If we want to be able to solve problems, we must be able to talk with others effectively.
- Research requires doing work to find credible sources and accurate information. Pay attention to details, and stay organized with how you take notes.
- Social media is not the best place to find accurate information.
- Finding fact-based information requires that you consider the source of information and separate the facts from opinion, and identify propaganda when you come across it.
- Government agencies and established non-governmental organizations can be good sources of fact-based information.
- Use the library for help in gathering credible reference information.
- Everyone has personal biases that can affect how they interact with the world. To mitigate this issue, use high-quality information, have others check your thinking for flaws, and look for information representing both sides of an issue.

Let's Check and See What We Know

Answer the following questions using the information provided in Lesson 10.2.

1. A broad problem that involves a dispute between two or more people is a(n):
   A. Issue
   B. Analysis
   C. Declaration
   D. Argument
2. The most reliable websites for information are:
   A. News outlets
   B. Social media
   C. Government websites
   D. Political organizations
3. *True or False?* Libraries are no longer good sources of reliable information.

4. One way to make reports free from bias is to:
   A. Use only one side of an issue in your report.
   B. Have someone read the report and offer feedback.
   C. Use only online information.
   D. Use information from both the internet and libraries.
5. *True or False?* If you cannot find the person or organization that created a website, it is not likely to be a reliable source of information.

10

Let's Apply What We Know

1. This lesson touched on a number of issues in WFNR. Can you think of any other issues from your community or recent news coverage? Do some research to find an issue having to do with wildlife, forestry, or natural resources that is a concern in your area. Can you explain what the controversy is? Look for different points of view.
2. It doesn't take a lot of effort to find unreliable information on social media. Look for news stories from reliable sources about misinformation in social media platforms. How can you tell the difference between what is real and what isn't?
3. Go to your school library and find ten reliable sources of information about WFNR issues. These can be books, publications, or online sources. How can you tell they are reliable?

Academic Activities

1. **Social Studies.** Read about current events in online news media. Find a topic related to wildlife, forestry, and natural resources. See if you can find the connection between the issue and its effect on humans.
2. **Language Arts.** Write a letter to the editor of your local newspaper. Express your opinion on a current event related to wildlife, forestry, and natural resources. Have your agriculture teacher read your letter and make suggestions for improvement.

Communicating about Agriculture

1. **Speaking.** Prepare an oral report based on the written report in this lesson's Hands-On Activity section. Present your information in your agriculture class.
2. **Reading.** Find a news report that connects an agricultural issue to a wildlife, forestry, and natural resources management issue. How are these issues related? How are they different?

SAE for ALL Opportunities

1. **Foundational SAE.** If you are in the process of developing your SAE program, consider using the information you learned while writing your report for this lesson. In particular, pay attention to the rules about selecting reliable sources of information. How can you use this knowledge to select useful information to help build your preferred SAE program?
2. **Foundational SAE.** While preparing your report on an issue related to wildlife, forestry, and natural resources, did you find any information that created an interest in a WFNR career? What information can you apply to the selection of your career of choice?

SAE for ALL Check-In

- How much time have you spent on your SAE this week?
- Have you logged your SAE hours?
- What challenges are you having with your SAE?
- How can your instructor help you?
- Do you have the equipment you need?

Ecosystems

Mongkolchon Akesin/Shutterstock.com

Learning Outcomes

By the end of this lesson, you should be able to:
- Define what makes an ecosystem. (LO 10.03-a)
- Describe the nutrient cycle, including decomposition, production, and consumption. (LO 10.03-b)
- Summarize the water cycle. (LO 10.03-c)
- Describe the process of succession in an ecosystem. (LO 10.03-d)
- Explain why biodiversity is important to an ecosystem. (LO 10.03-e)

Words to Know

aquifer	evaporation	pioneer species
condensation	groundwater	succession
decomposition	nutrient cycle	
ecosystem	omnivores	

Before You Begin

You might not believe that you live near wildlife, but you do. The next time you go outdoors, pause for a few minutes and take a look around. You might notice birds flying and perching on the branches of trees. You might see insects buzzing around. In the spring, you might see pollinator insects at work in flower beds and shrubs. Maybe there's a squirrel or two scampering on tree branches. If you wait long enough, you will observe plenty of wildlife species. What do you think you might see?

You share the space you live with many wildlife species. Just as we need certain things to live – a place to sleep, food to eat – the wildlife species that live among us do too. When you stop and look, you may not see a lot of wildlife at first, but there are millions of species of plants and animals. Nature is hard at work, day and night, creating a habitat for all of these creatures to live.

Przemyslaw Muszynski/Shutterstock.com

Figure 10.3.1. Black cherry (*Prunus serotina*) fruit feeds many bird species. Birds eat the fruit and then expel the undigested seeds as waste. These seeds sprout and grow into trees. In this way, birds help distribute tree seeds in the ecosystem.

What Is an Ecosystem?

An *ecosystem* is a community of living organisms that have a beneficial relationship. These organisms work together to provide for the community's needs: shelter, food, and safety.

An ecosystem depends on nonliving things. Soil nutrients, minerals, and water, as well as sunlight, weather, and climate, all play a role in the life of an ecosystem.

How does this work? A plant's seed falls into the soil, where nutrients and moisture help it to germinate and grow. It eventually produces a fruit containing seeds. A bird eats the fruit but cannot digest the hard seed. The seed passes through the bird's digestive system, where acids soften its hard coating. When the bird expels the seed, it is deposited back into the soil, ready to germinate and grow (**Figure 10.3.1**). In this example, soil, water, nutrients, seeds, and birds worked together to reproduce and spread a plant species in an ecosystem.

How an Ecosystem Works

The Nutrient Cycle

The *nutrient cycle* is a natural process of converting living organisms into organic matter and then into inorganic matter. This inorganic matter produces living things. Wow!

All organisms live and die; this is a fact of nature. Let's say that a squirrel dies from a disease. The squirrel falls to the ground in the forest and slowly decomposes among the leaves. Before long, the squirrel's body (formerly living organism, now organic matter) has decomposed down to carbon dioxide, oxygen, nitrogen, and minerals (inorganic matter). These compounds are essential nutrients for plants and trees, including oaks and hickories. Oak trees absorb this matter and produce acorns, and hickories do the same to produce nuts that are food for squirrels. The dead squirrel eventually provides the nutrients needed to produce food for other squirrels. This is the nutrient cycle.

Recycling is an essential part of an ecosystem. Water and nutrients are constantly recycled in an ecosystem. This process of continuous renewal protects the biodiversity within the ecosystem. Every living thing lives in an ecosystem, and every ecosystem is in the process of becoming something else. This constant change requires that ecosystems recycle the resources they need to keep change happening. The nutrient cycle is one example. Nature recycles nutrients back into the ecosystem so that living organisms can use them again, but this takes energy and effort. For nutrients to become available again, decomposition has to occur.

Decomposition happens when dead plants and animals break down into simpler matter such as water, sugars, and minerals. Living plants and animals take up the nutrients, water, and minerals from the decomposition process (**Figure 10.3.2**). Over time, plants lose leaves, and animals die. They become part of the organic litter in the soil. Decomposition occurs again, and the nutrients once again become available for use by other living organisms.

Let's look at this even more closely. Ecosystems can recycle everything within them completely. The living organisms that make up the nutrient recycling process are categorized as follows:

Producers. Green plants that use photosynthesis to make plant structures and sugars are producers. Photosynthesis produces plant sugars that serve as food for the next group of organisms in the nutrient cycle.

Consumers. Consumers consume the nutrients created by the producers. Consumers are animals. They cannot make their own food and have to rely on other consumers and producers for food. There are *primary, secondary,* and *tertiary consumers*, or top predators (**Figure 10.3.3**).

Perception of Reality/Shutterstock.com

Figure 10.3.2. This decaying log provides nutrients for ferns and mosses.

Figure 10.3.3. This energy pyramid explains the relationship of primary, secondary, and tertiary consumers. *Do you think people have a place on the energy pyramid?*

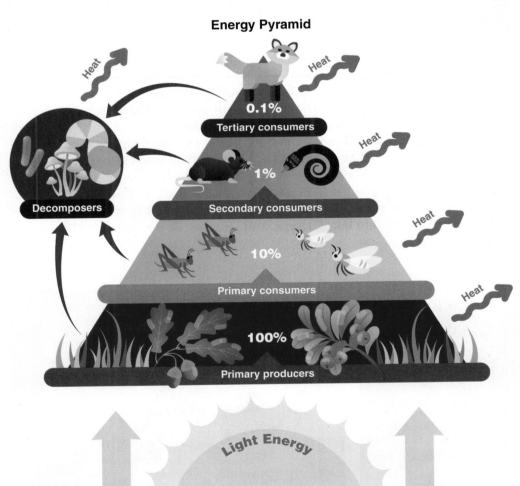

Energy Pyramid

Heat

Heat

0.1%
Tertiary consumers

Heat

Decomposers

1%
Secondary consumers

Heat

10%
Primary consumers

Heat

100%
Primary producers

Light Energy

VectorMine/Shutterstock.com

10

Figure 10.3.4. The turkey vulture is a consumer. *Would you guess it is a primary, secondary, or tertiary consumer? Why?*

hubert999/Shutterstock.com

Primary consumers are herbivores that only eat plant materials. Secondary consumers eat primary consumers (**Figure 10.3.4**). Tertiary consumers are top predators, or animals that eat other animals and have no natural enemies to prey on them. Secondary consumers and top predators are divided into carnivores and omnivores. Carnivores only eat other animals. **Omnivores** eat plants and animals.

The Water Cycle

Virtually all of the water on this planet has been here since it was formed. What does this mean? It means that Earth is very good at recycling water. All of the water that we have is recycled. Most of it is recycled naturally. The water cycle depends upon two major processes, *evaporation* and *condensation*. **Evaporation** is the process of water molecules changing from liquid to gas. **Condensation** is the reverse—the process of changing water vapor into a liquid.

Did You Know?

Scientists have discovered that a tiny bit of water may try to escape through the atmosphere into outer space. Ultraviolet light breaks down the water molecules, and only a tiny bit of hydrogen escapes into outer space.

In **Figure 10.3.5**, you see a lake on a sunny day. You see the water vapor rising up from the surface of the water. Sunlight heats the water, which converts it into water vapor. The water vapor enters the atmosphere and rises high enough to where it cools and condenses into very tiny droplets. These small droplets are so light that they don't fall immediately back to Earth. These droplets combine with other droplets to form clouds. Eventually, the water droplets in these clouds get heavy enough to fall back to Earth as precipitation in the form of rain, snow, or ice. Rainwater, once it reaches the soil surface, either runs off into lakes, rivers, and streams, or it soaks into the ground (**Figure 10.3.6**).

AlexReut/Shutterstock.com

Figure 10.3.5. *What is causing this mist to appear over this lake? Have you ever seen this in person before?*

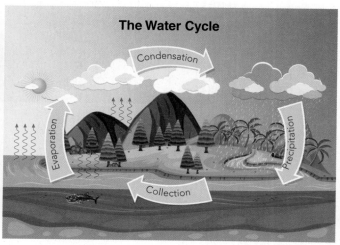

brgfx/Shutterstock.com

Figure 10.3.6. The water cycle is a process that is critically important to nature and healthy ecosystems.

Water that soaks into the ground is called **groundwater** (Figure 10.3.7). Concentrated quantities of groundwater deep beneath Earth's surface are called **aquifers**. An aquifer is like an underground lake, in terms of the amount of water it holds, but it is not the same as the lakes on Earth's surface. Instead, the water is held in beds of rock and sand. If you want to get a sense of what this might look like, take a glass jar filled with coarse sand and slowly add water until the sand is entirely saturated.

Groundwater and aquifers are essential. Groundwater supplies water to plants and trees. Aquifers can exist in arid areas. Well drillers sink wells that reach aquifers and provide water for cities and towns. In rural areas, this is still the primary water source for household use and farmland irrigation.

How do water quality technicians remove impurities from the water we drink? Cities and towns treat water with chlorine-based chemicals that kill bacteria and make the water safe to drink. Additionally, groundwater is filtered to remove solid impurities. For wells in rural areas, the water has gone through a natural filtering process underground. However, some impurities may exist in liquid form in the water. Water with too much iron in it tastes terrible and may eventually clog water pipes in your home. Soluble lead is a serious health concern that can lead to long-term problems. It is a good idea to test your water at home periodically. Lab tests can tell you what impurities are in the water and whether it is safe to drink. Water filtration systems can be added to your home's water supply to ensure its safety as well.

Groundwater

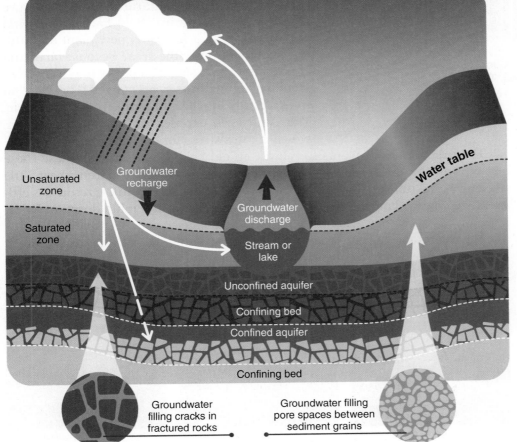

Figure 10.3.7. All of the water we use on a daily basis comes from groundwater. *Did you imagine that groundwater looked like this?*

VectorMine/Shutterstock.com

The Process of Ecological Succession

If you walked through a forest along the same path every day for a month, there would not be much observable change from one day to the next. But a forest is an ecosystem, and ecosystems are constantly changing. You may notice an ecosystem change from one season to the next. In a forest, trees lose their leaves, and grasses turn brown and go dormant. You may also see fewer forms of wildlife during the colder winter months, as they shelter from the cold weather or hibernate. Ecosystems are also always going through a much more gradual and long-term change. This change is called **succession**. This is an integral part of an ecosystem: the transition from one state to another.

Kazakova Maryia/Shutterstock.com

Figure 10.3.8. Over time, this pond filled in and became a marsh when left undisturbed. This is ecological succession at work.

Think of a city or town near you. You may have noticed over the past year that businesses open and businesses close. New houses are built, and old buildings are torn down and replaced by new buildings. New roads are constructed, and new traffic patterns are developed. This is how succession works in nature. Over time, a marshland may fill in with soil and organic matter and become dry land (**Figure 10.3.8**). A pine forest may develop into a hardwood forest. An old farm field that goes uncultivated will become covered with weeds, then woody plants, then trees, and then become a forest.

Sometimes a farmer or rancher will stop growing crops on a piece of land. They will let the field go back to its natural state. When nature converts an agricultural field back into a wild area, what habitat forms as new plant and animal species colonize the area? Because farmland is essentially bare soil, it's a pretty harsh environment for many plants to get started. There are no trees already in place for plants that need shade. The open area exposes soil to sun and wind, which creates stress for growing young plants.

The first species of plants to inhabit an old field are **pioneer species**. These are typically hardy organisms capable of surviving in times of extreme heat, little moisture, and full sun. Pioneer species are generally able to utilize resources efficiently. Pioneer plant species tend to have seeds that can remain dormant for an extended period. They also utilize nutrient and water resources rapidly, allowing them to reproduce quickly. You may be familiar with some of these hardy pioneer species, as they may be called weeds in other contexts! The first plants in some old fields are weeds, such as ragweed, crabgrass, and foxtail (**Figure 10.3.9**).

Dudaeva/Shutterstock.com

Roel Meijer/Shutterstock.com

Chris Hill/Shutterstock.com

Tatyana Parilova/Shutterstock.com

Figure 10.3.9. Hardy plants are required to take root in a bare field. *What do you think these plants have in common?*

Many of these early pioneer species are *annuals*. That is, they complete their life cycle in one year. After three or four years, annuals are replaced with *perennials*. These are plants have a life cycle that continues beyond one year. Some examples of perennial pioneer species would be bayberry, black cherry, and multiflora rose.

Eventually, tree seedlings begin to sprout among these weeds. Many of these tree species are shade tolerant and start to grow under the thick mat of weeds. However, as the seedlings grow, they become a mixture of shrubs and young trees, and the old field resembles a woodlot.

These trees will grow up and begin to shade out the sun-loving plants and shrubs. These will be replaced by shade-loving plants that can thrive in the understory of the forest. In the final stage of succession, the forest becomes divided into four layers: canopy, shrub, herbaceous, and litter.

The treetops are known as the *canopy layer*. In this layer, organisms that need plenty of sunlight make their homes (**Figure 10.3.10**).

The *shrub layer* is underneath the canopy. These trees are those smaller shade-loving plants that provide food and shelter for plant and animal species that thrive on the forest floor (**Figure 10.3.11**).

Beneath the shrub layer is the *herbaceous layer*. This is made up of smaller shade-tolerant species that can live without direct sunlight (**Figure 10.3.12**).

Finally, the *litter layer* is the layer of dead and decaying material on the forest floor. Like the other three layers, the litter layer attracts organisms that thrive in that environment (**Figure 10.3.13**).

Succession is a continuous process. The process is very gradual, and some areas of an old field may transition more quickly than others. If left undisturbed, mature forests become old-growth forests and develop the particular characteristics that separate them from younger forests. There is no final stage, though; succession means that transition is constantly occurring.

Andrej Safaric/Shutterstock.com

Figure 10.3.10. This is a unique walkway built at the height of the canopy to allow people to experience this layer of the forest firsthand.

Photodigitaal.nl/Shutterstock.com

Figure 10.3.11. The shrub layer is characterized by a variety of short, woody plant species.

Kuzmenko Viktoria photografer/Shutterstock.com

Figure 10.3.12. Wood anemone is a shade-loving plant that grows in the herbaceous layer of a forest.

Graham Corney/Shutterstock.com

Figure 10.3.13. Although this looks like merely rotting leaves, the litter layer contains insects, fungi, invertebrates, and other organisms that thrive in this environment.

Wirestock Creators/Shutterstock.com

Figure 10.3.14. The Sonoran desert in Arizona and Southern California looks desolate, but it is home to 39 plant and animal species, countless insects, and almost certainly other species we haven't discovered yet.

Biodiversity

Biodiversity is the enormous variety of life on Earth. In an ecosystem, biodiversity refers to every living thing in that ecosystem. There are more than 8 million species of plants and animals on Earth. So far, scientists have only been able to identify and name a little more than one million species.

Biodiversity is essential because it takes a variety of plants and animals to provide humans with food, clothing, and shelter. The more diverse the plant and animal life in an ecosystem, the healthier and more effective it is in producing oxygen and clean air for all living things (**Figure 10.3.14**). This diversity increases the capacity for pollinating plants; with more species of pollinators,

---

Making Connections STEM

Making a Solar Still

Most of the world's surface is covered by water, and yet many regions of Earth suffer from prolonged drought. Water is all around us in the air and soil, and we have to find a way to extract it for human use.

Water vapor condenses at a certain temperature. If we have a way to capture that condensed water vapor, we can use it as drinking water. In this activity, you will make a solar still to extract water from the soil and air.

For this activity, you will need:

- A cup or small pot
- A few medium-size stones
- A six-by-six foot sheet of clear plastic sheeting
- A shovel
- Space to work

Here are your instructions:

1. Dig a hole about four feet wide and three feet deep.
2. Place the cup or pot in the center of the bottom of the hole.
3. Cover the hole with the plastic sheet, allowing it to sag in the middle. Do not let it touch the cup or the bottom of the hole!
4. Use rocks to hold the edges of the plastic sheeting in place.

5. Place a small rock in the center of the sheet to weigh it down in the middle.

Depending upon the level of soil moisture, and the amount of water vapor present in the air, you may be able to collect as much as 32 ounces of water over two days.

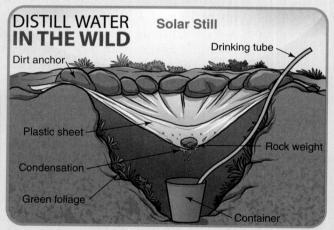
zombiu26/Shutterstock.com

Consider This

1. What would it take to create a solar still large enough to produce drinking water for a small town?
2. Would a solar still work in a desert setting? Why or why not?

---

more types of plants will be successfully pollinated and reproduce. Farmers, ranchers, and others working in the agricultural, food, and natural resources industry depend on the diversity of wildlife to sustain productivity.

We treasure the variety and diversity of life on Earth. Imagine what the world would be like if there were only a few different types of plants and animals. Living in this type of world would be very dull indeed. And what would happen if one or two of those plant species were to develop a disease that wiped it out? That would be a disaster for us! We have a great deal of biodiversity on this planet, and it is our job to understand it and protect it.

Hands-On Activity

Building a Mason Jar Terrarium

One quick and easy way to experience an ecosystem is to make a terrarium. A terrarium is a sealed glass container with soil and plants.

aon168/Shutterstock.com

To build a terrarium, you will need:

- 1 glass jar or container, such as a Mason jar
- Rocks (polished pebbles, sea glass, marbles, etc.)
- Sphagnum moss
- Potting soil
- Plants (You want plants that will stay small, like boxwood, croton, friendship plant, watermelon peperomia, cacti, and succulents.)
- Spoons
- Paper towels

Here are your directions:

1. Remove the lid and clean the inside of the glass container. If you are building a closed terrarium, do not punch or drill holes in the metal lid. A closed terrarium will not need watering, as the water vapor will condense and fall back on the plants. If you punch or drill holes in the metal terrarium lid, you have an "open terrarium" that occasionally requires watering the plants.
2. Place rocks in the container. This layer aids in drainage and aeration.
3. Soak dried sphagnum moss in water, then squeeze out the excess. Place this moss over the top of the rocks.

4. Carefully place several inches of soil on top of the moss. The plants you will be putting in the terrarium will determine the type of soil you use. The soil doesn't need to be flat.
5. Plant the plants in the soil layer, carefully pressing down on the soil to allow good contact between the soil and the roots.
6. Water the plants sparingly. You will need to water your terrarium lightly every two to four weeks if you have an open terrarium.
7. When you take your terrarium home, please place it in a location that will not receive direct sunlight. Too much sunlight will harm the plants.

Consider This

1. Once you have your terrarium in place, take note of it from time to time. How do you see the elements of an ecosystem working to keep your terrarium healthy?
2. How do you think it will look in a year?

10

What Did We Learn?

- An ecosystem is a community of living organisms working together to provide for the community's needs: shelter, food, and safety.
- The nutrient cycle converts living organisms into organic and then inorganic matter, which nourishes living things.
- Recycling is an essential part of an ecosystem. Water and nutrients are two essential items that are recycled in an ecosystem.
- Producers and consumers work together to recycle nutrients within an ecosystem.
- The water cycle consists of water vapor entering Earth's atmosphere and rising high enough to where it eventually cools and condenses into tiny droplets. These water droplets form together to become clouds, and then become heavy enough that they fall back to Earth as precipitation.
- Groundwater and aquifers are important to plants and people.
- Ecosystems go through constant, gradual, and long-term change called succession.
- Pioneer species are the first to populate fields and spaces returning to the wild.
- Biodiversity is the term we apply to the enormous variety of life on Earth.

Let's Check and See What We Know

Answer the following questions using the information provided in Lesson 10.3.

1. The gradual change an ecosystem goes through is called:
   A. Condensation
   B. Succession
   C. Biodiversity
   D. Ecostatic change

2. A community of living and nonliving organisms is called a(n):
   A. Bio-universe
   B. Biodiverse field
   C. Ecosystem
   D. Terrarium

3. The water cycle involves these actions in this order:
   A. Condensation, evaporation, precipitation
   B. Evaporation, precipitation, condensation
   C. Precipitation, condensation, evaporation
   D. Evaporation, condensation, precipitation

4. An underground pool (or "lake") of water is called a(n):
   A. Bio-aquifer
   B. Aquifer
   C. Ground lake
   D. Bio-groundwater

5. The first species usually found in an old field during succession are:
   A. Climbing vines
   B. Trees
   C. Priority plants
   D. Pioneer plants

Let's Apply What We Know

1. Examine a field near your school. Based on2 your understanding of ecological succession, what type of ecosystem does it appear to be?
2. When it rains at your home, where does the rainwater go? Does most of it soak into the ground? Does it run into a wastewater system? Do some research to explain what you find.

Academic Activities

1. **Geography.** Conduct some research in the school library on a geographic region of your choice. Pick a part of the world that is not similar to where you are currently living. What types of ecosystems exist there? How are they similar or different from your own?
2. **History.** Dig into the history of the place you live. What was the location like before it was inhabited? If you live in a town or suburban area, see if you can find out what it looked like before it became a town. Was it a forest? A grassland? A swamp? What type of ecosystem existed before your home location was inhabited?

Communicating about Agriculture

1. **Reading.** Search the news for reports of a hazardous waste leak or oil spill. Find out where the accident happened, and then see if you can guess how it might damage the ecosystem where the accident occurred.
2. **Writing.** Create a public service presentation that explains the water cycle and how humans can prevent contamination of water resources. Share your presentation with the other students in your agriculture class.

SAE for ALL Opportunities

1. **Foundational SAE.** As you begin your SAE, look for ways to use the power of the ecosystem to your advantage. If you have a garden or plant science program, how might you arrange plants so that they support each other's growth and development?
2. **Foundational SAE.** Search online for information on the types of science careers that focus on ecosystems. Examples include hydrologists, plant scientists, foresters, and oceanographers. What can you find? Do any of these careers interest you?

SAE for ALL Check-In

- How much time have you spent on your SAE this week?
- Have you logged your SAE hours?
- What challenges are you having with your SAE?
- How can your instructor help you?
- Do you have the equipment you need?

10

Creating Conservation Plans

Jacob Lund/Shutterstock.com

How does a conservation plan benefit an ecosystem?

Learning Outcomes

By the end of the lesson, you should be able to:
- Explain the purpose of conservation plans. (LO 10.04-a)
- Describe the steps of developing a conservation plan. (LO 10.04-b)
- Identify what makes a conservation plan successful. (LO 10.04-c)

Words to Know

conservation
conservation plans

estuaries
habitat restoration

indicator species
objectives

Before You Begin

Humans almost caused the extinction of the humpback whale. Humpbacks live in oceans around the world. They migrate more than 16,000 miles each year. They spend much of their time feeding in the polar seas and then move to warm tropical or subtropical waters to breed and give birth. They typically eat krill, a tiny shrimp-like organism, and small fish.

*Humpback whales were prized for their meat and fat in previous centuries, and were hunted extensively in the 1900s (**Figure 10.4.1**). By 1966, the population of humpback whales had fallen by 90 percent, placing them on the brink of extinction. Something had to be done to save the humpback whale. But how do you do that? What's involved in developing a plan that will save a species from extinction?*

Figure 10.4.1. The humpback whale was hunted almost to extinction before conservation efforts began.

Andrei Stepanov/Shutterstock.com

What Is Conservation?

So, what do we mean when we talk about conservation? **Conservation** is the careful use of resources to prevent unnecessary waste. It involves protecting species from extinction, protecting and restoring natural habitats, and doing what is necessary to protect natural ecosystems. Conservation protects the natural biological diversity of plants and animals.

It's important to remember that all species of plants and animals depend on others for their survival. If one plant or animal species disappears, that affects other plants and animals in the ecosystem. All things are connected in an ecosystem. When one element is in danger, it harms everything.

There may be particular reasons or relationships, though, that make conserving a species or habitat important. Whether it's rainforests, absorbing carbon dioxide from our atmosphere, or ice caps, keeping our planet's temperature down, whether it is preserving native plant species that support bird and insect life or protecting an endangered predator to keep a food web in place, there are plenty of reasons to conserve individual species. Once you understand what you're trying to save and why, you can communicate the value of action more clearly to other people.

A conservation plan is a way to communicate this value to others, and an effective way to take action. **Conservation plans** are put in place to protect ecosystems or species within those ecosystems.

Preparing a Conservation Plan

Identify the Problem

With any new conservation effort, the first step is to identify what we're trying to conserve, and why it needs our help. What is happening that is causing it to decline?

Let's look at monarch butterflies. The monarch butterfly is in severe decline in the United States. The main problem that monarch butterflies are facing is that the plants they need to live and reproduce are becoming scarcer. Monarch butterflies love native milkweed species. As milkweed disappears, monarch butterflies have to work harder to find a place to live and reproduce (**Figure 10.4.2**).

In this case, the first part of the conservation plan would be to identify the issue (monarch populations are declining because milkweed is scarce). We will take this information and move it forward to the next step.

Let's look at humpback whales again as another example. For decades, the fishing industry hunted whales for food and blubber, which was used to manufacture other products (candles, oil, soaps). Over time, the whaling industry wiped out a large percentage of the population. In more recent years, scientists have found that whales die from entanglement in fishing nets and being struck and injured by boats (**Figure 10.4.3**). When fewer whales were sighted, scientists concluded that the humpback whale population might be in trouble.

Sean Xu/Shutterstock.com

Figure 10.4.2. The monarch butterfly is becoming an endangered species due in part to the loss of habitat.

DesiDrewPhotography/Shutterstock.com

David Gonzalez Rebollo/Shutterstock.com

Figure 10.4.3. Humpback whales experience collisions with boat propellers (left), and this may lead to injury or death (right).

Scientists determined that humpback whales were injured or killed by boat strikes and entanglement in commercial fishing nets. They also thought that changing water temperatures might be affecting their food supply.

Did You Know? An international ban on whaling has helped to increase the population of whales. Two countries, Japan and Norway, still allow commercial whaling. In 2020, these two countries harvested 810 whales in the Atlantic and Pacific Oceans.

Develop Goals and Objectives

Once we have the problem identified, the next step is to develop the goals and objectives of a conservation plan. Take a look at the two goals below:

Goal 1: Increase the number of humpback whales in the Pacific Ocean.

Goal 2: By December 31, 2025, increase the number of humpback whales in the Pacific Ocean by 10 percent.

Which of these goals is easier to measure? If you selected Goal 2, you are right! Goal 2 has a deadline and a specific target. Setting specific goals makes it easier to know whether or not you achieved your goal (**Figure 10.4.4**).

Once you have identified the problem and set some goals to address it, you will create a list of objectives to accomplish along the way. ***Objectives*** are the intermediate steps for achieving goals. With objectives in place, you can begin to act.

Looking at the humpback whales again, a conservation plan may have included goals of increasing populations by 10 percent in the next two years. The objectives might be: *Identify issues in krill supply*, or *Reduce boat strikes*. Scientists then collect data on the availability and location of krill, and on the number, time, and location of boat strikes.

Figure 10.4.4. Conservation plan objectives need to resemble Smart Goals.

Mustafa Aydogan/Shutterstock.com

Inventory Resources

The third step in conservation plan development is to inventory resources. How many of these creatures exist? What is the situation we are working with? Biologists use sampling techniques and observations to estimate the populations of plants and animals considered in a conservation plan (**Figure 10.4.5**).

Analyze the Data

The fourth step is to analyze data. Once researchers have collected information about the situation, the next step is to determine what the data tells us.

Let's go back to our example of the humpback whale. Scientists gathered more information to test their hypothesis. They collected data on whale sightings in the oceans. They examined whale diets to determine what they were eating and whether or not whales were finding enough food. Scientists looked at their reproductive habits, and their travel patterns. They learned as much as they could.

Let's assume they found a high incidence of boat strikes in the area off the coast of Newport, Oregon. Why could this be? What could prevent this? Having specific information helps us to start thinking about how we might take steps to address the issue effectively.

Consider Alternative Solutions

Once you've analyzed all the data, the fifth step is to consider possible solutions. This part of a conservation plan looks a lot like what you would do in a research project. Once you have identified the problem and you have data that informs your thinking, you can brainstorm ways to solve the problem. We may try several different things because we're unsure what is necessary to get the job done (**Figure 10.4.6**).

Francescomoufotografo/Shutterstock.com

Lena_viridis/Shutterstock.com

Figure 10.4.5. These scientists are collecting information about animal populations. The biologist above is collecting samples of waterborne insects as part of a conservation plan inventory. The biologist in the lower photo is using motion detection cameras to collect data about animal populations and movements.

GLF Media/Shutterstock.com

Figure 10.4.6. This wildlife crossing bridge prevents animals from being killed crossing a major highway. This animal crossing is placed in this particular spot because analysts determined that this is the area where most animal accidents occur. *What do you think they did to figure this out?*

10

Do Wolves Change Rivers?

In 2014, a video appeared on YouTube called "How Wolves Change Rivers." The video hypothesized that the reintroduction of gray wolves into Yellowstone National Park significantly altered the flow of rivers and streams by controlling the elk and beaver populations. Elk and beaver populations were (supposedly) causing damage to the plants near rivers and streams. Scientists credited wolves with decreasing these populations and improving Yellowstone's water quality and availability. Did wolves have a positive and significant effect on the rivers and streams in Yellowstone? Some scientists disagree with the conclusions expressed in the video.

Don't look for a right or wrong answer. It is a complex issue, so look to gather as much good information as you can to inform your thoughts.

Bobs Creek Photography/Shutterstock.com

For this activity, you will need:

- Internet access
- Pen and paper

Here are your directions:

1. Find the video. Watch it, and take some notes.
2. Take a look at news articles and videos about wolves in Yellowstone. What do you learn about the effect wolves had on the ecosystem?
3. Take a look at news articles and videos about Yellowstone Park in general. What did you learn about the ecosystem and its water? Could any other animals or factors have affected the water supply?
4. As a class, have a discussion about what you found.

Consider This

1. What was interesting about this exercise? What was difficult?
2. What were the most persuasive arguments presented in your class? Do you agree with them?

Peter Johnsen, US Fish and Wildlife Service

Figure 10.4.7. The delta smelt is a tiny fish. Increases or decreases in its population provide insight into water quality in its native habitat.

The delta smelt is a tiny little fish that lives in estuaries in California (**Figure 10.4.7**). *Estuaries* are bodies of water where freshwater from inland sources meets saltwater from the ocean. They are notoriously sensitive habitats. This little fish is important because of what it tells us about the environment around it. The delta smelt is an *indicator species*. The increase or decline of this population serves as a metric to tell us about the environmental conditions in which it lives.

Through research, we have determined that the delta smelt population declines when pollution enters the estuary. The pollution of natural water bodies is not acceptable, and measures need to be taken to reduce it. Once scientists identify the problem, how do we protect the delta smelt?

They may try several things, including reducing the amount of agricultural runoff into natural bodies of water. They might install buffer zones along natural water bodies to prevent the flow of pesticides and chemicals into the water. They might try reducing the amount of sewage treatment water that enters the estuary. They might try reducing boat traffic and the pollution that comes with it. They might also consider *habitat restoration*, or putting time, money, and resources into proactively cleaning and fixing the area.

There are several alternatives to address the problem. Some of these may work, and some of them may not. The purpose here is to try things to see if they work to increase the delta smelt population in this area.

Once we have determined what alternatives might work, we test them. Those solutions that provide the best results become part of the conservation plan.

Get Community Buy-In

The sixth step in our plan is to make decisions about how to implement conservation efforts. In this step, we will consider the resources necessary to provide conservation measures. Is there enough money and expertise available to make a difference in this case? If not, what can we do about it? No conservation plan has a chance of success without the necessary resources to carry it out. State or local governments may get involved at this step if the expenditure of tax dollars is required to carry out the plan. This process usually means a period of public comment and input (**Figure 10.4.8**). The decision-making process can take a long time because people want and need to voice their concerns about how to best use public resources.

Barry Croom

Figure 10.4.8. Making decisions about a conservation plan requires the time and attention of many stakeholders.

Conservation Plan Implementation

Now we're on to the seventh step, which is implementation. The resources necessary to carry out the conservation plan are in place, the experts are in place to carry out the plan. All that's left is to dig in and get to work (**Figure 10.4.9**)!

Figure 10.4.9. Once the plan is agreed upon, it is time to set it in motion.

Rawpixel.com/Shutterstock.com

Evaluating the Conservation Plan

The final step in the conservation plan process is evaluation. We must determine whether or not the plan works. Biologists might go out into natural areas, collect samples, and perform analysis to determine if the plan worked. If it did, then perhaps the same plan can be reproduced in other areas where the problem exists. If the plan does not work, changes can be made based on the new data, and a new strategy can be created.

Wow! This seems like an awful lot to do just to save a single creature somewhere. Just remember (and it bears repeating): All things are dependent on each other in an ecosystem. When one element is in danger, it harms the whole ecosystem. That is why conservation planning is so important, even when it takes time and work.

Hands-On Activity

Saving the Greater Prairie Chicken

The greater prairie chicken is not currently an endangered species in the United States, but it soon may be. There are approximately 400,000 of these birds in the United States, but their population is declining rapidly. Prairie chickens live in native prairie grasses and open savannas in North Dakota, South Dakota, Nebraska, and Kansas. In recent years, pesticides and human encroachment into their habitats have resulted in declining populations.

Your task is to prepare a basic conservation plan for the greater prairie chicken. Let's go!

Danita Delimont/Shutterstock.com

Steps:

1. First, conduct research in your library or online to identify the issues leading to the greater prairie chicken population decline.
2. Prepare a list of objectives that would be necessary to increase prairie chicken numbers.
3. Identify the resources required to carry out the conservation plan. Do they need more space or habitat? Do they need to be protected from hunting?
4. Analyze any information you find about the greater prairie chicken, and formulate some solutions that might solve their problems. Once you have a list, rank the ideas from most likely to work to least likely to work. Then decide which ones you think might be most effective in increasing the greater prairie chicken population.
5. There is no need to implement or evaluate the plan at this point. However, include in your draft any information that you can find about alternatives that did work to improve the greater prairie chickens' populations in a given area.
6. Review your plan. Include maps, pictures, and charts that help to explain the ideas and information in your plan.

What Did We Learn?

- Conservation is the careful use of resources to prevent unnecessary waste. A conservation plan is a plan to protect natural ecosystems or species within those ecosystems.
- With any new conservation effort, the first thing to do is to identify what we're trying to conserve and develop a list of objectives that will lead you toward solving the problem.
- A conservation plan requires that researchers collect data about a problem, analyze it, and formulate solutions to solve the problem. These potential solutions may need to be approved by community members to secure approval or funding to carry them out.
- Once a conservation plan has been approved and implemented, the last step is to evaluate the results and make any necessary changes to the plan.

Let's Check and See What We Know

Answer the following questions using the information provided in Lesson 10.4.

1. What is the primary reason that humpback whales almost became extinct?
   A. Hunting
   B. Overeating
   C. Having to travel too far to reach food sources
   D. Climate change
2. Which of the following is *not* a conservation tactic?
   A. Installing buffer zones between agricultural land and streams
   B. Reusing old motor oil to repair roadways
   C. Collecting data about the migratory patterns of endangered species
   D. Planting native milkweed for monarch butterfly habitat
3. What is the best of these solutions for increasing monarch butterfly populations?
   A. Allow legal hunting of the birds that eat them.
   B. Build natural areas for the butterflies in cities and towns.
   C. Blanket the entire US with milkweed plants.
   D. Plant milkweed along the butterflies' migratory path.

4. *True or False?* An indicator species is a species for which the increase or decrease of its population tells us about the environmental conditions in which it lives.
5. Which of the following conservation goals is the *least* measurable of all choices listed?
   A. Increase Florida panther populations in the next few years.
   B. By December 2027, increase monarch butterfly populations by 15 percent in the Southeastern United States.
   C. Increase humpback whale populations in the North Atlantic Ocean by 8 percent before January 2030.
   D. Establish a no-fishing zone one mile off the coast of North Carolina by January 2025, to prevent the overfishing of Bluefin Tuna.

10

Let's Apply What We Know

1. What is an endangered species? How do they factor into conservation plans?
2. What are the causes for the declining population of greater prairie chickens? How does this compare to an endangered species in your state? How are they similar? How are they different?
3. What do we learn from the data we collect from animals, such as their location and travel patterns?
4. What is the purpose of developing several alternatives to solve a conservation problem?
5. What is the last step in developing and carrying out a conservation plan? Why is this final step necessary?

Academic Activities

1. **Social Studies.** Find and read articles about the northern spotted owl controversy. Protecting this species has implications on the local economies near its habitat. How did saving the northern spotted owl habitat affect the local economy? And how did protecting the owl's habitat affect the owl population?
2. **Language Arts.** Create a public service message, poster, or flyer that brings public attention to an at-risk species. Present this to the rest of the students in the agriculture class, and post it in the agriculture classroom.

Communicating about Agriculture

1. **Writing.** Prepare a report on an endangered or at-risk species of plant or animal in your community. Conduct research to determine the potential causes for the declining populations of this species.
2. **Speaking.** You completed a conservation plan draft about the greater prairie chicken in this lesson. Using that document, prepare a four-minute presentation that describes your findings and present your report in one of your classes at school. Include what you discovered about the conservation process. What did you learn from your research?

SAE for ALL Opportunities

1. **Foundational SAE.** Analyze your SAE to determine the quality and value of the natural resources you are using in it. For instance, if you are growing a garden as part of your SAE project, in what condition is the soil? Soil is an essential natural resource. Without adequate nutrients and water, the soil cannot provide the necessary support for growing plants. Complete a survey of the natural resources that you use in your SAE, and determine which ones require special attention in the area of conservation.
2. **Foundational SAE.** There are a wide variety of careers associated with conservation. Conduct an exploratory activity to determine if there are careers out there that interest you in the area of wildlife management and conservation or soil conservation.

SAE for ALL Check-In

- How much time have you spent on your SAE this week?
- Have you logged your SAE hours?
- What challenges are you having with your SAE?
- How can your instructor help you?
- Do you have the equipment you need?

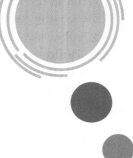

The Wildlife, Forestry, and Natural Resources Industry

Dudarev Mikhail/Shutterstock.com

Essential Question

What is the value of wildlife, forestry, and natural resources to the US economy?

Learning Outcome

By the end of this lesson, you should be able to:
- Identify the components of the wildlife, forestry, and natural resources industry. (LO 10.05-a)

Words to Know

British thermal unit (BTU)
carrying capacity
custodial management

manipulative management
nonrenewable
nuclear energy

renewable
silviculture

Before You Begin

Being outdoors is great for you! Taking a walk or going to a park helps boost your energy after a study session. Hiking in the woods is a form of exercise that helps build muscle and reduces soreness in muscles and joints. Sunlight allows your body to absorb vitamin D, a nutrient needed for strong bones.

There are plenty of things to see and do outdoors. Swimming, boating, camping, hunting, and fishing are just a few of the things that you can do. If you'd like to work outdoors, there are a ton of careers to choose from: forest ranger, fishing enforcement officer, wildlife biologist, camp counselor, and the list goes on. The wildlife, forestry, and natural resources industry employs 651,000 people, according to the US Bureau of Labor Statistics. And the pay is pretty good! People with jobs in WFNR make an average of $35.56 per hour.

So, what exactly is the wildlife, forestry, and natural resources industry? What does it do? Why does it matter? This lesson will give us a bird's-eye view of the industry. Let's take a look!

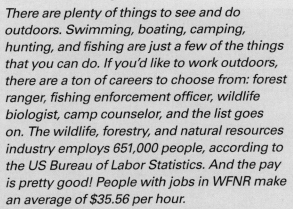

10

Nicole Glass Photography/Shutterstock.com

Figure 10.5.1. Wildlife management seeks to protect and conserve wildlife species, and this sometimes means setting boundaries on human action.

Vaclav Matous/Shutterstock.com

Figure 10.5.2. The red fox is a predator that, if left unmanaged, can quickly reduce the populations of other wildlife species.

Wildlife Management

Why do we have an industry devoted to wildlife and natural resources? Wildlife management is necessary to protect and conserve natural resources. Without wildlife management, the unchecked hunting of wildlife species could lead to extinction. Wildlife management requires that we understand animal biology, and how organisms live and grow. We need to know their preferred habitat. Scientists also study how wildlife interact with humans. Wildlife managers sometimes have to manage human behavior to protect and conserve nature (**Figure 10.5.1**). Wildlife management works for the benefit of all plants and animals, not just one wildlife species, which can require some complicated balancing. Wildlife populations have to be managed so that there are enough animals to continue to thrive, but not so many that they overrun an area (**Figure 10.5.2**). Wildlife managers ensure that the habitat supports the wildlife species in a place, including their needs for food, shelter, water, and space.

Approaches to Wildlife Management

There are two general types of wildlife management: *manipulative management* and *custodial management*. The carrying capacity of a given area determines which of the two types that wildlife managers use. The ***carrying capacity*** is the ability of the habitat to support wildlife (**Figure 10.5.3**).

Figure 10.5.3. Deer grazing in this meadow depend on its carrying capacity, or its ability to provide a sustainable food source.

oliver_schulz/Shutterstock.com

Manipulative management acts on a wildlife species by increasing or decreasing its population through changes to the food supply, changing the habitat, or managing predators. Approving a certain number of hunting licenses for a predatory animal per year, for example, is a form of manipulative management.

Custodial management prevents wildlife species from declining. The purpose is to minimize external factors such as human interaction or urban development. For example, it is appropriate in a national park to discourage the feeding of animals by visitors.

Habitat Restoration

Habitat for wildlife species includes food, shelter, water, and space to live and reproduce. Poor management, extreme weather, and climate change can negatively affect any of these four elements. When this happens, wildlife management specialists have to step in to improve the habitat and bring the elements back into balance.

Human intervention in a natural habitat can also reduce the amount of food, shelter, water, and space available to wildlife. Sometimes, wildlife management is a form of human management. We have to manage what we do as humans to protect and conserve the nature on this planet (**Figure 10.5.4**).

Jason Finn/Shutterstock.com

Figure 10.5.4. Habitat restoration often requires that areas be closed off to human traffic. *What do you think the consequences would be if people didn't abide by signs like this one?*

 **Did You Know?** Although many natural resources jobs require working outdoors in rugged conditions, people with disabilities can be successful in this career field by taking advantage of specialized equipment and training (**Figure 10.5.5**).

Figure 10.5.5. It's possible for those with different abilities to enjoy the outdoors and pursue a career in natural resources.

AndriyShevchuk/Shutterstock.com

10

Forestry and Forest Management

Forestry is the management and care of forests. Forestry is big business in the United States. Trees take water, soil nutrients, and carbon dioxide and create wood using the sun's energy. Wood is a renewable resource and one that we employ in many aspects of our lives. The US Forest Service reports that the production and sale of forest products generates $200 billion in annual revenue, and the industry employs about one million workers (**Figure 10.5.6**). The forestry industry not only produces wood products and paper, but it recycles as well: More than half of the paper produced in the US is recycled and used again.

Silviculture

Silviculture is a type of forest management that controls the establishment, growth, health, and quality of forests. The purpose of silviculture depends on the ultimate use of the woodland. For instance, some trees are grown to produce wooden timbers used in construction. Some tree species create pulpwood used to make paper products. Other forests are managed to produce chemicals or fruit. The end-use of the trees determines the method by which the forest is managed.

Silviculture involves tree production, including prescribed burning, wildlife habitat development, harvesting operations, replanting, and reseeding. Think of silviculture as a form of tree farming.

fotorobs/Shutterstock.com

Kletr/Shutterstock.com

Mark Agnor/Shutterstock.com

Jacob Lund/Shutterstock.com

Figure 10.5.6. Timber harvesting and transportation, paper production, and reforestation practices are all part of the forestry industry.

Forest Products

You might think that all we get from trees are wood and pulp for making paper. Not so! We produce hundreds of products from trees. We can create lumber for houses and furniture, toothpicks, baseball bats, musical instruments, toys, fences, decks, railroad ties, kitchen utensils, ladders, pencils, and matches. Trees also produce wood chips, or pulp. Pulp is used for making paper, paper bags, the books and notebooks that you use in your classes, and all kinds of other paper products (**Figure 10.5.7**). Everything from cardboard boxes to paper towels to wallpaper comes from wood pulp.

Mark Agnor/Shutterstock.com

Figure 10.5.7. Paper is one product derived from trees. This roll of paper will be used to make cardboard.

The value of trees does not just stop with wood and pulp; chemicals can be extracted from trees. Some trees have natural dyes and scented oils that are useful in producing other products. Turpentine is a product used in making paints. Cleaning products, insecticides, polishes, perfumes, plastics, and even personal deodorants come from the chemicals derived from trees. You may take an aspirin to relieve the pain when you have a headache. The active ingredient in aspirin is acetylsalicylic acid, which comes from certain willow trees.

If you take wood pulp and break it down into fibers, you can use those to make fabric, adhesive, floor tiles, and even luggage. Sometimes cellulose fibers are used to make helmets and hard hats. Products such as rope and twine come from the fibers of trees and other woody plants.

Trees may provide food for wildlife, but they also provide food for humans! We get a large variety of foods directly from trees. Think of all the fruit trees growing in your neighborhood—everything from almonds and apples to cherries and citrus. Pears, pecans, plums, tangerines, and walnuts are all grown on trees. Others provide food from the bark: Cinnamon is derived from the inner bark of trees from the genus *Cinnamomum* (**Figure 10.5.8**).

If you enjoy a piece of chocolate now and then, we can be grateful to the cacao tree for producing the berries that make chocolate. If you want a cup of coffee or tea from time to time, you are welcome to thank the coffee tree for producing the berries that can be dried and then ground to make coffee, or tea bushes for producing leaves that can be brewed (**Figure 10.5.9**). Hundreds of products come from trees, and scientists continue to study the secrets that trees hold to find more new and useful products.

Ciprian23/Shutterstock.com

Figure 10.5.8. This farmer is harvesting pieces of cinnamon tree bark on a plantation in Zanzibar, Tanzania. *Did you ever imagine that tree bark could be so delicious?*

Thomas Trompeter/Shutterstock.com

Figure 10.5.9. The tea we drink comes from the new green growth produced each year by tea plants like these grown near Charleston, South Carolina.

10

Energy Production

The energy we use to heat and light our homes, power our cars, and charge our smartphones comes from the environment. Energy sources can be either *renewable* or *nonrenewable*. A **renewable** energy source can quickly be replenished once it has been used. Wind energy, solar power, and geothermal energy are renewable energy sources. Hydroelectric dams use the power created by moving water to generate electricity, making water an essential renewable energy source.

Nonrenewable energy sources take a long time to replenish after being used. Coal, natural gas, and oil are examples of nonrenewable energy sources. Coal, natural gas, and oil are fossil fuels derived from the decomposition of organic matter in Earth's crust. These sources of energy took millions of years to form and can't be magically replaced.

Avigator Fortuner/Shutterstock.com

Figure 10.5.10. Refineries convert crude oil into fuel and lubricants. *Do you have any of these plants near you?*

According to the American Petroleum Institute, the oil and gas sector provides about 10 million jobs and accounts for about eight percent of the annual US economy ($110 billion). The United States produces about 11 million barrels of crude oil per day. This oil is refined into gasoline, diesel fuel, heating oil, and lubricating oil (**Figure 10.5.10**). Approximately 113 million cubic feet of natural gas is withdrawn from underground sources each day in this country and used for heating homes and businesses, cooking, and operating refrigeration and cooling equipment.

US coal production has declined in the last 30 years. Coal production is measured in **British Thermal Units**, or **BTUs**. A BTU is the amount of heat required to raise the temperature of one pound of water by one degree Fahrenheit. The amount of coal produced in the United States in recent years is about 13 quadrillion BTUs! Coal is used almost exclusively for heating, so it makes sense to measure its production in heat units. Interestingly, most coal goes into the creation of another energy source: electricity. Coal is burned to create heat that converts water to steam. The steam then passes through a turbine which operates a generator to produce electricity.

There are some problems with coal as an energy source. Burning coal creates sulfur dioxide, carbon dioxide, and other gases that harm the environment. Burning coal also releases toxic elements such as mercury and other heavy metals into the atmosphere. The coal industry is working to reduce these adverse environmental effects. Power plants use scrubbers to remove poisonous gases from coal smoke. Coal-fired power plants are also working on removing carbon dioxide from coal emissions and diverting them underground for permanent storage. While it may be impossible to eliminate the negative effects of coal emissions, the industry has reduced its pollution.

It is hard to place a price on renewable energy sources that never run out, but the value of the energy produced by solar power, wind, water, and geothermal energy tops $881 billion. All of these renewable energy sources produce about seven percent of the world's energy needs. The capacity of this industry sector is bound to grow in the future. As the world population increases, so will the demands for new and efficient energy sources.

Nuclear energy is a sector of the energy industry that sits somewhere between renewable and nonrenewable sources. **Nuclear energy** is a nonrenewable energy source that uses radioactive elements such as uranium to generate power. However, nuclear energy acts like a renewable energy source in that it produces no carbon emissions. There are 96 nuclear power plants in the United States, and those plants produce a little over 809 million kilowatt-hours of electrical energy each year. This is about 17 percent of the annual total electrical output of the United States (**Figure 10.5.11**). Careers and areas of study associated with nuclear energy include engineering, geology, chemistry, and mathematics.

hrui/Shutterstock.com

Figure 10.5.11. Nuclear energy uses the heat generated from splitting uranium atoms to heat water into steam which is then used to power a turbine. The cooling towers remove heat from the water to be recycled and used again in the energy production process.

Hands-On Activity

Harnessing Solar Energy for Good

Solar energy is easy to harness when you know what you're doing. Do you think you can do it? Let's find out!

Here are your materials:

- Empty pizza box
- Aluminum foil
- Plastic wrap
- Tape
- Scissors
- Ruler
- Paper plate
- Marshmallows
- Chocolate
- Graham crackers

Here are your instructions:

1. Cut a flap in the top of the pizza box, leaving a 1–2 inch border on the sides and front.
2. Wrap the flap in aluminum foil, shiny side out, so that the foil faces the inside of the box. Tape the foil in place.
3. Wrap the rest of the inside of the pizza box with aluminum foil in the same way, taping it in place.
4. On the outside of the pizza box, wrap the hole made by the flap with plastic wrap. Stretch it tightly and tape it in place.
5. Place marshmallows on the paper plate. Place the plate inside the box, propping the box open with a ruler.
6. Place your box in the sun, so that the sun is flowing in! Watch while the sun works its magic. It may take up to an hour to cook your marshmallows, so be patient!
7. Once the marshmallows are cooked (careful! They're hot!), feel free to eat them – with or without graham crackers and chocolate.

Consider This

1. How did the marshmallows cook? Did the results differ from using other methods?
2. Did the results of this experiment surprise you at all? Why or why not?
3. What else can you harness solar energy to do?

Outdoor Recreation

The outdoor recreation industry adds more than $374 billion to the US economy every year. Boating and fishing make up the largest share of this, generating almost $31 billion. The recreational vehicle sector is the second largest outdoor activity, based on its economic impact. RV owners contribute $19 billion to the US economy each year.

Do you like to go skiing? If you do, you are in good company. Americans like to ski a lot. This activity contributes $1.2 billion to the economy every year (**Figure 10.5.12**).

It takes a lot of gear to get Americans outdoors! People want outdoor clothing, fishing gear, skis, boats, tents, and all kinds of other retail items to enjoy the outdoors. Food, lodging, and travel services are also an essential part of the outdoor recreation industry.

Festivals and Outdoor Events

Agricultural education students participate in state and local fairs through the livestock show program and competitive agricultural events. These are indoor/outdoor agricultural activities. Fairs like this are a form of recreation. Fairs were historically gatherings of people who came together to celebrate the crop harvest and share farming achievements.

Artur Didyk/Shutterstock.com

yanik88/Shutterstock.com

AboutLife/Shutterstock.com

gorillaimages/Shutterstock.com

Figure 10.5.12. Snow sports are a significant outdoor recreation activity in the United States.

Festivals, outdoor art exhibitions, and concerts are all forms of outdoor recreation (**Figure 10.5.13**). These activities make up 25 percent of the outdoor recreation industry.

 **Did You Know?** During the COVID pandemic, people were looking for safe activities that allowed for social distancing, and many took the opportunity to rediscover outdoor activities. Interest in national parks, camping, hiking, and other outdoor activities jumped exponentially, leading in some cases to long lines, messy camping areas, and equipment shortages.

Leszek Glasner/Shutterstock.com

Figure 10.5.13. Outdoor musical concerts and festivals are also a part of outdoor recreation.

Outdoor recreation helps to improve human health, relieve stress, and provide a break from work. More than that, the outdoor recreation industry offers viable career options for people who like to work outdoors. There's a lot to like about outdoor recreation!

Making Connections #Language_Arts

Hiking Journal

You will go on a short hike for this activity. You will be doing more, though, than just walking in the woods. Take along a notebook and pen or pencil to record what you see and experience.

Before going on your outdoor walk, take appropriate safety precautions. Go with your class or a friend or group of friends. Make sure that your parents and teacher know where you are. Stay on the trail and avoid activities that could cause injury to you or the plants and animals that inhabit the forest or park.

For this activity, you will need:
- Shoes
- A notebook
- Pen or pencil

Steps:
1. Describe the plants and other living organisms you see.
2. Describe the weather, the temperature, wind, and cloud cover.
3. Write about the level of difficulty in walking along the trails you encounter. Is it easy or hard? Did you wear the proper clothing and footwear for a walk in the woods?
4. Share your observations with your teacher and classmates.

Consider This

1. Describe the best and worst parts of the experience.
2. What would you do differently the next time you go for a walk in the woods?
3. How did you feel about having a break from plugging in and using technology?

10

What Did We Learn?

- The wildlife, forestry, and natural resources industry consists of wildlife management, forestry and forest management, the energy industry, and outdoor recreation.
- Wildlife management is necessary to conserve wildlife and natural resources and prevent the excessive harvesting of wildlife species. It may include manipulative management, custodial management, and habitat restoration.
- Forestry and the forest management industry produce trees for wood and wood products, chemicals, and food.
- The energy industry produces renewable and nonrenewable energy from natural resources.
- Outdoor recreation consists of activities that Americans like to do outside, like skiing, boating, camping, and attending fairs and festivals.

Let's Check and See What We Know

Answer the following questions using the information provided in Lesson 10.5.

1. Which of the following is *not* produced from trees?
   A. Wallpaper   C. Tangerines
   B. Twine   D. Asphalt

2. Wildlife habitat includes all of the following *except*:
   A. Transportation   C. Water
   B. Shelter   D. Food

3. The largest component of the outdoor recreation industry by revenue is:
   A. Skiing
   B. RV travel
   C. Boating and fishing
   D. Fairs

4. Which of the following is a nonrenewable energy source?
   A. Wind power   C. Solar power
   B. Water power   D. Nuclear power

5. Why is coal production declining in the US?
   A. Coal emissions harm the environment
   B. The increase of renewable energy sources
   C. Burning coal releases heavy metals such as mercury into the environment
   D. All of these play some role.

Let's Apply What We Know

1. Why is wildlife management important? What sorts of wildlife management activities happen in your community?
2. Make a list of the products you've used since you got up this morning that come from wood. Are there substitutes for those products? Why or why not?
3. How do renewable and nonrenewable energy sources differ? What are the pros and cons of each?

4. How does outdoor recreation affect the US population? What impact does it have on the economy? On our resources? On our health? Find some details to support your answer.

Academic Activities

1. **Reading.** Visit a reputable online newspaper or magazine and read about advances in energy production. Prepare a brief issue paper that discusses the energy sources of the future, and predict which energy sources have the potential for the most growth and development.
2. **Writing.** Design a poster that describes the four elements that wildlife need to survive: food, water, shelter, and space. Highlight at least one wildlife species in your poster.

Communicating about Agriculture

1. **Writing.** Pretend that you are hiring summer staffers at your outdoor recreation business. Create an advertisement for your business— what it does, where it is, what activities are offered—and write a description of what you need summer staff to do for you. Illustrate it with photographs or images you draw or find online. How can you get people excited to come and work with you?

SAE for ALL Opportunities

1. **Foundational SAE.** For many students, WFNR is a bit of a surprise. There is a lot to explore here! What topics are you most interested in? Brainstorm options for integrating your interests in wildlife, forestry, and natural resources into an SAE. Talk to your agriculture teacher if you need some direction.

SAE for ALL Check-In

- How much time have you spent on your SAE this week?
- Have you logged your SAE hours?
- What challenges are you having with your SAE?
- How can your instructor help you?
- Do you have the equipment you need?

10

United States Forest Regions

Learning Outcomes

By the end of this lesson, you should be able to:
- Identify the three major forest regions in the United States. (LO 10.06-a)
- Describe the characteristics of the primary forest regions of the United States. (LO 10.06-b)
- Explain why different species thrive in different regions. (LO 10.06-c)

Words to Know

boreal forest

rainforest

taiga

temperate forest

tropical forest

Before You Begin

When the first people came to the area that is now the Southeastern United States, they probably saw a landscape that looked like **Figure 10.6.1**. *Almost 90 million acres of the American Southeast were covered by longleaf pine and savannah grassland species, according to the United States Forest Service (USFS). Now only five million acres of longleaf pine forest remain.*

What caused this decline? One reason is that national forest fire policy once required that all forest fires be immediately extinguished. This policy prevented the natural fires that longleaf pines need to control competing species. Urban development, agriculture, and unrestrained timber harvesting reduced this type of forest to a small fraction of what it was. What does this mean for our forests? Let's read on together and find out.

US Forest Service

Figure 10.6.1. Longleaf pines, like these, were once prevalent in the Southeast. *What happened to these trees?*

There are three major types of forests in the United States: temperate, boreal, and tropical. Each brings unique benefits to the US.

Temperate Forests

Temperate forests are located primarily in the Eastern United States. These forests experience four distinct seasons each year. They exist in moderate climates and have a growing season of between 140 to 200 days, with an average of six frost-free months per year. The temperature range in a temperate forest is more varied than in the other types of forests found in the US. During the winter, temperatures in a temperate forest can reach 20 degrees below zero Fahrenheit. During the warmer months, temperatures can average as much as 86 degrees Fahrenheit. Rainfall in a temperate forest is less than that of a tropical forest. As a result of lesser rainfall and the widely ranging seasonal temperatures, there are fewer plant species in a temperate forest than in others. You may find as few as three to four tree species per square mile.

The major broadleaf species within a temperate forest fall into one of these categories: oaks, beeches, maples, and birches. Let's examine these species more carefully.

Oaks

Oak trees are one of the most predominant deciduous trees in a temperate broadleaf forest. There are about 500 species of oaks across the world, but in the United States, oaks fall into two major categories: red oaks and white oaks. While most oaks look similar, there are differences between red oaks and white oaks (**Figure 10.6.2**). One important difference is that white oak species have unique structures in their xylem tissue called *tyloses*. These are like small valves in the xylem tissue. The tyloses close off the xylem's water-carrying pathways in the trunk when the tree is harvested. This helps to make the lumber waterproof. Wooden barrels are made of white oak lumber because they will not leak like barrels made from red oak and other tree species.

Beeches

Beech trees are large with smooth bark. The leaves are glossy and produce a nut-type fruit. Trees in the beech family have simple leaves with a main vein running down the middle. The bark of the American beech is smooth and light gray, although the bark can have a rough appearance on other species. Beech bark is easily damaged. In parks and public areas, you can see the names of people carved into the bark of these trees, causing permanent damage to the tree.

Bildagentur Zoonar GmbH/Shutterstock.com

Hintau Aliaksei/Shutterstock.com

Figure 10.6.2. An obvious difference between white and red oak trees is in the leaves. White oak leaves (left) have rounded lobes; red oak leaves have pointed lobes.

10

The beech family is essential in the temperate forest. Beech nuts provide food for many species. The commercial value of the tree is in its suitability for furniture-making. Beech trees are also widely used in plywood, lumber, and log cabin construction. Beechwood is burned to smoke meats and cheeses in the food-processing industry.

Maples

There are more than 125 species of maple trees, and nearly all of them are native to Asia. Maple trees are easy to identify in the landscape because of their hand-shaped leaves; winged seed pods; and distinctive red, orange, and yellow coloring in the fall. In the United States, red maple is the most common species. Red maples turn a brilliant red in the fall. Maples produce a winged fruit called a *samara*. These fruits are made in the spring and early spring breezes carry them great distances.

Maple wood is hard and has many industrial uses. Maple is used to make bowling pins because of its hardness. It is also used to make baseball bats and musical instruments (**Figure 10.6.3**). Maple has a beautiful and distinctive *grain* (wood pattern) when stained in furniture-making.

Birches

Birches are another major species in temperate forests. Birch trees are medium-size and short-lived. They are pioneer species that appear first in newly disturbed areas, such as old fields. Birch trees typically have a bark that resembles tattered paper. Birches have simple, pinnate leaves with serrated margins. They produce a samara-type fruit. Birch is used in making fine furniture and plywood. The wood is strong and flexible (**Figure 10.6.4**).

Mixed Temperate Forests

In a mixed temperate forest, you might find a few evergreen and conifer species mixed in with the deciduous trees we have discussed. Pines, cedars, firs, spruces, and other evergreen species are part of the mixed temperate forest biome.

Depending on where you live, a mixed temperate forest will include a variety of trees best suited for that habitat. A mixed temperate forest in the Southeastern United States would likely include oaks, hickories, beeches, and birches, along with loblolly pine, eastern red cedar, magnolia, and eastern white pine.

Did You Know?

Maple syrup is one of the products derived from the sweet sap of sugar maples. It takes about 42 quarts of maple tree sap to make a single quart of maple syrup. That may explain why real maple syrup is expensive in most grocery stores. It is expensive to produce!

optimarc/Shutterstock.com

Figure 10.6.3. Electric guitars are often constructed from maple because the wood carries sound well.

AlexLMX/Shutterstock.com

Figure 10.6.4. Birchwood is an excellent choice for skateboards because of its strength and flexibility.

In Oregon, a mixed temperate forest would include Douglas fir (**Figure 10.6.5**), western red cedar, and big leaf maple.

Boreal Forests

Boreal forests, or *taiga*, are found in Alaska and across Canada. These are forests where the temperatures remain very low throughout the year, and about half of the precipitation is snow. The soil is typically fragile and low in nutrients. Evergreen tree species especially suited for life in the northernmost biomes make up the boreal forests. Conifer trees, such as pine, spruce, and fir, are more plentiful here. Alaskan forests are also populated with aspen, larch, poplar, and birch trees.

Hugh K Telleria/Shutterstock.com

Figure 10.6.5. Douglas fir is a species of evergreen that is prevalent in the Western United States. This important lumber tree can reach 200 feet tall.

Did You Know? Forests are a valuable source of food for humans. Forest plants produce fruits, nuts, berries, and seeds, all of which are high in protein and minerals needed by humans.

Spruces

Spruces are an essential tree species in boreal forests. There are about 35 species of spruce that inhabit the taiga. Most spruces grow to a large size, with whorled branches, and have a conical tree shape. Their leaves look like needles. Spruce trees have four-sided needles attached to a small peg-like structure. Cones hang downward after they are pollinated (**Figure 10.6.6**).

Spruce trees are useful for construction lumber. Spruce is also useful as pulpwood for paper-making. The wood fibers in the spruce pulp have characteristics that make the paper stronger. Like maple, spruce is good for constructing musical instruments. Spruces are also used as ornamental trees during the holiday season.

Alexa.Sha/Shutterstock.com

Figure 10.6.6. *What characteristics of this tree identify it as a spruce species?*

Larches

Larches are a species of pine tree. Larches are one of the dominant plants in the boreal forest, and they typically grow to 150 feet in height. The western larch is the tallest within the species and can reach almost 200 feet.

Larches are one of the most critical timber-producing species in the Western United States (**Figure 10.6.7**). The wood is dense, waterproof, and decay-resistant. This makes this tree useful for indoor and outdoor construction. As firewood, it has a high heating capacity when burned. Native Americans used the sap and gum from the western larch to treat cuts and bruises, colds, and sore throats.

Sean-ONeill/Shutterstock.com

Figure 10.6.7. The western larch is easy to identify in the boreal forest. *Have you seen a blazing yellow pine tree before?*

10

BMJ/Shutterstock.com

Figure 10.6.8. Boreal forests provide beautiful scenery, and a home to many species of wildlife, like these caribou.

Reuben Jolley Photography/Shutterstock.com

Figure 10.6.9. All of these trees are connected. They are a small part of the Pando Clone, the largest and heaviest living organism on Earth.

Aspens

Aspen trees are a medium-size deciduous tree with a mature height between 20 and 80 feet. The bark is smooth with a greenish-gray or whitish-gray color.

Aspens live up to 150 years. They grow best in moist soils but can thrive in almost any soil type. They are considered pioneer species and quickly sprout in bare soil and areas burned by wildfire. According to the US Forest Service, quaking aspen is the most widely distributed tree in the United States and grows mainly at elevations over 5,000 feet.

The aspen provides a habitat for various wildlife, including bears, deer, elk, and moose, and smaller mammals. Aspen forest stands are also home to grouse and migratory birds. Aspen is a visually appealing species and attracts tourists interested in viewing its brilliant fall colors (**Figure 10.6.8**). Aspen can reproduce by sprouts coming from the roots of existing trees. This allows the species to fill in disturbed areas as a pioneer species.

Did You Know? The Pando Clone, located in Fishlake National Forest in Utah, is a forest stand of interconnected aspen trees. It is the largest and heaviest single living organism on Earth (**Figure 10.6.9**).

Birches

As in the temperate forests, birches are an essential part of the boreal forest biome. Birches are relatively short-lived trees, yet they are essential to the boreal forest because of their value to wildlife. Birds like to feed on the sap, seeds, and flowering buds. Deer, moose, and the snowshoe hare feed on green twigs and tender new growth.

Did You Know? Nearly one-fourth of all medicines used by humans come from forest plants, and nearly two-thirds of all drugs used to combat cancer are derived from forest plants.

Tropical Forests

There are many wildlife and plant species in the *tropical forest*. The winter months are very mild, and the days are usually 12 hours or more. Temperatures average between 65 and 80 degrees Fahrenheit throughout the year. Typically, tropical forests receive precipitation throughout the year, averaging more than 78 inches of rainfall. In this environment, plant material decomposes rapidly and is taken up quickly by other plants. The soil is usually relatively nutrient-poor and acidic. Tropical forests are very dense and have canopies far overhead, allowing very little sunlight to penetrate the forest floor. There are many plant species in a tropical forest because of the favorable growing conditions. In some cases, you may find as many as 100 different tree species in a one-square-mile block of forest land. This diversity of plant material draws in an assortment of wildlife, including a wide variety of birds, mammals, and insects.

Not all tropical forests are rainforests. A *rainforest* is any forest that receives more than 100 inches of rainfall each year. The number and size of tropical and subtropical forests in the United States are tiny compared to temperate and boreal forests. The climate does not support the growth of tropical plants in many places other than parts of Florida and Hawaii. You will also find tropical and subtropical forests in Puerto Rico, Guam, and other US territories in the South Pacific.

In tropical ecosystems, trees are vital for several reasons. Trees provide a windbreak from ocean winds and protect coastal areas from excessive erosion when storms occur (**Figure 10.6.10**). Trees provide shelter, shade, and habitat for a wide variety of other plants and animals.

Francisco Blanco/Shutterstock.com

Figure 10.6.10. This mangrove forest in Florida's Biscayne Bay protects land from erosion, and helps to rebuild land areas. *How do you suppose mangroves accomplish this?*

Hands-On Activity

Tree Biography

In your history class, you might have to write a report on a famous person or event in history. These short reports help you develop an understanding of a person or event.

After reading this lesson, prepare a "tree biography." What is that? You will prepare a report on a tree species that grows naturally in your community. This report will describe the natural history of the tree – a description of the tree, its growth habits, and the role it plays in the forestry industry.

Steps:

1. In your "tree biography," write about where the tree prefers to grow and the specific characteristics that make it similar to or different from other trees.
   - Include the common name of the tree species and the scientific name.
   - Describe the best growing environment for the tree species.
   - Describe the tree's leaves, bark, trunk, and the shape of the tree's crown. For example, are the branches upright and spreading, or does the tree crown have a dense ball-shaped appearance?
   - Describe how the tree benefits wildlife. Does it produce a fruit that wildlife need as a food source? Does the tree provide shelter for wildlife?
2. Prepare your tree biography as a written report and give it to your instructor. Prepare a presentation to share your essay results with other students in the class. If possible, bring leaf and branch samples to show your classmates.

Consider This

1. What did you find interesting about the other students' presentations?
2. If you had to do this activity over again, what would you do differently?

Why Do Trees Grow Where They Do?

What causes one species of tree to grow where it does, while a different species grows elsewhere? Just as you prefer certain living conditions, so do trees. Each tree species has its own specific needs. Some trees prefer moist soil to grow in, while others prefer dryer soils. Some tree species, like the sabal palm, prefer a long, warm growing season. Other species, such as the red spruce, prefer cooler climates. Loblolly pines prefer moist soils in lowlands and swampy areas. Trees such as the poplar and aspen can grow at high altitudes and are adapted to rocky soils in cooler climates.

Is it possible for a tree to grow outside of its preferred environment? Yes, some trees can grow outside of their normal range. However, they are not likely to grow as quickly or get as large as they would in their preferred environment.

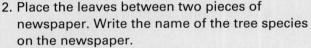

Making Connections #STEM

Building a Leaf Collection

A leaf collection is an excellent way to identify trees and other forest plant species and helps you understand how trees grow and develop.

You can collect leaves in any season of the year. In the spring and summer, leaves are easily plucked from the trees. In the fall and winter, you may have to look around on the ground to find leaves. Evergreen tree species will always have some leaves available.

For this activity, you will need:

- Clean white card stock or paper
- Heavy books
- Old newspapers
- 1 three-ring binder
- Clear tape
- A tree identification book or guide for identifying and labeling leaves

Here are your steps:

1. To begin, collect leaves from various tree species around your school or home. Before collecting leaves, try to identify the tree species. It is easier to identify the leaves before removing them from the tree.

2. Place the leaves between two pieces of newspaper. Write the name of the tree species on the newspaper.
3. Place the leaves between heavy books to flatten them out. You will need to leave them for several days as they dry out.
4. Once the leaves are flat and dry, tape each one to a sheet of white paper or card stock. Write the name of the tree species in the lower right-hand corner of the paper. If you have plastic sheet protectors, you can place the leaf samples in them before storing them in a three-ring binder.

Consider This

1. Sketch some of the leaves in your collection. Can you see the veins in the leaves as you draw them? Is there a pattern to how the leaf veins run throughout the leaf?
2. Why do you suppose the leaves are shaped the way they are?

Review and Assessment

What Did We Learn?

- There are three major forest regions in the US: temperate, boreal, and tropical.
- Temperate forests experience all four seasons, and have a broader temperature range than other forest regions. Oaks, beeches, maples, and birches are the most prevalent types of trees in the temperate forest.
- Mixed temperate forests are a subtype of temperate forests that include some evergreen and conifer species.
- Boreal forests are found in the coldest climates and are mostly comprised of evergreens and conifers. Spruces, larches, aspens, and birches are the most prevalent types of trees in the boreal forest.
- Tropical forests exist in frost-free areas where the climate is favorable for the year-round growth of plants.
- Trees grow in different places because they need and want different conditions. Some prefer wet climates, and others like dry. Differences in altitude, temperature, and precipitation make a difference in what species are most likely to be successful.

Let's Check and See What We Know

Answer the following questions using the information provided in Lesson 10.6.

1. What is the likely cause of the decline in longleaf pine forests?
   A. Air pollution
   B. Overhunting
   C. Urban development
   D. Too much government regulation

2. Another name for a boreal forest is _____.
   A. temperate forest
   B. subtropical forest
   C. Pando Clone
   D. taiga

3. Which of the following species has excellent waterproof and outdoor capabilities?
   A. Red spruce
   B. Western larch
   C. Douglas fir
   D. Loblolly pine

4. The Pando Clone is the largest living organism in the world and is comprised of the _____ species.
   A. loblolly pine
   B. Norway spruce
   C. sabal palm
   D. quaking aspen

5. Which of the following species is a tropical species?
   A. Coconut palm
   B. Yellow birch
   C. White oak
   D. Longleaf pine

Let's Apply What We Know

1. What are the most common tree species in your community? Look up the growing conditions preferred by these species. What does this tell you about the climate and state of the soil in your community?

2. Look at a map of the United States, focusing on your state and those surrounding it. What type(s) of forests would you expect to find in your community? Why?

Academic Activities

1. **Reading.** Go to your library and find books written by the botanist William Bartram. Bartram traveled around the American colonies in the 1700s, making notes of the plants and animals he observed. Potential colonists in Europe read his journals and learned much about what they could expect should they come to North America. This was especially useful because travelers to the new colonies could prepare more effectively for living and working there. You may find it interesting that Bartram was able to travel as far as he did and observe as much as he did. He learned much of what he learned by walking through the forests and wilderness. What do you think about his writings? What can you learn from him?

2. **Reading.** John Muir was a famous botanist and conservationist. He wrote several books about his travels around America and living in the wilderness. One of the books he wrote was titled *A Thousand Mile Walk to the Gulf*. Read excerpts from this book under the direction of your agriculture teacher. What do you think about what you read? What did you learn by reading a book written more than a hundred years ago? What does it teach you about the wilderness and the need for conservation?

Communicating about Agriculture

1. **Visual Communication.** Create a map of the United States, showing the different types of forests and where they are prevalent. Use colors, symbols, illustrations, and figures to communicate your findings to your classmates.

SAE for ALL Opportunities

1. **Foundational SAE.** If you are just beginning to develop a Supervised Agricultural Experience, consider creating one related to forestry. This career path would examine all aspects of tree growth and development, including harvesting and lumber production. Foresters work in the public and private sectors, managing public and private forest lands. Conduct some basic research in the library to determine if forestry is a potential career path for you.

2. **Foundational SAE.** This lesson defined the major forest regions in the United States and addressed other aspects of these forests. Are you interested in a career in wildlife management? Perhaps you might be interested in a career in the United States Fish and Wildlife Service? Conduct some basic research to determine what other careers might be available to you.

SAE for ALL Check-In

- How much time have you spent on your SAE this week?
- Have you logged your SAE hours?
- What challenges are you having with your SAE?
- How can your instructor help you?
- Do you have the equipment you need?

Forests and Forest Products

Learning Outcomes

By the end of this lesson, you should be able to:

- Explain why forests and trees are important to people and the planet. (LO 10.07-a)
- Describe the uses of and products produced from forests managed for consumption. (LO 10.07-b)

Words to Know

pitch
plywood
pulpwood

sawtimber
superfund site
tar

turpentine

Before You Begin

*In the days of wooden sailing ships, forest management was crucial. Ship construction required straight and curved oak timbers. Live oak trees were cultivated and harvested primarily because of their ability to grow the curved limbs needed for the ribs of a wooden ship (**Figure 10.7.1**). During the colonial days, agents of the British king would apply a small sign to trees reserved for the particular use of building ships for Great Britain. These trees were so important that the theft or damage of one of the king's trees would result in harsh penalties.*

*Trees produce more than just wood. Trees also produce chemical products such as tar, pitch, and turpentine. **Tar** is a thick sticky substance derived from tree sap and used to waterproof rope, the sails on sailing ships, and wooden-hulled ships. A thicker version of tar is called **pitch**. It was used as a waterproofing agent to caulk the seams in wooden ships. Pitch is currently used to polish high-quality glass lenses for scientific use. **Turpentine** is an oil distilled from pine trees' sap and is used in paints and medications.*

Tar, pitch, and turpentine were so important to the construction of sailing ships that they were known collectively as naval stores. The naval stores industry was critically important in the colonial days of the United States, and there is still some use for naval stores in modern manufacturing. This lesson will dig into the forest products industry to show how trees affect almost every aspect of modern society.

Figure 10.7.1. Live oaks were important in producing the large, curved timbers needed in shipbuilding. *What about oak trees would make them well-suited to this task?*

The Importance of Forests

What do trees do? It may just seem like trees are standing there, taking up space. They don't seem to be doing much of anything! We know better, though. Science tells us that trees are critical to human health. Forests of trees are beneficial because they combine and magnify the benefits offered by single trees.

Human Health and Air Quality

Trees produce oxygen. Humans need oxygen to live, and life on Earth would be difficult without the oxygen-producing capabilities of trees. The US Forest Service estimates that a small leafy tree can produce the same amount of oxygen in one season as ten people can inhale in one year. Photosynthesis uses carbon, oxygen, and hydrogen to create plant sugars and starches essential for plant growth. Through photosynthesis, trees serve as giant filters to clean the air we breathe.

Trees can absorb airborne particles and pollutants, such as carbon monoxide, sulfur dioxide, and nitrogen dioxide. Trees filter these pollutants out of the atmosphere and replace them with oxygen and carbon dioxide. Small openings in the underside of leaves allow gas pollutants to enter the leaf. Once those gases enter the leaf, they remain until the leaves or the tree itself dies and falls to the forest floor. The US Forest Service estimates that trees remove one-third of all the airborne pollutants annually produced by cars and fossil-fueled power plants. In Chicago alone, the US Forest Service estimated that trees in city parks and along city streets remove 18,000 tons of air pollution annually. That is the weight equivalent of 1,500 fully loaded school buses!

Carbon dioxide is a greenhouse gas. Carbon dioxide in the atmosphere can absorb and emit heat from the sun, causing the planet's surface to heat up. Trees absorb and lock away carbon dioxide in their plant tissues. Trees store this carbon in their wood, preventing it from entering the atmosphere as a greenhouse gas.

Clean Soil and Water

Trees also help to clean the soil. Trees absorb chemicals and other pollutants that have entered the ground. Trees can either hold these pollutants inside their tissues or change them into less harmful forms. Trees filter sewage, farm chemicals, and animal waste. They can also clean or filter the water that runs off into streams. Scientists have created a special species of poplar tree that can break down common environmental pollutants. Some of these chemicals are very dangerous to humans. Trichloroethylene, benzene, carbon tetrachloride, and vinyl chloride are common pollutants at superfund sites in the United States. A *superfund site* is an area where hazardous chemicals have become a significant and immediate danger to humans and the environment. Superfund sites require extensive remediation to remove and dispose of toxic waste. According to the National Institutes of Health, these hybrid poplars can remove dangerous pollutants from the soil and air.

Trees' root systems can slow down surface water and prevent erosion. During a storm or flash flood, water flows across slopes and creates significant soil erosion and contamination of lakes and streams. The root systems of trees slow down stormwater runoff.

Shade, Insulation, and Noise Prevention

Trees provide shade and cooling for humans and wildlife on hot days. The shade from trees can reduce air conditioning usage in some urban and suburban areas in the summer. This helps to conserve electrical energy. The US Forest Service has determined that shade trees can reduce ground-level outside temperatures by as much as 12 degrees.

During windy seasons and winter storms, trees serve as windbreaks (**Figure 10.7.2**). In this way, a tree can help a homeowner reduce their heating bill by as much as 30 percent, according to the US Forest Service. Trees make homes and outdoor areas cooler in the summer and warmer in the winter.

Forests are also capable of reducing noise pollution. Trees planted along urban streets can capture and buffer noise from traffic (**Figure 10.7.3**). Trees planted as a buffer zone along major highways can also reduce the noise from freeways and airports.

Wildlife Habitat

According to the World Wildlife Fund, more than 300 million people live in forests. More than 1.6 billion people depend on forests for their livelihoods. Beyond that, forests are essential in providing habitat for various plants and animals. Forests are home to more than 80 percent of the biodiversity on our planet. *Biodiversity* is the variety of life in an ecosystem.

Figure 10.7.2. Trees provide a windbreak to prevent wind erosion of soil. *How does this provide benefits to humans?*

Figure 10.7.3. Urban trees provide shade, habitat for wildlife, and reduce traffic noises in neighborhoods. *What neighborhoods in your area provide the best shade trees?*

While tropical rainforests have the most biodiversity, all forests play a critical role in providing a place for wildlife to live and thrive. Forests are home to wildlife, including insects, reptiles, amphibians, fish, mammals, and birds. All of these creatures take advantage of the characteristics of trees to survive. You may wonder how a fish swimming in a stream has any use for a tree growing on the stream bank. Let's think about it! Trees can provide shade, keeping stream areas cool and suitable for fish growth and development.

The Importance of Forest Products

Timber Production

Some forests are managed to produce timber for human consumption. People use timber for all types of construction projects, including homes, businesses, and even extraordinary things like railroad bridges and highway bridges. According to the US Forest Service, the United States produces about 16 billion cubic feet of timber per year. The production of sawtimber continues to increase.

Sawtimber is wood used for lumber (**Figure 10.7.4**). It comes from trees large enough to be sawed into lumber. This lumber is used in the construction of buildings. Some sawtimber is used to make poles to support electrical wires and telephone wires. Trees too small to make into lumber produce ***pulpwood***, to be made into paper products and wood chips for landscaping, and fuel pellets for power plants.

Lumber mill operators use some trees to make plywood. ***Plywood*** is a versatile construction item used as the foundation for flooring, the roofs of homes, and even a structural load-bearing unit in home construction. Making plywood is a complicated process. Essentially, the process involves peeling sections of a log and then flattening them into sheets that can be glued together and cut to specific lengths and widths (**Figure 10.7.5**).

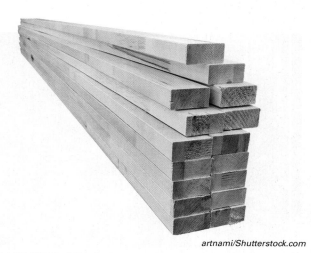

artnami/Shutterstock.com

Figure 10.7.4. Sawtimber is used to make dimension lumber, such as these two-by-fours. *Why do you think this is called dimension lumber?*

Stock image/Shutterstock.com

Figure 10.7.5. Plywood production involves preparing the logs, then peeling off thin sheets. These sheets are then glued together, pressed, and sawed into the appropriate lengths and widths.

Making Connections #|STEM

Recycling in Your Community

Recycling paper makes good sense. It reduces the amount of waste that goes into landfills, creates jobs in the recycling industry, and reduces the need to harvest lots of trees. But how easy is it to recycle paper in your community? Let's do a little research to see how your school and community measure up concerning paper recycling! Here are some questions to guide your research:

At School:

1. Are there recycling containers conveniently located around your school?
2. How do teachers reduce the amount of paper used in class?
3. Does the school lunchroom have separate waste bins for food, plastics, and paper?

In Your Community:

1. Does the local garbage removal company provide recycling bins for customers?
2. Are there local community collection sites for recyclable materials?

In the News:

1. Are there stories in the news about issues related to recycling?
2. Are community groups featured in the news for their recycling efforts?

Consider This

1. What did you find in your research?
2. What conclusions can you draw from your findings?
3. Were you surprised by the information you discovered?
4. What are you doing to reduce the amount of waste you create?

Food Products

Trees also produce food for wildlife and humans. Did you know that the United States produces more than $25 billion annually in fruits, tree nuts, and berries? Just three states—California, Washington, and Florida—produce almost three-quarters of the US's annual crop of fruits and nuts (**Figure 10.7.6**). Around the globe, trees are a crucial food source for humans and wildlife alike. Citrus—oranges, lemons, grapefruit, tangerines, and others—account for 14 percent of all fresh fruit produced for Americans.

Figure 10.7.6. The Columbia River flows past an apple orchard in Washington. *Why might Washington's climate work well for growing apples?*

Gestalt Imagery/Shutterstock.com

Bob Pool/Shutterstock.com

Figure 10.7.7. More than 98 percent of all hazelnuts produced in the US are grown in the Willamette Valley in Oregon. *Have you ever tried hazelnuts?*

Citrus trees require warm temperatures, so you will likely find them in the southernmost and warmest states. Apples, peaches, plums, pears, pecans, cherries, and other tree fruits are grown across the United States (**Figure 10.7.7**). How many fruit and tree nut species can you name?

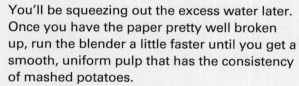

Hands-On Activity

Making Paper

One of the best ways to understand the processes behind producing forest products is to make one of those products. For this activity, you will be making paper.

For this exercise, you will need the following items:

- Newspaper print sheets
- Water
- Blender
- Window screen
- Plastic tub
- Towel
- Cloth or rag

We will not be making paper directly from wood chips or pulp as you might find in a modern manufacturing facility. We're going to skip forward just a few steps and use preprocessed pulp in the form of old newspaper. The best paper for this exercise would have a rough surface. Newspaper print is good for this.

Here are your directions:

1. Take the newspaper and tear it into tiny strips about two inches long.
2. Add the paper strips to the blender and cover them with water. Run the blender at low speed. If the blender appears to be struggling to mash up the paper, add more water.

You'll be squeezing out the excess water later. Once you have the paper pretty well broken up, run the blender a little faster until you get a smooth, uniform pulp that has the consistency of mashed potatoes.

3. Take the window screen and place it across the top of the plastic tub.
4. Spread the pulp thinly and uniformly across the screen. Place a towel over the top of the paper pulp and press down to push the water out of the pulp and allow it to drain into the tub.
5. Take the screen over to a table that can handle getting wet. With assistance from one of your classmates, flip the screen onto the tabletop.
6. Remove the screen and press down with a cloth on the pulp. Allow the paper to dry undisturbed for about a day. If your paper pulp is very thick, it will take longer to dry.

Consider This

1. How does this process differ from the complete process of making paper? Do some research.
2. Was this more or less difficult to do than you thought? What does this make you think about recycling?

What Did We Learn?

- Forests are valuable to humans in many ways. They provide oxygen, remove carbon dioxide from the atmosphere, and absorb air pollution. Forests clean soil and water, and provide shade, insulation, and absorb noise pollution.
- Forest products are important to people also. Forests can be managed to produce sawtimber, pulpwood, and plywood for construction. Managed forests also produce fruit and nuts for people to eat.

Let's Check and See What We Know

Answer the following questions using the information provided in Lesson 10.7.

1. *True or False?* Some tree species can remove toxic waste from the environment.
2. *True or False?* Trees planted in the appropriate location can reduce air-conditioning costs in the summer and heating costs in the winter.
3. Carbon dioxide is a type of:
   A. Mineral
   B. Substitute for oxygen
   C. Greenhouse gas
   D. Fertilizer
4. Tar pitch and turpentine are:
   A. Naval stores
   B. Pollutants
   C. Forms of tree sap
   D. Essential in modern automotive repair
5. A wood product that comes from trees large enough to saw into lumber is called:
   A. Plywood   C. Paper wood
   B. Pulpwood   D. Sawtimber

Let's Apply What We Know

1. Conduct an online search to learn about superfund sites. Are there any near you? What caused the government to designate it as a superfund site, and what was necessary to clean up the hazardous waste? If you look at photos of the superfund site now, are trees used as part of the landscape? What purpose do you think the trees are serving?
2. As we mentioned in this lesson, forest trees can reduce noise pollution. Look to see how trees are used in your local neighborhood or the nearest city or town. Are trees used to reduce noise? Are they planted on city streets to provide shade for sidewalks, and do they dampen the sound of traffic on the road?

10

Academic Activities

1. **Writing.** Prepare a poster that describes the process of making paper or cardboard. Use pictures from online sources to help you demonstrate each step of the papermaking process. Share your poster with your classmates.
2. **Reading.** Using online and library resources, read about the naval stores industry. Analyze what you read. What stood out to you as being the most exciting information?

Communicating about Agriculture

1. **Writing.** Create a flyer to be posted around your community describing how important trees are to the area. Use pictures and precise language to get your message across clearly and concisely.
2. **Writing.** Research US pulpwood production. What trees are involved? Where are these forests grown? How many are harvested each year? What jobs are available in this industry? Create a three-paragraph essay describing your findings.

SAE for ALL Opportunities

1. **Foundational SAE.** Some SAE projects require construction. With the assistance of your parents or your agriculture teacher, visit a hardware store or lumber store and learn about the types of lumber available to you. Which kinds of lumber are most expensive, and which ones are most useful for the project you are considering?

SAE for ALL Check-In

- How much time have you spent on your SAE this week?
- Have you logged your SAE hours?
- What challenges are you having with your SAE?
- How can your instructor help you?
- Do you have the equipment you need?

Szymon Bartosz/Shutterstock.com

Forest Management Practices

Learning Objectives

By the end of this lesson, you should be able to:

- Describe the history of forest management in the United States. (LO 10.08-a)
- Explain the purpose of modern forest management. (LO 10.08-b)
- Name the items necessary for a forest management plan. (LO 10.08-c)
- Identify the elements of good forest management practices. (LO 10.08-d)
- Summarize the process of forest management, from planning to harvest. (LO 10.08-e)
- Explain how wildfires are a part of forest management. (LO 10.08-f)

Words to Know

appraisal	prescribed fire	surveyor
clear-cutting	selection cutting	sustainable forest management
forest management	shelterwood cutting	windthrow
inventory	slash	
log deck	stream buffer zone	

Before You Begin

Have you ever listened to news coverage of a forest fire before? What did the news say about it? Do you think fires are bad or do you think they can be helpful? Talk to a classmate and brainstorm positive and negative results of a forest fire.

10

The History of US Forest Management

Forestry has existed for thousands of years. Roman emperors protected forests for use in national defense, but extensive artificial reforestation did not begin until the 18th century. Forestry grew most rapidly and extensively in Germany. Other countries soon followed Germany's lead, and the French National Forest School opened in 1825. English foresters who were preparing to serve in the forests of India trained at the school.

In 1817, the United States Congress enacted legislation to preserve public land in the Southeast to produce naval stores. In 1873, the United States government set aside $2,000 to study forest conditions in the United States and its territories. In 1876, the United States Department of Agriculture surveyed forest resources in the United States. In 1887, Congress passed the Timber-Culture Act, requiring the planting and cultivating of a certain number of trees on public land. In 1891, President Benjamin Harrison protected 2.4 million acres of woodland by creating the Yellowstone National Park Timber Land Reserve in Wyoming and the White River Plateau Timber Land Reserve in Colorado. More than 17.5 million acres were subsequently set aside for Yellowstone, Yosemite, Sequoia, and General Grant Parks. The forestry movement gathered momentum in the United States as it moved into the 19th century.

The Biltmore Estate and Forest

Forestry as a science got its start in the United States through the support of George Vanderbilt, the grandson of Cornelius Vanderbilt, the 19th century railroad baron. Vanderbilt often vacationed in the mountains and valleys near Asheville, North Carolina, and developed an appreciation for the natural beauty of the southern highlands. The weather in Asheville is temperate and pleasant year-round, making it an excellent location for a summer home. Vanderbilt purchased 125,000 acres of forestland along the French Broad River south of Asheville, and renamed it Pisgah Forest.

Vanderbilt carved out a section of this newly purchased forest to establish the residence, formal gardens, and farms for his new estate: Biltmore (**Figure 10.8.1**). Most of the estate was forestland. Vanderbilt had no idea how to manage a large forest, and it was clear that he would need expert assistance. In 1891, Vanderbilt hired Gifford Pinchot to take over these duties.

Pinchot described the forests surrounding the Biltmore Estate as "deplorable in the extreme." Area farmers had harvested the best trees for lumber and firewood without attempting to reseed. Cattle grazed on the cutover areas, destroying young trees and seedlings. Frequent burnings by farmers cleared land for pasture, destroyed valuable tree seedlings, and removed shrub vegetation that protected the mountain slopes from erosion. Loggers could not access the area for fear of causing severe soil erosion. Old trees were preventing the growth of younger trees by taking their water, sunlight, and nutrients.

ZakZeinert/Shutterstock.com

Figure 10.8.1. The Biltmore Estate remains a popular tourist attraction today.

Pinchot fell into his work with a passion. He combined the roles of forester and lumberman to develop a practical approach to forest management. He began improving cuttings and newly graded roads designed to curb soil erosion. He managed selective cuttings of low-grade species to produce cordwood and firewood while protecting the more valuable trees. He constructed a sawmill to make construction-grade lumber.

The improvements undertaken by Pinchot significantly improved the timber stand and accelerated favorable tree growth. Pinchot's scientific forestry had improved the overall health of the forest. Because of the innovative work at Biltmore Forest, men interested in studying forestry were applying to Pinchot for jobs. Biltmore had acquired the reputation as the center of forestry in America.

Pinchot left Biltmore in 1895 to work in the forestry division in the United States Department of Agriculture. Before leaving, he sought advice on a replacement to continue the work at Biltmore. His mentor, Dietrich Brandis, had just the person in mind: a young German by the name of Carl Alwin Schenck.

Schenck arrived in the United States on April 8, 1895. At Schenck's arrival, Biltmore Forest consisted of a conglomeration of farms situated on nutrient-depleted soils. Although Pinchot's efforts to manage the forest had improved tree stands and erosion control, the long-time lack of management left a lot of work to be done. Soil erosion continued to be a problem, and the lumbermen in the region refused to adopt Pinchot's new management program. Schenck started reforestation efforts, road building in Biltmore Forest, and a series of logging and sawmill operations.

The Biltmore Forest School

In his first year at Biltmore, Schenck began to give lectures to forestry apprentices on rainy days when they could not work in the forests. The forest division office served as a makeshift classroom (**Figure 10.8.2**). Schenck saw this as a way to improve the quality of forestry.

The school followed a plan involving lectures in the morning and fieldwork in the afternoon. Schenck often convened "school" on horseback or the forest floor. Schenck found that experience was the best teacher and encouraged his students to spend as much time as practical in the woods or the lumberyards. Students had to learn all aspects of the industry during the one-year course and six-month internship.

From November to April, Students marked trees for culling from the forest, graded firewood, estimated the costs of cutting and hauling lumber to market, made seed beds, planted seeds and seedlings, and learned to operate the sawmill machinery needed to convert logs into saleable lumber. Students learned how to estimate standing timber, conduct logging and milling operations, and practiced surveying.

The Biltmore Forest School graduated 400 students between 1898 and 1913. Many of these "Biltmore Immortals," as Schenck called them, became prominent foresters in government agencies and private industry. By October 1913, more than 70 percent of forestry school graduates in the United States had graduated from the Biltmore Forest School. As the Biltmore Forest School grew, so did competition from other schools. By 1897, more than 20 universities included forestry education in the curriculum.

Library of Congress

Figure 10.8.2. Biltmore rangers studied forestry concepts in this school building in the mornings and then worked in the forest in the afternoons.

10

Modern Forest Management

The purpose of ***forest management*** is to protect and enhance the health of the forest by eliminating or reducing factors that harm trees. Insect damage, improper harvesting practices, poor replanting practices, soil erosion, and disease can destroy the value of forests.

Forest management has, at its core, the purpose of producing timber, but it can work to achieve other goals as well. Forests are more than just trees growing together; a forest is an ecosystem. Many organisms live within a forest that depend on each other for life. Managing a forest well can benefit all of them.

LeManna/Shutterstock.com

Figure 10.8.3. Camping and hiking are excellent recreational uses of a forest. *What would you like to do in a forest?*

F-Focus by Mati Kose/Shutterstock.com

Figure 10.8.4. Forest management requires the removal of mature or diseased trees so that healthy trees can grow without competition for water and nutrients.

Wildlife management and recreation management are essential to the forester because they provide additional value and income while the trees grow into a valuable product. It can take a long time for trees to reach the optimal lumber value. What is a forester supposed to do in the meantime? Some privately held forests are excellent places for hunting, camping, fishing, hiking, horseback riding, and other outdoor recreational pursuits (**Figure 10.8.3**).

Private owners own more than 35 percent of all forested lands in the United States. Although there are many national and state forests, private owners make up the single largest group of forest owners. Because private owners are free to choose their specific management plan, there are different ways to approach the management of forests. This lesson will focus on ***sustainable forest management***: the techniques and practices that protect soil and water resources.

If there is one thing that tree farmers must have, it is patience. Timber production is often a once-in-a-lifetime activity for forest owners. Imagine that you are a tree farmer that has to spend an entire career producing one or two crops of trees. It takes Douglas fir trees seven to ten years to grow to a size suitable for Christmas trees. It may take 20 or 30 years for the forest of Douglas fir to be big enough for timber harvest. In the meantime, a tree farmer has to manage the forest: removing dead or diseased trees, controlling pests, and thinning out smaller trees so that the best trees can grow (**Figure 10.8.4**).

Did You Know? More than two and a half *billion* trees are planted every year in the United States, according to the American Forest and Paper Association. That's about 6.8 million trees per day!

Forest Management Plans

Regardless of the forest management approach a tree farmer uses, all must have a forest management plan in place. Every forest management plan includes:

1. A set of goals. What is the purpose of the timber stand? Some forest owners grow trees for pulpwood. Others produce trees for sawtimber. Some forest owners sell ornamental or seasonal trees, such as the traditional Christmas tree (**Figure 10.8.5**). The goals of the forest owner dictate the type of management plan needed.

2. A contract with a lumber company. Forest owners rarely have the equipment and expertise to harvest trees from their forests (**Figure 10.8.6**). They need a timber company to harvest trees and replant seedlings for the next generation of trees in the woods. The contract tells loggers how, when, and where to harvest timber.

3. A consulting forester. While forest owners may have a lot of knowledge about the trees in their forest, they often need the advice of a well-trained forester. Foresters understand the growth and development of trees and can often spot problems early.

Tomasz Majchrowicz/Shutterstock.com

Figure 10.8.5. Shearing trees is an essential part of managing a Christmas tree farm. Tree farmers trim tree branches and foliage to create an attractive tree shape.

Best Management Practices

The best management practices involve three essential elements: conservation, protection, and enhancement.

Conservation

Conserving forests for timber production involves many resources, including water, soil, and energy. Water and soil are essential for the healthy growth of forests. Power is needed to maintain and harvest trees. Good management plans conserve soil, water, and energy to make forest management a sustainable enterprise.

Did You Know?

While some areas of the world have seen declines in forest acreage, the total number of trees on the planet has increased by 20 percent over the last 40 years.

Sheryl Watson/Shutterstock.com

Figure 10.8.6. Lumber companies have specialized equipment for cutting and removing logs from a forest. *What skills would be required to handle this equipment?*

10

Joanne Bell/Shutterstock.com

Figure 10.8.7. The spongy moth is one of the most destructive pests in the Eastern United States. *Can you name any other pests that attack trees?*

Protection

Forests are constantly under the threat of damage from pests and poor management practices. Good management plans protect forests: a pest management plan (**Figure 10.8.7**), a prescribed burn plan, and a wildfire prevention plan. Forests need protection from people, too. People can do serious damage to soil and water resources. Unrestricted all-terrain vehicle activity, unauthorized logging, or unregulated hunting can injure the health of forests.

Enhancement

Best management practices involve careful attention to the health of the forest, soil, and water resources. As climate change threatens the health of ecosystems, we must find ways to enhance and improve conservation and protection efforts.

The Forest Management Cycle

Planning

Planning is essential to the health and value of a forest. Planning requires attention to detail, observation of the growing conditions in the woods, and a desire to conserve, protect, and enhance the growth of trees for the best possible outcome (**Figure 10.8.8**).

One of the most important things for a forest owner to know before timber harvest is where their property boundaries are located. *Surveyors* define property boundaries to ensure that logging operations do not infringe on others' property. Surveyors create property maps that help landowners estimate property tax expenses, and these maps are the starting point for an inventory.

beeboys/Shutterstock.com

Figure 10.8.8. Planning is essential to the overall health and value of a forest. *What might this forester be looking for?*

Inventory

Before cutting timber, a landowner should conduct a forest *inventory*, noting the tree species, soil type(s), and wildlife in a forest. This information will help the landowner to create a smart plan to maximize the profitability of harvest.

Part of this inventory is to figure out what tree species exist on the land. Some species are more valuable than others, depending on their intended use. For example, oak lumber is generally more expensive than pine lumber and makes high-value products such as furniture and cabinets. On the other hand, it takes a long time for oak trees to reach maturity, so there will be more time between harvests than there might be for a pine forest. The species of trees in a forest largely determine the value of the forest.

Soil contains the nutrients and water needed for the survival and growth of trees. Soils that are dry or infertile will not support tree growth. Soils that are too wet encourage trees to put down shallow roots, and root rot may develop.

Trees with shallow or damaged roots may experience **windthrow**, when trees fall over in high winds.

Landowners will also inventory the wildlife living in their forests. They can do this by trapping, using surveillance cameras, examining tracks, and looking at other wildlife signs. Endangered species living in a forest need special protection that may limit the timber harvest. Some wildlife species thrive in a forest environment, and sport hunting may be a lucrative option for landowners to pursue while they wait for the trees to mature.

The Timber Sale

The landowner will conduct a timber sale when the lumber is at its most valuable. How does a landowner go about doing this? The first step in a timber sale is to determine which part of the forest will be harvested and sold. A forester will help the landowner identify the trees to be cut and appraise their value. An **appraisal** is an estimate of the worth of something. Foresters appraise timber stands to determine their value.

Forest owners and loggers conduct a walkthrough of the site prior to harvest to make sure they know the terrain where they will be working. A site visit helps the logging crew understand the slopes and grounds they will have to work around when cutting timber. Loggers and forest owners also mark property boundaries at this point.

Streams and waterways are highly susceptible to pollution from logging operations. The equipment used by loggers to cut timber and skid logs churns up topsoil and may cause runoff into bodies of water. Loggers mark **stream buffer zones**, or areas next to streams where logging operations are prohibited (**Figure 10.8.9**). Logging equipment stays out of these zones so that standing plants and ground cover capture and prevent soil runoff.

The **log deck** is a central location in a timber cutting operation where log skidders bring cut timber for trimming and stacking. Stacks of logs are loaded onto tractor-trailer trucks for transportation to sawmills. The log deck area should be carefully selected and cleared to allow adequate space to operate equipment and temporarily store logs awaiting transport (**Figure 10.8.10**). Log decks should be on flat ground to prevent soil erosion and runoff, and for equipment's safe and efficient operation.

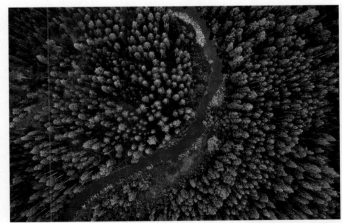

nblx/Shutterstock.com

Figure 10.8.9 Loggers will set aside a buffer zone next to this stream to keep soil runoff from entering and polluting it.

fotorobs/Shutterstock.com

Figure 10.8.10. Log decks are where cut trees are trimmed, and logs are stacked until they are transported out of the forest. *Why is it important to get the right spot identified for this work?*

Michael Leslie/Shutterstock.com

Figure 10.8.11. Signs like this one warn motorists of slow-moving logging trucks and equipment that may be on the roadway ahead. *Why is it important to give drivers this warning?*

Slash is a by-product of a logging operation. **Slash** is the parts of the tree left in the woods after a tree is harvested, like the leftover branches, twigs, leaves, bark, and other bits. As it rots, this organic material returns nutrients to the soil and helps to prevent erosion. Slash also protects seedlings and provides habitat for wildlife as it moves back into a logged area. Slash should be spread as evenly as practical in a logged area.

It is essential to mark the place where logging trucks will enter public roads. Loaded log trucks are heavy and long, and move slowly when they first enter a public roadway from a logging road. Loggers will often place signs on public roads warning motorists of nearby logging operations (**Figure 10.8.11**).

According to the US Department of Labor, logging is one of the most dangerous occupations in the United States. The tools and equipment used in logging operations can cause injury or death if not used safely and correctly. Loggers have to consider the massive weights of trees and the effects of wind and terrain on trees as they fall during the cutting operation. Extreme heat and cold, wind, rain, snow, and lightning make the work difficult. Logging operations often occur miles away from emergency medical services and towns where medical treatment is available. Loggers and forest owners are responsible for making sure that logging operations are conducted in the safest possible manner.

Tree Harvesting

There are three primary methods of whole-tree harvest: clear-cutting, shelterwood cutting, and selection cutting.

Clear-Cutting

Clear-cutting cuts down all the trees in a given area, resulting in ample open space that is now ready for reseeding or transplanting young trees (**Figure 10.8.12**).

Figure 10.8.12. Clear-cutting operations can result in severe soil erosion problems unless managed carefully. *When do you think it would make sense to use clear-cutting?*

Karin Hildebrand Lau/Shutterstock.com

Shelterwood Cutting

Shelterwood cutting takes place in several stages. The first stage in shelterwood cutting is to remove undesirable tree species. The next phase is to remove enough trees to allow light to reach the forest floor, encouraging new seedlings to grow. As the growing seedlings need more light, loggers harvest taller overstory trees. The standing trees serve as a shelter to provide seedlings optimal conditions for growth.

Selection Cutting

Selection cutting involves picking and choosing specific trees to remove from the forestland. For instance, low-value trees may be removed so that trees with a higher value will have room to grow and less competition for water, nutrients, and sunlight. Trees may be selectively harvested as they reach marketable value.

Forest Fire Management

Wildfire is one of the most destructive forces in a forest. Wildfire destroys trees, habitats, and even homes. The US Forest Service reports that more than 73,000 wildfires burn almost 7 million acres of American forest land every year. Fighting forest fires is a complicated process involving firefighters from local, state, and federal agencies.

Hands-On Activity

Forest Management in Practice

Once your agriculture teacher divides you into groups, consider this photo depicting a 1,200-acre forest. The landowner intends to harvest trees from this forest next year. Using the management techniques described in this lesson, describe how your group would manage this forest based on the following questions:

1. What are the first steps you need to take to develop your management plan?
2. What questions do you need answered in order to get a good plan in place?
3. What cutting practice would you choose to use? Why?
4. How would you manage this habitat for wildlife?
5. How could you protect and conserve soil and water resources?

Once you have completed this activity, share your results with the rest of the class.

Atstock Productions/Shutterstock.com

Consider This

1. What was the hardest part of this activity?
2. What additional information did you need?

10

Figure 10.8.13. This forest is recovering from a fire. *What work will foresters do to help restore the forest in this area?*

In the past, whenever a wildfire was discovered burning in a forest, the first response was to assemble a team of firefighters and go put it out. Since then, we have found that the best way to control forest fire is to allow some fires to burn in a controlled way. We call these *prescribed fires*. A **prescribed fire** is a small fire in a monitored area that burns up dead leaves and matter on the forest floor. By getting rid of this fuel source, the risk of a major wildfire diminishes.

Forest managers also use fire forecasting methods to anticipate when the risk for wildfire is most significant. They can then assemble the resources and equipment necessary to respond in the event of a wildfire. Part of the work of the forest service and private forest management companies is in the rehabilitation and restoration of areas after a wildfire (**Figure 10.8.13**).

Career Corner FORESTER

It takes a great deal of management and planning to protect the health and vitality of forests in the United States. **Foresters** *are professionals who manage forested lands for recreation, conservation, and economic uses.*

What Do They Do? Foresters spend a lot of time outdoors inventorying the type, number, and location of trees in the forest. Foresters appraise the value of the timber in a forest and help private landowners sell their timber. After harvest, foresters help plant and grow new trees and monitor those trees for healthy growth.

MAYA LAB/Shutterstock.com

What Education Do I Need? Becoming a professional forester requires a four-year degree in forestry from a college or university. However, some foresters start as forest technicians, a position that only requires a two-year associate degree from a community college or technical forestry program.

How Much Money Can I Make? The average forester in the United States makes about $67,710 each year.

How Can I Get Started? You can build an SAE around your interest in forestry. Research how to estimate timber from trees, and create a lesson teaching others how to do it. Work to create a sustainable management plan for a local forested area. Research what trees grow in your area, and what those trees can be used for. You can also interview or job shadow a forester.

What Did We Learn?

- The science of forest management in the US developed at the Biltmore Forest School, under the guidance of Gifford Pinchot and Carl Alwin Schenck.
- The main goal of modern forest management is to maximize the value of timber harvests while also managing resources. Sustainable forest management protects soil and water resources and wildlife habitat while managing forests.
- Forest management plans include goals, a contract with a lumber company, and a consulting forester.
- Best management practices for forests include conservation, protection, and enhancement.
- The forest management cycle includes planning, inventory, sale, and harvest. Trees may be harvested in several ways, including clear-cutting, shelterwood cutting, and selection cutting.
- Fire management is part of forest management. Foresters work to prevent, identify, and coordinate fighting fires, and then to reseed and restore forest following fires. Prescribed fires are a tool that foresters use to prevent large fires.

Let's Check and See What We Know

Answer the following questions using the information provided in Lesson 10.8.

1. The primary purpose of forest management is:
   A. Timber production
   B. Deer habitat
   C. Soil conservation
   D. Water improvement

2. More than _____ percent of forests in the United States are owned by private owners.
   A. 25   C. 45
   B. 35   D. 55

3. It takes between _____ years to produce a Douglas fir Christmas tree.
   A. two and five   C. seven and ten
   B. four and seven   D. 10 and 20

4. A forest management plan includes all of the following *except*:
   A. A set of goals
   B. A contract with a lumber company
   C. A consulting forester
   D. An air quality plan

5. In the harvesting phase of the forest management plan, it is crucial to remove the _____ trees first.
   A. youngest   C. fastest-growing
   B. shortest   D. mature

10

Let's Apply What We Know

1. With the guidance of your agriculture teacher, survey your school's property. Many schools have undeveloped or natural areas. Develop a management plan for this natural area that includes conserving soil and water and protecting wildlife habitat. Using resources provided by your teacher, see if you can identify between five and ten plants in this natural area.
2. Conduct research about the public forests in your area. See if you can locate a forest management plan for one of these forests online. Do these plans conserve, protect, and enhance natural resources?

Academic Activities

1. **Math.** How do foresters calculate the volume of lumber in a tree? Do some research to find out how this works. Once you have the information, measure a tree in your yard, a local park, or at your school, as best you are able. How much wood is in that tree?

Communicating about Agriculture

1. **Reading.** Visit your local library or reliable online sources to discover what you can about Gifford Pinchot. He was the first person to lead the US Forest Service. Read about his efforts to conserve and protect national forests. His career focused on developing good forest management policies. In your own SAE project, do you have an interest in developing a new way of doing things? Pinchot used his career interest to establish his leadership ability. How can you be a leader in the career path you are pursuing through your SAE program? Make some notes to yourself.

SAE for ALL Opportunities

1. **Foundational SAE.** Research the types of jobs and careers associated with forest management planning. We have identified the role of the forester in forest management, but what about other occupations that may be of interest? For instance, a botanist or biologist might be called to survey endangered species in a forest. What careers did you find? Do any of these career areas interest you?

SAE for ALL Check-In

- How much time have you spent on your SAE this week?
- Have you logged your SAE hours?
- What challenges are you having with your SAE?
- How can your instructor help you?
- Do you have the equipment you need?

Anatomy of a Tree

Learning Outcomes

By the end of this lesson, you should be able to:
- Describe the four parts that all trees have. (LO 10.09-a)
- Explain what purpose leaves serve for trees. (LO 10.09-b)
- Describe what purpose stems serve for trees. (LO 10.09-c)
- Explain why roots are important to trees. (LO 10.09-d)
- Describe how tree flowers are pollinated. (LO 10.09-e)
- List the four types of fruit that trees produce, and give examples of each. (LO 10.09-f)
- Describe how trees disperse their seeds. (LO 10.09-g)

Words to Know

apical meristem	heartwood	pome
bark	lateral meristem	sapwood
berry	margin	
drupe	nut	

Before You Begin

Take a moment and look at your fingerprints. What do you notice about them? Can you recognize any patterns? Compare the fingerprints on your right hand to the ones on your left hand. Are they similar or completely different?

Did you know trees also have their own set of "fingerprints"? Trees have rings on the inside that look a little bit similar to fingerprints. No two trees have the same ring patterns. What do you think causes the rings in each tree to be different? Jot down a few of your ideas on a piece of paper.

Tree Physiology

All trees have leaves, stems, roots, and flowers. Let's take each in turn, and explain their functions.

Leaves

We see leaves on trees, plants in our yards, and houseplants in our homes. What do leaves do?

The primary functions of a leaf are to perform photosynthesis and respiration. *Photosynthesis* is the process by which a plant converts energy and nutrients into sugars that it uses for growth and reproduction. The green color of most leaves indicates the presence of *chlorophyll*. Plants use this chemical compound to make plant nutrients in the presence of sunlight.

Leaves are essential for trees and other plants. Unlike humans, mammals, and insects, plants have to make their own food. Most leaves have a flat surface, making it easy to capture sunlight and make the plant sugars needed to grow. Leaves also draw water from the roots to maintain plant cell health. In trees, as in other plants, tubes called *xylem* move water through the plant, and other tubes called *phloem* move nutrients created during photosynthesis downward from the leaves to be stored until needed.

Leaves pull carbon dioxide out of the air to carry out photosynthesis. They absorb carbon dioxide through small openings on the underside of the leaf called *stomata*. Leaves release oxygen as they complete the process of photosynthesis.

Respiration, which occurs at night, is the reverse of photosynthesis (**Figure 10.9.1**). Plants absorb oxygen at night and release carbon dioxide back into the atmosphere. Plants emit much less carbon dioxide than oxygen, so there is a net gain in oxygen production from plants.

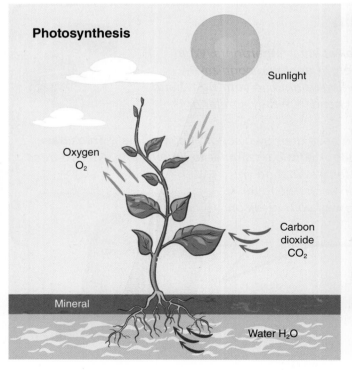
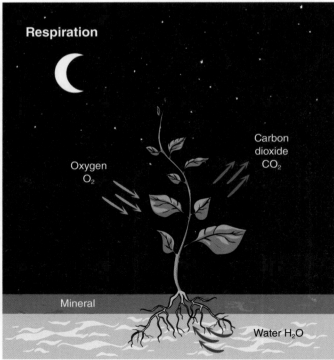

Figure 10.9.1. Photosynthesis and respiration are commonly described as opposite processes. *Why is this?*

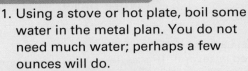

Making Connections #STEM

Fun with Chlorophyll

Chlorophyll is a chemical that helps plants convert sunlight energy into glucose, a type of suger used by plants to make new leaves and other plant tissues. Most plants have green leaves as a result of the presence of chlorophyll, but what would a leaf look like without it? This experiment breaks down the cell walls within the leaf to allow the chlorophyll to be extracted.

You will need the following materials for this experiment:

- A hot plate or stove
- A metal pan that can be used to heat water
- Tweezers
- 1 8- to 10-ounce drinking glass or glass jar
- 70% isopropyl rubbing alcohol, about 2 ounces
- A green leaf from a deciduous tree or bush
- Water

Here are your directions:

1. Using a stove or hot plate, boil some water in the metal plan. You do not need much water; perhaps a few ounces will do.
2. Once the water boils, turn off the stove and place the leaf in the hot water for 30 to 60 seconds.
3. Use tweezers to remove the leaf from the hot water and place it in the glass.
4. Pour enough alcohol into the glass or jar to fully cover the leaf.
5. Set the glass in the pan of warm water and leave it for about an hour.
6. While you wait, clean up your work area.

Consider This

1. At the end of the hour, what did you notice when you came back to look at the leaf specimen?
2. What color was the leaf, and what color was the liquid in the glass?

Leaf Arrangement, Venation, and Shape

Not all leaves are the same. They have different shapes and sizes and are arranged in unique ways on different species of plants. Compare the two leaf shapes in the **Figure 10.9.2**. The species of the plant determines the shape and size of its leaves.

We classify leaves based on three elements: leaf arrangement, leaf venation, and leaf shape.

LiuSol/Shutterstock.com

yilmazsavaskandag/Shutterstock.com

Figure 10.9.2. Compare the maple leaf (left) with the persimmon leaf (right). *How are the leaves different? How are they similar?*

Leaf Arrangement. Let's start with leaf arrangement. Leaves are arranged on the stem in one of three ways: opposite, alternate, or whorled (**Figure 10.9.3**).

Leaf Venation. Leaves can also be classified and identified by their vein patterns. These veins carry water and nutrients to and from the leaf. There are three basic leaf venation patterns: pinnate, parallel, and palmate, as shown in **Figure 10.9.4**.

Figure 10.9.3. Leaves can be arranged in several different ways on a stem or branch. *Can you find examples of these arrangements in your own yard or on school grounds?*

Leaf Arrangements

Arrangement	Appearance	Examples
Opposite: Leaves are arranged on the stem directly opposite other leaves.	*ChunPicture/Shutterstock.com*	Ashes; maples; olives
Alternate: Leaves are arranged in alternating locations on the stem.	*Mikhnyuk Galina/Shutterstock.com*	Oaks; beeches; hickories
Whorled: Leaves arise directly out of the stem and appear to be wrapped around the stem.	*Ihor Hvozdetskyi/Shutterstock.com*	Mountain laurel; southern catalpa

Goodheart-Willcox Publisher

Theeraya Nanta/Shutterstock.com

drpnncpptak/Shutterstock.com

Maliwan kittidacha/Shutterstock.com

Figure 10.9.4. These leaves share a similar color, but the vein patterns in them differ. *How would you describe pinnate (left), parallel (center), and palmate (right) venation to someone else?*

Leaf Shape. Leaves come in many shapes and sizes. Some look like bird feathers, and others are heart-shaped. Some have smooth edges, and others have rough or sharp edges. The tips of leaves can be pointed, rounded, or somewhere in between. The edge of a leaf is called a *margin*. Leaf margins can be smooth, serrated, toothed, or lobed. **Figure 10.9.5** shows some of the different types of leaf shapes, including the tips and bases of each leaf.

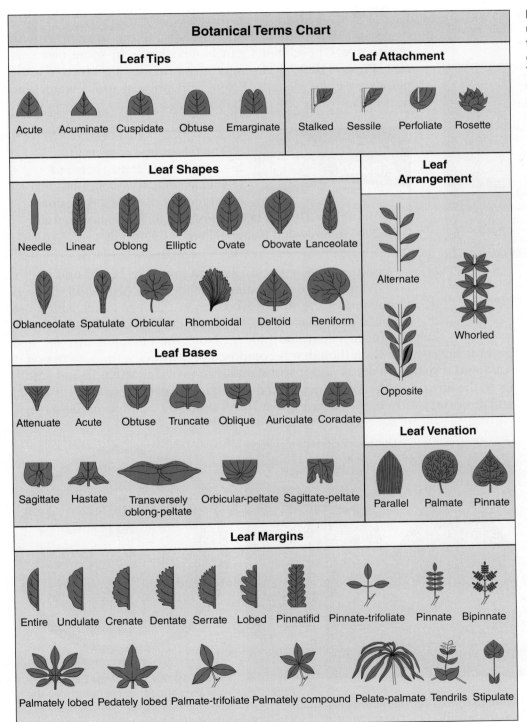

Figure 10.9.5. Arborists use charts like this to help them correctly identify different tree species. They use each trees unique characteristics to narrow down the correct identification.

prodepran/Shutterstock.com

Figure 10.9.6. Note the presence of a bud at the base of this compound leaf from the mountain ash. That tells us this is a compound leaf, rather than multiple leaves.

Figure 10.9.7. The sequoia is a large tree in the Western US. Its stem can reach 30 feet across.

Simple and Compound Leaves

Some trees have simple leaves, and some have compound leaves. Take a look at **Figure 10.9.6**. There appear to be many leaves on each of these stems, but what you are looking at are single leaves with many leaflets. You might wonder how you can tell whether something is a simple leaf or a compound leaf. It can be hard to tell these two apart. Remember that a leaf is that part of the plant immediately beneath the bud on a stem. Look for the bud on a branch, and you will find where the leaf connects.

Stems

As in other plants, stems provide for the upright growth of the tree. These stems can be enormous in some tree species (**Figure 10.9.7**)! The stem also provides a pathway for the plant's vascular system. Xylem and phloem move water and nutrients up and down the tree.

Did You Know? The tallest tree species in the world is the coast redwood (*Sequoia sempervirens*). One coast redwood, named Hyperion, is more than 379 feet tall (**Figure 10.9.8**). It was discovered in 2006 in Redwood National Park in California.

As a tree grows, the main stem grows in height and width. The living part of the stem is called the **sapwood**. This is comprised of the newly grown xylem and phloem, which are replaced each year. These sit immediately under the tree's bark. The older xylem and phloem from the previous year die and become heartwood. Even though it is composed of dead cells, the **heartwood** is solid and resistant to decay and insect damage. It provides strength and stability to the growing tree. As the sapwood dies each year, new sapwood replaces it, and dark rings form in the stem. **Figure 10.9.9** shows an example of these rings.

Figure 10.9.8. This coast redwood (*Sequoia sempervirens*) is named "Hyperion." It is the tallest tree in the world. *How do you think that scientists were able to measure this tree?*

Figure 10.9.9. Tree rings can help scientists and foresters determine the approximate age of trees because rings are created every year, on average, by the growth of new xylem and phloem.

On average, trees create a single new set of these rings each year. By examining the number of rings in a tree stem, scientists can estimate the age of a tree. Tree rings also serve as a type of "forest history book." By comparing the rings of several trees in an area, foresters can date a previous forest fire, insect infestation, or other historical forest event that might have affected tree growth that year.

Creative bee Maja/Shutterstock.com

Figure 10.9.10. Although this bristlecone pine (named Methuselah) doesn't have a lot of leaves on it, it is still living. Methuselah is more than 48 centuries old!

> **Did You Know?** Trees are the oldest living organisms on Earth. In the White Mountains of California, there is a Great Basin bristlecone pine (*Pinus longaeva*) that is estimated to be more than 4,853 years old (**Figure 10.9.10**)! How do you suppose that a tree species can live this long?

At the top of each growing stem is a specialized area of fast-growing cells called the ***apical meristem***. The growing tip of the stem produces new cells that elongate the stem.

Like other plants, trees have secondary stems that branch from the main stem. At the tip of each of these branches is a ***lateral meristem***. This area has specialized growing cells that allow tree branches to grow in length and diameter. The tree's ***bark*** protects its main stem. The bark of a tree is old phloem tissue that has died and been replaced with new phloem tissue. The bark is essential to the tree's health. It protects the growing sapwood from wind and sun damage. It also helps to prevent injury from animals, insects, and diseases. Each tree species has unique bark, making it useful as a tool to identify certain tree species (**Figure 10.9.11**).

Roots

Roots absorb water and nutrients for the tree from the soil. They also provide trees with a stable foundation to grow to a considerable height. Additionally, roots can store sugars and proteins.

Flowers

For a tree species to survive, it must have a way to reproduce. Trees may reproduce through sexual or asexual means. *Sexual reproduction* involves the creation of a fruit or seed through pollination and fertilization. *Asexual reproduction* occurs when plants reproduce through vegetative shoots or stem cuttings (**Figure 10.9.12**).

Promsangtong Thai/Shutterstock.com

Figure 10.9.12. Bamboo is a type of tree that reproduces through below-ground stems called *rhizomes*.

mjurik/Shutterstock.com   Mircea Costina/Shutterstock.com

Figure 10.9.11. The red oak species on the left has a different style of bark than the American beech species on the right.

10

Trees have flowers like any other plant. Flowers are the reproductive organ of the tree. Some trees' flowers have both the male and female parts required for fertilization. In other tree species, some trees have male flowers and others have female flowers (**Figure 10.9.13**).

Pollination

Some tree species are wind-pollinated. Wind picks up the pollen spores from male flowers, or male flower parts, and transports them to the female flower or flower parts. Wind-pollinated tree species include oaks, pine, maples, and hickories. Wind-pollinated flowers are easy to spot because they often lack the showy petals or bright colors needed to attract insect or animal pollinators. The stigmas and stamens of the flowers are exposed to the wind currents (**Figure 10.9.14**). These flowers produce a large amount of pollen.

Fruit Types

Trees produce fruit as part of the reproduction process. Fruits come from the ovaries of flowers, and are nature's way of providing the seeds produced by the plant with the best possible chance of survival. We may think of oranges and apples when we think of fruit, but some fruits are far from delicious to humans. The fruit of the Osage orange tree may taste good to squirrels, but humans find it tastes awful. Native Americans used oak acorns as a food source, but we rarely use them today for this purpose.

Figure 10.9.13. The yellow poplar (*liriodendron tulipifera*) on the left has a flower with both the male and female components needed for fertilization. This showy flower attracts insect pollinators. Mulberry trees (shown on the right) have male flowers and female flowers on separate trees.

Nick Pecker/Shutterstock.com

Skrypnykov Dmytro/Shutterstock.com

Figure 10.9.14. These cones represent the male flower (left) and the female flower (right) of a pine tree. *How can you tell that these are flowers?*

samray/Shutterstock.com

Nikolay Kurzenko/Shutterstock.com

There are four types of fruits: drupes, berries, pomes, and nuts.

Drupe. *Drupes* are fleshy fruit with one seed in their centers. Examples include peaches, plums, cherries, and olives. A hard coating called the *endocarp* surrounds the seed. This is covered by a thick fleshy layer called the *mesocarp*. All of this is covered by the skin called the *exocarp*. A *pedicel* is a stem that holds the fruit to the tree (**Figure 10.9.15**).

Berry. *Berries* are similar to drupes but have many seeds in the center instead of just one seed. Bananas are a type of berry because there are many dark seeds in the center of the fruit. Other trees that produce berries include persimmon, Eastern red cedar, pomegranate, and mulberry.

Pome. *Pomes* are similar to berries, but the seeds are grouped in the core of the fruit, rather than spread through. Pome fruits include the flower's ovary, but they also may consist of part of the stem. Examples of trees that produce pomes are apples, pears, mayhaws, and quinces (**Figure 10.9.16**).

Nut. A *nut* is a fruit with an edible seed enclosed in a hard shell coating. The hard layer is the wall of the ovary of the flower. It becomes very hard as the seed matures. Examples of trees that produce nuts are oak, chestnut, beech, and hazelnut.

Seed Dispersal

As seeds do not move on their own, trees rely on others to spread their seeds. One way that seeds are dispersed is by wind. Many seeds are very light or have tiny wings that float short distances in the wind current. Maples, ashes, and elms produce these types of seeds. They are called *samaras* (**Figure 10.9.17**). When pine cones open, the seeds disperse. Each pine seed is attached to a papery wing that helps the seed travel in the wind.

Gravity is also a method by which trees disperse seeds, especially for tree species that require some shade while growing. Oak acorns are too heavy to be carried by wind currents, so they fall to the ground underneath the parent tree, where they can grow with some protection from the harsh sunlight. Apples and oaks are good examples of tree species with gravity-dispersed seeds.

Animals also disperse seeds. They eat the fruit and expel the seeds during digestion. In some cases, animal stomach acids soften the seed coat and prepare the seed to germinate.

Still other tree species rely on water to disperse seeds. The water carries them to places along the water's edge where they can germinate (**Figure 10.9.18**).

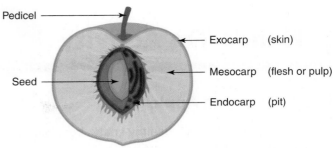

Kazakova Maryia/Shutterstock.com

Figure 10.9.15. Drupes contain many different parts. *Which part of the drupe fruit do we find most edible?*

PIXbank CZ/Shutterstock.com

Figure 10.9.16. The edible part of the fruit from the pear tree is specialized stem tissue surrounding the flower ovary.

Robson90/Shutterstock.com

Figure 10.9.17. The red maple reproduces through winged seeds called *samaras*. *What advantages do these seeds have?*

Ethan Daniels/Shutterstock.com

Figure 10.9.18. The fruit from a coconut palm drifts into a shallow part of a calm lagoon in the tropical Western Pacific. Coconuts can disperse thousands of miles by floating across the sea before being washed ashore.

In the drupe diagram: Pedicel, Seed, Exocarp (skin), Mesocarp (flesh or pulp), Endocarp (pit)

10

Dissecting an Apple

Apples are a favorite healthy snack, but have you ever wondered what you are eating? The fruit of an apple does not look like an orange or blueberry. Why is an apple so different from other fruits?

For this activity, you will need these items:

- 2 apples
- 1 small kitchen knife
- Paper plates
- Paper towels or rags
- 1 magnifying lens
- Notebook
- Pen or pencil

Note on knife safety: You will be using a sharp knife to cut the two apples. Follow your instructor's safety procedures precisely. Do not pick up or use the blade until your instructor tells you to do so. Knives can cause serious injury to you and others! Use care when handling the knife at your lab station.

Here are your directions:

1. Examine the whole apple. Sketch the outline of the apple. Note the location of the small stem where it was attached to the tree. Then look at the opposite end of the fruit. Are those tiny leaves? Use your magnifying glass for a closer look. Those small leaves are the sepals initially found at the base of the apple flower. Compare the sepals in the diagram below with those on the fruit. How are they similar or different?

Anatomy of an Apple Flower

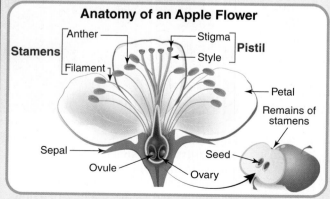

Designua/Shutterstock.com

2. Place one apple on the paper plate and, using the knife, carefully slice the apple into two halves, from the stem to the sepals. Draw the cross-section of the apple in your notebook. Make a note of the stem's location, sepals, and seeds. In the center of the apple half, you should be able to see a cross-section of the ovaries of the apple flower. The ovary of the flower is where the seeds form.

3. Set aside the two apple halves from Step 2 and place the other whole apple on the paper plate. Carefully slice the apple in half the other way, just like you see here.

PixaHub/Shutterstock.com

You should see the ovary in the center of the apple slice and perhaps a few of the seeds. But look closely at the white fleshy part of the fruit. Do you see greenish dots spaced at regular intervals in a circle around the ovary? Those dots are a cross-sectional view of xylem and phloem bundles. You may remember that the xylem carries water throughout the plant, and the phloem carries nutrients downward inside the plant. Make detailed drawings of the cross-sectional pieces of fruit in your notebook, and label each part of the flower and stem tissue.

Consider This

1. What part of the apple tree is the apple? How can you tell? Does this surprise you?

What Did We Learn?

- All trees have leaves, stems, roots, and flowers.
- The primary functions of a leaf are to perform photosynthesis and respiration, producing energy and food for the tree.
- Leaves can be classified based on three elements: leaf arrangement, leaf venation, and leaf shape.
- Tree stems provide for upright growth and move water and nutrients through the tree. Tree stems get much larger than those of other plants.
- Tree roots absorb water and nutrients from soil and provide a solid foundation for the tree's growth.
- Trees use flowers to reproduce. Some are wind-pollinated, and others require the assistance of insects or birds.
- Trees produce different types of fruit protect their seeds: drupes, berries, pomes, and nuts.
- Gravity, wind, animals, and water help to disperse tree seeds.

Let's Check and See What We Know

Answer the following questions using the information provided in Lesson 10.9.

1. Apples are what type of fruit?
   - A. Drupe
   - B. Berry
   - C. Pome
   - D. Nut
2. Stems provide a pathway for _____.
   - A. xylem and phloem
   - B. plant oils
   - C. pollen
   - D. samaras
3. The leaf is the plant part located on the stem directly beneath the _____.
   - A. root
   - B. trunk
   - C. leaflet
   - D. bud
4. Palmate vein patterns in a leaf resemble:
   - A. The rings in a tree trunk
   - B. A bird's feather
   - C. The palm of a human hand
   - D. Compound leaves
5. Tree bark is actually old _____ tissue.
   - A. phloem
   - B. xylem
   - C. dried wood
   - D. heartwood

Let's Apply What We Know

1. Compare two leaves: a simple leaf and a compound leaf. How are they similar? How are they different?
2. Under the supervision of your agriculture teacher, go out and collect leaves from several tree species at your school or in a nearby wooded area. Please be sure that you do not injure the plants. Bring the leaf samples into the classroom and describe them using the terms you learned in this lesson.

Academic Activities

1. **Engineering.** What role do roots play in tree growth and stability? Using building toys or toothpicks and Play-Doh, build a structure that resembles a tree – trunk and branches. How does the structure of the tree's roots affect its stability? Experiment with shallow roots, deep roots, narrow roots, wide roots. Try balancing each in a dish of soil, grass, or other non-sticky matter. What do you find? What can the roots tell you about the health of the tree?

Communicating about Agriculture

1. **Writing.** Design a one-page "cheat sheet" of all the key terms used in this lesson. Use this sheet outdoors with your agriculture class to describe plant characteristics and if you are trying to assemble a leaf collection or identify trees.

SAE for ALL Opportunities

1. **Foundational SAE.** Suppose you are beginning an SAE related to plant sciences. Learn how to describe plants using the terms in this lesson. When you describe plants to others, you must use the correct terminology.

SAE for ALL Check-In

- How much time have you spent on your SAE this week?
- Have you logged your SAE hours?
- What challenges are you having with your SAE?
- How can your instructor help you?
- Do you have the equipment you need?

LESSON
10.10

Inga Linder/Shutterstock.com

Commercially Important American Trees

Learning Outcomes

By the end of this lesson, you should be able to:
- Explain how trees are classified as hardwood or softwood. (LO 10.10-a)
- Identify species of commercially important hardwood trees in the United States. (LO 10.10-b)
- Identify species of commercially important softwood trees in the United States. (LO 10.10-c)

Words to Know

board foot   hardwood   softwood

Before You Begin

Flip through this lesson and look at the photographs. Do you recognize any of these trees? If so, where have you seen them? If you do not recognize any of these as an exact match, do you see similar trees in your area?

10

Maren Winter/Shutterstock.com

Hardwoods and Softwoods

Industry professionals classify trees into two major categories: hardwoods and softwoods (**Figure 10.10.1**). *Hardwoods* have seeds with a covering. These are called *angiosperms*. Oak, apple, and walnut species are angiosperms. These trees are usually deciduous, meaning that they lose their leaves during cold weather. Their wood is typically dense and strong, giving hardwood its name.

Softwoods are trees with seeds that do not have a protective coating, such as pines and spruces. These tree species are known as *gymnosperms*. Softwood trees are usually evergreen, and typically have softer, spongier wood than hardwood trees.

Commercially Important Tree Species

According to the US Forest Service, annual lumber consumption in the United States is 54.7 billion board feet. A *board foot* is a measure used extensively in the lumber industry. It represents a board that is one square foot and one inch thick. Lumber is bought and sold based on board feet. Approximately two-thirds of this consumption is softwood and one-third is hardwood.

As we introduce you to some of the most widely produced trees for commercial purposes in this lesson, bear in mind that these tree species are representative of the types of commercially essential trees in the United States, but are not the only trees of economic value. There are hundreds of species with marketable value—far too many to include here. **Figure 10.10.2** covers the ten most-grown species by volume—an estimation of the board feet of lumber produced.

Rob Jump/Shutterstock.com

Figure 10.10.1. The northern red oak on the top is a hardwood tree. The loblolly pine (bottom) is a softwood tree. *Can you see any of the differences between the two?*

Top 10 Species by Live Volume in 2002

Common Name	Genus	Species	Volume in Cubic Feet	Percent of All Volume
Douglas Fir	*Pseudotsuga*	*menziesii*	114,757,096,586	12.8%
Loblolly Pine	*Pinus*	*taeda*	59,017,744,741	6.6%
Ponderosa Pine	*Pinus*	*ponderosa*	36,468,201,905	4.1%
Red Maple	*Acer*	*rubrum*	35,335,862,859	3.9%
Western Hemlock	*Tsuga*	*heterophylla*	31,976,022,390	3.6%
Lodgepole Pine	*Pinus*	*contorta*	28,724,432,949	3.2%
White Oak	*Quercus*	*alba*	28,653,130,630	3.2%
Sugar Maple	*Acer*	*saccharum*	23,871,411,502	2.7%
Yellow Poplar	*Liriodendron*	*tulipifera*	23,203,250,041	2.6%
Northern Red Oak	*Quercus*	*rubra*	21,303,062,900	2.4%

US Forest Service

Figure 10.10.2. The most widely grown and valuable commercial trees in the United States are listed above. *Does anything here surprise you?*

Commercially Important Hardwood Tree Species

Cottonwood

Cottonwood is a fast-growing tree species, reaching a maximum height of 196 feet. It prefers moist soils from floodplains up to 9,000 feet on mountain slopes, and is found throughout most of the Western US (**Figure 10.10.3**). The black cottonwood is the largest native cottonwood in the United States and is grown for lumber and pulpwood production.

Most cottonwoods have *lanceolate*, or pointed oval, leaves that are dark green on top and grayish underneath, with fine teeth along the leaf margin. The cottonwood has a straight, large trunk with furrows and broad ridges. It is named after the cottonlike seeds it produces in early summer (**Figure 10.10.4**).

YegoroV/Shutterstock.com

Figure 10.10.3. Cottonwood trees are found throughout the US, but are especially prevalent in the West.

Aspen

Aspens grow to about 100 feet tall. This native species can survive in diverse habitats, from streamsides to mountainsides. Aspens grow in Northern states across the US. Aspen wood is light and soft but strong, and is used for lumber and pulpwood.

The aspen has oval leaves that are dark green on top and whitish underneath, with finely toothed margins. The bark is smooth and yellow to yellow-white. Aspen trees are beautiful, and their leaves produce remarkable color in the fall (**Figure 10.10.5**).

Ann Cantelow/Shutterstock.com

Figure 10.10.4. The cottonwood is named for its seeds.

California Black Oak

The California black oak grows to about 80 feet in height. It is a prime example of the types of oaks that make up the Western hardwood industry (**Figure 10.10.6**). As the name suggests, it grows on the West Coast of the United States, primarily in California. This tree species is grown primarily for lumber.

SNEHIT PHOTO/Shutterstock.com

Figure 10.10.5. Aspen trees make a striking sight in the fall. *Do you think this is due to the color of their bark or fall foliage?*

Richard Thornton/Shutterstock.com

Figure 10.10.6. The California black oak tree is important in the Western lumber trade. *What qualities does this tree have that make it suitable to lumber production?*

10

Martin Fowler/Shutterstock.com

Figure 10.10.7. Ash trees have compound leaves and winged samaras to disperse its seeds.

R_Johnson/Shutterstock.com

Figure 10.10.8. Walnut trees have large compound leaves and hard, round fruit, which help in identifying them.

Ash

Most species of ash grow 45 to 50 feet tall. Their wood is hard and shock-resistant, making it an excellent choice for baseball bats and boat paddles, in addition to lumber uses.

All ash trees have compound leaves arranged opposite each other on the stem (**Figure 10.10.7**). The bark is gray with interlaced furrows.

Did You Know? Baseball bats are made from maple, hickory, ash, and bamboo woods.

Walnut

Walnut species live all over the United States, from Maine to Georgia, south into Texas, and west into Arizona and California. Walnuts are a prized fruit, and cabinetmakers use walnut wood in fine furniture construction. It is easy to cut and sand, and the unique patterns of the wood grain make it an attractive wood for furniture-making.

Walnut trees can reach 150 feet tall. The compound leaf of the deciduous tree alternates on the stem (**Figure 10.10.8**). Leaflets are dark green and paler green underneath.

Maple

Maples are a predominantly Northern species, but can live almost anywhere in the United States. Maples are deciduous trees with palmate leaves: three large lobes with irregular serrated leaf margins (**Figure 10.10.9**).

Sugar maples in the New England states yield a sugary sap used in making syrups and food products (**Figure 10.10.10**). Cabinetmakers use maple species in furniture construction. Maple wood has a fine texture, and cabinetmakers can machine it easily, making it a good choice for furniture-making.

ArTDi101/Shutterstock.com

Figure 10.10.9. Maple leaves are distinctive and easily identifiable, even to tree novices. *Where have you seen maple leaves before?*

Studio Light and Shade/Shutterstock.com

Figure 10.10.10. Sugar maples produce a sweet sap used to make syrups.

Basswood

Basswood trees grow to 100 feet and do not have branches on the lower half of the main trunk (**Figure 10.10.11**). They grow all across the US. Basswood is soft, light, and easy to cut and finish. Cabinet-makers use basswood for lumber, veneer, plywood, wood-carving, and ornamental work. Artisans and musical instrument makers may use it to make the wood body of electric guitars.

The American basswood (also called the American linden) has an unusual leaf structure. The ovate leaf has serrated margins. The top side of the leaf is dark green, and the underside is light green with tiny hairs. A unique small leaf called a *bract* is attached to the flower cluster's stalk. Flowers are yellow and droop downward (**Figure 10.10.12**).

Pecan

The pecan tree is native to the United States and grown for its delicious nut-type fruit. The natural range of the pecan tree includes Florida, Louisiana, Georgia, most of Arkansas, and parts of Texas, Oklahoma, Mississippi, Tennessee, Illinois, Missouri, and Indiana. However, the value of this species has caused foresters and landowners to plant it in most Eastern and Midwestern states to farm. Pecans are grown primarily for their fruit, but the hard and strong wood makes good tool handles and flooring (**Figure 10.10.13**).

Pecans can grow to a height of 150 feet. The compound leaves are arranged alternatively on the stem, with 9 to 17 leaflets.

Did You Know? More than 100-million pounds of pecans are commercially produced in the United States each year, according to the United States Department of Agriculture.

Bezbod/Shutterstock.com

Figure 10.10.11. Basswood (also called American linden) trees grow a unique, pyramid-shaped tree crown.

Gerry Bishop/Shutterstock.com

Figure 10.10.12. Basswood trees have a unique droopy leaf called a *bract* that connects to the flower stalk.

Aaron Nystrom/Shutterstock.com

tayloradempsey/Shutterstock.com

Figure 10.10.13. Pecan trees are raised throughout the South and East US to produce fruit.

nnattalli/Shutterstock.com

Figure 10.10.14. Northern red oak trees have long, multilobed, pointy leaves.

Oleksandr Lytvynenko/Shutterstock.com

Figure 10.10.15. White oak leaves have round lobes. The acorns are long and slender. *How do these look different than red oak leaves?*

Red Oak

Many commercial red oak species live in all regions of the United States. The northern red oak range includes the eastern half of the United States, from Mexico to the Canadian border. The southern red oak lives in almost the same area as the northern red oak.

The northern red oak grows in well-drained soils. It grows well in full sun and is a dominant tree species in many natural forest communities. Northern red oak is heavy and hard, making it a good choice for flooring, cabinetmaking, paneling, and as handles for agricultural hand tools.

Red oaks have dark bark and leaves with pointed lobes (**Figure 10.10.14**). The mature leaves are seven to eight inches long and have between seven and eleven pointed lobes. Trees can grow to heights of more than 100 feet.

White Oak

White oak tree species are both water-resistant and rot-resistant. White oaks live in the Northern and Southern US east of the Mississippi River. Industry professionals use it in construction, home wood flooring, and agricultural hand tools and implements. White oaks grow slowly, but cabinetmakers prize them for their strength and durability.

White oak leaves are five to nine inches long and have rounded lobes arranged alternately on the stem (**Figure 10.10.15**). The bark is light gray.

Did You Know?

The USS Constitution is the oldest ship in the United States Navy. The white oak timbers add strength to the vessel, which explains why the ship's nickname is "Old Ironsides."

Zack Frank/Shutterstock.com

Gabriela Beres/Shutterstock.com

Figure 10.10.16. The yellow poplar's flower serves to attract the insects that are needed to pollinate it.

Yellow Poplar

The yellow poplar, or tulip tree, is a fast-growing species native to the Eastern United States. It gets its name from the yellow tulip-like flowers it produces in the late spring (**Figure 10.10.16**). Yellow poplar lumber is relatively weak but easy to cut and finish. It is used as a veneer and to make cabinets and furniture.

The leaves are pinnate with four distinct lobes. The bark is brown and furrowed. The yellow flowers attract insects.

What Happened to the Chestnut Tree?

In the early 1900s, the American chestnut (*Castanea dentata*) was a commercially important tree in the United States. The tree produced a nut-type fruit prized by both humans and wildlife. The tree is now listed as critically endangered by the International Union for Conservation of Nature.

Your task is to figure out what happened to the American chestnut. What happened to cause this tree species, once widespread, to die in large numbers?

Here are some questions to guide your research:
1. When did the American chestnut begin to die off?
2. What diseases, if any, caused the American chestnut population to decrease?
3. What effect did the rapid decrease in the American chestnut population have on humans and wildlife?
4. What methods are being explored to help increase the American chestnut population?

Commercially Important Softwood Tree Species

Douglas Fir

Douglas fir is an essential lumber species in the Western United States. It grows west of the Mississippi River, and in some other Northern states: Minnesota, New York, Pennsylvania, and New Jersey. Douglas fir lumber is excellent for construction, and the lumber industry makes a variety of veneers from it. Douglas fir is also ideal for use in making wooden telephone and power poles, pier supports, and pilings. Douglas fir lumber that has been treated with a preservative is quite durable. Douglas fir is also used for Christmas trees, and their seeds provide food for many bird species and small mammals.

This evergreen member of the pine family has a round crown once it matures and can reach heights greater than 180 feet. The bark is dark brown, thick, and furrowed. The leaves are one to two inches long. The seed cones of the Douglas fir can mature in the first season of growth and are approximately two to four inches long (**Figure 10.10.17**).

Hugh K Telleria/Shutterstock.com

Sandra Standbridge/Shutterstock.com

Figure 10.10.17. Douglas fir trees are especially prevalent in the North and West US, where they grow tall and produce lots of timber.

Western Larch

The western larch grows in the Northwestern United States. This evergreen can reach heights of 200 feet. The western larch is a hardy lumber tree used in framing and finishing during the construction process. It also makes good pulpwood and firewood.

The needles of the western larch are less than two inches long, but they come in bundles much like those of pine. These needles turn a golden yellow in the fall, making the tree easy to identify. The bark is gray or brown and flaky until the tree reaches maturity, when it develops deeper furrows (**Figure 10.10.18**).

Coast Redwood

The coast redwood is one of the most important lumber species on the West coast. It lives almost exclusively in California and Oregon. Coast redwood lumber is resistant to decay, insects, and disease, making it a vital lumber source for exterior construction. This species grows very tall, and at least one specimen tree grew to a height of 377 feet. That is 77 feet longer than a football field!

The leaves of this evergreen are one-fourth to one-half of an inch long and are dark green on top, with two blue-white bands below. The leaves are arranged in a spiral around the stem. The seed cones are small: 9/16 to 1 ¼ inches long.

This species is unusual because it needs fog for healthy growth. Because these trees grow very tall, it is difficult for the tree to "pump" water through its vascular system to the upper branches of the trees. The leaves absorb fog droplets at the canopy height of the tree (**Figure 10.10.19**).

Ian Dewar Photography/Shutterstock.com

Grigorii Pisotsckii/Shutterstock.com

Figure 10.10.18. Western larch trees are recognizable in fall for their stunning yellow color, and in the spring for their unique blooms.

hkalkan/Shutterstock.com

Figure 10.10.19. The leaves of the coast redwood absorb fog droplets, providing the water necessary for healthy growth.

Sitka Spruce

The Sitka spruce is a large evergreen tree found along the West Coast in California, Oregon, and Washington, growing to 200 feet. The Sitka spruce is a versatile species. Native Americans used its sap and leaves for medicinal purposes, and the tree produces high-grade lumber for framing and construction, plywood, and specialty items such as ladders and scaffolding.

Sitka spruce needles are yellow-green on top and blue-green underneath, and about one inch long. The white lines on top of the leaf are rows of stomata. The cones of this evergreen species are one to four inches long and hang down. The bark is gray and smooth (**Figure 10.10.20**).

Figure 10.10.20. Sitka spruce needles are yellow-green on top and blue-green on the undersides.

Fraser Fir

The Fraser fir is grown in four Southern states: Virginia, Tennessee, North Carolina, and Georgia. It is native to the Appalachian Mountains. The Fraser fir is a small evergreen tree grown primarily for Christmas trees (**Figure 10.10.21**).

The needle-like leaves are in two rows that spiral around the stem. They are slightly less than one inch long and are dark green on the topside and light green underneath. The cones are erect and are between one and three inches in length.

Did You Know?

Fraser firs frequently serve as the official White House Christmas tree.

Red Pine

Red pines are native to the Northeastern United States. Red pine lumber is hard and strong, making it a good choice for utility poles and pulpwood. This species is also planted as a windbreak to protect livestock from cold winter winds and protect crops from wind erosion.

Red pines grow to 80 feet tall. The dark green needles of this evergreen species are four to six inches long, in bundles of two (**Figure 10.10.22**). The cones are ovoid and grow to just over two inches.

Figure 10.10.21. The Fraser fir is a preferred species for Christmas trees. *How long have these trees had to grow to reach this size?*

Loblolly Pine

The loblolly pine is native to the Southeastern US and is the principal lumber tree in that region. The loblolly pine grows rapidly, reaching between 60 and 100 feet. This species loses its limbs as the tree ages. This limb-shedding process, along with thick, scaly bark, provides some measure of protection against fire damage (**Figure 10.10.23**). The dark green needles are six to ten inches long.

Figure 10.10.22. Red pine needles grow in bundles of two.

Figure 10.10.23. Loblolly pines are grown in rows on plantations. These trees will become pulpwood. *How do these pines differ from other evergreen trees?*

NataVilman/Shutterstock.com

Figure 10.10.24. Eastern red cedar is easy to identify, when you know what you're looking for. *What distinctive elements can you identify in this photo?*

Eastern Red Cedar

The eastern red cedar is native to the Eastern US, but is cultivated in several Western states as a field border to minimize wind erosion in agricultural fields. The wood is fragrant and colorful when finished. It is knotty and unsuitable for lumber production but has multiple uses in furniture-making and cabinetry.

The tree grows to a mature height between 10 and 40 feet, forming a pyramid shape. The leaves are opposite and scale-like. Branches and trunks are reddish-brown, and its distinctive fruits are light blue (**Figure 10.10.24**).

Did You Know?

Are you using a wooden pencil? If so, chances are that it is made with graphite (the lead) encased in cedar wood. Cedar species are most commonly used to make pencils because it is easy to shape and form the barrel of the pencil.

Hands-On Activity

Tree Scavenger Hunt

Your task is to identify local tree species using a tree identification guide. Using that guide, examine the leaf structure and arrangement on the tree. The way the leaves are arranged can be a clue to the tree species. Gather as many clues as you are able to help with the identification!

While you are searching for the proper name of the tree, collect the some additional samples. You may collect these plant specimens or take a picture of the specimen with your mobile phone, according to your teacher's preference.

Collect the following:
- Pinnate leaf
- Lobed leaf
- Palmate leaf
- Compound leaf
- The fruit, nut, or seed from a tree species

There are some great resources available to you online. The USDA plant database describes many tree species and provides photos for comparing to your plant specimen. iNaturalist is a mobile phone app that allows you to take a picture of a plant and have it automatically identified by a community of citizen scientists. The Arnold Arboretum at Harvard University provides online information to assist with tree identification.

Consider This

1. Were you able to identify your tree specimen? If so, what was the most helpful characteristic you looked at? If not, what was the characteristic you got hung up on?
2. Were you able to find all of the types of leaves and seeds that were part of the scavenger hunt? What was the easiest to find? The hardest? Do you know what types of trees any of those came from?

Review and Assessment

What Did We Learn?

- Wood is classified as hardwood or softwood. Hardwood comes from deciduous trees with covered seeds, or angiosperms. Softwood comes from evergreen trees with uncovered seeds, or gymnosperms.
- Lumber production is measured in board feet.
- Commercially important hardwood tree species include, but are not limited to, cottonwood, aspen, California black oak, ash, walnut, maple, basswood, pecan, red oak, white oak, and yellow poplar.
- Commercially important softwood tree species include, but are not limited to, Douglas fir, western larch, coast redwood, Sitka spruce, Fraser fir, red pine, loblolly pine, and eastern red cedar.

Let's Check and See What We Know

Answer the following questions using the information provided in Lesson 10.10.

1. Trees that produce seed with a seed coating are classified as:
   A. Angiosperms   C. Alternate
   B. Gymnosperms   D. Lanceolate

2. *True or False?* Most softwoods are evergreens.

3. A board foot is a unit of measure that represents a board 1 foot square and ____ inch(es) thick.
   A. 1   C. 100
   B. 12   D. 144

4. Of the following tree species, which one is produced in the largest volume?
   A. Red maple
   B. Loblolly pine
   C. Douglas fir
   D. Northern red oak

5. Which tree species is known for the flavorful nut-type fruit it produces?
   A. Osage-orange
   B. Northern red oak
   C. Sugar maple
   D. Pecan

Let's Apply What We Know

1. You identified some of the trees in your community at the beginning of this lesson. Based on your lesson work, try to identify some of those trees that you could not identify earlier. Are you able to do this?

Academic Activities

1. **History.** Go to the library and research one of the tree species mentioned in this lesson. Prepare a report that explains how the timber industry uses this tree. Report on historical uses as well. For instance, white oak trees were once used in wooden shipbuilding. Tell the story of the tree, including its uses and natural history.

Communicating about Agriculture

1. **Presenting Information.** Using a poster board, create a large table comparing and contrasting the different trees covered in this lesson. Include information about the tree's height, habitat, bark, leaves, fruit, seeds, and uses, as well as any fun facts you can find. Hang your poster in the classroom. Do additional research as needed to make your poster stand out!

SAE for ALL Opportunities

1. **Foundational SAE.** Research careers associated with the scientific study of trees and plant biology. Which ones are interesting to you? What career opportunities exist in these fields?

SAE for ALL Check-In

- How much time have you spent on your SAE this week?
- Have you logged your SAE hours?
- What challenges are you having with your SAE?
- How can your instructor help you?
- Do you have the equipment you need?

Natural Resources Management Practices in Agriculture

Pictureguy/Shutterstock.com

Learning Outcomes

By the end of this lesson, you should be able to:
- Explain why it is important to conserve water and soil resources. (LO 10.11-a)
- Describe common soil and water conservation practices in agriculture. (LO 10.11-b)

Words to Know

buffer strip	cover crops	no-till
conservation cover	crop rotation	strip tillage

Before You Begin

What do you think of when you think of conservation? Take two minutes and make a list of every word, practice, or idea that pops into your head. Share your thoughts with your class. Did anyone think of things you didn't? What do you suppose conservation practices accomplish?

10

Soil and Water Conservation

Why is it important to conserve soil and water resources? It seems like soil and water are everywhere when we go outside. As we have covered in other lessons, Earth's resources, while plentiful, are not endless.Wherever and however we can, using resources in a responsible, sustainable way allows us to continue to have access to them in the future.

Only about one percent of Earth's water is usable for drinking. It doesn't make sense for us to leave the water running all day in our homes; why would that be okay in an agricultural field? Sustainably using water prevents depletion and protects water quality. Usable water is scarce in many areas of the world. Using only what we need and doing so in a thoughtful way helps to ensure there will be water for others.

Like other plants, crops need fertile topsoil to grow and thrive. Topsoil is the layer of soil that is nearest the soil surface. It contains more nutrients than any other layer of soil. Not all soil has the necessary nutrients to feed plants; it can take years to develop fertile topsoil. If it blows away in the wind, is washed away in a flood, or all of its nutrients are quickly used up through poor management, it may be impossible to replace.

Common Natural Resources Management Practices

Cover Crops and Buffer Strips

Farmers and ranchers use *cover crops* to prevent soil erosion and maintain soil moisture (**Figure 10.11.1**). Farmers use cover crops between their regular crop rotations in the field. Here's an example: A farmer harvests her soybeans in the fall and plans to plant cotton in that same field late next spring. If she leaves the area bare of any vegetation, wind and water erosion will carry away topsoil and reduce the soil's ability to grow plants.

To prevent erosion, she plants a cover crop of ryegrass. This ryegrass will hold the soil in place during the winter months. In early spring, she may harvest the ryegrass as feed for cattle, or she may plow it under and let it become organic matter in the soil to feed the next crop.

Buffer strips are planted in fields when a crop is growing. A *buffer strip* is a narrow swath of plants that put space between the crop and other areas or streams and ponds. This strip prevents water and nutrients from agricultural fields from running into nearby streams and causing water pollution (**Figure 10.11.2**).

Antsipava Volha/Shutterstock.com

Figure 10.11.1. Cover crops prevent soil erosion and soil moisture loss. *How does this work?*

Roel Meijer/Shutterstock.com

Figure 10.11.2. This buffer strip prevents runoff from the crop field from entering wild areas. *What benefits are to be found here?*

Mark Baldwin/Shutterstock.com

Figure 10.11.3. Conservation cover differs from standard cover crops because of its emphasis on attracting pollinators and supporting wildlife. *What benefits do farmers see from this?*

Conservation Cover

A *conservation cover* is similar to a cover crop, with some added benefits. If farmland is going to be out of production for a long time, the farmer may plant a conservation cover crop. Conservation cover is usually a mixture of native plants and wildflowers. These plants provide habitat for wildlife, food for deer and other herbivores, and a place for beneficial insects such as honey bees and butterflies to gather pollen. By attracting bees and other pollinators, farmers may increase the population of these essential pollinators (**Figure 10.11.3**).

Crop Rotation

Crop rotation is the practice of planting crops on an alternating basis. For instance, a farmer grows corn in a field in the spring. The following spring, he produces soybeans. Corn and soybeans have different nutrient needs, and each crop has specific pest problems (**Figure 10.11.4**). Planting the same crop repeatedly in the same field depletes the soil of nutrients and allows pests to establish a foothold. Rotating crops cuts down on the depletion of nutrients and reduces the presence of pests.

Strip Tillage and No-Tillage

Strip tillage is a method by which a seedbed is tilled without disturbing the soil on either side (**Figure 10.11.5**). This method exposes the soil to sunlight, which can dry out overly wet soils and warm the seedbed to the optimum germination temperature. This reduces the time and energy needed to produce a crop while protecting the earth from over-tillage.

Tillage changes how soil particles stick together. In a natural environment, soil particles cling to each other. This makes it easy for the soil to hold water and nutrients close to the surface at the depth where crop seeds are planted. Excessive tillage breaks down the ability of soil particles to stick together. This makes the soil surface dryer and reduces the nutrients available to help crop seeds germinate and grow.

Sometimes the best method for protecting and conserving soil and water is to avoid tillage altogether. Tillage uses mechanical tools to turn and mix the topsoil layer to prepare a suitable bed for seeds to germinate while removing weeds and other pests. *No-till* leaves the soil in place from year to year. This allows a layer of organic matter to build up, which helps preserve soil moisture and maintains the structure of the soil particles.

Kelly Marken/Shutterstock.com

Figure 10.11.4. Soybeans can convert nitrogen in the air into a usable form of nitrogen fertilizer. Part of that nitrogen fertilizer is available to the crop growing in the same field for the next growing season.

Helga_foto/Shutterstock.com

Figure 10.11.5. These soybeans are planted directly into the organic matter from last year's crop, without tillage of any kind. *How does this benefit the soil?*

Something to remember about no-till: No-till positively affects *soil, oil,* and *toil.*

Soil: No-till protects soil, preventing damage to soil structure and its ability to hold water.

Oil: This refers to the fuel needed to operate the equipment. By reducing the energy necessary to till the soil, farmers save on fuel costs.

Toil: Reducing the number of times a farmer has to till the soil leaves time for other essential tasks.

Figure 10.11.6 describes common soil conservation practices to prevent soil erosion. Soil conservation methods avoid soil erosion and reduce the time and effort necessary to produce crops.

Did You Know? Charles Darwin is best known for developing the theory of evolution, but did you know that he was also a soil scientist? Darwin studied the role of earthworms in the decomposition of organic matter in soils.

Soil Conservation Practices

Practices	Example
Strip Tillage Farmers use plows to remove strips of plant residue from the field. Farmers till strips about six inches wide and four to eight inches deep. The farmer may inject fertilizer into the strip before planting. Strip tillage is helpful for warming and drying the soil to optimum levels for planting.	*Natural Resources Conservation Service*
No-Till Farmers use no-till to prevent wind and water erosion in their fields. This method requires that the farmer plant directly through a layer of organic matter. There is no plowing the soil surface. This method decreases erosion, increases soil moisture, and increases the number and variety of beneficial soil organisms that break down organic matter into plant nutrients.	*Jair Ferreira Belafacce/Shutterstock.com*
Contour Farming Contour farming requires placing rows to follow the slope contour around a hill, rather than up and down. This process forms rows that serve as hundreds of small dams that slow down surface water runoff and prevent erosion.	*Joseph Sohm/Shutterstock.com*
Terracing Farmers use terraces to level steep slopes on hillsides. Terraces provide a flat surface for planting, which helps prevent surface water runoff.	*Travel mania/Shutterstock.com*

Figure 10.11.6. These popular methods of conserving soil and water resources are used throughout agriculture. *If you live in an agricultural community, which of these soil conservation methods have you seen in practice?*

Forages

Sometimes crop choice can be used to conserve natural resources. Forages are plants that are used to feed livestock. Examples of forage crops include tall fescue, red clover, Kentucky bluegrass, and common lespedeza. Planting fields with these crops may reduce issues with erosion, nurture the soil, and provide food for livestock (**Figure 10.11.7**).

majeczka/Shutterstock.com

Figure 10.11.7. These sheep are grazing on forage grasses. Livestock must be moved from one pasture to another to allow the forage to recover. *What benefits do forage grasses offer as cover crops?*

Hands-On Activity

Erosion Explosion

This activity will show how quickly erosion can damage a farm field.

For this activity, you will need:

- A stream table
- Soil
- Water
- A glass or cup
- Rocks
- A bucket
- Notebook
- Pen or pencil

This activity can create a messy lab environment. Ensure that you keep your work area clean and keep any spilled water off the floor to prevent slip hazards.

Here are your directions:

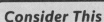

1. Set up the stream table to have a slope of about 10 percent. Your agriculture teacher can help you measure this. Set your bucket under the drainage hole at the end.
2. Fill the upper two-thirds of the stream table with the soil your agriculture teacher gives you. Spread the soil evenly and flatten it by pressing on it with your hands.

3. Pour a prescribed amount of water out of your cup onto the soil at the top edge of the stream table and record the effects of this water on the soil.
4. Record the results of your experiment. What happened to your soil? Did any erosion occur?
5. Allow the water to drain fully from the soil and prepare for your next experiment. Reduce the slope from 10 percent to 5 percent. Your agriculture teacher can indicate the appropriate angle of the slope for you. Repeat Steps two, three, and four and record the results. Compare the erosion from this experiment to the previous investigation. Did more or less erosion occur?
6. Set up again, this time placing rocks on top of the soil. When you place the stones in the appropriate locations as determined by your agriculture teacher, repeat Steps two, three, and four again, and record the results in your notebook. Compare the results from the first experiment with the second and third experiments.

Consider This

1. How were the results similar in all three trials? How were they different?
2. How does the slope of a land affect its susceptibility to erosion?

10

Managing Water Resources

Agricultural production uses a significant amount of the nation's water resources. According to the USDA, agricultural irrigation uses about 42 percent of all freshwater withdrawals from groundwater and reservoirs. Farmers and ranchers irrigate almost 60 million acres of farmlands every year.

How can we protect this water resource when it is critically needed for agricultural production? Actually, farmers are leading the charge when it comes to water conservation by applying some pretty cool methods:

- **Drip Irrigation.** A drip irrigation system delivers water directly to a plant's roots, where it is most needed (**Figure 10.11.8**).
- **Capturing and Storing Water.** Farmers and ranchers capture rainwater and store it for use later when water is needed but rain isn't in the forecast.
- **Scheduling Irrigation.** Farmers schedule irrigation during cooler times of the day when evaporation is less likely to result in water loss.
- **Drought-Tolerant Crops.** Scientists have developed crop varieties that are tolerant of dry weather. These crops need less irrigation than other crops.
- **Rotational Grazing.** Farmers and ranchers move livestock from one pasture to another to prevent overgrazing. This reduces stress on forage grasses and helps them to survive periods of dry weather.

Floki/Shutterstock.com

Figure 10.11.8. Drip irrigation places water at the soil level where it can be easily taken up by plant roots. *How does this save water?*

- **Composting and Mulch.** Farmers and ranchers compost plant materials and use it as a mulch that helps the soil hold water longer during dry weather periods.
- **Livestock Manure Management.** The average beef cow produces more than 65 pounds of manure *each day*. According to the US Department of Agriculture, a large cattle feedlot can produce more than 172,000 tons of waste every year. Farmers and ranchers store and distribute both liquid and solid manure to be used to fertilize growing crops. Animal manures are rich in nutrients and helpful in growing plants. Animal waste is toxic in high quantities, so spreading it out on farmland puts its nutrients to work to benefit crop plants while reducing its toxicity. Livestock waste also holds a lot of water that can be returned to the soil through effective manure management.

What Did We Learn?

- Conserving resources is important to make sure we have access to them in the future.
- Cover crops help to protect soil from wind and water erosion between crop rotations.
- Buffer strips protect natural areas and water from agricultural runoff.
- Conservation cover serves the same functions as a cover crop, but provides food and habitat for wildlife and pollinators.
- Crop rotation conserves resources by preventing nutrient depletion in soil and keeping pests from getting a foothold.
- Strip tillage is tilling without disturbing soil on either side of the seedbed, which protects the soil. No-till protects soil, oil, and toil by leaving the soil alone.
- Forages are plants that are used to feed livestock. This prevents erosion and maintains soil moisture.
- Proper water conservation practices reduce the amount of water needed to produce agricultural crops and livestock.

Let's Check and See What We Know

Answer the following questions using the information from Lesson 10.11.

1. Planting into a narrow, prepared seedbed in a no-till field is called:
   A. Advanced no-tillage
   B. Strip-tillage
   C. Contour tillage
   D. Terracing
2. *True or False?* Land on a slope is more likely to be susceptible to water erosion.
3. Terraces help to prevent ____.
   A. insect infestations
   B. plant disease
   C. soil erosion
   D. excessive soil moisture

4. Soybeans can add nitrogen to the soil with the help of root nodules containing a special kind of ____.
   A. nitrogen   C. mineral
   B. bacteria   D. air
5. A farmer uses a ____ crop to prevent soil erosion and maintain soil moisture between crops.
   A. temporary
   B. buffer
   C. cover
   D. pollinator

10

Let's Apply What We Know

1. Take a field trip to a local farm or nature park with your class. Examine the growing plants and determine the conservation methods that the farmer or horticulturist is using. Which ones do you recognize from the lesson?

Academic Activities

1. **Math.** In the hands-on activity, we worked with slope, which is rise divided by run. Here is the formula in practice: Let's say that the slope of a field rises 10 feet for every 100 feet of travel. This is a 10-foot rise divided by a 100-foot run. The equation then is 10/100 or 0.10. This is equal to a 10 percent slope. Now, you try it. Suppose that there is a field with a slope that rises 10 feet over a distance of 200 feet. Calculate the slope of the field.

Communicating about Agriculture

1. **Writing.** Create a flyer describing soil and water conservation practices that can be used in agriculture. Be sure to communicate the benefits of each approach to the soil or water, and to farmers. Describe what situations each might be used in.

SAE for ALL Opportunities

1. **Foundational SAE.** Using the resources in your library and those on the internet, search for careers in agriculture related to soil and water conservation. Look up the position of soil and water technician. What type of education is required, and what kind of work does this technician do? Does this career interest you?

SAE for ALL Check-In

- How much time have you spent on your SAE this week?
- Have you logged your SAE hours?
- What challenges are you having with your SAE?
- How can your instructor help you?
- Do you have the equipment you need?

Tom Reichner/Shutterstock.com

Wildlife and Game Management

Learning Outcomes

By the end of this lesson, you should be able to:
- Explain why it is important to protect wildlife populations. (LO 10.12-a)
- Describe the methods that wildlife biologists use to study wildlife. (LO 10.12-b)
- Describe the basic characteristics of different categories of wildlife. (LO 10.12-c)
- Recognize common species of United States game birds, fish, and mammals. (LO 10.12-d)

Words to Know

game species	passerine	radio telemetry
metamorphosis	preening	
ornithology	quadruped	

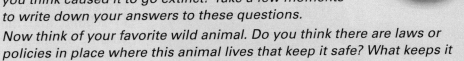

Before You Begin

Can you think of an animal that is now extinct? What about the woolly mammoth? How long do you think it took for the woolly mammoth to go extinct? What do you think caused it to go extinct? Take a few moments to write down your answers to these questions.

Now think of your favorite wild animal. Do you think there are laws or policies in place where this animal lives that keep it safe? What keeps it from going extinct?

In this lesson, we are going to learn about wildlife management and how it protects wildlife from going extinct.

10

The Importance of Wildlife

The conservationist Aldo Leopold once said, "There are some who can live without wild things and some who cannot." Historical studies and archaeological efforts tell us otherwise! Humans have been around for millions of years, and we have had relationships of some sort with animals the entire time. Ancient art shows humans hunting animals. Ancient burial sites have yielded animal artifacts. The survival of Native Americans depended on wildlife as a source of food and clothing. Early settlers to the colonies in North America found it necessary to hunt and fish to survive until agriculture was well established.

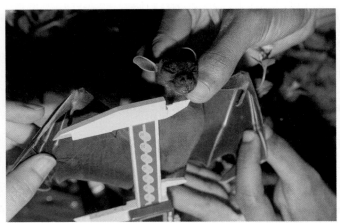

Chokniti-Studio/Shutterstock.com

Figure 10.12.1. Biologists measure the wingspan of this brown bat. *Why might they be doing this?*

Nicole Helgason/Shutterstock.com

Figure 10.12.2. The study of wildlife can take you to all kinds of places. This marine biologist is studying black spiny sea urchins in the Caribbean Sea off the coast of Honduras.

Modern food production methods have replaced the need to hunt and fish as a primary source of nutrition for most Americans. However, modern food production has not reduced the importance of animals to human life. Wildlife species are part of our ecosystems. They have relationships with plants, soil, water, and other species. Those interconnected relationships are connected to humans, too. Our health depends on fully functional ecosystems.

The United States has long recognized the need to maintain and protect wildlife habitat, but this is an expensive endeavor. It costs a lot of money to maintain and support healthy habitats. The United States Congress has passed two laws that provide for the funding of habitat restoration efforts. These funds come about through *excise taxes*, or taxes on specific goods and services for a particular purpose.

The Federal Firearms and Ammunition Excise Tax is imposed on the sale of hunting and fishing equipment. This tax supports habitat restoration for wildlife. The Federal Aid in Sportfish Restoration Act also uses funds generated from taxes on the sale of fishing and boating equipment to rebuild and maintain fisheries and aquatic wildlife habitat.

Wildlife Research

Scientists can monitor the health of our planet by watching the plants and animals that coexist with humans. Scientists conduct research that examines the health of wildlife, plants, and other organisms in an ecosystem. Wildlife biologists use the scientific method to design and conduct studies of wildlife species (**Figure 10.12.1**).

Scientists often need to examine wildlife up close in their research. This involves collecting samples of wildlife species or the safe capture of wildlife. Studying wildlife up close helps us understand how other species live within their ecosystem (**Figure 10.12.2**).

Population Surveys

Scientists use population surveys to determine the quantity of a particular wildlife species in a given area. Wildlife moves around to get the best access to food and shelter. Some species migrate from one location to another during the winter and spring seasons.

What are scientists looking for when they conduct a population survey? It is more complicated than just counting the number of animals. Scientists use population surveys to find deviations from expected wildlife behavior. For example, let's assume that a herd of elk usually travels from high mountain elevations to a local valley each winter. Recently, the elk herd changed its destination. Scientists would be curious why the elk started venturing into this new valley. Does it have to do with a change to plant life? An evolving predator population? Changes to water? Wildlife biologists are interested in investigating and answering these questions.

Capturing and Tracking Wildlife

Scientists may need to examine wildlife up close to determine the health and vitality of the species. To do this, scientists use safe techniques to trap or capture animals for study. After examination, the animal can be released safely back into the wild (**Figure 10.12.3**).

Tracking animal movements helps scientists research wildlife species' migration and travel patterns. One of the best ways to track and monitor wildlife is *radio telemetry*. Let's say that an animal is captured and fitted with a radio frequency identification (RFID) device. This device has specific identifiers detected by an antenna and radio receiver. The scientists use the radio receiver and antennae to communicate with the RFID device attached to the animal (**Figure 10.12.4**). This allows scientists to track specific animals without interfering with their movements in the wild. Small animals have the RFID device glued to their backs, while larger animals are fitted with collars embedded with RFID (**Figure 10.12.5**). The antennae and receiver pick up the electromagnetic signal from the RFID and use it to pinpoint the animal's location.

Colin Seddon/Shutterstock.com

Figure 10.12.3. Scientists released this pine marten back into the wild after being studied and cared for at a wildlife rescue center.

Nancy Bauer/Shutterstock.com

Figure 10.12.4. This biologist uses a radio telemetry antenna to track a pack of gray wolves in a forest.

Tom Reichner/Shutterstock.com

Figure 10.12.5. Biologists are using RFID to track this California bighorn sheep in the Cascade Mountain foothills near Yakima, Washington.

SERGEI PRIMAKOV/Shutterstock.com

1Roman Makedonsky/Shutterstock.com

Figure 10.12.6. This deer is ready for her close-up! Scientists use cameras to capture images of wildlife moving through the forest. *What are the benefits and drawbacks of this technology?*

Trail cameras are another way to collect population data on large mammals. Scientists set up surveillance cameras in an area. Whenever this camera detects motion, it takes a picture (**Figure 10.12.6**). Scientists retrieve the data card from the camera and examine the photographs for the presence of wildlife.

Categories of Wildlife

Wildlife management in the US is primarily concerned with manging game species. *Game species* are those species of animals that may be legally harvested through hunting, trapping, or fishing. Non-game species are those animals that are not approved for legal harvest, often because they are endangered or threatened with extinction. This lesson differentiates between game and non-game species, but you should always consult your local game laws and regulations before hunting and fishing.

Birds

They may not look like it, but birds evolved from dinosaurs. Over the last 60 million years, birds have developed into a species of vertebrates with feathers and the ability to lay hard-shelled eggs. The smallest known bird species is the bee hummingbird, which grows to two inches. The largest known bird is the ostrich, growing as tall as nine feet. There are more than 18,000 bird species worldwide, and the United States is home to about 1,100 of these.

Most birds are *passerines*, meaning they can perch on surfaces such as limbs and twigs (**Figure 10.12.7**). Other birds, such as penguins, some seabirds, and waterbirds, are best adapted for swimming. They may have webbed feet especially suited for walking or swimming.

Figure 10.12.7. This female cardinal (left, brown feathers) and male cardinal (right, red feathers) are passerines, perching on a tree limb.

Bonnie Taylor Barry/Shutterstock.com

Birds live in diverse environments and can be found on all seven continents. Bird skeletons contain lightweight bones filled with air cavities that connect to the respiratory system. These bones are lightweight and strong.

Most birds can fly. There are some birds that cannot fly, like penguins, kiwis, and ostriches (**Figure 10.12.8**). Still others, such as wild turkeys, cannot fly for long distances.

Birds eat a varied diet of fruit, seeds, and nectar from flowers. Some species, such as the turkey vulture, dine on carrion. Other birds, like hawks and eagles, feed on small birds, mammals, fish, and reptiles (**Figure 10.12.9**). Birds do not have teeth. Instead, they swallow food ground up in a special stomach compartment called a *gizzard*. Birds may swallow small stones to help grind food for digestion.

Figure 10.12.8. Penguins cannot fly, but they can make up for it by swimming in search of their food.

Perhaps the most distinguishing characteristic of birds is their feathers. Feathers are a critical part of the bird's survival because they provide warmth, camouflage from predators, and are essential to flight. In nature, you may notice that birds clean their feathers of mites and lice. This is the process of **preening**. Preening removes dirt and helps to apply a waxy substance secreted from a gland near the bird's tail to the feathers (**Figure 10.12.10**). This protects the feathers and makes them waterproof.

Birds communicate with each other primarily through visual signals and by singing. Bird songs are an effective means of communication. Birds sing to attract mates, to communicate danger, and to claim territory. Birds may also raise and lower their wings and ruffle their plumage to scare away predators or attract a mate.

Many birds travel in large flocks. Flocks provide some measure of safety and help to increase foraging efficiency. It is challenging for a predator to sneak up on a large flock of birds without being detected. Some other bird species live in small family groups.

There is an entire branch of zoology that concerns itself with the study of birds, called **ornithology**. By studying birds, scientists can determine a lot about the health of habitats and ecosystems for other animals as well.

Did You Know?

Turkey vultures use the sun to bake harmful parasites off of their feathers as part of their preening process.

10

Figure 10.12.9. This western osprey dines on fish. *What makes it suited to hunting and catching fish?*

Figure 10.12.10. This adult bald eagle is preening its tail feathers. *Why is preening important?*

Some populations of birds are specially managed as game species. Game wardens track the populations of these animals, and manage the numbers that can be hunted. Some of the most common game birds in the US are introduced in **Figure 10.12.11**.

Common Species of Game Birds

Description	Appearance
Wild Turkey. The wild turkey is native to North America and inhabits much of the eastern half of the United States. Wild turkeys prefer to live in mixed hardwood-conifer forests, although they will forage for food in open pastures and fields. Despite being a large bird that spends much of its time on the ground, turkeys can fly for short distances and often roost for the night to avoid predators. Turkeys are omnivorous, eating seeds, nuts, berries, and insects. They also eat tiny frogs and reptiles on occasion. Predators of wild turkeys include raccoons, foxes, and raptors such as the bald eagle.	 *Todd Boland/Shutterstock.com*
Common Pheasant. The common pheasant, or ring-necked pheasant, is one of the most popular game birds worldwide. In the United States, the common pheasant inhabits farmland, woodland, and wetlands. The common pheasant was introduced in the United States in 1773 from Asia and Europe. Pheasant hunting is a popular sport, but in the last thirty years, the population in the United States has decreased.	 *WildMedia/Shutterstock.com*
Northern Bobwhite. The northern bobwhite, or bobwhite quail, is native to the Eastern United States. This species is omnivorous, enjoying small insects, berries, seeds, and agricultural grains in its diet. Although it is a game bird species in the US, it has been classified as a near threatened species by the International Union for the Conservation of Nature. Habitat loss is the primary reason for the declining numbers of the northern bobwhite.	 *Dennis W Donohue/Shutterstock.com*
Ruffed Grouse. The ruffed grouse is a medium-size game bird that lives in forested areas in the northern regions of the United States. The ruffed grouse is an omnivore and eats insects and various berries, seeds, and vegetation.	 *Kevin Cass/Shutterstock.com*
Mourning Dove. The mourning dove is one of the most widespread bird species in the United States and is a popular game species. Mourning doves live in lightly wooded areas and grasslands, avoiding thick and heavy forest areas. Their diet consists of seeds. The mourning dove is an abundant game species, but there are concerns that its population is declining due to habitat degradation.	 *Matthieu Moingt/Shutterstock.com*

Figure 10.12.11. These game birds are commonly found in the US. *Do any live in your area?*

Hands-On Activity

Building Bird's Nests

For this activity, you will need to perform some research and use your imagination. Your task will be to build a bird nest using the same materials a specific bird species would use. Some bird species build their nests from grass leaves and pine needles, such as the nest portrayed in **Figure A**.

Aguadeluna/Shutterstock.com

Figure A. A bird's nest made from pine needles and grass leaves.

Other bird species may use a mixture of mud and plant matter to create their nests, such as the swallows in **Figure B**.

Olexandr Panchenko/Shutterstock.com

Figure B. This swallow's nest is made of mud, stones, twigs, and grass leaves. The mud helps the nest stick to a smooth surface.

Before building your nest, research a bird species native to your community and build the nest using materials you gather outside. Remember, only use materials that a bird would be able to pick. Do not damage growing plants by removing limbs or twigs. That's not something a bird would do! Design and build your nest so that it holds together and looks like one that a bird would make.

In addition to the bird's nest, prepare a short report on the bird species that builds the chosen nest. Include in your essay the feeding and nesting habits of your bird species.

Share your bird nest design and report with your classmates.

Lost Mountain Studio/Shutterstock.com

Consider This

1. Was it difficult to build the nest? What was the biggest challenge?
2. What materials were the most commonly used in your class? What were the most unique? What does this tell you about birds in your area?

10

Shane Myers Photography/Shutterstock.com

Figure 10.12.12. Modern sharks have not changed much from their 100 million-year-old ancestors.

Kevin Cass/Shutterstock.com

Figure 10.12.13. Chinook salmon travel back to the same stream each year to spawn.

Fish

Almost everyone can identify a fish when they see it, but what do we know about them? There are more than 24,000 species of fish on the planet, and many of those are ancient. Fossil records indicate that the earliest fish species existed more than 455 million years ago! Some of the shark species swimming in the world's oceans today have changed little over the last 100 million years (**Figure 10.12.12**).

Fish live in almost every body of water. Some fish species live in the deepest regions of the world's oceans, and some have been found in lakes and ponds at elevations at or above 17,000 feet.

Fish are cold-blooded. They can thrive in temperatures as cold as 28 degrees Fahrenheit and as warm as 108 degrees Fahrenheit. Most fish species can see, hear, and smell. Some species can even feel or detect water movement.

Some fish species are known to migrate long distances. This migration is triggered by the need to reproduce, reach key feeding areas, or seek refuge from predators (**Figure 10.12.13**).

Fish feed on plants, plankton, or other fish. Some fish swim in schools, while others prefer to forage for food alone.

The freshwater fishing industry is big business in the United States. The large number of game fish species makes it impossible to list them in this lesson, but some of the most common game fish are largemouth bass, crappie, bluegill, smallmouth bass, catfish, walleye, and muskellunge (**Figure 10.12.14**). Trout, salmon, and sturgeon are also popular game fish species. More than 55 million Americans go fishing each year, generating $10.5 billion for fishing equipment and licenses. More than 57,000 businesses in the United States sell fishing equipment.

More than 8.5 million people each year try their luck at fishing for saltwater game fish species. Of the estimated 956 million fish caught annually, anglers release more than 64 percent back into the ocean. Spotted seatrout, Atlantic croaker, striped bass, black sea bass, and bluefish are some of the most popular game fish (**Figure 10.12.15**).

Maclane Parker/Shutterstock.com

Figure 10.12.14. The largemouth bass is a popular game fish.

Pelow Media/Shutterstock.com

Figure 10.12.15. Striped bass live in the Atlantic Ocean and the Gulf of Mexico. This vital game species also lives in inland lakes and reservoirs.

Mammals

There are more than 6,400 mammal species across the world. Most mammals are **quadrupeds**. They can use arms and legs, or front and rear legs, for moving from place to place.

Mammals have a highly developed brain and communication skills. They can see, hear, touch, smell, and taste. Several mammal species, such as swine, cattle, and sheep, have been domesticated for human use. Mammals have a coat of hair that provides warmth in cooler temperatures and camouflage as protection against predators.

Some mammals, like bats, can fly. Others, like whales, can swim. Others can crawl, run, walk, and leap (**Figure 10.12.16**).

Mammals live in numerous habitats: in the trees, underground, on grassland, and in water. Some mammals have adapted to living in more than one habitat (**Figure 10.12.17**).

Mammals are warm-blooded creatures. It takes a lot of energy to maintain a warm body temperature and requires that mammals spend a lot of time seeking food. Some mammals are omnivorous and can live on a diet of plant material and meat (**Figure 10.12.18**). Some mammals are carnivorous and possess teeth for ripping and tearing flesh from prey.

Stu Porter/Shutterstock.com

Figure 10.12.16. The fastest land animal in the world is the cheetah, which can reach speeds of nearly 70 miles per hour.

IgorCheri/Shutterstock.com

Figure 10.12.17. Brown bats can adapt to many habitats but prefer dark places to reside during the day.

BGSmith/Shutterstock.com

Figure 10.12.18. Bears are omnivorous, enjoying a diet of plants and meat. *What other omnivorous animals can you name?*

Many mammals in the US are managed as game animals. Some of the most common game animals are discussed in **Figure 10.12.19**.

Common Large Game Mammals

Description	Appearance
Caribou. Caribou, or reindeer, live in the far northern regions of North America. They are ruminants and enjoy a plant-based diet. The natural predators of reindeer include wolves, wolverines, and black, brown, and polar bears.	 *Ghost Bear/Shutterstock.com*
Wild Boar. Wild boars came to the United States from Asia and Africa. Wild boars can reproduce quickly, making them a serious problem in feral populations. The wild boar is omnivorous. The species requires about 4,000 calories per day to survive, and its intensive foraging damages fragile ecosystems.	 *WildMedia/Shutterstock.com*
Elk. Elk is one of the largest species in the deer family. Male elk can weigh up to 1,100 pounds. Female elk weigh between 370 and 650 pounds. Elk live in a wide range of habitats, from forest to grasslands. They are ruminants and eat a plant-based diet. You can find elk in most Northwestern and Northern states in the United States.	 *Patrick van Asselt/Shutterstock.com*
White-Tailed Deer. The white-tailed deer is the most widely distributed deer species in the United States. An adult male deer weighs between 150 and 300 pounds, and the female adult deer is a bit smaller. White-tailed deer adapt to a wide range of habitats and are herbivorous ruminants. They consume legumes and other plants, including crops. Wolves, cougars, American alligators, and jaguars all prey on white-tailed deer.	 *Tony Campbell /Shutterstock.com*
Bighorn Sheep. Bighorn sheep are native to the cooler mountain regions of the Western United States. Bighorn sheep get their name from the males' large curved horns. They travel in large herds and can climb steep slopes in search of food and shelter. Bighorn sheep are ruminants who graze on grasses and shrubs. Their natural predators include black bears, grizzly bears, wolves, and mountain lions.	 *bunlee /Shutterstock.com*
American Black Bear. The American black bear is native to North America. Although it is smaller than other bear species found here, it is the most widely distributed of all bear species. Black bears typically inhabit forests, so you are most likely to find them in the most heavily forested regions of the United States. Their diet consists primarily of vegetation. Unlike other mammalian species, black bears hibernate for long periods during the winter, when food supplies are scarce.	 *Dennis W Donohue/Shutterstock.com*

Figure 10.12.19. These animals are managed as game species in the US. *Can you name others?*

Reptiles and Amphibians

There are 394 species of reptiles and 304 species of amphibians known to exist in the United States. The varied climate within the United States makes it easy for a wide variety of these species to live.

Reptiles and amphibians are vertebrates, so they have backbones that serve as a foundation for their skeleton. Reptiles and amphibians are cold-blooded. Their blood temperature can range from near freezing to the current air temperature. Some reptiles and amphibians are able to fertilize and lay eggs as part of the reproductive cycle (**Figure 10.12.20**).

Reptiles and amphibians are not the same. Reptiles have a skin layer covered with scales. Amphibians have smooth or slimy skin, which helps them to breathe through their skin. Reptiles cannot do this and must rely on lung capacity alone for breathing.

At birth, reptiles appear to be miniature versions of adults. Amphibians have a more complex life cycle because they start in a larval stage.

What do reptiles and amphibians eat? Most reptiles and amphibians are carnivorous (**Figure 10.12.21**). However, some species, like turtles, will also eat plants.

Reptiles

Crocodiles, snakes, turtles, and lizards are all reptiles. The only member of the crocodile family native to the United States is the American alligator (**Figure 10.12.22**). This species is typically found in the Southeastern United States.

SunflowerMomma/Shutterstock.com

Figure 10.12.20. The loggerhead sea turtle lays its eggs in beach sand. These eggs hatch, and the turtles head for the ocean surf.

cynoclub/Shutterstock.com

Figure 10.12.21. Snakes and lizards are carnivorous and eat insects and small mammals, amphibians, and other reptiles.

SunflowerMomma/Shutterstock.com

Figure 10.12.22. The American alligator is the largest reptile found in the US. Male alligators grow up to 15 feet long and weigh up to 1,000 pounds. Female alligators can reach a length of 10 feet. Alligators consume fish, birds, mammals, and other reptiles.

There are more than 2,900 species of snakes worldwide. Most species are harmless to humans. Snakes are closely related to lizards. Snakes vary in length from 4 inches up to 25 feet! They are carnivorous and can flex their jaws to consume small mammals and birds.

Many types of lizards are also found in the United States. Depending on the species, they may be herbivorous or carnivorous. Most lizards enjoy a diet of plants, insects, and small mammals.

Turtles are reptiles with a hard protective shell composed of bone. Many turtles live in water. They're primarily omnivorous, enjoying a diet of insects, grasses, and fruit.

Amphibians

Frogs, toads, and salamanders make up the bulk of amphibian species. Amphibians have front legs and hind legs with appendages, or fingers, on each foot. These allow for grasping and climbing. For frogs and toads, the front and hind legs make swimming very easy.

Amphibians go through a life cycle where the young grow from eggs into adults, with significant body structure changes. This process is called *metamorphosis*. Amphibians lay eggs that hatch into a larval stage. Over time, the larvae develop into adults (**Figure 10.12.23**).

Figure 10.12.23. The life cycle of a frog doesn't look much like the life cycle of a turtle. *What differences can you identify?*

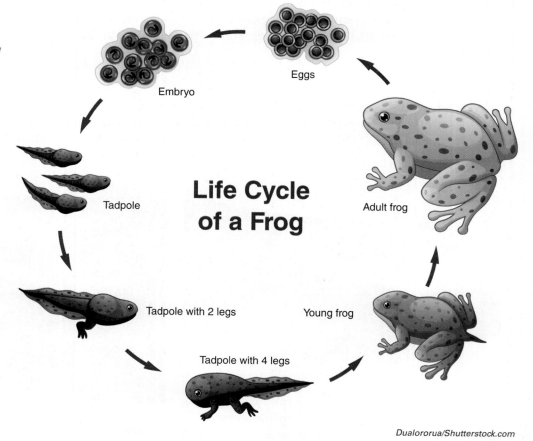

Life Cycle of a Frog

Eggs

Embryo

Tadpole

Tadpole with 2 legs

Tadpole with 4 legs

Young frog

Adult frog

Dualororua/Shutterstock.com

Career Corner ▸ WILDLIFE BIOLOGIST

> *If all of this sounds interesting to you, perhaps you're cut out to be a* **wildlife biologist***!*

What Do They Do? Wildlife biologists study the health of animal populations; collect and analyze data; organize research projects; develop land and water use plans; track, tag, and move animals; study human interactions with and effects on animal populations; interact with fish and game wardens and wildlife rehabilitators; and write academic papers.

What Do I Need to Be Good at This Job? Wildlife biologists should be observant, good problem-solvers, and have strong communication and critical-thinking skills. You should also enjoy spending time outside and time with animals!

koonsiri boonnak/Shutterstock.com

What Education Do I Need? Wildlife biologists are required to have a four-year degree in animal behavior, zoology, wildlife conservation, biology, or wildlife and environmental law. Masters or doctoral degrees may be preferred for some positions.

How Much Money Can I Make? The median annual salary for a wildlife biologist is about $62,000.

How Can I Get Started? Design an SAE to study an animal population in your area. Research how human actions in your area have affected wildlife populations, and what evidence you see of that. Find a local wildlife biologist and attend a talk they give to the public, or set up an informational interview about their work.

10

What Did We Learn?

- Wildlife management is important, as humans and animals share habitats and environments, and complex interdependencies. Wildlife management can provide clues about changes to our shared environment.
- Scientists use tools such as population surveys, wildlife capture, radio telemetry, and trail cameras to monitor wildlife populations.
- Categories of wildlife to be managed include birds, fish, mammals, reptiles, and amphibians.
- Game species are species of animals that may be legally harvested through hunting, trapping, or fishing.
- Birds are warm-blooded vertebrates descended from dinosaurs. They hatch from eggs. Most birds can fly, but some cannot. Birds do not have teeth, but have a gizzard to help them grind and digest their food. Birds' feathers keep them warm and provide camouflage.
- Fish are cold-blooded vertebrates that live in bodies of water. They may migrate long distances. Fishing is big business, and many species are managed for consumption.
- Mammals are warm-blooded vertebrates that produce milk for their young. They have highly developed brains and communication skills. Hair provides them warmth in cool temperatures. Mammals are diverse: some fly; some swim; they live in all types of environments.
- Reptiles and amphibians are similar in many ways. They are cold-blooded vertebrates. Many of them lay and hatch from eggs. Reptiles have scaly, thick skin, breathe through lungs, and newly hatched young resemble small adults. Amphibians have slimy skin, breathe through their skin, go through a larval stage, and are able to grip things with their toes.

Let's Check and See What We Know

Answer the following questions using the information provided in Lesson 10.12.

1. Which of the following is *not* a reptile?
   - A. Frog
   - B. Snake
   - C. Lizard
   - D. Alligator
2. *True or False?* Reptiles can breathe through their skin.
3. Which of the following is a passerine bird?
   - A. Cardinal
   - B. Penguin
   - C. Mallard duck
   - D. Kiwi

4. To grind up their food during digestion, birds have a unique digestive organ called a(n):
   - A. Cloaca
   - B. Crop
   - C. Gizzard
   - D. Aorta
5. Which is the following is a warm-blooded species?
   - A. Tree frog
   - B. Alligator
   - C. Salmon
   - D. Black bear

Let's Apply What We Know

1. Conduct some research to determine the species of reptiles and amphibians that live in your community. Are any of them poisonous to humans? What benefits do they provide for humans? For instance, some snakes can help control rodents around your home.
2. Conduct some research to determine which species in each group of animals mentioned in this lesson are endangered. What is causing the decreased population of the species?

Academic Activities

1. **Reading.** Prepare a report on a wildlife species that interests you. Select one familiar to you. Your essay should describe the animal's range and habitat, its food sources, and how to identify the species. Share your findings with your classmates.

Communicating about Agriculture

1. **Social Media.** Create a mock social media profile for an animal native to your area. Where does it live? What does it eat? What does it like to do? What activities might you find it doing?
2. **Reading and Writing.** Research animals classified as game in your state. Where do they live? How are they managed?

SAE for ALL Opportunities

1. **Foundational SAE.** Conduct some research related to the topic of this lesson to see what types of careers would be available to you. Are you interested in learning more about animals? Would you be interested in becoming a research scientist to study these animals more closely?
2. **Foundational SAE.** There are business opportunities in wildlife management. There are several careers associated with recreational hunting and with wildlife sightseeing. Are there any careers or job opportunities available to you in your community related to wildlife and recreation? If so, what are they? If not, what is the potential for a business related to wildlife and recreation?

SAE for ALL Check-In

- How much time have you spent on your SAE this week?
- Have you logged your SAE hours?
- What challenges are you having with your SAE?
- How can your instructor help you?
- Do you have the equipment you need?

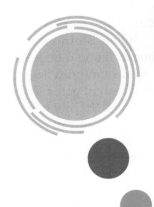

UNIT 11
Agricultural Engineering

SAE for ALL *Profile*
Building on the Family Business

For five generations, welding and steel fabrication has been Stormy Knight's family livelihood. His dad owned a business called Advanced Fabrications, where Stormy held his first job sweeping the floors in middle school. Soon, he would be apprenticing under workers and eventually earn a place at his own machine on the floor.

In addition to working at his dad's shop, Stormy also developed his talents in his school's agricultural mechanics shop. He was an active FFA member, serving his chapter for two years as the treasurer and president. Stormy's SAE project was a machine of his own design, the Gravity Deer Feeder. Stormy competed in the Georgia National Fair with his project, earning a blue ribbon. He qualified as a State Proficiency Finalist and Finalist for Central Region State Star in Agribusiness, and his passion for the family industry was confirmed.

Toward the end of his time in high school, Stormy started working at a local business, Georgia High Tech. They did an abundance of metal fabrication work, but specialized in aerospace parts, which require a special certification. Stormy saw a chance to gain experience that he couldn't get with his family business, and took on the challenge. He worked and studied with workers at the shop, stayed late to practice, and eventually took the aerospace test. When he didn't pass the test on the first try, Stormy didn't give up; he went back and passed it.

After high school, Stormy pursued a welding degree at his local technical college. When he finished, he took a job at Plant Vogtle, a Georgia energy plant where his dad had started decades earlier. He spent months practicing different weld tests, and in June 2019 he passed them. He was now employed through the electrician's union as an electrician welder. Stormy stayed on at Plant Vogtle for two more years, before returning home to help his dad start a new family business, Oconee Welding and Fabrication.

Today, Stormy never knows what will come through the shop doors. It could be a 5060E John Deere tractor needing a custom cage, aluminum float tubes that hold airplane fuel, or 16-foot ornamental gates for a house down the street. He loves what he does. Stormy has been granted so many opportunities, and it all started with his SAE project. Stormy says, "The best way to better yourself is to push yourself to the places you think are unreachable. Hard work and self-discipline pay off."

- How did Stormy's SAE prepare him for career success?
- What opportunities do you see to improve or build on an existing business in your area? What skills might you need?

Top: Stormy Knight
Bottom: Stormy Knight

DedMityay/Shutterstock.com

Safe Agricultural Work Practices

Tyler Olson/Shutterstock.com

Learning Outcomes

By the end of this lesson, you should be able to:
- Describe the attitudes and attributes students can develop to maximize safety. (LO 11.01-a)
- Identify potential hazards on an agricultural site. (LO 11.01-b)
- Classify injuries as nonimpact, impact, and contact injuries. (LO 11.01-c)
- Describe measures that can be taken to maintain a safe environment on an agricultural site. (LO 11.01-d)
- Identify basic personal protective equipment (PPE) and describe its importance to staying safe. (LO 11.01-e)
- Explain how reaction time relates to safety. (LO 11.01-f)
- Describe safe practices in the agricultural laboratory. (LO 11.01-g)

Words to Know

accident	first aid	reaction time
decibel	material safety data sheet (MSDS)	

Before You Begin

Have you ever cut your finger when you weren't paying attention? Maybe it was with a pair of scissors, maybe a knife in the kitchen or garage, or maybe it was with something else. Why did that happen?
Turn to your neighbor to tell them about the incident. What could you have done to prevent it?

Many agricultural workers do work that puts them at high risk for injuries. Agricultural machinery, chemicals, loud noises, unpredictable animals, and prolonged sun exposure all pose a risk of injury (**Figure 11.1.1**). These risks require increased awareness and effort to avoid accidents. *Accidents* happen unexpectedly and unintentionally, and they often result in damage or injury. Maintaining a safe work environment is a primary goal in every agricultural workplace and in the agricultural laboratory.

Developing Safe Habits

Safety begins with you. You must take responsibility for your safety and the safety of others. Have a positive attitude about safety and encourage others to do the same. Pay attention during safety lessons or instructions from your teacher. Always take safety seriously by staying focused on what you are doing and keeping alert while you are working. Help your classmates or coworkers to stay safe by doing your tasks carefully and according to instruction, and speaking up when you see issues. Safe work environments require everyone to be accountable.

Potential Hazards

Identifying potential hazards will help you to stay safe and avoid accidents. Be familiar with your work environment. Pay attention to your surroundings and recognize when something is not right or out of place.

Machinery and Equipment

Many types of machinery and equipment are used around a farm (**Figure 11.1.2**) and in agricultural laboratory settings, from tractors to table saws, portable saws, welders, and hand tools. Each of these items can cause injury to people who are not paying attention or are not using them properly.

Sodel Vladyslav/Shutterstock.com

Figure 11.1.1. Working with large equipment can pose hazards to safety. *What hazards can you identify in this photo? List ways to address the safety hazards that you identify.*

Did You Know?

A little over 5,000 workers in the US die on the job per year. That averages to more than 100 deaths per week, or about 15 deaths per day. About 20 percent of those deaths are in construction-related activities.

Figure 11.1.2. These large machines are common on a farm or in an agricultural setting. *What type of training do you need to be able to handle these machines properly?*

royal_indiana/Shutterstock.com

11

SpeedKingz/Shutterstock.com

Figure 11.1.3. This worker doesn't look like he's very happy right now. *What might be causing his pain? What could he do to address it?*

Knowledge of how to use and maintain each piece of equipment is critical to safe operation and the safe completion of work. Equipment manufacturers provide manuals detailing operations that should be followed. If you are not familiar with how to use a tool properly, you shouldn't be using it.

Noise

Noise is measured in decibels (dB). A *decibel* is a unit for expressing the relative intensity of sounds. Decibels range from zero, for barely perceptible sounds, to 130, where the sound alone causes pain (**Figure 11.1.3**).

Gas-powered lawn equipment, motorcycles, car horns, large machinery, and other equipment can make sounds that measure greater than 100 decibels. This can cause hearing damage or hearing loss with extended exposure. Once hearing loss occurs, you cannot gain it back. If you are going to use loud equipment or be around it, you need to protect your ears.

Air Quality

The air quality of a location is measured by the Environmental Protection Agency and is expressed as the Air Quality Index (AQI). The US Air Quality Index is a color-coded scale that ranges from 0 to 500, with lower numbers representing cleaner air (**Figure 11.1.4**). Pollution can increase the AQI of an area. Higher numbers are typically located near large cities or industrial areas.

Air quality is a consideration indoors as well. Classrooms, laboratories, or workshops can have air quality issues affected by the kind of work being performed. Many chemicals, paints, or finishes require ventilation from fans or air-conditioning equipment to keep the air in the space clean and breathable. Dust from sawing or sanding wood, chemicals released while welding, and animal waste require proper ventilation to protect workers' (or students') lungs. It is important to understand the type of work being performed and if it requires ventilation or outside work to maintain healthy air quality.

Figure 11.1.4. *Have you seen this Air Quality Index information before? If so, where? Where would you look to find your community's AQI?*

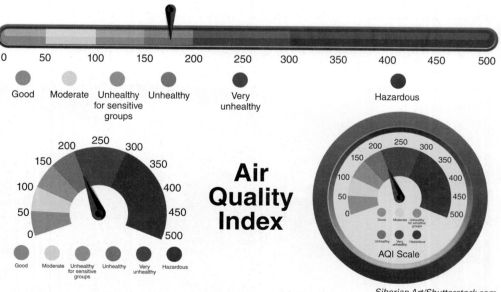

Siberian Art/Shutterstock.com

Chemicals

Many chemicals are used in agricultural applications. Pesticides, paint, finishes, fuels, and lubricants are common in agricultural work, and it is important to understand how to handle them safely. Get familiar with the potential hazards posed by any substance before you use it. This information can be found in a material safety data sheet. *Material safety data sheets (MSDS)* (**Figure 11.1.5**) contain information about the properties of hazardous substances. These are required on many work sites by the Occupational Safety and Health Administration (OSHA), to provide information to employees or emergency personnel with safety procedures for handling or working with potentially hazardous substances. Material safety data sheets are mandated to be kept on-site where the chemicals are used.

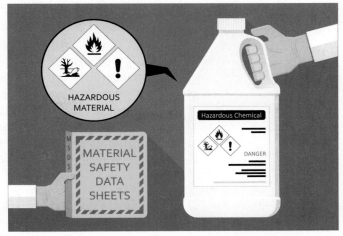

Pepermpron/Shutterstock.com

Figure 11.1.5. Material safety data sheets are critically important to maintaining lab safety. *Why?*

Proper storage of chemicals is another important part of chemical safety in the lab. Chemical and flame-resistant cabinets and containers are to be used for the storage of hazardous chemicals to protect workers, students, and the environment. Spill containment materials should be readily available. Chemicals should be stored in their original containers, labels attached, and should not be stored above or with materials that may be part of a chemical reaction in case of a spill.

Injuries

Injuries occur when someone is hurt. While this is not an exhaustive list, some of the most common injuries fall into three categories: nonimpact injuries, impact injuries, or injuries resulting from contact with objects.

Nonimpact injuries are a result of excessive physical effort from lifting, pushing, pulling, turning, throwing, or other bodily effort (**Figure 11.1.6**). They can also result from non-strenuous, repetitive motions like lifting and turning that are repeated, often over long periods of time. The most common nonimpact injuries are back injuries, though repetitive use injuries can happen to many other parts of your body.

Figure 11.1.6. *What injury risks do you see in this photo?*

Photographee.eu/Shutterstock.com

Impact injuries occur when your body sustains the impact of a blow. Falling is the most common cause of impact injuries and happens more commonly to older adults. Slipping and falling or bumping into another object is another source of impact injuries. Jumping down from a height can also cause an impact injury. Most impact injuries result in muscle sprains, strains, or tears.

Contact injuries occur when your body contacts equipment or other objects. This might include bumping your head, being struck by equipment, getting caught between objects, touching something hot, or touching something sharp. The most common types of contact injuries include cuts, lacerations, and punctures. The most common body parts affected are the hands and fingers, head, and eyes.

Staying Safe in an Agricultural Environment

Personal Protective Equipment

Personal protective equipment (PPE) is equipment worn to minimize exposure to hazards that cause injury or illness. Familiar examples of PPE include safety glasses, gloves, earplugs, hard hats, masks, and body suits. PPE should be safely designed and constructed to fit comfortably to encourage use. An improper fit could cause dangerous exposure. Anyone using personal protective equipment should be trained to know when it is necessary, what kind is necessary, how to wear it, how to adjust it, and how to remove it. Wearers should understand the limitations of the equipment, its proper care and maintenance, and how to dispose of the equipment once its useful life is complete.

Personal protective equipment (**Figure 11.1.7**) may include items to protect your:

- **Eyes.** Safety glasses, goggles, welding helmets, and face shields
- **Head.** Hard hats and head covers
- **Hearing.** Earplugs and earmuffs
- **Hands and Fingers.** Leather gloves, welding gloves, and rubber gloves
- **Back.** Back braces and support straps
- **Breathing.** Respirators and face masks
- **Body.** Coveralls, overalls, full body suits, and aprons
- **Feet.** Steel-toed shoes or boots, splash guards, and rubber boots

CHUYKO SERGEY/Shutterstock.com

Figure 11.1.7. Personal protective equipment, or PPE, comes in many shapes and forms. *What types of PPE do you use in your life?*

First Aid

The attention given to a person suffering from an illness or injury to preserve life, prevent the condition from worsening, or to promote recovery is called ***first aid***. First aid is generally performed by someone with basic medical training (**Figure 11.1.8**). First aid training is widely given and may be required in certain occupations. First aid often involves the treatment of minor cuts, burns, abrasions, or punctures to reduce or stop bleeding. Basic first aid kits should be available in agricultural workplaces, including the agricultural mechanics laboratory.

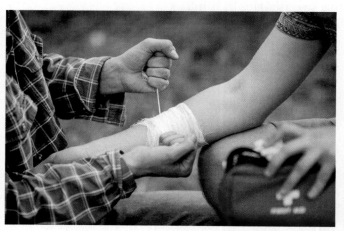

PRESSLAB/Shutterstock.com

Figure 11.1.8. Basic first aid is performed regularly, at home or in an agricultural setting. *Why might you need first aid kits in your ag classroom?*

Reaction Time

Reaction time is the time between a stimulus (the cause of a reaction) and your body's response. Reaction time is a simple form of speed and it depends on your nervous system (**Figure 11.1.9**). The initial signal from a stimulus is recognized by the nervous system. The reaction develops in your brain, then runs through the spinal cord and nerves to the muscles, resulting in a reaction. An understanding of average human reaction times is important to preventing accidents and developing a safe workplace. You might think you're pretty fast, but are you faster than the table saw you're working with? The answer might surprise you!

Simple reaction time is the reaction time to a single stimulus. Think of how you might react to a loud barking dog surprising you, or touching a hot pan. Your simple reaction time is determined by genetics and age. Simple reaction time is impaired by being tired or fatigued. It can also be improved with training, but only to a point.

Complex reaction time is the reaction to one stimulus chosen from many available stimuli and it increases reaction time. For example, imagine being presented with something that scares you, like a snake. In a simple reaction situation, you would jump away, and that would be that. But what if your back is against a wall? Now you are presented with a choice. Do you jump to the right or do you jump to the left? Having to make additional decisions increases your reaction time, which slows your response.

Many workplace or laboratory situations that require a reaction are going to be complex. The best way to prepare for a safe response is with proper training.

Maxisport/Shutterstock.com

Figure 11.1.9. Elite athletes have exceptionally fast reaction times. *How does your reaction time affect safety practices around an agricultural setting?*

Hands-On Activity

Reaction Time!

Superman was faster than a speeding bullet. How fast are you? Olympic sprinters have a reaction time of about 1/10th of a second. By the time they even begin to move, an object thrown by the blade of a lawn mower has already traveled 25 feet! Use the provided yardstick and charts to determine your reaction time in seconds. When you're done, compare your times to the provided consequence chart to see just how slow you are in comparison to agricultural machinery.

For this activity, each group will need the following:

- Yardstick
- Reaction Time Chart (Distance in inches converted to time in seconds)
- Consequences Chart

Here are your instructions:

1. Get into groups of two or three students.
2. The first student will place their hand flat on the edge of a table or desk.
3. The second student will hold a yardstick vertically one foot away from the edge of the desk.
4. The bottom of the yardstick should be held level with the edge of the desk.
5. The first student will drop the yardstick without any indication or warning while the second student tries to catch the yardstick as fast as possible, moving in a horizontal direction.
6. Read the number of inches that the yardstick has fallen above the first student's hand.
7. Use the provided Reaction Time Chart to determine the reaction time in seconds.

Continued

11

Reaction Time! *Continued*

8. Use the provided Consequence Chart to find out just how quickly an accident can happen.
9. Swap students until all have a chance to test their reaction time.
10. Repeat the process to see just how quickly you can respond.
11. Try the process again using a warning signal to see if reaction time is improved.

Distance in Inches	Reaction Time in Seconds
32	0.4082
33	0.4146
34	0.4208
35	0.427
36	0.433

Reaction Time Chart

Distance in Inches	Reaction Time in Seconds
1	0.0721
2	0.1020
3	0.125
4	0.1443
5	0.1613
6	0.1767
7	0.1909
8	0.2041
9	0.2165
10	0.2281
11	0.2394
12	0.25
13	0.2602
14	0.2700
15	0.2795
16	0.2887
17	0.2976
18	0.3061
19	0.3146
20	0.3227
21	0.3307
22	0.3385
23	0.3461
24	0.3536
25	0.3608
26	0.368
27	0.375
28	0.3819
29	0.3886
30	0.3953
31	0.4018

Consequences of Reaction Time

Agricultural machinery moves fast – faster than you do! In the time it takes you to react to noticing your shirt caught in a machine, or that something is flying at you, or that the tractor is out of control, you may already be in trouble!

If a student records a reaction time of 0.5 seconds, their clothing could be wrapped 4.5 times around a rotating hydraulic engine before they could respond, or an object thrown by a mower could travel 125 feet before they could try to get out of the way.

Reaction Time in Seconds	PTO @ 540 rpm (rotations)	Tractor @ 20 mph (dist. in ft)	Object Thrown by Lawn Mower (dist. in ft)
.1	.9	2.93	25
.2	1.8	5.86	50
.3	2.7	8.79	75
.4	3.6	11.72	100
.5	4.5	14.66	125

Consider This

1. What does the relationship between reaction time and consequences tell you about the importance of safety in an agriculture lab or on an agricultural job?
2. What best practice rules could you put in place to help prevent accidents, taking reaction time into account?

Laboratory Safety

The agricultural laboratory is like many well-equipped facilities used by contractors or farmers. These facilities include basic tools, equipment, and storage based on their intended use. Agricultural laboratories are valuable for training purposes and should be clean, organized, and well-maintained. Your lab should also have a safety plan in place. This safety plan is important when planning activities and procedures to take place in the facility. Adequate storage and proper safety equipment, safety training, and organization reduce the chances of accidents and make the workplace and training environment more productive.

Tools, Equipment, and Machinery

There are many tools and pieces of machinery that are used in agricultural applications. These items may be large and stationary or small and portable. They may be powered by fuel, or by electrical power from a plug or battery. Regardless of the type, size, or kind of tool or equipment, workers (and students) must be familiar with the safe operation and use of each item before they operate it.

All tools and equipment should be well-maintained and in good working order. Tools with blades or bits must be sharp to function properly. The extra pressure needed to make a dull tool work can cause a tool to bind or slip, making an accident more likely.

Tool and equipment manufacturers typically provide instructions for the safe operation, use, and care for their specific equipment. It is important to read and understand all operational and safety instructions and to be aware of specific personal protection equipment (PPE) required while using the tool or equipment.

Organization and Layout

"Everything has its place" should be the motto of agricultural laboratories and facilities. A clean and orderly workspace increases comfort and convenience while improving learning efficiency and increasing personal safety (**Figure 11.1.10**). All students or workers should have a clear understanding of what an organized workspace looks like.

Large equipment and stationary machinery should have clear areas designated as safety zones around each machine. Only equipment operators should be allowed in this area while the machine is in use. The location of large equipment should be carefully planned. Manufacturer recommendations should be followed for installation, use, adjustment, and repair. Keep all guards and shields in place and in proper working condition. All blades, knives, or bits should be sharp and clean.

Portable equipment and hand tools should be returned to their proper storage location after use. Tools and equipment kept in their proper place are easily found when needed, increasing workplace efficiency. All portable tools should be maintained following the manufacturer's recommendations for use, adjustment, and repair when needed.

Daleen Loest/Shutterstock.com

Figure 11.1.10. Organization is a key to laboratory safety. *How does everything having a place make the lab safer?*

TheBlueHydrangea/Shutterstock.com

Figure 11.1.11. This cabinet is specially designed to hold and contain chemicals. *What special properties might it have?*

General guidelines for an orderly laboratory include:
- All students or workers help with cleanup.
- Specific organization and cleanup jobs are assigned to all to increase cooperation and knowledge.
- Machinery and equipment are clean and well-maintained.
- All portable equipment and tools are cleaned and placed in their proper storage locations.
- Floor is clean and all trash is picked up and placed in appropriate containers.
- Fuels, solvents, paints, cleaners, greases, etc. are all in approved containers and safely stored (**Figure 11.1.11**).
- All benches and countertops are cleared and cleaned after use.

General Classroom/Laboratory Safety Rules

Specific safety rules may be required for the use of specialized equipment, tools, or materials. Any specific rule should align with the recommendations from the manufacturer and should be addressed by the instructor prior to the use of such equipment, tools, and materials. The following general safety rules are a good start toward developing safe behavior to help avoid accidents in the agricultural laboratory:
- Know locations of first aid kits, fire extinguishers, safety showers, and eyewash stations. Review how to use them at the start of a new term.
- Know emergency exit and evacuation routes.
- Know the location of material safety data sheets (MSDS) for any chemicals or hazardous materials used in the laboratory.
- No horseplay will be tolerated.
- Safety glasses should always be worn while working in the laboratory.
- Wear appropriate personal protective equipment when needed (**Figure 11.1.12**).
- Avoid distracting or startling anyone working in the laboratory.
- Use tools and equipment only for their designed purpose.
- Long hair and loose clothing must be properly secured to avoid entanglement.
- Only perform work in the laboratory under the supervision of your instructor.
- Immediately notify your instructor of any accident that occurs.
- Regularly inspect all equipment for wear or damage and repair as needed according to the manufacturer's recommendation.

Figure 11.1.12. This turbine engineer is wearing personal protective equipment to keep him safe on the job. *What elements of PPE do you see here? What might those items indicate about the risks of the job?*

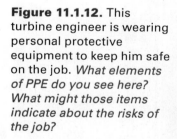

Oil and Gas Photographer/Shutterstock.com

Careers in Safety

The United States Department of Labor oversees the Occupational Safety and Health (OSHA) administration. If safety in the workplace is important to you, there are many potential careers that allow you to be a part of taking care of others.

Examples of OSHA-related job positions include:

- Safety and Occupational Health Specialist
- Chemist
- Program Analyst
- Writer
- Industrial Hygienist
- Electrical Engineer
- Manager

Each of these jobs have competitive salaries, great retirement benefits, flexible work hours, and are located across the United States.

LESSON 11.1 Review and Assessment

What Did We Learn?

- Safety in the agricultural environment starts with you. It is your responsibility to have a positive attitude about safety, to listen to instructions, to stay focused and alert, and speak up when you see issues.
- Agricultural sites are full of potential hazards, including powerful machinery and equipment, noise, poor air quality, and chemicals.
- Injuries occur when someone is hurt. Injuries can be nonimpact, impact, or contact injuries.
- Personal protective equipment is worn to protect individuals from exposure to harm. Appropriate PPE varies depending on the job or task at hand.
- Reaction time is the time between a stimulus and a response. Reaction times can be simple or complex.
- Laboratory safety practices include understanding how to properly use tools, equipment, and machinery and keeping tools in working order. Labs should be organized and clean. Safety zones keep people away from large or dangerous machines. Chemicals should be stored properly.

11

Let's Check and See What We Know

Answer the following questions using the information provided in Lesson 11.1.

1. An unfortunate event that happens unexpectedly and unintentionally, resulting in damage or injury, is called a(n):
   A. Improper attitude
   B. Accident
   C. Maintenance problem
   D. Safety procedure

2. *True or False?* Hearing protection should be worn when operating gas-powered lawn equipment for extended periods of time.

3. Extended exposure to decibel levels _____ can cause hearing damage.
   A. higher than 85
   B. lower than 85
   C. between 30 and 130
   D. at any level

4. Which of the following would be the best source of information about the proper procedures for operation, maintenance, and repair of machinery or equipment used in the laboratory?
   A. Manufacturer's operator or repair manual
   B. Google
   C. Your best judgment
   D. Textbook

5. _____ provide chemical safety and hazard information to students and workers, and must be kept on site.
   A. Product labels
   B. Containers
   C. Material safety data sheets
   D. Operator's manuals

6. Appropriate PPE in an agricultural setting does *not* include:
   A. Safety goggles
   B. Bandages
   C. Steel-toed boots
   D. Respirator

7. Slipping and falling on a wet surface and hurting your hip is an example of a(n):
   A. non-contact injury
   B. contact injury
   C. impact injury
   D. nonimpact injury

8. The attention given to someone suffering from an illness or injury to preserve life, prevent the condition from worsening, or to promote recovery is called _____.
   A. cardiopulmonary resuscitation
   B. band-aids
   C. tourniquets
   D. first aid

9. Which of the following factors affect reaction time?
   A. Age
   B. Fatigue
   C. Genetics
   D. Age, fatigue, and genetics

10. *True or False?* A clean, orderly, well-organized laboratory or workplace is *not* generally recognized to be any safer than a cluttered, disorganized one.

Let's Apply What We Know

1. How does having a healthy attitude about safety and paying attention make a workplace safer?
2. How does reaction time relate to safety in an agricultural setting?
3. Pick an agricultural workplace. Do some research, and identify potential safety hazards that might be found in that workplace.

Academic Activities

1. **Social Sciences.** What impact has the emphasis on workplace safety had on the American agriculture industry over the last 50 years? How does the US compare to other countries? Does increased workplace safety have an effect on productivity in the workplace?

2. **Mathematics.** There are many statistics that focus on workplace safety and accidents. Find one of these statistics pertaining to an agricultural workplace (www.osha.gov may be a good place to start). What does your statistic mean? How does the collection of data and statistics allow companies to increase workplace safety and reduce accidents?

3. **Science.** How can the scientific method be applied to identify methods of improving workplace safety? Is scientific research a useful tool to reduce workplace hazards?

4. **Language Arts.** How important is it to communicate safe practices effectively? Are warning labels really needed? How can an unclear set of instructions cause workplace accidents?

Communicating about Agriculture

1. **Social Media.** Work with your class to create a social media account promoting safe practices in ag classrooms (or classrooms generally), and other agricultural education spaces. How can you promote good habits and safety to others?

SAE for ALL Opportunities

1. **Foundational SAE.** Interview three local farmers or producers and have them give you a list of injuries that have occurred at their operations. What were the similarities and differences among the types of injuries? Did any of the injuries cause a loss of work time or money? Could they have been prevented? What is being done to prevent them in the future? Create a presentation report to share with your class that summarizes what you found out.

2. **Research SAE.** Work with your agriculture teacher to identify potential hazards in your agricultural lab space. What opportunities do you see to improve student safety? Write a set of detailed recommendations to improve the lab environment and present them to your class.

SAE for ALL Check-In

- How much time have you spent on your SAE this week?
- Have you logged your SAE hours?
- What challenges are you having with your SAE?
- How can your instructor help you?
- Do you have the equipment you need?

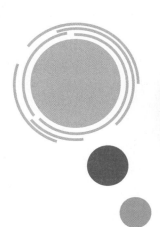

11

Simple Machines and Common Tools

ifong/Shutterstock.com

Learning Outcomes

By the end of this lesson, you should be able to:
- Identify the basic simple machines and describe how they work. (LO 11.02-a)
- Define the main categories of tools and give examples of each. (LO 11.02-b)

Words to Know

fulcrum	pneumatic tools	tool
hand tools	power tools	wedge
inclined plane	pulley	wheel and axle
lever	screw	work
mechanical advantage	simple machines	

Before You Begin

What is the first thing that comes to mind when you hear the word tool? For many of you, it might be a screwdriver or a hammer. Did you know a computer or your phone can be a tool? Have you ever used your phone to complete a difficult task, like getting directions to a place you have never been before or helping you figure out how large a tip to leave at a restaurant?

Make a list of anything that might be considered a tool. Then, put a star next to the items that you listed that might be found in a shop or garage.

A **_tool_** is a device that is used to perform a specific type of function or task. Tools make it easier to complete difficult or labor-intensive tasks. Although some animals have been observed using objects as tools, only humans use tools to create other tools (**Figure 11.2.1**). People have created and modified numerous tools to serve specific purposes. Many tools use simple machines individually or in combination to get things done.

Simple Machines

Simple machines are devices that have few or no moving parts and are used to modify motion and decrease the effort required to perform work. Simple machines use mechanical advantage to increase force. **_Mechanical advantage_** is the ratio of the force that performs useful work to the force applied. Think about it this way: Mechanical advantage means that, by putting some effort into using the tool, the tool magnifies that effort. It does work, so that you put less effort in. **Work** here is defined as a measurement of the energy transfer that occurs when an object is moved over a distance by an external force. It takes a certain amount of work to move a heavy rock 10 feet. When you use a tool like a wheelbarrow to help do it, the work done is the same – but the tool does some of it, so you put in less effort to get it done.

The simple machines are the inclined plane, lever, wedge, wheel and axle, pulley, and screw.

Inclined Plane

An **_inclined plane_** has a sloped (or inclined) surface (**Figure 11.2.2**) and is used for lifting heavy objects. The force required to move an object up the incline is less than the weight of the object being raised. The steeper the incline, the closer the required force is to the object's full weight. Examples of inclined planes in use include loading ramps, wheelchair ramps, and switchback roads up steep hills.

Gorodenkoff/Shutterstock.com

Figure 11.2.1. The use of tools is one of the things that separates humans from animals. *Have you heard of animals using primitive tools before?*

Did You Know?

The earliest hand tools were chiseled from rock using rocks! There is debate over how long ago that was, but it wasn't until about 3,000 BC (5,000 years ago) that the first tools were made from metal (bronze).

Did You Know? The plow is a simple machine consisting of an inclined plane. When driven by a motor and combined with others, this inclined plane can do a lot of work!

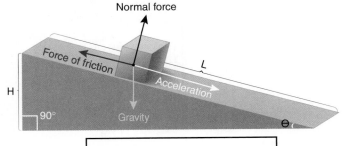

Normal force

Force of friction

Acceleration

H

90°

Gravity

L

Θ

| Mechanical advantage (MA) | = | Lenth of the slope (L) / Height of the slope (H) |

Figure 11.2.2. An inclined plane makes it easier to lift heavy items. *Can you think of an example of this machine in action?*

11

Designua/Shutterstock.com

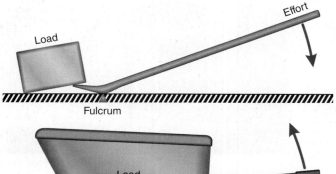

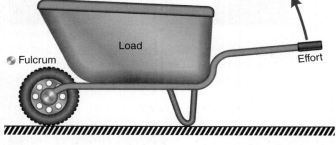

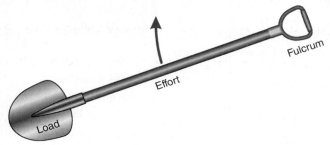

Fouad A. Saad/Shutterstock.com

Figure 11.2.3. Levers are hidden in many popular and useful tools. *Can you think of another tool that utilizes levers?*

Lever

A *lever* is a long bar that rests on a support called a *fulcrum* (**Figure 11.2.3**). When a downward force is applied to one end (typically the longer) the force can be transferred and increased in an upward direction on the other end. Levers can be used to move or lift heavy objects, even when their weight is greater than the weight of the person trying to move it. Many tools use leverage to accomplish work. Examples include crowbars, pliers, and the claw of a hammer.

Wedge

A *wedge* is an object that tapers to a thin edge. Pushing the wedge in one direction creates a perpendicular or sideways force (**Figure 11.2.4**). A wedge is usually made of a hard material such as wood or metal and is used for splitting, tightening, or lifting objects. An axe is a common example of a wedge where the downward swing of the handle exerts a force on the wedge of the axe head that is useful in splitting firewood.

Wheel and Axle

A *wheel and axle* is made up of a circular object (wheel) that rotates or revolves on or around a smaller shaft or rod (axle) (**Figure 11.2.5**). One of the earliest uses was probably for raising a bucket of water from a well with a rope. It requires less force to turn the wheel and lift the water than it would take to lift the bucket itself.

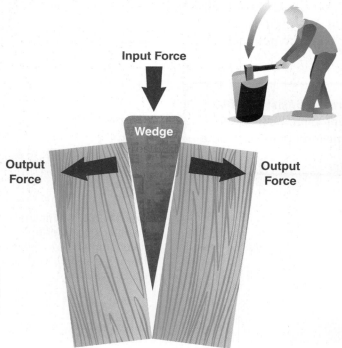

VectorMine/Shutterstock.com

Figure 11.2.4. A wedge is another simple machine that is hidden in many everyday tools. *Can you think of another tool using the wedge?*

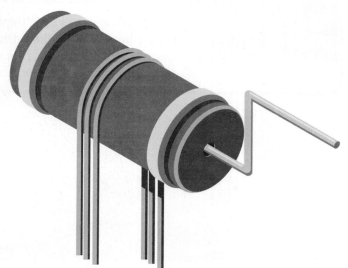

vectorOK/Shutterstock.com

Figure 11.2.5. The wheel and axle is a simple machine that you may not be familiar with. *Can you think of a time this might come in handy?*

Pulley

A **pulley** is a wheel that carries a rope, chain, belt, or other flexible cord on its rim. Pulleys can be used singly or in combination (**Figure 11.2.6**) to provide mechanical advantage to transmit force to motion. Combinations of pulleys and ropes were referred to as block and tackle and were historically used on ships to raise and lower sails.

Screw

A **screw** is a cylinder wrapped with a continuous helical rib. It can be thought of as a wedge wrapped around a cylinder. We often think of screws as fasteners, and we'll discuss them in another lesson. For the purposes of this lesson, screws are also simple machines used as motion modifiers. As an example, a grain auger is used to transfer grain from one container to another (**Figure 11.2.7**).

Tools and Equipment

There are many ways to classify the tools and equipment found in the agricultural mechanics laboratory. Many tools are grouped based on who uses them. Carpenters, mechanics, masons, metalworkers, plumbers, and many other types of tradespeople use tools specific to their trade or work. The challenge with identifying tools this way is that many of them are used in multiple trades. An easier method to sort tools is by how they are used. Tools can be grouped in one of two main categories: *hand tools* or *power tools*. Power tools can be further described as either *portable power tools* or *stationary power tools*.

Hand tools are tools used by someone for manual activities such as cutting, chopping, sawing, or hammering, where the power is provided by the person. **Power tools** are used to perform many of the same operations as hand tools, but with power supplied by an electric motor or engine. Portable power tools are designed for a wide variety of uses and are easily transported to a jobsite for work. Portable power tools may be battery-operated for convenient use when electricity is not readily available. Stationary power tools are usually large and hard to move, and stay in one place.

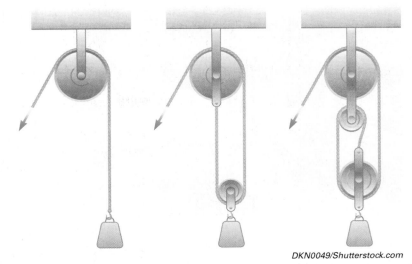

DKN0049/Shutterstock.com

Figure 11.2.6. Pulleys can be built of various complexity, to lift items of all sizes. *When do you think a pulley might come in handy?*

Lutsenko_Oleksandr/Shutterstock.com

Denton Rumsey/Shutterstock.com

Figure 11.2.7. The interior chute of the grain augur is a screw that moves grain up the chute to the other end.

Andrey_Samorodov/Shutterstock.com

Figure 11.2.8. Hand tools include many of the most common and useful tools. *How many of these tools can you identify? Where have you seen them before?*

Hand Tools

Hand tools (**Figure 11.2.8**) are valuable for completing work in design, construction, maintenance, and repair in a variety of industrial trades. From building houses to the assembly or repair of a tractor, hand tools make work faster and more efficient. Hand tools can be used for layout and measurement, cutting, sawing, boring, drilling, fastening, prying, striking, clamping, gripping, and digging. The referenced hand tools in this section are shown in a table at the end of this lesson.

Layout, Measurement, and Marking

A *layout tool* is used to measure or mark material before it is cut or shaped for use. These tools increase precision and efficiency by reducing mistakes and waste, and are valuable when planning, designing, and building projects.

Common layout tools include:
- **Squares and Rules**
  - Combination square
  - Framing square
  - Marking gauge
  - Quick (speed) square
  - Steel rule
  - T-bevel
  - Try square
- **Tapes and Lines**
  - Chalk line
  - Plumb bob
  - Tape measure
- **Levels**
  - Line level
  - Post level
  - Torpedo level
- **Marking**
  - Carpenter's pencil
  - Punch
  - Scratch awl
  - Soapstone

Cutting, Sawing, and Boring

These tools are used to cut or bore material such as wood or metal to reduce its size or modify its shape to a specific dimension. These tools may cut by removing a small portion of the material or separating the material. The cutting edge of the tool must be designed so that it is harder than the material being cut.

Common cutting, sawing, or boring tools include:
- **Sawing**
  - Backsaw
  - Coping saw
  - Crosscut handsaw
  - Hacksaw
  - Keyhole saw
  - Miter saw
  - Ripping handsaw
- **Cutting**
  - Aviation cutters
  - Diagonal (side) cutters
  - Tin snips
  - Utility (razor) knife
  - Wire strippers

- **Boring**
  - Brad point bit
  - Countersink bit
  - Forstner bit
  - Spade bore bit
  - Step bit
  - Twist bit

Fastening, Prying, and Impact

Fastening tools include screwdrivers, wrenches, and hammers that are used to turn or drive fasteners into materials to hold them together. Hammers are also impact tools and prying tools, using the basics of mechanical advantage to increase force and complete work more efficiently. This grouping of tools includes a wide variety used in many different applications.

Common fastening, prying, and impact tools include:

- **Fastening**
  - Allen wrench
  - Nut runner and driver
  - Screwdriver
  - Socket and socket handle
  - Wrench
- **Prying**
  - Crowbar
  - Flat wrecking bar
  - Nail puller
- **Impact**
  - Ball peen hammer
  - Chisel
  - Curved or straight claw hammer
  - Dead blow hammer
  - Punch
  - Rubber mallet
  - Wood mallet

Clamping and Gripping

Tools in this category are useful for holding things securely. They can be used to grip or hold wood, metal, pipe, wire, bolts, and many other items. They are commonly used in combination with other tools to hold materials so they can be cut, bent, or otherwise modified.

Common clamping and gripping tools include:

- **Clamping**
  - Bar clamp
  - "C" clamp
  - Corner clamp
  - Spring clamp
  - Vise
  - Wooden hand screw clamp
- **Gripping**
  - Groove joint pliers
  - Lineman's pliers
  - Locking pliers
  - Long nose pliers
  - Slip joint pliers

Digging

Digging tools are used to work the soil or other ground material by loosening or removing it. Digging ditches, digging postholes, removing manure from stalls, moving feedstuffs, and many other activities are completed with the mechanical advantage provided by digging tools.

Common digging tools include:

- Fork
- Posthole digger
- Rake
- Round point shovel
- Spade
- Square point shovel

Hands-On Activity

Driving a Nail

Driving nails is a common construction activity. While it is a simple task if you have done it before, it can be quite the challenge if you have never attempted to do it. Driving a nail requires some strength, but it also requires coordination and technique. This activity will give students practice at completing one of the most basic woodworking skills.

For this activity, you will need:

- Safety glasses
- 16-oz curved or straight claw hammer
- 2 to 5 16d nails
- Board (preferably 4 × 4 × 8 or similar, as it is thick enough to take all the nails)
- Clamp or vise
- Woodworking table, bench, or sawhorses

Setup:

1. Put on safety glasses.
2. Secure the board to the table, bench, sawhorses, or other appropriate base using a vise, clamps, or other means of clamping.
   A. The board should be secured at approximately waist height for students (28 to 36 inches).
3. Get into groups of two or four students.
4. Watch your teacher's demonstration of proper technique with the hammer. Listen carefully to any safety instructions.

Procedure:

1. Grip the hammer correctly.
   A. Like shaking hands.
   B. Grip closer to the head when starting the nail.
   C. Grip near the end of the handle when driving the nail for more power.

2. Start the nail.
   A. Hold the nail between your thumb and forefinger using your nondominant hand.
   B. Place your fingers near the top of the nail.
   C. Grip the hammer closer to the head (about the middle) and tap the nailhead lightly until the nail has sunk into the wood to a depth that it can stand on its own.
3. Drive the nail.
   A. Focus on the nailhead, not the hammer. With practice, your brain coordinates the strike so that you hit what you are looking at.
   B. Swing from the elbow for maximum efficiency and strength. When you need more control, swing from the wrist. A common mistake is to only swing from the wrist.
   C. Let the hammer's weight do most of the work. You do not need to exert all your strength when hammering. This leads to fatigue, which causes wild swings, bent nails, and damaged work.
4. Finish the drive.
   A. Once the nail is driven nearly all the way, slow your swing for the final distance to avoid damaging the wood.

Consider This

1. What was the most difficult part of driving a nail?
2. Would you have been able to push the nail into the board without the hammer? If not, what does this tell you about the hammer?
3. Think about the simple machines we learned about in this lesson. Which one is a hammer? How can you tell?

Power Tools

Many power tools are designed to operate using an electric motor. The electric motors that run power tools can be battery-operated or plug directly into a power outlet. Combining the mechanical advantage of simple machines with the power of electrical motors allows for more work to be completed or more force to be exerted. Power tools may be used in woodworking and metalworking applications in the agricultural mechanics laboratory.

Other sources of power for power tools include engines fueled by gasoline or diesel, and pneumatic power tools. ***Pneumatic tools*** are powered by compressed air turning a shaft, similar to the way gasoline engines and electric motors operate (**Figure 11.2.9**).

Power tools can be categorized based on how they are used, such as woodworking power tools or metalworking power tools. Many power tools can be used in different job applications. Power tools are often grouped into two major categories: portable and stationary. The referenced power tools in this section are shown in a table at the end of this lesson.

Portable Power Tools

Portable power tools are smaller, lighter, and often less expensive than stationary versions, making them easier to transport to the jobsite. They can be operated by alternating current (AC) or direct current (DC) electric motors. Portable electric generators are sometimes used to provide electricity in areas where it is not readily available. Common portable power tools include:

- **Electric AC- or DC-operated tools (Figure 11.2.10)**
  - Drill
  - Grinder
  - Portable circular saw
  - Reciprocating saw
  - Router
  - Sander
- **Pneumatic tools (Figure 11.2.11)**
  - Drill
  - Grinder
  - Impact chisel
  - Nail gun
  - Paint sprayer
  - Ratchet
  - Sander
  - Saw
- **Gasoline- or diesel-powered tools (Figure 11.2.12)**
  - Chain saw
  - Hedge trimmer
  - Weed trimmer

MossStudio/Shutterstock.com

Figure 11.2.9. These pneumatic tools use compressed air to supply the pressure and force they need to work. *What precautions might a user need to take in working with these tools?*

Chepko Danil Vitalevich/Shutterstock.com

Figure 11.2.10. Electric power tools work by being plugged into a wall outlet. *Can you think of a scenario in which this would be a problem?*

frantic00/Shutterstock.com

Figure 11.2.11. This cabinetmaker is using a pneumatic nail gun to fasten two boards together. *Why would this be a good tool to use for this task?*

Mike Pellinni/Shutterstock.com

Figure 11.2.12. Many outdoor power tools are gasoline-powered. *What advantages can you think of to working with gas-powered tools?*

Dobrovizcki/Shutterstock.com

Figure 11.2.13. Many of the tools in this woodworking shop are large, heavy, and difficult to move—the kind of tools that can't efficiently or safely be moved from site to site. *What is another name for this category of tools?*

Stationary Power Tools

Stationary power tools are often larger and heavier than portable tools, requiring them to be used at a permanent location (**Figure 11.2.13**). Work is usually completed at the laboratory or shop location and then transported to the jobsite for further modification or assembly. Stationary power tools are useful in mass production applications where numerous, repetitive tasks are completed at a single location. Most stationary power tools have a large electric motor and are permanently mounted to the floor. Those that are not permanently mounted are heavy enough to remain stable while working with material.

Common stationary power tools include:
- Band saw
- Cold saw
- Ironworker
- Jointer
- Panel saw
- Planer
- Radial arm saw
- Stationary sander
- Table saw

Hand Tools

The following are tools and equipment that you may be expected to use in an agricultural occupation. It is important to know the correct names of the tools in addition to their intended uses. You should use the proper tool for the task at hand. It is essential that you wear the necessary personal protective equipment, such as safety glasses, when using hand and power tools.

Boring

Brad-point bit

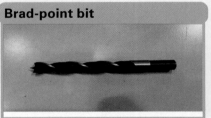
Kevin Jump

Countersink bit

Kevin Jump

Forstner bit

Kevin Jump

Spade-bore bit

Kevin Jump

Step bit

Kevin Jump

Twist bit

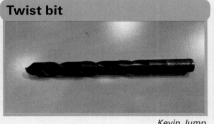
Kevin Jump

Clamping

Bar clamp
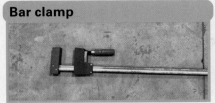
Kevin Jump

"C" clamp

Kevin Jump

Corner clamp
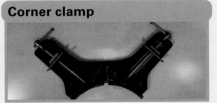
Kevin Jump

Spring clamp
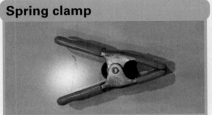
Kevin Jump

Vise

Kevin Jump

Wooden hand-screw clamp

Kevin Jump

Cutters

Aviation cutters

Kevin Jump

Diagonal cutters

Kevin Jump

Tin snips

Kevin Jump

Utility knife

Kevin Jump

Wire strippers

Kevin Jump

Digging

Fork

Kevin Jump

Posthole digger

Kevin Jump

Rake

Kevin Jump

Round point shovel

Kevin Jump

Spade

Kevin Jump

Square point shovel

Kevin Jump

Fastening

Allen wrenches

Kevin Jump

Nut runner and driver

Kevin Jump

Screwdrivers

Kevin Jump

11

Fastening (continued)

Sockets and socket handle

Kevin Jump

Wrenches

Kevin Jump

Gripping

Groove-joint pliers
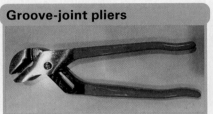
Kevin Jump

Lineman's pliers
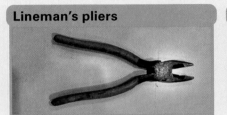
Kevin Jump

Locking pliers

Kevin Jump

Long nose pliers
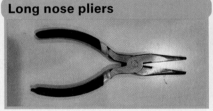
Kevin Jump

Slip-joint pliers
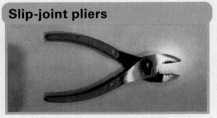
Kevin Jump

Impact

Ball peen hammer

Kevin Jump

Chisel
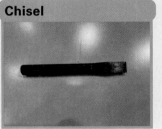
Kevin Jump

Curved/Straight claw hammer

Kevin Jump

Dead blow hammer
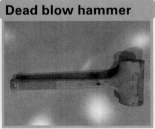
Kevin Jump

Punch
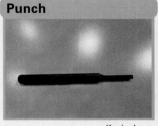
Kevin Jump

Rubber mallet

Kevin Jump

Sledgehammer

Kevin Jump

Wood mallet

Kevin Jump

Layout

Carpenter's pencil
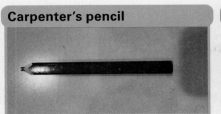
Kevin Jump

Chalk line
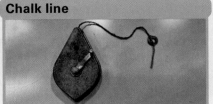
Kevin Jump

Combination square

Kevin Jump

Framing square

Kevin Jump

Line level
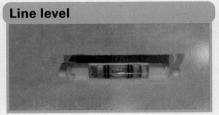
Kevin Jump

Marking gauge

Kevin Jump

Plumb bob

Kevin Jump

Post level

Kevin Jump

Punch
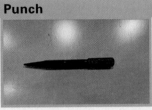
Kevin Jump

Quick (speed) square

Kevin Jump

Scratch awl

Kevin Jump

Soapstone
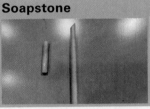
Kevin Jump

Steel rule

Kevin Jump

T-bevel
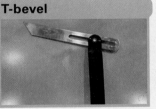
Kevin Jump

Tape measure

Kevin Jump

Torpedo level
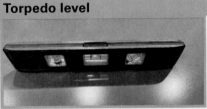
Kevin Jump

Try square
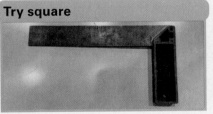
Kevin Jump

Prying

Crowbar

Kevin Jump

Flat wrecking bar

Goodheart-Willcox

Nail puller
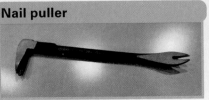
Kevin Jump

Saws

Backsaw

Kevin Jump

Coping saw

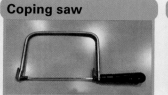

Kevin Jump

Crosscut handsaw

Kevin Jump

Hacksaw

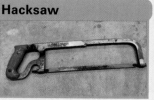

Kevin Jump

Keyhole saw

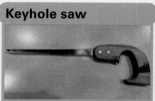

Kevin Jump

Miter saw

Kevin Jump

Ripping handsaw

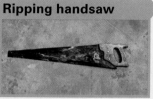

Kevin Jump

Power Tools

Gasoline or Diesel Tools

Chain saw

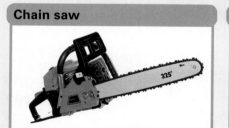

yevgeniy11/Shutterstock.com

Hedge trimmer

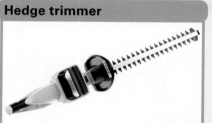

STILLFX/Shutterstock.com

Weed trimmer

Libor Fousek/Shutterstock.com

Pneumatic Tools

Drill

Shutter Baby photo/Shutterstock.com

Grinder

JITTAKORN KAEO-EAIM/Shutterstock.com

Impact chisel

natatravel/Shutterstock.com

Nail gun

ericlefrancais/Shutterstock.com

Paint sprayer

AlexLMX/Shutterstock.com

Ratchet

pooooom/Shutterstock.com

Sander

Yanas/Shutterstock.com

Portable Electric Tools

Drills

kasarp studio/Shutterstock.com

Grinder

kasarp studio/Shutterstock.com

Portable circular saw

MNI/Shutterstock.com

Reciprocating saw

HappyPictures/Shutterstock.com

Routers

mihalec/Shutterstock.com

Sander

Krysja/Shutterstock.com

Stationary Power Tools

Band saw

Matveev Aleksandr/Shutterstock.com

Cold saw

Goodheart-Willcox Publisher

Ironworker

Goodheart-Willcox Publisher

Jointer

Goodheart-Willcox Publisher

Panel saw

Supermaw/Shutterstock.com

Planer

Miriam Doerr Martin

Radial arm saw

mavo/Shutterstock.com

Stationary sander

FOTOGRIN/Shutterstock.com

Table saw

Nakornthai/Shutterstock.com

11

Maintaining Tools and Equipment

Laboratory or shop tools and equipment represent a large investment. They must be well-maintained and adjusted to deliver the best return. Poorly adjusted equipment does not operate efficiently and can cause problems with work quality and safety. Properly maintained equipment lasts longer, works more efficiently, and is much safer to operate than poorly maintained equipment. Owners and workers must follow the recommended procedures for adjustments, operations, sharpening, and repairing all equipment. Be sure to completely and thoroughly read the manual that accompanies the equipment and familiarize yourself with all procedures before attempting to make any adjustments to the machinery.

If something seems off to you with one of the tools in your lab, or you aren't sure what to do with it, go straight to your instructor to let them know. It is unsafe to use a tool that may be broken or have issues, and it is always a good idea to get extra help when using tools or equipment that is new to you.

Career Corner ▷ TRACTOR AND EQUIPMENT SERVICE TECHNICIAN

Interested in working with tools? I might have a job for you!

Tractor and equipment service technicians *work for tractor and equipment companies. Technicians can either work at a dealership or travel around with a truck and tools to assist farmers and other equipment operators on-site with any problems they have.*

What Do They Do? Tractor and equipment service technicians fix broken equipment! They use their tools and knowledge of the equipment to help farmers and others solve equipment problems. If you like solving problems, providing customer service, and working with tools, this might be a good job for you.

What Education Do I Need? Many companies will provide internships, with a guaranteed job after graduation from a training program.

How Much Money Can I Make? Service technicians make an average of $44,000 per year.

ndoeljindoel/Shutterstock.com

How Can I Get Started? Design an SAE to work at a local hardware store and learn as much as you can about the purposes of different tools. Take an agricultural mechanics class, and spend time with your instructor in the lab, practicing your skills. Seek out equipment mechanics to job shadow, interview, or assist with a project.

What Did We Learn?

- Tools help us to perform specific functions or tasks.
- Simple machines are devices with few or no moving parts that decrease the effort required to perform work. They use mechanical advantage to increase force.
- The simple machines are the inclined plane, lever, wedge, wheel and axle, pulley, and screw. Many tools are made from these machines, either alone or in combination.
- Tools are often classified as hand tools or power tools. Power tools are classified as portable or stationary.
- Hand tools are used to complete work in all types of industrial trades. Different tools are used for layout, cutting, sawing, boring, fastening, prying, impact, clamping, gripping, and digging.
- Power tools are designed to operate using a motor to amplify force. Some are electric, others are gasoline- or diesel-powered, and others are pneumatic, or powered by compressed air. Portable power tools can be transported easily to jobsites, but stationary power tools are too large or heavy to move.

Let's Check and See What We Know

Answer the following questions using the information provided in Lesson 11.2.

1. A _____ is a device that is used to perform a specific function or task.
   - A. fastener
   - B. nail
   - C. glove
   - D. tool

2. Which of the following items is *not* a simple machine?
   - A. Lever
   - B. Pencil
   - C. Pulley
   - D. Wedge

3. Identify the best example of a wedge from the list below.
   - A. Hammer
   - B. Tire
   - C. Axe
   - D. Mallet

4. *True or False?* Some tools are a combination of simple machines used together.

5. Identify the tool from the list below that is the best example of a hand tool.
   - A. Portable circular saw
   - B. Electric drill
   - C. Hammer
   - D. Chain saw

6. A framing square would be best categorized as a(n):
   - A. Impact tool
   - B. Cutting tool
   - C. Gripping tool
   - D. Layout tool

7. A wedge wrapped around a cylinder is a good description of a:
   - A. Screw
   - B. Nail
   - C. Staple
   - D. Pulley

11

8. Pneumatic tools are operated and driven by:
   A. Electricity
   B. Compressed air
   C. Batteries
   D. Human muscle

9. A chain saw would be best categorized as a:
   A. Stationary power tool
   B. Pneumatic power tool
   C. Hand tool
   D. Portable power tool

10. Which of the following action items should be completed first by a student who notices a problem with a tool in the agricultural mechanics laboratory?
    A. Notify the instructor immediately.
    B. Leave the instructor a note attached to the tool.
    C. Throw the tool in the trash.
    D. Continue using the tool as long as it is safe.

Let's Apply What We Know

1. How many simple machines have you used today? Think through what you have done and what tools you have used since you got out of bed this morning. Have any of them been based on the concepts in this lesson?

2. What advantages do hand tools have over power tools? When might you choose to use them? Now, switch it around. What advantages do power tools have? When might you choose to use them?

Academic Activities

1. **Social Sciences.** How has the availability of tools been a benefit to agriculture in our country in recent decades? How has that experience been different in other parts of the world?

2. **Science.** What is the difference between the metal used to make a nail and the metal used to make a hammer? Why are they different? What would happen if they were the same?

Communicating about Agriculture

1. **Writing.** What is the importance of effective communication when using tools? Pick a tool and write a short description of that tool without using its common name. Give your description to a classmate and see if they can identify it. What challenges would we have if there were no common names of tools?

SAE for ALL Opportunities

1. **Foundational SAE.** Visit a local tool supply store and interview the owner or manager. Ask them about the tools that they sell. Which tool is sold most often? Which tool is least often sold? Who or what kind of company purchases the most tools? What items are sold at the store besides tools? What opportunities are available for part time work for a future Placement SAE?

SAE for ALL Check-In

- How much time have you spent on your SAE this week?
- Have you logged your SAE hours?
- What challenges are you having with your SAE?
- How can your instructor help you?
- Do you have the equipment you need?

Measurement and Layout

Bannafarsai_Stock/Shutterstock.com

Learning Outcomes

By the end of this lesson, you should be able to:
- Explain why standard weights and measures are important in agriculture. (LO 11.03-a)
- Describe the differences between the metric and US customary systems. (LO 11.03-b)
- Name the main units of measurement in the metric and US customary systems. (LO 11.03-c)
- Explain how accurate measurements and specialized tools improve project outcomes. (LO 11.03-d)

Words to Know

angle	layout	time
area	length	United States customary system
bushel	mass	
commodity	measure	volume
computer-aided design (CAD)	metric system	weight
	temperature	

Before You Begin

Have you ever had a time in your life when measuring or knowing the correct measurement was really important? Several years ago, I was in the football game of my life! It was me and my friends playing against the kids from a rival school. We were running out of time and I caught the pass and went running for the touchdown. I was tackled before I made it to the end zone, but I knew we still had a chance as long as I made it to a first down, which is 10 yards on the football field. I am pretty sure I held my breath while the referees were measuring. How do you think they determined if I made a first down or not? Can you think of some measuring devices that might have been used?

Create a list of five tools you have used to take measurements before. Beside each tool, give examples of things we measure with that tool. And in case you were wondering, measuring was on my side and we won that football game!

Weights and measures have long been a necessary part of human life. Many early civilizations developed standards for measurement and tools for measuring (**Figure 11.3.1**). Early historical documents, including the Bible, gave instructions in their text for the need for common and fair measurements. To **measure** is to determine the size, amount, or degree of something by using an instrument or device marked in standard units. Stones, containers, and amounts that animals (or humans) could lift or move served as the basis of many early weight measurements.

Agricultural Measurements

Why is it important that we learn about measurement in an agriculture class? Let's start by talking about commodities. A **commodity** is a good or product that is interchangeable with other goods of the same type and meets specific minimum standards. Agricultural commodities include crops and animals that are produced and raised on a farm to provide food for people and other animals. These commodities may be traded in large quantities, all over the world. We have talked elsewhere about how the crops produced in one area may be different than those produced elsewhere because of differences in climate, soil, or other available resources. To have access to all kinds of crops, countries need to be able to trade with others. But how can they do that in a way that is consistent and fair? Measurement is the answer. By establishing standards for weights, measures, and conversions, different countries are able to trade their products. The United States Department of Agriculture (USDA) produces a document that gives the weights, measures, and conversion factors for agricultural commodities.

Measurement Systems

The two primary systems of measurement are the *metric system* and the *United States customary system*. Only three countries (Burma, Liberia, and the United States) have not adopted the metric system as their official system of measurement. The use of the metric system has been sanctioned by law in the United States since 1866 but has not displaced the US customary system in wide use. The US armed forces, scientific, and medical communities exclusively use metric measurements.

Figure 11.3.1. If you look carefully at this image, you may see indications of measurement as a part of early Egyptian society. *What hints about measurement do you see?*

matrioshka/Shutterstock.com

History

Early civilizations required measurement for construction and trading agricultural products. Records from early civilizations indicate measurement was important, even if it varied from region to region and changed over time.

- **Ancient Egypt.** The Egyptian *cubit* was the most widespread standard length measurement in the ancient world. The cubit was based on the length of the arm from the elbow to extended fingertips. It was standardized by a royal master cubit of black granite, by which all other cubits were measured. The Great Pyramid of Giza was built using the cubit as a base unit of measurement and is a testament to the accuracy of Egyptian measurement.
- **Babylon.** The Babylonian *mina* is one of the earliest known weights and is estimated to weigh between 23 and 34 ounces.
- **Greece and Rome.** A basic unit of Greek measurement was the *finger*, which was about ¾ inch long. Roman length measurements were based on the Roman standard foot called the *pes*. This unit was divided into 16 digits or 12 inches. Larger linear units were expressed in units of the Roman foot. The furlong was 625 Roman feet, the mile was 5,000 Roman feet, and the league was 7,500 Roman feet.
- **Ancient China.** The ancient Chinese measurement system was separate from the Mediterranean and European systems, but it also used parts of the body as of the guide for length units. There was great variation in their measurements until 221 BC when the first emperor of China, Shihuangdi, set regulations for basic units. The ancient Chinese were known for high accuracy and quality measuring instruments.

Did You Know?

The foot, which is 12 inches in length, was based on the length of a man's foot. How many inches is your foot?

Metric System

Today, most nations use the metric system of measurement. The ***metric system*** is a system of measurement that uses the meter, liter, and gram as the basic units of length, capacity, and weight. The first suggestion of what would eventually become the metric system was around 1670 by Gabriel Mouton, the vicar of St. Paul's Church in Lyon, France. The metric system was approved in June 1799.

Metric units are straightforward because the system is based on powers of ten with standard prefixes used to distinguish between units of different size. These prefixes come from Latin and Greek terms (**Figure 11.3.2**).

Prefix	Symbol	Number	Power	Example (length)
Giga-	G	1 000 000 000	10^9	Gigameter
Mega-	M	1 000 000	10^6	Megameter
Kilo-	k	1 000	10^3	Kilometer
Hecto-	h	100	10^2	Hectometer
Deka-	da	10	10^1	Dekameter
–	base unit	1	10^0	Meter
Deci-	d	0.1	10^{-1}	Decimeter
Centi-	c	0.01	10^{-2}	Centimeter
Milli-	m	0.001	10^{-3}	Millimeter
Micro-	μ	0.000 001	10^{-6}	Micrometer
Nano-	n	0.000 000 001	10^{-9}	Nanometer

Figure 11.3.3. Common metric prefixes. *Do you know these prefixes from any other words?*

Goodheart-Willcox Publisher

11

These prefixes are applied to the seven base units identified by the International System of Units (SI) **(Figure 11.3.3)**. The seven base units are:

- Meter (length)
- Second (time)
- Mole (amount of substance)
- Ampere (electric current)
- Kelvin (temperature)
- Candela (luminous intensity)
- Kilogram (mass)

United States Customary System

In 1790, George Washington addressed Congress stating the need for "uniformity in currency, weights, and measures." Currency was quickly settled in decimal form, but weights and measures proved a more difficult challenge. Thomas Jefferson and John Quincy Adams were in favor of the metric system; however, the United States at that point primarily used weights and measures similar to the old English system.

Standards were set by the Office of Standard Weights and Measures under the Treasury Department. These were a combination of modern and older British weights and measures. The standards were finalized, and new states were presented with these sets of standards as they were admitted to the union. These standards became known as the *United States customary system.*

The Office of Standard Weights and Measures became the National Bureau of Standards on July 1, 1901 by an act of Congress. The National Bureau of Standards sponsored a national conference on weights and measures with the purpose of coordinating standards among the states. In 1988, the National Bureau of Standards became the National Institute of Standards and Technology as a part of the Commerce Department.

The US government remains unique among nations in refraining from exercising control of measurement standards at the national level, with the exception of the Metric Act of 1866, which permitted the use of the metric system in the United States.

Figure 11.3.3. These are the seven base units for the International System of Units. *Which are you most familiar with? Which are you least familiar with?*

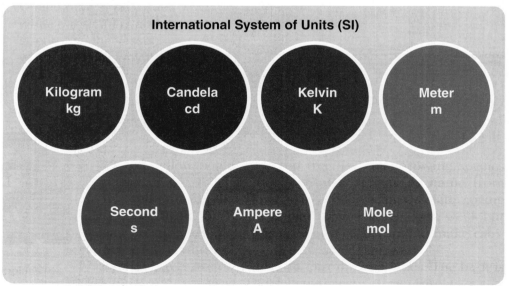

aydngvn/Shutterstock.com

Measurement Units

Standard units of measure are important to be able to buy and sell products consistently and fairly. Scientists and researchers in medicine, agriculture, and engineering depend on standard units to effectively communicate with one another and to document accurate findings for research. There are four bases of measurement that are found in everyday use: weight, length, time, and temperature.

Weight and Mass

Weight and mass are similar but are not the same. *Weight* is the quantity of matter contained by an object as measured by the downward force of Earth's gravity (**Figure 11.3.4**). The *mass* of an object is the measure of the amount of matter it contains. Mass is consistent, but weight may change, depending on where it is measured. An object that weighs six pounds on Earth would weigh one pound on the moon, yet its mass would remain the same. The moon's gravitational pull is about one-sixth that of Earth's, reducing the object's weight but not the amount of matter that it contains.

Basic Units of Weight

US Customary	Metric
1 pound = 16 ounces	1 gram = 1,000 milligrams
1 ton = 2,000 pounds	1 kilogram = 1,000 grams

Weight is the common measure used to buy and sell standard amounts of agricultural products. Standard weights must be used for a fair market value to be given to a product. Agricultural commodities such as corn, soybeans, and oats are measured by the bushel. A *bushel* is a measure of capacity (volume) equal to 64 US pints (35.2 liters). Different products are assigned a standard weight per bushel, based on how much they typically weigh. For example, a bushel of shelled corn has an assigned weight of 56 pounds and a bushel of wheat has an assigned weight of 60 pounds (**Figure 11.3.5**). The wheat has a higher assigned weight because more wheat fits in the bushel, and produces a heavier bushel than corn.

Common Agricultural Products	Bushel Weight
Wheat, white potatoes, soybeans	60 pounds
Shelled corn, rye, sorghum grain, flaxseed	56 pounds
Barley, buckwheat, apples	48 pounds
Oats	38 pounds

Weight and Mass

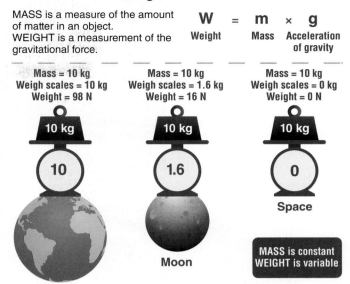

MASS is a measure of the amount of matter in an object.
WEIGHT is a measurement of the gravitational force.

$$W = m \times g$$

Weight = Mass × Acceleration of gravity

Mass = 10 kg
Weigh scales = 10 kg
Weight = 98 N

Mass = 10 kg
Weigh scales = 1.6 kg
Weight = 16 N

Mass = 10 kg
Weigh scales = 0 kg
Weight = 0 N

Earth
Moon
Space

MASS is constant
WEIGHT is variable

VectorMine/Shutterstock.com

Figure 11.3.4. The difference between weight and mass can be difficult to understand, given that the concepts are related and the words are used casually in much the same way. *How would you explain the difference to a classmate?*

Ken Felepchuk/Shutterstock.com

Figure 11.3.5. These bushels of shelled corn all weigh the same amount, so they can be traded fairly. *Would a bushel of green beans weigh more, less, or the same? Why?*

11

Length

Length is a measure of distance. Length units are used in measuring items such as boards, fasteners, the height of trees, and land. Length measurements are also the basis of other measurements, including angles, area, and cubic volume.

Basic Units of Length

US Customary	Metric
1 foot = 12 inches	1 centimeter – 10 millimeters
1 yard = 3 feet	1 meter = 100 centimeters
1 mile = 5,280 feet	1 kilometer = 1,000 meters

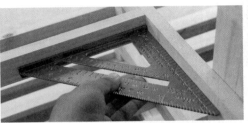

AMLbox/Shutterstock.com

Figure 11.3.6. Measuring angles is an important skill in carpentry. *Why is this?*

Triff/Shutterstock.com

Figure 11.3.7. The needle of a compass always points north. *How can this help you find your location?*

Figure 11.3.8. Measurements help a building fit together properly. *What would happen if the measurements were off?*

Angles

An **angle** is the figure formed by two lines extending from the same point. Angle measurements are useful when determining if a structure is square, for land surveying, and for navigation (**Figure 11.3.6**).

For land survey and navigation, direction from a known point (usually north) is measured in degrees of a circle with a compass. The magnetic pull of Earth draws the needle of a compass in a consistent direction, giving a basis to work from (**Figure 11.3.7**). The circle of the compass is divided into 360 equal units called *degrees*. Each degree is further divided into 60 minutes and each minute is divided into seconds. Using this method, accurate directions from north (or south, if you are in the Southern Hemisphere) can be given in degrees, minutes, and seconds.

Architects, engineers, and construction workers use angle measurements to check the stability and precision of a structure (**Figure 11.3.8**). The layout of a floor system needs to be consistent for all the parts of a building to fit correctly. Precise length and angle measurements allow that to happen. Walls need to be straight up and down, or *plumb*. They are typically perpendicular to the floor.

Stephen Coburn/Shutterstock.com

Area

Area is the amount of space inside a boundary of a flat (two-dimensional) object or shape (**Figure 11.3.9**). Area measurements are determined by multiplying two perpendicular lengths (area = length × width) to determine the surface area in square units. Surface area calculations are useful in determining the amount of material needed to cover a structure. Land area is measured in square units called acres (US customary) or hectares (metric).

Basic Units of Area

US Customary	Metric
1 square foot = 144 square inches	1 square centimeter = 100 square millimeters
1 square yard = 9 square feet	1 square decimeter = 100 square centimeters
43,560 square feet = 1 acre	1 square meter = 100 square decimeters
1 square mile = 640 acres	1 hectare = 10,000 square meters

Volume

Volume is the amount of space occupied by a cubic (three-dimensional) object as measured in cubic units. The volume of an object is determined by multiplying the three-dimensional lengths (volume = length × width × height) together to arrive at cubic units. Volume calculations are useful for determining the capacity of containers, the volume of concrete for a foundation, or the volume of a farm pond (**Figure 11.3.10**).

Basic Units of Volume

US Customary	Metric
1 cubic foot = 1,728 cubic inches	1 square centimeter = 100 square millimeters
1 cubic yard = 27 cubic feet	1 square decimeter = 100 square centimeters
	1 square meter = 100 square decimeters
	1 hectare = 10,000 square meters
1 cup = 8 ounces	1 cubic centimeter = 1 milliliter
1 pint = 16 ounces, or 2 cups	1 milliliter = 0.001 liters
1 quart = 32 ounces, or 2 pints	1 liter = 1,000 milliliters
1 gallon = 4 quarts	1 kiloliter = 1,000 liters

DifferR/Shutterstock.com

Figure 11.3.9. The area of each of these fields is the same. *Why might it be important to use a standard unit of measure in planting?*

Marvpix/Shutterstock.com

Figure 11.3.10. Measurements allow a contractor to calculate exactly how much concrete is needed to pour a foundation. *What happens if those measurements are incorrect?*

Inc/Shutterstock.com

Figure 11.3.11. Understanding the flow rate of chemicals helps farmers know how much pesticide they need to treat their fields. *Why would farmers do these calculations? Would it be more efficient to go out into the fields and just start spraying?*

Time

Time is the measurable period during which an action, process, or condition exists or continues. Time measurement is commonly used in conjunction with other measurements. Farmers commonly use volume over time to calculate flow rates for applying herbicides, pesticides, fertilizers, and other agricultural chemicals, expressed in a rate of gallons per minute (US customary) or liters per minute (metric) (**Figure 11.3.11**). This makes it possible to apply the precise amount needed. Another example is the calculation of miles per hour (US customary) or kilometers per hour (metric) to determine speed.

Units of Time

1 year = 365 days
1 day = 24 hours
1 hour = 60 minutes
1 minute = 60 seconds

Temperature

Temperature is a measure of heat or cold. There are many scales to measure temperature, but the main two scales used worldwide are Fahrenheit and Celsius. Daniel Gabriel Fahrenheit was the first man to make two thermometers that showed the same reading. In 1724, he developed a scale to measure temperature. His method spread throughout the British empire and is still used in many countries today. Shortly after, Anders Celsius developed a system of measuring temperature that uses an even 100-degree difference between the freezing and boiling points of water.

Units of Temperature

Event	Fahrenheit	Celsius
Water Freezes	32 degrees	0 degrees
Normal Body Temperature	98.6 degrees	37 degrees
Water Boils	212 degrees	100 degrees

Making Accurate Measurements

Tools used for measurement and layout vary based on how they are used, what they are used for, what they are made from, and the measurement system of units that they use. *Layout* is the process of preparing a pattern or method in detail for future work. Layout and measurement tools are used to guide workers as they prepare a site or materials for construction. Careful planning during the design, measurement, and layout process makes work more efficient, and decreases the chance of mistakes.

Accurate measuring tools are important for the layout of any project. The level of accuracy needed depends on the type of work being completed. A higher level of accuracy is needed for building the internal parts of an engine compared to the construction of a house. If a house is off square by a sixteenth of an inch, it isn't perfect, but it is much less noticeable than it would be in an engine. The engine would not be able to operate with that level of inaccuracy. The measurement tools used in the design and layout of an engine part must be much more accurate than those used in building construction.

Hands-On Activity

Marking a Board to Cut

Wood construction projects involve cutting lumber to a specified length. Lumber can be purchased in a variety of thicknesses, widths, and lengths. The standard lengths that are purchased are often cut to build different types of projects, from picnic tables to houses. It is very important to make accurate cuts in terms of length and angle. Square (90-degree) cuts are the most common.

slawomir.gawryluk/Shutterstock.com

For this activity, you will need the following:

- Board (1 × 4 × 8 will work well)
- Square
- Pencil
- Tape measure
- Sawhorse or worktable

Here are your instructions:

1. Place the board to be marked on the table or sawhorse.
2. Measure the desired length of the board and make a small mark to indicate the length. The mark should be small, thin, and perpendicular to the edge of the board (see figure).
3. Place the framing square firmly against the edge of the board.

4. Position the blade of the square so that it is over the length mark that you previously made.
5. Use the pencil to draw a line across the board and through the mark.
6. Use the tape measure to recheck the length of the board to the marked line.
7. The board is now ready to be sawed to length.

Consider This

1. Make sure your pencil is sharp. Dull points make wider marks which can increase error. What could you use to mark a board if a pencil is not available?

11

Solcan Design/Shutterstock.com

Figure 11.3.12. CAD software is the basis of modern design, and provides exact measurements for manufacturing. *How do you think this innovation changed the industry?*

Computer-Aided Design (CAD)

Modern computers and computer design software are used throughout the construction industries (**Figure 11.3.12**). The development of *computer-aided design (CAD)* software to assist in product design has made significant advances in recent years. CAD software has grown from two-dimensional design to three-dimensional design and complex modeling with the kind of precision that has enabled it to become the dominant design tool for architects and engineers around the world.

Career Corner   LAND SURVEYOR

If you love to measure things, being a **land surveyor** *may be the job for you!*

What Do They Do? Land surveyors measure property boundaries by taking precise measurements using specialized equipment. The job involves office work and field work, including time spent working outdoors in potentially hazardous or challenging conditions.

What Education Do I Need? Most surveyors have an associate or bachelor's degree, and they must be licensed before they can certify legal documents in their state or local areas.

How Much Money Can I Make? Land surveyors make an average of $68,000 per year.

Dmitry Kalinovsky/Shutterstock.com

How Can I Get Started? You can start an SAE learning how to use specialized measurement equipment. You can work with a local government to see how they measure and mark land for use in your area. You can work with conservation groups or other local organizations to help measure and mark off protected areas, or you can job shadow a land surveyor in your area.

What Did We Learn?

- Standard weights and measures help to promote the fair trade of agricultural commodities.
- The two primary systems of measurement used are the metric system and United States customary system. All but three countries (including the United States) have adopted the metric system as their official system of measurement. It has been sanctioned by law here, but has been slow to displace the customary system.
- Standard units of measurement exist for measuring many things. For the purposes of this unit, the most important of these are weight and mass, length (including angles, area, and volume), time, and temperature.
- Mass is the measure of the amount of matter in an object, and is consistent in all places. Weight is a measure of the amount of matter an object contains, as measured by the force gravity exerts on the object. Weight is subject to change, depending on the gravitational pull where the measurement is taking place. Common units of weight include the pound and ton (US customary) and gram and kilogram (metric).
- Length is a measure of distance, which can also be used to help calculate angles, area, and volume. Common units of length include the foot, yard, and mile (US customary) and the centimeter, meter, and kilometer (metric).
- Time is the measurable period during which an action, process, or condition exists or continues. It is typically measured in seconds, minutes, and hours.
- Temperature is a measure of heat or cold. It is measured in degrees Fahrenheit or degrees Celsius.
- Accurate measurements are important to making projects work, whether in keeping commodity trades fair and consistent, making sure a house can stand, or that an engine can run. Computer-aided design is a tool that has become an industry standard because it increases the precision and accuracy of planning and manufacture.

11

Let's Check and See What We Know

Answer the following questions using the information provided in Lesson 11.3.

1. ____ served as one basis of measurement in early civilizations.
   A. Animal strength
   B. The metric system
   C. The US customary system
   D. The wheel

2. The Egyptian ____ was the most widespread standard of length measurement in the ancient world.
   A. foot
   B. hand
   C. rod
   D. cubit

3. A key advantage of the metric system is that it is:
   A. Based on half units
   B. Easily divided by units of 10
   C. Known for not using decimals
   D. The oldest system of measurement in existence

4. The basic metric unit for length is the:
   A. Foot
   B. Kilogram
   C. Meter
   D. Liter

5. The quantity of matter contained by a unit and measured by the downward force of gravity is:
   A. Mass
   B. Weight
   C. Volume
   D. Time

6. The circle is divided into ____ units called degrees.
   A. 360
   B. 90
   C. 60
   D. 4

7. How many square feet can be found in one square yard?
   A. 3
   B. 6
   C. 9
   D. 27

8. *True or False?* Weights and measurements need to be standardized and consistent around the world for fair commerce and trade.

9. *True or False?* Careful planning during the design, measurement, and layout process allows for more efficient completion of planned work.

10. A measurement of ¼ of an inch can be found in the:
    A. Metric system
    B. US customary system
    C. Fahrenheit scale
    D. Celsius scale

Let's Apply What We Know

1. In this lesson, we learned about measuring weight, length, angles, area, volume, time, and temperature. Give an example of how each of these measurements can be used in an agricultural context.
2. In your own words, why is it important to have a standard system of measurement?

Academic Activities

1. **Social Sciences.** How do countries that use different measurement systems work together when compiling data that requires measurement? Is there a need for common measurements? What would happen if each country developed and used their own system of measurement? Would there be any problems or challenges?
2. **Mathematics.** Some areas of agriculture involve precision measurements while others seem to work well with close estimates. Give an example of an agricultural activity where a precise measurement would be required. Give an example where an estimated measurement would be acceptable. Why is it important to have precise measurements in some situations?

3. **Science.** Is there a standard system of measurement for scientific practices around the world? Why is it important to have a standard? What would happen if there were no standard system of measurement in the world's scientific community?

Communicating about Agriculture

1. **Persuasive Writing.** The US is one of only three countries that does not use the metric system as its official measurement system. Is this a good thing? Why or why not? Write a persuasive essay arguing for or against the US's measurement system. Do additional research, if necessary. Present your arguments to the class.

SAE for ALL Opportunities

1. **Foundational SAE.** Interview an engineer in your community. Ask them how important measurements are in their career field. Be prepared to share the findings of your interview with your teacher and class.
2. **Research SAE.** Compare different feed types for animals. For example, you could raise two groups of baby chickens and use a different feed source for each group. Use scales to measure and weigh the amount of feed given to each group and then weigh the chicks to determine which group has the greatest rate of gain over a specific period of time.

SAE for ALL Check-In

- How much time have you spent on your SAE this week?
- Have you logged your SAE hours?
- What challenges are you having with your SAE?
- How can your instructor help you?
- Do you have the equipment you need?

Common Fasteners and Materials

Momizi/Shutterstock.com

Essential Question

Why is it important to know the properties of materials used in agricultural projects?

Learning Outcomes

By the end of this lesson, you should be able to:
- Describe why it is important to understand the properties of materials used in agricultural projects. (LO 11.04-a)
- Explain what fasteners do and describe the three main categories of fasteners. (LO 11.04-b)
- Describe the sources, advantages, and disadvantages of common construction materials. (LO 11.04-c)

Words to Know

alloy	lumber	plastic
bolt	metal	thermoplastic
concrete	nail	thermoset
fastener	nut	washer
ferrous	oriented strand board (OSB)	

Before You Begin

Have you ever made a gingerbread house around the holidays? Did you have trouble getting it to remain standing? Did the wall always fall over? Or maybe you couldn't get the roof to stay on? Do you think if you had different or better materials, the job would have been easier?

The materials we use in construction matter. It is important to be familiar with the materials that are best suited for the project at hand, whether that is a gingerbread house or a real building.

I n an agricultural setting, there are always projects to be done. Knowing what materials are suited to different projects and purposes will increase your chances of a successful build.

Fasteners

There are many different types of fasteners— so many, in fact, that many hardware stores dedicate an entire section to them. **Fasteners** are devices that are used to connect and secure one object to another. These connections may be permanent or not (**Figure 11.4.1**). Although fasteners vary by size, shape, and use, most can be grouped as a nail, screw, or bolt.

Mikbiz/Shutterstock.com

Figure 11.4.1. *Would you consider these nuts and bolts to be permanent or impermanent fasteners?*

Nails

Nails are slender and straight fasteners made from metal and composed of three parts: the head, shank, and point (**Figure 11.4.2**). The *head* is the wider top of the nail, where force is applied to drive it into a material. The head of a nail is usually flat. The *shank* is the long, slender stem that makes up the main portion of the nail. The *tip* of the nail is usually pointed to easily penetrate the material. Nails are classified based on the structure of their parts, materials they are made from, and size (diameter and length). Most nails are designed to be driven with a hammer, while others can be driven with power tools like the pneumatic nail gun. Nails made from materials like aluminum, stainless steel, and galvanized steel are used in exterior applications because of their resistance to corrosion. Some nails with deformed shanks (ring, spiral, or barbs) have greater holding power than those with smooth shanks. There are nails specifically designed for roofing applications, finish applications, masonry and concrete, and many other situations with unique requirements.

Nail sizes are expressed as a number and the letter "d." The *gauge*, or diameter, of nails depends on the type and length of the nail. As the gauge increases, the diameter and length increases. A 6d nail is two inches long, an 8d nail is two and a half inches long, and a 16d nail is three and a half inches long (**Figure 11.4.3**).

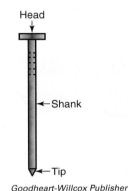

Goodheart-Willcox Publisher

Figure 11.4.2. *What are the purposes of the different parts of the nail?*

Nail Gauge Size Chart

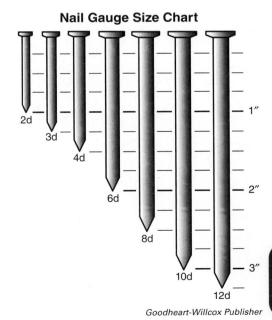

Goodheart-Willcox Publisher

Figure 11.4.3. Nails come in different gauges. This may be confusing, so a visual cue may help keep things straight!

Did You Know? The letter "d" in nail sizes comes from the Roman term *denarius*, for a small coin or penny. The penny system of nail sizes began in England centuries ago, where nails were priced by size and their cost (in pennies) for one hundred nails. 6d nails would have cost six pennies per hundred; 8d nails would have cost eight pennies per hundred.

11

Screws

Some situations require greater holding strength than nails can provide. *Screws* are slender, straight fasteners with a raised helical thread running around the shank and a slotted head for turning to grip materials more tightly. The thread of the screw cuts into the material when the screw is turned, creating an internal thread that pulls fastened materials together to provide greater holding power. Screw threads may be coarse or fine, depending on the specific application.

Screw sizes are based on diameter and length. The gauge of a screw represents the diameter, with higher gauge numbers indicating a thicker diameter. Screw length typically ranges from ¼ inch to five inches. Like a nail, the parts of the screw are the head, shank, and tip. The screw has three basic head types: flat, oval, and round. Screws also have recessed slots for driving. These driving slots are shaped to fit different screwdrivers, including flat, Phillips, square, and star (**Figure 11.4.4**). The shank of a screw can be partially threaded or entirely threaded down its length. Screws may be driven using hand tools such as a screwdriver or power tools such as an electric drill.

Bolts

A **bolt** is an externally threaded fastener designed to be inserted through holes in materials or assembled parts with the use of a **nut** on the opposite side to tighten the materials together. A bolt consists of a head and shank (**Figure 11.4.5**). The shank may be partially or entirely threaded. Nuts and bolts can also be used with washers. **Washers** are flat, round pieces of metal placed under the head of a bolt or nut to distribute the load placed when tightening occurs. Bolts are usually classified by the shape of their head, their diameter or length, their material, or by their thread characteristics.

It can be difficult to tell the difference between a screw and a bolt. A general distinction is that a bolt passes through material and locks it in place using a nut, whereas a screw threads directly into the material as an anchor.

Did You Know?

Bolts are classified by the hardness of the metal, also referred to as *bolt grade*. The higher the number, the harder the metal and the stronger the bolt.

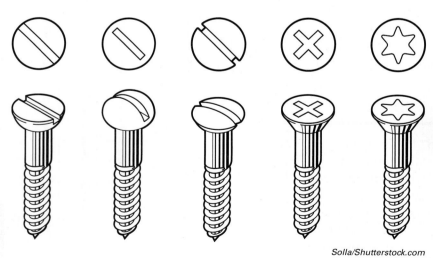

Solla/Shutterstock.com

Figure 11.4.4. Different screw slots require different types of screwdrivers. *Which of these have you seen and worked with previously?*

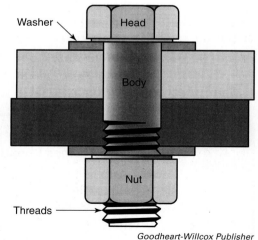

Goodheart-Willcox Publisher

Figure 11.4.5. The parts of a bolt are like those of a screw or nail. *How is the structure of this setup different than those fasteners?*

Common Construction Materials

Many different materials are used in agricultural construction, but the primary materials are wood, metal, plastic, and concrete. Wood and wood products are used extensively to build structures like barns, sheds, and fences. Metal is the primary material for most machinery and agricultural equipment. Metal is also used in fences and roofing. Plastics can be found in plumbing, irrigation, food processing, and other structural components. Concrete is used in structural footings, foundations, and walls. It is valuable to have a basic knowledge of common materials and their uses when planning, designing, and building agricultural structures and equipment.

Wood

Wood and wood products come from trees. ***Lumber*** is a wood product that has been cut to specific sizes for building and construction purposes. Lumber and other wood products come from a number of tree species but are generally divided into softwoods and hardwoods. In general, hardwoods are denser and harder than softwoods. Most lumber used in structures is made from softwoods because it is easier to work with and less expensive to produce (**Figure 11.4.6**). Some softwoods are resistant to rot and can be used in exterior applications where moisture is a problem. Other softwoods are chemically treated to resist rot (**Figure 11.4.7**). Common softwoods used include pine, spruce, Douglas fir, hemlock, cypress, redwood, and cedar.

Did You Know? Wood from some trees is more resistant to rot than others. Cypress, for example, is highly resistant to rot and often used as exterior siding or in other areas that come in contact with outside elements.

Kenneth D Carpenter/Shutterstock.com

Figure 11.4.6. Evergreen forests and trees, like these, provide many of the softwood building materials used in construction across the country.

Chris Hill/Shutterstock.com

Figure 11.4.7. Softwood forests produce lumber used in common projects, like fencing. *What characteristics make a material good for fencing?*

11

Hands-On Activity

Driving Screws into Wood

Wood projects are often held together using nails, screws, and glue (or a combination of all three). Screws are commonly used to fasten boards together. They provide a strong bond and can be removed if necessary.

For this activity, you will need:

- A variety of screwdrivers and screws
  - Slotted screwdriver and slotted wood screws
  - Phillips-head screwdriver and Phillips-head screws
  - Square driver and square drive screws
  - Torx® (star) driver and Torx® (star) drive screws
- Board (2 × 4 × 8 will work well)
- Sawhorse or worktable
- Safety glasses

Here are your instructions:

1. **Secure the board.** Put on your safety glasses. Have a partner hold the board or clamp it to a sawhorse or table so that it will not move during the activity.

2. **Start the screw.** As you begin, make sure that the screw is at a 90° angle and will go straight into the board. Make sure you have the correct driver for the screw that you intend to drive.
3. **Maintain correct alignment and pressure.** Keep the screwdriver aligned with the direction of travel and maintain a constant pressure as you turn clockwise to drive the screw into the board.
4. **Finish flush.** Continue until the screw head is flush (even) with the surface of the board.

Consider This

1. How easy or difficult was it to drive the screw? Why do you think this was the case?
2. Were some screw types (coarse thread vs. fine thread) easier to drive than others?
3. How did using different types of screwdrivers and screw heads affect the results?
4. Which of the types of screwdrivers did you prefer? Can you think of an application where one type of screwdriver and screw head would work better than others?

Colorshadow/Shutterstock.com

Figure 11.4.8. Plywood is an important material used in building projects. *What types of projects might plywood be used in?*

Other wood products used in agricultural construction include plywood, oriented strand board (OSB), fiberboard, poles, and fence posts. *Plywood* is a wood building material made from veneers (thin wood layers) bonded together with an adhesive (**Figure 11.4.8**). Plywood can be made from softwood or hardwood but is mostly made from softwood. Plywood is used for siding, sheathing, roof decking, concrete form boards, floors, and containers. ***Oriented strand board (OSB)*** and other particle board products are like plywood in that they are made from thin strips of wood bonded together with a strong adhesive. In OSB, the wood particles are oriented to increase strength and durability. OSB is often less expensive to make than plywood, making it more popular in some applications. Fence posts and poles are made from trees. The requirements for a tree to make a quality pole are stricter than for other products, making poles more expensive to produce.

Metal

Metal is a solid material composed of one or more chemical elements that have a crystalline structure with high thermal and electrical conductivity. Metals are hard, shiny, *malleable* (can be shaped or formed), *fusible* (easily melted or fused), and *ductile* (able to be drawn into a wire). Metals can be pure (consisting of a single element) or an **alloy** (consisting of two or more elements).

Alloyed metals are classified as *ferrous* or *nonferrous*. **Ferrous** metals have iron as the primary alloying element and are magnetic. Examples of ferrous metals include cast iron, carbon steel, and galvanized steel. Cast iron is an alloy of iron and carbon that is widely used for pipe, machine tools, engine blocks, and equipment components. Cast iron can be easily machined and highly resistant to wear. Carbon steel contains small amounts of carbon and other materials. It is classified as low, medium, or high carbon depending on how much carbon it contains. Galvanized steel is steel that is coated with a thin layer of zinc to prevent rust and corrosion.

Nonferrous metals do not contain iron. These are typically softer than ferrous metals and are not magnetic. Examples of nonferrous metals include aluminum, copper, magnesium, and brass. Aluminum is lightweight and noncorrosive with good thermal and electrical conductivity. Aluminum is commonly used in alloy form for additional strength. It can be found in siding, roofing, building components, electrical wire, and many other applications. It is the most common nonferrous metal used in structural, mechanical, and agricultural applications.

Copper is highly ductile and malleable with impressive flexibility and strength. These properties make it a popular choice for electrical conduction (**Figure 11.4.9**). Copper is commonly alloyed to increase its strength and can be found in electrical components, plumbing pipe, and hand tools where sparking is a problem. Copper is alloyed with zinc to form brass and with tin to form bronze. Brass and bronze were some of the earliest known alloys.

Plastic

Plastics are synthetic and semisynthetic building materials. Plastics are grouped into two main categories: *thermoplastics* and *thermosets*. **Thermoplastics** soften when heated and reform into a solid when cooled. Thermoplastics are recyclable and common examples include acrylic, polyethylene, polypropylene, and polyvinyl chloride (PVC).

A **thermoset** is chemically changed during initial processing but does not soften with additional applications of heat. These plastics typically cannot be recycled. Examples of thermoset plastics are silicone, epoxy, and polyester. Thermoset plastics are often combined with fiberglass to increase strength and are used in automobile panels, tractor body panels, building panels, and many other applications.

Plastics are low-cost products that are resistant to corrosion and moisture and make excellent insulators. They can be flexible, easily colored, and made into any shape. Plastics, especially PVC, are widely used in plumbing applications, plastic sheeting, harvesting, and processing equipment (**Figure 11.4.10**).

Love Silhouette/Shutterstock.com

Figure 11.4.9. When you peel back the coating and look at the actual wires in electrical wire, you are likely to find copper. *What about copper makes it good for conducting electricity?*

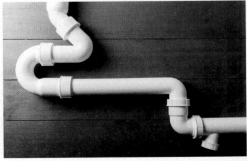

ronstik/Shutterstock.com

Figure 11.4.10. *Have you seen a pipe like this snaking through your house? What do these pipes do? Why is plastic a good choice of material for this purpose?*

11

Figure 11.4.11. Concrete can be formed into a variety of complicated, beautiful, or decorative shapes. *Where else might concrete be used?*

4Max/Shutterstock.com

Concrete

Concrete is a durable material known for its strength and its resistance to fire and decay. It is widely used in construction for foundations, retaining walls, feed bunks, block walls, posts, and poles. Concrete is a mixture of cement, gravel, sand, and water. When first mixed and before it hardens, concrete can be formed into almost any shape, making it a versatile construction component (**Figure 11.4.11**).

Concrete is often made from one part cement, two parts sand, and three parts gravel. Water is then slowly added to the mixture until it becomes workable. Cement forms a paste when mixed with water. As the water dries, this paste acts as a bonding agent to hold all the components together, forming an extremely hard and durable structure. The strength and durability of concrete relies primarily on the type of cement used. Concrete can be mixed by hand in a small container, by machine (like a portable mixer), or in large quantities and shipped by truck to a jobsite for large construction projects.

Career Corner ▶ PORTABLE SAWMILL OPERATOR

If you're interested in producing construction materials, it might make sense for you to own your own business and **run a portable sawmill**. *Let's learn more!*

What Do They Do? Portable sawmills make it possible to cut trees and saw lumber on-site without having to move the trees to a lumberyard. Entrepreneurs who want to run their own business buy the equipment and schedule visits to lumber-growing sites.

What Education Do I Need? While no formal education is required, you will benefit from training in how to run the equipment properly, and from knowing how to run a business. On-site training and an associate degree in business (or equivalent on-site training) would be helpful.

How Much Money Can I Make? The average portable sawmill operator makes about $40,000 per year.

Fexel/Shutterstock.com

How Can I Get Started? Reach out to a sawmill operator in your area. Talk to them about safety and training, or arrange to follow them for a day. Put together an SAE about promoting safety on a jobsite.

What Did We Learn?

- Not all materials are appropriate for all projects. Understanding the properties of the materials used help you to choose the best materials for the job.
- The three main types of fasteners are nails, screws, and bolts. Nails penetrate material to keep pieces connected. Screws have a thread that increases the strength of their bond. Bolts use nuts and washers to fasten pieces together.
- The most common construction materials include wood, metal, plastic, and concrete.
- Wood is used to create lumber, plywood, fiberboard, poles, and fence posts. Softwood is used more frequently than hardwood, as it is easier to manage, cheaper to grow, and more resistant to rot.
- Metals are solid materials composed of one or more chemical elements that have a crystalline structure and high thermal and electrical conductivity. Metals can be pure or alloyed, ferrous or nonferrous. Metals are used to conduct electricity, create siding, roofing, machine tools, and engine blocks.
- Plastics are synthetic and semisynthetic building materials. Plastics are inexpensive and easy to mold, and resistant to corrosion and moisture. They are often used in plumbing, harvesting, and processing equipment.
- Concrete is known for its strength and resistance to fire and decay. Concrete is a mix of cement, gravel, sand, and water.

Let's Check and See What We Know

Answer the following questions using the information provided in Lesson 11.4.

1. _____ are devices that are used to connect or secure one object to another.
   A. Fasteners
   B. Nails
   C. Bolts
   D. Nails, bolts, and screws are all examples of fasteners.

2. Which of the following type of nail shanks has the least amount of holding power?
   A. ring
   B. spiral
   C. barb
   D. smooth

3. Which of the following nail materials would be least effective in exterior applications because of corrosion?
   A. Aluminum
   C. Galvanized steel
   B. Iron
   D. Stainless steel

4. A nut is used with a:
   A. Bolt
   C. Nail
   B. Screw
   D. Staple

5. _____ are used under the head of a bolt to distribute the load placed when tightening occurs.
   A. Screws
   C. Hammers
   B. Nails
   D. Washers

6. *True or False?* Softwoods are often harder than hardwoods.

11

7. Which of these trees is *not* a softwood?
   A. Pine
   B. Oak
   C. Cypress
   D. Cedar

8. _____ is a solid material composed of one or more chemical elements with a crystalline structure and high thermal and electrical conductivity.
   A. Plastic
   B. Wood
   C. Metal
   D. Concrete

9. Galvanized steel is steel that is coated with a thin layer of _____ to prevent rust and corrosion.
   A. aluminum
   B. zinc
   C. plastic
   D. copper

10. *True or False?* Nonferrous metals are known for their strong magnetic attraction and ability to insulate against the flow of electricity.

Let's Apply What We Know

1. Not all fasteners are the same. When would you choose to use nails? Screws? Bolts? How are the circumstances different?
2. How do ferrous and nonferrous metals differ? When would it be better to use one instead of the other?

Academic Activities

1. **Social Sciences.** Certain species of trees are grown for use as lumber and building materials. Crops like corn, soybeans, and cotton are planted, grown, and harvested in a single season. Pine trees can take 30 years or longer to reach maturity. What challenges are associated with such a long growing period?
2. **Mathematics.** Trees are often measured as they grow to get an estimate of the volume of wood they contain so a value can be appraised. The diameter and height of the tree are measured and volume tables are used to calculate the amount of wood. How are tree diameters measured? How are tree heights measured? What equipment/tools are used to measure trees? What formulas are used to estimate the tree volumes?

Communicating about Agriculture

1. **Research and Writing.** Pick a building material mentioned in this lesson – a type of wood, a type of metal, a type of plastic, concrete. Research how this material is used in agricultural contexts. What are its advantages and disadvantages? Write an essay explaining your findings.

SAE for ALL Opportunities

1. **Research SAE.** Compare and contrast wood materials used in agricultural construction applications. Wood samples can be exposed to harsh environmental conditions and then observed to determine how well they hold up. Create an experiment using different species of wood and then explain your results.

SAE for ALL Check-In

- How much time have you spent on your SAE this week?
- Have you logged your SAE hours?
- What challenges are you having with your SAE?
- How can your instructor help you?
- Do you have the equipment you need?

Irrigation and Plumbing

Floki/Shutterstock.com

Learning Outcomes

By the end of this lesson, you should be able to:
- Explain what irrigation is and why it is needed. (LO 11.05-a)
- Describe the different types of surface and subsurface irrigation. (LO 11.05-b)
- Describe water supply and drainage plumbing systems. (LO 11.05-c)
- List the advantages and disadvantages of copper, galvanized steel, PVC, and PEX piping. (LO 11.05-d)

Words to Know

active rainwater harvesting

passive rainwater harvesting

plumbing

potable

reclaimed water

reservoir

sanitary drainage system

septic tank

sewage

subsurface irrigation

surface irrigation

water main

Before You Begin

Take a minute to get up and walk around your house or classroom. Do you see pipes? What do they look like? Where do you see them? Why do you think they are there? Pipes are an important part of the infrastructure of the world we live in. How might understanding how and why these pipes work change your understanding of the building you're in?

11

PHOTO JUNCTION/Shutterstock.com

Figure 11.5.1. Two-thirds of Earth's surface is covered with water, but that water isn't usable for human consumption. *Why is it important to conserve usable water?*

NDAB Creativity/Shutterstock.com

Figure 11.5.2. These sprinklers are applying water to these plants at a nursery. *Why is it important to control the amount of water used?*

Humans, animals, and plants all require water for survival, as our bodies are made up mostly of water. Only a small fraction of the water on Earth is drinkable, as most water is contained in salty oceans (**Figure 11.5.1**). So, how can we access the water we need?

Early civilizations figured out how to dig wells, divert water from lakes and rivers, and capture rainfall. The early Egyptians diverted water from the Nile river to water their crops, and some of their diversion techniques are still used today. Modern agriculturalists use these and other methods to collect and use available water to provide for livestock and crops.

Irrigation

As the world's population has increased, so has the demand for food. Highly efficient methods of growing crops have been developed, including irrigation. *Irrigation* is the process of applying controlled amounts of water to the land to assist in crop production (**Figure 11.5.2**). Irrigation also helps maintain landscapes in dry areas and during periods of inadequate rainfall.

Water Sources

The two main sources of water are *surface water* and *groundwater*. Surface water is found on Earth's surface, in lakes, rivers, and reservoirs. A ***reservoir*** is a lake where water is stored for human use (**Figure 11.5.3**). Groundwater lies under the surface of the land, filling openings in layers of soil and rocks. These underground, water-filled rock openings are called *aquifers*. Groundwater must be pumped from an aquifer to the surface for use.

Figure 11.5.3. This reservoir was artificially created to provide water for the surrounding area. *How are the reservoirs in your area managed? Where do they come from?*

Kletr/Shutterstock.com

Surface water can be collected and stored many ways (**Figure 11.5.4**). Rainwater is often collected from runoff and stored in tanks for future use. This is called ***active rainwater harvesting***. This allows stored water to be used as needed, but storage capacity is limited to the size of the tank or barrel. ***Passive rainwater harvesting*** involves earthworks or other structures that divert water to needed locations. These systems are efficient, but water is only available for a limited time after rainfall occurs.

Groundwater is usually accessed by pumping water from a well that accesses an aquifer. The depth of wells varies by location. Some have shallow, easily accessed aquifers, while other locations have aquifers deep underground. About 14 percent of the United States population relies on well water, with most of these wells found in rural areas.

Most farmers rely on rainfall to provide the water needed for crops to grow. Farmers who irrigate their crops access water from reservoirs or wells. Varying amounts of water are needed for irrigation depending on the needs of the crop, amount of rainfall, regional location, and temperature.

Did You Know? If the amount of water provided from irrigation and rainfall are equal, the rainfall is better for the plants! The clouds from rain cool the environment on a hot day and increase moisture in the air, allowing better absorption of water through the plant. Raindrops fall faster than irrigation droplets and are more evenly distributed through the soil, allowing the plants to take up water more effectively.

Rainwater Harvesting

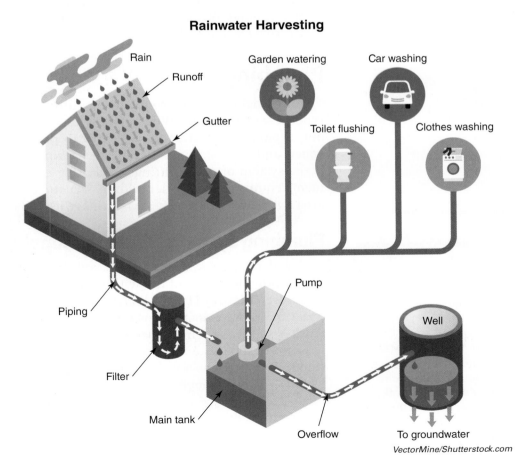

Figure 11.5.4. This home is collecting rainwater for use. *Where does it end up? How does it get there?*

VectorMine/Shutterstock.com

Methods of Irrigation

Surface and subsurface irrigation are the two main types of irrigation. **Surface irrigation** is irrigation that happens aboveground. It can involve flooding a field, using a sprinkler system, or using a drip system that applies water from above. **Subsurface irrigation** consists of methods where irrigation water is applied below the soil surface. Let's look at some of these types of irrigation in more detail:

Wanderlust Media/Shutterstock.com

Figure 11.5.5. This center-pivot irrigation system is designed to be moved across fields. *How can you tell? What advantages does this system have for farmers?*

- **Surface irrigation.** Flooding, sprinkler, and drip systems are the most common, though others exist.
  - **Flooding.** Water is distributed over the surface of the soil by the flow of gravity. Water is introduced into level or graded furrows or basins and allowed to advance across the field.
  - **Sprinkler.** Water is distributed over the surface by sprinkler systems. These systems may be permanently or temporarily mounted and portable. They may be stationary or moved mechanically across the field by hose or a center-pivot system (**Figure 11.5.5**).
  - **Drip.** Water is applied to the soil surface in small drops or streams. The discharge rate of these systems is very low, allowing them to be used on all soil types.
- **Subsurface irrigation.** This type of irrigation consists of methods where water is applied below the soil surface. In some applications, trickle emitters and pipes can be buried below the surface in the plant root zone.

Plumbing

All agricultural environments require some type of water supply. Whether it is indoors or out, a plumbing system is necessary to provide water and drainage for wastewater. **Plumbing** is the system of pipes, fittings, and other equipment required to supply a building, structure, or area with the water supply and drainage it needs.

Plumbing Systems for Water Supply

A water supply system carries water from the supply point to points of use within a building or structure (**Figure 11.5.6**). Water comes from a water main. A **water main** is the primary line in a water supply system. It may come from a public source, such as a city water supply, or a private source, such as a well. Public water sources usually travel through a meter and payment is made based on the amount of water used. Private sources do not charge for water use, except for the electricity or fuel required to power the pump for the well.

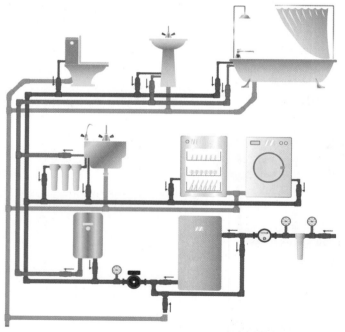

Artem Novosad/Shutterstock.com

Figure 11.5.6. The water supply system for this house carries water to all the points in the house that require it. The blue lines represent cold water supply lines, the red lines indicate hot water supply lines, and the brown lines represent drain lines. *Are there other points in your house that require water and aren't shown in this diagram?*

Water supplied for human consumption is called *potable* water. It must be free from impurities that could cause disease or other negative effects. The quality of potable water must meet the requirements of state health agencies. Public supply systems are regulated for water quality, but the quality of a private water supply is the responsibility of the owner.

Livestock watering systems do not have the same strict requirements as water for human consumption. These systems may use water for livestock watering (drinking), dairy operations, mixing with chemicals, or other agricultural operations. Water used in livestock systems may be non-potable. Some water can even be used more than one time before it passes back into the natural water cycle. This *reclaimed water* may be initially used for watering livestock and the excess can be piped into storage containers for rinsing waste from dairy barns before being drained into the sanitation or septic system for treatment. Agricultural facilities that have piping for potable and non-potable water must have the piping clearly identified by specifically colored labels and arrows indicating the direction of flow (**Figure 11.5.7**).

Plumbing Systems for Drainage

A *sanitary drainage system* carries wastewater from plumbing fixtures and appliances to a sanitary sewer. The sanitary drainage system must be properly vented to ensure adequate removal of sewage and for sewer gases to escape into the atmosphere. *Sewage* is waste material, such as human urine or feces, that is carried away from homes or other buildings in a system of drainage pipes. The sanitary drainage system carries the waste material into a sewer system or septic tank. *Septic tanks* are underground tanks made of concrete, fiberglass, or plastic through which domestic wastewater flows for basic treatment (**Figure 11.5.8**). Septic tanks provide sewage treatment in rural areas where public treatment facilities are not available.

Photographs by Michael/Shutterstock.com

Figure 11.5.7. These labels indicate both the direction of water flow and what kind of water may be carried in these pipes.

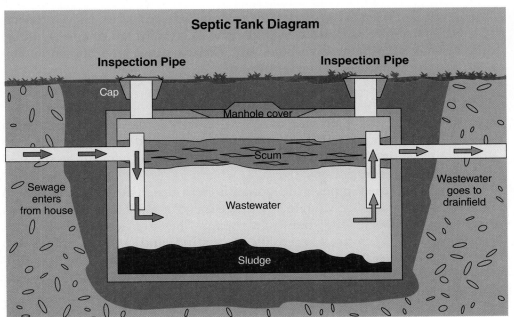

Slave SPB/Shutterstock.com

Figure 11.5.8. Many Americans have septic tanks to provide sewage treatment, rather than pumping all waste to a treatment plant. *Do you have a treatment plant in your area, or a septic system?*

11

The parts of a sanitary drainage system include:

- **Waste pipe.** A waste pipe carries liquid waste from a sink, bathtub, or shower to a drain.
- **Soil pipe.** A soil pipe carries water and fecal matter from a toilet to a drain or sewer.
- **Vent pipe.** A vent pipe allows air into the drainage system.
- **Stack.** A stack is a vertical line of waste, soil, or vent pipes extending through one or more stories.
- **Fixture Drainage Trap.** Fixture drainage traps are devices installed in the drainage line to prevent the passage of sewer gas into the building or structure.
- **Cleanout.** A cleanout is a fitting with a removable plug installed into a sanitary drainage pipe to allow clear access to the pipe for removing stoppages or obstructions.

OlegDoroshin/Shutterstock.com

Figure 11.5.9. Copper pipes have long been popular for use in plumbing. *What makes them well-suited to the job?*

Plumbing Materials and Components

Water is conveyed to individual fixtures (sinks, faucets) through distribution pipes. These distribution pipes are made from copper, steel, or plastic.

Copper

Copper is one of the most common materials used for plumbing pipes (**Figure 11.5.9**). The two main types of copper pipes are *rigid* and *flexible*. Rigid (hard) copper is used throughout a structure for the water supply. Flexible (soft) copper tubing is malleable and is used in short runs where space is tight, and flexibility is needed.

Copper pipes have many advantages. They are long-lasting, having a life span of at least 50 years. They are durable and temperature-resistant, so they can carry hot or cold water. Bacteria do not live well in copper, and copper does not leach into water, making it a safe option for carrying potable water. On the downside, copper pipes are expensive, often costing many times what other materials would cost. There are also environmental concerns with the process of copper mining.

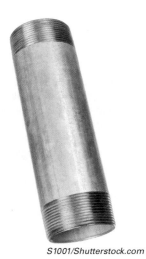

S1001/Shutterstock.com

Figure 11.5.10. The threaded ends of galvanized pipes were one of the advantages that made this material popular for plumbing. *Why is it no longer in favor?*

Galvanized Steel

Galvanized steel pipe was once the preferred material in residential plumbing (**Figure 11.5.10**). The steel pipes are coated in a layer of zinc to prevent corrosion and the pipe ends are threaded so they can be screwed into fittings for connections.

Galvanized steel pipes are strong and less expensive than copper, so they're a great choice... right? Not necessarily. Galvanized steel pipes have a much shorter life span than copper. They are heavy and difficult to work with. They are susceptible to rusting, and corroded pipes can leach lead into water supplies, which is not safe for human health. Galvanized steel is not used to conduct potable water supplies anymore for that reason.

Plastic

Polyvinyl chloride, or PVC, pipes are popular for plumbing (**Figure 11.5.11**). PVC is a thermoplastic made from a combination of plastic and vinyl. PVC pipes are rigid and usually white. They work well with highly pressurized water and are great for main supply lines. PVC can be used as a supply line for potable water and for sanitary drain lines. PVC pipes come in varying thicknesses and diameters, making them useful in a variety of applications.

PVC pipes last for a long time. They are resistant to corrosion and do not rust, meaning they can last far longer than any metal pipes. They are light and easy to use and install. PVC pipes are also relatively cheap, making them an economical choice. On the downside, PVC pipes can warp in hot water, and the quality of the pipes is degraded in sunlight. PVC must be used inside.

Andy Dean Photography/Shutterstock.com

Figure 11.5.11. PVC pipes are popular because they are inexpensive, light, and easy to work with.

Cross-linked polyethylene (PEX) is another popular plastic material used for piping (**Figure 11.5.12**). PEX is new to the market but is extensively used in a variety of plumbing applications. PEX pipes are flexible and can be used in long continuous runs without needing the fittings that PVC pipe requires. PEX is easy to cut and join, making installation quick. PEX piping can come in hard lengths or long rolls.

PEX, like PVC, is resistant to rust and corrosion, so it can last a long time. PEX is extremely flexible, minimizing the use of fittings and making it ideal for replacement jobs. PEX is easy to install. It is considerably cheaper than the most expensive options and can handle hot and cold water easily. PEX is not without downsides, however. It can be damaged by UV radiation and can't be exposed to sunlight. Although PEX has met the US' stringent safety standards, there are some who don't entirely trust it to be safe. PEX can also affect the taste and odor of water, particularly if it sits in a pipe.

Figure 11.5.12. PEX piping is growing in popularity now. *What advantages do you see on display in this photo?*

David Papazian/Shutterstock.com

Hands-On Activity

Building a Portable Sprinkler

Portable sprinklers are useful. This design connects to a hose and can be moved about a garden or lawn. It is relatively inexpensive to build and can be sold as a fundraiser for class projects or FFA activities.

Kevin Jump

Materials List:

Item Description	Quantity
½" PVC pipe (see below for cutting list)	10 feet (commonly found in 20-foot sections)
½" PVC elbow	7
½" PVC tee	4
½" PVC 4-way	1
½" PVC adapter (slip to MNPT)	1
½" PVC adapter (slip to FNPT)	1
½" PVC FNPT to ⅝" hose adapter	1
Sprinkler head with ½" MNPT	1
PVC solvent	1 can
Teflon tape	1 roll

PVC Pipe Cutting List:

Description	Quantity
½" PVC pipe – 8 inches long	14
½" PVC pipe – 6 inches long	1
½" PVC pipe – 2 inches long	1

Pipe Fitting Acronyms:

- **NPT** = National Pipe Thread = American standard thread type
- **MNPT** = NPT with male threads
- **FNPT** = NPT with female threads
- **Slip** = Connection with no threads designed to use PVC solvent

Connection Methods:

- **PVC solvent** – PVC solvent is used to connect PVC pipe and various PVC slip joint fittings using a chemical compound that welds the connections together. This welded connection, when performed properly, is stronger than the pipe itself.
- **Teflon tape** – Teflon tape is used to provide a watertight seal on threaded connections by wrapping the tape around the male (MNPT) threads and then connecting them to the female (FNPT) threaded fitting.

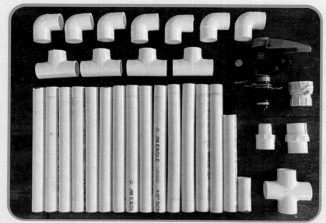

Kevin Jump

Here are your instructions:

1. Before beginning this activity, put on your safety glasses.
2. Use the cutting list as a guide to cut the ½" PVC pipe into the specified pieces.
   A. There are several methods and tools used to cut PVC pipe, from power tools to hand saws.
   B. One of the most simple and effective methods is by using a ratcheting cutter that works like pruning shears. They have a straight blade, are simple to use, cut with no waste, and are accurate.

Continued

Building a Portable Sprinkler *Continued*

3. Begin by assembling the base using **Figure A** below as a guide.
   A. Make sure to dry fit (no solvent) all the parts first to make sure they fit and the lengths are correct.
   B. When applying solvent, make sure that each fitting is facing the correct direction. Solvent is permanent.
4. Assemble the upper portion of the sprinkler assembly using **Figure B** as a guide.
   A. Pay attention to fitting locations and make sure each fitting is rotated to face the correct direction when solvent is applied.

B. The top fitting is the ½" PVC slip to FNPT connection. This is where the sprinkler head (**Figure C**) will attach.
5. Connect the sprinkler head to the top of the sprinkler frame. Remember to use Teflon tape on the threads to avoid leaks.
6. Connect the Hose End connection to the MNPT adapter fitting (**Figure D**) on the base. Remember to use Teflon tape on the threads to avoid leaks.

Kevin Jump
Figure A

Kevin Jump
Figure B

Kevin Jump
Figure C

Kevin Jump
Figure D

Career Corner · AGRICULTURAL IRRIGATION SYSTEM SPECIALIST

If this is interesting to you, you might explore being an **agricultural irrigation system specialist**!

What Do They Do? Agricultural irrigation system specialists provide system planning, design, and repair for large-acreage irrigation systems in production agriculture. Modern agricultural irrigation systems can be automated and monitored through an app on your phone! Specialists need to be able to measure land, assemble large equipment, install wells, understand the technological components of a system, and troubleshoot those same systems quickly and efficiently.

What Education Do I Need? You will need an associate degree or higher and a certification in irrigation systems.

How Much Money Can I Make? The average agricultural irrigation systems specialist makes about $50,000 per year.

Attasit saentep/Shutterstock.com

How Can I Get Started? Design an SAE to create and test your own irrigation system. Research what types of irrigation systems are used in different settings (farms, nurseries, gardens) in your area. Meet with an agricultural irrigation systems specialist to interview them or job shadow them for the day.

What Did We Learn?

- Only a small fraction of the water on Earth is usable, so using it responsibly is important. All crops and life require water, and getting water to the places that need it is important and often challenging.
- Irrigation is the process of applying controlled amounts of water to the land to assist in crop production.
- Surface water and groundwater are the main sources of usable water. Surface water includes rainwater, and water found in lakes, rivers, and reservoirs. Groundwater comes from underground aquifers that are tapped and pumped to the surface for use.
- Surface and subsurface are the two main types of irrigation. Surface irrigation techniques include flooding, sprinkler, and drip irrigation. Subsurface irrigation includes systems set up beneath the soil at the root level.
- Plumbing is a system of pipes, fittings, and other equipment required to supply a building, structure, or area with the water supply and drainage it needs. Water supply comes to a building through a water main and is transported around the structure through distribution pipes. Human water supplies need to be potable; livestock watering systems do not have that requirement.
- The sanitary drainage system carries wastewater from fixtures and appliances to the sanitary sewer. The sanitary drainage system includes waste pipes, soil pipes, vent pipes, and fixture drainage traps.
- Plumbing pipes are typically made from copper, steel, or plastic. Each material has advantages and disadvantages to its use.

Let's Check and See What We Know

Answer the following questions using the information provided in Lesson 11.5.

1. _____ is (are) necessary to all life on Earth.
   A. Trees
   B. Coal
   C. Petroleum
   D. Water

2. Groundwater is sourced from:
   A. Rivers
   B. Aquifers
   C. Lakes
   D. Oceans

3. Collecting rainwater runoff from the roof of a structure is known as:
   A. Plumbing
   B. Passive rainwater harvesting
   C. Active rainwater harvesting
   D. Irrigation

4. Flooding a field is a method of:
   A. Drip irrigation
   B. Subsurface irrigation
   C. Sprinkler irrigation
   D. Surface irrigation

5. Water fit for human consumption is known as:
   A. Non-potable water
   B. Sewer water
   C. Septic water
   D. Potable water

6. Fixture drainage traps are devices installed in the drainage line and to prevent the passage of _____ into a building or structure.
   A. backflow
   B. waste
   C. water
   D. sewer gas

7. *True or False?* Copper is the least expensive material used in the construction of plumbing pipes.

8. Which of the following pipe materials would be risky to use to carry water?
   A. PEX
   B. PVC
   C. Copper
   D. Galvanized steel

9. Copper pipe can have a life span of ____ years or longer.

A. 5
B. 15
C. 25
D. 50

10. Sanitary drain systems include ____ pipes to allow sewer gases to escape.

A. waste
B. soil
C. vent
D. cleanout

Let's Apply What We Know

1. What materials are best suited to indoor plumbing? What materials are best suited to outdoor uses? Why?
2. What are the advantages of different methods of irrigation? What are the disadvantages?

Academic Activities

1. **Social Sciences.** Does water use create conflict in certain areas of the world? What challenges do locations face with the use of water for irrigation? How can these challenges be addressed?
2. **Mathematics.** How do you determine the size of a well needed to provide irrigation water for a crop? How do you determine land area? How do you calculate the amount of water that is produced by an irrigation system over a large acreage in an hour?
3. **Science.** Is there a difference between rainwater and irrigation water? Is one healthier for a plant? What happens to water that is not used by a plant after irrigation? How does irrigation water fit into the water cycle?
4. **Language Arts.** Analyze a brochure for irrigation products. What kind of wording is used to market the product? How can the brochure be improved? Do you see any information that may be left out?

Communicating about Agriculture

1. **Writing.** Create a table showing the advantages and disadvantages of different plumbing materials. Use clear and precise language. Add additional research.

SAE for ALL Opportunities

1. **Research SAE.** How are plants affected by the amount of water available? Select 12 tomato seedlings and transplant them to individual containers. Divide them into three groups of four plants. Give one group of plants no water. Give another group a restricted amount of water. Give the last group adequate amounts of water. Observe and record the results of your observations daily. What differences do you see after a day, a week, and a month? Take pictures and prepare a report to explain the results to your class and teacher.
2. **Entrepreneurship SAE.** Design and construct a drip irrigation system for a container garden that can be attached to a garden hose and a timer. Put together a kit to be sold with the greenhouse plants from your school greenhouse.

SAE for ALL Check-In

- How much time have you spent on your SAE this week?
- Have you logged your SAE hours?
- What challenges are you having with your SAE?
- How can your instructor help you?
- Do you have the equipment you need?

11

Fundamentals of Electricity

arturnichiporenko/Shutterstock.com

Learning Outcomes

By the end of this lesson, you should be able to:
- Explain the role of atoms in producing electricity. (LO 11.06-a)
- Describe the main sources of electricity production. (LO 11.06-b)
- Describe how electricity is measured. (LO 11.06-c)
- List the parts necessary to complete an electric circuit. (LO 11.06-d)

Words to Know

alternating current (AC)
alternator
amperage
atom
battery
compound
conductors
current
direct current (DC)

electricity
electromagnetic
  induction
electromotive force
electron
element
free electrons
generator
insulator

load
neutron
nucleus
parallel circuit
proton
series circuit
voltage
wattage

Before You Begin

Do you think electricity has an effect on your daily life? Go through an average day and make a list of each time you use electricity. What devices do you use most? Can you imagine your life without electricity? What is electricity anyway, and how is it created? In this lesson, we'll find out!

E lectricity is used to power tools, equipment, and machinery for production, processing, heating, cooling, lighting, and other agricultural applications (**Figure 11.6.1**). It is important for agricultural workers to understand the principles of electricity to work with it.

Where Does Electricity Come From?

Elements are the building blocks that make up all matter. They cannot be broken down into other materials. Some matter may contain only a single element, such as pure aluminum. Two or more elements joined together at the atomic level are called *compounds*. For example, water is a compound formed from two hydrogen atoms and one oxygen atom (**Figure 11.6.2**).

An *atom* is the smallest particle showing the properties of a single element. An atom has three parts: protons, neutrons, and electrons. *Protons* are dense, positively charged particles found in an atom's center, or *nucleus*. *Neutrons* are dense, neutrally charged particles also found in the nucleus. *Electrons* are negatively charged particles that orbit around the nucleus. Electrons are attracted to protons. This attraction keeps the electrons from flying away, much like Earth's gravity keeps the moon in orbit. The movement of electrons around the nucleus is so fast that their position is difficult to locate at any specific moment. You can think of them as a cloud of negatively charged particles rotating around the nucleus (**Figure 11.6.3**).

Electrons are generally positioned in layers about the nucleus. Each layer contains a specific number of electrons. The first layer holds a maximum of two electrons. Each additional layer holds eight or more electrons, with the outer layer containing a maximum of eight. If the outer layer does not have eight at any point in time, it will try to find enough electrons to get to eight. This is called *the octet rule*. To satisfy the octet rule, atoms can gain, lose, or share electrons with other atoms.

GLF Media/Shutterstock.com

Figure 11.6.1. This solar panel is gathering electricity for a farm. *How might it be used?*

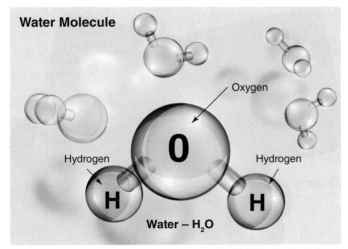

Double Brain/Shutterstock.com

Figure 11.6.2. Water contains two different elements (hydrogen and oxygen), bonded together in a consistent way.

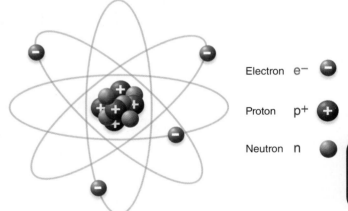

Lookiepixie/Shutterstock.com

Figure 11.6.3. Electrons move in a cloud around the nucleus of protons and neutrons. The difference in charge between the electrons and protons keeps the electrons from flying off to another molecule.

Electron Movement

The movement of electrons between charged particles results in *electricity*. Electrons that move easily between atoms are called *free electrons*. Materials that allow free electron movement are called *conductors* (**Figure 11.6.4**). Pure metals like silver, gold, and copper are excellent conductors of electricity. Materials that resist the movement of electrons are called *insulators*. Materials such as glass, plastic, rubber, pure water, and air are insulators. Metals conduct electricity well because they have many free electrons, while insulators have few, if any, free electrons.

As we saw with the octet rule, atoms are most stable when they have a full outer shell, and the atom has a neutral charge. Electrons will flow from negatively charged locations (more electrons than protons) to positively charged locations (more protons) to find stability. This flow of electrons is called *current.* It will flow until a neutral state is reached.

There are two types of electrical current. *Direct current (DC)* flows in one direction and can be found in batteries and battery-operated devices. *Alternating current (AC)* changes direction at regular intervals and is used to power devices that plug into the electrical power grid. Unlike DC current, AC current can be transmitted at high voltages over great distances, making it more useful for powering homes and businesses.

Figure 11.6.4. Electrical conductors and electrical insulators have different properties. *What do the materials in each of these categories have in common?*

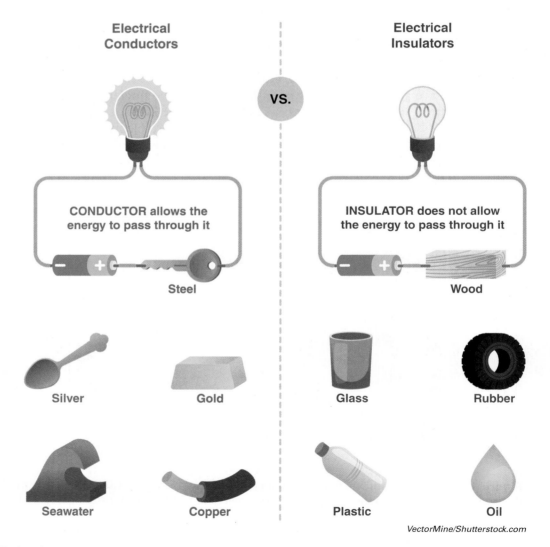

Electrical Conductors

VS.

Electrical Insulators

CONDUCTOR allows the energy to pass through it

Steel

INSULATOR does not allow the energy to pass through it

Wood

Silver

Gold

Glass

Rubber

Seawater

Copper

Plastic

Oil

VectorMine/Shutterstock.com

How Do We Produce Electricity?

Electron flow must be maintained to produce electricity. Electrical energy can be produced chemically, as in a battery, or mechanically, by another source powering a generator. A battery produces electricity by harnessing a chemical reaction between two electrodes. An electric generator produces electricity by creating a difference in electric potential between two electrical poles.

Batteries

A *battery* is a device made up of chemical cells that produce a stream of free electrons (**Figure 11.6.5**). Batteries are used to power devices that require direct current electricity. A battery cell is a source of direct current that has limited energy potential. Separate cells are connected in a single unit to raise the total energy output. This group, or battery, of cells gives the battery its name. The direct current electricity that batteries provide is storable, making it preferred power for portable devices.

Batteries can be rechargeable or not. Non-rechargeable batteries (also called primary batteries) can be used once. Once their stored power is used, they need to be replaced with new batteries. Alkaline and lithium batteries are both non-rechargeable batteries. Rechargeable batteries (also called secondary batteries) can be recharged and reused once their initial equilibrium is reached. Lead-acid and lithium-ion batteries are examples of rechargeable batteries (**Figure 11.6.6**).

Generator

A *generator* moves electrons from one terminal (charged point) to another through a process called electromagnetic induction. *Electromagnetic induction* is the production of electromotive force across an electrical conductor in a changing magnetic field. Generators can produce direct or alternating current. AC generators are called *alternators* and are built so that opposite moving current is produced with every half turn. The speed of rotation determines the frequency. In the United States, the alternating current produced for household use changes direction 60 times per second. This directional change is called a *cycle* and the frequency of change is called a *hertz* (or cycles per second). The frequency of electricity produced in the United States is 60 hertz.

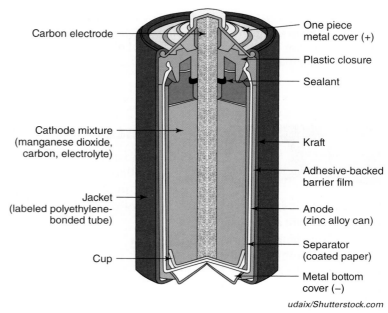

Carbon electrode — One piece metal cover (+) — Plastic closure — Sealant — Cathode mixture (manganese dioxide, carbon, electrolyte) — Kraft — Adhesive-backed barrier film — Jacket (labeled polyethylene-bonded tube) — Anode (zinc alloy can) — Cup — Separator (coated paper) — Metal bottom cover (–)

udaix/Shutterstock.com

Figure 11.6.5. A battery produces energy through managed chemical reactions. *Did you imagine that batteries had so much inside?*

13_Phunkod/Shutterstock.com

Figure 11.6.6. *Did you know that car batteries are constantly recharging while the engine is running? How is this possible?*

Anton Mislawsky/Shutterstock.com

Figure 11.6.7. Electricity is used in many agricultural facilities to keep things moving cleanly, quickly, and efficiently. Being able to manage the flow of electricity with precision makes this work possible.

Measuring Electricity

Some appliances or processes require small amounts of electricity, and others use large amounts. To meet this varied demand, electricity must be measured and controlled for accurate distribution (**Figure 11.6.7**). This will determine the speed, volume, and force as electricity flows through a conductor. Electricity is measured in units of *amperage, resistance, voltage,* and *wattage.*

Amperage

The flow of electricity is called *current.* The rate of the current flow is ***amperage.*** The ampere (A), or amp, is the unit used to measure current flow. An ampere is the quantity of electrons that pass a given point in one second. The volume of electrons conveyed in one second at a current of one ampere is called a *coulomb.*

Resistance

The opposition of electrical flow is called ***resistance*** (R) and is measured in units called ohms (Ω). *Ohms* are units expressing the electrical resistance in a circuit producing a current of one ampere when subjected to a potential difference of one volt.

Voltage

The pressure (electrical potential) that moves electrons through a conductor is called ***voltage*** or ***electromotive force (E).*** The unit of measure of this pressure is called the volt (V). Most residential appliances and equipment are 120-volt, with some 240-volt appliances. Some industry applications can have equipment rated at higher voltages.

Wattage

Power is the rate at which work is completed. In electricity, this is measured in ***wattage.*** A *watt* (W) is equal to one joule per second, corresponding to the power in an electric circuit in which the potential difference is one volt and the current is one ampere. A *joule* is a unit of work required to move an electric charge of one coulomb through an electrical potential difference of one volt. A watt of power is a very small unit and is usually expressed in kilowatts (1,000 watts) and measured over time (kilowatt-hours).

Electrical Circuits

An electrical circuit requires a power source, a conductor pathway, a load to consume power to produce work, and a path to ground to be complete. There are two basic types of circuits: series and parallel. Each circuit type has specific uses.

Did You Know?

Light bulbs produce light and heat. The efficiency of a light bulb is measured as a comparison of how much light it produces compared to how much heat it produces. Incandescent bulbs produce 10 to 20 percent light and 80 to 90 percent heat. By comparison, an LED bulb produces 40 to 50 percent light and 50 to 60 percent heat. LED bulbs are about twice as efficient as incandescent bulbs!

Hands-On Activity

Direct Current (DC) Voltage Measurement

Alkaline batteries have a direct current and a 1.5-volt rating no matter the size: AAA, AA, C, and D batteries have the same voltage. You can determine the charge of an alkaline battery by measuring its voltage using a digital multimeter. You can interpret the multimeter's readings like this:

- 1.5 volts or higher means that the battery is fully charged.
- 1.3 volts or more means that the battery still has some charge.
- 1.2 volts or fewer means that the battery is discharged and can be replaced.

For this activity, you will need the following:

- Several alkaline batteries per group (vary sizes and charge levels)
- Digital multimeter
- Safety glasses
- Rubber gloves (optional)

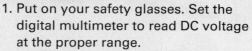

Here are your instructions:

1. Put on your safety glasses. Set the digital multimeter to read DC voltage at the proper range.
2. Connect the multimeter leads to the battery terminals, with the red lead to the (+) terminal and the black lead to the (-) terminal.
3. Record the voltage reading of each battery and the rated size of the battery.
4. Stack two batteries together (positive of one battery to the negative of the other) and record the voltage reading by placing the multimeter leads on the outside terminals.
5. Add a third battery to the stack in the same way and record the voltage reading.

Consider This

1. What happens to the voltage when batteries are stacked together?
2. What happens when voltage is reduced in a battery?
3. Why do you think a battery is no longer useful at 1.2 volts? Why wouldn't it have half of its power left and register 0.75 volts?

Sources of Power

Power plants or electrical generating stations provide most of the power for home, industry, and agricultural applications. These stations use natural resources such as coal, natural gas, nuclear material, wind, water, or solar power to operate large generators that send electricity into an attached power grid by supply lines. Small generators, solar cells, and batteries may be capable of generating enough electrical power for an individual residence or other small purpose.

Conductor Pathway

The flow of electrons requires a pathway to move. This is typically provided by copper, aluminum, or steel wires. There are other metals, such as silver and gold, that provide better conduction but are too expensive to use. Copper is a better conductor than aluminum but costs more to produce. Copper is used in residential and other on-site lines, while aluminum is used in longer runs of transmission lines from power plants to residential areas. The aluminum transmission lines are usually reinforced with steel. The aluminum provides good conduction while the steel mainly increases strength.

11

zDigital Genetics/Shutterstock.com

Figure 11.6.8. All of these are examples of load in an electrical circuit. *What examples do you see when you look around your classroom?*

Load

An electrical *load* is any device that uses electricity to produce work. These devices (**Figure 11.6.8**) can be lights, ovens, air-conditioning units, washing machines, vacuum cleaners, power tools, and any other appliances that connect to an electrical outlet. The work produced by an electrical load converts the electrical energy into another form such as light, heat, or mechanical energy (**Figure 11.6.9**).

Path to Ground (Bonding)

In an electrical circuit, the path to ground is a direct physical connection to the earth that provides a common return path for electric current and is a reference point from which voltages are measured. This physical connection to the earth connects exposed conductive parts (such as metal boxes) directly to the ground, so that any failures of insulated wires or other materials that create dangerous voltages will cause protective devices to turn off the power in the circuit. A ground connection also limits the build-up of static electricity which can be dangerous in areas with flammable materials.

Figure 11.6.9. Energy converts into different forms through work. *Can you think of an example of each of these transformations?*

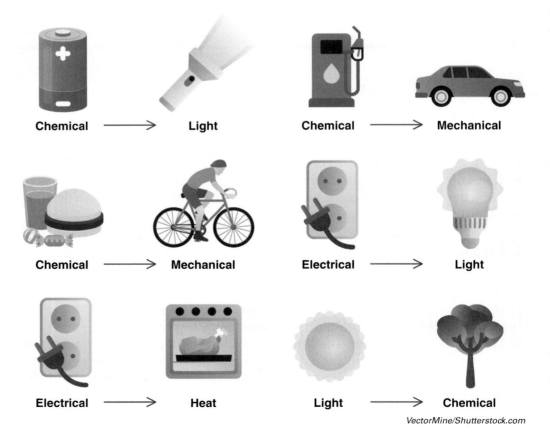

Chemical	⟶	Light
Chemical	⟶	Mechanical
Chemical	⟶	Mechanical
Electrical	⟶	Light
Electrical	⟶	Heat
Light	⟶	Chemical

VectorMine/Shutterstock.com

Series Circuits

A *series circuit* has only a single path for electrical current to flow (**Figure 11.6.10**). The electricity in a series circuit flows *through* every device, or load. If any single device, or load, stops working, the entire circuit is disabled. The current draw of any load in the series circuit also affects the remaining loads in the circuit. Series circuits are not commonly used in residential applications. They can be found in electronic devices and other applications where it is necessary for current to stop if any part of the circuit stops working. Common modern applications of series circuits include the wiring within appliances where a portion of the appliance circuit may be protected by a fuse. The fuse is wired in series so that if it fails, it will shut off power to the appliance to protect it from potential damage.

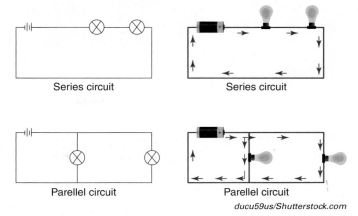

Series circuit

Series circuit

Parellel circuit

Parellel circuit

ducu59us/Shutterstock.com

Figure 11.6.10. Series circuits require all the devices on the circuit to work; if one is out, the whole thing goes down. *Why can't all circuits work this way?*

Parallel Circuits

A *parallel circuit* has more than one path for the electric current to flow, with each path having the ability to operate loads independently. If something happens to one of the loads in the circuit, the remaining loads will continue to operate. In a parallel circuit, each individual load draws from the total voltage available. Residential, industry, and agricultural wiring is completed using parallel circuits to provide power to appliances and equipment. Some components of a parallel circuit (switches, fuses, circuit breakers) are wired in series to protect the circuit.

Career Corner   ELECTRICIAN

If you think you'd like to work with electricity, then being an **electrician** *might be right for you!*

What Do They Do? Electricians have many job opportunities. They can work at electrical power plants, for construction companies, or as private contractors making installations in any kind of location, from farm production systems to private homes.

What Education Do I Need? Training varies based on the job performed, but most electricians require training after high school and are required to complete a certification program. In addition, they must continue training throughout their career to keep up to date with changing technology.

LightField Studios/Shutterstock.com

How Much Money Can I Make? The average electrician makes about $59,000 per year.

How Can I Get Started? Design an SAE studying how electrical circuits work. Design and build a simple electrical device. Work in your school's agricultural mechanics lab to learn as much as you can about working with electricity. Meet with an electrician and conduct an informational interview or job shadow.

What Did We Learn?

- Electricity is produced by the flow of free electrons from negatively charged locations to positively charged locations.
- Direct current flows in one direction. It is portable, and can be produced by the chemical reaction in batteries. Alternating current changes direction at regular intervals. Alternating current is used to power devices that plug into the electrical power grid.
- The distribution of electricity must be measured and controlled to meet the needs of different appliances. Electricity is measured in terms of amperage (current), resistance (opposition to flow), voltage (pressure), and wattage (power).
- Electrical circuits require a power source, a conductor pathway, a load, and a path to ground to be complete. Circuits can be series circuits or parallel circuits.

Let's Check and See What We Know

Answer the following questions using the information provided in Lesson 11.6.

1. _____ are the building blocks that make up all matter and cannot be broken down into other materials.
   A. Particles   C. Solids
   B. Compounds   D. Elements

2. A(n) _____ is the smallest recognized particle that exhibits the specific properties of an individual element.
   A. element   C. neutron
   B. atom   D. proton

3. _____ are negatively charged particles that orbit around the nucleus of an atom.
   A. Protons   C. Gases
   B. Neutrons   D. Electrons

4. The movement of electrons between charged particles results in energy called _____.
   A. heat   C. electricity
   B. light   D. physics

5. Materials that resist the movement of electrons are called _____.
   A. insulators   C. atoms
   B. conductors   D. neutrons

6. Which of the following materials would be a good example of a conductor?
   A. Rubber   C. Glass
   B. Copper   D. Plastic

7. *True or False?* Appliances that plug into a wall are powered by direct current.

8. Batteries are known to produce _____ current.
   A. alternating   C. parallel
   B. series   D. direct

9. The rate of the flow of electricity is measured in _____.
   A. volts   C. watts
   B. ohms   D. amps

10. In an electrical circuit, the path to _____ is a direct physical connection to the earth that provides a common return path for electric current.
   A. neutral   C. least resistance
   B. ground   D. source

Let's Apply What We Know

1. Describe how the movement of electrons produces electricity.
2. Why do batteries produce direct current only?
3. What are the differences between a parallel circuit and a series circuit? Why would an electrician choose to use one over another?

Academic Activities

1. **Social Sciences.** How long has electricity been available to homes in the United States? What US president is credited with making electricity widely available? Are there any countries that do not have widespread electricity? What are the challenges of living in such a location? How can electricity be made available to remote locations?
2. **Mathematics.** How is electricity measured? How much electricity is used by the average home in the United States? What is the average cost to provide electricity to a home?
3. **Science.** What is a conductor? What is an insulator? Why is it important for electricity to be able to flow without hindrance? Which element is the best conductor of electricity? Why is the best conductor of electricity not used to distribute electricity to homes?

Communicating about Agriculture

1. **Language Arts.** Write a paragraph to explain what electricity is to someone who may have never seen it before. If you could travel back in time 300 years and speak with a person from that period, do you think they would believe you when you explained what electricity is and how it is used today? What kind of advancements might we expect 100 years from now?

SAE for ALL Opportunities

1. **Foundational SAE.** Interview local electricians or electrical supply store owners or managers. What jobs are available in your local area for someone with no training? What kind of training is needed to have a job in your local electrical industry? Be prepared to share your findings with your teacher or in class.

SAE for ALL Check-In

- How much time have you spent on your SAE this week?
- Have you logged your SAE hours?
- What challenges are you having with your SAE?
- How can your instructor help you?
- Do you have the equipment you need?

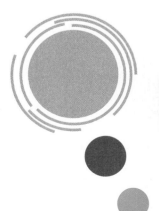

11

Small Engine Basics and Maintenance

Essential Question

Why is it important to know how to care for small engines?

Learning Outcomes

By the end of this lesson, you should be able to:
- Name the key parts of a small engine. (LO 11.07-a)
- Describe the four processes of the engine operation cycle. (LO 11.07-b)
- Describe the systems that keep an engine running. (LO 11.07-c)
- Explain what basic maintenance tasks are required to maintain a small engine. (LO 11.07-d)

Words to Know

camshaft	engine block	spark plug
carburetor	flywheel	stroke
combustion	four-stroke engine	two-stroke engine
crankshaft	piston assembly	valve train
cycle	small engines	

Before You Begin

Think about the gadgets, tools, and appliances found around your house. What do you have to do to keep them in working order? How does taking care of things affect their lifespan? Turn to a classmate and share two things that you must do around the house to help take care of your things. Share your most surprising or creative example with your class.

Engine Basics

Small engines are commonly used to power lawn mowers, generators, tractors, pressure washers, water pumps, and other equipment. They save time by completing tasks faster than people could on their own.

Combustion, or burning fuel in the presence of oxygen, creates powerful reactions. These reactions power engines. ***Small engines*** are internal combustion engines rated at 25 horsepower (hp) or less, used to convert energy into mechanical energy. Most small engines are air-cooled, single cylinder, and gasoline-powered, though this does not describe all of them.

Parts of a Small Engine

No matter the size, the basic components of engines are similar. Small engines are made up of a few basic parts: the engine block, crankshaft, piston assembly, valve train, camshaft, and flywheel. (**Figure 11.7.1**). There are a lot of new terms in describing the engine parts and how they work, and it might seem a bit confusing to start. Hang in there—we'll explain everything along the way!

Engine Block

The ***engine block*** is usually a single piece of machined cast iron or aluminum that contains the crankshaft, piston assembly, valve train, and camshaft (**Figure 11.7.2**). The flywheel and carburetor are attached to the outside of the block. It must be strong enough to contain these moving parts and also contain the power developed by expanding gases during combustion.

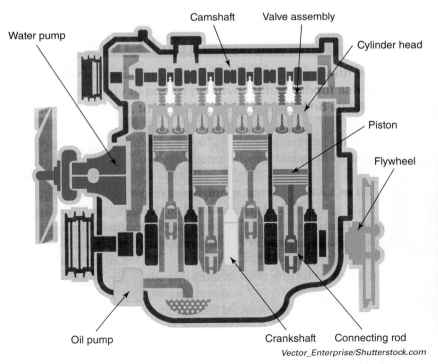

Vector_Enterprise/Shutterstock.com

Figure 11.7.1. Engines are complicated machines, so it may help to see how these pieces fit together. *Have you looked inside an engine before?*

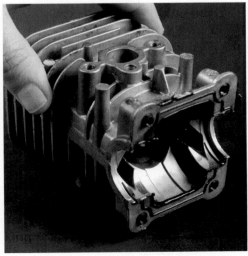

Albert Lozano/Shutterstock.com

Figure 11.7.2. The engine block of a small engine protects the workings inside. *Why is it important for it to be durable?*

11

Jose M. Peral Photography/Shutterstock.com

Figure 11.7.3. The crankshaft is a crucial part of any small engine.

Crankshaft

The *crankshaft* of a small engine converts the up and down motion of pistons into circular or rotating motion (**Figure 11.7.3**). The rotation of the crankshaft turns the camshaft, powering the motion of the valve train, and turns the flywheel, powering the ignition system and providing moving air for the cooling system.

Piston Assembly

The *piston assembly* includes the piston, piston rings, piston pin, and the connecting rod (**Figure 11.7.4**). The piston is located at the top end of the assembly and is in direct contact with the heat and force of combustion. Pistons are commonly made from aluminum or steel. They travel up and down inside of the cylinder of the engine block and are connected to the crankshaft by the piston pin and the connecting rod. Most pistons have three piston rings that run around the outside to create a seal between the top of the piston and the crankcase. This seal allows for compression of the air and fuel mixture to happen. Without these rings, gases created by combustion would leak past the piston and the engine would not generate the intended amount of power.

Valve Train

The *valve train* is typically composed of intake valves, exhaust valves, and the camshaft (**Figure 11.7.5**). The crankshaft turns the camshaft by the connection of a gear. The *camshaft* opens the intake and exhaust valves at exactly the right times to allow a mixture of fuel and air into the combustion chamber and exhaust gases to exit. The camshaft gear is twice the size of the crankshaft gear, allowing the crankshaft to rotate twice for every single rotation of the camshaft. The movement of the piston, crankshaft gear, camshaft gear, and valves must be perfectly timed for proper engine operation. If the timing is off, the engine will not function properly or will operate at reduced power.

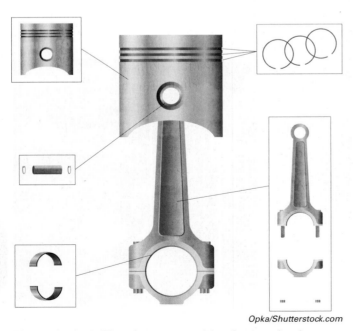

Opka/Shutterstock.com

Figure 11.7.4. The piston assembly of an engine is made up of many small parts, all of which are critical to engine function.

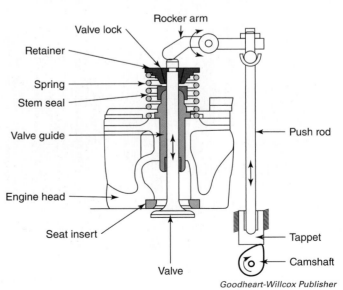

Goodheart-Willcox Publisher

Figure 11.7.5. The engine valve train has many components. *Why is the valve train's work important?*

Flywheel

A *flywheel* is a heavy revolving wheel that increases the engine's momentum and provides a reserve of available power during interruptions in the delivery of power to the machine. In a small engine, the flywheel is attached to the magneto end of the crankshaft and contains magnets that produce ignition current while rotating (**Figure 11.7.6**). Its momentum keeps the crankshaft rotating and smooths engine operation.

Carburetor

Small gasoline engines cannot ignite liquid fuel. The fuel must be vaporized and mixed with air to be ignited by an ignition spark. A *carburetor* produces the proper mixture of fuel and air to ignite and operate the engine (**Figure 11.7.7**). Air is drawn into one end of the carburetor and mixed with the liquid fuel. This mixture of fuel and air is then forced into the cylinder for compression and ignition.

Engine Operation

Basic engine operation involves four basic processes that occur in sequence and then repeat. Each process occurs during a stroke of the piston. A *stroke* is a piston's movement from one end of the cylinder to the other. A *cycle* is the completion of the processes of intake, compression, power, and exhaust. In the *intake stroke*, the fuel and air mixture is drawn into the cylinder. In the *compression stroke*, the mixture is compressed. In the *power stroke*, the fuel and air mixture is ignited and burns rapidly, producing power. In the *exhaust stroke*, the gases given off in combustion are forced out of the cylinder in preparation for the cycle to start again.

There are two basic types of small engines, categorized by the number of piston strokes required to complete a cycle. Engines that complete a cycle of these four events in four piston movements are called *four-stroke engines*, and those that complete a cycle in two piston movements are called *two-stroke engines*.

Nordroden/Shutterstock.com

Figure 11.7.6. Flywheels are crucial to engine performance. *How do they help an engine run smoothly?*

GS23/Shutterstock.com

Figure 11.7.7. A carburetor mixes fuel and air to provide the right mix of fuel to the engine. *Why do some newer engines not include a carburetor?*

11

Figure 11.7.8. The cycle of a four-stroke engine is complex, and the timing of the engine's parts is critical. *What could throw this process off?*

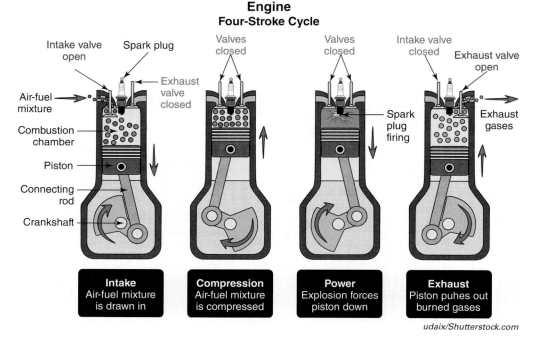

Engine
Four-Stroke Cycle

Intake valve open | Spark plug
Exhaust valve closed
Air-fuel mixture
Combustion chamber
Piston
Connecting rod
Crankshaft

Valves closed
Valves closed
Intake valve closed | Exhaust valve open
Spark plug firing
Exhaust gases

| **Intake** Air-fuel mixture is drawn in | **Compression** Air-fuel mixture is compressed | **Power** Explosion forces piston down | **Exhaust** Piston puhes out burned gases |

udaix/Shutterstock.com

In a ***four-stroke engine***, four strokes are needed to complete the cycle. Each revolution of the crankshaft produces two strokes of the piston. As a result, four-stroke engines require two revolutions of the crankshaft to complete a single operating cycle (**Figure 11.7.8**).

The ***two-stroke engine*** completes a cycle of all four events with only two strokes of the piston and a single revolution of the crankshaft (**Figure 11.7.9**).

The major differences between four-stroke and two-stroke engines are shown in **Figure 11.7.10**.

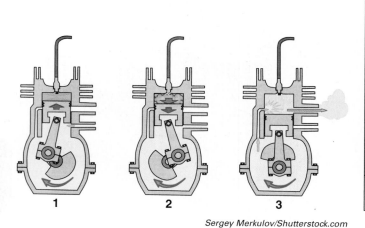

1   2   3

Sergey Merkulov/Shutterstock.com

Figure 11.7.9. A two-stroke engine cycles through in fewer turns of the crankshaft. *How does this affect its efficiency?*

Four-Stroke Engine	Two-Stroke Engine
Four strokes per cycle	Two strokes per cycle
More moving parts	Fewer moving parts
Larger size and weight (same horsepower)	Smaller size and weight (same horsepower)
Lubrication occurs from oil in a sump	Lubrication occurs from oil mixed with fuel
More fuel efficient	Less fuel efficient
Smoother, quieter, slower operation (rpm)	More erratic, louder, faster operation (rpm)
More expensive	Less expensive
Greater maintenance required	Less maintenance required
Longer lasting and more durable	Less durable

Goodheart-Willcox Publisher

Figure 11.7.10. Four-stroke engines and two-stroke engines operate in similar ways but are not the same. *What are the most important differences in choosing one or the other for use?*

Engine Systems

The major functions and operations of an engine are carried out by specific engine systems. Each system must work properly and coordinate with other systems for the engine to operate efficiently. A breakdown or malfunction of any system part can cause all systems to stop working. An engine malfunction is usually caused by one part of a single system. The major engine systems are the fuel, compression, cooling, lubrication, and electrical systems.

Fuel System

The fuel system of a small engine consists of a supply tank, fuel lines, air filter, and carburetor (**Figure 11.7.11**). The carburetor provides fuel and air to the cylinder for combustion. As the piston moves down during the intake stroke, it creates a vacuum that draws air through the air filter to remove foreign particles (dust). The air is then pulled through the throat of the carburetor. This passage, called the *venturi*, is narrow in the center. The air moves more quickly through the venturi's low-pressure zone. Liquid fuel is released into the airflow and is vaporized as it moves along the intake passageway into the cylinder (**Figure 11.7.12**).

Many modern engines no longer use carburetors. Instead, they use an injector pump and fuel injection system that injects vaporized fuel directly into the combustion chamber.

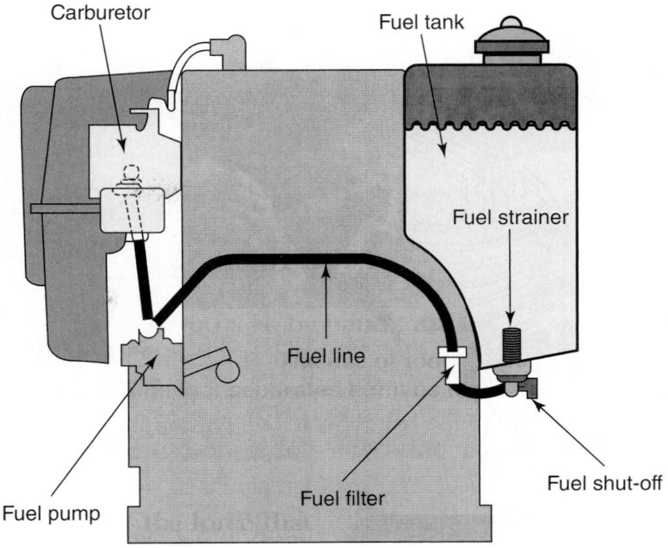

Compression System

The compression system consists of the cylinder, piston assembly, the cylinder head and head gasket, and the valves. The upward movement of the piston during the compression stroke compresses the fuel mixture at the top of the chamber. The downward movement during the power stroke caused by the rapid expansion during combustion creates the power that drives the engine. It is important for the cylinder to be tightly closed during both strokes. If there is a compression leak, the fuel mixture will not be compressed tightly enough for proper ignition and power will be lost during the power stroke. A loss of compression can usually be traced to a piston that doesn't fit properly in the cylinder, a bad head-gasket, or valves that do not close or seal properly.

Cooling System

A small engine operating at a high rate of speed can ignite the fuel and air mixture more than 2,000 times per minute. This combustion process can generate heat as high as 3,000 degrees Fahrenheit. If the heat gets too high, moving engine parts can expand to the point of locking up or seizing.

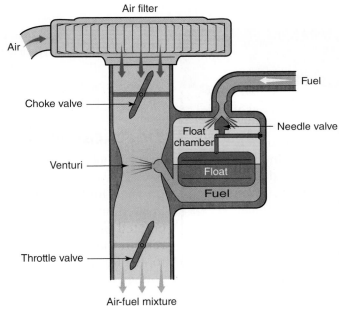

Goodheart-Willcox Publisher

Figure 11.7.11. The fuel system of an engine prepares the fuel and transports it to the piston assembly in the engine to be ignited.

EreborMountain/Shutterstock.com

Figure 11.7.12. The carburetor is a key piece of an engine's ability to work. *What happens to the engine if the carburetor malfunctions?*

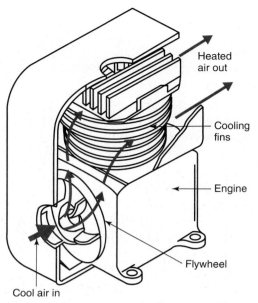

Heated
air out

Cooling
fins

Engine

Flywheel

Cool air in

Goodheart-Willcox Publisher

Figure 11.7.13. The cooling system of the engine needs to be working to keep the engine from overheating. *What happens when an engine overheats?*

Did You Know?

When gasoline is consumed and burned by an engine, it is considered chemical energy. That chemical energy is converted to mechanical energy to provide work. Only about 15 percent of the chemical energy from gasoline is converted into mechanical energy. The rest is lost to heat and friction.

To prevent this from happening, the heat must be dispersed. Most small engines are air-cooled. The cooling system of a small engine includes the flywheel, the shroud, and the cooling fins on the outside of the cylinder block (**Figure 11.7.13**). The shroud of a small engine is a cover that is designed to allow air to be pulled into and across the fins of the cylinder by the rotation of the flywheel. The cooling fins of the cylinder block create more surface area for heat distribution and the movement of air across these fins removes heat from the surface. It is important to keep debris from building up inside of these fins for proper cooling. If the shroud is in place and the cooling fins are clean, the engine should operate at the proper temperature.

Lubrication System

The rapid, repetitive movement of the metal parts inside the engine creates friction, causing wear and producing heat. It is important for these moving parts to have proper lubrication to reduce friction, heat, and wear. In most four-stroke engines, oil is held in a reservoir inside the crankcase, called a *sump*. When the engine is not in operation, the oil is in the sump. When the engine is operating, devices such as an oil dipper or oil slinger push oil throughout the inside of the crankcase as the crankshaft turns. Two-stroke engines have oil mixed in the fuel. This fuel and oil mixture flows through the crankcase providing necessary lubrication. The oil in a small engine provides lubrication, cleans carbon deposits from the engine compartment, cools the engine, and helps provide a seal between the piston, piston rings, and cylinder wall.

Electrical System

The fuel and air mixture in the cylinder is ignited by a spark. Small engines have a small electric generator called a magneto containing a magnet that provides high-voltage pulses as it passes a coil of conductor. When the magnet, located on the flywheel, passes by the armature coil, a low-voltage current is created in the primary circuit of the coil and converted to a high-voltage current in the secondary circuit of the coil. This current, which can produce up to 40,000 volts of electricity, travels along a wire to the ***spark plug***. The high voltage of this current causes the electrical charge to cross the air gap at the end of a spark plug, generating a spark that ignites the compressed fuel and air mixture at the precise moment necessary to create the power stroke.

Small Engine Maintenance

Small engines are known for their durability and dependability. Basic maintenance tasks, when properly performed, will keep engines running for many years.

Cleaning

Small engine cooling systems need to be clean and free from buildup of debris to function properly. A dirty engine can overheat, causing excessive wear and other problems. Engines that operate in dirty environments, such as lawn mowers, need to be cleaned more often. It is a good practice to clean an engine after every use. It does not require much time to clean an engine if it is done often.

Soap, water, and a rag can be used to wash and wipe down a dirty engine to remove dirt, debris, or other buildup. Remember to allow the engine to cool before washing with cold water. Cold water that is suddenly placed on a hot engine block can cause it to crack.

Checking and Changing Oil

The oil level in the crankcase should be checked prior to use. Check the oil level each time you add fuel. Engine manufacturers typically provide a dipstick to check the oil level in the sump.

Engine manufacturers provide specific recommendations for changing oil. The oil itself does not wear out but becomes contaminated with particles that prevent proper lubrication.

Air Filter

The air that enters the carburetor and combustion chamber needs to be free from dust and other particles that can cause excessive wear in the cylinder. A clogged filter prevents the proper amount of air from entering the carburetor and causes the engine to run poorly. Some filter types can be cleaned with soap and water, while others are designed to be thrown away and replaced with a new one. The air filter should be cleaned or replaced at regular intervals.

Spark Plug

Spark plugs need to be periodically checked and cleaned or replaced. A dirty plug, or one with the gap at the wrong setting, can cause ignition problems and engine failure. Observe the tip of the spark plug for excessive wear, carbon buildup, oil, or other problems. The spark plug electrode should be clean, dry, and free from wear. Check the spark plug gap with a wire feeler gauge (**Figure 11.7.14**).

Dmytro Mykhailov/Shutterstock.com

Figure 11.7.14. Spark plugs may need to be replaced to keep an engine in good shape.

Hands-On Activity

Testing the Spark Plug on a Small Engine

A common cause of small engine failures is a faulty spark plug. It is inexpensive and easy to fix. It is also relatively simple to check.

For this activity, you will need the following:

- Small gasoline engine (3- to 5-horsepower)
- Spark gap tester
- Table
- Clamps or other means of securing the engine to a table

Here are your instructions:

1. Securely clamp the engine to the table.
2. Remove the spark plug from the cylinder head, so the engine won't accidentally start.
3. Connect the tester lead to the plug wire.
4. Attach the opposite end of the tester to the engine block.
5. Pull the starter rope quickly and observe the tester view window to see the spark.

Consider This

1. What did you observe when you did this?
2. How might this test indicate that something is wrong with the spark plug?

11

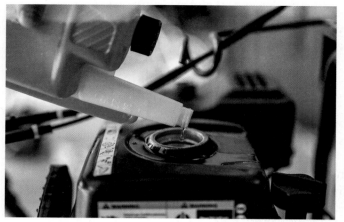

Figure 11.7.15. Preparing a small engine for proper storage will help keep it running year after year. *Why is it important to do this?*

Lost_in_the_Midwest/Shutterstock.com

Adjust the gap setting as needed based on manufacturer's recommendations and replace in the cylinder head. Make sure when you replace the spark plug that it is tightened to the proper torque specifications.

Preparation for Storage

Some small engines may be used seasonally and must be stored for extended periods of time between uses. Proper storage helps an engine to last by preventing corrosion and moisture damage from condensation or bad fuel (**Figure 11.7.15**). If an engine needs to be stored for an extended period, clean it thoroughly. Add fuel stabilizer to the fuel and fill the tank to the top. If you don't have fuel stabilizer, make sure that no fuel remains in the tank. Drain any oil from the crankcase. Remove the spark plug, add a small amount of oil to the cylinder, rotate the cylinder, and replace the spark plug. Do not attach the plug wire. Rotate the crankshaft to make sure that the cylinder and valves are closed.

When using an engine for the first time after storage, refill the crankcase with oil. Remove the spark plug and rotate the engine to disperse oil. Reattach the spark plug and fill the fuel tank. Allow the engine time to get to proper operating temperature before running it.

Career Corner — SMALL ENGINE REPAIR TECHNICIAN

Small engines are found in many applications around our home and community. With such a large population of engines, local **repair technicians** can be very busy.

What Do They Do? Small engine repair technicians repair machines with small motors. They diagnose problems, replace parts, test fixes, and teach their customers about proper maintenance.

What Education Do I Need? Most maintenance and repair jobs for small engines are relatively simple and with basic training can be completed quickly. Advanced training is also provided by some companies.

How Much Money Can I Make? A small engine repair technician can make $40,000 per year.

How Can I Get Started? Design an SAE working with engines. Build a small engine, or start a repair business for small engines in your neighborhood. Spend time in your ag mechanics lab learning about engines and how to work with and care for them. Interview an engine repair technician, or shadow them on their job.

alexkich/Shutterstock.com

What Did We Learn?

- Small engines are internal combustion engines rated at 25 horsepower or less, used to convert energy into mechanical energy. Most small engines are air-cooled, single cylinder, and gasoline-powered.
- The key parts of an engine include the engine block, which houses the moving pieces of the engine; the crankshaft, which converts the motion of the pistons into circular motion; the piston assembly, where combustion takes place; the valve train, which contains the intake and exhaust valves and the apparatus to open them; the flywheel, which provides power during interruptions and produces current to ignite the spark plug; and the carburetor, which mixes fuel and air in proper proportion for ignition.
- An engine operation involves four basic processes that occur in a consistent cycle: intake, compression, power, and exhaust. In the intake stroke, the fuel and air mixture is drawn into a cylinder. In the compression stroke, the mixture is compressed into a smaller space. In the power stroke, the compressed fuel is ignited and burned rapidly. In the exhaust stroke, gases given off in combustion are forced out of the cylinder.
- Engines run because several systems work together to keep things running. Every engine has a fuel system, which manages preparing fuel for ignition and moving it into the piston system. The compression system is home to the compression and combustion of fuel, producing energy to make the engine go. The cooling system prevents the engine from overheating and locking. The lubrication system keeps the engine from overheating and protects the moving parts from wear. The electrical system provides the energy that ignites the fuel and keeps the engine running smoothly.
- Regular engine maintenance includes keeping an engine clean, checking and changing the oil, changing the air filter, replacing spark plugs, and preparing engines for storage.

Let's Check and See What We Know

Answer the following questions using the information provided in Lesson 11.7.

1. Small engines are internal combustion engines rated at _____ horsepower (hp) or less.
   A. 5
   B. 10
   C. 25
   D. 50

2. Which of the following equipment would *not* be operated by a small engine?
   A. Weed trimmer
   B. Lawn mower
   C. Pressure washer
   D. Farm truck

3. The _____ of a small engine has the purpose of converting the up and down motion of the piston into circular or rotating motion.
   A. crankshaft
   B. camshaft
   C. valve
   D. cylinder

4. Most pistons have _____ piston ring(s) located in grooves that run around the outside of the piston.
   A. one
   B. two
   C. three
   D. four

5. Which of the following lists of engine processes is in the correct order of operation?
   A. Intake, power, compression, exhaust
   B. Intake, compression, power, exhaust
   C. Compression, exhaust, power, intake
   D. Power, intake, compression, exhaust
6. Combustion occurs during the _____ stroke.
   A. intake
   B. exhaust
   C. compression
   D. power
7. *True or False?* The two-stroke engine accelerates faster and has a higher horsepower to weight ratio than a four-stroke engine.
8. *True or False?* The camshaft rotates twice as fast as the crankshaft.
9. During the compression stroke in a four-stroke engine:
   A. The exhaust valve is closed, and the intake valve is open.
   B. The exhaust valve is open, and the intake valve is closed.
   C. Both valves are open.
   D. Both valves are closed.
10. The magneto of a small engine can produce as much as _____ volts of electricity.
   A. 12
   B. 120
   C. 600
   D. 40,000

Let's Apply What We Know

1. How do two-stroke and four-stroke engines differ?
2. How does each of an engine's key systems help it to run?
3. What are the key parts of an engine? What does each of these parts do?
4. Why is engine maintenance important?

Academic Activities

1. **Social Sciences.** What are the advantages of using small engines to complete tasks? How can food be grown more efficiently by using the power provided by an engine? Are there locations in the world that do not have access to engines? How is food grown in those locations?
2. **Science.** What fuels are available to power engines? What are the advantages/disadvantages of engines powered by different fuel sources?

Communicating about Agriculture

1. **Media.** How do you care for an engine? Create a flyer, poster, or video describing how to care for a small engine, and why it is important to do so. Share your message with your class.

SAE for ALL Opportunities

1. **Entrepreneurship SAE.** One of the most important parts of small engine care involves basic maintenance of the engine by replacing the engine oil, oil filters, air filters, and other components. Design an SAE to provide that service to people who live near you.

SAE for ALL Check-In

- How much time have you spent on your SAE this week?
- Have you logged your SAE hours?
- What challenges are you having with your SAE?
- How can your instructor help you?
- Do you have the equipment you need?

Project Planning, Design, and Construction

Learning Outcomes

By the end of this lesson, you should be able to:
- List the considerations you need to make in planning a project. (LO 11.08-a)
- Explain how the design process works with the planning process. (LO 11.08-b)
- List the items that should be included in a bill of materials. (LO 11.08-c)
- Summarize the steps in the construction phase of project. (LO 11.08-d)
- Describe the steps of a project, from planning to assembly. (LO 11.08-e)

Words to Know

bill of materials	crosscut	kerf
budget	cutting list	rip cut
construction	finish	scale drawing

Essential Question

How does planning affect project outcomes?

Before You Begin

Think about the last project you completed for school. Did you picture it in your head before you started it? Did the final product turn out differently than what you imagined? Or did you follow through with your original plan and not make any changes? What would you do differently next time? Project planning is a learned skill, and a very useful one to have. As you think about your previous project, how do you think that knowing more about project planning and execution could have helped you?

11

Planning

A project may be as complicated as the layout of a farming operation or as simple as a fence repair. Whatever the project, a well-developed plan is essential to its success. Careful thought and consideration must be given to the details. Most agricultural mechanics projects can begin by considering the project's use, materials and tools needed to complete the project, and the budget and time frame required.

Project Use

What will the project be used to accomplish? Does it fill a need or a want? Does it solve a problem? Is it functional or decorative? Will it be for personal use or will it be sold to generate income? All these questions (and others) need to be considered and answered during the initial project planning phase.

Materials Needed

There are many types of materials that can be used in project construction. Once the use of the project is established, materials are selected based on availability, suitability, and cost.

Many structures in the Southeastern United States use pine boards for framing, as it is readily available. However, pine is not always appropriate. If used outside in humid climates, it must be treated to prevent rot from moisture or ground contact. Other types of wood may be more resistant to rot but may not be readily available in that area. Cedar and cypress provide greater rot resistance but also cost more.

olrat/Shutterstock.com

Figure 11.8.1. These cables are wrapped with steel to keep the cost of building transmission towers low while still providing necessary protection to the wires underneath.

The cost of a material is a factor that must be considered when selecting materials. Copper is a much better conductor of electricity than aluminum, but aluminum is much cheaper than copper. Long runs of outside transmission lines use aluminum-wrapped steel to keep construction costs manageable (**Figure 11.8.1**).

Tools Required

Many options are available when selecting tools for use, and certain jobs require specific tools to complete. When planning the construction of a project, it is important to consider what tools are available and what tools will be required to complete the task. If you do not own all the tools required, you will have to purchase, rent, or borrow those items (**Figure 11.8.2**).

Figure 11.8.2. Major construction equipment can be very expensive to own! Many people rent the equipment they need to keep costs down.

Jim Lambert/Shutterstock.com

Budget and Time

A *budget* is an estimate of income and expenses for a specific time or project. A budget helps you keep track of what money you can spend, what money you have already spent, and what you believe the remaining items will cost. An important consideration for any project is its cost. The total cost of a project includes materials, labor, and tools. Other costs may include power supply, permits, rentals, or safety considerations. If the estimated cost of a project is greater than the money available, then actions need to be taken to reduce expenses.

A major factor associated with the budget is time. The value of your time, employee time, or other hired labor is usually calculated in dollars per hour (**Figure 11.8.3**). It may be easier to estimate the costs associated with small projects. Larger or more complicated projects may be more difficult to plan. Delays because of weather, mistakes, or supply chain issues can cause unexpected expenses and delays and should be considered during planning.

Design

When the initial planning is complete, you can transition to the design phase of the project. With any project, it is impossible to predict every detail with complete accuracy, but taking the time and effort to try will save time and money. During design, you may find that a little more planning is needed. This may require a little back and forth between planning and design before construction begins. Careful planning and design will allow the construction phase of the project to run more efficiently.

Purpose

While the planning phase explores the potential uses of a project, the design phase locks in on its intended purpose. The purpose served by the project will form the basis of the design. If the purpose of something is entirely for looks, then strength is a lesser priority. A project may solve a problem. A project may be a complete unit, such as a birdhouse, or it can be a part of something else, like a lean-to addition on a barn. Once the purpose is identified, design decisions can be made to guarantee the project's purpose is fulfilled (**Figure 11.8.4**).

Iurii Stepanov/Shutterstock.com

Figure 11.8.3. Because you don't see people's work in the end product, you might forget to plan for it. You have to pay people for their time and help!

bogdanhoda/Shutterstock.com

Figure 11.8.4. Brick is an appealing material option for many construction projects. *When would brick suit the purpose of a project? When would it be inappropriate?*

Did You Know? There are structures built during the Middle Ages that are still standing today. Rather than using nails or fasteners, they were designed from large timbers cut to fit together, making them very sturdy and strong.

11

J.J. Gouin/Shutterstock.com

Figure 11.8.5. This bulldozer is idle while it waits in the snow. While snow may not be a deal breaker for such a powerful machine, weather delays can cause real problems in construction projects.

Did You Know?

The average house in the United States could be constructed in six weeks or fewer if there are no delays or complications in the building process. In reality, it usually takes between four and six months to complete a build. Why do you suppose this is?

Schedule

A construction project schedule is produced based on prior experience of how long it takes to complete certain tasks. Many factors can affect the schedule and add to the time. Outdoor projects can be affected by weather (**Figure 11.8.5**). A detailed schedule for a small project that only takes a few hours to complete is not as important as a schedule for fencing a 50-acre pasture. What materials are needed and are they available? Will delivery of some items take longer than others? What will you do if equipment breaks down and must be repaired? What will the weather be like? Careful planning considers potential challenges and builds a schedule to allow for delays or problems. You cannot predict all the problems you might encounter, but careful planning allows challenges to be managed more effectively.

Even the smallest construction projects benefit from having a planned order of construction. Identify the materials, hardware, and tools necessary to complete all tasks. Some advanced preparation of materials, such as cutting boards to length, may be necessary before assembly begins. Some project parts may need to be assembled before others. Experience helps in planning.

Drawing

A project sketch or drawing can provide a visual representation of a project that is useful for planning, design, and assembly. The drawing can represent the complete project or a smaller portion of the project to give instruction for assembly. A drawing can be as simple as a hand sketch using pencil and paper to give an idea of what the finished product will look like (**Figure 11.8.6**). More complicated or larger projects may require a technical drawing completed to scale. *Scale drawings* are drawings where each component in the drawing is completed in proportion to the intended output (**Figure 11.8.7**). A ratio of 1 to 12 (1:12) can represent one inch equaling one foot and a small project such as a picnic table can be drawn on a single sheet of paper to represent the full-size item.

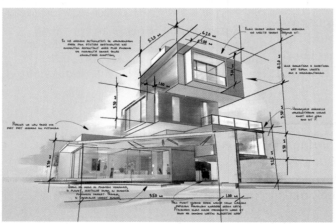

Franck Boston/Shutterstock.com

Figure 11.8.6. Drawings like this can give a good sense of what the project should look and feel like upon completion. *Does looking at this give you a sense of what the builder is looking to do?*

Chaosamran_Studio/Shutterstock.com

Figure 11.8.7. Scale drawings are carefully drawn and designed to be exact, miniature versions of projects. *Why might this be helpful to have?*

A completed drawing may contain several different views of the project to help with construction. The drawing may be completed by hand using drawing instruments or it may be completed using a computer. Computer-assisted design (CAD) programs can be used to complete complicated two-dimensional or three-dimensional drawings quickly and efficiently.

Bill of Materials

Once planning and design are complete, a list of all needed materials should be compiled. This ***bill of materials*** is a complete list of all materials, fasteners, and hardware needed to complete a project. This list can also include tools and assembly instructions. A bill of materials may include a ***cutting list*** that shows the exact dimensions of certain parts of a project. The bill of materials list is useful for making purchases and for calculating the material cost. Once a total cost is determined for materials, the estimate for labor cost can be determined using the schedule and a total project cost can be determined.

Construction

Once planning and design are complete and a bill of materials is in place, assembly, or ***construction***, can begin. It is important to pay attention to details during the construction phase of a project. Take notes along the way to help understand the process. Even though we consider the planning and design complete, the reality is that plans may be revisited and designs may be revised during construction to help make future projects even better (**Figure 11.8.8**).

Construction consists of several processes: material selection, preparation, assembly, and finish.

MBI/Shutterstock.com

Figure 11.8.8. Part of learning from building a project is going back and reviewing what went well (or didn't) and refining design elements or plans for next time.

Material Selection

The final selection of materials to use will be based on location, use, availability, quality, and cost. If the project will be made of metal parts, selection is a little easier because those components are made in a uniform and consistent manner (**Figure 11.8.9**). In contrast, lumber is a natural product and may contain defects that can have a negative effect on the finished project. It is important to take the time to check all of the materials for their quality and suitability (**Figure 11.8.10**).

SimoneN/Shutterstock.com

Figure 11.8.9. These mass-produced metal tubes are identical and can be ordered and used without much worry about individual pieces.

Jupiter Candy/Shutterstock.com

Figure 11.8.10. Wood, as a natural material, has more differentiation between pieces. It is important to pay attention to make sure you get the quality and type of wood you need.

Figure 11.8.11. Knots are the most common natural wood defect. *Why wouldn't this board be desirable for construction?*

Recognizing potential defects will help you select wood that will be well-suited to the project. It is also important to know what materials are available in your area. If a plan calls for a material that must be shipped to your location, costs will go up. It might save money to use materials that are more readily available.

Lumber defects are either natural, due to warping from the drying process, or a result of physical damage during the manufacturing process. Common natural lumber defects include knots and splits (**Figure 11.8.11**). These defects can affect the look of the wood, as well as its strength. Warping, or improper drying, can also cause damage to boards. If you see lumber that is bowed, twisted, or otherwise isn't flat, that is due to warp. Warped wood can be difficult to work with and should be avoided (**Figure 11.8.12**).

Lumber can also be damaged during the manufacturing process. These defects might include teeth marks from blades, burns, knife marks, chipped wood, or other damaged grain. Wood damage can also be caused by fungi, termites, or other insects that eat wood. Lumber should be stored so that it remains flat, straight, and dry to prevent warping or other defects.

Preparation

Once materials and hardware are gathered, they must be organized, marked, and/or cut for assembly. It is best to select material that does not require a lot of preparation before assembly but that is not always possible. Some preparation usually must be done prior to assembly and this process of measuring and marking lines for cut is called *layout*. Layout needs to be accurate for assembly to be correct. Having and knowing how to use the proper measuring and layout tools will make this process much easier.

Use the dimensions in the project plan to identify where to measure and mark each individual piece. Careful attention to measurement and layout will reduce waste material and increase accuracy. Remember when cuts are made to allow for the width of the saw blade. This blade width is known as the *kerf* and is the gap made by the removal of material by the saw blade. When cutting lumber to length, remember to check and make sure that the first end that you measure from is square. If it is not, use a saw to square the end first by removing as little material as possible (**Figure 11.8.13**).

Figure 11.8.12. This lumber is warped, which means it didn't dry evenly. *Why would it be difficult to use warped lumber in a project?*

Figure 11.8.13. The more skilled the carpenter at cutting and shaping lumber, the less lumber is wasted in the construction process.

Once you have the first end square, you can proceed to measure and mark the length of the board with greater accuracy. Tools like table saws can be set to specific dimensions to cut lumber without marking it. This is common when ripping lumber to specific widths. Ripping, or making a *rip cut,* is cutting a piece of wood parallel to the grain. By comparison, a *crosscut* divides a piece of wood perpendicular to the grain. Crosscuts are used when cutting a board to length and rip cuts are used to reduce the width of a board.

It is important to note that softwood lumber purchased for construction is referenced based on the nominal sizes of the boards. The actual size of each piece will be different after surfacing (**Figure 11.8.14**). Softwood lumber is also categorized by different quality grades (**Figure 11.8.15**).

Assembly

Once the materials have been selected and cut to size, the construction of the project can begin. Metal pieces can be welded together using a variety of machines and methods. There are epoxies and adhesives that work well when joining thin metal materials. Sheet metal screws may be used in some applications.

Wooden projects have many options when it comes to fasteners, adhesives, tools, and methods of joinery. Nails, screws, wooden joints, and glue work well to hold projects together depending on the size and desired use of the product.

When following a plan, there is usually a construction sequence that has been developed to save time. A step-by-step guide of instructions for layout, cutting, and assembly is typically included in purchased plans. If you design your own plan, put some thought into what a logical sequence of construction would look like and make notes of all the steps necessary to complete the project. Think about what the final project will look like and organize the process to complete assembly. Evaluate each step during the construction process and make notes about anything that you would modify or change. This will allow future assembly of the same project to be more efficient.

Finish

Applying a *finish* is the final step in the construction process. It gives the project an attractive appearance and protects it from corrosion. Paint gives metal and wood surfaces a finished appearance and prevents rusting or rotting. Wood can also be stained or given an oil finish to bring out the natural beauty of the wood. Concrete products can be painted or stained for appearance and protection from corrosion.

Nominal vs. Actual Lumber Dimensions

Nominal Size	Actual Size
1″ × 2″	¾″ × 1 ½″
1″ × 4″	¾″ × 2 ½″
2″ × 2″	1 ½″ × 1 ½″
2″ × 4″	1 ½″ × 3 ½″
4″ × 4″	3 ½″ × 3 ½″

Goodheart-Willcox Publisher

Figure 11.8.14. Nominal and actual lumber dimensions are different. It's important to understand this in planning, to make sure you get what you need!

Softwood Grades

Grades	Characteristics
Clear	Has no knots.
Select or select structural	Very high-quality wood. Broken down into Nos. 1–3 or grades A–D; the lower grades will have more knots.
No. 2 common	Has tight knots, no major blemishes; good for shelving.
No. 3 common	Some knots may be loose, often blemished or damaged.
Construction or standard	Good strength; used for general framing.
Utility	Economy grade; used for rough framing.

Goodheart-Willcox Publisher

Figure 11.8.15. Softwoods may be graded, giving you a sense of the quality and appropriate uses for an individual piece of wood.

11

Figure 11.8.16. This paint is presumably designed to be used in an outdoor application. *Why is it important to verify that this is true?*

Surface preparation is necessary before applying a finish to the project. Sanding the surface removes unwanted material and smooths the surface, helping paint adhere more effectively. Sometimes a primer or surface sealer is necessary before the paint is applied to fill in small spaces and prepare the surface to accept the paint. Some finishes are designed to be used in exterior applications to protect surfaces from weather damage (**Figure 11.8.16**). Whatever finish you decide to use, be sure that you follow the manufacturer's instructions for proper application.

Hands-On Activity

Picnic Table Construction

In this activity, you and your classmates will be constructing picnic tables from the following materials and plans.

For this activity, you will need:

- Bar clamps
- Cordless or electric drill
- $^{13}/_{32}''$ diameter drill bit
- Drill bits (square or star to match screws)
- Marking tool
- Measuring tape
- Safety glasses
- Saw (hand- or power saw)
- Square (combination, framing)

Materials List

Quantity	Description	Size (nominal)
13	Pine boards (pressure-treated optional)	2″ × 4″ × 8′
2	Pine boards (pressure-treated optional)	2″ × 6″ × 8′
1	Pine boards (pressure-treated optional)	2″ × 6″ × 10′
2	Pine boards (pressure-treated optional)	2″ × 12″ × 8′
32	Carriage bolts (galvanized optional)	3 ½″ × ⅜″
32	Flat washers	⅜″
32	Nuts (nylon lock nuts optional)	⅜″
100	Deck screws (square or star drive)	#10 × 2 ½″

Cutting List

Quantity	Description	Size (nominal)
8	Tabletop	2″ × 4″ × 8′
3	Tabletop cross braces	2″ × 4″ × 28″
4	Legs	2″ × 6″ × 33 ½″ (allow extra length for angle)
2	Seat cross boards	2″ × 6″ × 5′
2	Diagonal brace	2″ × 6″ × 41 ½″ (allow extra length for angle)
2	Seat brace	2″ × 4″ × 85″

Here are your instructions:

Tabletop

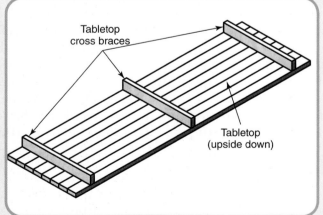

Tabletop cross braces

Tabletop (upside down)

1. Select eight 2 × 4 boards and cut them to exactly eight feet long.

Continued

Picnic Table Construction *Continued*

2. Cut three tabletop cross braces from a 2 × 4. Each brace should be 28 inches long to be flush with the edges.
3. Center one brace and place the other two braces six inches from each end. Attach the braces using 2 ½ inch wood screws. Bar clamps may be used to hold the boards together.

Table Legs

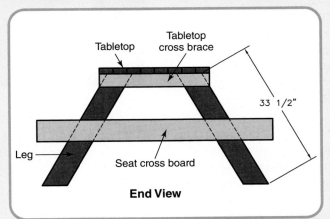

End View

Goodheart-Willcox Publisher

1. Flip the completed tabletop face down to mount the legs, seat cross boards, and leg braces.
2. Cut each identical leg from the 2 × 6 boards with a 60-degree angle on each end. The angles will be parallel to each other. Each leg will be 33 ½ inches long.
3. Mount each leg to the top two cross braces on each end two inches from the end of the table. Mount them on the inside of the cross braces using the carriage bolts, washers, and nuts. Use two bolts per leg.

4. Mount the seat cross boards so the top of each is 17 inches from the ground and centered on the legs. Attach the cross boards using carriage bolts, washers, and nuts. Use two bolts per leg.
5. The legs and seat cross boards can be temporarily secured using wood screws to hold them in place while you drill holes and mount the carriage bolts.

Diagonal Braces

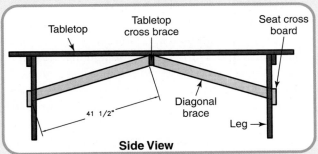

Side View

Goodheart-Willcox Publisher

1. Keep the table upside down. Cut two diagonal braces from 2 × 4s; make each 41 ½ inches long. The angle on each end should be about 17 degrees and they will be parallel to each other.
2. Attach the diagonal braces with wood screws.

Consider This

1. What was the most fun part of this exercise? What was the most difficult?
2. What could you do to make it easier the next time you try this project?

Making Connections # Careers

Building Agricultural Shelters

Equipment used in agricultural applications lasts longer if it is protected from the elements. The design and construction of agricultural shelters provides a great job. You can work in this industry with any kind of education, from a high school diploma to an advanced degree if you want to be a mechanical engineer.

What Did We Learn?

- During project planning, you should ask questions about what your project will be used for, what materials are needed, what tools are required, what budget is available, and what time is required to complete the project.
- The design phase of the project includes finalizing the purpose of the project, building a schedule, and creating a project sketch or scale drawing. Planning and design are closely connected, and sometimes planning questions may be revisited during design.
- A bill of materials includes all the materials, fasteners, and hardware that will be needed to complete a project. A bill of materials may also include a cutting list.
- The construction, or assembly, phase of the project follows design. Final materials will be selected, based on their cost, availability, and suitability to the purpose. Once materials are gathered, some will need to be prepared: measured, cut, and organized. Once the materials are ready, they will be assembled according to a plan. Applying a finish is the final step of the project, which improves the appearance of the project while protecting the final project from rust, rot, or other damage.

Let's Check and See What We Know

Answer the following questions using the information provided in Lesson 11.8.

1. A well-developed _____ is essential to any project.
   A. summary
   B. plan
   C. tool supply
   D. materials list

2. Which of the following factors would be least important to consider when planning a project?
   A. Use
   B. Materials
   C. Budget
   D. Project title

3. *True or False?* Lumber used outside or in contact with the ground needs to be protected from rot and insect damage.

4. Which of the following tools would most likely be a rental tool for a construction project?
   A. Hammer
   B. Portable circular saw
   C. Ladder
   D. Tractor

5. Which of the following costs is *not* visible when looking at a completed project?
   A. Labor
   B. Nails
   C. Lumber
   D. Paint

6. Which of the following items would *not* be included in a budget for a construction project?
   A. Time/labor costs
   B. Lumber costs
   C. Nail costs
   D. Paint color choice

7. A _____ is a complete list of all materials, fasteners, and hardware needed to complete a project.
   A. cutting list
   B. bill of materials
   C. budget
   D. schematic drawing

8. A project _____ can give a visual representation of a project that is useful for planning, design, and assembly.
   A. budget
   B. drawing
   C. cutting list
   D. tool list

9. A _____ is the most common natural lumber defect.
   A. warp
   B. twist
   C. knot
   D. check

10. A _____ is the thickness of material removed by the cutting of the sawblade and must be considered in measurement and location of cut.
    A. crosscut
    B. rip cut
    C. kerf
    D. dimension

Let's Apply What We Know

1. What are the questions you need to answer before you start a project? Why is it important to ask questions before starting?
2. What considerations need to be made regarding materials used in a project?
3. What can affect a project's schedule? Why is it important to consider those factors?

Academic Activities

1. **Social Sciences.** How does the growing world population affect the need for shelter and housing and the job market for home construction? How can design engineers make such activities more efficient?
2. **Mathematics.** Calculate the cost of a small construction project, like a birdhouse. What materials are needed to complete the project? What tools are needed? Why is it important to be able to accurately measure materials and calculate cost?

Communicating about Agriculture

1. **Writing.** One of the most important parts of a project is reviewing how it went upon completion. Think through a recent project you've completed, either for school or an activity. How did it go? What would you do differently next time? Write a short essay describing your project, what you learned, and what you would do differently next time.

SAE for ALL Opportunities

1. **Entrepreneurship SAE.** Design and construction of small woodworking projects can be a profitable SAE project with a minimal starting investment in tools. Picnic tables, storage buildings, and even fences can be designed and constructed for sale and allow flexible hours for a student. How might you get started?

SAE for ALL Check-In

- How much time have you spent on your SAE this week?
- Have you logged your SAE hours?
- What challenges are you having with your SAE?
- How can your instructor help you?
- Do you have the equipment you need?

Connecting Producers and Consumers through Ag Marketing

SAE for ALL Profile
Sticking with Her Roots

Sherita Barnes owes her love of agriculture to her family's business. "Every summer, you could find me learning how to grow and harvest vegetables and running our vegetable stand. I loved the hustle and bustle, the rhythms of passing seasons, and connecting with people," said Sherita.

Sherita grew up in Marshall, Michigan and attended the local high school, where she became involved in agricultural education and FFA. Says Sherita, "My agriculture teacher showed me how to take my love for gardening and working with customers to the next level." Sherita used her SAE to expand the fall agritourism activities at her family's business, including building a corn maze, growing pumpkins, and launching a fall festival for the community. During her senior year, she received recognition in the FFA Proficiency Award program and earned the state FFA degree.

That was great, but even better was how the experience made her feel. "That whole experience taught me so much about agriculture, but what I really loved was seeing our community find a new love for agriculture. I loved that I could use my project to connect with my friends and neighbors."

As a student at Michigan State University, Sherita studied agricultural economics and marketing. Her senior project involved developing marketing resources for small farms. "Agriculture is big business in the United States, but there is a need for all farmers—big and small. Developing resources that could help small farms and small businesses, like my family's, was a rewarding experience."

As a marketing specialist with the United States Department of Agriculture's Agricultural Marketing Service, Sherita works to create marketing to support American farmers. "I love my job, and I am thankful that my family invested the time and effort to teach me the value of work and connection with community, and the importance of a safe and wholesome food supply."

When she isn't at her USDA office, Sherita and her husband spend time each summer in the garden behind their home in College Park, Maryland, teaching their two children how to raise vegetables. "I may be a long way from my home in Michigan," said Sherita, "But you cannot get too far away from your roots."

- How do you see agriculture connecting with your community?
- How could you help agriculture become more integrated into your community?

Top: Rido/Shutterstock.com
Bottom: FrimuFilms/Shutterstock.com

eddie-hernandez.com/Shutterstock.com

What Is Marketing?

Essential Question

Why is marketing an essential part of running a successful agricultural business?

Learning Outcomes

By the end of this lesson, you should be able to:
- Describe role of marketing in a successful business. (LO 12.01-a)
- Explain how agricultural marketing differs from other types of marketing. (LO 12.01-b)
- Describe the types of agricultural markets. (LO 12.01-c)

Words to Know

agricultural marketing
Community Supported Agriculture (CSA)

farmers market
integrator
marketing

processor
retail market
wholesale market

Before You Begin

Class! I have the best idea for a new business, and I am so excited! Fashion jewelry for livestock animals is guaranteed to be a hit, don't you think? I've already come up with a name: Bovine Beautiful. I am just not sure how to get started. Does anyone have any suggestions?

If you were starting a business in the agriculture industry, what things do you think would be important? Can you come up with a list?

What Is Marketing?

Marketing includes all the activities involved in transferring goods and services from producer to consumer. Marketing's purpose is to encourage and facilitate exchange from one business or individual to another. Marketing is not just the final financial transaction; a lot goes into marketing a product. Buying supplies, renting equipment, paying workers, advertising, processing, and selling are all part of marketing a product. Marketing begins as products are in the early stages of development.

Marketing involves many different people and takes a lot of planning and time. It is not a one-time activity, but rather an ongoing strategy that helps businesses remain connected to consumers. A business that understands the importance of marketing knows how to maintain long-lasting relationships with their customers (**Figure 12.1.1**).

What Is Agricultural Marketing?

Agricultural marketing covers the processes and services involved in moving an agricultural product from the farm to the consumer. These services include handling agricultural produce and other raw products in a way that satisfies farmers, processors, and consumers. There are many steps involved in marketing agricultural goods: planning, growing and harvesting, grading, packing and packaging, storage, distribution, advertising, and sales. Marketing nonfood products does not include as many steps, as those products do not need to be grown, harvested, or graded. Additionally, nonfood products do not usually require the tight time constraints for distribution that fresh, perishable final products do (**Figure 12.1.2**).

Producers and Consumers

In agriculture, producers raise the animals or grow the crops. Producers typically sell what they produce to ***processors*** who change the products into a different (perhaps more desirable) form, such as raw potatoes into potato chips or apples into applesauce (**Figure 12.1.3**).

Nixx Photography/Shutterstock.com

GagliardiPhotography/Shutterstock.com

The Image Party/Shutterstock.com

Figure 12.1.1. These businesses and brands are highly recognizable and have loyal customers. *What do McDonalds, Starbucks, and Frito Lay do to maintain a connection with their customers?*

Shutter B Photo/Shutterstock.com

Figure 12.1.2. Agricultural marketing ensures that there is a fresh, clean, and safe food supply available year-round. *How is this part of marketing?*

Zoia Kostina/Shutterstock.com

Figure 12.1.3. Cotton has few uses in its raw, harvested state. Processing cotton into more usable forms like clothing or linens is part of the marketing process.

12

Hands-On Activity

Consumer Preference Poll

Consumers often have different preferences. These may relate to price, taste, color, size, nutritional value, or other factors. Have you ever wondered how your consumer choices are different from your peers? Let's take an anonymous survey to determine the consumer preferences of your classmates.

For this activity, you will need:

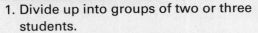

- Blank paper or poster board
- Pencil or pen
- Envelopes
- Images of products
- Magazines or device for internet research

Here are your directions:

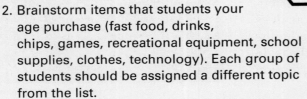

1. Divide up into groups of two or three students.
2. Brainstorm items that students your age purchase (fast food, drinks, chips, games, recreational equipment, school supplies, clothes, technology). Each group of students should be assigned a different topic from the list.
3. Each group will find images of different items that belong in their category. Print out examples of the images or cut them out of a magazine. Glue the images to a paper or poster board, labeling each image with a letter (A, B, C, etc.). For example, if you were doing types of potato chips, plain would be A, sour cream and onion might be B, ridged could be C, and barbecue could be D.
4. Each group will set up a station in the classroom. Each station should include:
   - Poster board or paper featuring pictures of the options from each category
   - Paper and marker
   - Envelope
5. Move around the room from station to station selecting your preferred option from the images given. Write your choice at each station on a piece of paper and place it in the envelope.

6. Once everyone has completed making their choices, each group should open the envelope at their station and view the responses.
7. Create a graph showing your class's consumer preferences for your category and share it with the rest of the class.

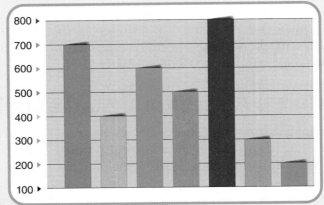

Kubko/Shutterstock.com

Sheila Fitzgerald/Shutterstock.com

Consider This

1. How did your personal preferences compare to the rest of the class? Where were your choices in the majority? Where did they differ from others' preferences?
2. Why is it important for producers to meet the needs of all consumers? How can producers meet these different needs?

A consumer is an individual, group of individuals, or business that uses the goods or services created by the producer. Everyone is a consumer! When you buy a school lunch, you are a consumer. When you stop at a convenience store for snacks or a drink, you are a consumer. When you shop at a grocery store, mall, or farmers market, you are a consumer. All consumers have specific needs and wants. Producers work to create products that meet that need or want (**Figure 12.1.4**). Consumers communicate their preferences, needs, or wants to producers with their purchasing decisions.

Chuck Wagner/Shutterstock.com

Figure 12.1.4. *Have you ever noticed how many different flavors of sports drinks are available? Why do you think this is? As a consumer, how do you let producers know your preferences?*

Types of Agricultural Markets

A producer has many different pathways to connect with their customers, especially in agriculture. How a producer connects with their customers often depends on the size of their operation. If a producer has a small operation where they produce one crop, they will market their product differently than a producer growing thousands of acres of multiple crops. Because each producer is different, there are many ways to market agricultural goods.

Farmers Markets

Farmers markets and farm stands are public, recurring events that allow farmers to sell fresh products directly to the consumer. Farmers markets can be located on the farm, on the side of the road, in a designated location inside a city or town, or even at a state-run operation. Farmers markets create personal connections and mutual benefits between farmers, shoppers, and the local community (**Figure 12.1.5**). Typically, farmers receive more money for their crops when selling at a farmers market, and consumers receive the freshest and highest-quality local food. The number of farmers markets in the United States has grown rapidly in recent years, from under 2,000 in 1994 to more than 8,600 markets in 2022.

Did You Know?

Farmers reap the benefit at farmers markets, pocketing upwards of 90 cents for each dollar of sales. Why do you think this is?

Figure 12.1.5. Farmers markets are a great way for farmers to connect with their customers. *Have you been to a farmers market? What did you notice about that experience?*

Rawpixel.com/Shutterstock.com

12

Jasmine Sahin/Shutterstock.com

Figure 12.1.6. CSAs create a partnership between the consumer and the farmer. The consumer signs up in advance to receive a box at regular intervals and in return gets a wide variety of fresh goods delivered locally. *What benefits do you see to this approach?*

Lori Sparkia/Shutterstock.com

Figure 12.1.7. Christmas tree farms are an example of a recreational marketing approach. This type of operation allows families to come to the farm and harvest their own Christmas tree for a fee. *Why might an agricultural business pursue this type of approach?*

Community Supported Agriculture (CSA)

Community Supported Agriculture (CSA) is a partnership between farmers and consumers that allows consumers to buy small portions of a farm's harvest in advance. Consumers pay an agreed-upon price annually, monthly, or weekly, starting at the beginning of the growing season. By paying at the beginning of the season, CSA members share in the risks of production and relieve the farmer of the time needed to market their produce in other ways, essentially giving the farmer more time to concentrate on growing high-quality food. This up-front payment also provides the farmer with immediate income to help pay for seed and other needed items at the beginning of season. In return, CSA members receive a variety of freshly picked vegetables every week, every two weeks, or once a month (**Figure 12.1.6**). CSAs usually offer fruits, vegetables, herbs, eggs, and limited meat and dairy products.

Recreational Marketing

Recreational marketing approaches are built around providing entertainment and amusement to consumers to encourage them to visit and spend their money with agricultural producers. These are usually tailored to a family atmosphere and include seasonal events. Recreational approaches can include things like festivals that encourage the public to come directly to the farm and pick up seasonal produce and other items while enjoying entertainment. In many areas of the country, fall festivals are particularly popular, featuring hayrides, apple cider, music, pumpkin picking, food, and entertainment for kids. Pick-your-own operations are another recreational approach. These allow families to go into the fields and harvest their own fresh produce for a small fee. Recreational marketing approaches do not make up a large percentage of the money spent on agricultural products in the country and don't necessarily move a large amount of goods, but they create a strong connection between the producer and the consumer (**Figure 12.1.7**).

Did You Know? Pick-your-own farms originated near London in Victorian market gardens during the 19th century.

Retail Markets

Retail markets allow farmers to sell directly to stores that then sell the product to the public. Selling to individual retail stores, such as grocery stores, allows farmers to earn more for their produce than selling wholesale. However, some retail stores are restricted from buying from local growers because they have contracts with wholesale producers. Additionally, the farmer must be able to produce guaranteed quantities on a set delivery schedule and meet standard harvesting and packing requirements to sell on the retail market.

Restaurants

In some cases, farmers can get a higher price selling a fresh product directly to restaurants, though the amount sold through restaurants is usually smaller than other marketing options.

Wholesale Markets

Wholesale markets allow farmers to work through a second party that collects, assembles, and distributes their product. A wholesale distributor can balance supply and demand by transporting products from multiple areas to one location. Wholesale distributors often work with multiple farmers at once to collect, process, and move produce to provide large quantities of products. Wholesale distributors allow farmers to move farm produce quickly at a fair price.

Making Connections #STEM

Product and Service Selection

A manufacturer listens to its customers before making production decisions. Companies analyze consumer buying trends, conduct market research surveys, and study competitors' product sales experiences to learn what consumers desire or need.

For example, a convenience food company may expand its frozen dinner offerings after learning that time-squeezed customers want more quick-fix options. A cosmetics maker may create a new line of organic cosmetics after discovering that consumers are concerned about synthetic additives in makeup and skincare products.

With a group of your classmates, brainstorm an opportunity to expand an existing product. Think about items you use every day. Is there a change that could make one of them better?

Think about foods or drinks you regularly consume. Is there a new flavor or change that could be made to improve the product? A piece of technology that needs additional features? A clothing item that doesn't exist yet, but should? Build a poster promoting your new and improved product.

Consider This

1. How can you ensure the success of your new product?
2. How would you market your new product to the public?
3. Why did you think this change would be successful? What data led you to propose your product?

12

branislavpudar/Shutterstock.com

Figure 12.1.8. Processing plants allow farmers to sell large quantities of produce at one time. While this type of marketing would not work for smaller producers, it allows larger growers to receive a fair price for their product before it is processed into a different product. *What might the corn on this assembly line be processed into?*

Processors

Farmers can also sell directly to processing plants, usually at a wholesale price. The producer may sign a contract with the processing facility ahead of production to ensure outlets for crops once they have been produced (**Figure 12.1.8**). All crops produced must meet the standards set by the processing facility. Selling directly to a processing facility allows farmers to move large quantities of products at one time.

Integrators

An *integrator* is a business involved in two or more stages of production. Typically, there is a signed agreement between the farmer and the integrator. The agreement explains what will be produced, how it will be produced, and what the producer and farmer will receive. In the poultry industry, the integrator provides baby chicks, feed, and other materials to the farmer. The farmer is responsible for raising the birds according to the integrator's rules and regulations, and then the integrator later processes and distributes the chickens.

Career Corner — MARKETING SPECIALIST

A **marketing specialist** *ensures that the appropriate messages and mediums are used to meet sales targets.*

What Do They Do? Marketing specialists plan, develop and direct distribution of product, develop and execute marketing programs, and develop a marketing strategy for the organization. They create incentive programs for salespeople and suppliers, monitor brand performance and use information to recommend action, conduct market research, and manage a marketing budget.

What Education Do I Need? A bachelor's degree in agricultural business, marketing, journalism, communications, education, or business administration is required.

Desizned/Shutterstock.com

How Much Money Can I Make? Marketing specialists make an average of $50,000 per year.

How Can I Get Started? You can build an SAE around conducting market research for your school's agriculture program. What do people think agriculture is, and what do they want from agriculture classes? Help your agriculture program figure out how to raise money through community interactions, whether those are sales or services. You can always meet with a marketing specialist to learn about what they do!

What Did We Learn?

- Marketing includes all activities involved in transferring goods and services from producer to consumer. Marketing's purpose is to encourage and facilitate exchanges between one business or individual and another.
- Agricultural marketing covers the processes and services involved in moving an agricultural product from the farm to the consumer. These services include handling of agricultural produce and other raw products in a way that satisfies farmers, processors, and consumers.
- Agricultural producers raise animals or grow crops. Consumers use the goods or services created by the producer. Everyone is a consumer! Consumers give producers information about their preferences by choosing what to buy.
- A consumer is an individual, group of individuals, or other businesses that use the goods or services created by the producer.
- An agricultural producer has many different pathways to connect with their customers. How a producer connects with their customers often depends on what they are selling and how much of it they must sell.
- Farmers may market their products through farmers markets, Community Supported Agriculture (CSAs), recreational marketing, retail markets, restaurants, wholesale markets, processors, and integrators.

Let's Check and See What We Know

Answer the following questions using the information provided in Lesson 12.1.

1. *True or False?* Marketing consists of all the activities involved in transferring goods from a consumer to a producer.
2. *True or False?* Marketing should *not* be considered until a producer has a final, finished product that is ready to sell.
3. Agricultural marketing is different because most of the sellable products come directly from the ____.
   - A. city
   - B. farm
   - C. factory
   - D. warehouse
4. A(n) ____ uses the goods or services created by the producer.
   - A. consumer
   - B. farmer
   - C. client
   - D. agent
5. *True or False?* How a producer connects with their customers often depends on the size of their operation and what they are selling.
6. ____ are public, recurring events that allow farmers to sell their fresh products directly to the consumer.
   - A. Community Supported Agriculture (CSA)
   - B. Retail markets
   - C. Wholesale markets
   - D. Farmers markets
7. *True or False?* Community Supported Agriculture is a partnership between the farmer and the consumer that allows consumers to buy a small portion of a farm's harvest in advance.

12

8. ____ allow farmers to work through second parties to collect, assemble, and distribute their products.
   A. Community Supported Agriculture (CSA)
   B. Retail markets
   C. Wholesale markets
   D. Farmers markets

9. ____ make contracts with farmers to buy their products, provided they meet quality standards, and use them to manufacture new products.
   A. Producers   C. Integrators
   B. Processors   D. Wholesalers

10. *True or False?* Integrators are only involved in the end stage of production.

Let's Apply What We Know

1. Explain the difference between marketing and agricultural marketing.
2. Farm producers have many options for marketing their raw products. Discuss at least three options available.

Academic Activities

1. **Math.** Research the average price a farmer would receive for a bushel of corn sold through a retail market, a wholesale market, and a farmers market. Discuss the pros and cons of each option for the farmer.
2. **History.** How were agricultural products marketed 50 years ago? Conduct some research to determine how marketing methods have changed in the last half-century. You may wish to interview an older friend or relative who remembers what it was like to buy or sell products in the mid to late 20th century.

Communicating about Agriculture

1. **Social Media.** How do local agriculture businesses promote themselves on social media? How do they use social media to connect and build relationships with customers? Craft three sample social media posts that your own, made-up agricultural business might use. Share them with your class.

SAE for ALL Opportunities

1. **Foundational SAE.** Talk to the producers at a local farmers market to learn how they connect with their customers. Write up a report of your findings and present it to the class, along with recommendations about how the producers might improve their connection to the community.

SAE for ALL Check-In

- How much time have you spent on your SAE this week?
- Have you logged your SAE hours?
- What challenges are you having with your SAE?
- How can your instructor help you?
- Do you have the equipment you need?

USDA Standards and Grades

Dusan Petkovic/Shutterstock.com

Essential Question

How does the government ensure the quality, freshness, and safety of the public food supply?

Learning Outcomes

By the end of this lesson, you should be able to:
- Describe how the US government keeps the food supply safe. (LO 12.02-a)
- Explain the difference between standards and grades. (LO 12.02-b)
- Discuss different USDA grades. (LO 12.02-c)
- Explain what other common labels might be found on food products and what they mean. (LO 12.02-d)

Words to Know

grading   organic   standard

Before You Begin

Do you know anything about buying a steak? The other day, I was shopping at the grocery store and picked up a steak that said "Prime" on the label. What do you think that means? The United States Department of Agriculture (USDA) could give us a hint or two. Find a partner and take a moment for a quick search about different grades of steak. Write down two or three things you learn from the USDA website (www.usda.gov).

allensima/Shutterstock.com

Figure 12.2.1. Food labels can be confusing. Grade A, USDA Choice, organic: *What do these symbols and words mean about the food we eat every day?*

I f you have recently visited a grocery store meat or dairy department, you might find yourself with a few questions about what the labels on your food mean. It's likely that you've seen cuts of different meats or cartons of eggs labeled with phrases like *USDA Choice* or *Grade A*. Those phrases sound good, but what do they mean? These labels can tell you a lot about the specific package you are considering and whether it's the right choice for you. Not only can understanding these labels help you determine the differences between different cuts of meat and other agricultural products, but they can make it easier for you to choose items best suited for your purpose (**Figure 12.2.1**).

Regulating a Quality Food Supply

Product labeling is a contract of trust between consumers and producers. This is especially true for the foods we eat and the companies that sell them. The responsibility of regulating and monitoring food labels is shared between many federal agencies, including the Food and Drug Administration (FDA) and USDA. Government regulations and standards over our food supply and other agricultural goods have not always existed; in fact, food safety standards have existed for fewer than 100 years.

Before there were federal laws, individual commodities had their own sets of standards, and few of them had their own formal regulations. These individual commodity standards often varied across state lines. This meant that consumers had no guarantee of the quality and freshness of their food supply (**Figure 12.2.2**). There were no unified standards covering all states and commodities until the Federal Food, Drug, and Cosmetic Act of 1938.

Figure 12.2.2. Before 1938, there was little regulation over how food was grown or processed in the United States. *What issues might have arisen from this lack of oversight?*

Morphart Creation/Shutterstock.com

In 1938, Congress passed the Federal Food, Drug, and Cosmetic Act, which gave the Food and Drug Administration (FDA) the authority to issue food safety standards. In 1946, the Agricultural Marketing Act (AMA) was passed, which expanded the scope of inspection and provided USDA the authority to inspect, certify, and identify the class, quality, and condition of agricultural products (**Figure 12.2.3**). Grading and quality identification activities were separated from inspection activities and assigned to USDA's Agricultural Marketing Service in 1981. The standards set by USDA on producers and processors continue to evolve as new technologies emerge and more research is done to keep our food supply safe.

Standards and Grades

Standards and grades are two separate programs within the United States Department of Agriculture. All agricultural products go through inspections to ensure that producers and processors are following the guidelines and standards set by the USDA. *Standards* are guidelines used to measure product quality to an acceptable level. The USDA has specific standards that relate directly to each commodity category. These categories include cotton, dairy products, vegetables, grain, beef, fish and seafood, and fruits (**Figure 12.2.4**). These products are inspected for freshness, cleanliness, and purity. Companies must participate in standard inspections. These inspections ensure safety in the food supply. They are paid for out of the USDA's budget, from American tax dollars.

Government regulations and standards ensure that we have a safe, healthy food supply. These standards also give consumers peace of mind when they shop, knowing that producers are acting as good stewards of the land they are using to grow our food and that livestock animals are being treated in a humane manner. Standards also ensure that processors are maintaining clean and safe facilities that prevent the spread of foodborne illnesses or zoonotic diseases.

Jasen Wright/Shutterstock.com

Figure 12.2.3. All food-processing plants must follow strict rules and regulations under the guidelines set by the USDA. This ensures that all food grown, raised, and processed meets the same strict quality standards.

Sorokina Viktoryia/Shutterstock.com

Figure 12.2.4. Food products are not the only items inspected by the USDA. Plant material, crops, and seeds also have strict standards. *Did you know all seed packages have an expiration date, just like food?*

12

Figure 12.2.5. Producers can hire certified graders to check the overall quality of their product for marketing purposes. These pork carcasses are checked for last rib backfat thickness, degree of muscling, and quality, and assigned a number, 1–4, indicating the grader's assessment of its overall quality. *How do producers benefit from this?*

RossHelen/Shutterstock.com

After the products are inspected according to USDA standards, producers and processors may request to have the products graded for quality by a federal grader. **Grading** is a postharvest practice wherein commodities are sorted into categories according to an expert's judgment of their overall quality (**Figure 12.2.5**). The USDA's Agricultural Marketing Service is responsible for grading agricultural products. Producers who choose to go through the grading process must pay for the service themselves.

Large-volume buyers such as grocery stores, restaurants, and even foreign governments use the USDA grading system to smooth business transactions; it is easier to make a deal when everyone knows and understands exactly what they are getting. USDA grades assure consumers that the products they buy have gone through a rigorous review process.

Making Connections # Technology

Metal Detectors and X-Ray Machines in Food Safety Inspections

When you are eating, you expect your food to be safe. All the food that is sold, processed, or consumed in the US goes through some type of inspection to ensure it meets the standards set by the USDA. One of the standards for food processing plants is they must check for metal and other foreign objects in the food before it is in its final packaging. You might be wondering how metal could get in your food in the first place, but what if a very small bolt or screw fell from a piece of machinery or factory equipment and went unnoticed?

Manufacturers are required to identify hazards, determine preventive controls to eliminate or reduce hazards, and determine ways to control hazards in the case of accidents. Hazards can be biological, chemical, and physical. Physical hazards include things like bone fragments, stone, metal fragments, wood, glass, and plastic. Food manufacturers have two lines of defense against foreign object contamination: metal detectors and X-ray equipment. How does this equipment work?

Metal detectors find small particles of metal using magnetic fields. When a particle of metal passes through a metal detector, the field is disturbed, creating a measurable change in energy. This output is used to detect metal.

X-ray inspection systems are based on the density of the product and the contaminant. X-rays are invisible light waves. As an X-ray penetrates a food product, it loses some of its energy. A dense area, such as a contaminant, will reduce the energy even further. As the X-ray exits the product, it reaches a sensor. The sensor converts the energy signal into an image of the interior of the food product. Foreign matter appears darker in the images, helping to identify foreign contaminants.

Baloncici/Shutterstock.com

Understanding USDA Grade Labels

Quality grading is an important postharvest practice for farmers who care about their market reputation. Farmers who practice quality grading can adjust to market demands more easily. Though there is a fee for having products graded, these products can usually be sold for a higher price than ungraded products. Quality grading can help farmers achieve their sales goals and assure their customers they are receiving a high-quality item in return.

Different commodities may be graded on:

- Size or weight
- Shape
- Color
- Texture
- Stage of maturity
- The presence of disease or physical deformation

There are USDA grade labels that producers and processors can attach to their product to let consumers know that their product has been graded. USDA quality grade marks are found on beef, lamb, chicken, turkey, butter, and eggs. For many other products, such as fresh and processed fruits and vegetables, the grade mark isn't found on the retail product. For these commodities, the grading service is used by wholesalers, and the final retail packaging may not include the grade mark. Quality grades are widely used as a common standard when buying and selling agricultural products.

Beef grades include Prime, which is an indication of the highest quality, followed by Choice, Select, and Standard, which is the lowest grade given (**Figure 12.2.6**). When grading poultry sold for meat, Grade A is the most common grade sold in supermarkets. What makes poultry products qualify for Grade A depends on the absence of "defects," such as discoloration. As poultry is graded, it either meets Grade A criteria for quality or it is downgraded to lesser grades (B and C), depending on the number of defects. Eggs are categorized into one of three consumer grades: AA, A, or B (**Figure 12.2.7**).

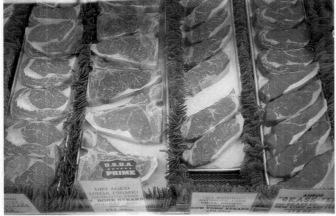

Elliott Cowand Jr/Shutterstock.com

Figure 12.2.6. Grading beef products is optional, but lets consumers know the quality of the meat they are buying. *How can this be an effective marketing tool?*

Keith Homan/Shutterstock.com

Figure 12.2.7. These eggs are Grade A, which means they are very high-quality. *Have you seen eggs with a different grade at the store?*

Hands-On Activity

Check Quality and Standard Labels

Quality labels can often be found on food product packaging. Have you ever looked closely to see what kinds of information it tells you about your food?

For this activity, you will need:

- Several empty food containers. It's best to have mostly unprocessed foods such as cheese, yogurt, milk, frozen meat, vegetables, or fruits.
- Paper and pencil
- Device for internet research

Here are your instructions:

1. Divide up into groups of two or three students.
2. Pick three food containers from different food groups to examine.
3. Examine the empty packaging for labels, seals, grades, or other symbols that you think might be important about the food product.
4. Write or draw each symbol on a piece of paper, leaving space to include information that you collect about that symbol.
5. Use an approved device to research what this symbol means and write a brief summary on your paper alongside the symbol. Explain what the symbol means and what the processor must go through to earn that grade or certification.

Consider This

1. Were there any symbols or words that surprised you? Why or why not?
2. Do you think producers are doing a good job of communicating to consumers what these symbols and grades mean? How would you improve upon that process?

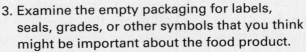

Other Common Food Labels

Certified Organic

Organic is a labeling term that indicates that the food or other agricultural product has been produced according to approved methods. Organic products must be produced using production practices that promote ecological balance, maintain or improve soil and water quality, minimize the use of synthetic materials, and conserve biodiversity. Organic products are typically grown using fewer chemicals, but the crops may exhibit lower yields as a result. The National Organic Program (NOP) develops the rules and regulations for the production, handling, labeling, and enforcement of all USDA-certified organic products. The organic standards describe the specific requirements that must be verified by a USDA-accredited agent before products can be labeled USDA Organic (**Figure 12.2.8**).

Tada Images/Shutterstock.com

Figure 12.2.8. Producers must follow strict production standards and undergo an inspection before they can label their products as USDA Organic.

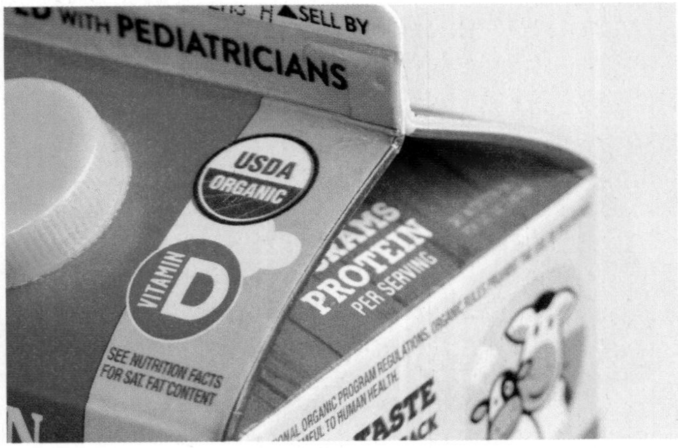

There are four different approved organic labels: 100% Organic, Organic, Made with Organic, or specific ingredients listed as organic. *100% Organic* is used to label any product that contains only purely organic ingredients.

Organic is the appropriate label for any product that contains at least 95-percent organic ingredients. *Made with Organic* is the label for a product that contains at least 70-percent organically produced ingredients. Specific organic ingredients listed on a nutrition label may be listed on products containing less than 70-percent organic contents.

Did You Know? Organic milk and organically raised broiler chickens are the top two organic US commodities, each with more than $1 billion in annual sales.

USDA Process Verified Program

The USDA Process Verified Program is a verification service that offers producers a unique way to market their products to consumers using approved marketing claims. Producers with an approved USDA Process Verified Program may develop promotional materials associated with their process verified points, use the USDA PVP shield in accordance with program requirements, and market themselves as "USDA Process Verified" (**Figure 12.2.9**).

Andriy Blokhin/Shutterstock.com

Figure 12.2.9. Producers must go through the Process Verified Program (PVP) before they can add additional marketing labels to their products. The eggs pictured here are labeled as *No Antibiotics, No Added Hormones,* and *Certified Humane.* All of these claims must be approved through the PVP before they can appear on a package.

Career Corner | BRAND MANAGER

Agricultural **brand managers** *utilize market research to promote specific products to consumers and ensure sales goals are met.*

What Do They Do? Brand managers control the strategy for entire brands, from idea to execution. They drive new product innovation and line extensions, oversee packaging design, differentiate from the competition, tailor presentation targets to key accounts, and review complaints related to their line.

What Education Do I Need? A bachelor's degree in business administration or marketing is required.

How Much Money Can I Make? Brand managers make an average of $110,000 per year.

wee dezign/Shutterstock.com

How Can I Get Started? You can build an SAE around creating a marketing plan for your school fundraiser, conduct an experiment on customer preference over different marketing techniques, or research different agricultural brands to determine how they made their brand successful. You can always interview a brand manager or set up a virtual field trip for your class.

12

What Did We Learn?

- The responsibility of regulating and monitoring food labels is shared between many federal agencies, including the Food and Drug Administration (FDA) and USDA.
- In 1938, Congress passed the Federal Food, Drug, and Cosmetic Act, which gave the Food and Drug Administration (FDA) the authority to issue food safety standards.
- All agriculture products go through inspections to ensure that producers and processors are following the guidelines and standards set by the USDA. Standards are guidelines used to measure acceptable product quality.
- Grading is an optional postharvest practice that consists of sorting a commodity into categories according to an expert's judgment of their overall quality. Not all products are graded, but those that are may be able to charge a premium.
- USDA labels include Prime, Choice, Select, and Standard for beef; Grades A, B, and C for chicken; and Grades AA, A, and B for eggs.
- Other common labels on products include Certified Organic, which signifies that the USDA has verified the products were wholly or partially produced according to organic standards. Other marketing claims are regulated according to the USDA Process Verified Program.

Let's Check to See What We Know

Answer the following questions using the information provided in Lesson 12.2.

1. *True or False?* In 1938, Congress passed the Federal Food, Drug, and Cosmetic Act, which gave the FDA authority to issue food safety standards.

2. In 1946, the AMA was passed to provide the USDA the authority to inspect, certify, and identify the class, quality, and condition of agricultural products. The AMA stands for the ____.
   A. American Marketing Act
   B. American Management Act
   C. Agricultural Marketing Act
   D. Agricultural Management Act

3. *True or False?* The entire country did *not* have unified standards that covered almost all agriculture commodities until the Federal Food, Drug, and Cosmetic Act.

4. ____ is/are an optional postharvest practice that consists of classifying a commodity based on observed quality.
   A. Standards   C. Organic
   B. Grading   D. Certified

5. *True or False?* The USDA's Agricultural Marketing Service is the agency responsible for grading agricultural products.

6. The highest beef grade is ____.
   A. Standard   C. Select
   B. Choice   D. Prime

7. The lowest grade of eggs as rated by the USDA is: ____.
   A. Grade AA   C. Grade B
   B. Grade A   D. Grade C

8. _____ is the label given to products grown under standards that promote ecological balance, improve soil and water quality, and conserve biodiversity.
   A. Standards
   B. Grading
   C. Organic
   D. Certified

9. *True or False?* Products with the word "organic" on the label are made with 100-percent organic ingredients.

10. The USDA Process _____ Program is a service that offers producers approval to use additional marketing claims such as "no hormones" or "antibiotic free."
    A. Standards
    B. Grading
    C. Organic
    D. Verified

Let's Apply What We Know

1. Explain the difference between standards and grades.
2. Discuss the benefits of grading and what the different commodity grades represent.
3. Explain what an organic label means, and how the four approved labels featuring the word *organic* differ.

Academic Activities

1. **Math.** Research the retail cost of graded versus ungraded food products, such as steaks or pork. Determine the cost to the producers to undergo the grading process. Is this a profitable decision?

Communicating about Agriculture

1. **Communicating Information.** Many consumers do not know what the symbols on their food packaging mean. Create a poster, infographic, or social media post to explain these labels clearly to other people. Share with your class, and get some feedback about how to make it even clearer.

SAE for ALL Opportunities

1. **Research SAE.** Spend time researching the products that are eligible for USDA grading. Prepare a report describing what products are graded, what the grades mean, and what role grading plays in the agriculture industry. Find data and statistics to help make your explanation clear.

SAE for ALL Check-In

- How much time have you spent on your SAE this week?
- Have you logged your SAE hours?
- What challenges are you having with your SAE?
- How can your instructor help you?
- Do you have the equipment you need?

12

Determining Price Points

Prostock-studio/Shutterstock.com

Prostock-studio/Shutterstock.com

Essential Question

What factors affect the retail price someone pays for a product?

Learning Outcomes

By the end of this lesson, you should be able to:
- Describe the relationship between supply and demand. (LO 12.03-a)
- Explain why market prices change. (LO 12.03-b)
- Show how price points are determined. (LO 12.03-c)

Words to Know

market price

price point

Before You Begin

Who loves a good deal? Have you ever bought something you didn't really need just because it was a good price? I am guilty of that myself, but I just can't resist a bargain. Have you ever wondered why producers can't sell products at the sale price all the time? Do you think producers and processors lose money when they put things on sale? Jot down on a piece of paper factors that you think affect the price of an item. We will check back after the end of the lesson to see how accurate our first thoughts were.

Have you ever been at a restaurant and noticed there was no price listed next to certain menu items, but rather the words "market price" (**Figure 12.3.1**)? How do you know the market price? Knowing the market price for products and services is key to knowing how producers can increase sales and grow their business, and how consumers can get a good deal. Market prices can vary depending on many factors.

Supply, Demand, and Market Price

Market price is the amount a product or service can be bought or sold for. A buyer wants to purchase an item for as little as possible. A seller wants to make a profit and get as much money as possible. The market price is where these two numbers meet—where the seller can make a profit at a price that a buyer is willing to pay.

The two factors that have the greatest impact on the market price are supply and demand. Supply is the amount of a good or service that is offered for sale. There are many different factors that affect supply. Think of crop yields: a strong yield results in a greater supply, whereas a poor yield leads to lesser supply (**Figure 12.3.2**). Government policies can also affect supply. Transportation costs, employment rates, product quality, and production costs all affect supply.

Demand is the quantity of a good that consumers are willing and able to purchase at a particular price during a given period. The demand for a product typically decreases as the price of the item increases. In turn, as the price of a product decreases, demand typically increases (**Figure 12.3.3**). Demand can be affected by interest rates, consumer opinion, and wage rates.

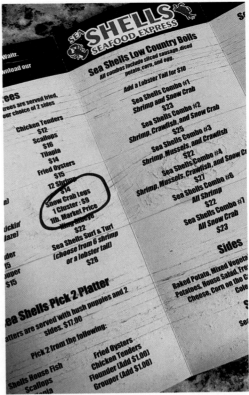

Melissa Riley

Figure 12.3.1. Prices of items can vary depending on the market. You might see items on a menu listed as "market price" instead of giving a dollar amount. This allows the price to change depending on the market without the restaurant having to adjust their printed menus.

Earl D. Walker/Shutterstock.com

Figure 12.3.2. A poor crop yield can decrease supply. *How do producers deal with this?*

Nelson Antoine/Shutterstock.com

Figure 12.3.3. There are typically items in high demand each holiday season. Consumers are willing to pay more for these items. *What does this have to do with supply and demand?*

12

Why Market Prices Change

There are two main reasons market prices change (**Figure 12.3.4**). These are:

1. A decrease in product or service availability
2. An increase in product or service availability

When there's a decrease in the availability of a good or service, consumers tend to agree to pay more because it's harder to find (**Figure 12.3.5**). These items become more valuable to consumers.

The opposite occurs when products or services are more readily available. When products and services are easy to obtain, consumers refuse to pay high prices for them. If consumers know that supply is plentiful, they are likely to walk away from high prices to find the product elsewhere at a lower cost.

Factors That Affect Market Price

Although the main drivers of change in market price are supply and demand, there are other factors that can also affect market price. Some of these are controllable, while others are out of the suppliers' hands. Factors that impact market price include:

- Natural disasters
- World events
- Wages paid to workers
- Decrease or increase in employment
- Pricing of luxury items versus necessities

Natural disasters or other world events (such as wars, terrorist attacks, or pandemics) can limit supplies available to manufacturers. Decreases in necessary supplies can slow down the production of goods or a business's ability to offer services. And if there's a shortage in products or services, demand can shoot up quickly.

Did You Know? After Hurricane Irma in 2017, up to 70 percent of the supply of cut foliage was harmed. This caused a shortage to florists, which resulted in higher prices for flower arrangements and other floral designs.

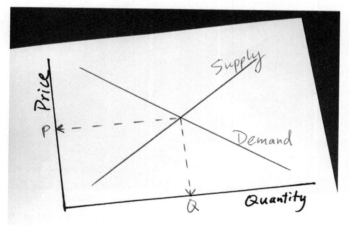

JohnKwan/Shutterstock.com

Figure 12.3.4. Market price falls where supply and demand meet. As supply changes, so will demand. *Why might supply change?*

CGN089/Shutterstock.com

Figure 12.3.5. During the COVID-19 pandemic, there were shortages of bathroom tissue and cleaning products. Customers were paying two and three times the normal price of these products. *Why?*

Determining Market Price through Supply and Demand

How much would the other students in your class be willing to pay for your exclusive t-shirt design? Let's find out!

For this activity, you will need:

- Pen, pencil, or markers
- Paper

Here are your instructions:

1. Get into groups of three. You will have four to six minutes to create a design for a t-shirt. Draw a quick sketch of your design on a piece of paper. Post the paper up on a wall, desk, or table at your teacher's direction.

2. Take five minutes to survey your classmates about how much they would be willing to pay for your t-shirt design. On a new piece of paper, write the numbers $0–$50 in increments of $5. Show your design to others and ask what price they would be willing to pay for the shirt. Record the price indicated by each student with a tally mark.

3. Plot your results on a line graph.

USBFCO/Shutterstock.com

HomeArt/Shutterstock.com

Consider This

1. How many shirts would you "sell" if you gave them away for free?
2. At what price would you sell no shirts? What price is higher than anyone is willing to pay?
3. What price would bring in the most total revenue?
4. What would be the market price for your shirt? Why?

Employment and the wages paid to workers can also affect the market price. A decrease in employment or wages may cause consumers to be more mindful of how they are spending their money. They might not be able to afford to pay the same prices as before, or might cut back on spending overall. Likewise, increases in employment opportunities and wages result in consumers being willing and able to pay more, leading to higher market prices.

Market prices of luxury items don't follow the same rules of supply and demand. Although the demand for luxury items is smaller, the prices are almost always high. Some consumers are willing to pay more for name brands and perceived quality or status (**Figure 12.3.6**).

Beauty and Sports/Shutterstock.com

Figure 12.3.6. Some people are always willing to pay more for luxury items such as sports cars or designer clothes. These items break the rules that typically apply to supply and demand. *How can they do this?*

Guiderom/Shutterstock.com

Figure 12.3.7. Determining price points can be tricky for retailers. You don't want to price yourself too high and be beat out by the competition, but you don't want to price yourself too low and lose money. *What factors do you think could affect the price point of a retail item?*

What Is a Price Point?

A *price point* is a point on a scale of possible prices at which something might be marketed. Price points are the retail prices that keep a product in high demand. It is the point where retailers are making the most money and consumers are willing to purchase (**Figure 12.3.7**). Price points are a balancing act because the retailer risks asking too little money when customers are ready to pay more, losing out on potential revenue. On the other hand, retailers do not want to set the price too high and not sell any items.

Hands-On Activity

Determine the Price Point

Price points can be tricky. Retailers are trying to get the most money for their product, while consumers are trying to get the best deal. Name brand and designer products usually sell for more than the store brand, but how much is too much or too little? The consumer will decide.

In this activity, you will be taking on the role of consumer or producer, and then assigned a level (for consumers: high, medium, or low amount of money; for producers: designer brand, name brand, or store brand). Consumers want to get the best item available to them for the least money. Producers want to sell the most product for the most money. How will you make this work?

Also, there's a twist: You are not allowed to talk or share your identity! You don't know who you are "talking" to, what other people are "selling," or how much money everyone else has.

For this activity, you will need:

- Fake money

Here are your instructions:

1. Get into six equal groups:
   - Three of the groups will serve as the consumers.
   - Three of the groups will serve as the producers.

2. The consumer groups will be assigned as a low-income, middle-income, and high-income group. Each group should be given fake money in accordance with its role in the game.
3. The producer groups will be assigned roles as designer brand, name brand, and store brand retailers. Each group will decide what sort of good they want to sell based on their assigned role.
4. Without talking, each consumer student will move around the room and try to complete a sale.
5. Once you have completed a transaction, you will take a seat, so that others know you are sold out or out of money.

Consider This

1. Did all the producers with the same product sell their product for the same price? What was the highest and lowest price received for each item?
2. When price points are determined on the retail market are they done blindly, as in this activity? What research can be done before prices are determined?

How Price Points Work

To determine the price and make it maximally profitable, retailers must test their prices and adjust as needed. There are a range of elements to think about while deciding on the correct price point, such as:

- The amount that consumers can pay
- The position of the company
- What the competition charges
- The supply volume
- How popular a product is compared to competitors (**Figure 12.3.8**)

Determining the right price point can be challenging and time-consuming, but it is worth it. Taking the time to research the market and determine the best price for a given product pays off for companies in the long run. Research has shown that smart pricing tactics allow companies to meet their revenue objectives.

Sheila Fitzgerald/Shutterstock.com

Figure 12.3.8. How popular a company's product is compared to its competitors can affect its price point. Both of these pasta noodles are the same, but one box is name brand and one box is store brand. The store can charge more for the name brand pasta because they are a more sought-after product than the store brand.

Career Corner ▶ PRICING COORDINATOR

If this is interesting to you, perhaps you might try being a **pricing coordinator**!

What Do They Do? Pricing coordinators serve as intermediaries between commodity traders, commodity marketers, and company administration. They lead design and management of price tests and tailor pricing actions based on the expected behavior of different customer segments. They ensure that accurate pricing is provided and collaborate to make sure that customers and internal stakeholders are getting the information they need.

Prostock-studio/Shutterstock.com

What Education Do I Need? An associate or bachelor's degree in agricultural business, finance, or accounting is required to be a pricing coordinator.

How Much Money Can I Make? Pricing coordinators make an average of $55,000 per year.

How Can I Get Started? You can build an SAE around researching prices of different agricultural goods at different local stores. Conduct an experiment about customer knowledge and preferences when it comes to food pricing or pricing of other agricultural products. Research the steps needed to set a final price on a product and present your findings to the class, or interview a pricing coordinator at a local business.

12

What Did We Learn?

- Market price is the amount a product or service can be bought or sold for.
- The two factors that have the greatest impact on market price are supply and demand. Supply is the amount of a good or service that is offered for sale. Demand is the quantity of a good that consumers are willing and able to purchase at various prices during a given period. Market price is where those lines cross.
- There are two main reasons why market price might change: a decrease in product or service availability or an increase in product or service availability.
- A price point is a point on a scale of possible prices at which something might be marketed. Businesses conduct research to set the correct price point for their goods.

Let's Check and See What We Know

Answer the following questions using the information provided in Lesson 12.3.

1. *True or False?* Market prices stay the same year-round and never change.
2. _____ is the amount of a product or service that is offered for sale.
   A. Supply
   B. Demand
   C. Market price
   D. Price point
3. The _____ is the quantity of a good that consumers are willing and able to purchase at various prices during a given period.
   A. supply
   B. demand
   C. market price
   D. price point
4. *True or False?* If the availability of a normal product decreases, the amount consumers will pay increases.
5. *True or False?* When products are easier to obtain, the price of the product decreases.

6. Which of the following does *not* impact market prices?
   A. Natural disasters
   B. World events
   C. Prices of luxury items
   D. Cost of advertising
7. A point on a scale of possible prices at which something might be marketed is the _____.
   A. supply
   B. demand
   C. market price
   D. price point
8. *True or False?* How much the competition charges has no effect on the price point of an item.
9. *True or False?* Luxury goods behave according to the laws of supply and demand.
10. Market prices are variable because _____.
    A. conditions in the market change
    B. market conditions do not change
    C. producers may decide to charge more
    D. producers may decide to charge less

Let's Apply What We Know

1. Explain the difference between supply and demand.
2. Discuss why market prices vary throughout the year.
3. Explain how price points work and factors that affect them.

Academic Activities

1. **Math.** Research the market price of at least three different agricultural commodities during the last five years. Create a graph to show how the market price of each item has changed over the last five years.
2. **History.** This lesson was focused on the marketing of current products and services. What about products that people no longer want or need? Interview three older adults to ask them for a list of products or services that people no longer want or need, For example, home milk delivery is virtually a thing of the past. What other products or services are no longer available for purchase, and why?

Communicating about Agriculture

1. **Writing.** Learn as much as you can about the supply and demand curves for a favorite food or drink. What affects the supply? What affects the demand? How does it compare to its competitors (if there are any)? Write a short essay about what you learned.

SAE for ALL Opportunities

1. **School-Based Enterprise SAE.** Use data and research to help your school's FFA organization figure out what to sell, and what to charge, at its fundraiser to maximize revenue. Conduct research into consumer needs and wants in your area, and what consumers might be willing to pay. Study potential competition. Present your plan to your organization, along with any supporting evidence for your idea.

SAE for ALL Check-In

- How much time have you spent on your SAE this week?
- Have you logged your SAE hours?
- What challenges are you having with your SAE?
- How can your instructor help you?
- Do you have the equipment you need?

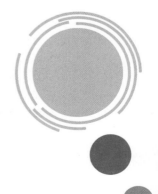

Rawpixel.com/Shutterstock.com

Marketing Strategies and Plans

Learning Outcomes

By the end of this lesson, you should be able to:
- Describe marketing strategies. (LO 12.04-a)
- Explain the purposes and benefits of having a marketing plan. (LO 12.04-b)
- List the steps to create a marketing plan. (LO 12.04-c)

Words to Know

market research marketing strategy
marketing plan target audience

Before You Begin

You have probably bought something that you really didn't need to purchase, but have you ever sold something you weren't planning on selling? Like at lunch, when you have all the good snacks, and your friends offer to buy some from you. Or when you have a cool trading card that someone needs and you like it, but someone makes you an offer you can't refuse. Has this ever happened to you? Tell your neighbor about a time you or a family member sold something you weren't planning to sell.

Selling and marketing are not the same thing. Sometimes selling something can take very little effort. Other times, selling something can be difficult and time-consuming. The difference between selling and marketing is that marketing requires a plan (**Figure 12.4.1**). Often, business owners confuse the act of selling their products to customers as marketing, when selling is only a small portion of the entire marketing plan.

Concentrating only on selling will not help you develop a long-term customer base. Marketing is about discovering and satisfying your customers' needs and wants while earning a profit. This involves attracting and retaining customers. You attract customers by promising and delivering better value than your competitors. You retain customers by continuing to deliver satisfaction. In other words, marketing brings the buyer and the seller together.

Marketing Strategy

A *marketing strategy* refers to a business's overall plan for reaching consumers and turning them into customers. A marketing strategy contains the company's value statement, key brand messaging, data on target customer demographics, and other high-level elements. Marketing strategies are important to help a business understand their target audience. A *target audience* is a group of people identified as being likely customers of a business (**Figure 12.4.2**). A thorough marketing strategy covers "the four Ps" of marketing: product, price, place, and promotion.

Marketing Plans

The marketing strategy is outlined in the *marketing plan*, which is a document that details the specific marketing activities a company plans to conduct, along with timetables for rolling out these initiatives. All businesses, big or small, need a marketing plan to determine which products should be sold, how they should be sold, and what type of financial gain is expected from the sale. Rather than assuming your product is perfect for everyone, the plan focuses on key people who are most likely to buy the product. Marketing plans are important because they make sales easier for any business owner. When the ideal customer is targeted in a smarter way, marketing costs are reduced and more leads are turned into sales.

Lisa F. Young/Shutterstock.com

Figure 12.4.1. *Have you ever tried to sell or trade an item and had a difficult time finding a buyer? What do you think makes some trades easier than others?*

quinky/Shutterstock.com

Figure 12.4.2. Marketing strategies are important to help a business understand their target audience. A target audience is a group of people identified as being likely customers of a business. *How can you find a target audience?*

12

Makistock/Shutterstock.com

Figure 12.4.3. The purpose of a marketing plan is to decide how you will sell your products or services to consumers. *How do marketing plans help businesses find success?*

Marketing plans lay out a specific set of objectives and specify a time frame (usually a calendar year) within which to achieve these objectives. A marketing plan helps a business focus their efforts on maximizing profits. Marketing plans help to clearly define the product or service being offered, identify customers and competitors, outline a strategy for attracting and retaining customers, and anticipate changes in the marketplace (**Figure 12.4.3**). The marketing plan also contains a detailed budget for the funds and resources required to carry out planned activities. The assignment of tasks and responsibilities to team members is also usually outlined in the marketing plan.

Purpose of a Marketing Plan

The purpose of a marketing plan is to:
- Reach a business's target audience
- Grow customer base
- Increase the amount of sales
- Obtain financing
- Set clear, realistic, measurable objectives
- Focus the total marketing efforts

Making Connections   # Creative_Writing

Writing Product Descriptions

Product descriptions are a key component of a marketing plan. They provide a lot of information about the product and present it in a way that is appealing. The best product descriptions connect to the customer and make the customer want to buy the product.

Product descriptions include:
- Informative content that highlights the benefits of the product
- A clear description of what the product looks like, what it's made from, what it does
- Expressive and engaging language

For this activity, write a product description for your favorite food or candy. Make sure to use correct grammar and punctuation.

VaLiza/Shutterstock.com

Consider This

1. How does your product description compare to your classmates' descriptions?
2. Which of your classmates' product descriptions were your favorites? Why?
3. Is there a common theme among your favorite product descriptions that made them memorable or more appealing than others?

Creating a Marketing Plan

Preparing a marketing plan is not as difficult as it might sound, though it does require some market research to get things going. *Market research* is the process of gathering information about consumers' needs and preferences. When conducting market research, businesses are usually trying to answer questions like:

- What market(s) do they serve?
- What unique features distinguish their product/service from others?
- What is the best way to get their product/service into the hands of targeted consumers?
- Who are their customers?
- How should they price their product/service?
- How should they promote or advertise their product/service?
- How is the target market changing or likely to change?

This information can be obtained in a variety of ways, including hiring consultants to conduct the research for the company. You can also search for information online, use social media to network, communicate with others in the industry, visit similar businesses, attend trade shows and conferences, or join professional organizations (**Figure 12.4.4**). Once sufficient information has been collected, the marketing plan can be created.

Sergey Ryzhov/Shutterstock.com

Figure 12.4.4. Conference trade shows are a great place for businesses to conduct market research. Trade shows give businesses the opportunity to interact with their existing and potential clients.

There are multiple ways that businesses create marketing plans, but generally, marketing plans follow these steps:

1. *Create a market objective.* Examples might include increasing awareness of the product or service, increasing sales and revenues by a certain percentage, or increasing the number of customers.
2. *Identify the target audience.* Examples include age, gender, profession, income level, individual or business, etc.
3. *Identify your competition.* Identify other businesses that are offering similar products or services to the target audience within the same price range.
4. *Describe the product or service.* Product descriptions are what sell a product and should include many details to help your customer make a buying decision.
5. *Define the distribution strategy.* Where will you sell your product? Refer to **Lesson 12.1** to recall different marketing locations.
6. *Develop a pricing strategy.* The price should be based on the current market price and production costs. It should include a profit. Refer to **Lesson 12.3** for more information about selecting a market price.
7. *Create a promotion strategy. Promotion* refers to the set of activities that inform consumers about a product or service. Promotion is how a business gets people to know about the products or services they provide. It deals with how and what they want to communicate to potential customers. We will cover more about this in **Lesson 12.5**.
8. *Create a marketing budget.* Identify how much time and money the business wants to allocate to marketing.

As businesses expand and markets change, a business can update any part of the marketing plan. Marketing plans are not permanent. They need to be routinely checked and revised to meet the business's current needs.

Career Corner — FOOD STYLIST

A **food stylist** *designs, prepares, and styles food for photography or on-air demonstration. If you like marketing and food, perhaps this is a career for you. Yum!*

What Do They Do? A food stylist makes food items appealing for photography and videography; goes through recipes step-by-step for on-air or video demonstration; maintains and cleans test kitchen or demonstration areas, cooking equipment, and utensils; and consults with marketing and food production staff about which foods or recipes should be presented.

What Education Do I Need? No formal training is required, but an associate degree in culinary arts, baking arts, nutrition, or a related field is helpful.

How Much Money Can I Make? Food stylists make an average of $35,000 per year.

How Can I Get Started? You can build an SAE around trying out recipes and creating online videos and pictures, research tips on how to photograph food and other products, research the steps needed to start a photography business, and interview a food stylist or ask them to be a class speaker.

New Africa/Shutterstock.com

Hands-On Activity

Create a Basic Marketing Plan

You own a marketing firm that specializes in developing marketing plans for a variety of products. You've been hired by your local school system to develop a marketing plan for a restaurant students can go to for lunch instead of the school cafeteria. You are creating your own new restaurant lunch option, but you can incorporate items already being served in the cafeteria as well.

For this activity, you will need to divide into small groups to create and present a basic introductory marketing plan. The goal of this activity is to get you thinking through the first steps in starting a successful business.

Here are your instructions:

1. Research the lunch options available at your school and decide on a new menu or menu items for your restaurant.
2. Develop a restaurant name and a slogan or tagline for the new food options. Be creative.
3. Identify your target market. Who are you trying to get to come to your restaurant? All students? Only a certain grade? Teachers?
4. Explain your price strategies and decisions.
5. Create a general promotion strategy.
   - What's your message and how will you communicate it?
   - How and in what format will you reach your customers?
   - Describe any advertisements.

Remember, the more you define your target market, the easier it will be to figure out how to reach them. All marketing decisions should be made with the target market in mind.

Consider This

1. What made creating a marketing plan challenging?
2. If you were conducting this plan for your company, what timeline would you need to complete all the steps of your plan?

Rawpixel.com/Shutterstock.com

12

What Did We Learn?

- The difference between selling and marketing is that marketing requires a plan.
- A marketing strategy refers to a business's overall plan for reaching consumers and turning them into customers. A thorough marketing strategy covers product, price, place, and promotion.
- The marketing strategy is outlined in the marketing plan, which is a document that details the specific marketing activities a company plans to conduct along with timetables for rolling out these initiatives.
- The purposes of a marketing plan include reaching a target audience, growing a customer base, increasing the amount of sales, obtaining financing, setting measurable objectives, and focusing the marketing efforts.
- Marketing plans require market research. Market research is the process of gathering information about consumers' needs and preferences.
- There are eight main steps to creating a marketing plan: creating an objective, identifying the target audience, identifying the competition, describing the product or service, defining the distribution strategy, developing a pricing strategy, creating a promotion strategy, and creating a marketing budget.

Let's Check and See What We Know

Answer the following questions using the information provided in Lesson 12.4.

1. *True or False?* Selling is only a small portion of marketing.

2. A group of people identified as being likely customers of a business is known as a ____.
   A. marketing strategy
   B. marketing plan
   C. target audience
   D. market research

3. A document that details the specific types of marketing activities a company conducts and contains timetables for rolling out marketing initiatives is a ____.
   A. marketing strategy
   B. marketing plan
   C. target audience
   D. market research

4. ____ occurs when businesses gather information about consumers' needs and preferences.
   A. Marketing strategy
   B. Marketing plan
   C. Target audience
   D. Market research

5. Which of the following is *not* a step to creating a marketing plan?
   A. Create a market objective.
   B. Develop a pricing strategy.
   C. Create a board of directors to oversee the marketing plan.
   D. Create a promotion strategy.

Let's Apply What We Know

1. Explain the difference between a marketing strategy and a marketing plan.
2. Discuss how market research is necessary before creating a marketing plan.
3. Describe the eight steps to creating a marketing plan.

Academic Activities

1. **Social Sciences.** Think of a public company – a famous national corporation or a local business. What do you think that business's marketing plan looks like? Create a mock-up of a basic marketing plan for your chosen company.
2. **Statistics.** Marketing plans are supposed to offer many benefits to the businesses that create them. How could you track the effects of your marketing plan? What data could you gather?

Communicating about Agriculture

1. **Speaking.** Create a short (two- to three-minute) speech describing the purposes and benefits of marketing plans. Have a friend film your speech, and watch it back. Are your points clear? Is your argument persuasive?

SAE for ALL Opportunities

1. **School-Based Enterprise SAE.** Create a basic marketing plan for your school's FFA program. Who is your target audience? What message do you want to get to them? Do some market research to support your plans.

SAE for ALL Check-In

- How much time have you spent on your SAE this week?
- Have you logged your SAE hours?
- What challenges are you having with your SAE?
- How can your instructor help you?
- Do you have the equipment you need?

Promoting and Advertising Agricultural Products

Appolinaria/Shutterstock.com

Appolinaria/Shutterstock.com

Essential Question

What role does promotion play in marketing agricultural products and services?

Learning Outcomes

By the end of this lesson, you should be able to:
- Explain what promotions are and what purpose they serve. (LO 12.05-a)
- Describe the benefits of promotional efforts. (LO 12.05-b)
- List popular methods that businesses may use to promote their products and services. (LO 12.05-c)
- Define the relationship between advertising and promotion. (LO 12.05-d)

Words to Know

advertising
advertising plan

brand
promotion

Before You Begin

*What does your favorite TV commercial look like? Mine is a car commercial where a family of dogs acts out situations as if they were people. It's so funny—have you seen it? The best TV commercials stick with us thanks to a catchy phrase or slogan, or maybe a song that gets stuck in your head (**Figure 12.5.1**). Why do you think businesses spend so much money on commercials? With your neighbor, make a list of five reasons why you think advertising is important for a business.*

sitthiphong/Shutterstock.com

Figure 12.5.1. TV commercials are one of the most recognized forms of promotion. *What TV commercials do you remember? What makes them effective?*

Business owners must stay on top of profits and losses and have a marketing plan, but did you know they also have to promote and expand their business? Businesses must keep growing and stay relevant to be competitive in the market. How do businesses know where, when, and how to promote their products?

Promotions

A ***promotion*** is any communication that influences people to buy products or services. Businesses promote their products and services to their target audience. Promotions can refer to an *effort* (like an ad), a *concept* (like a temporary price reduction), or an *item* (like a branded t-shirt). A poster at a bus stop and a billboard on the side of the highway are forms of promotion. Promotions can also be attractive limited-term deals that a business offers to prospective customers.

Promotion is a vital aspect of any business. Without at least some level of promotion, a business can't get customers, and without customers, it's only a matter of time before the business will have to close. While all businesses need promotion, no two businesses will have the same promotional needs. Promotional tactics also vary significantly among industries. Promoting food and agricultural goods is quite different than promoting a new piece of technology, like a cell phone. A corner store in a small town might only need a sign that can be seen from the sidewalk to tell customers about an existing sale. Other businesses may need to buy ad time on a streaming service or social media to reach their audience.

New businesses may have to go through a period of trial-and-error while they experiment with different promotional styles before they find the mix that is best suited to them. Established businesses may experiment with new promotional strategies also, in an attempt to reach new customers (**Figure 12.5.2**).

Benefits of Promotions

You might think of promotion as the voice of your company carrying your brand's message to the audience. A ***brand*** is a marketing or business concept that helps people identify a company, product, or individual.

travelview/Shutterstock.com

Figure 12.5.2. *How many promotions can you spot in this image of New York City's Times Square?* In a typical year, nearly 360,000 people pass through here each day. That is a lot of exposure for a business and people will pay top dollar to promote their business there.

Rawpixel.com/Shutterstock.com

Figure 12.5.3. Businesses offering a free sample of their product is a promotional marketing plan that increases customer traffic to their shop, stall, or table. *Why is this?*

Without marketing promotions, the business brand would not gain the attention of new or existing customers. Promotions help businesses in many ways.

- *Promotions increase brand awareness.* Businesses use promotion to spread information about their brand and company. This helps potential customers find out more about the business, investigate available products, and make purchases.
- *Promotions provide accurate information.* Have you ever purchased something online only to find it wasn't quite what you expected? Did this experience cause you to think negatively about the company? Providing target audiences with reliable information through promotions will gain new clients and keep current customers coming back.
- *Promotions increase customer traffic.* Have you ever visited a grocery store when they were passing out free samples of a product? This type of promotion allows customers to try a new product without any commitment (**Figure 12.5.3**). This familiarity may lead to more interest in the product being sampled, which could turn into sales.

Making Connections # Creative_Writing

Promoting Yourself

At some point in life, we all find ourselves in situations where we must promote ourselves. Maybe this is to our guardians when we are asking permission to do something or go somewhere with friends. When you try out for a team, audition for the school play, or run for an officer position in a club or organization at school, you must promote your strengths, characteristics, experience, and what you have to offer to the group. Let's try it out!

1. Write a half page marketing promotion about yourself.
2. Answer the following questions in your write-up:
   - What message do you communicate to others? What image do you portray?
   - What makes you awesome and unique? What makes you a great friend and classmate? What are examples of some of your successes?
   - What message would you like to communicate about yourself? What would you like to change about your image? How could you do that?

Shift Drive/Shutterstock.com

Tie in personal stories to let readers know the real you and allow them a chance to see how you became the awesome person you are today.

Consider This

1. How difficult was this writing assignment? Why?
2. If you were promoting yourself, who would be your target audience?

Business Promotion Methods

Businesses can promote their brand, products, and services in many ways. Here are examples of different promotional tactics businesses use:

- *Word of Mouth.* Word of mouth is one of the most effective ways to promote a business and, best of all, it is free! Businesses providing excellent customer service benefit the most from word of mouth. Happy customers are more likely to mention a business, product, or service to a friend or family member, and those referrals are more likely to lead to new business than just about any other type of promotion. People listen to those they trust!
- *Website.* A website allows a business to keep customers up to date about what products are available and provide information about current specials. Websites also keep customers up to date on locations, hours of business, contact numbers, and other important details.
- *Social Media.* Social media can be a free or inexpensive way to promote a business. Having a presence on multiple social media platforms like Facebook, Instagram, Twitter, or YouTube allows businesses to reach a wide variety of customers through targeted ads (**Figure 12.5.4**).
- *Business Cards and Branded Promotional Materials.* Even in the digital age, business cards are still relevant, and are often exchanged at meetings or conferences to easily keep track of the connections made. Additionally, promotional items branded with a company logo, like pencils, pens, hats, golf balls, or shirts spread awareness of the company brand.
- *Charity Sponsorship.* You may see businesses getting involved in local charitable events by hosting, sponsoring, or attending. These charitable efforts must be publicized in some way to qualify as a promotion. Being associated with a local charity is a great way to create a positive feeling about the business within the community.

Did You Know?

It takes between five and seven impressions for a person to remember a brand, and use of color improves brand recognition by 80 percent.

Did You Know?

The ritual of exchanging *meishi*, which is Japanese for 'business card,' is a highly valued practice in Japan.

Slam Stock/Shutterstock.com

Figure 12.5.4. Social media is a great way for agribusinesses to connect with consumers of all ages. Many agribusinesses use social media to advertise events, promote new products, and interact with their customers. *What are the risks and benefits of this approach?*

12

articular/Shutterstock.com

Figure 12.5.5. Colorful posters and flyers are ways that companies can advertise their products and services. *Have you seen similar promotions for agribusinesses in your area?*

Advertising

The words "promotion" and "advertising" are often used interchangeably, but they are not the same thing. *Advertising* is a type of promotion, referring to a specific action taken to reach the people most likely to buy the company's products. The goal of advertising is to entice these consumers to buy. Advertisements are paid for by the business and are intended to inform or influence the people who see them.

Types of Advertising

There are many options for advertising the products or services that a business has to offer. The best choices will vary from business to business and depend on their customers, budget, industry, and business goals (**Figure 12.5.5**). A few different ways that agribusinesses advertise might include:

- Flyers or pamphlets
- TV
- Radio
- Internet ads
- Journals or blogs
- Giveaways
- Face-to-face sales
- Emails
- Catalogs or magazines

Creating an Advertising Plan

An *advertising plan* is an outline for how a business will use advertising to promote their products and reach new customers. Having a plan ensures that the money a business spends on advertising reaches the correct audience. It also establishes benchmarks that a business can use to assess whether their strategy is effective at reaching customers within the limits of the marketing budget.

The advertising plan should include:

- Advertising budget
- Target audience
- Key marketing messages
- Preferred advertising platforms
- Marketing and revenue goals

Did You Know?

Highway billboards have been around for more than 100 years. They let travelers know about a business or experience located nearby. Billboard advertising prices vary depending on the location and type of billboard (stationary, digital). Cost can range from $250 per month to more than $10,000 per month.

melissamn/Shutterstock.com

Hands-On Activity

Create an Advertising Plan

You and your teammates own a marketing firm that specializes in developing advertising plans for local businesses. You are focusing your efforts on a local school project or event (existing or made-up) and need to create an advertising plan to help promote it.

For this activity, you will need to get into small groups to create and present an advertising plan. You and your team can use an existing event taking place at your school (such as a school dance) or create your own school activity.

Here are your instructions:

1. Decide on a school event. Create a name for your event. Be creative!
2. Develop a slogan, tagline, or hashtag for the event.

3. Identify your target audience.
4. Determine the best advertising method to reach your target audience.
5. Assuming that budget is not an issue, outline the best advertising plan to reach your target audience and promote your school function. This plan should include a time line and advertising platforms.

Consider This

1. How did your advertising plan compare to your classmates'?
2. How do advertising methods change if you want to reach different age groups? How would you reach parents? Teachers? Grandparents?

Career Corner   GRAPHIC DESIGNER

If you're interested in promotions or advertising, **graphic designer** *might be a career for you to check out. Let's do it!*

What Do They Do? A graphic designer creates marketing content and media using computer software. They produce many types of promotional materials: print, video, social media, presentations, advertising, blogs, newsletters, posters, and flyers.

What Education Do I Need? A graphic designer should obtain an associate or bachelor's degree in graphic design, digital marketing, or a related field.

How Much Money Can I Make? Graphic designers make an average of $45,000 per year.

Andrey_Popov/Shutterstock.com

How Can I Get Started? You can build an SAE researching how graphic designers work in the agriculture industry, conduct a social science experiment on consumer preference on graphic brands or logos common in the agriculture industry, or volunteer at an art center to learn more about design work. Of course, you can always interview a graphic designer or ask them to be a class speaker.

What Did We Learn?

- A promotion is any communication that influences people to buy products or services.
- Promotion is the voice of your company carrying your brand's message to the audience. A brand is a marketing or business concept that helps people identify a company, product, or individual.
- Promotions help businesses in many ways, including increasing brand awareness, providing accurate information, and increasing customer traffic.
- Businesses can promote their brand, products, and services in many ways, including word of mouth, social media, and websites.
- Promotion and advertising are not the same thing. Advertising is a type of promotion, referring to a specific action taken to reach the people most likely to buy the company's products.
- An advertising plan is an outline for how a business will use advertising to promote their products and reach new customers.

Let's Check and See What We Know

Answer the following questions using the information provided in Lesson 12.5.

1. *True or False?* A pop-up advertisement during an online game is a type of business promotion.
2. A marketing or business concept that helps people identify a company, product, or individual is a company _____.
   A. advertisement
   B. marketing plan
   C. brand
   D. target audience
3. Which of the following is *not* a benefit of having a business promotional plan?
   A. Increases brand awareness
   B. Creates new products
   C. Provides accurate information
   D. Increases customer traffic
4. *True or False?* Word of mouth is a successful promotional tactic.
5. *True or False?* Supporting local charity events is *not* a good way to promote a business.

6. _____ is one specific action a business takes to promote a product or service.
   A. Advertising
   B. A marketing plan
   C. A brand
   D. A target audience
7. *True or False?* Advertisements are a type of business promotion.
8. Which of the following is considered a type of advertisement?
   A. Giveaway
   B. Flyer
   C. TV commercial
   D. All of these would be considered forms of advertisement.
9. Which of these methods is *not* a popular way for businesses to promote their products?
   A. Word of mouth
   B. Social media
   C. Business cards
   D. Skywriting

10. Promotions provide _____ information to consumers.
    A. detailed
    B. accurate
    C. confident
    D. false

Let's Apply What We Know

1. Explain the difference between promoting and advertising.
2. Discuss why knowing your target audience is important for any business when developing a promotional plan.
3. Explain the steps to create an advertising plan.

Academic Activities

1. **Social Science.** Pick a historical figure that you wish to learn more about. Do some basic research about this figure. Come up with four ways that you can promote this person and their achievements to your class. Use at least three different methods to get the word out about this figure. Be creative!

Communicating about Agriculture

1. **Speaking.** Present a plan to promote agricultural education classes to the other students at your school. Talk about the messages you would use, your target audience, and the methods you would use to reach them.
2. **Personal Communication.** Word of mouth advertising is a powerful but underrated method for spreading the word about products and services. Make a list of five to ten agricultural products like vegetables, fruits, meats, etc. Interview three adults and ask them for the best place to purchase these products. Compare their answers. What did you discover as a result of this activity?

SAE for ALL Opportunities

1. **Research SAE.** Reach out to a local agricultural business to learn about their promotional activities. Work with them to come up with a plan to improve their efforts, and present it to your class.

SAE for ALL Check-In

- How much time have you spent on your SAE this week?
- Have you logged your SAE hours?
- What challenges are you having with your SAE?
- How can your instructor help you?
- Do you have the equipment you need?

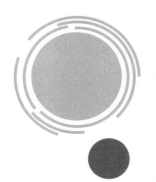

12

Technology in Marketing and Agribusiness

Zapp2Photo/Shutterstock.com

Learning Outcomes

By the end of this lesson, you should be able to:
- Describe how agribusinesses use technology to address challenges. (LO 12.06-a)
- Explain how technology is used to manage customer relationships. (LO 12.06-b)
- Describe how technology is used to manage data to benefit businesses. (LO 12.06-c)
- Explain how technology increases the efficiency of agricultural transportation. (LO 12.06-d)

Words to Know

analytics	cookie	supply chain
automation	remarketing	
conversion optimization	search engine	

Before You Begin

Have you ever been shopping online, only to see an item you were looking at appear in advertisements on other websites? Have you ever wondered how those advertisements seem to find you? The next time you go online, notice how many advertisements you see. Are they directed toward things that interest you? Modern advertising techniques use technology to place marketing ads in front of consumers. Take two minutes and make a list of the advertisements you have seen recently on a computer or mobile phone while using the internet. Are these advertisements similar to products or services you recently searched?

Challenges of Modern Agricultural Businesses

No one said that running a business was easy, and agribusinesses are no different. Agribusinesses face many challenges as they try to produce quality products, connect with their customers, and complete sales to meet revenue targets.

Accurate and Timely Information

One of the most pressing needs for agribusiness owners is prompt and accurate information. Farmers and ranchers need up-to-the-minute weather information to make decisions. Bad weather can cause serious damage to an agricultural operation (**Figure 12.6.1**).

Modern technology places reliable news sources at the fingertips of anyone operating agricultural businesses. The internet allows news organizations to post news or information as fast as it can be researched and written. Take the National Weather Service (NWS) website as an example. The NWS provides current information about weather patterns starting a few days in advance. Farmers and ranchers appreciate having current information on weather patterns because the weather may determine the success or failure of crops and livestock operations.

Aleksandr Ozerov/Shutterstock.com

Figure 12.6.1. Hail damage can destroy a farmer's chances of a successful growing season. *How does fast and accurate information help to avoid this outcome?*

Farmers and ranchers need to be aware of current commodity prices to determine the best time to sell their products. People in agricultural sales and services need to have current pricing and inventory information for their products. If someone comes into a small engine repair shop and asks for a replacement part, they expect to get a quick response from the person behind the counter about the price and availability of the part (**Figure 12.6.2**).

Tracking Market Prices and Economic Indicators

In a previous lesson, we discussed the concepts of supply and demand, and how they affect the cost of goods sold. Depending on consumer demand and the supply of products available, the price of a particular product will go up or down. Farmers try to time the sale of agricultural products when the price is high, so that they can generate the most income. This means that farmers have to pay attention to the news so they have a sense of what market prices are doing (**Figure 12.6.3**).

Nejron Photo/Shutterstock.com

Figure 12.6.2. Customers seeking repair parts want a fast and efficient response from service technicians. *How does this help a business?*

Artie Medvedev/Shutterstock.com

Figure 12.6.3. If a grain price is low and likely to rise, farmers may choose to store harvested grain until market prices improve.

Joseph Sohm/Shutterstock.com

Figure 12.6.4. Professional brokers are monitoring the prices of agricultural commodities such as corn and soybeans at the Chicago Mercantile Exchange. *Why would this information be helpful to producers?*

Farmers and ranchers use technology to gather information about the market price for the goods and services they produce. Both general and agricultural news outlets provide information about the health of the economy. One important resource for agricultural producers and agribusinesses is the Chicago Mercantile Exchange (CME) (**Figure 12.6.4**). The Chicago Mercantile Exchange tracks the sales and prices of agricultural commodities like soybeans, cattle, and grain. Farmers, ranchers, and agricultural businesses monitor the major stock market indexes and the CME to determine how the market is reacting to their products. They must get up-to-the-minute information to know precisely when to buy and sell products to get the best possible price. Technology has changed the way farmers and ranchers produce agricultural products. Today's agricultural decision-making depends on these computerized systems and instantaneous information.

The US economy depends on the internet to trade goods and services. According to the US Bureau of Economic Analysis, online sales and services account for 9.6 percent of all US services and goods sold, or more than $2.1 trillion. The internet helps businesses manage sales. It would be difficult (if not impossible) for modern agribusinesses to conduct sales, manage inventory, and keep customers without the internet.

Managing the Supply Chain

Agribusinesses depend on the supply chain to run their business smoothly. *Supply chain* is a term used to explain how raw materials move from one company to another (**Figure 12.6.5**).

Agribusinesses depend on the steady flow of materials through the supply chain. Technology helps agribusinesses purchase raw materials from suppliers. Manufacturers convert raw materials into finished products. This process is managed through automated technology, robotics, and computerized systems.

Figure 12.6.5. The supply chain moves raw materials to a manufacturing facility that makes finished products out of these materials. These products are then shipped to distribution points and then to retail stores for consumer purchase. *Why is it important to have a predictable supply chain?*

tele52/Shutterstock.com

Once the finished products leave the manufacturer, they travel through a sophisticated transportation network to the point of sale, perhaps making multiple stops along the way (**Figure 12.6.6**). The retailer then has to have a system for storing goods and retrieving them for display and sale.

Using Technology in Agribusiness

Now that we have covered some of the challenges that modern businesses can face (and how technology can help), let's take a closer look at some of the ways that technology is used in modern agribusiness.

donvictorio/Shutterstock.com

Figure 12.6.6. Container ships can carry up to 20,000 tractor-trailer loads at once. It takes a sophisticated system to manage the transportation and distribution of the products shipped in these containers.

Technology and Customer Relationship Management

Customer relationships are essential to agribusiness. The more a customer trusts and receives positive feedback from a business, the more likely it is the customer will return. Building customer relationships takes time and effort, yet this relationship can be damaged or destroyed by one bad experience. Customers appreciate receiving quality goods and services promptly. Today, agribusinesses use the internet to maintain good relationships with customers.

Analytics

Agribusinesses use analytics to determine if advertising is working (**Figure 12.6.7**). *Analytics* are computer software programs that track websites and products that customers are viewing and purchasing. Almost every aspect of an online sale is trackable: how long a customer stays on a website, the customer's location, the types of products they prefer. This information allows agribusinesses to customize the shopping experience for customers.

Here's another example of analytics technology at work. When you go to a grocery store, have you ever noticed how the products are placed on the shelves? The most popular and expensive items are placed at eye level on the shelves as you walk down each aisle. Why is this? In a modern supermarket, product placement is driven by technology. A lot of research goes into placing the products at the right location so that consumers can easily find and purchase them. Supermarkets conduct studies using cameras and sensors to determine how long a shopper spends in a particular area looking at a specific product.

There are even more analytics at a grocery store! When shoppers get to the checkout counter, the cashier scans their loyalty card (**Figure 12.6.8**). This may provide the shopper with discounts on purchased items, encouraging them to come back.

boonchoke/Shutterstock.com

Figure 12.6.7. Agricultural consultants use technology to help producers improve their business operations. Looking at analytics is an important part of this process.

Robert Kneschke/Shutterstock.com

Figure 12.6.8. Customer discount or loyalty cards provide shoppers with incentives to shop. This supermarket uses the data gathered to make decisions about what products to offer. *What are the benefits and drawbacks of this practice?*

New Africa/Shutterstock.com

Figure 12.6.9. Conversion optimization encourages customers to stay longer on a website to view information, shop, and make purchases. *What conversion optimization features have you noticed on websites?*

Did You Know?

The Harvard Business Review reports that only three percent of customers complete online purchases on their first visit to a site.

Did You Know?

In 2020, more than 306 billion emails were sent and received each day.

Perhaps more importantly, though, it also gives the supermarket information about consumer product preferences. Every time a loyalty card is scanned, the customer is providing the supermarket with the date and time of purchase, and the types of products purchased. This information helps the supermarket deliver the goods and services that people want at the right time and the right place.

Conversion Optimization

Advertising aims to get people to come to a website and stay long enough to buy something. ***Conversion optimization*** is what businesses do to get customers to stay longer on a website and purchase a product or service. Let's say a company has an online catalog. The web page for each item has an "add to cart" or "add to wish list" button to encourage customers to make a purchase or return later. A business wants to maximize its conversion optimization rate to attract buying customers (**Figure 12.6.9**).

There are some drawbacks to conversion optimization. Too many pop-up pages can serve as a distraction. Asking too many questions or pushing too many requests to add products or services to the purchase slows the checkout process and can frustrate customers.

Email

Email is a valuable tool for communicating with customers. Email allows agribusinesses to send customers newsletters, coupons, and other information about products and services. For example, a small engine repair shop can send customers confirmation emails regarding service appointments. Invoices and receipts for service are emailed to customers.

Search Engine Marketing

If you have used an internet search engine, you have likely seen advertisements pop onto your computer screen. These advertisements appear because the search engine takes the words you typed into the search box and conducts an automated search for products and services related to your question. A ***search engine*** is a computer software program that is used to find information on the internet. Google, Bing, Yahoo, and DuckDuckGo (among others) are all search engines.

Agribusinesses want to be found online, so they may submit their company information to search engines. The search engines store this data. When a consumer looks for a product or service sold by a particular company, that company's information appears on the consumer's computer screen.

Remarketing

Online marketing allows a business to put information about its goods and services on the internet for consumers to find. Sometimes a consumer will buy a product from the business through its online sales portal, but often the customer leaves the website without purchasing anything. To remind customers of their recent visit to the site, businesses collect information about a customer's visit using a software program called a cookie. A ***cookie*** is a collection of information about a customer that is stored in their web browser. Whenever you go online, your web browser "remembers" which websites you visited and may show you advertisements from those websites. This is known as ***remarketing.***

Sending promotions to customers based on their previous internet use is an effective marketing strategy. Websites use cookies to personalize the customer experience. Cookies can be a benefit for customers because they help to focus their shopping experience.

Mobile Phones

The Pew Research Center reports that more than 85 percent of Americans own a smartphone. More than 37 percent of adults say that they only use their smartphones to access the internet. Agribusinesses have developed websites to be easily viewable on a smartphone screen. Websites optimized for viewing on a mobile phone allow consumers to utilize most functions of a website, including purchasing goods and services.

Automation

Consumers visit an agribusiness web page to complete an online form to purchase goods or services. The customer's online purchase triggers the website to automatically create an emailed receipt for online purchases and send a message to staff to prepare the product for shipping. For agribusinesses that sell services, a customer's online request triggers an email confirmation to schedule the service. How is this possible? *Automation* allows the agribusiness to send and receive customer messages without continual human input. How might this benefit an agribusiness?

Data Management and Analysis

Agribusiness owners and managers need to make good decisions based on data. The more reliable the information available to an agribusiness owner is, the more informed decisions they can make about business operations.

Companies use technology to maintain records with up-to-the-minute accuracy, including virtually every aspect of the agribusiness:

- Number of customers and when they visited
- The most frequent goods or services purchased
- The frequency and type of customer complaints
- Employee work records and time worked
- Records of sales and expenses

Monitoring Systems

Agribusinesses use modern tools to monitor the growth and health of the business. Video surveillance provides security and prevents theft or inventory loss. Business operators use automated systems to monitor environmental controls in cold storage facilities. Greenhouse operators use computerized systems to control light and irrigation systems in plant growth facilities.

Tracking Inventory

Agricultural businesses use the internet to keep track of the inventory needed to conduct business (**Figure 12.5.10**). Technology allows companies to track shipments and predict when shortages will occur. Some large retail chain stores allow customers to check the availability of products in all stores in a region. Using technology, the store personnel can recommend where a customer can get an item not available in the local store.

Gorodenkoff/Shutterstock.com

Figure 12.6.10. Agribusinesses use a computerized system that tracks every item in an inventory. *Why would this information be helpful?*

12

Hands-On Activity

Reviewing Analytics for Bark N Suds

For your SAE, you work as a pet grooming technician for Bark N Suds, your family's mobile pet grooming service. You wash and groom pets in one of two mobile pet grooming salons. You also manage the business website, where customers can schedule appointments and buy pet care products.

Each week, you prepare reports to help you analyze customer traffic flow on the Bark N Suds website. View the three charts below. **Chart A** describes the amount and type of products sold by the business over two weeks.

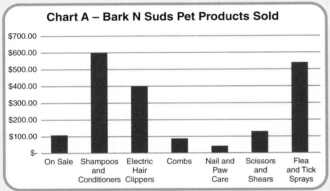

Chart A – Bark N Suds Pet Products Sold

Goodheart-Willcox Publisher

1. Which products sold well, and which products did not?
2. How might you modify the Bark N Suds website to encourage more purchasing?

Chart B shows the pages visited by customers making grooming appointments over the last two weeks.

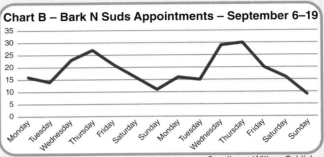

Chart B – Bark N Suds Appointments – September 6–19

Goodheart-Willcox Publisher

3. Which days of the week experienced the most customer traffic on the website?
4. What are some things you might try to encourage more customer traffic during the less busy days?

Chart C shows the internet sales income for the last two weeks.

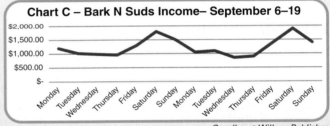

Chart C – Bark N Suds Income– September 6–19

Goodheart-Willcox Publisher

5. What might account for the increased sales on certain days?
6. How can you increase sales on the days when customer traffic is low?

Share your answers to these questions with your instructor and classmates.

Consider This

1. What did looking at the analytics tell you about your business that you might not have noticed or seen in person?
2. How can analytics help you operate a business?

Transportation of Agricultural Products

Agricultural businesses rely on the efficient and effective transportation of their products. Technology has improved transportation systems significantly over the last decade. Global positioning systems (GPS) provide information to manufacturers and shippers about the location of product shipments.

A modern agricultural business can locate almost any product in transit by consulting GPS (**Figure 12.6.11**). Tractor trailers, ships, and airplanes all utilize GPS tracking.

It can be expensive to produce and transport agricultural products. Agricultural businesses are always looking for ways to reduce this cost. As in addressing many other problems, technology is the answer! One of the ways that companies can reduce expenses is by reducing the amount of energy consumed in this process. Farmers and ranchers use solar power to provide electricity in remote areas. Solar energy can be used to generate heat and electricity. Advances in solar energy technology have made this energy reliable and economical.

The transportation industry is developing new ways to use electricity to power cars and machines. Small electrical cars have been around for some time, but the large scale use of electrically powered vehicles to transport agricultural goods is a relatively new concept. Automobile manufacturers are exploring the development of electrical tractor trailers (**Figure 12.6.12**).

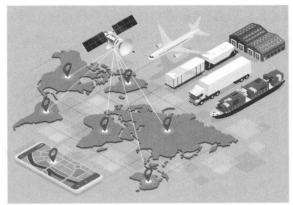

Golden Sikorka/Shutterstock.com

Figure 12.6.11. Global positioning systems (GPS) link transportation systems together to more efficiently track the location of agricultural products in transit.

Andrus Ciprian/Shutterstock.com

Figure 12.6.12. This is an artist's concept of an electrical tractor-trailer that can be used to transport agricultural goods. *What benefits would this have?*

What Did We Learn?

- Modern agricultural businesses face many challenges. Agribusinesses rely on technology for accurate and timely information, up-to-the-minute market prices, and supply chain management.
- Agribusinesses use technology to manage customer relationships, using tools and techniques like analytics, conversion optimization, email, search engine marketing, remarketing, mobile phones, and automation.
- Agribusinesses use technology to collect and analyze data. This data can be used to monitor production and security systems and track inventory.
- Technology can be used to transport agricultural products efficiently. Innovations in this area include GPS, solar energy, and electrical vehicles.

12

Let's Check and See What We Know

Answer the following questions using the information provided in Lesson 12.6.

1. *True or False?* Agricultural businesses need up-to-date information to make the best production decisions.

2. The ____ tracks the current price of agricultural commodities traded in the US.
   A. Chicago Mercantile Exchange
   B. Dow Jones Industrial Average
   C. Agricultural Trading Desk
   D. NASDAQ

3. An agribusiness owner uses a computer program to determine how many customers visit the company's online sales catalog. This is an example of the use of ____.
   A. analytics
   B. conversion optimization
   C. inventory management
   D. stock remarketing

4. An agribusiness's online web store seeks to have a ____ conversion optimization rate to attract more customers to buy products.
   A. low   C. high
   B. medium   D. steady

5. An agribusiness owner wants to send coupons to existing customers to encourage them to shop on the store website. The most efficient way to send the coupons to customers is by:
   A. Sending the coupons to customers through the US mail.
   B. Handing out the coupons to customers shopping in the store.
   C. Sending the coupons to customers by email.
   D. Calling the customers on the phone and telling them where to pick up the coupon.

6. Online businesses use software called ____ to gather basic information about a customer when they visit a website.
   A. site documents   C. URLs
   B. cookies   D. HTTPs

7. The technology that makes it possible to track a shipment anywhere in the world is:
   A. Satellites
   B. Global positioning systems (GPS)
   C. Chicago Mercantile Exchange (CME)
   D. Cookies

8. Supermarket loyalty cards are designed to:
   A. Speed up the checkout process
   B. Determine if foods are fresh when purchasing
   C. Allow entry into the store
   D. Collect basic information about customer purchases

9. More than ____ percent of adults access the internet solely via their phone.
   A. 30   C. 90
   B. 60   D. 120

10. A(n) ____ is a tool that people use to search for information on the internet.
    A. internet service provider
    B. search engine
    C. uniform resource locator
    D. hypertext markup language

Let's Apply What We Know

1. What are some ways that technology has improved the way customers buy goods and services?

2. What are some ways that technology has helped agribusinesses become more efficient?

3. Conduct a web search for local agricultural businesses. Do they have websites designed for customers to buy products and services?

4. As you read about conversion optimization and remarketing, did you agree with its use? Why or why not?

5. Interview an older adult in your family or someone in your community about their experiences shopping for goods and services using technology. Ask them to explain how technology has changed the way they make purchases.

Academic Activities

1. **Social Science.** What do you think about the techniques businesses use to encourage people to shop and buy products online? Are you comfortable with internet-based businesses knowing your shopping habits? How has technology changed the way that people interact while shopping?
2. **Reading.** Visit the school library and find a textbook or other book written before 1990. Compare the amount of text in the book with a text written in 2015 or later. What differences do you notice? Is the reading more accessible or more challenging in the older text? How much reading do you do in an average week? How has technology changed your reading habits?

Communicating about Agriculture

1. **Writing.** Find a website that uses conversion optimization to encourage customers to stay longer on the website. What is your opinion about this technology from the standpoint of the agricultural business? Write a short essay. Include the pros and cons of the agribusiness' use of the technology.
2. **Speaking.** Prepare a three-minute oral presentation on one of the technology tools described in this lesson (remarketing, email, analytics, etc.). Please explain how to use the tool and whether or not you think it is a good idea.

SAE for ALL Opportunities

1. **Foundational SAE.** In your SAE program, what are some ways to market your product or service online? Please note that it isn't always necessary to sell something online, but you could use the internet to provide information about your SAE. Consider developing a website that highlights features of your SAE. Make sure that you have parental permission before beginning.

SAE for ALL Check-In

- How much time have you spent on your SAE this week?
- Have you logged your SAE hours?
- What challenges are you having with your SAE?
- How can your instructor help you?
- Do you have the equipment you need?

Glossary

A

abattoir. An animal processing facility. (9.5)

accident. An unfortunate event that happens unexpectedly and unintentionally, typically resulting in damage or injury. (11.1)

accountability. Answering for one's actions or decisions. (2.2)

active listening. The act of concentrating on the person speaking, understanding their message, and responding appropriately. (2.4)

active rainwater harvesting. Collecting rainwater from runoff to be stored in tanks for future use. (11.5)

additives. Chemicals added to foods to preserve them by inhibiting the growth of microbes. (6.3)

adventitious roots. Roots that grow from the stem of a plant. (8.3)

advertising. One specific action a business takes to promote a product or service. (12.5)

advertising plan. An outline for how a business will use advertising to promote their products and reach new customers. (12.5)

agenda. The order in which items should be presented in a meeting Also known as the *order of business*. (1.6)

agribusiness. A business that provides agricultural goods or services. (1.2)

agricultural education. The teaching of agriculture, natural resources, and land management. A total program consists of experiential learning through classroom and laboratory instruction, Supervised Agricultural Experiences outside the classroom setting, and leadership development through involvement in the FFA. (1.1)

agricultural engineering. A career path focused on innovations in electronics, power, hydraulics, and engineering. (1.2)

agricultural literacy. A basic understanding of agricultural systems, how food is grown, and the effect of agriculture on the economy and environment. (5.2)

agricultural marketing. The processes and services involved in moving an agricultural product from the farm to the consumer. (12.1)

agriculture. The science or practice of farming, including cultivation of the soil for the growing of crops, and the rearing of animals to provide food, fiber, and other products. (1.1)

agronomist. An expert in the science of soil management and crop production. (1.1)

algae. A soil organism that can make its own nutrients through photosynthesis. (7.1)

alloy. Metals consisting of two or more elements. (11.4)

alternating current (AC). Electric current that changes direction at regular intervals. (11.6)

alternator. A generator that produces alternating current. (11.6)

amendment. A change to allowed portions of a motion. (1.6)

amino acids. Structures made up of carbon, hydrogen, oxygen, and nitrogen. (9.7)

amperage. The rate of current flow. (11.6)

analytics. Computer software programs that track web pages and products that customers are viewing and purchasing. (12.6)

angiosperm. A seed-bearing plant that protects its seed in an ovary. (8.2)

angle. A figure formed by two lines extending from the same point. (11.3)

animal science. The study of the care, health, and safety of animals and the products they provide. (1.2)

annual. A plant that completes its life cycle in one growing season or a single year. (8.2)

antibiotics. Drugs that are designed to kill a specific strain of bacteria. (6.2)

apiaries. Areas where large numbers of beehives are placed for active management and commercial breeding. (9.10)

apical meristem. The growing tip of a stem. (10.9)

apiculture. The care and management of honey bees. (9.10)

appraisal. An estimate of the worth of something. (10.8)

aquaculture. The production of aquatic organisms under controlled conditions. (9.10)

aquifer. An underground lake with the water held in beds of rock and sand. (10.3)

Note: The number in parentheses following each definition indicates the lesson in which the term can be found.

area. The amount of space inside a boundary of a flat (two-dimensional) object or shape. (11.3)

artificial intelligence. The ability of computers to "think" and make decisions without human input. (5.3)

asexual propagation. Reproduction of plants from the vegetative parts of a plant. (8.8)

atom. The smallest particle showing the properties of an individual element. (11.6)

automation. Technology that allows businesses to manage tasks without human input. (12.6)

B

bacteria. Single-celled organisms. (7.1)

balanced ration. A ration in which all the nutrients needed by an animal are provided and correctly proportioned. (9.7)

bark. Old phloem tissue that has died and hardened on the outside of a tree, protecting the new phloem tissue. (10.9)

battery. A device made up of chemical cells that produce a stream of free electrons. (11.6)

berry. Fruit that has many seeds in the center instead of just one seed; examples include bananas, persimmons, and pomegranates. (10.9)

bias. An unreasoned outlook or opinion about something. (10.2)

biennial. A plant that requires two growing seasons or two years to complete its life cycle. (8.2)

big data. Large amounts of information that are collected and analyzed to detect patterns or make predictions. (5.3)

bill of materials. A complete list of all materials, fasteners, and hardware needed to complete a project. (11.8)

biodiversity. The amount and variety of life on Earth. (5.2)

biological pest management. The use of living organisms to reduce the population of pests. (8.9)

biotechnology. Scientific modification to the genetic makeup of living cells to produce new functions in those cells. (1.2)

bison. Species of large mammal similar to cattle that are native to North America. (9.10)

blacksmith. A skilled worker who creates metal objects out of iron and steel using tools to heat, bend, and shape the metal. (5.1)

board foot. A unit of measure in timber production that equals one piece of wood, one foot square, and one inch thick. (10.10)

boll weevil. A beetle that feeds on cotton buds and flowers. (3.2)

bolt. An externally threaded fastener designed to be inserted through holes in materials or assembled parts to tighten materials together. (11.4)

boreal forest. Forests where the temperatures remain very low throughout the year, and about half of the precipitation is snow. The soil is typically fragile and low in nutrients. Also called a *taiga*. (10.6)

Bos indicus. Species of cattle that originated in South Asia; are suited to withstand hot climates. (9.2)

Bos taurus. Species of cattle that originated in Europe, Northeast Asia, and parts of Africa; are suited to withstand cool climates. (9.2)

brand. A marketing or business concept that helps people identify a company, product, or individual. (12.5)

breed. Group of animals within the same species that share specific, distinctive characteristics, such as size, color, and shape. (9.3)

breeding animal. Livestock raised for reproduction. (9.4)

British thermal unit (BTU). The amount of heat required to raise the temperature of one pound of water by one degree Fahrenheit. (10.5)

broilers. Chicken bred to produce meat. (9.4)

bud. A stem's growing point that contains undeveloped leaves or flowers. (8.3)

budget. An estimate of the income and expenses for a specific time or project. (11.8)

buffer strip. A narrow swath of plants that serve as a barrier zone between a crop and other fields, streams, and ponds. (10.11)

bushel. A measure of capacity (volume) equal to 64 US pints. (11.3)

by-product. A secondary product made in the manufacturing of something else. (7.4)

C

caloric value. The heat produced by food when burned or metabolized by the body. (6.4)

camelid. Even-toed ruminant mammals with a three-compartment stomach. (9.10)

camshaft. The portion of the valve train that opens the intake and exhaust valves within the engine at the proper times. (11.7)

canning. The process of preserving cooked food in cans or jars through heat and pressure. (6.3)

carbohydrates. The main energy nutrient found in livestock rations; made up of sugars, starches, cellulose, and lignin. (9.7)

carburetor. The portion of the engine that produces the mix of fuel and air to ignite the engine. (11.7)

cardiac muscle. An involuntary muscle which contracts to pump blood and maintain blood pressure throughout the body. (9.6)

career and technical education (CTE). The practice of teaching and learning through hands-on experiences in skill-based career areas. (1.1)

Career Development Event (CDE). An FFA event in which students demonstrate knowledge and skills they have learned in the agriculture classroom in a competitive atmosphere. These events encourage teamwork and individual achievement. (1.4)

career pathway. A small group of career opportunities within a larger career cluster. Careers in a pathway involve similar knowledge and skill sets. (1.2)

career success. The process of continuously demonstrating the qualities, attributes, and skills necessary to succeed in or further prepare for a chosen profession. (1.4)

carnivore. An organism that feeds on animal tissues. (9.1)

carrying capacity. The ability of a habitat to support wildlife. (10.5)

cartilage. Tough and flexible connective tissue that provides support between the bones at joints and forms structures like the nose, ears, and trachea. (9.6)

cartography. The creation of maps using global positioning systems. (10.1)

Center for Veterinary Medicine (CVM). A US government organization that approves safe food additives and manages the Food and Drug Administration's medicated feed programs. They also monitor for animal feed contaminants. (9.7)

cereals. Any grass grown for the edible parts of its grain. (3.3)

chair. The person responsible for conducting a meeting. The term *chairwoman* or *chairman* may be used in place of chair. (1.6)

character. The attributes and traits that make up and distinguish our individual nature. (2.1)

chemical pest management. Pesticides used to control harmful pests. (8.9)

chlorophyll. A pigment found in plant cells that converts light energy into food. (8.5)

chloroplasts. Specialized structures within individual leaf cells where photosynthesis occurs. (8.5)

chromosomes. Long threads of DNA that carry the genetic information of a cell. (5.4)

circulatory system. A bodily system that moves blood and nutrients throughout the body. (9.6)

classroom and laboratory instruction. Learning that takes place with a teacher in a classroom, laboratory, greenhouse, or outdoor setting. (1.3)

clay. The smallest soil particle. (7.1)

clear-cutting. Cutting down all the trees in an area, resulting in ample open space. (10.8)

climate. The measure of the average temperature, precipitation, wind, and humidity of an area over a long period of time. (3.4)

climate change. The average long-term changes to Earth's atmosphere caused by the release of greenhouse gases into the atmosphere. (3.4)

collagen. Structural protein found in some animal tissues, like bone and skin. (9.5)

combustion. Burning fuel in the presence of oxygen to produce energy. (11.7)

commodity. A good or product that is interchangeable with other goods of the same type and meets specific minimum standards. (11.3)

communication. The process of exchanging thoughts, ideas, messages, or information between two or more people. (2.4)

Community Supported Agriculture (CSA). A partnership between farmers and consumers that allows consumers to buy small portions of a farm's harvest in advance. (12.1)

companion animal. A domesticated animal kept for pleasure or companionship. (9.11)

complete fertilizer. A substance that contains the three primary macronutrients: nitrogen, phosphorus, and potassium. (7.3)

complete flower. A flower that has all four major flower parts: pistil, stamen, petals, and sepals. (8.6)

complex carbohydrates. Cellulose and lignin; difficult substances to digest. Also called *fiber*. (9.7)

compost. Decomposed organic matter that can be added to the soil to help plants grow. (7.4)

compound. Two or more elements joined together at the atomic level. (11.6)

computer-aided design (CAD). The use of computer software to draw or model structures or parts of structures. (11.3)

concentrates. Livestock feeds that contain less than 18 percent crude fiber. (9.7)

concrete. A durable material known for its strength and its resistance to fire and decay. It is a mixture of cement, gravel, sand, and water. (11.4)

condensation. The process of water molecules changing from gas to liquid. (10.3)

conductors. Materials that allow free electron movement. (11.6)

conformation. Quality muscle shape and structure. (9.4)

conservation. The careful use of resources to prevent unnecessary waste. (10.4)

conservation cover. A mixture of native plants and wildflowers that provide habitat for wildlife, food for deer and other herbivores, and a place for beneficial insects to gather pollen. (10.11)

conservation plans. A plan to protect a species or habitat. (10.4)

conservation tillage. A soil-tilling method which improves the quality and health of soil by reducing tillage operations. (5.2)

construction. The creation or building of a structure or project. (11.8)

constructive. Helpful to your ability to get better. (2.4)

consumer. A person who purchases goods and services for personal use. (3.1)

contagious. Easily transmitted from one organism to another. (9.8)

conversion optimization. Activities companies engage in to get people to stay longer on a website to achieve higher engagement or purchase. (12.6)

cookie. A collection of information about a customer that is stored in their web browser. (12.6)

cover crops. Crops used between regular crop rotations to prevent soil erosion and maintain soil moisture. (10.11)

crankshaft. An engine part that converts the up and down motion of the pistons into circular or rotating motion. (11.7)

creativity. The capacity to use our imagination to generate or recognize new ideas and possibilities or create something original that is of value. (2.5)

creed. A set of values and beliefs that guide our thoughts and behaviors. (1.4)

critical thinking. Our ability to think deeply in a clear and informed way. (2.5)

crop rotation. The practice of planting crops on an alternating basis each year. (10.11)

crosscut. A type of woodcut made perpendicular to the grain. (11.8)

cross-contamination. The unintentional transfer of harmful bacteria to food from other foods, cutting boards, and utensils if not handled properly. (6.2)

cross pollination. When the pollen of one plant travels to and pollinates the flower of another plant. (8.6)

cull. To remove an animal from the herd. (9.4)

cultivars. A group of plants from the same species grown for their desirable traits that keep these traits when reproduced. (8.2)

cultural pest management. Growing and environmental management techniques used to control pests. (8.9)

curiosity. The strong desire to know or learn something new. (2.5)

current. The movement of electrons from negatively charged locations (more electrons than protons) to positively charged locations (more protons). (11.6)

custodial management. Minimizing external factors such as human interaction or urban development to prevent wildlife species from declining. (10.5)

cutting list. A list of project materials that includes the exact dimensions required. (11.8)

cuttings. Detached parts of a parent plant such as the leaves, stems, or roots that grow into new plants. (8.8)

cycle. The process of intake, compression, power, and exhaust. (11.7)

D

dam. Female parent. (9.4)

data. Pieces of information collected through scientific observation. (2.5)

decibel. A unit for expressing the relative intensity of sounds. (11.1)

deciduous plants. Plants that are leafless during a portion of the year. (8.2)

decomposition. The process of dead plants and animals breaking down into organic matter such as water, sugars, and minerals. (10.3)

dehydration. The process of reducing the moisture in food to low levels to increase shelf life. (6.3)

demand. Consumers' desire for a particular product. (4.3)

demeanor. The way you present your emotions in your actions. (9.9)

dendrology. The scientific study of trees. (10.1)

dicots. Plants characterized by two embryonic leaves in their seedling stage; leaf vines that resemble veins or lines crossing; and flower parts in multiples of four or five. (8.2)

digestion. The process of breaking down feed into simple substances that can be absorbed by the body for energy and nutrient value. (9.6)

direct current (DC). Electrical current that flows in one direction. (11.6)

direct seeding. Planting seeds directly into the soil outdoors. (8.7)

Discovery Degree. An award that recognizes students enrolled in seventh and eighth grade agricultural science classes who are making steps toward productive engagement in FFA. (1.4)

disease. A condition that negatively affects structure or function of a living creature with specific signs or symptoms. (9.8)

division. Separating parts of a parent plant into sections that will grow into new plants. (8.8)

DNA. Short for deoxyribonucleic acid. A material that contains all of an organism's genetic information. (5.4)

domesticate. To tame and breed plants or animals. (1.1)

dormancy. A period of slow or inactive plant growth. (8.2)

drake. A male duck. (9.3)

drought. A long period with little or no rainfall. (5.1)

drupe. Fleshy fruit with one seed in its center; examples include peaches, plums, and cherries. (10.9)

Dust Bowl. Massive dust storms over the Midwest in the 1930s caused by eroding soil. (3.2)

E

E. coli. A bacterium commonly found in the intestines of humans and other animals, some strains of which can cause severe food poisoning. (6.2)

ecological management. Career path dedicated to conserving and protecting natural ecosystems. (10.1)

economy. The management and use of money, resources, and labor. (4.1)

ecosystem. A community of living organisms that have a beneficial relationship. (10.3)

edible. Able to be eaten. (9.5)

effective communication. Ensuring clear, concise, and consistent information is shared, received, and understood by all involved. (2.4)

electricity. The movement of electrons between charged particles. (11.6)

electromagnetic induction. The production of electromotive force across an electrical conductor in a changing magnetic field. (11.6)

electromotive force. Pressure (electrical potential) that moves electrons through a conductor. Also called *voltage*. (11.6)

electron. Negatively charged particles that orbit around the nucleus of an atom. (11.6)

element. The building blocks that make up all matter. They cannot be broken down into other materials. (11.6)

emblem. A representational image or series of images that serves as a distinctive symbol of a nation, organization, or family. (1.4)

embryo. An immature plant. (8.3)

emissions. The gases and particles which are pushed into the air by production processes and transportation. (3.4)

empathy. The ability to understand and share the feelings of another person. (2.1)

endosperm. Specialized tissue in a seed that stores food to aid in growth and development. (8.3)

engine block. A single piece of machined cast iron or aluminum which contains all of the internal components of an engine within its housing. (11.7)

engineer. A scientist who designs and builds machines and structures. (5.1)

environmental management. Career path that brings nature and people into a relationship where both can benefit. (10.1)

Environmental Protection Agency (EPA). The United States governmental agency responsible for creating standards and laws promoting the health of individuals and the environment. (3.4)

environmental science. The study of human activities and their related impacts on the environment. (1.2)

erosion. Physical wearing away of Earth's surface (7.2)

estuaries. Bodies of water where freshwater from inland sources meets saltwater from the ocean. (10.4)

ethanol. A corn-based fuel that is often added to gasoline. (3.1)

evaporation. The process of water molecules changing from liquid to gas. (10.3)

evergreen plants. Plants that stay green and keep their leaves year-round. (8.2)

experience. Practical contact with and observation of facts or events. (9.9)

experiential learning. A learning style that provides hands-on opportunities to learn from and reflect on experiences. (1.1)

extrinsic reward. Outside reward or motivator related to someone else, or a physical, tangible reward. (2.3)

F

fact. Objective information that is widely known to be accurate. (10.2)

fairness. The ability to act in a way that is free of judgment, bias, or favoritism. (2.1)

farmers markets. Public places for farmers to sell their fresh products directly to the consumer. (12.1)

fastener. A device that is used to connect and secure one object to another. (11.4)

feed additives. Materials incorporated into animal food to improve feed efficiency, promote faster growth, and treat diseases and parasites. (9.7)

feedback. Any information from others about your performance. (2.4)

feed conversion. The ability to efficiently convert feed rations into a desired output, such as meat or milk. (9.1)

feed grain. Any grain used as the main energy ingredient in livestock feed. (3.3)

fermentation. The process of converting the carbohydrates in foods to alcohol or organic acids using microbes like yeast and bacteria. (6.3)

ferrous. Adjective describing metal that has iron as the primary alloying element and is magnetic. (11.4)

fertilized egg. Eggs that are fertilized by roosters, and can produce chicks. (9.5)

fertilizer. A substance that is spread on the ground or mixed in the soil to help plants grow by providing nutrients required for plant growth and health. (7.3)

fiber. 1. A thread or filament used to create woven or composite material. (3.1) **2.** Cellulose and lignin; difficult substance to digest. Also called *complex carbohydrates*. (9.7)

fiber crops. Field crops grown for their fibers which are used to make paper, cloth, or rope. (3.3)

fibrous roots. Very shallow root system with many roots that spread out throughout the soil. (8.3)

finish. The final step in the construction process; gives the product an attractive appearance while providing any protective measures required to preserve the project for the future. (11.8)

finishing. Feeding an animal an energy-dense diet to promote rapid growth and add meat and fat to their frame in preparation for slaughter. (9.2)

first aid. The attention given to a person suffering from an illness or injury to preserve life, prevent the condition from worsening, or to promote recovery. (11.1)

first impression. The thoughts and opinions people have about each other after a first meeting. (2.3)

flight zone. Areas of personal buffer space that herd animals need to feel calm and safe. (9.9)

floriculture. The study, cultivation, and marketing of flowers and ornamental plants. (8.1)

flowers. The reproductive part of a plant. (8.3)

flywheel. A heavy, rotating wheel used to increase momentum and provide a reserve of power to the engine during interruptions of power. (11.7)

fodder. Something fed to domestic animals, especially coarse food for cattle, horses, or sheep. (9.1)

followership. The ability to follow a leader effectively and add value. (2.2)

Food and Drug Administration (FDA). A US government organization responsible for protecting and promoting public health through the control and supervision of food safety, vaccines, cosmetics, tobacco products, and pharmaceuticals. (6.1)

foodborne illness. An illness that develops from consuming a food contaminated with a disease-causing agent. (6.2)

food desert. An area, typically inside a city or town, where the citizens have limited access to affordable and nutritious food. (5.2)

food deterioration. The process of a food product breaking down to a point where it is no longer healthy to eat. (6.3)

food insecurity. Condition experienced by a household that is uncertain of having or unable to acquire enough food to meet the needs of every family member. (6.1)

food loss. Any edible food that goes uneaten at any stage in the supply chain. (6.1)

food nutrition label. Information listed on food products to describe a food in terms of caloric and nutritional value for consumers. (6.4)

food preservation. All food processing methods used to slow down or stop the rate of spoilage. (6.3)

food safety. The conditions and practices used to preserve the quality of food to prevent contamination and foodborne illnesses. (6.1)

food science. The study of the physical, biological, and chemical makeup of food, the causes of food deterioration, and the nature of food processing. (1.2)

food security. The condition of household having access to enough food for an active and healthy lifestyle at all times. (6.1)

food system. All processes and infrastructure involved in the production, processing, transportation, and sale of food before it reaches your plate. (6.1)

food waste. All available food considered to be of good quality that is intended for human consumption but instead is discarded by supermarkets or consumers. (6.1)

forest management. A plan for managing a forest to produce timber while also managing wildlife and recreational interests. (10.8)

forestry. The science of managing, conserving, and creating forests. (10.1)

Foundational SAE. A program that allows beginning agriculture students to familiarize themselves with agriculture careers and concepts before selecting an Immersion SAE program. (1.5)

four-stroke engine. An engine in which four strokes of the piston are needed to complete the engine cycle. (11.7)

fowl. A bird of any kind. (9.2)

free electrons. Electrons that move easily between atoms. (11.6)

freeze drying. The process of removing moisture from a food product at very low or freezing temperatures. (6.3)

freeze injury. A condition in which plants are damaged when low temperatures freeze the water in their plant tissue. (8.4)

fruit. A protective tissue around the seed which also aids in seed dispersal. (8.3)

fulcrum. The support that a lever rests on. (11.2)

fungi. Small spore-bearing organisms that form parasitic or symbiotic relationships with plants or animals. (8.9)

G

game species. Animal species that may be harvested through hunting, trapping, or fishing. (10.12)

gavel. A hammer-shaped mallet that the chair uses to direct the membership. (1.6)

generally recognized as safe (GRAS). Individual substances or chemicals added to food that are considered safe and are not subject to regulation. (9.7)

generator. A device that moves electrons from one terminal (charged point) to another through a process called electromagnetic induction. (11.6)

genes. Segments of chromosomes that contain specific information about an organism's traits. (5.4)

genetic engineering. The process of cutting apart DNA strands and removing or inserting new genes. (5.4)

genetically modified organism (GMO). An organism that has had its DNA modified using genetic engineering techniques. (5.4)

genetics. The study of characteristics that plants and animals inherit from a parent plant or animal. (5.1)

geology. The study of the rocks that make up Earth. (10.1)

germinate. To sprout from dormancy. (8.7)

grading. A postharvest practice that consists of classifying a commodity into categories separated by overall quality. (12.2)

grafting. The process of merging two different plants or plant parts together so that they will connect and continue to grow as one, forming a new plant. (8.8)

grains. Cereals suitable as food for people. (3.3)

gratitude. The quality of being thankful or appreciative. (2.1)

greenhouse. A structure made of plastic or glass that provides the ideal growing environment for plants to thrive. (8.1)

greenhouse effect. The trapping of the sun's solar energy in Earth's atmosphere resulting in a warming atmosphere. (3.4)

grit. A term used to describe multiple areas of character including persistence, willingness to fail, and the ability to have the strength and courage to move forward and try again. (2.1)

gross domestic product (GDP). The value of all goods and services that are created in a country. (4.1)

groundwater. Water present underneath Earth's surface layer of soil. (10.3)

gymnosperm. A plant that carries its seed in a cone. (8.2)

H

habitat restoration. Putting time, money, and resources into cleaning and rehabilitating a natural area. (10.4)

halter. A rope or strap for leading or tying an animal. (9.9)

hand. The unit of measure used to find the size of horses; equal to 4 inches. (9.3)

hands-on learning. Instruction with activities that apply your knowledge, such as interacting with materials, objects, equipment, machinery, and animals. (1.3)

hand tools. Tools used for manual activities such as cutting, chopping, sawing, or hammering where the power is provided by a person. (11.2)

hardiness. Ability of a plant to withstand cold temperatures typical during winter months. (8.2)

hardscapes. Constructed areas within a landscape such as walkways, patios, arbors, and retaining walls. (8.1)

hardwood. The wood of deciduous angiosperm trees. (10.10)

heartwood. Tree stem tissue that is solid and resistant to decay and insect damage. It provides strength and stability to a growing tree. (10.9)

heliciculture. Snail farming. (9.10)

herbaceous plants. Plants with stems that are soft and die back to the ground each year. (8.2)

herbivore. An organism that feeds on plants. (9.1)

honesty. The quality of speaking and acting truthfully. (2.1)

horticulture. The production, processing, and sale of plants for food, comfort, and beauty. (8.1)

horticulturist. A person who studies and specializes in raising horticultural crops to full maturity. (8.1)

host. An organism that parasites feed from. (9.8)

humidity. The amount of water vapor in the air. (8.4)

humus. Organic matter that is formed by the decomposition of plant material by microorganisms in the soil. (7.1)

hydrology. The study of the movement of water on Earth's surface. (10.1)

hydroponics. The growing of plants with their roots in a nutrient solution instead of soil. (8.1)

hypoallergenic. Unlikely to cause an allergic reaction. (9.10)

hypothesis. An informed guess or explanation based on what you already know about a topic. (2.5)

I

imperfect flower. A flower that lacks either the female (pistil) or male (stamen) part of the flower. (8.6)

incidental motion. A motion related to parliamentary procedure rules and procedures and not with the business at hand. (1.6)

inclined plane. A sloped surface used for lifting heavy objects. (11.2)

incomplete fertilizer. A fertilizer that is missing one or more of the primary macronutrients. (7.3)

incomplete flower. A flower that lacks one or more of the major parts. (8.6)

indicator species. Organisms whose population increases or decreases serve as a metric to tell scientists about the environmental conditions in which they live. (10.4)

indirect seeding. Process in which seeds are planted in a place that is different from where the plants will grow to full maturity. (8.7)

infectious. Able to cause harm to or infect another organism. (9.8)

infrared radiation. A type of light that is invisible to human eyes but can be felt as heat radiating from objects. (5.2)

ingredients. The individual foods or substances combined to create a food product or recipe. (6.4)

innovation. The creation and adoption of ideas that make improvements to existing products and processes. (5.1)

inorganic fertilizer. A nutrient compound that is derived from materials other than plants and animals, such as mineral salts. (7.4)

insect. A small anthropoid animal that has six legs, three distinct body parts, and generally one or two pairs of wings. (8.9)

insulator. Material that resists the movement of electrons. (11.6)

integrated pest management (IPM). The use of a combination of biological, cultural, physical, and chemical tools to reduce the damage caused by pests. (5.2)

integrator. A business involved in two or more stages of production. (12.1)

interiorscaping. The design, installation, and maintenance of foliage plants inside buildings. (8.1)

internet of things (IoT). Machines that communicate with each other with minimal human interaction. (5.3)

internode. The space between each node on a stem. (8.3)

intrinsic reward. A reward in the form of pride in your work, satisfaction in your abilities, or the pleasure of a sense of responsibility; additionally, basic needs, such as food, water, clothing, and shelter. (2.3)

inventory. Goods or supplies that a business maintains in stock. (10.8)

irrigation. The process of adding water to soil. (8.4)

issue. A problem that typically involves a dispute between two or more parties. (10.2)

K

kerf. The blade width of the saw used to cut materials. (11.8)

kindness. The sincere act of being gentle, caring, and helpful toward others. (2.1)

L

landscapers. Workers that design plans for plant layout, install plant material, and maintain the plants in the landscape environment. (8.1)

lateral meristem. Specialized growing cells that allow tree branches to grow in length and diameter. (10.9)

law of supply and demand. The principle that as the consumer desire for a product goes up, the amount of it available for sale goes down. (4.3)

layers. Chickens bred to produce eggs. (9.4)

layout. The process of measuring and marking lines for cut. (11.3)

leader. A person who guides or directs a group or organization. (2.2)

leadership. The personal qualities related to being able to guide or direct others. (2.2)

Leadership Development Event (LDE). An FFA event in which students use critical-thinking skills, effective decision-making skills, and communication skills while appreciating the benefit of fair competition and individual achievement. (1.4)

leaf cuttings. Reproduction of plants from the leaf of a parent plant. (8.8)

leaves. Part of a plant attached to the stem that captures sunlight for the plant to carry out photosynthesis. (8.3)

legumes. A type of plant with nodules on the roots that contain specialized, nitrogen-producing bacteria. (9.7)

length. A measure of distance. (11.3)

lever. A bar that rests on a support called a fulcrum and is used to lift heavy materials. (11.2)

ligament. Tissue that connects bones to other bones. (9.6)

light quality. The intensity or brightness of light. (8.4)

light quantity. The amount and duration of light emitted by a light source. (8.4)

lipids. Fats and oils made up of carbon, hydrogen, and oxygen. (9.7)

livestock. Domesticated animals bred and raised for the purposes of producing food or products. (1.1)

livestock confinement housing. Housing where large populations of animals are kept in small areas. (9.9)

load. Any device that uses electricity to produce work. (11.6)

loam. Soil made with a balance of the three main components: sand, silt, and clay. (7.1)

log deck. A central location in a timber cutting operation where log skidders bring cut timber for trimming and stacking. (10.8)

long-term goal. A task you want to accomplish more than a year into the future. (2.2)

lumber. A wood product that has been cut to specific size for building and construction purposes. (11.4)

M

macronutrients. Nutrients that plants need in the largest quantities: nitrogen, phosphorus, and potassium. (7.3)

main motion. A motion that brings a new topic or idea before the membership. (1.6)

malnutrition. A condition in which a person does not receive the necessary nutrients in their food to maintain a healthy and well-functioning body. (4.2)

manipulative management. Actions taken to increase or decrease the population of a wildlife species such as changing the food supply, changing the habitat, or managing predators. (10.5)

manure. A combination of animal feces and bedding material such as straw or wood shavings. (7.4)

marbled. Having a desirable ratio of meat to fat; a popular quality in meat, but particularly in beef. (9.3)

margin. The edge of a leaf. (10.9)

marine aquaculture. The growth and harvesting of saltwater species. (9.10)

market animal. Livestock raised for food consumption. (9.4)

market price. The amount a product or service can be bought or sold for. (12.3)

market research. The process of gathering information about consumers' needs and preferences. (12.4)

marketing. The activities involved in transferring goods and services from producer to consumer. (12.1)

marketing plan. A document that details the specific types of marketing activities a company conducts and contains timetables for rolling out various marketing initiatives. (12.4).

marketing strategy. A business's overall plan for reaching prospective consumers and turning them into customers of their products or services. (12.4)

marrow. The soft center of the bone. (9.6)

mass. The measure of the amount of matter an object contains. (11.3)

material safety data sheets (MSDS). Reference materials that contain information about the properties of hazardous substances. (11.1)

maternal. Known for mothering; one of the two main qualities, along with muscling, sought in breeds of swine. (9.3)

measure. Determining the size, amount, or degree of something by using an instrument or device marked in standard units. (11.3)

mechanical advantage. The ratio of the force that performs useful work to the force applied. (11.2)

mechanical pest management. Using tools or equipment to control pests. (8.9)

metal. A solid material composed of one or more chemical elements that have a crystalline structure with high thermal and electrical conductivity. (11.4)

metamorphosis. The life cycle process of growing from eggs into adults with significant body structure changes. (10.12)

meteorology. The study of Earth's weather patterns. (10.1)

metric system. A system of measurement that uses the meter, liter, and gram as the basic units of length, capacity, and weight. (11.3)

micronutrients. Nutrients needed in small or limited amounts that are essential for plant growth and development. (7.3)

minerals. Inorganic particles that make up the largest component of soil. (7.1)

minutes. A complete, written record of the events of a meeting. (1.6)

mission statement. An explanation of what an organization strives to provide for each of its members; a summary of its aims and values. (1.4)

mohair. A type of long, silky goat hair used to create soft wool products. (9.3)

mole. Small mammal that burrows through the soil, eating roots and uprooting plants as it digs. (7.1)

monocots. Plants characterized by one embryonic leaf in their seedling stage; parallel veined leaves; and flower parts arranged in multiples of three. (8.2)

monogastric. Having a single, nonruminant stomach for digestion. (9.6)

motion. A formal proposal by a member, which if agreed upon by the membership, will result in action. (1.6)

motivation. The influence that makes us want to do things and finish them, even if those things are difficult; a key to staying focused. (2.3)

motto. A short sentence or phrase that captures the beliefs or ideals of an organization. (1.4)

muscular system. A body system composed of cells or fibers that contract to produce movement. (9.6)

mutton. Sheep meat. (9.3)

N

nail. Slender and straight fasteners made from metal and composed of three parts: the head, shank, and point. (11.4)

National FFA Organization (FFA). A national student organization that promotes the relevancy of agricultural education lessons through realistic, hands-on opportunities and competitions outside of the classroom. (1.3)

natural resources. Renewable or nonrenewable materials originally provided to us by nature. (1.1)

nematode. Multicellular organism that resembles a worm and feeds on bacteria. (7.1)

nervous system. A body system responsible for transferring information throughout the body. (9.6)

neutron. Dense, neutrally charged particles found in an atom's center or nucleus. (11.6)

node. An area of the stem where leaves develop. (8.3)

nomad. A person with no fixed residence who moves from place to place, usually seasonally and within a well-defined territory. (9.1)

nonrenewable. Taking a long time to replenish after being used. (10.5)

nonverbal communication. The exchange of information without the use of words. (2.2)

no-till. A method by which crops are planted in fields without the use of tillage equipment. (10.11)

nuclear energy. A nonrenewable energy source that uses radioactive elements such as uranium to generate power. (10.5)

nucleus. An atom's center. (11.6)

nursery. A place where shrubs, ornamental trees, and plants are grown for transplanting into landscape areas. (8.1)

nut. Fruit with an edible seed enclosed in a hard shell coating; examples include oaks, chestnuts, and hazelnuts. (10.9)

nut. A device used on the opposite side of a bolt to tighten materials together. (11.4)

nutrient cycle. A natural process of converting living organisms into organic matter and inorganic matter. (10.3)

nutrients. Substances that provide nourishment for growth and the maintenance of life. (9.7)

O

objectives. The intermediate steps for achieving goals. (10.4)

observation. A process used to gain information through detailed examination and experimentation. (2.5)

offal. Animal parts left over from the meat production process. (9.5)

oilseeds. Grains that are valuable for the oil content they produce. (3.3)

olericulture. The science, cultivation, processing, storage, and marketing of herbs and vegetables. (8.1)

omnivores. Animals that eat plants and other animals. (10.3)

opinions. Thoughts based upon emotions and feelings. (10.2)

optimism. The act of being hopeful and confident about the future. (2.1)

order of business. The order in which items should be presented in a meeting. Also known as an *agenda*. (1.6)

organic. A labeling term that indicates that the food or other agricultural product has been produced through specific, verified methods. (12.2)

organic farming. The production of food that integrates cultural, biological, and mechanical practices that recycle resources, and promote ecological balance while conserving biotechnology. (5.2)

organic fertilizer. Naturally occurring nutrient material that originates from plants or animals. (7.4)

oriented strand board (OSB). A particle board product with the wood particles oriented to increase strength and durability. (11.4)

ornamental horticulture. The production of plants for their beauty. (8.1)

ornithology. The scientific study of birds. (10.12)

Ownership SAE. A program in which students create, own, and operate a business that provides goods and/or services to the marketplace. Also called an *Entrepreneurship SAE*. (1.5)

P

palatable. The quality of tasting good to an animal. (9.7)

parallel circuit. A circuit with more than one path for the electric current to flow, with each path having the ability to operate loads independently. (11.6)

parasites. Organisms that live in, with, or on another and deprive it of nutrients. (9.8)

parliamentary procedure. A set of rules and regulations for properly conducting a meeting. (1.6)

passerine. A species of bird that can perch. (10.12)

passive rainwater harvesting. Using earthworks or other structures to divert rainwater to needed locations. (11.5)

pasteurization. Heat sterilization. (9.5)

pathogen. Any microorganism or virus that can cause a disease. (8.9)

Percent Daily Value. Value located on the Nutrition Facts label of a food product as a guide to the nutrients in one serving of food. (6.4)

perennial. Plants that have an unspecified life span, meaning they can live for many years. (8.2)

perfect flower. A flower that contains both male and female flower parts. (8.6)

performance records. Production data collected on livestock animals; used to make decisions about herd management. (9.4)

perishable. The quality of spoiling or deteriorating if not stored properly and moved to market in a timely manner; used to describe food. (4.1)

personal growth. The result of continual effort to improve yourself intellectually, morally, and physically. (1.4)

personal leadership plan. A strategy to accomplish your goals. (2.2)

personal protective equipment (PPE). Equipment worn to reduce exposure to hazards. (9.9)

pest. A living organism that can cause loss or damage to a plant. (8.9)

pesticide. Any chemical used to control pests. (8.9)

petals. Brightly colored modified leaves that surround the male and female organs of a plant and serve to attract pollinators. (8.6)

phloem. Tissue in a plant stem that transports the food made in the leaves to the rest of the plant. (8.3)

photoperiodism. Plants' growth responses based on the number of hours of light they receive each day. (8.4)

photosynthesis. A series of chemical reactions in which plants take carbon dioxide from the atmosphere, add water, and use the energy from sunlight to produce sugar. (8.5)

phototropism. The bending of a plant toward or away from a light source. (8.4)

physiology. The branch of biology that deals with the normal functions of living organisms and their parts. (9.6)

pioneer species. The first species of plants and animals to inhabit an area. (10.3)

pistil. Female organs of a flower. (8.6)

piston assembly. The part of the engine that includes that piston, piston rings, piston pin, and connecting rod. (11.7)

pitch. A thicker version of tar. (10.7)

Placement SAE. A program that gives students the chance to get knowledge and understanding in a chosen agricultural field as paid employees or unpaid volunteers. Also called an *Internship SAE*. (1.5)

plantations. Large acreage properties on which crops are grown by resident labor. (3.2)

plant diseases. Irregular conditions in plants that affect their appearance, growth, and fruit or flower production. (8.9)

plant propagation. Process of producing a new plant. (8.7)

plant science. The study of plants and how they affect the world. (1.2)

plastic. A synthetic or semisynthetic building material. (11.4)

plumbing. The system of pipes, fittings, and other equipment required to supply a building, structure, or area with the water supply and drainage it needs. (11.5)

plywood. A versatile construction item made of peeled, flattened sections of logs that are then glued together and flattened into sheets. (10.7)

pneumatic tools. Tools powered by compressed air turning a shaft. (11.2)

polled. Without horns. (9.3)

pollen. Male sex cells of a plant. (8.6)

pollination. Reproductive process that takes place when pollen grains from the anther transfer to the sticky surface of the stigma. (8.6)

pome. Fruit with the seeds grouped in the core, rather than spread through; examples include apples and pears. (10.9)

pomology. The cultivation, processing, storing, and marketing of fruits and nuts. (8.1)

potable. Water suitable for human consumption. (11.5)

power tools. Tools used to perform many of the same operations as hand tools, but with power supplied by an electric motor or engine. (11.2)

precision agriculture. The method of managing farmland with the assistance of computers or satellite information. (3.2)

preening. The action a bird takes to remove dirt from its feathers. It allows a waxy substance secreted from a gland near the bird's tail to be applied to the feathers. (10.12)

premier leadership. Leadership using skills that are developed and improved for one's entire life, promoting self-awareness, focus, and accountability in all areas of life. (1.4)

prescribed fire. A small fire in a monitored area that burns up dead leaves and matter on the forest floor. (10.8)

prevention. Stopping something before it happens. (9.8)

price point. A point on a scale of possible prices at which something might be sold. (12.3)

privileged motion. A motion that pertains to the rights and needs of the organization. (1.6)

problem-solving. The process used when searching for a solution to a problem. (2.5)

process. A series of actions or steps followed when seeking answers to a problem. (2.5)

processing. Changing raw agricultural products into more useable forms. (4.1)

processor. Business or factory that changes raw products into a different (perhaps more desirable) form, such as raw potatoes into potato chips or apples into applesauce. (12.1)

producers. Workers such as farmers or ranchers that produce raw agriculture products. (3.1)

professionalism. How we conduct ourselves in a professional or work setting; goes beyond what we wear to include our behavior and attitudes. (2.3)

promotion. The entire set of activities that inform people about a product or service. (12.5)

propaganda. Communication that attempts to use emotions and feelings to lead you to think or feel a certain way about something. (10.2)

proteins. Compounds made up of amino acids that serve as the structural components of body tissues like muscle, hair, and ligaments. (9.7)

proton. Dense, positively charged particles found in an atom's center or nucleus. (11.6)

protozoa. Single-celled organisms that feed on bacteria. (7.1)

pseudo-ruminant. A classification of animals that are similar to ruminants, in that they have a multi-part stomach and chew their cud, but do not have a rumen. (9.10)

pulley. A wheel that is carries a rope, chain, belt, or other flexible cord on its rim; used to multiply force when lifting items. (11.2)

pulpwood. A product from small trees that is used to make wood chips in the manufacture of paper and other products. (10.7)

punctuality. The characteristic of completing a task or meeting an obligation ahead of a previously designated time. (2.3)

Pure Food and Drug Act. Legislation that prevents the manufacture, sale, or transportation of misbranded or poisonous foods, drugs, or medicines. (3.2)

Q

quadruped. Organisms that can use arms and legs, or front and rear legs (four total), for moving from place to place. (10.12)

quarantine. To isolate or separate sick animals from others as a means of protecting the larger group. (9.8)

quorum. The number of members that must be present in a meeting for business to occur. (1.6)

R

radio frequency identification (RFID). A radio transponder used to send messages and data to a computer. (5.3)

radio telemetry. The use of a radio receiver and antennae to communicate with a radio frequency identification (RFID) device and track specific animals without interfering with their movements in the wild. (10.12)

rainforest. Any forest that receives more than 100 inches of rainfall each year. (10.6)

ration. The amount of feed given to livestock over a twenty-four-hour period to meet its needs. (9.7)

reaction time. The time between a stimulus (the cause of a reaction) and your body's response. (11.1)

reclaimed water. Water that can be used more than one time before it passes back into the natural water cycle. (11.5)

recordkeeping. The written history of one's activities compiled by collecting and entering data. (1.5)

regenerative agriculture. A type of conservation tillage that is combined with other processes to increase biodiversity, reduce soil erosion, and improve the water cycle. (5.2)

regurgitate. To force food back into the mouth from the rumen; part of ruminant's digestive process. (9.6)

remarketing. Technique of sending advertisements to customers based on their previous browsing history. (12.6)

renewable. The ability to be quickly replenished after being used; used to refer to energy sources. (10.5)

Research SAE. A program in which students ask a question and work through the scientific method to learn new information or gather information that supports existing research. (1.5)

reservoir. A lake where water is stored for human use. (11.5)

resistance. Opposition to electrical flow. (11.6)

respect. The positive way we treat and give attention to others, especially when considering their feelings, rights, wishes, and traditions. (2.1)

respiration. The process of sugars made in photosynthesis combining with oxygen to produce energy in a form that can be used by plants. (8.5)

respiratory system. Anatomical system that manages oxygen intake and absorption. (9.6)

responsibility. Being reliable or dependable for something in one's control or management. (2.2)

retail. Selling directly to consumers through grocery stores and restaurants. (4.1)

retail markets. Farmers selling directly to stores that in turn sell that product to the public. (12.1)

rip cut. A type of woodcut made along the grain. (11.8)

robotics. The design and construction of machines that do repetitive work for humans. (5.1)

role model. A person looked to as an example for their integrity and good behavior. (2.1)

root. A plant structure that anchors the plant in the ground and takes up water and nutrients. (8.3)

roughages. Livestock feeds that contain more than 18 percent crude fiber. (9.7)

ruminant. Even-toed mammal that chews the cud regurgitated from the rumen, such as a cow. (9.6)

Rural Electrification Administration (REA). A US government organization tasked with building electrical power plants and building transmission lines to rural areas that previously had no access to electricity. (5.2)

S

salmonella. A foodborne bacterial pathogen sometimes found in the intestines of chickens. (6.2)

sand. The largest soil particle. (7.1)

sanitary drainage system. A system that carries wastewater from plumbing fixtures and appliances to a sanitary sewer. (11.5)

sapwood. The living part of a tree stem, comprised of newly-grown xylem and phloem. (10.9)

sawtimber. A construction product from large trees used to make dimension lumber for constructing buildings. (10.7)

scale drawing. A sketch, drawing, or model completed in a specified proportion to the final product. (11.8)

scarification. Process of soaking or scratching the hard seed coat, allowing the seed to sprout. (8.7)

School-Based Enterprise SAE. A program in which students lead business enterprises that take place on school campus using school facilities and provide goods or services. (1.5)

scientific method. A step-by-step method of thinking about, experimenting with, and making sense of complex questions. (2.5)

scientific name. A Latin, two-part name consisting of the genus and species. (8.2)

screw. A slender, straight fastener with a raised helical thread running around the shank and a slotted head for turning to grip materials more tightly. (11.2)

search engine. A computer software program that is used to find information on the internet. (12.6)

second. When a second member expresses that they also would like to discuss a motion with the membership. (1.6)

seed. The reproductive unit of plants. (8.3)

seed coat. Protective covering for the developing seed. (8.3)

seedlings. Plants grown from seeds. (8.7)

selection. Breeding with the hope of maintaining desired characteristics. (9.3)

selection cutting. Picking and choosing specific trees to remove from the forestland. (10.8)

self-discipline. The ability to stay focused on a task or goal without being easily side-tracked. (2.3)

sepals. The green leaflike structures beneath the petals that serve as a protective covering of the flower before it opens. (8.6)

separation. When vegetative parts are completely removed from the parent plant and replanted to grow on their own. (8.8)

septic tank. An underground tank made of concrete, fiberglass, or plastic through which domestic wastewater flows for basic treatment. (11.5)

series circuit. A circuit with only a single path for electrical current to flow. (11.6)

service animals. Animals that serve as aides and helpers to their owners. (9.11)

Service Learning SAE. A program in which students plan, conduct, and evaluate a project that is designed to provide a service to the community. (1.5)

serving size. An amount used to describe one portion of food. (6.4)

sewage. Waste material, such as human urine or feces, that is carried away from homes or other buildings in a system of drainage pipes. (11.5)

sexual propagation. Reproduction of plants from seed. (8.7)

sharecropping. A method of agricultural labor in which the landowner provided the use of farmland to a farmer or tenant in exchange for a share of the crop produced. (3.2)

shelf life. The period a food product will remain safe to eat. (6.3)

shelterwood cutting. Removing enough undesirable tree species to allow light to reach the forest floor, encouraging new seedlings to grow. (10.8)

short-term goal. A task you want to accomplish in less than a year. (2.2)

silt. A type of soil particle that is smaller than sand but larger than clay. (7.1)

silviculture. A type of forest management that works to control the establishment, growth, health, and quality of forests. (10.5)

simple carbohydrates. Sugars and starches that are easily digested. (9.7)

simple machines. Devices that have few or no moving parts and are used to modify motion and decrease the effort required to perform work. (11.2)

sire. Male parent. (9.4)

skeletal muscles. Voluntary muscles attached to bones or skin to provide movement. (9.6)

skeletal system. System comprised of bones, joints, cartilage, and teeth; provides structure, protection, and support for the body. (9.6)

slash. The parts of the tree left in the woods after harvest. (10.8)

small engines. A type of internal combustion engine rated at 25 horsepower (hp) or less. (11.7)

Smith-Hughes Act. A government act that provides federal aid to states, allowing them to offer courses to high school students to study agriculture, industry, and consumer and family sciences. (1.3)

smooth muscles. Involuntary muscles in the walls of hollow organs like the stomach or bladder; also found in blood vessels and the respiratory tract. (9.6)

social media. Electronic communication between members of an online social network created to share information, ideas, messages, and other content. (10.2)

softwood. The wood of evergreen gymnosperm trees. (10.10)

soil. A complex combination of minerals, gases, liquids, living organisms, and organic matter that support the growth and development of plants and animals. (7.1)

soil biodiversity. A variety of life that exists with the soil, including bacteria, fungi, earthworms, and other organisms. (7.2)

soil conservation practice. A method of farming that prevents the erosion of soil. (7.2)

soil deposition. Soil particles settled in a new location. (7.2)

soil detachment. Soil removed by a type of force. (7.2)

soil horizon. A layer parallel to the soil surface whose physical, chemical, and biological characteristics differ from the layers above and beneath. (7.1)

soil pH. An indication of the hydrogen ion activity in the soil or growing medium. (7.3)

soil profile. A diagram of a vertical section of soil depicting the horizons. (7.1)

soil test. The calculation of plant nutrients available in the soil at a given time. (7.3)

soil transport. Soil particles moved from one location to another. (7.2)

spark plug. A device that generates a spark and ignites the compressed fuel and air mixture at the precise moment necessary to create a power stroke in an engine. (11.7)

specialty animals. Livestock raised in small numbers in small operations, often for a niche market. (9.10)

stamen. Male organs of the flower. (8.6)

standards. Levels or guidelines used to measure product quality defined as average or acceptable. (12.2)

stem. Part of a plant that moves nutrients and food throughout the plant. (8.3)

stem cuttings. Reproduction of plants from the stem of a parent plant. (8.8)

sterilization. The controlled heating of food products to eliminate living microorganisms in food. (6.3)

stomata. Tiny openings on the surface of leaves. (8.5)

stratification. Moist, cold, extended period of rest to allow for seed germination. (8.7)

stream buffer zone. Areas next to streams where logging operations are prohibited. (10.8)

strip tillage. A method of planting crops in a narrow band of tilled soil, without disturbing the surface of the field. (10.11)

stroke. The movement of the cylinder of the piston from one end to the other. (11.7)

subsidiary motion. A motion that deals with managing other motions. (1.6)

sub-surface irrigation. Methods of applying water to crops below the soil surface. (11.5)

succession. The gradual and long-term changes occurring in an ecosystem as it transitions from one state to another. (10.3)

superfund site. A hazardous waste site requiring a significant cleanup to reduce immediate danger to humans and the environment. (10.7)

Supervised Agricultural Experience (SAE). Practical, hands-on agricultural activities that are completed by students outside of their regular class time. (1.3)

supply. The amount of a product currently available. (4.3)

supply chain. The way that materials move from one company to another. (12.6)

surface irrigation. Methods of applying water to crops aboveground. (11.5)

surveyors. Workers who create property boundaries to ensure that logging operations do not infringe on others' property. They also create property maps that help landowners estimate property tax expenses. (10.8)

sustainable agriculture. New ways of farming and ranching, so that natural resources are not used up at a faster rate than they can be replenished. (5.1)

sustainable forest management. Techniques and practices used in managing forested lands that protect soil and water resources. (10.8)

switch. The tip of an animal's tail. (9.3)

symptoms. Signs of illness. (9.8)

T

taiga. Forests where the temperatures remain very low throughout the year, and about half of the precipitation is snow. The soil is typically fragile and low in nutrients. Also called *boreal forests*. (10.6)

taproot. One large root that grows directly downward and has a few small roots that branch off the main root. (8.3)

tar. A thick, sticky substance derived from tree sap and used for waterproofing. (10.7)

target audience. A group of people identified as being likely customers of a business. (12.4).

tariff. A tax on incoming goods from another country to make a domestic product more competitive in the market. (4.3)

team. A group of people who come together to achieve a common goal. (2.2)

technology. The application of science, engineering, and mathematics to solve real-world, practical problems. (5.2)

temperament. An animal's demeanor or natural behavioral and emotional state. (9.1)

temperate forest. Forests that experience four distinct seasons each year.

temperature. A measure of heat or cold. (11.3)

tendon. Strong, fibrous collagen tissue that attaches muscles to the skeleton. (9.6)

terminal. Known for good carcass quality and muscling; one of the two main qualities, along with maternal characteristics, sought in swine. (9.3)

terminal bud. The primary growth point of the stem; usually forms the central leader or main trunk of the plant. (8.3)

theory. A set of statements that attempt to explain what we see or experience in the natural world. (5.1)

thermoplastic. A recyclable material that softens when heated and reforms into a solid when cooled. Common examples include acrylic, polyethylene, polypropylene, and polyvinyl chloride (PVC). (11.4)

thermoset. A material that is chemically changed during initial processing but does not soften with additional applications of heat. Examples include silicone, epoxy, and polyester. (11.4)

tillage. The process of preparing soil for planting crops. (5.2)

time. The measurable period during which an action, process, or condition exists or continues. (11.3)

time management. The process of organizing and planning your day by dividing your time between activities. (2.3)

ton-miles. The movement of 2,000 pounds (one ton) a distance of 5,280 feet (one mile). (4.1)

tool. A device that is used to perform a specific type of function or task. (11.2)

trade. The exchange of desired goods. (4.2)

traits. Inherited characteristics. (9.1)

transpiration. The process of leaves releasing water vapor into the air. (8.5)

tropical forest. Forests that exist in frost-free areas where climate is favorable for the year-round growth of plants. (10.6)

turf grass A collection of green plants that form a ground cover. (8.1)

turpentine. An oil distilled from pine trees' sap, used in paints and medications. (10.7)

two-stroke engine. An engine in which two strokes of the piston are needed to complete the engine cycle. (11.7)

U

udder. Milk-producing organ of cattle. (9.4)

unfertilized egg. Eggs that are laid without being fertilized by roosters, and so cannot produce chicks. These are produced for human consumption. (9.5)

United States Centers for Disease Control (CDC). A US government agency tasked with protecting public health by preventing and controlling the spread of disease. (6.2)

United States customary system. A system of measurement that uses the foot, gallon, and pound as the basic units of length, capacity, and weight. (11.3)

United States Department of Agriculture. The United States' executive department responsible for overseeing federal laws related to farming, forestry, rural development, and food. (3.1)

urban agriculture. The production and distribution of food, fiber, and natural resources in cities and towns. (5.2)

V

vaccination. The use of a vaccine (dead or mild virus) to stimulate an organism's immune system so that it recognizes a disease and offers protection from future infection. (9.8)

valve train. The portion of the engine that is composed of intake valves, exhaust valves, and camshaft. (11.7)

variety meats. Intestines and organ meats, ground up and mixed for human consumption. (9.5)

verbal communication. The use of words to exchange information. (2.2)

vertical farming. The production of food on vertically aligned surfaces, like stacks of planters along a wall. (5.2)

virus. Small, protein-coated organism that invades living cells and replicates in the host it infects. (8.9)

visual communication. Nonverbal communication that uses signs, symbols, or pictures. (2.4)

vitamin deficiency. A lack of vitamins that leads to many growth and disease problems. (9.7)

viticulture. The cultivation of grapes to be eaten fresh or to be used in the making of juice, raisins, jams, jellies, and wines. (8.1)

voltage. Pressure (electrical potential) that moves electrons through a conductor. Also called *electromotive force*. (11.6)

volume. The amount of space occupied by a cubic (three-dimensional) object as measured in cubic units. (11.3)

W

washer. Flat, round piece of metal placed under the head of a bolt or nut to distribute the load when tightening occurs. (11.4)

water main. The primary line in a water supply system. (11.5)

wattage. The rate at which electrical work is completed. (11.6)

weather. The measure of the current temperature, precipitation, wind, and humidity of an area. (3.4)

wedge. A simple machine that tapers to a thin edge, used to apply force and split things apart. (11.2)

weed. A plant growing out of place or an unwanted plant. (8.9)

weight. The quantity of matter contained by an object as measured by the downward force of Earth's gravity. (11.3)

wheel and axle. An assembly made up of a circular object that rotates or revolves on or around a smaller shaft or rod. (11.2)

wholesale. Products that are sold to grocery stores and restaurants for resale. (4.1)

wholesale markets. Second parties that collect, assemble, and distribute farmers' or producers' products. (12.1)

wilderness area. An area where the impact of human activities is minimal or nonexistent. (10.1)

wildlife refuge. A protected place for wildlife. (10.1)

windthrow. Trees falling over in high winds. (10.8)

withdrawal time. The amount of time from injection until a medicine is no longer detected in an animal's body. (9.8)

withers. The ridge between a horse's shoulders. (9.3)

woody plant. Plant with a hard stem and buds that survives through the winter months above ground. (8.2)

wool. The hair-like material that covers sheep. (3.1)

work. A measurement of the energy transfer that occurs when an object is moved over a distance by an external force. (11.2)

work ethic. The values, attitudes, and behaviors toward every job or task you complete. (2.3)

written communication. The use of written language to convey messages. (2.4)

X

xylem. The tissue in a plant stem that transports water and nutrients from the roots to all other parts of the plant. (8.3)

Z

zoonotic diseases. Diseases that can be transferred between humans and animals. (9.8)

Index

area, 705
artificial intelligence, 242
asexual propagation, 398–403
    cuttings, 399–400
    definition, 399
    grafting, 403
    separation and division,
        400–402
aspens, 596
assembly, 759
atom, 733
attitudes and habits, 95–100
    first impression, 96–97
    motivation, 100
    work ethic, 98–99
automation, 813

B

bacteria, 300
bacteria growth, 272–273
    moisture, 273
    nutrients, 272
    pH, 273
    time and temperature, 272
balanced ration, 489
barbed wire, 137
bark, 627
battery, 735
beeches, 593–594
beef cattle, 427, 436–437
    Angus, 436
    Brahman, 436–437
    Charolais, 437
    Hereford, 437
    Simmental, 437
berry, 629
bias, 558
biennial, 350
big data, 239
bill of materials, 757
Biltmore Estate and Forest, 610–611
Biltmore Forest School, 611
biodiversity, 223, 568–569
biological pest management, 412
biotechnology, 17, 142, 247–250
    chromosomes, 249
    DNA, 249
    genes, 250
    genetic engineering, 250
    history, 247–248
birches, 594, 596
birds, 656–659
bison, 518–519
blacksmith, 215
board foot, 634
boll weevil, 137
bolt, 714

boreal forest, 595–596
    aspens, 596
    birches, 596
    larches, 595
    spruces, 595
Bos indicus, 427
Bos taurus, 427
brand, 801
breed, 436
breeding animal, 451
British thermal unit (BTU), 586
broilers, 457
bud, 356
budget, 755
buffer strip, 646
bushel, 703
business. *See* agribusiness
by-product, 328, 463

C

caloric value, 290
camelids, 519–520
    alpacas, 519
    camels, 520
    llamas, 519
camshaft, 744
canning, 282
capturing and tracking, 655–656
carbohydrates, 485, 489
carburetor, 745
cardiac muscle, 474
career and technical education (CTE), 8
Career Development Event (CDE), 42
career pathway, 15
career readiness, 548
careers
    ecosystem management, 543
    fish and wildlife management,
        544–545
    forestry and forest management, 544
    minerals and geology, 545
    natural resources engineering, 544
    outdoor recreation, 545
    pathways in wildlife and natural
        resources, 543–545
    wildlife and natural resources,
        540–548
career skills
    attitudes and habits, 95–100
    character, 74–79
    communication, 103–108
    curiosity and creativity, 111–117
    leadership, 82–90
career success, 42
carnivore, 421
carrying capacity, 582
cartilage, 473

cartography, 545
cats, 528
cattle, 453
Center for Veterinary Medicine (CVM),
    488
cereals, 147
chair, 60
character, 75
character skills, 74–79
    cornerstones, 76–77
    cultivating, 75–76
    growth opportunities, 78–79
    models, 79
chemical pest management, 412
chemicals, safety, 673
chlorophyll, 375
chloroplasts, 375
chromosomes, 249
circulatory system, 479
classroom and laboratory instruction, 29
clay, 302
clear-cutting, 616
climate, 156–163
    agriculture effect on climate
        change, 159–162
    climate change, 157–159
    definition, 157
    effect of climate change on
        agriculture, 163
    role of agriculture in addressing
        climate change, 163
climate change, 157–159
clothing, 126–127
collagen, 467
combustion, 743
commercial crops, 146–153
    fiber crops, 151–153
    grain crops, 147–150
    peanuts, 150
    soybeans, 150–151
commodity, 700
commodity prices, 199–200
communication, 103–108
    constructive feedback, 107–108
    effective, 104–106
    listening strategies, 106
    types, 105–106
Community Supported Agriculture
    (CSA), 770
companion animals, 431–432, 525–532
    agriculture and, 526
    cats, 528
    choosing, 530–531
    dogs, 526–527
    other types, 528–529
    responsible pet ownership, 531–532
    service animals, 529
    wild animals, 529

complete fertilizer, 321
complete flower, 385
complex carbohydrates, 485
compost, 328
compound, 733
compression system, 747
computer-aided design (CAD), 708
concentrates, 487
concrete, 718
condensation, 564
conductor pathway, 737
conductors, 734
conflict resolution, 89
conformation, 451
conservation, 573, 613
conservation cover, 647
conservation plans, 572–578
    alternative solutions, 575–577
    community buy-in, 577
    data analysis, 575
    definition, 573
    evaluation, 578
    goals and objectives, 574
    implementation, 577
    problem identification, 573–574
    resource inventory, 575
conservation tillage, 223
construction
    definition, 757
    See also project construction
construction materials, 715–718
    concrete, 718
    metal, 717
    plastic, 717
    wood, 715–716
constructive, 107
consumer, 128
contagious, 497
conversion optimization, 812
cookie, 812
cooling system, 747–748
copper, 726
corn, 147
cotton gin, 134–135
cover crops, 646
crankshaft, 744
creativity, 113
creed, 36, 75
critical thinking, 115–116
crop rotation, 647
cross-contamination, 274
crosscut, 759
cross pollination, 384
cull, 451
cultivar, 346
cultural pest management, 413
curiosity, 112
curiosity and creativity, 111–117

current, 734
custodial management, 583
cutting list, 757
cuttings, 399–400
cycle, 745

D

dairy, 464
dairy cattle, 428, 438
    Brown Swiss, 438
    Holstein, 438
    Jersey, 438
dam, 458
data, 113
data management, 813–814
decibel, 672
deciduous plant, 350
decomposition, 563
dehydration, 284
demand, 197
demeanor, 505
dendrology, 547
design. *See* project design
dicot, 348
digestion, 475
digestive system, 475–478
    avian, 478
    monogastric, 476
    ruminant, 477–478
direct current (DC), 734
direct seeding, 392–393
Discovery Degree, 43
disease, 497–500
    control, 500
    improper nutrition, 498
    infectious, 497
    parasites, 497–49
division, 400–402
DNA, 249
dogs, 526–527
domesticate, 5
domestication, 420–421
    breeding, 421
    eating and growth, 421
    temperament, 420
dormancy, 350
drainage systems, 725–726
drake, 444
drought, 213
drupe, 629
Dust Bowl, 138

E

E. coli, 275
ecological management, 543

economy, 170–178, 196–201
    banking and finance, 176
    commodity prices, 199–200
    definition, 171
    food manufacturing and sales, 173
    food processing industry, 174
    historical supply and demand, 198
    jobs in agriculture, 172–178
    law of supply and demand, 197
    natural resources industry, 175
    policy and trade, 200–201
    price, 197–198
    transportation of agricultural products, 177–178
    value of agriculture, 171
ecosystem, 561–569
    biodiversity, 568–569
    definition, 562
    nutrient cycle, 562–564
    process of succession, 566–567
    water cycle, 564–565
ecosystem management, 543
edible, 465
education, 546–548
effective communication, 104
electrical circuits, 736–739
    conductor pathway, 737
    load, 738
    parallel circuits, 739
    path to ground, 738
    power sources, 737
    series circuits, 739
electrical system, 748
electricity, 732–739
    amperage, 736
    batteries, 735
    definition, 734
    electrical circuits, 736–739
    electron movement, 734
    generator, 735
    measurement, 736
    resistance, 736
    source, 733–734
    voltage, 736
    wattage, 736
electromagnetic induction, 735
electromotive force, 736
electron, 733
electron movement, 734
element, 733
emblem, 36
embryo, 361
emerging technologies, 234–242
    agricultural revolution, 236–237
    artificial intelligence, 242
    big data, 239
    Internet of Things (IoT), 239–241

integrated pest management (IPM), 226–227
integrator, 772
interiorscaping, 339
Internet of Things (IoT), 239–241
internode, 356
Internship SAE, 48–49
intrinsic reward, 100
inventory, 614
irrigation, 370, 721–724
    methods, 724
    water sources, 722–723
issues, 551–558
    definition, 552
    northern spotted owl, 553–554
    research skills, 554–558

K

kerf, 758
kindness, 77

L

laboratory safety, 677–679
    organization and layout, 677
    tools, equipment, machinery, 677
landscapers, 338
larches, 595
lateral meristem, 627
law of supply and demand, 197
layers, 457
layout, 706
leader, 83
leadership, 83
Leadership Development Event (LDE), 40
leadership skills, 82–90
    accountability, 87
    characteristics, 84
    conflict resolution, 89
    effective communicators, 86–87
    focus, 88
    followership, 89–90
    goals, 84
    listening skills, 85–86
    opportunities, 90
    passion, 85
    personal, 83–84
    personal traits, 85
    responsibility, 87–88
    team leadership, 88–89
leaf cuttings, 400
leaves, 359, 622–626
    arrangement, 624
    shape, 625
    simple and compound, 626
    venation, 624

legumes, 487
length, 704
lever, 684
ligament, 473
light, plant growth environment, 365–366
light quality, 365
light quantity, 366
lipids, 485
listening strategies, 106
livestock, 5
livestock breeds, 435–445
    beef cattle, 436–437
    dairy cattle, 438
    goat breeds, 442–443
    horse breeds, 439–440
    poultry breeds, 443–445
    sheep breeds, 441–442
    swine breeds, 440–441
livestock by-products, 465–469
    edible, 466–468
    inedible, 468–469
livestock confinement housing, 509
livestock facilities, 509–511
    equipment, 511
    fire safety, 510
    grain storage, 509
livestock feeding and nutrition, 483–490
    carbohydrates, 489
    fats and oils, 489
    feed additives, 488
    feed classifications, 487–488
    feed labels, 488–489
    minerals, 487
    proteins, 485
    rations, 489–490
    vitamins, 486
    water, 484
livestock health, 494–500
    diseases, 497–500
    health management, 495–496
livestock industry
    beef cattle, 427
    companion animals, 431–432
    dairy cattle, 428
    goats, 430
    horses, 428
    poultry, 430
    sheep, 429
    swine, 429
livestock physiology, 472–480
    circulatory system, 479
    digestive system, 475–478
    muscular system, 474
    nervous system, 480
    respiratory system, 478–479
    skeletal system, 473

livestock production, specialty, 514–522
livestock products, 463–465
    dairy, 464
    eggs, 465
    fiber, 464–465
    meat, 463–464
livestock safety, 503–511
    animal factors, 505–507
    facilities, 509–511
    human factors, 504–505
    importance, 504
    personal protective equipment (PPE), 508
    safe chemical use, 508
livestock selection, 450–458
    animal fiber production, 452
    breeding, 451
    cattle, 453
    considerations, 453–457
    goat, 454
    horse, 455
    market, 451
    milk production, 452
    performance records, 458
    poultry, 456–457
    sheep, 454
    swine, 454
livestock waste reduction, 465–469
load, 738
loam, 302
log deck, 615
long-term goals, 84
lubrication system, 748
lumber, 715

M

machinery, safety, 671–672, 677
macronutrients, 316
main motion, 62
major US commercial crops. See commercial crops
malnutrition, 184
mammals, 661–662
manipulative management, 583
manures, 328
maples, 594
marbled, 436
margin, 625
marine aquaculture, 517
market animal, 451
marketing, 766–772
    agricultural marketing, 767–772
    definition, 767
    strategies and plans, 792–797
marketing plans, 793–797
    creating, 795–797
    purpose, 794

planning. *See* project planning
plantations, 134
plant classification
    foliage retention, 350
    names, 346–347
    plant life cycles, 350
    reproduction, 348
    stem type, 349
plant disease, 410
    bacterial, 410
    fungal, 410
    viral, 411
plant growth environment, 364–370
    air, 368
    light, 365–366
    temperature, 366–368
    water, 368–370
plant life cycles, 350
plant nutrition, 316–317
    micronutrients, 317
    primary macronutrients, 316
    secondary macronutrients, 316
plant parts, 354–361
    flowers, 360
    fruits and seeds, 361
    leaves, 359
    roots, 355–356
    stems, 356–358
plant propagation, 390
plant reproduction, 348
plant reproductive parts, 381–386
    flower parts, 382
    pollination, 383–384
    types of flowers, 385–386
plant science, 20
    See also horticulture
plastic, 717, 727
plumbing, 724–729
    definition, 724
    drainage, 725–726
    materials and components, 726–729
    water supply, 724–725
plywood, 604
pneumatic tools, 689
polled, 436
pollen, 382
pollination, 383–384, 628
pome, 629
pomology, 341
population surveys, 655
potable, 725
poultry, 430, 456–457, 520–521
poultry breeds, 443–445
    chickens, 443
    ducks, 444
    emus, 520
    geese, 444–445
    ostriches, 520

    quails, 521
    rheas, 521
    turkeys, 445
power sources, 737
power tools, 688–690, 694–695
    definition, 685
precision agriculture, 141, 224–226
    decision-making and response, 226
    measuring, 226
    observing, 225–226
preening, 657
premier leadership, 39
prescribed fire, 618
preservation. *See* food preservation
prevention, 495
price point, 788–789
price point determination, 784–789
    market price, 785–787
    price points, 788–789
    supply and demand, 785
privileged motion, 61
problem-solving, 112
process, 112
processing, 171
processor, 767, 772
producer, 128
professionalism, 98
Progressive Era
    boll weevil, 137–138
    George Washington Carver, 138
    Pure Food and Drug Act, 137
project construction, 757–761
    assembly, 759
    finish, 759–760
    material selection, 757–758
    preparation, 758–759
project design, 755–757
    bill of materials, 757
    drawing, 756–757
    purpose, 755
    schedule, 756
project planning, 754–755
    budget and time, 755
    materials, 754
    tools, 754
promotions, 801–803
    benefits, 801–802
    definition, 801
    methods, 803
propaganda, 556
protection, 614
proteins, 485
proton, 733
protozoa, 300
pseudo-ruminant, 519
pulley, 685
pulpwood, 604

punctuality, 98
Pure Food and Drug Act, 137

Q

quadruped, 661
quarantine, 500
quorum, 60

R

rabbits, 521
Rachel Carson, 211
radio frequency identification (RFID), 242
radio telemetry, 655
rainforest, 597
rations, 489–490
reaction time, 675–676
reclaimed water, 725
Reconstruction period, 136–137
    barbed wire, 137
    sharecropping, 136
    steam tractor, 136
recordkeeping, 53
recreational marketing, 770
regenerative agriculture, 223–224
regurgitate, 478
remarketing, 812
renewable, 586
reptiles, 663–664
Research SAE, 50
research skills, 554–558
    fact-based sources, 556
    internet sources, 557
    library, 557
    personal bias, 558
    social media, 555
reservoir, 722
resistance, 736
respect, 76
respiration, 375
respiratory system, 478–479
responsibility, 87–88
restaurants, 771
retail, 172
retail market, 771
rice, 148
rip cut, 759
robotics, 214, 222
role model, 79
root, 355–356, 627
roughages, 487
ruminant, 477
Rural Electrification Administration (REA), 221

S

safety, 670–679
    developing habits, 671
    first aid, 674
    hazards, 671–673
    injuries, 673–674
    laboratory safety, 677–679
    personal protective equipment (PPE), 674
    reaction time, 675–676
salmonella, 274
sand, 302
sanitary drainage system, 725
sapwood, 626
sawtimber, 604
scale drawing, 756
scarification, 391
School-Based Enterprise SAE, 50
science, 208–217
    agricultural mechanization, 214
    crop production and management, 213
    famous agricultural scientists, 211–213
    improving agriculture and natural resources, 213–215
    innovation in agriculture, 215–217
    livestock and poultry production, 213
    scientific method, 209–210
    sustainable agriculture, 215
scientific method, 114
    elements of reasoning, 209–210
scientific name, 346
scientists
    George Washington Carver, 212
    modern leaders, 213
    Norman Borlaug, 212
    Rachel Carson, 211
screw, 685, 714
search engine, 812
second, 63
seed coat, 361
seed dispersal, 629
seedlings, 393
seeds, 361, 391–394
    direct seeding, 392–393
    indirect seeding, 393–394
selection, 436
selection cutting, 617
self-discipline, 98
sepals, 382
separation, 400–402
septic tank, 725
series circuit, 739
service animals, 529
Service Learning SAE, 51

serving size, 290
sewage, 725
sexual propagation, 389–394
    definition, 390
    plant propagation, 390
    seeds, 391–394
shade, 603
sharecropping, 136
sheep, 429, 454
sheep breeds, 441–442
    Dorset, 441
    Hampshire, 441
    Katahdin, 441
    Suffolk, 441
shelf life, 281
shelterwood cutting, 617
short-term goal, 84
silt, 302
silviculture, 584
simple carbohydrates, 485
simple machines, 683–685
    definition, 683
    inclined plane, 683
    lever, 684
    pulley, 685
    screw, 685
    wedge, 684
    wheel and axle, 684
sire, 458
skeletal muscles, 474
skeletal system, 473
slash, 616
small engine, 742–750
    definition, 743
    engine operation, 745–746
    engine systems, 747–748
    maintenance, 748–750
    parts, 743–745
small engine maintenance, 748–750
    air filter, 749
    checking and changing oil, 749
    cleaning, 748–749
    preparation for storage, 750
    spark plug, 749–750
Smith-Hughes Act, 26
smooth muscles, 474
social media, 555
softwoods, 634
    coast redwood, 640
    Douglas fir, 639
    eastern red cedar, 642
    Fraser fir, 641
    loblolly pine, 641
    red pine, 641
    Sitka spruce, 641
    western larch, 640

soil
    components, 298–303
    conservation, 307–312
    definition, 299
    fertilizers, 320–322
    fertilizer sources, 325–331
    importance of, 299
    inhabitants, 300–302
    pH, 317–320
    plant nutrition, 316–317
soil biodiversity, 311
soil components, 298–303
    soil particles, 302
    soil profile, 303
soil conservation, 307–312, 646
    conservation practice, 312
    soil erosion, 309–310
    soil loss, 308
    soil preservation, 311–312
soil conservation practice, 312
soil deposition, 309
soil detachment, 309
soil horizon, 303
soil pH, 317–320
soil profile, 303
soil test, 320
soil transport, 309
sorghum, 148
soybeans, 150–151
spark plug, 748–750
specialty animals, 515
specialty livestock production, 514–522
    apiculture, 515
    aquaculture, 515–518
    bison, 518–519
    camelids, 519–520
    poultry, 520–521
    rabbits, 521
spruces, 595
stamens, 382
standard, 777
steam tractor, 136
steel plow, 135
stem, 356–358, 626–627
stem cuttings, 400
stem type, 349
sterilization, 282
stomata, 377
stratification, 391
stream buffer zone, 615
strip tillage, 647
stroke, 745
subsidiary motion, 62
subsurface irrigation, 724
succession, 566
superfund site, 603